Freeman Laboratory Separates in Biology

The exercises listed below are available as Freeman Laboratory Separates. Each Separate is derived from the experiments in the corresponding laboratory manual. The Separates are self-bound, self-contained exercises. They are 8½ inches by 11 inches in size, and are punched for a three-ring notebook. They can be ordered in any assortment or quantity. When ordering Separates, please use both the Separate title and the ISBN.

LABORATORY SEPARATES BY HELMS, HELMS, KOSINSKI, AND CUMMINGS

From *Biology in the Laboratory,* Third Edition
1998, 928 pages, ISBN 0-7167-3146-0

0-7167-9303-2	Science—A Process and Appendixes I, II, and III
0-7167-9304-0	Observations and Measurements: Microscope
0-7167-9305-9	Observations and Measurements: Measuring Techniques
0-7167-9306-7	pH and Buffers
0-7167-9307-5	Using the Spectrophotometer
0-7167-9308-3	Organic Molecules
0-7167-9309-1	Prokaryotic Cells
0-7167-9310-5	Eukaryotic Cells
0-7167-9311-3	Osmosis and Diffusion
0-7167-9312-1	Mitosis
0-7167-9313-X	Enzymes
0-7167-9314-8	Energetics, Fermentation, and Respiration
0-7167-9315-6	Photosynthesis
0-7167-9316-4	Meiosis: Independent Assortment and Segregation
0-7167-9317-2	Genes and Chromosomes: Chromosome Mapping
0-7167-9318-0	Human Genetic Traits
0-7167-9319-9	DNA Isolation
0-7167-9320-2	DNA—The Genetic Material: Replication, Transcription, and Translation
0-7167-9321-0	Molecular Genetics: Recombinant DNA
0-7167-9322-9	Genetic Control of Development and Immune Defenses
0-7167-9323-7	The Genetic Basis of Evolution
0-7167-9324-5	Genetic Basis of Evolution II—Diversity
0-7167-9325-3	Diversity—Kingdoms Enbacteria, Archaebacteria, and Protista
0-7167-9326-1	Diversity—Fungi and the Nontracheophytes
0-7167-9327-X	Diversity—The tracheophytes (Vascular Land Plants)
0-7167-9328-8	Diversity—Porifera, Cnidaria, and Wormlike Invertebrates
0-7167-9329-6	Diversity—Mollusks, Arthropods, and Echinoderms
0-7167-9330-X	Diversity—Phylum Chordata
0-7167-9331-8	Plant Anatomy—Roots, Stems, and Leaves
0-7167-9332-6	Angiosperm Development—Fruits, Seeds, Meristems, and Secondary Growth
0-7167-9333-4	Water Movement and Mineral Nutrition in Plants
0-7167-9334-2	Plant Responses to Stimuli
0-7167-9335-0	Animal Tissues
0-7167-9336-9	Introduction to the Study of Anatomy, and the External Anatomy and Integument of Representative Vertebrates

0-7167-9337-7	The Anatomy of Representative Vertebrates: Behavioral Systems
0-7167-9338-5	The Anatomy of Representative Vertebrates: Digestive and Respiratory Systems
0-7167-9339-3	The Anatomy of Representative Vertebrates: Circulatory and Urogenital Systems
0-7167-9340-7	The Basics of Animal Form: Skin, Bones, and Muscles
0-7167-9341-5	Physiology of Circulation
0-7167-9342-3	Gas Exchange and Respiratory Systems
0-7167-9343-1	The Digestive, Excretory, and Reproductive Systems
0-7167-9344-X	Control—The Nervous System
0-7167-9345-8	Behavior
0-7167-9346-6	Communities and Ecosystems
0-7167-9347-4	Predator—Prey Relations
0-7167-9348-2	Productivity in an Aquatic Ecosystem

LABORATORY SEPARATES BY WALKER AND HOMBERGER

From *Anatomy and Dissection of the Fetal Pig,* Fifth Edition
1998, 120 pages, ISBN 0-7167-2637-8

0-7167-9296-6	External Anatomy, Skin, and Skeleton
0-7167-9297-4	Muscles
0-7167-9298-2	Digestive and Respiratory Systems
0-7167-9299-0	Circulatory System
0-7167-9300-8	Urogenital System
0-7167-9301-6	Nervous Coordination: Sense Organs
0-7167-9302-4	Nervous Coordination: Nervous System
0-7167-9288-5	Glossary of Vertebrate Anatomical Terms

From *Anatomy and Dissection of the Rat,* Third Edition
1998, 120 pages, ISBN 0-7167-2635-1

0-7167-9289-3	External Anatomy, Skin, and Skeleton
0-7167-9290-7	Muscles
0-7167-9291-5	Digestive and Respiratory Systems
0-7167-9292-3	Circulatory System
0-7167-9293-1	Urogenital System
0-7167-9294-X	Nervous Coordination: Nervous System
0-7167-9295-8	Nervous Coordination: Sense Organs
0-7167-9288-5	Glossary of Vertebrate Anatomical Terms

W. H. FREEMAN AND COMPANY
41 Madison Avenue, New York, NY 10010
Houndmills, Basingstoke, RG21 6XS, England

From *Dissection of the Frog,* Second Edition
1998, 120 pages, ISBN 0-7167-2636-X

0-7167-9218-4	External Anatomy, Skin, and Skeleton
0-7167-9221-4	Circulatory System
0-7167-9220-6	Digestive and Respiratory Systems
0-7167-9219-2	Muscles
0-7167-9223-0	Nervous Coordination: Nervous System
0-7167-9224-9	Nervous Coordination: Sense Organs
0-7167-9222-2	Urogenital System
0-7167-9288-5	Glossary of Vertebrate Anatomical Terms

LABORATORY SEPARATES BY ABRAMOFF/THOMSON

From *Laboratory Outlines in Biology VI*
1994, 526 pages, ISBN 0-7167-2633-5

0-7167-9081-5	Biologically Important Molecules: Proteins, Carbohydrates, Lipids, and Nucleic Acids
0-7167-9082-3	Light Microscopy
0-7167-9083-1	Cell Structure and Function
0-7167-9084-X	Subcellular Structure and Function
0-7167-9085-8	Cellular Reproduction
0-7167-9086-6	Movement of Materials Through Plasma Membranes
0-7167-9087-4	Enzymes
0-7167-9088-2	Cellular Respiration
0-7167-9089-0	Photosynthesis
0-7167-9090-4	Mendelian Genetics
0-7167-9091-2	Chromosomal Basis of Heredity
0-7167-9092-0	Human Genetics
0-7167-9093-9	Expression of Gene Activity
0-7167-9094-7	Kingdom Monera
0-7167-9095-5	Kingdom Protista I: Algae and Slime Molds
0-7167-9096-3	Kingdom Protista II: Protozoa
0-7167-9097-1	Kingdom Fungi
0-7167-9098-X	Kingdom Plantae: Division Bryophyta
0-7167-9099-8	Kingdom Plantae: The Vascular Plants
0-7167-9100-5	Plant Anatomy: Roots, Stems, and Leaves
0-7167-9101-3	Flowers and Fruits
0-7167-9117-X	Transport and Coordination in Plants
0-7167-9118-8	Plant Growth and Development
0-7167-9119-6	Plant Development: Hormonal Regulation
0-7167-9102-1	Kingdom Animalia: Phyla Porifera, Cnidaria, and Ctenophora
0-7167-9120-X	Kingdom Animalia: Acoelomates (Phylum Platyhelminthes) and Pseudocoelomates (Phyla Nematoda and Rotifera)
0-7167-9103-X	Kingdom Animalia: Phylum Mollusca
0-7167-9104-8	Kingdom Animalia: Phylum Annelida
0-7167-9105-6	Kingdom Animalia: Phylum Onychophora and Arthropoda
0-7167-9106-4	Kingdom Animalia: Phylum Echinodermata
0-7167-9121-8	Kingdom Animalia: Phyla Hermichordata and Chordata
0-7167-9107-2	Vertebrate Anatomy: External Anatomy, Skeleton, and Muscles
0-7167-9108-0	Vertebrate Anatomy: Digestive, Respiratory, Circulatory, and Urogenital Systems
0-7167-9109-9	Biological Coordination in Animals
0-7167-9110-2	Nervous System Physiology
0-7167-9111-0	Fertilization and Early Development of the Sea Urchin
0-7167-9112-9	Fertilization and Early Development of the Frog
0-7167-9113-7	Early Development of the Chick
0-7167-9114-5	Analysis of Surface Water Pollution by Microorganisms
0-7167-9115-3	Analysis of Solids in Water

LABORATORY SEPARATES BY WISCHNITZER

From *Atlas and Dissection Guide for Comparative Anatomy,* Fifth Edition
1993, 291 pages, ISBN 0-7167-2374-3

Anatomy of the Dogfish Shark

0-7167-9240-0	External Morphology
0-7167-9241-9	Skeletal System
0-7167-9242-7	Muscular System
0-7167-9243-5	Digestive and Respiratory Systems
0-7167-9244-3	Circulatory System
0-7167-9245-1	Urogenital System
0-7167-9246-X	Sense Organs
0-7167-9247-8	Nervous System

Anatomy of the Protochordates

0-7167-9239-7	Morphology

Anatomy of the Lamprey

0-7167-9263-X	Morphology

Anatomy of the Mud Puppy Necturus

0-7167-9248-6	External Morphology
0-7167-9249-4	Skeletal System
0-7167-9250-8	Muscular System
0-7167-9251-6	Digestive and Respiratory Systems
0-7167-9252-4	Circulatory System
0-7167-9253-2	Urogenital System
0-7167-9254-0	Sense Organs and Nervous System

Anatomy of the Cat

0-7167-9255-9	External Morphology
0-7167-9256-7	Skeletal System
0-7167-9257-5	Muscular System
0-7167-9258-3	Digestive and Respiratory Systems
0-7167-9259-1	Circulatory System
0-7167-9260-5	Urogenital System
0-7167-9261-3	Endocrine System and Sense Organs
0-7167-9262-1	Nervous System

Biology in the Laboratory

THIRD EDITION

Doris R. Helms

Carl W. Helms

Robert J. Kosinski

John R. Cummings
Clemson University

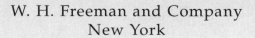

W. H. Freeman and Company
New York

ACQUISITIONS EDITOR: *Patrick Shriner*
ASSOCIATE EDITOR: *Debra Siegel*
PROJECT EDITORS: *ESNE, Inc., Erica Seifert*
DEVELOPMENT EDITOR: *Judith Wilson*
I'RODUCTION COORDINATOR: *Paul W. Rohloff*
ADMINISTRATIVE ASSISTANT: *Ceserina Pugliese*
TEXT AND COVER DESIGNER: *Vicki Tomaselli*
INTERIOR ILLUSTRATION COORDINATOR: *Bill Page*
INTERIOR ILLUSTRATIONS: *Fine Line Illustrations*
COMPOSITION: *York Graphic Services*
MANUFACTURING: *Vail-Ballou Press*

ISBN: 0-7167-3146-0 (EAN: 9780716731467)

Printed in the United States of America

Eighth printing

BRIEF CONTENTS

The following is the contents for the complete version of Biology in the Laboratory, *Third Edition. Refer to the first page for a complete listing of W. H. Freeman's separates in biology, including the separates for this manual.*

CONTENTS

The following is the contents for the complete version of Biology in the Laboratory, Third Edition.
The symbol 👁 *indicates a guided investigation;* ✔ *indicates a verification exercise (see Preface).*

PREFACE

It is a pleasure to bring *Biology in the Laboratory*, 3/e, to your classroom. In this new edition, we have combined popular, innovative laboratory exercises and the most effective pedagogical features to foster observation, experimentation, critical thought, and application of scientific principles. We are also pleased to include a new *Biobytes* **CD-ROM** with the laboratory manual to support and extend student investigation.

Biology in the Laboratory accompanies *Life: The Science of Biology*, 5/e, by William K. Purves, Gordon H. Orions, H. Craig Heller, and David Sadava; however, the manual is designed to support all general biology texts and courses. We hope you enjoy using *Biology in the Laboratory*.

To the Student

The Study of Biology is changing rapidly. As scientists discover more and more about our world, fascinating questions multiply as do the experimental means for exploring them. This edition of Biology in the Laboratory provides you with new ways to explore, apply, and connect concepts and guides you through techniques you will need for your own investigations. A special feature, called **Extending Your Investigation** provides you with an opportunity to discover the excitement, and, yes, even the frustration of being a scientist: slime molds may leave the confines of the Petri dishes, bacterial cultures may stink, and organisms are usually unpredictable. Nevertheless, you are likely to find the process of investigation challenging, illuminating, fascinating, and, most of all, fun!

To the Instructor

Content

In preparing *Biology in the Laboratory*, 3/e, we incorporated the best laboratories and most popular features of *Biology in the Laboratory*, 2/e, and *More Biology in the Laboratory*. This edition provides a choice of 46 laboratory topics and more than 200 experiments. A diversity of instructional approaches, including simple guided inquiries, more complex experimental designs, and original student investigations, presents many opportunities for students to gain skills, experience, and confidence with the tools and techniques of the scientific method. Students can participate in investigations emphasizing current biological explorations in recombinant DNA, the genetic basis of evolution, the genetic control of development and immunity, and animal behavior.

A new Laboratory I explores methods of scientific inquiry and sets the tone for a more investigative approach throughout the manual. This laboratory makes use of a reaction time computer program included on the new *Biobytes* CD-ROM that accompanies this manual. In the course of conducting an original reaction-time investigation, students learn how to form hypotheses, make predictions and observations, and gather, interpret, and report their data. Statistical treatment of data is emphasized and can be easily accomplished using the *Biobytes* CD-ROM.

Laboratories on eukaryotic cells, the genetic basis of evolution, predator-prey relations, and the aquatic ecosystem are all new. In addition, the genetics sequence, Laboratories 13-18, has been revised to include restriction mapping of chromosomes, forensic DNA studies, up-to-date considerations of protein synthesis, and recombinant DNA techniques.

The diversity laboratories, 22-27, have been modified to reflect the three-domain, six-kingdom classification of life. Major reorganization of laboratories on osmosis and diffusion (Laboratory 8) population genetics

(Laboratory 20), and plant mineral nutrition (Laboratory 30) allow students to compile multiple sets of data from class experiments and analyze them statistically.

An integrated series of dissections (Laboratories 33 through 36) explores evolutionary relationships among vertebrates by comparing the major organ systems of the shark, frog, turtle, and rat. (You may find that your students learn a great deal about evolutionary trends by simply reading through these laboratories if time or resources preclude performing the dissections.) The emphasis is on concepts and comparisons, rather than on the details of anatomy.

Laboratories 37 through 41 emphasize structure, function, and phylogenetic relationships in vertebrate physiology.

The appendices equip students with some basic scientific skills—using statistics, reporting data, writing laboratory reports, using the pH meter and spectrophotometer, and understanding the meaning of the roots, prefixes, and suffixes commonly used in biology.

ııııı Pedagogy ııı

The goal of today's biology educators is nothing less than the transformation of science education. We want our classrooms to be student-centered, to emphasize the formulation of questions above the search for the "right" answers. Investigations should emphasize problem-solving and should provide opportunities for students to think critically. Hands-on science is central to what students must know and be able to do.

Our experience has shown that students and instructors are best served by laboratories that encourage the development of the technical, observational, and intellectual skills required to take full advantage of an investigative approach to learning. To this end, exercises specifically designed to build proficiency in each area have been included. Beginning with the first laboratory and continuing through this manual, students are assisted with the design and execution of investigations as well as statistical analyses of results. These skills are further developed by **Extending Your Investigation,** a new feature included in most of the 46 laboratories. Many extensions are designed to be carried out at home or in the dormitory, although some may require access to equipment available only in the laboratory. We have had great success with simple investigations in which students work on their own. Science becomes not just a two- or three-hour scheduled class, but an everyday experience. All **Extending Your Investigation** exercises are identified in a shaded box.

Some exercises are intended to sharpen observational skills and develop the ability to distinguish observations from inferences. These Verification Exercises, designated by the symbol ✔, help students to understand and apply specific concepts. Other exercises guide students through experimental protocols, teaching them to use laboratory techniques proficiently and appropriately and to understand the reasoning behind them. These Guided Investigations are designated by the symbol 👁.

Some questions have been added throughout to help students gain as much as possible from their observations. Our goal is that they understand not just the "hows" but also the "whys" of the techniques they are using, that they integrate new knowledge and deepen their understanding of concepts. The questions at the end of each laboratory are designed to prompt recall of pertinent facts, develop logical thought, encourage problem solving, and highlight the connections between new learning and previous knowledge.

Every effort has been made to provide complete, concise directions that enable students to work on their own, whether engaged in open, guided, or verification investigations. Many exercises require the collection and analysis of numerical data and the creation of graphs and charts—important scientific tools.

In providing a wide variety of exercises that can be tailored to strengthen students' skills, we have tried to create a student-centered, question-centered laboratory approach that will be practical—and enjoyable—for students and instructors.

‖‖‖‖ The *Biobytes* CD-ROM ‖‖‖‖‖‖‖‖‖‖‖‖‖‖‖‖‖‖‖‖‖‖‖‖‖‖‖‖‖‖‖‖‖‖‖‖‖‖

We are enthusiastic about the addition of four interactive CD-ROM versions of Biobytes designed by Robert J. Kosinski. These programs run on both PC and Macintosh platforms. Most of the Biobytes programs can be used either as data generators or as games (simulations) to augment laboratory studies. SEEDLING explores the effects of varying environmental conditions on the growth of plants. ALIEN investigates cardiopulmonary physiology in humans and fictitious alien beings subjected to different types and levels of physical stress. CYCLE simulates some of the physiological variables associated with human fertility, including the menstrual cycle and pregnancy. DUELING ALLELES allows the user to manipulate the frequencies of various alleles in a population in order to explore the principles of population genetics. We think you will find that Biobytes offers a wide variety of flexible, fun-filled learning experiences.

‖‖‖‖ Format ‖‖‖‖‖‖‖‖‖‖‖‖‖‖‖‖‖‖‖‖‖‖‖‖‖‖‖‖‖‖‖‖‖‖‖

The laboratories in this manual may be used in any sequence to complement a diversity of approaches in the lecture course. Although the order of exercises within a laboratory is designed to develop a particular topic in a logical manner, all exercises can stand alone. Exercises from different laboratories can be used together to suit your emphases and requirements. Time guidelines in the Preparator's Guide can be used to select exercises that fit your laboratory period.

We have used a popular, perforated page format because it offers maximum flexibility. Some illustrations and charts are useful in more than one laboratory and can easily be moved to be included in a loose-leaf laboratory notebook. Supplementary exercises, quiz papers, drawings, and laboratory data can always be added. Students have the convenient option of bringing to the laboratory only the pages required for the day's work.

‖‖‖‖ Instructor's Manual—Materials, Preparation, and Answer Guidelines ‖‖‖‖‖‖‖‖‖‖‖‖‖‖‖‖‖‖‖‖‖‖‖‖‖‖‖‖‖‖

We believe that learning in the laboratory need not depend on expensive, elaborately furnished facilities. Requirements for materials and equipment have been kept to a minimum. The accompanying Preparator's Guide provides complete instructions for the preparation and ordering of all materials and is arranged for ease in tailoring preparations to the size of the class and the amount of time available. Also included is information about possible substitutions, alternative preparations, and how to avoid pitfalls.

An Answer Guide provides examples of actual experimental and class data and answers for all experiments and questions, including suggested hypotheses.

‖‖‖‖ Acknowledgments ‖‖‖‖‖‖‖‖‖‖‖‖‖‖‖‖‖‖‖‖‖‖‖‖‖‖‖‖‖‖‖‖‖

Renewed attention to the importance of hands-on learning and student-centered laboratory instruction has fostered a sharing of ideas and materials among laboratory instructors at all educational levels. This wealth of creative work has allowed us to present a broad range of topics in a variety of ways. We have enjoyed working with the many instructors and investigators who have helped to bring this edition of Biology in the Laboratory to your classroom.

We would especially like to thank all of the instructors we have heard from over the years. Their comments and suggestions have helped to improve *Biology in the Laboratory*. Please join our ranks and e-mail your suggestions to Doris R. Helms at BIOL110@Clemson.edu or call for assistance 864-656-2418.

Exercises on observations and measurements (Laboratories 1 and 2) were adapted from *More Biology in the Laboratory* and the work of Jean Dickey, Clemson University, a valued colleague and friend.

The human pedigree exercises (Laboratory 15) were based on the materials and suggestions provided by Nina Caris, Texas A&M University, College Station, Texas.

Peggy O'Neill Skinner, Bush School, Seattle, Washington, contributed to the DNA isolation exercises (Laboratory 16) and deserves a great deal of thanks for her patience and inspiration.

The recombinant DNA exercises (Laboratories 14 and 16) were adapted from exercises made available by Carolina Biological Supply Co. and developed by Greg Freyer, Columbia University, College of Physicians and Surgeons, and David Miklos, DNA Learning Center, Cold Spring Harbor Laboratory, New York, and also from exercises made available by Edvotek and developed by Jack Chirikjian, Georgetown University, School of Medicine, Washington, DC.

Ken House, a former high school teacher and an AP Biology friend, is wonderful to work with and deserves credit for "woolly worms," the basis for an experiment on selection pressure in Laboratory 20.

James Colacino, Clemson University, assisted in designing an exercise that explores the properties of hemoglobin (Laboratory 39).

Fred Stutzenberger and Jzuen-Rong Tzeng, both of Clemson University, developed the laboratory on the immune response (Laboratory 19).

Exercise B in Laboratory 21, Electrophoretic Analysis of LDH in Ungulate Mammals, was adapted from the work of John N. Anderson, Purdue University and Modern Biology, Inc., West Lafayette, IN.

Barbara Speziale, Clemson University, contributed many hours to the development of the plant diversity and plant anatomy and physiology laboratories (Laboratories 22, 23, 24, and 28 through 31).

Elizabeth Godrick, Boston College, contributed to the extensive list of Greek and Latin roots, prefixes and suffixes in Appendix VII.

Ray Gladden, Carolina Biological Supply Company, was instrumental in the development of the Preparator's Guide.

We would like to thank our copyeditor, Linda Strange, who has assisted us with multiple editions of *Biology in the Laboratory*, seeing to the minor adjustments that make such a difference. Also, without the leadership of Patrick Shriner of W. H. Freeman, the *Biobytes* CD-ROM would still be a wish rather than reality. And without the able assistance of Erica Seifert of W. H. Freeman and Susannah Noel of ESNE, Inc., we would never have completed this edition.

Most of all, we would like to thank our editor and special friend, Judith Wilson, who joins us in bringing you this edition of *Biology in the Laboratory*. Judith's assistance is of the type usually given by a coauthor. Her knowledge of biology and her quest to make it understandable for students add an important element to the development of this multidimensional endeavor. Without Judith's dedication, understanding, and uncanny feel for what is needed and what can be left out, this laboratory manual would never have been possible.

We hope that *Biology in the Laboratory* offers you many hours of enjoyable learning and contributes to an ever increasing fascination with the study of our living world.

Doris R. Helms
Carl W. Helms
Robert J. Kosinski
John R. Cummings

September, 1997
Clemson, South Carolina

Biology in the Laboratory

Science—A Process

OVERVIEW

Science is a way of examining and finding order in the natural world. It is a dynamic process of asking questions and then seeking answers. Observations lead us to formulate questions and, with our limited knowledge, we may offer tentative explanations or make educated guesses about the answers to our questions.

Scientists call a tentative explanation a **hypothesis.** Experimentation follows, providing information that may support or refute a hypothesis. From data, often reinforced by statistics, conclusions can be made about what it is we wish to know.

In this laboratory, you will investigate reaction time—the length of time it takes to react to a stimulus. If available, a computer-generated reaction time program will be used. A less complex series of tests can be conducted without the use of a computer. By formulating hypotheses, designing experiments, and analyzing data, you will engage in the process of science.

> **Please Note:** During your laboratory course, you will often be required to formulate hypotheses and design investigations in the Procedure section of an exercise. You will also have the opportunity to design your own experiments to apply what you have learned. These experiments are designated as Extending Your Investigation. Some of these are designed as class experiments; others are intended to be carried out at home.

Laboratory I provides you with the basic tools to carry out scientific investigations and the statistical tools to evaluate your results.

STUDENT PREPARATION

Before coming to the laboratory, complete Exercise A and steps 1 and 2 of Exercise F.

EXERCISE A   The Scientific Method

The scientific approach is a powerful method for understanding the natural world because it is based on observations of how the world works. However, not just any observations will do: the observations must be systematic and objective. The method for making these observations is sometimes broken down into a series of steps referred to as the **scientific method.**

The scientific method of inquiry is an important part of everything we do in our daily lives; we simply do not recognize the steps because we are so used to them. Scientific inquiry involves the steps outlined below.

Problem You want to find out whether a combination of anti-cholesterol drugs X and Y is more effective in reducing high cholesterol levels than either of the drugs given separately. You might proceed as follows.

Step 1 Make **observations** that lead to the formulation of a question.

> You observe that drugs X and Y, used independently, lower LDL (low-density lipoprotein) levels in the blood. You question whether drug X plus drug Y, given in combination, would be even more effective.

Step 2 The question leads to a tentative explanation or educated guess—a **hypothesis.** Prior knowledge or research or even intuition can contribute to the formulation of a hypothesis. The hypothesis must be tested in a way *that allows it to be proven false.* We can never prove that a hypothesis is true, but we can support the hypothesis if repeated experiments do not falsify it.

> You formulate the following hypothesis: If individuals with high cholesterol levels are treated with a combination of drugs X and Y, their cholesterol levels will be lowered more than for similar groups treated with drug X alone or drug Y alone.

For statistical reasons (to be covered later), you also devise a **null hypothesis** or prediction of what would happen if the experimental treatment has *no* effect.

> The null hypothesis is that cholesterol will be lowered by the same amount with all three drug treatments.

Step 3 Make **predictions** about the results you would expect if the hypothesis is correct. In this way, scientists begin to formulate an experimental design. Hypotheses are often stated in the form of predictions.

> You predict that treating individuals with drugs X and Y in combination will be more effective than treatment with either drug alone. As part of your experimental design, you know that you will need to compare at least three experimental groups, and you may begin to plan how to identify participants for your study.

Step 4 Clearly define the experiment's independent, dependent, and standardized **variables.** The **independent variable** is the factor that is being manipulated in the current experiment.

> The independent variable is the type of drug treatment. (In the experiments you will be performing in this laboratory, you will usually deal with only one independent variable at a time.)

The **dependent variable** is the aspect of the system that is showing some response to the manipulations of the independent variable.

> The dependent variable could be any of the many aspects of an individual's condition that define the difference between life-threatening high cholesterol levels and lower levels typical of healthy individuals.

The **standardized variables,** or **controlled variables,** are all the variables that are held constant between the treatments.

> The way the drugs or drug mixture are administered, the frequency with which the subjects are checked, and the average ages and general health characteristics of the individuals assigned to the treatments are all standardized variables.

Step 5 Define the **experimental treatments.** A treatment is a test group of individuals that are subjected to the same levels of the independent variable.

> The group that gets drug X alone is one treatment, the group that gets drug Y alone is another treatment, and the group that gets both drug X and drug Y is the third treatment.

Step 6 Select materials and identify experimental methods and methods of data collection and analysis, as part of a well-planned **experimental design.** These are incorporated into a **procedure** that tests whether the predicted results occur.

You identify a large group of individuals who have high cholesterol levels but are otherwise healthy. You randomly assign them to groups that will get drug X alone, drug Y alone, and drugs X and Y. You make sure that all the treatments have a fair chance against one another by ensuring that they all have subjects with a similar range of ages, health, previous treatment histories, and so forth. If you do not do this and one treatment ends up with most of the younger, healthier patients and the other with most of the older, sicker ones, it will be impossible to say whether the results of the experiment are due to the drugs or to the biased selection of people entering each treatment.

The experiment should also have a fourth **control treatment.**

In the control group, individuals with high cholesterol get *no* drug treatment but are given placebos and are held to the same standardized variables.

Step 7 Perform experiments and collect data.

You make sure that all participants who will be taking the drugs understand the drug dosages and administration conditions that the experiment demands, that the subjects are checked frequently, and that all medical personnel are using the same definition of high cholesterol. You should pre-plan procedures for dealing with inevitable problems such as patients who miss drug treatments or drop out of the program.

Step 8 Analyze **results** (data) from the experiments that test the hypothesis, using statistics when necessary, and interpret these results to determine whether the hypothesis is supported or falsified. This leads to a **conclusion.**

Step 2 included the statement of a null hypothesis, or prediction of what would happen if no treatment effect occurred. While the null hypothesis may seem negative and uninteresting, it is important because most statistical techniques can only test a null hypothesis. Therefore, you will probably end up concluding either that **the data allowed us to reject the null hypothesis** (meaning that there was a treatment effect), or that **the data did not allow us to reject the null hypothesis** (meaning that there was no evidence of a treatment effect). You *cannot* say, "We proved there was an effect" or "We proved there was no effect." Although it is common to talk about "experimental proof," experiments do not *prove* anything. **Experiments can only offer evidence that either supports, or fails to support, hypotheses.**

Step 9 Repeat the process, using a more refined question about the system.

Assume that the combination of drugs was more effective than either drug taken individually. Next, you might ask if the best results are obtained when the two drugs are given in equal or in unequal amounts. Or, you might ask if a combination of drugs X, Y, and Z is more effective than a combination of just X and Y.

The usual result of an experiment is more questions.

ııııı Objectives ıı

☐ Recognize the stages of scientific inquiry as it applies to everyday experiences.

☐ Make observations, formulate hypotheses, make predictions, and design experiments to test hypotheses.

ııııı Procedure ıı

You have just received a grade on your first major examination in biology, and it is not as high as you had hoped. You wanted an "A" and you earned a "C." You talk to your professor and do a lot of thinking about what might have gone wrong. Perhaps your mistake was reviewing your notes only before the exam rather than every night. Or perhaps you made some other mistakes.

   1. Using this scenario, apply the steps of the scientific method to identify what you might do to raise your grade on the next exam.

Step 1 Make observations that lead to a question. _____

_____

Step 2 Formulate a hypothesis. _____

_____

Step 3 Make predictions based on this hypothesis. _____

_____

Step 4 Define the independent and dependent variables. _____

_____

Step 5 Define the experimental treatments. _____

_____

Step 6 Design your experiment. _____

_____

Step 7 Perform the experiment and collect data. _____

_____

Step 8 Analyze your data. _____

_____

Step 9 Plan a more refined experiment. _____

_____

2. Now, suppose your grade on the next examination remains low. *How could this cause you to revise your conclusions about your original hypothesis?* _____

_____

EXERCISE B　　Reaction Time Experiments: Making Observations

One of the basic features of life is that living things react to stimuli from the environment. This may be as obvious as a frog hopping away as you approach, or as subtle as a plant changing its pattern of hormone secretion in response to increasing day length in the spring.

Reaction time is the length of time it takes to begin a response to a stimulus. Any number of stimuli can evoke specific responses. For instance, seeing movement out of the corner of your eye could cause you to turn your head. Another example might be slamming on the brakes when an animal runs out in front of your car. From your normal day-to-day life, list some stimuli and their associated, observed responses. Be careful not to extend your observations into inferences (*inferences* are explanations or interpretations of observations).

In this experiment, you will use a computer program to test your reaction time. The program presents a stimulus and then uses the computer's internal clock to measure the time it takes you to respond to the

Stimulus	Observed Response

stimulus by pressing the spacebar. If the computer program is not available, you will test your ability to catch a ruler dropped between two of your fingers—can you react fast enough to catch it?

IIIII **Objectives** IIIIIIIIIIIIIIIIIIIIIIIIIIIIIIIIIIIIIIIII

☐ Make observations about responses to stimuli.

☐ Make observations about differences in reaction times.

Note: If you are *not* using a computer for this exercise, skip to Exercise B, Part 2.

PART I   **Making Observations (*for Laboratories Using Computers*)**

IIIII **Procedure** IIIIIIIIIIIIIIIIIIIIIIIIIIIIIIIIIIIIIIIIIIIII

1. Your instructor will provide you with directions for loading the reaction time program.

2. After starting the program, choose *Collect some reaction time data.*

3. Choose *Use just the reaction time program.*

4. The reaction time program will present you with a stimulus and then ask you to respond by pressing the spacebar. The stimuli can be either visual or auditory, and simple or complex. Possible reaction time tests include:

   • **X at a known location** An "X" appears in the middle of the screen.

   • **Spot the dot** A period appears somewhere on the screen. The period may be either high-contrast (white on a blue background) or low-contrast (black on a blue background). The dot may appear on a small area of the screen or anywhere on the whole screen.

   • **Symbol recognition** You are given a list of 1 to 10 letters. Then letters appear in the middle of the screen and you press the spacebar *only* if the letter is on your list. For example, if your list is "A G B," you press the spacebar if you see an A, or G, or a B, but *not* if you see any other letter.

   • **Reaction to sound** The computer sounds a tone.

   • **Tone recognition** The computer sounds either a low tone or a high tone. You press the spacebar only if the *high* tone sounds.

5. Take some time to become familiar with what each test requires.

6. In the space below, record observations about your reaction time on the different tests, possible environmental effects on your reaction time, or differences in reaction times among individuals.

These observations will serve as information for formulating a hypothesis. You will work in a laboratory group of eight (four pairs), so share your observations with others in your laboratory group. Once you are familiar with the program, you and the other members of your lab group will devise a reaction time hypothesis (Exercise C) and an experiment to test your hypothesis (Exercise D). You will then collect and analyze your data (Exercises E–G).

PART 2

Making Observations (*for Laboratories Not Using Computers*)

||||| **Procedure** |||||||||||||||||||||||||||||||||||||||||

1. Work in pairs. Place a ruler (preferably stainless steel) in the crevice formed by the second (index) and third fingers of your partner's preferred hand (right, if right-handed; left, if left-handed). Orient the ruler so that the 1-cm mark is downward (Figure IB-1). Now ask your partner to open the two fingers as wide as possible.

2. Drop the ruler. When caught, record the millimeter marking at the top surface of the fingers.

Figure IB-1   *Positioning of ruler for reaction time experiment.*

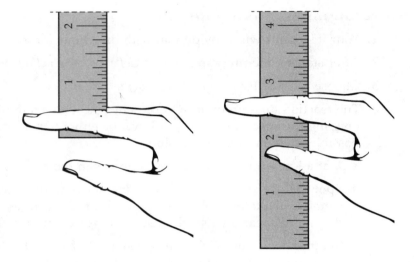

3. Try this technique several times. Make sure that the original position of the ruler, before dropping (step 2), is the same each time. In the space below, record observations about the mechanics and "times" of reactions (measured as number of millimeters) for different partners. Are there any environmental or individual differences that might be affecting reaction times?

These observations will serve as information for formulating a hypothesis. You will work in a laboratory group of eight (four pairs), so share your observations with others at your laboratory table. Once you are familiar with the mechanics of the exercise, you and the other members of your lab team will devise a reaction time hypothesis (Exercise C) and an experiment to test your hypothesis (Exercise D). You will then collect and analyze your data (Exercises E–G).

✔ **EXERCISE C**   **The Reaction Time Experiments: Formulating a Hypothesis**

Use of our general and perhaps nonsystematic observations often leads to devising a question about the observed system. A **hypothesis** is a question, often stated in the form of an educated guess or possible answer (tentative explanation) to a question. Hypotheses lead to predictions—indeed, hypotheses are often stated as predictions. Two additional criteria must be met by a hypothesis:

1. We must have a hypothesis that can be **falsified** (refuted or proven false). If there is no possibility of proving that a hypothesis is false, then it cannot be tested. Why? Because we can never "prove" that a hypothesis is "true." We can only add to the body of evidence that **supports** a hypothesis.

2. The hypothesis must focus on a limited, specific, **well-defined** problem. A hypothesis that is too broad often entails consideration of multiple variables that confuse our interpretation of results.

Formal testing of a hypothesis requires that we distinguish between two alternative possibilities:

1. The variable being manipulated has an effect.

2. The variable being manipulated has *no* effect.

The second possibility ("no effect") is often referred to as the **null hypothesis** because it states the alternative possibility—that no effect occurs. A null hypothesis is a statement of "no relationship."

A null hypothesis must also be falsifiable. If we can falsify the null hypothesis, our results support the hypothesis. If, in testing our hypothesis, we find that the data support the null hypothesis, we need to evaluate both the design of the experiment and the usefulness of the hypothesis. Scientists generally accept the null hypothesis unless they have evidence that causes them to reject it.

ⅠⅠⅠⅠⅠ Objects ⅠⅠ

☐ Formulate a hypothesis and null hypothesis that can be used to investigate differences in reaction times.

Note: If you are *not* using a computer for this exercise, skip to Exercise C, Part 2.

PART I   Formulating a Hypothesis (*for Laboratories Using Computers*)

In this exercise, you will devise and test a hypothesis involving reaction time. For instance, you may wish to investigate whether there is a difference in response time to different stimuli or whether response times differ between two or more groups (male vs. female, athlete vs. nonathlete, etc.).

Example Seated around a laboratory table are six band members. Three of the students are woodwind players and the other three are percussionists. Both groups believe they have the fastest responses. They decide to settle the dispute like scientists, so they develop a simple investigation.

Observation: More and faster finger movement is required to play a woodwind instrument than a percussion instrument.

Hypothesis: Woodwind players will have faster reaction times than percussionists in the test "X at a known location."

Null hypothesis: There is no difference in the reaction times of woodwind players and percussionists.

ⅠⅠⅠⅠⅠ Procedure ⅠⅠ

With others in your group, use your observations from Exercise B to develop a reaction time hypothesis and null hypothesis. After agreement has been reached, supply the information below.

Stimulus used: _____

Observation: _____

Hypothesis: _____

_____

Null hypothesis: _____

_____

👁 **PART 2** **Formulating a Hypothesis** (*for Laboratories Not Using Computers*)

In this exercise, you will devise and test a hypothesis involving reaction time. For instance, you may wish to investigate whether response times differ between two or more groups (male vs. female, athlete vs. nonathlete, etc.).

Example Seated around a laboratory table are six band members. Three of the students are woodwind players and the other three are percussionists. Both groups believe they have the fastest responses. They decide to settle the dispute like scientists, so they develop a simple investigation.

> *Observation:* More and faster finger movement is required to play a woodwind instrument than a percussion instrument.

> *Hypothesis:* Woodwind players will have faster reaction times than percussionists for catching a ruler.

> *Null hypothesis:* There is no difference in the reaction times of woodwind players and percussionists.

⊪⊪⊪ **Procedure** ⊪⊪⊪⊪⊪⊪⊪⊪⊪⊪⊪⊪⊪⊪⊪⊪⊪⊪⊪⊪⊪⊪⊪⊪⊪⊪⊪⊪⊪⊪⊪⊪

With others in your group, use your observations from Exercise B to develop a reaction time hypothesis and null hypothesis. After agreement has been reached, supply the information below.

Stimulus used: _____

Observation: _____

Hypothesis: _____

_____

Null hypothesis: _____

_____

👁 **EXERCISE D** **The Reaction Time Experiments: Developing an Experimental Design**

To test a hypothesis, you have to design an experiment. The design process often begins when you make predictions from your hypothesis. Next, you must identify your experimental variables and treatments. You must also determine how to control all other factors that might influence your results. Finally, you should consider how you want to collect and analyze data.

👁 **PART 1** **Identifying Variables** (*for All Laboratories, Using or Not Using Computers*)

Developing a good experimental design requires the experimenter to define what factors will be varied or held constant during the test. The **independent variable** can be anything the experimenter decides to manipulate (vary)—density, temperature, light intensity, altitude, concentration, to name a few of the hundreds of possibilities. Probably the most commonly used independent variable is time: the experimenter takes measurements once a day, every 2 hours, every 2 minutes, or whatever time period is appropriate for the experiment.

Example Suppose you want to test the effect of time of day on reaction time. Do you react more quickly at 7 A.M. or at noon? The independent variable is time of day.

a. The independent variable in an experiment on the effect of time of day is _____

The **dependent variable** is the aspect of the system that is showing response to the manipulations of the independent variable. It is what is measured (or counted or somehow recorded) at different "settings" or levels of the independent variable. An experiment may have several dependent variables—all of the things that are affected and can be measured as you manipulate the independent variable. In any experiment, however, you should always strive to have only one independent variable at a time.

b. The dependent variable in an experiment on the effect of time of day is ＿＿＿＿＿＿＿＿＿＿＿＿＿

Since an experiment should have only one independent variable, all other things that could vary during the experiment must be kept constant or be controlled. These are the **controlled variables** or **standardized variables.**

c. What would happen if there were more than one independent variable in an experiment on the effect of time of day? (For example, suppose all your morning readings were just after breakfast, but all your noon readings were just before lunch?) ＿＿＿＿＿＿＿＿＿＿＿＿＿＿＿＿＿＿＿＿＿＿＿＿＿＿＿＿＿＿＿＿＿＿

＿＿＿

d. What conditions would you want to keep constant in an experiment designed to measure the effect of time of day on reaction time? ＿＿＿＿＿＿＿＿＿＿＿＿＿＿＿＿＿＿＿＿＿＿＿＿＿＿＿＿＿＿＿

＿＿＿

ııııı Objectives ıııııııııııııııııııııııııııııııııııı

☐ Identify and distinguish between dependent, independent, and controlled variables.

ııııı Procedure ıııııııııııııııııııııııııııııııııııııı

Identify the dependent, independent, and controlled variables for your reaction time experiment.

Dependent variable(s): ＿＿＿＿＿＿＿＿＿＿＿＿＿＿＿＿＿＿＿＿＿＿＿＿＿＿＿＿＿＿＿＿＿＿＿＿

Independent variable: ＿＿＿＿＿＿＿＿＿＿＿＿＿＿＿＿＿＿＿＿＿＿＿＿＿＿＿＿＿＿＿＿＿＿＿＿＿＿

Controlled (standardized) variables: ＿＿＿＿＿＿＿＿＿＿＿＿＿＿＿＿＿＿＿＿＿＿＿＿＿＿＿＿＿

PART 2   Defining Experimental Treatments (*for All Laboratories, Using or Not Using Computers*)

A **treatment** is a test group (or treatment group) of individuals that are subjected to the same levels of independent variable. For instance, you might make the following comparisons:

Same group of subjects given two different tests:

Treatment 1 Women given test A.

Treatment 2 Women given test B.

Two different groups of subjects given the same test:

Treatment 1 Men given test A.

Treatment 2 Women given test A.

Same group of subjects given same test at two different times or under two different conditions:

Treatment 1 Women given test A before lunch.

Treatment 2 Women given test A after lunch.

It is also possible to have more than two treatments:

Treatment 1 Women given test A before lunch.

Treatment 2 Women given test A 1 hour after lunch.

Treatment 3 Women given test A 2 hours after lunch.

Treatment 4 Women given test A 3 hours after lunch.

In the experiments listed above, all subjects in each treatment or test group (treatment group) experience the same manipulation or level of independent variable. But how can the experimenter assure that the outcome is in fact due to manipulation of the independent variable? This is usually done by running a **control** for the experiment.

When comparing two treatments, one treatment group may serve as the *control treatment* and the other as the *experimental treatment* exposed to the independent variable. In before-and-after tests, the "before" conditions often serve as the control treatment. It *is* possible to have no true control group (for example, when you compare the reaction times of men and women).

ⅠⅠⅠⅠⅠ Objectives ⅠⅠⅠ

☐ Identify treatment groups in an experiment.

ⅠⅠⅠⅠⅠ Procedure ⅠⅠⅠ

For your reaction time experiment, define the treatment groups (try to limit your experiment to two treatment groups).

Treatment 1: _____

Treatment 2: _____

a. Do you have a "control" group for your experiment? _____

_____

b. Explain why or why not. _____

PART 3   Defining Data Collection and Analysis Procedures (*for All Laboratories, Using or Not Using Computers*)

Once you know what your variables are and have identified your treatment groups, you should list the steps you will follow to test your hypothesis. You should also consider how you want to collect data; the steps you take to collect data often determine (and limit) the ways in which you can analyze the data.

For the reaction time experiments, you will need to decide what kind of analysis (for statistical purposes) you want to perform on your data. You have a choice of a **paired** or **unpaired** analysis. The following rules will help you decide this question:

> **Paired analysis** The same individuals are used for each treatment (both treatment groups contain the same individuals)—typical of before-and-after experiments.

> **Unpaired analysis** Different individuals are used for each treatment (each treatment group is a different group of individuals).

Example Suppose you want to look at before-and-after treatments. You are working in a group of four students: A, B, C, and D. An unpaired test would randomize the four individuals and treat them as two different treatment groups, comparing all observations in treatment 1 with all observations in treatment 2. It would be better to treat this as a paired test, comparing before and after for the same individuals (compare first observation in treatment 1 with first observation in treatment 2). For a paired test, be sure to test individuals in the same order! (See Appendix I, Part B.)

Unpaired		Paired	
Treatment 1	Treatment 2	Treatment 1	Treatment 2
A	C	A	A
B	A	B	B
C ← random →	D	C ← paired →	C
D	B	D	D

ıııı **Objectives** ıııııııııııııııııııııııııııııııııııııı

☐ Design an experimental procedure.

☐ Determine the type of analysis appropriate for an experimental design.

ıııı **Procedure** ıııııııııııııııııııııııııııııııııııı

1. In the space below, list the steps of your experimental procedure.

2. Determine whether your data analysis should be conducted as a paired or unpaired test.

   Type of analysis: _____

   a. *Why did you choose this form of data analysis?* _____

   _____

There is no substitute for a good experimental design: formulation of a clear, falsifiable hypothesis and null hypothesis, predicting outcomes, defining variables, and identifying treatments. The following suggestions will help you design valid experiments.

Suggestions for Designing Experiments

1. **Use adequate replication.** If you want to generalize your results to a larger group, you should have at least 10 different people in each treatment (when testing one group against another), or at least 10 people (when doing a before-and-after test on the same individuals). This will probably require you to go outside your lab group to get enough people. If you *cannot* get 10 people, you will have to note that small sample size has weakened the conclusions of your experiment. Perhaps you will only be able to draw conclusions about the individuals tested, not broader groups such as all men or all women.

   Regardless of the number of people, collect at least 10 reaction times per person per treatment. For example, if you are testing men versus women, each person should do 10 reaction times. If you are testing effects of caffeine, each person should do 10 reaction times before caffeine and 10 after.

2. **Do not try to make up for a small number of individuals by having each person perform more tests.** All this will yield is increasingly precise estimates of the reaction times of these particular individuals. If Bob is the only man in a men-versus-women experiment and Bob does 100 reaction time tests, this doesn't make Bob any more representative of men as a group.

3. **Avoid bias.** Bias occurs when one treatment has an advantage or disadvantage that has nothing to do with the independent variable. For example, in a men-versus-women experiment, let's say that all the men are athletes, and athletes have faster reaction times. If the results show that men have faster reaction times, is this because these individuals are male or because they are athletes?

   A common kind of bias relates to the time the tests are done. For example, say that in a men-versus-women study, all the women do the test first, then all the men. If the men watch the women, they may learn tricks that will improve their own performance. One way to combat this problem would be to have men and women alternate as they do the tests.

4. **Do not allow fatigue to become a factor.** It's better to have each individual do several short series of reaction time tests rather than complete all tests at one long sitting. This will also help with the "who goes first" problem. For example, say a group of students is doing a men-versus-women experiment and wants to have each individual log 10 reaction times. It would be better to have men and women alternate doing five tests at each sitting rather than have each person try to complete all 10 tests at once.

EXERCISE E   The Reaction Time Experiments: Conducting the Test

Once the experimental design and procedures have been identified and materials have been gathered to conduct your experiments, you are ready to test your hypothesis. Observations and data must be collected in a systematic way. A laboratory notebook or log is a must.

Another important aspect of experimental design is replicating the experiments. If you do something only once, you cannot be sure your results are valid.

a. What is the advantage of increasing the number of replications? _____

_____

You should *not* expect results from test to be the same—you cannot control everything. Some variation will always be present. Most experimental data give us an average result of many experiments carried out under identical—or as nearly identical as possible—conditions.

|||||| Objectives ||||||||||||||||||||||||||||||||||||||

- ☐ Conduct a reaction time test.
- ☐ Collect data in a systematic fashion.

|||||| Procedure *(for All Laboratories, Using or Not Using Computers)* |||||||||||||||||||||||||||||||||

Identify and describe all parts of your reaction time experiment.

Hypothesis: _____

_____

Null hypothesis: _____

_____

Prediction: _____

_____

Independent variable: _____

Dependent variable: _____

Standardized variable(s): _____

Treatment 1: _____

Treatment 2: _____

Procedure: _____

_____

Type of data analysis (paired or unpaired): _____

Note: If you are *not* using a computer for this exercise, skip to Exercise E, Part 2.

PART I   Conducting the Reaction Time Experiments *(for Laboratories Using Computers)*

|||||| Procedure ||||||||||||||||||||||||||||||||||||||||||

1. The directions that follow present some steps for using the computer to study reaction times. Your instructor will provide you with additional information for your specific computer.

2. After starting the program, choose *Collect some reaction time data.*

3. Since you have already worked with a selection of tests and have formulated your hypothesis and designed your procedure, you are ready to conduct your experiment to analyze the data.

(You will do this using a statistical package that is part of the reaction time program.) Choose *Use the reaction time program and then perform an immediate statistical analysis.*

4. Enter how many treatments you are using (usually two, but certainly more than one) and press ENTER.

5. If you want to repeatedly use *exactly* the same reaction time test, answer yes (Y) to the question asking about this. If you want to make the slightest alteration (like using high-contrast spot-the-dot one time and low-contrast another time, or using one-letter symbol recognition one time and three-letter another time), answer no (N) to the question.

6. For the purpose of data analysis, you must decide which kind of statistical analysis you want—a *paired* or an *unpaired* analysis. Indicate which you are going to use. You should use a paired analysis *only* if you have two treatments and will be using the same individuals in the same order, with the same number of observations in each treatment.

7. Choose the reaction time experiment you would like to do. As you set up each experiment, you will be asked to what treatment this group of observations belongs. There is no need to enter treatment 1 first and then treatment 2. If you want to do treatment 2 first, type in that the treatment is 2 as you set up your first experiment.

8. As you finish each set of tests in your experiment, you will be asked if you want to add the observations to the treatment. If the data are valid, indicate yes (Y). Then the next group of tests will begin. There is no need to write down the results of each test. The computer will keep track of the data for you.

9. *Do not* indicate that this is the end of the experiment unless all students (all treatment groups) have completed all of their tests and you want the final statistical analysis.

10. When all data have been collected, the program will send them for statistical analysis. Record your results in the tables below. (You must have at least two treatments.) Chi-square and probability data should be recorded for use in Exercise G.

Treatment	Average

For an Unpaired Analysis

Treatment	Below Median	Above Median

Chi-square = _____

Probability = _____

For a Paired Analysis

Treatment 1 Higher	Treatment 2 Higher

Chi-square = _____

Probability = _____

11. After completing the data tables, continue to Exercise F and Exercise G, Part 1. **You will use the data recorded above for these exercises.**

 PART 2 **Conducting the Reaction Time Experiments (*for Laboratories Not Using Computers*)**

IIIII **Procedure** IIIIIIIIIIIIIIIIIIIIIIIIIIIIIIIIIIIIIII

1. Conduct your "drop the ruler" reaction time experiment, collecting at least 10 reaction times (trials) per person or subject. Remember, if you are conducting a before-and-after paired test, collect 10 values for "before" and 10 for "after" for each individual, keeping the individuals in the same order.

2. Record your data for treatments 1 and 2 in the tables below.

Treatment 1 (Treatment Group 1)

Trial	Subject							
	A	B	C	D	E	F	G	H
1								
2								
3								
4								
5								
6								
7								
8								
9								
10								
Mean								

Treatment 2 (Treatment Group 2)

Trial	Subject							
	A	B	C	D	E	F	G	H
1								
2								
3								
4								
5								
6								
7								
8								
9								
10								
Mean								

3. After completing the data tables, continue to Exercise F and Exercise G, Part 2. **You will use the data recorded above for these exercises.**

EXERCISE F | Presenting Experimental Data

Data consist of information that can be measured or counted and recorded. Various types of tables or graphs are used for presenting data. You must decide how the data are best presented so that your results can be readily communicated to others. Scientists must be able to accurately graph data collected from their experiments, and must be able to interpret graphs showing results from experiments done by other people.

IIIII **Objectives** IIIIIIIIIIIIIIIIIIIIIIIIIIIIIIIIIIIIIIIII

☐ Present experimental data using appropriate formats.

IIIII **Procedure** IIIIIIIIIIIIIIIIIIIIIIIIIIIIIIIIIIIIIIIII

Complete steps 1 and 2 before coming to the laboratory. Step 3 requires data collected during the laboratory period.

1. Data are usually presented visually in the form of **tables, line graphs,** or **bar graphs.** To determine which type of presentation is appropriate for your data, you must first decide whether your experimental variables (dependent, independent, and standardized) are **continuous** or **discrete.** Knowing this, you can present your data in the most effective way. Review Appendix I, Part A, before proceeding to step 2.

2. The data in Table IF-1 are results from an experiment on reaction time. Subjects were exposed to various volumes (decibels) of three types of background music and, while listening, were asked to respond to observing an "X" that would appear at random intervals in the center of their computer screen (see Exercise B, Part 1, step 4).

Table IF-1 Reaction Time (in milliseconds) to "X at a Known Location" Test for Male and Female Subjects with Different Levels of Background Noise

Decibel Level	Country Music		Classical Music		Rock Music	
	Male	Female	Male	Female	Male	Female
60	226	241	259	252	245	238
	243	239	276	250	262	236
	292	237	325	248	311	234
70	218	232	251	243	241	234
	222	219	255	230	245	220
	220	224	253	235	243	225
80	187	224	220	235	215	230
	224	221	257	232	252	227
	221	223	254	234	249	229
90	233	232	266	243	264	264
	235	234	268	245	266	243
	233	241	269	252	267	250
100	246	239	279	250	286	257
	223	221	256	232	263	239
	229	271	262	282	269	289

Many different hypotheses could be tested using the above experiment. After each of the following hypotheses (*a* through *e*), write the null hypothesis and identify the format (table, bar graph, or line graph) you would use to present the data.

    a. Women will have a faster reaction time than men when country music is being played.

      Null: _____

      Presentation format: _____

    b. Increasing the volume of rock music will affect the time it takes to react to the stimulus.

      Null: _____

      Presentation format: _____

    c. Men will be significantly affected by an increase in noise volume from 70 to 80 decibels, but women will not.

      Null: _____

      Presentation format: _____

    d. Musical style affects men's reaction time, but not women's.

      Null: _____

      Presentation format: _____

    e. An increasing volume of rock music will affect a woman's reaction time, but not a man's.

      Null: _____

      Presentation format: _____

3. Prepare a table, bar graph, or line graph for *your* experimental data from Exercise E. Use a separate sheet of graph paper if needed. You will use this information as part of preparing a laboratory report (see Appendix II) if requested by your instructor.

EXERCISE G | **Interpretation: Conducting Statistical Analyses and Forming a Conclusion**

A distinctive quality of scientific thinking is that a hypothesis can never be considered to be "proven" as a result of experiments or observations. Scientists accept ideas, theories, and hypotheses when they cannot show them to be false. They accept the null (no treatment effect) hypothesis unless they have evidence to falsify it.

    Scientists often employ statistics to interpret experimental results and to determine whether the results meet a minimal level of acceptance or should be rejected. This allows us to come to a **conclusion** about a hypothesis. We might then suggest other hypotheses and test them against the null (no-effect) hypothesis.

    The experimental method, designed to be replicated or repeated by others, is the process that precedes the scientific acceptance of a hypothesis. Such acceptance is always provisional. We cannot prove that a hypothesis is true because we can never be certain that we examined all the evidence or considered all possible alternative hypotheses.

    Testing hypotheses leads to more questions. In this way, evidence accrues in support of a hypothesis, which may, after years of testing, gain the status of a theory. A **theory** is a generalization (based on many observations and experiments) that forms a basis for further studies.

    Treating data statistically allows us to decide whether to accept or reject our null and alternative hypotheses. Say that you took the same reaction time test 10 times, then another 10 times, and then another 10 times. There is no reason to believe that your reaction time is changing, but it would be very unusual if all three average (mean) reaction times came out exactly the same. That is, even in the absence of any true treatment, successive groups of trials will have different means due simply to *random variation*. How do we distinguish between random variation and variation caused by true treatment differences? This is the purpose of experimental statistics.

    Statistical tests allow us to generalize from a subset sampled during experimentation to a larger population, based on the probability that chance alone caused the difference observed between treatment groups or samples. Recall that the null hypothesis for any experiment is that there is no difference between the treatment groups or samples. The null hypothesis is useful because it is easily testable by statistics. Statistics give the probability that the results are due to chance and not some real treatment difference between the treatment groups. Arbitrarily, we can define the level at which chance is not affecting our decision. For biologists, this is usually a probability level of $p = 0.05$. This means that we have only 5 chances in 100 of drawing the wrong conclusion for the entire population based on our sample data.

- If the computed probability is low (usually less than 5%) that the difference between two samples or treatment groups is caused by chance, *we can reject the null hypothesis* and accept the alternative hypothesis that there *is* a treatment effect.

- If the computed probability is *high* (usually greater than 5%), *we cannot reject the null hypothesis;* we accept the null hypothesis (no difference) because we do not have enough evidence to declare that there is a treatment effect.

These principles can be seen in the statistical test you will use in this lab, the **chi-square test.** The chi-square test gives us a χ^2 value which, when compared with critical values in a χ^2 table (see Table AI-2 in Appendix I, Part A), allows us to determine the probability that our assumption of no difference or a difference due to chance only (null hypothesis) is a safe assumption. Depending on experimental design, you will use chi-square to perform either an unpaired median test or a paired median test.

If you used the computer reaction time program, the computer has performed the chi-square test for you. If you performed the "drop the ruler" test, you will have to calculate χ^2 using your calculator or a computer made available by your instructor. The reaction time computer program will allow you to enter data and will perform statistical tests for you even though you did not use the computer for the reaction time experiment. (See Appendix I, Part B, for a complete description of the chi-square test.)

IIIII Objectives IIIIIIIIIIIIIIIIIIIIIIIIIIIIIIIIIIIIIIIII

☐ Use chi-square and probability values to determine whether to accept or reject a hypothesis.

Note: If you did *not* use a computer for the reaction time test, skip to Exercise G, Part 2.

PART I   Analyzing Chi-Square Results (*for Laboratories Using Computers*)

IIIII Procedure IIIIIIIIIIIIIIIIIIIIIIIIIIIIIIIIIIIIIIIII

1. After reviewing the chi-square median test (Appendix I, Part B), return to your data tables in Exercise E, Part 1. Record below the chi-square and probability values for your experiment.

    Chi-square = _____

    Probability = _____

2. Determine the degrees of freedom (*df*) for your experiment (see Appendix I, Part B).

    Degrees of freedom = _____

    a. Is the computed value for probability above or below the 5% probability level ($p \geq 0.05$)?

    _____

    b. Do you accept or reject your null hypothesis? _____ Why?

    _____

    _____

3. Examine the chi-square table in Appendix I, Part A (Table AI-2). If your χ^2 value is above that associated with the $p = 0.05$ level, it will be found to the right in Table AI-2; if it is below, it will be to the left. Remember that you are looking at the probability of being *wrong* if you accept the null hypothesis. So, if the probability of being wrong is higher, your χ^2 value will be *smaller* than at $p = 0.05$. In this case, you would accept the null hypothesis because the risk

of being wrong if you reject it is too great. If the probability of being wrong is lower, your χ^2 value is *higher* than at $p = 0.05$. You would reject the null hypothesis, with less than a 5% chance of being wrong to do so.

4. Continue with Exercise G, Part 3.

PART 2   Analyzing Chi-Square Results (*for Laboratories Not Using Computers*)

Procedure ||||||||||||||||||||||||||||||||||||||||||||

1. Using your data from Exercise E, Part 2, compute your χ^2 value. This can be done using a calculator or the reaction time computer program if made available by your instructor. Written directions for calculating chi-square are summarized in Appendix I, Part B. Your instructor will provide you with a worksheet (included in the *Instructor's Guide and Preparation Manual*) for performing the χ^2 calculation appropriate to the type of test (paired or unpaired) you performed.

   Chi-square = _____

2. Determine the degrees of freedom (*df*) for your experiment (see Appendix I, Part B).

   Degrees of freedom = _____

3. Using this *df* value, determine the probability (*p*) from Table AI-2 (Appendix I, Part B).

   Probability = _____ to _____

   a. *Is your χ^2 value above or below the critical value in Table AI-2 (Appendix I, Part B) at the* $p = 0.05$ *level?* _____

   b. *Do you accept or reject your null hypothesis?* _____ *Why?*

   _____

   _____

4. Examine the chi-square table (Table AI-2). If your χ^2 value is above that associated with the $p = 0.05$ level, it will be found to the right in Table AI-2; if it is below, it will be to the left. Remember that you are looking at the probability of being *wrong* if you accept the null hypothesis. So, if the probability of being wrong is higher, your χ^2 value will be *smaller* than at $p = 0.05$. In this case, you would accept the null hypothesis because the risk of being wrong if you reject it is too great. If the probability of being wrong is lower, your χ^2 value is *higher* than at $p = 0.05$. You would reject the null hypothesis, with less than a 5% chance of being wrong to do so.

PART 3   Forming Conclusions

When drawing conclusions from experimental data, it is important not to "overstate" your findings. Suppose a man and a woman try the same reaction time test, and the man has a faster reaction time. Does this mean the experimenters have the evidence to reject the null hypothesis, that they should conclude that men have faster reaction times than women?

Certainly not. This particular man might be faster than the average man, and this particular woman might be slower than the average woman. Furthermore, especially if the difference in reaction times is small, tomorrow the results might be reversed. The experimenters might be able to draw conclusions about these two individuals, but drawing conclusions about men in general and women in general goes way beyond these data.

Objectives ||||||||||||||||||||||||||||||||||||||||||||

☐ Draw a conclusion from your reaction time experiment.

IIIII **Procedure** IIIIIIIIIIIIIIIIIIIIIIIIIIIIIIIIIIIIIIIIIIIIIIIII

Consider your data table or graph, the students you tested, and the computed statistical probability that your results were or were not due to chance alone. In the space below, record what you conclude from your reaction time experiment.

 EXERCISE H | **Writing a Report**

You now have all of the information necessary to write a laboratory report. Directions for writing a report and an example of a student lab report are included in Appendix II.

Laboratory Review Questions and Problems

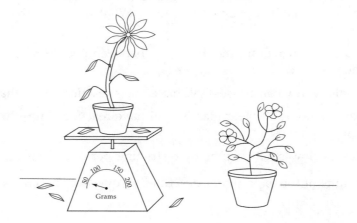

1. For each of the following statements about the situation depicted above, indicate whether it is an observation or an inference.

   a. The total mass of the plant pot and soil is 50 g.

   b. The plant on the balance is not as healthy as the plant on the table.

   c. The larger plant on the table has a greater mass than the one on the balance.

   d. The plant on the balance has three leaves.

   e. The plant on the balance is smaller than the plant on the table.

   f. The plant on the balance has lost some leaves.

2. From the following graph, provide the information requested below.

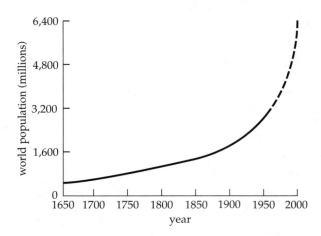

What was the world population in 1900? _____

Title of graph: _____

Dependent variable: _____

Independent variable: _____

Description of information provided by graph: _____

3. For a reaction time experiment, a group hypothesizes that people who wear glasses will have slower reactions than people who don't. What would be the best null hypothesis for this group?

  a. There is no reaction time difference between people who wear glasses and people who don't.

  b. People who wear glasses will have *faster* reaction times than people who don't.

  c. People who don't wear glasses will have reaction times that are half those of people who do wear glasses.

  d. Glasses slow reaction time by restricting peripheral vision.

4. The results of the experiment described in question 3 were as follows.

Treatment	Average
Glasses	0.225
No glasses	0.287

Treatment	Below Median	Above Median
Glasses	14	6
No glasses	6	14

Chi-square = 6.4

Probability = 1%

"Probability = 1%" means there is a 1% chance that:

  a. there is a reaction time difference between people who wear glasses and people who don't.

    b. glasses slow down a person's reaction time.

    c. a difference this large between the glasses and no-glasses groups would have arisen by chance even if the two groups had the same reaction times.

    d. it would be correct to reject the null hypothesis.

5. What would be the best conclusion for the group performing the experiment described in questions 3 and 4?

    a. They must reject their null hypothesis and report that glasses slow reaction times.

    b. The null hypothesis cannot be rejected.

    c. Glasses seem to have no effect on reaction time.

    d. The null hypothesis has been falsified, but the results still do not support the group's original hypothesis.

Observations and Measurements: The Microscope

OVERVIEW

The success of a scientific experiment depends upon several factors: the problem must be well defined, the variables must be identified, and the experimental techniques and equipment used must be appropriate for the method of inquiry.

Understanding and properly applying the methods of scientific inquiry require that you become proficient at observing and recording data accurately. To do this, you need to be familiar with the types of instruments used for experimental work and with proper sampling techniques. During this laboratory period you will learn about the use and care of the **compound microscope** and the **dissecting microscope.** You will prepare living materials for observation and you will learn to use the microscope to measure the size of cells.

STUDENT PREPARATION

Prepare for this laboratory by reading the text pages indicated by your instructor. Familiarizing yourself in advance with the information and procedures covered in this laboratory will give you a better understanding of the material and improve your efficiency. Review Laboratory I, "Science—A Process."

✔ | EXERCISE A | Identifying the Parts of the Compound Microscope

The simplest example of a microscope is a double convex lens of the type that is used as a magnifying glass. In the late 1500s, two Dutch spectacle makers developed the compound microscope. Their device had two convex lenses placed at either end of a tube and was capable of magnifying an object to 10 times (10×) its actual size. Today, developments in microscopy provide scientists with a wide selection of instruments with which to view the smallest organisms and even the components of individual cells. These microscopes range in complexity from the relatively simple models you will use in the laboratory today to highly sophisticated scanning and transmission electron microscopes.

IIIII **Objectives** IIIIIIIIIIIIIIIIIIIIIIIIIIIIIIIIIIIIIII

☐ Locate the optical and mechanical parts of the compound microscope.

☐ Discuss the function of each part of the compound microscope.

IIIII **Procedure** IIIIIIIIIIIIIIIIIIIIIIIIIIIIIIIIIIIIIII

To use the microscope properly, you must first be familiar with the care of this expensive and delicate instrument. Keep the following precautions in mind:

- Always carry the microscope in an upright position. Use one hand to grasp the arm of the microscope; use the other to support the base. The eyepiece (ocular lens) slides into the body tube and could fall out if the microscope is tilted.

- Never place the microscope close to the edge of the lab table or counter. Be sure to place the electrical cord out of the way and not in a position where it could catch and drag the microscope to the floor.

- Use only lens paper for cleaning the lenses. Using your fingers, handkerchief, or other materials could smudge or damage the lenses.

- When you are finished with your observations, turn off the illuminator and rotate the low-power objective into viewing position. Never put a microscope away with the high-power objective in the viewing position.

1. Obtain a compound microscope from your instructor.

2. Study the **optical** system of the microscope, familiarizing yourself with the location and function of each part. The letters in Figure 1A-1 correspond to the parts described below. After locating each part on the diagram, write the name of the part in the lettered space on Figure 1A-1, and identify that part on your own microscope.

Figure 1A-1 *Parts of the compound microscope. Write the names of the parts in the spaces provided.*

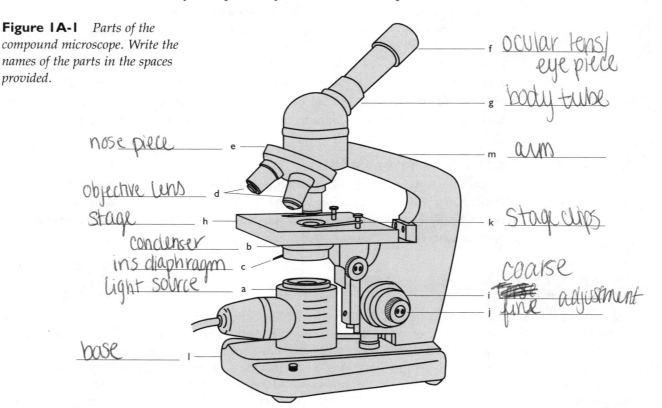

f ocular lens/ eye piece
g body tube
m arm
k stage clips
i coarse adjustment
j fine adjustment

e nose piece
d objective lens
h stage
b condenser
c iris diaphragm
a light source
l base

a Light source May be built into the base with a lens that focuses light onto the lower condenser lens or may be a separate light that is focused onto the condenser lens by a mirror.

b Condenser Contains a system of lenses that focuses light on the object (Figure 1A-2). Some microscopes may not have a condenser, particularly if they do not have a built-in light source. Others have either a movable or a fixed condenser. If your microscope is equipped with a movable condenser, locate the knob that raises and lowers the condenser and circle it on the diagram.

Mechanical stage not shown.

Figure 1A-2 *The microscope condenser focuses light onto the specimen on the microscope stage. Observe the point at which the dotted lines cross.*

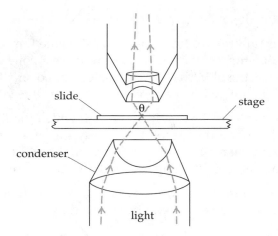

c Iris diaphragm Used to adjust the amount of light striking the object. It can be opened or closed using the lever on the side of the condenser. On some microscopes this function is accomplished by a disk-aperture diaphragm. (Different-sized holes in the diaphragm are used to view objects at different magnifications.)

d Objective lenses Mounted on a revolving **nosepiece** or **turret e.** Most new microscopes are **parfocal;** that is, when an object is in focus with one lens, the lenses can be changed without completely losing focus. Each objective contains a complex lens system. The lens closest to the specimen produces the **magnification.** Magnification is indicated on the side of the objective (Figure 1A-3). The nosepiece usually holds the following objectives (check to see which of these are present on your microscope): scanning objective (4× magnification), low-power objective (10× magnification), high-power objective (40×, 43×, or 45× magnification), and oil-immersion objective (100 × magnification).

Figure 1A-3 *The power or magnification of an objective lens is engraved on the side of the objective.*

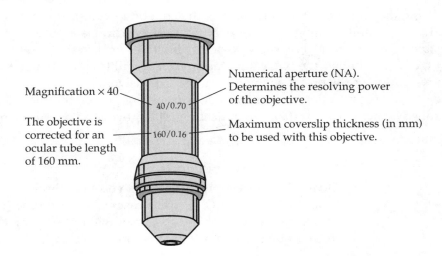

Magnification × 40

The objective is corrected for an ocular tube length of 160 mm.

40/0.70

160/0.16

Numerical aperture (NA). Determines the resolving power of the objective.

Maximum coverslip thickness (in mm) to be used with this objective.

A second lens within the objective is responsible for limiting its **resolving power:** the ability to reveal detail, to distinguish two closely spaced objects as being two rather than one. The smaller the distance between two objects that can be distinguished from one another, the better is the resolving power of the instrument used to view the objects. The unaided human eye can distinguish (resolve) two objects when they are at least 0.1 mm apart, whereas with the light microscope, the human eye can distinguish two objects as separate when they are up to 1,000 times closer than that!

Resolving power (R) is dependent on three factors:

Angular aperture (θ) Examine Figure 1A-2. Note the cone of light entering the objective. The optimum value for θ is the angle that produces a cone of light whose diameter just matches the diameter of the objective. When angle θ is too small, the resolution is poor. One of the functions of the microscope's condenser is to produce the ideal angle of the cone of light.

Refractive index (n) The medium through which the light must travel will affect the shape of the cone of light and thus resolution. Air has a refractive index of $n = 1$. Oil has a greater refractive index than air ($n = 1.5$) and is often used to increase resolution of the microscope at higher powers by increasing the angle (θ) of the cone of light that passes into the objective.

Wavelength of light (λ) The shorter the wavelength of light, the greater is the resolution of the objective. The value of λ can be changed by using colored filters.

The value of R (resolving power) can be determined by the expression

$$R = \frac{\lambda}{2\,[n \sin (\tfrac{1}{2}\theta)]}$$

where

R = resolving power

λ = wavelength of light used

n = refractive index of the medium between the lens and object

θ = angular aperture of light cone

The expression $[n \sin (\tfrac{1}{2}\theta)]$ in the above equation is known as the **numerical aperture** (NA), and the equation can be rewritten as

$$R = \frac{\lambda}{2\,\text{NA}}$$

Note that since λ is the only term in the equation expressed in units, the value of R will be expressed in the same units: nanometers (nm). Numerical aperture is a pure number: it is unitless. The numerical aperture is engraved on the side of all objectives next to the number indicating magnification (Figure 1A-3). The higher the NA value, the smaller R will be. And remember, the smaller R, the better will be the resolution of the objective (and the more expensive!).

As magnification increases, so does resolving power, but the relationship is not linear: resolution always increases less than magnification. Magnification without increased resolution is not advantageous for studying specimens. *a. Why?* _____

_____

f **Ocular lens** or **eyepiece** The lens you look through. It will usually magnify objects to 10 times their size (10×). In some cases, the body tube **g** can be rotated, making it easier for someone else to view the specimen without moving the entire microscope. If your microscope has one ocular, it is mononuclear. If there are two oculars, it is binocular.

3. Study the **mechanical system** in the same manner as you studied the optical system. Letters in Figure 1A-1 correspond to the parts described below. Continue to label the diagram and locate each part on your own microscope.

h **Stage** Holds the slide to be viewed. The stage can be moved vertically by turning the **coarse adjustment knob i** and the **fine adjustment knob j.** These are located in different places on different types of microscopes, either separately or together. Coarse adjustment is used for initial focusing of specimens at low power. Fine adjustment makes very slight changes, allowing precision focusing at higher power.

k Stage clips Hold the slide so that it can be moved by hand. If your microscope does not have stage clips, it will be equipped with a mechanical stage (not shown in Figure 1A-1). Adjustment knobs are used to move the slide in the horizontal plane, that is, side to side and toward and away from you.

l Base and **arm m** Important support parts of the microscope; these also allow for easy carrying.

EXERCISE B | Using the Compound Microscope

Objectives

☐ Learn to use the compound microscope properly.

☐ Learn to position specimens properly and adjust the microscope for optimum use.

☐ Calculate the magnification of a specimen.

☐ Define "resolution" and explain how it applies to obtaining maximum clarity of image.

Procedure

To get maximum performance from the microscope, you will need to adjust the illumination and focus properly.

1. Place the microscope on the table with the ocular pointing toward you. If your microscope has a binocular head (two eyepieces), you will need to adjust the distance between the two oculars to match the distance between your eyes. There is an adjustment dial between the oculars. Turn the dial to change the interocular distance until the oculars are positioned in front of your eyes. When the adjustment is correct, you can comfortably see a single, round field of view.

2. Revolve the nosepiece until the scanning objective (4×) is in line with the body tube. You will hear (or feel) a click when the objective is properly engaged (otherwise you will see only your eyelashes against a black background).

3. Turn on the light switch and adjust the illumination. Higher-power objectives require more illumination. Adjust the *condenser* if it is movable. When using the 4× or 10× low-power objective, the upper lens of the condenser should be about 5 mm below the slide, but when using a 40× (high-power) objective lens, the upper lens of the condenser should be slightly below its uppermost position or about 2 mm below the slide.

   The best way to adjust the condenser is to begin by placing the sharpened tip of a pencil directly on top of the light source. Then, using the condenser adjustment knob, move the condenser up and down until the tip of the pencil is clearly in focus. The condenser should be readjusted with each objective, but the amount of adjustment needed will be slight.

   The iris diaphragm or disk-aperture diaphragm must also be adjusted to obtain the proper balance between contrast and resolution. The amount of light that enters the microscope affects both **contrast** (ability to distinguish something from its background) and **resolution** (ability to distinguish two points as separate). Best observation with the microscope occurs when neither contrast nor resolution is maximized. To adjust the iris diaphragm, remove the ocular lens (or one of the lenses, if the microscope is binocular) and use the diaphragm lever to adjust its leaves until they are just at the edge of the circle of light seen through the open ocular. Then, adjust the leaves so that they are approximately one-third of the way in toward the center of the circle of light. At this setting, the best balance exists between contrast and resolution. Replace the ocular lens before proceeding.

4. Use the coarse adjustment knob to adjust the stage *downward* before placing the slide on the microscope stage. (*Note:* If your microscope has a movable body tube rather than a movable stage, move the body tube *upward*.)

a. Which way did you have to turn the coarse adjustment knob (toward you or away from you) to increase the distance between the objective and the stage? _____

5. Obtain a slide of the letter **e** (or make a wet-mount slide: cut the letter **e** from an old newspaper, add a drop of water, and cover with a coverslip). Place the slide on the microscope stage, making sure that the coverslip is facing upward toward the objective. Hold the slide in place with stage clips or the mechanical stage. If your microscope is equipped with a mechanical stage, do *not* try to place the slide beneath the arms of the slide carrier. Pinch together the two metal extensions of the carrier and its arms will open to allow for insertion of the slide. Now position the slide so that the specimen is directly over the hole in the stage. You will need to do this by hand if your microscope is equipped with stage clips. To move the mechanical stage, use the two knobs beneath the right side of the stage. The upper knob moves the stage backward and forward. The lower knob moves the stage from side to side.

6. To adjust the focus, you will need to move the objective lens using the coarse and fine adjustment knobs. Watch from the side as you adjust the stage upward (or the body tube downward) until the objective lens is almost touching the slide. While doing this, observe the direction in which you are turning the coarse adjustment knob—toward you or away from you. With experience, you will develop a feel for which way to turn the knob. Take care not to bump the slide with the objective lens because this could damage both the objective lens and the slide.

7. While looking through the eyepiece, use the coarse adjustment knob to slowly move the stage downward away from the objective lens (or move the body tube upward away from the stage). When the specimen becomes visible, turn the fine adjustment knob slowly to sharpen the focus. If you are using a binocular microscope, you may have to adjust the focus for each of your eyes. Sometimes the visual acuity of our two eyes is different. After you think you have the letter **e** in focus, close your left eye and fine-tune the focus by turning either the coarse or fine adjustment knob. Once the picture has been sharpened, open your left eye and close your right eye. Turn the binocular focus knob of the left ocular tube until the object is in sharp focus. You can determine how much your specimen is magnified by multiplying the power of the objective lens by the power of the ocular lens. If you are using a binocular microscope, consider the power of only one of the ocular lenses.

| Total magnification = ocular magnification × objective magnification |
---

8. Move the slide to the right.

*b. Which way does the letter **e** move (as viewed through the microscope)?* _____ *c. Is it upside down?* _____ *Is it backwards?* _____ *(It is important to remember these spatial relations.)*

9. Now obtain a prepared slide of *Oscillatoria.* Place the slide on your microscope stage and repeat steps 1 through 8.

10. Rotate the nosepiece so that the 10× objective clicks into place. Again, check the space between the objective lens and your slide. The **working distance** (the space between the objective lens and the slide) decreases with higher-power objectives and increased magnification (Figure 1B-1). If your microscope is parfocal, you will probably be able to bring your specimen into focus by using only the fine adjustment. If you do need to use the coarse adjustment, remember: never turn the coarse adjustment while looking through the ocular. Always view from the side.

   The size of the **field of view** (the area you can see) varies inversely with magnification. The greater the magnification, the smaller is the field of view (Figure 1B-1).

11. Examine a slide of *Spirogyra* or *Oedogonium.* Count the number of cells you can see in one strand at 4× and 10×.

Figure 1B-1 *The field of view and the working distance change with magnification. (When each of these magnifications is used with a 10× ocular, the magnification is multiplied by 10.)*

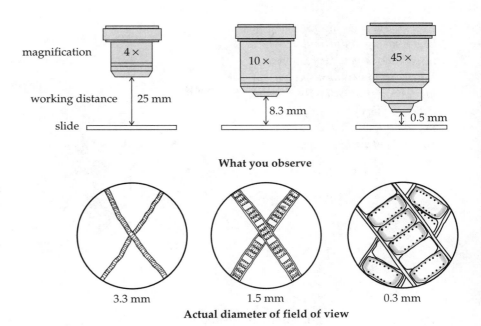

d. What happened to your field of view as you changed from the 4× objective to the 10× objective?

_____

e. Examine Figure 1B-1. What can you say about the relationship between magnification and field

of view? _____

12. Practice moving the slide until you can move it smoothly. Locate something in the field of view that can be followed as the slide is shifted. Attempt to move the object completely around the edge of the field without losing it (manipulations of this type will be particularly important when you view live organisms). Rotate the fine adjustment knob as you move the slide.

f. Do some parts of the specimen appear to be in focus when other parts are not? _____

If you said yes, you are probably correct. Your specimen is not of equal thickness in all places and the depth to which your microscope can focus (called the **depth of field**) is limited. The higher the power or magnification, the shallower is the depth of field. You can study thick objects by continually changing the fine focus, thereby bringing into focus different planes through the specimen. This allows you to "optically section" some materials (Figure 1B-2).

13. When you can operate the microscope successfully using 10× magnification, change to the 40× high-power objective. When using high power, the object to be viewed must be at the center of the field because the high-power objective magnifies only a small portion of the field of view observed under low power (Figure 1B-1).

14. Again, to avoid hitting the slide with the objective lens, watch from the side as you switch to the high-power objective. You may need to use the fine adjustment knob to focus the specimen. Remember: never use the coarse adjustment knob with higher power.

g. What is the total magnification of your specimen with the 40× objective in place? _____

15. On a separate sheet of paper, accurately draw what you observe at this magnification. Drawings should always be done in pencil and be labeled to indicate the name of the specimen and the total magnification.

Figure 1B-2 *This schematic shows how focusing at different depths reveals the structure of an object. When viewing a thick specimen at a higher power (40× or 100×), you may use this optical sectioning process to examine the specimen's structure at various depths.*

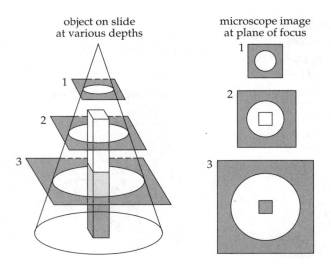

object on slide at various depths

microscope image at plane of focus

Below are a few additional suggestions for using a microscope.

- Be certain that you have the proper amount of light.
- Try to keep both eyes open. This will take some practice on your part, but will be less tiring for your eyes.
- Always clean ocular and objective lenses with lens paper before use.
- If you wear glasses, remove them.
- Always begin by using a low-power objective lens to find the specimen. You can then turn to a higher power to make your observations.
- If you are having difficulty locating the specimen, use a systematic pattern to search the slide.
- If all else fails, ask your instructor for help.

👁 EXERCISE C │ Preparing a Wet-Mount Slide

A wet-mount slide is a temporary preparation. Specimens are mounted in a drop of liquid and are covered with a coverslip.

IIIII Objectives IIIIIIIIIIIIIIIIIIIIIIIIIIIIIIIIIIIIIIIIIIIII

☐ Learn the proper technique for making a wet-mount slide.

IIIII Procedure IIIIIIIIIIIIIIIIIIIIIIIIIIIIIIIIIIIIIIIIIIII

1. Place a drop of water in the center of a clean microscope slide. To the drop, add a torn piece of an *Elodea* leaf.

2. Place one edge of a coverslip at the edge of the water drop and gently lower it so that the water containing the specimen completely spreads out under the coverslip (Figure 1C-1). Take care not to trap bubbles under or around the specimen. Do not press down on the coverslip. If there is too much water, draw off the excess by touching the corner of a paper towel to the edge of the coverslip.

3. Examine your wet mount under low power and then under high power. Remember, care must be taken to avoid touching the coverslip with the objective lens. Not only will it move your specimen, it might break the coverslip or scratch the objective lens.

Figure 1C-1   *Technique for*
preparing a wet-mount slide.

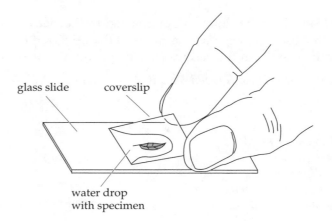

glass slide   coverslip

water drop
with specimen

a. Give three reasons for using a coverslip when preparing a wet mount.

_____

4. Use the 10× objective to focus the torn edge of the *Elodea* leaf. Notice that only part of the whole thickness of the leaf is in focus. Using the fine adjustment knob, focus on various planes throughout the thickness of the leaf. Note that the leaf is more than one cell layer thick.

5. Now switch to your 40× objective. You should notice that even less of the leaf's thickness is in focus at any one setting. What you have just accomplished is a demonstration of a principle of magnification that was discussed earlier: the higher the magnification, the shallower is the depth of field (depth of the area that is clearly focused).

6. With the *Elodea* leaf still in position, notice the movement of the little green bodies inside each cell (Figure 1C-2). These are chloroplasts, organelles responsible for photosynthesis in plant cells. The movement you observe is called **cyclosis,** or cytoplasmic streaming. As the cytoplasm moves around the large central vacuole, it carries with it dissolved substances as well as suspended organelles. Does cyclosis occur in the same direction in all cells?

Figure 1C-2   *A cell of an* Elodea
leaf. The small round bodies in the cell
are chloroplasts.

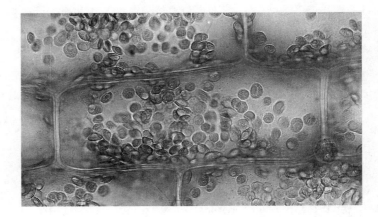

b. What might be the practical advantage of cyclosis to the cells of the leaf?

_____

7. Now replace the liquid around your specimen with a concentrated salt solution (Figure 1C-3). Place a drop of the concentrated salt solution at the edge of your *Elodea* wet mount. Draw the

solution under the coverslip by placing the corner of a paper towel at the opposite edge. Watch closely to see what happens to the cells of the leaf. You will probably see the cytoplasm shrink away from the cell walls. This phenomenon is called **plasmolysis.**

c. *Does cyclosis continue or does it stop?* _____

Figure 1C-3 *The solution under a coverslip can be changed without removing the coverslip by using a piece of torn paper towel to absorb the fluid at one corner edge of the coverslip, and introducing new solution with a pipette or eyedropper near the opposite corner.*

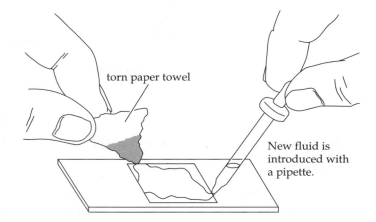

torn paper towel

New fluid is introduced with a pipette.

EXTENDING YOUR INVESTIGATION: MEASURING CYCLOSIS

In steps 6 and 7 of this exercise, you observed cyclosis. Do you think that environmental conditions can affect natural functions of cells, including such processes as cyclosis? *Does temperature affect cyclosis? Does light affect cyclosis?* Choose one of these questions and formulate a hypothesis to explore it. (Refer to Laboratory I, Science—A Process.)

HYPOTHESIS:

NULL HYPOTHESIS:

What do you **predict** will happen to the rate of cyclosis for the conditions you have chosen to investigate?

Identify the **independent variable** in this experiment.

Identify the **dependent variable** in this experiment.

Design an experimental procedure to test your hypothesis. (How might you measure cyclosis? You might consider the rate of chloroplast movement as a measure of cyclosis activity.)

PROCEDURE:

Now, determine the rate of cyclosis. (How many measurements should you make?)

RESULTS:

From your results, describe what happened to the rate of cyclosis given the environmental conditions you chose. (Graphing your data may make your results easier to interpret.)

Do your results support your hypothesis?

Your null hypothesis?

Was your prediction correct?

What do you **conclude** about the effects of varied environmental conditions on cyclosis?

8. If pond water, algal, or protozoan cultures are available, make wet mounts of these by placing one drop of culture medium on a clean slide and applying a coverslip. Some of the unicellular organisms you will see are protozoans, which may move very rapidly across the microscope's field. It may be necessary to slow these organisms by adding a drop of methyl cellulose (Protoslo) to the culture drop (mix with a toothpick).

On a separate piece of paper, draw, in as much detail as possible, one or more of the organisms you observe and insert your drawing into the laboratory manual.

9. Turn off the microscope light, coil the cord as directed by your instructor, and return your specimens and microscope to the proper place.

✔ **EXERCISE D**   **Measuring the Size of Objects Using the Compound Microscope**

The microscope can be used as a tool to gather quantitative data in addition to serving as an instrument for making qualitative observations.

||||| **Objectives** |||||||||||||||||||||||||||||||||||||||

☐ Determine the diameter of the field of view for microscope objectives.

☐ Estimate the size of an object from the diameter of the field of view.

☐ Accurately determine the size of an object using an ocular micrometer.

||||| **Procedure** ||||||||||||||||||||||||||||||||||||||||||

The size of objects viewed with the compound microscope can be estimated by first determining the diameter of the field of view (see Exercise B) for a particular microscope objective and then estimating the size of the specimen by comparing it with the total diameter of the field of view.

1. Place a transparent ruler across the field of view under scanning power and record the diameter in millimeters: _____ What is the diameter in micrometers? _____ The diameter of the field of view using the scanning objective (A) can be used to calculate the

diameter using any other objective (B) (*Recall:* total magnification = objective magnification × ocular magnification):

$$\frac{\text{Total magnification A}}{\text{Total magnification B}} \times \text{diameter A } (\mu m) = \text{diameter B } (\mu m)$$

2. Calculate the diameters of the fields of view using the other objectives on your microscope.

Objective	Diameter (μm)

3. Obtain a prepared slide of *Nostoc*, a type of cyanobacterium. Estimate the length of one cell in micrometers (*Hint:* Use the diameter of the field of view to determine the length of a strand of cells, then divide by the number of cells): _____ μm

   a. *If a cell measures 10 μm at 100×, what is the length at 200×?* _____

4. To determine the diameter of the field of view more accurately, you can use an **ocular micrometer,** a glass disk with a scale etched into it. It is placed in the ocular of the microscope and is visible when you look through the ocular. The ocular micrometer appears as in Figure 1D-1a. Notice there are numbered divisions, but no units per division.

   Due to variations among microscopes, an ocular micrometer must be **calibrated** for the microscope with which it will be used. A **stage micrometer** is used to calibrate an ocular micrometer. The stage micrometer (Figure 1D-1b) looks like a microscope slide but has a standard scale etched into it. The smallest divisions are 0.01 mm in length. It is just like a tiny ruler!

Figure 1D-1   *Ocular micrometer and stage micrometer.*

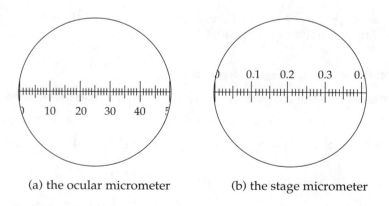

(a) the ocular micrometer   (b) the stage micrometer

You can calibrate an ocular micrometer as follows. Using your scanning objective (4×), look through the ocular containing the micrometer and focus on the stage micrometer. Now move your objective to low power (10×) and refocus on the micrometer using the fine adjustment knob. The two scales should appear to be superimposed on one another. Move the stage micrometer to match up its left end with the left end of the ocular micrometer

Figure 1D-2 *Ocular and stage micrometers are aligned for calibration. (For clarity, the stage and ocular micrometers are shown separated here; as you look through the ocular, the two scales should appear superimposed.)*

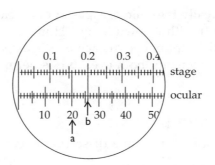

(Figure 1D-2). The actual distance between the lines of the ocular micrometer can be calculated by finding a line on the ocular scale that aligns exactly with a line on the stage micrometer scale.

Example   In Figure 1D-2 observe that 20 spaces (**a**) on the ocular micrometer equal 0.16 mm and that 26 spaces (**b**) equal 0.2 mm. Therefore, 1 space on the ocular scale equals:

$$\frac{0.16}{20} \text{ or } \frac{0.20}{26} = 0.008 \text{ mm} = 8 \text{ } \mu m$$

Your instructor has already calibrated the ocular micrometer (using the 10× objective) in one of the microscopes on demonstration. Note that the value for a division varies for different objectives. *b. Why?* _____

You will need to recalibrate the ocular micrometer for each objective.

c. 10× objective: 1 division on the ocular micrometer = _____ mm = _____ μm

d. 43× objective: 1 division on the ocular micrometer = _____ mm = _____ μm

It is now possible to measure the size of a specimen by viewing it through an ocular containing an ocular micrometer that has been calibrated.

5. Remove the stage micrometer. Using low power, locate a filament of *Nostoc* and bring it into clear focus. Switch to high power and, as accurately as possible, count the number of ocular micrometer divisions from one end of a cell to the other or the number of cells that reach from one division to the next. (Each cell is approximately the same size.) For example, if you had a cell that fit between the 0 and the 25 on your ocular, you could then determine its length. If you had calibrated the ocular at 43× and each division was equal to 2 μm, then to determine the length of the cell you would multiply the number of divisions the cell covered by the measurement of each division: 25 × 2 μm = 50 μm.

   You determined that each division of the ocular micrometer equals _____ μm.

   Therefore, the length of one *Nostoc* cell = _____ μm. Show your work in the space below.

✔ **EXERCISE E**   **The Stereoscopic Dissecting Microscope**

Most of the biological specimens you observe through the compound microscope are very thin and can almost be considered two-dimensional. Larger specimens, including whole organisms, can be viewed more easily using the **stereoscopic microscope**. Dissection is often done using this microscope. Thus, it is sometimes called a **dissecting microscope**.

The dissecting microscope is really two microscopes in one—each eyepiece has its own objective and ocular lenses. Each eye is presented with its own image. Human beings ordinarily see in three dimensions by virtue of receiving a slightly different image in each eye. This possibility is lost in the compound microscope with its single objective lens for both eyes, but is permitted by the design of the dissecting microscope.

ⅠⅠⅠⅠⅠ Objectives ⅠⅠⅠⅠⅠⅠⅠⅠⅠⅠⅠⅠⅠⅠⅠⅠⅠⅠⅠⅠⅠⅠⅠⅠⅠⅠⅠⅠⅠⅠⅠⅠⅠⅠ

☐ Locate and describe the function of each part of a dissecting microscope.

☐ Learn to use the dissecting microscope properly.

ⅠⅠⅠⅠⅠ Procedure ⅠⅠⅠⅠⅠⅠⅠⅠⅠⅠⅠⅠⅠⅠⅠⅠⅠⅠⅠⅠⅠⅠⅠⅠⅠⅠⅠⅠⅠⅠⅠⅠⅠⅠⅠⅠⅠ

1. Use your knowledge of the compound microscope to identify the parts of the dissecting microscope. Write the names of the parts on the lines provided in Figure 1E-1: **eyepiece** or **ocular, focus knob, body tube, arm, objective, stage,** and **base.**

Figure 1E-1   *The stereoscopic, or dissecting, microscope.*

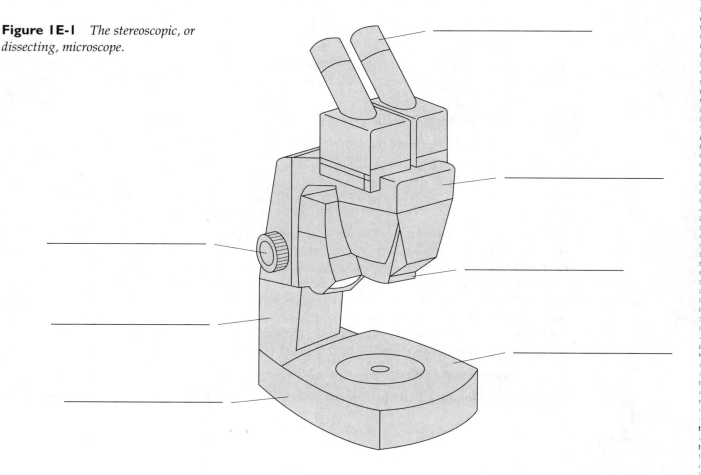

2. Find the **magnification changer** on your microscope. Most, but not all, dissecting microscopes have this feature. If present, it will be located on the side or top of the microscope. Draw the magnification changer on Figure 1E-1. What final magnification can your dissecting microscope achieve?

3. The **light source** for your microscope may be a separate light. Light can be **transmitted** through a specimen on the microscope stage by proper adjustment of the mirror below the

stage. Light can be **reflected** from the surface of a specimen by positioning the light source above the specimen so that it shines downward onto the microscope stage. If your illuminator is built into the microscope, light will be adjusted by two switches, the **transmitted light** switch and the **reflected light** switch. Familiarize yourself with each of these and add them to Figure 1E-1.

4. Select a three-dimensional specimen from those provided. Center your specimen on the stage plate.

5. Grasp both ocular housings and move them together or apart to adjust the spacing until you are comfortable using both eyes simultaneously to view a single field.

6. Try using both reflected and transmitted light to illuminate your specimen.

   a. *Which works better?* _____ *Why?* _____

   _____

7. Use the coarse adjustment knob to focus. There is no fine adjustment knob on the stereoscopic microscope.

8. If possible, change to a higher magnification and refocus as necessary to view different portions of your specimen.

9. Sketch your specimen in the space below.

10. Turn off the microscope light, coil the cord as directed by your instructor, and return your specimens and microscope (and its light) to the proper place.

Laboratory Review Questions and Problems

1. What is the resultant total magnification of an object as seen through a microscope with 10× oculars and each of the following objectives? *a.* 4× _____ *b.* 10× _____ *c.* 43× _____

2. Describe the difference between magnification and resolution.

3. Which part of the microscope is most important in determining its resolving power? Why?

4. You are interested in purchasing a microscope. The salesperson shows you several microscopes, each with one of the following objectives. Which would you want to purchase?

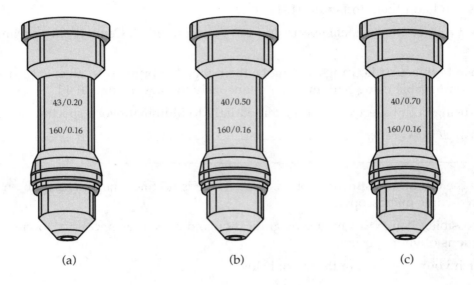

(a)   (b)   (c)

5. Briefly describe the function of each of the following parts of the microscope.
Objective

Ocular

Iris diaphragm

6. In which direction do you turn the coarse focus adjustment knob to move the objective *away* from your slide? _____ In which direction do you turn the fine adjustment? _____

7. What is the purpose of the condenser on a microscope? Relate this to the angle θ and to the numerical aperture (NA) of the objective.

8. The resolution of a microscope objective depends on numerical aperture. Is the resolution better with a high NA or low NA?

9. You are using a microscope with a 100× objective that has a numerical aperture of 0.71. You place a dot of oil ($n = 2$) on the coverslip of the slide you want to observe. What is the resolving power of the 100× objective if you are using light of wavelength 560 nm?

Which is preferable: a large or small R? _____ Why?

10. Suppose your 10× objective has an NA of 0.25 and you are using a green filter that limits the wavelength of light striking the objective to 560 nm. What is the resolving power for this objective? _____ What type of dimensions does your answer have? _____ This is the theoretical value for *R*. Suppose $\theta = 26°$. What would be the true resolution for your system? _____

11. What is meant by *field of view?* What will happen to the field of view for each resultant magnification as you change objectives from 4× to 10× to 43×?

12. Using your 4× objective, you measure the diameter of the field of view to be 3 mm. You notice that the length of a large letter **E** takes up half the width of the diameter of your field at 10×. The oculars on your microscope are 10×. What is the size of the letter **E**? Draw what this letter would look like through the microscope.

13. You use an ocular micrometer to measure the size of a cell, as seen below. The stage micrometer is divided into 0.1-mm units. Which of the following accurately represents the size of the cell? *a.* 3.2 μm; *b.* 350μm; *c.* 35 mm; *d.* 0.0035 mm.

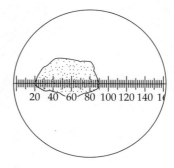

Observations and Measurements: Measuring Techniques

OVERVIEW

We learn about the world by making observations and comparing them with other observations. The major difference between the types of comparisons we make in everyday life and in the discipline known as science is that science requires a more rigorous, focused, and systematic approach. By using a common system of units of measurement and by making precise and accurate measurements, scientists ensure that procedures and results can be reported and repeated anywhere in the world.

In this laboratory you will become familiar with techniques used to make accurate measurements.

STUDENT PREPARATION

Prepare for this laboratory by reading the text pages indicated by your instructor. Familiarizing yourself in advance with the information and procedures covered in this laboratory will give you a better understanding of the material and improve your efficiency.

✔ EXERCISE A   The Metric System

The two systems of measurement in common use are the English system and the metric system. Most scientific data are expressed in units of the metric system. The modern metric system is called the **International System of Units (SI).**

||||| Objectives |||||||||||||||||||||||||||||||||||||||||||||

☐ Identify the units of the metric system and make conversions among these units.

☐ Express measurements in scientific notation.

☐ Distinguish between accuracy and precision.

☐ Report data in significant figures.

||||| Procedure |||||||||||||||||||||||||||||||||||||||||||||

In the metric system, the basic units are the **meter (m)** for length, the **liter (l)*** for volume, and the **gram (g)** for mass.

---

*This has been changed to L in the SI system of measurements, but l is still the notation biologists use most frequently.

Table 2A-1   Commonly Used Metric Units

Prefix	Meaning	Size Relative to Base Unit	Length	Volume	Mass
nano-	billionth	10^{-9}	nanometer (nm)	nanoliter (nl)	nanogram (ng)
micro-	millionth	10^{-6}	micrometer (μm)	microliter (μl)	microgram (μg)
milli-	thousandth	10^{-3}	millimeter (mm)	milliliter (ml)	milligram (mg)
centi-	hundredth	10^{-2}	centimeter (cm)		
kilo-	thousand	10^{3}	kilometer (km)		kilogram (kg)
mega-	million	10^{6}			
giga-	billion	10^{9}			

1. The prefixes used with the basic metric system units indicate either a fraction of a unit or a multiple of a unit, depending, in part, on the size of what is being measured and the degree of accuracy of the measurement. (See Exercise C, Part 4.) Metric prefixes always express a power of 10 by which the basic unit has been multiplied; see Table 2A-1.

    Convert the following measurements from one unit of metric measurement to another as indicated.

    1 nm = $\underline{1.0 \times 10^{-9}}$ m   25 μm = $\underline{.0025}$ cm

    1 cm = $\underline{10}$ mm   10 cm = $\underline{100}$ mm

    1 mm = $\underline{1,000,000}$ nm   12 l = $\underline{12,000}$ ml

    1 m = $\underline{1,000,000}$ μm   22 μl = $\underline{.022}$ ml

    1 μm = $\underline{.0001}$ cm   250 μm = $\underline{250,000}$ nm

    1 mg = $\underline{1000}$ μg   10 μg = $\underline{10,000}$ ng

    1 mg = $\underline{1.0 \times 10^{-6}}$ kg   9 μm = $\underline{9.0 \times 10^{-6}}$ m

    1 l = $\underline{1,000,000}$ μl   50 g = $\underline{50,000}$ mg

2. Compare metric units with the English units given in Table 2A-2.

Table 2A-2   English and Metric Equivalents

English	Metric
1 inch (in)	2.54 cm
0.4 in	1 cm
39.37 in	1 m
0.62 mile (mi)	1 km
1 mi	1.6 km
1 fluid ounce (fl oz)	29.57 ml
1 quart (qt)	946.4 ml
1.05 qt	1 l
1 ounce (oz)	28.53 g
1 pound (lb)	453.6 g
2.2 lb	1 kg

$5 \text{kg} \times \dfrac{2.2 \text{lb}}{1 \text{kg}} = 11 \text{ lbs}$

$2 \text{oz} \times \dfrac{28.53 \text{g}}{1 \text{oz}} = 57.06 \text{ grams}$

a. If you were dieting, would you rather lose 10 pounds or 5 kilograms? $\underline{5 \text{kg}}$

b. Would you rather have 50 grams or 2 ounces of chocolate (assuming that you love chocolate and are not dieting)? $\underline{2 \text{oz}}$

c. *Would you be likely to get a speeding ticket if you were driving 85 km/hr in a 55 mi/hr speed zone?*
NO, 52.7 mi/hr

d. *If 2 qt or 2,500 ml of orange juice costs $1.79, which package would be the better bargain?* 2500 mL OJ

e. *If you have a cup that holds 50 ml of coffee, would it be full if you poured 50,000 μl of coffee into the cup?* YES, = 50,000 uL also

3. When measuring the number, mass, length, or volume of objects, scientists write numbers in **scientific notation.** A number is expressed in scientific notation when it is written as a product of a decimal number between 1 and 9 and the number 10 raised to the proper power. For instance, the number 100 can be written as 1×10^2, 34,698.5 can be written as 3.46985×10^4, and 0.0069 as 6.9×10^{-3}. (*Note:* When a number is larger than 10, the decimal point must be moved left. When the decimal point is moved left, the power of 10 is positive and equal to the number of decimal places moved; for example, $20 = 2.0 \times 10^1$. When the number is smaller than 1, the decimal point must be moved right, and the power of 10 is negative and equal to the number of decimal places moved; for example, $0.01379 = 1.379 \times 10^{-2}$.) Convert the numbers in the following table to scientific notation, and then convert to the metric units indicated in the second column.

	Scientific Notation	Metric Unit Conversion
0.00013 g	1.3×10^{-4} g	.13 mg
0.00000625 1	6.25×10^{-6} 1	.00625 ml
2,323,000 m	2.3×10^6 m	2.3×10^{12} µm
10 µg	1.0×10^1 µg	1.0×10^{-8} kg
1,654 km	1.6×10^3 km	1.6×10^6 m

4. When reporting measurements, it is permissible to estimate one digit beyond the precision of the measuring device. For example, the "ruler" shown in Figure 2A-1 is marked in centimeters. The length of line A could be reported as 7.5 cm by estimating the distance covered beyond the 7-cm mark. This measurement has two significant figures, one certain (7) and one estimated (uncertain) (0.5). All certain digits, plus the next estimated digit, are called **significant figures.** Line B is exactly 7 cm long. This may be reported as 7.0 cm, again using one certain (7) and one estimated (0) digit. To report the length of line B as 7.00 would *not* be correct because only one estimated digit can be reported.

f. *How many significant figures would be used in reporting the length of line C?* 2 *Are both digits in front of the decimal point significant?* yes

Figure 2A-1

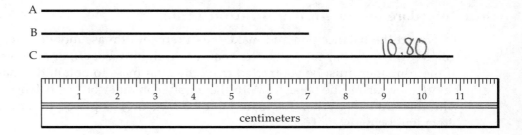

A

B

C   10.80

centimeters

    g. *Look at a ruler. If the length of an object is to be reported in centimeters, what is the maximum number of significant figures to the right of the decimal point in any measurement made with this instrument?* _____

    h. *If a balance can measure mass to an accuracy of ±0.01 g, what is the maximum number of significant figures if the object being measured is slightly heavier than 100 g?* _____ *If, using the same balance, you find that an object has a mass of 100 mg, how many figures in this measurement can be considered significant?* _____

5. Typically, length is measured with a ruler or meter stick. Your instructor will provide a set of objects to measure in centimeters. Use the maximum number of significant figures for each measurement and record your data in the following table. When you have completed each measurement, convert it to millimeters and meters.

	Length				Length		
Object	cm	mm	m	Object	cm	mm	m
1 pump		18.6		6			
2 ball		13.1		7			
3 calcula-tor		18.2		8			
4 book		27.7		9			
5				10			

At the end of this laboratory, you will use these data to construct a histogram.

✔ **EXERCISE B** **Measuring Mass**

Mass is measured by using a balance or scale. Most measuring of mass (commonly called weighing) that you will do in this course will be in the range of tenths of a gram through tens of grams.

    The **gram** (**g**) is the standard unit of mass. It is defined as 1/1,000th of the mass of a 1-kg block of platinum-iridium alloy. The commonly used subdivisions of the gram are the **milligram** (**mg**) (1,000 mg = 1 g) and the **microgram** (μ**g**) (1,000,000 μg = 1 g). Measuring devices sensitive enough to measure milligram or microgram amounts may not be available in your laboratory.

||||| **Objectives** ||||||||||||||||||||||||||||||||||||||

☐ Learn to use a balance.

☐ Explain what is meant by "taring" the balance.

||||| **Procedure** |||||||||||||||||||||||||||||||||||||||

To protect the balance pan and make measurement of mass more convenient, a piece of paper (usually called "weighing paper"), a plastic boat (or "weighing boat"), or another container may be used. The mass of the container must be subtracted from the total mass to obtain the mass of the material itself. This can be done by **taring** the balance. With the empty container on the balance pan, the balance is set to zero with the tare knob. When the material to be measured is placed in the container, the mass of the material alone will be registered.

When using balances without a tare knob, you must first measure and record the mass of the paper or container and add it to the mass of the material you want to measure. You then add material to the container until you obtain the total amount calculated. Alternatively, subtract the mass of the container from the total mass of the container plus material.

Example The paper has a mass of 0.2 g and you wish to measure 2 g of NaCl. You must add NaCl until the balance reads 2.2 g.

Example You are given a beaker of NaCl and asked to determine the mass of the NaCl. First determine the mass of the beaker with the NaCl in it, then remove the NaCl, determine the mass of the beaker, and subtract the mass of the beaker from the first measurement.

1. Examine the balance available in your laboratory.

    a. What is the maximum mass that can be measured? _____ *b. What is the maximum number of significant figures that can be used to report data?* _____

2. Place a piece of paper or a plastic boat on the balance pan and read the result. Tare the balance, or record this mass if no tare knob is present.

3. Obtain three rubber stoppers (or other objects provided by your instructor). Determine and record the mass in grams of each stopper. The manufacturer's specifications indicate the accuracy of the balance you are using. Record the possible error with each of your measurements in Table 2B-1.

Table 2B-1   Measuring Mass

Object	Mass	Possible Error	Potential Range of Values
1			
2			
3			

EXERCISE C   Measuring Volume

Life, as you know, depends on water and the types and concentrations of particles dissolved in it. To determine the concentration of dissolved substances, you must be able to measure volume accurately. A number of devices are available to measure volume, but the most common are the pipette, graduated cylinder, and volumetric flask.

Objectives

☐ Measure volume and report data accurately and precisely.

☐ Select and use appropriate measuring devices.

PART I   The Pipette

Pipettes are usually used to measure liquid volumes of 10 ml or less, although some types of pipettes can measure up to 100 ml or more. Pipettes are of two types: **to deliver** (TD) and **to contain** (TC). The label near the upper end of the pipette shaft indicates the type of pipette. Most pipettes are the TD type. There are two varieties of TD pipettes: "delivery" and "blow out." Examples are shown in Figure 2C-1. Notice that on the delivery pipette (Figure 2C-1a), the scale stops before the pipette narrows. When using this type of pipette, it is important to deliver liquid only down to this line. The blow-out pipette (Figure 2C-1b) is graduated to the tip, so all of the liquid must be delivered, usually by blowing out (**to**

Figure 2C-1   *Types of pipettes: (a) to deliver (TD) delivery pipette, (b) to deliver (TD) blow-out pipette.*

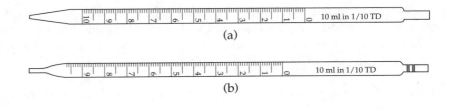

(a)

(b)

Figure 2C-2   *To contain (TC) pipette (λ = μl).*

deliver/blow-out pipettes are usually distinguished from other types by two frosted rings on the upper part of the pipette shaft).

   When using a to contain or TC pipette, simply allow the liquid to flow out of the pipette until the flow stops. The pipette is *not* blown out and a small amount of liquid will remain in the end of the pipette, but this has been taken into account in calibrating the pipette. Some micropipettes (Figure 2C-2) are calibrated in this manner.

   Before using any type of pipette, you need to determine how the pipette is calibrated. Both the largest and smallest volumes that a pipette can measure are marked on the upper end of the pipette. The first or larger number gives the total volume and the second, smaller number indicates how the pipette is calibrated into smaller dimensions—much like a centimeter ruler. For example, "5 ml in 1/10" means that the pipette measures 5 ml from the line marked "5" to the line marked "0" (delivery pipette) or from the line marked "5" to the tip (blow-out pipette). Each large line marks 1 ml and each smaller division represents 1/10 ml.

ⅢⅢ Procedure ⅢⅢⅢⅢⅢⅢⅢⅢⅢⅢⅢⅢⅢⅢⅢⅢⅢⅢⅢⅢⅢⅢⅢ

The **propipette** is a mechanical device used to draw liquids into a pipette. (It is particularly useful when working with hazardous materials.) There are two types: the bulb type (Figure 2C-3a) and the plunger type (Figure 2C-3b). Your instructor will demonstrate the use of one or both types. *Note:* In this laboratory you will **NOT** pipette by mouth. You will always use a propipette or rubber bulb.

Figure 2C-3   *Types of propipettes. (a) Bulb type:* Ⓢ *is pinched to suck up liquid and* Ⓔ *is pinched to release the liquid. If a blow-out pipette is used,* Ⓐ *must also be squeezed to remove all liquid. (b) Plunger type in which a wheel is turned to suck up or release liquid; a plunger can be used for rapid dispensing.*

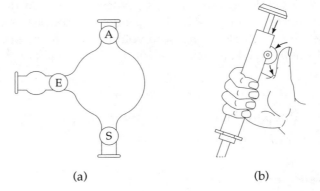

(a)   (b)

1. Examine the pipettes on demonstration in your laboratory room. Notice the variety of types and scales. <u>Before using a pipette, always examine the tip and read the markings on the upper part of the pipette shaft.</u> **DO NOT TOUCH THE TIP WITH YOUR FINGERS!** If you are not familiar with the markings, you will end up with your pipette in a solution and have no idea of how much liquid to pull up or release.

2. Obtain a 10-ml disposable pipette and a beaker of red liquid. Determine whether the pipette is designed for delivery or blow-out. Attach the propipette.

3. Measure out 8 ml: carefully suck the red liquid up into the pipette to a point slightly above the 0 level.

4. Keep the pipette in the liquid with the tip pressing against the side of the container and slowly adjust the level to the 0 line. To read the level correctly, your eye should be directly in line with the meniscus. The bottom of the meniscus marks the correct volume. (*Note:* No meniscus is formed in plastic labware.)

5. Move the pipette to the container into which you wish to dispense the liquid and release the liquid in the proper manner, depending on the type of pipette you are using.

6. Pipette 8 ml of the red liquid into a 100-ml graduated cylinder. Repeat four more times.

   a. *What volume of red liquid should the cylinder contain?* _____ *What volume of liquid does it*

   actually contain? _____ *b. If you found a difference in volume, how do you explain it?*

   _____

   _____

7. Practice using the propipette by measuring out 3.2 ml, 4.8 ml, and 7.3 ml of the red-dye solution.

   | **When you have finished, place your pipettes in the pipette washer, tips upward!** |

8. Now obtain a **microcapillary pipette.** This is a simple glass capillary tube used for measuring microvolumes. It is calibrated in microliters (μl), often with multiple graduations. A wire plunger is used to deliver liquid from the pipette. Alternatively, a smaller version of the propipette (Accropet) can be used to fill or empty the microcapillary tube. This is done by rotating the wheel on the end of the Accropet (Figure 2C-4).

Figure 2C-4 *Types of glass microcapillary pipettes: (a) microcapillary pipette with plunger, (b) microcapillary pipette with Accropet attached.*

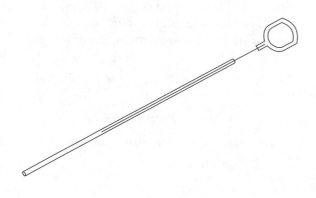

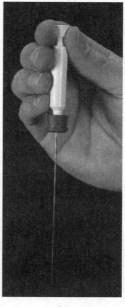

(a)   (b)

9. Place a drop of red-dye solution on a piece of Parafilm or wax paper. Now use a 50-μl microcapillary pipette to draw up the drop of liquid. How much liquid do you estimate is in the drop? _____ Dispense the liquid carefully onto the Parafilm at a different location.

  c. *If you wanted to deliver 0.1 ml of a solution, you could use a regular 0.1-ml pipette or a*

  _____-μl *microcapillary pipette.*

10. Research technicians in biotechnology laboratories need to measure microvolumes accurately and repeatedly. To do this, **digital micropipettes** are used (Figure 2C-5). These pipettes usually have a continuous range of 1 μl to 100 μl, and even fractional volumes can be selected. A digital display allows you to set the pipette to a desired amount by turning the handle of the digital micropipette. A plastic tip (which can be sterilized if needed) is used to hold the liquid and the tip can be ejected easily. Depending on the brand of digital pipette used, there are "stops" on the plunger that allow you to fill and dispense and to eject the tip. Your lab assistant will demonstrate how to use the type of digital micropipette available in your laboratory.

Figure 2C-5 *Digital micropipettes. Disposable plastic tips of different sizes fit on the ends of the pipettes.*

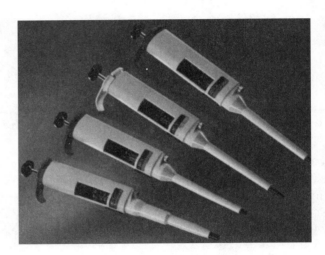

11. Use the digital micropipette to transfer 25.5 μl of red-dye solution onto a piece of Parafilm or wax paper. After dispensing the dye, pick up the drop with the same pipette and transfer it to another location on the Parafilm.

PART 2 The Graduated Cylinder

For measuring larger amounts of fluid, graduated cylinders are used. Two types are in common use: glass cylinders and cylinders made from organic polymers such as polycarbonates. Water adheres more strongly to glass than to polymers, and thus the water at the top of a column will "climb" the sides of a glass cylinder to form a bowl-like meniscus. To obtain a precise measurement, volume *must* be read at the bottom of the "bowl."

Procedure

1. Partially fill a 100-ml graduated cylinder with water and set it on a table.

2. View the cylinder and estimate volume: (a) from a standing position _____, (b) at eye level _____, and (c) from below _____ (see Figure 2C-6).

  a. *How do volumes estimated from these various positions compare?* _____

  _____

Figure 2C-6 *Reading the volume in a graduated cylinder.*

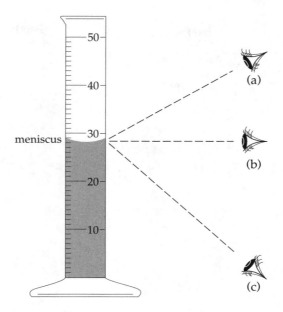

3. Repeat the procedure using a graduated cylinder made from an organic polymer. You will notice that there is much less of a meniscus and that the volume can be read more precisely from any eye level.

4. Add water to the 10-ml level in a 100-ml glass graduated cylinder. Pour the 10 ml of water into a 10-ml graduated cylinder. What does the measurement read on this cylinder?

   b. *Which cylinder would be more accurate for measuring 5 ml of solution: the 100-ml cylinder or the*

   10-ml cylinder? _____

EXTENDING YOUR INVESTIGATION: COMPARING MEASURING DEVICES

There are two additional types of measuring devices on your laboratory table: a beaker and an Erlenmeyer flask. If you wanted to measure 50 ml of water, do you think the beaker or the flask would be as accurate as the graduated cylinder? Formulate a hypothesis to explore this question.

HYPOTHESIS:

NULL HYPOTHESIS:

Which measuring device do you **predict** will be the most accurate?

Design a procedure to test your hypothesis.

PROCEDURE:

Identify the **independent variable** in this experiment:

Identify the **dependent variable** in this experiment:

Now, accurately measure 50 ml of water using the three measuring devices available. (How many measurements should you make for each measuring device?)

RESULTS:

From your results, describe the accuracy of the beaker and the Erlenmeyer flask.

Do your results support your hypothesis?

Your null hypothesis?

Was your prediction correct?

What do you **conclude** about the type of measuring device that should be used to measure liquids?

 PART 3   **The Volumetric Flask**

Volumetric measuring devices do not have graduations. Volumetric equipment is calibrated at the factory to measure a single volume—for example, a 25-ml flask, a 10-ml pipette, or a 200-ml flask—and should be used *only* to measure that volume—25, 10, or 200 ml, respectively. Volumetric equipment is more accurate than graduated equipment.

ⅠⅠⅠⅠⅠ **Procedure** ⅠⅠⅠⅠⅠⅠⅠⅠⅠⅠⅠⅠⅠⅠⅠⅠⅠⅠⅠⅠⅠⅠⅠⅠⅠⅠⅠⅠⅠⅠⅠⅠⅠ

1. Place 100 ml of water into a graduated cylinder. Measure this as accurately as possible.

2. Pour the 100 ml of water into a volumetric flask. The water should be exactly at the line on the neck of the volumetric flask.

   a. Is it? _____ *A volumetric flask is a very accurate measuring device. What does this mean in terms of when you should measure with a graduated cylinder and when you should measure with a volumetric flask?* _____

   The greater accuracy of the volumetric flask is due to the narrowness of its neck, which extends the last few milliliters of solution over a longer distance than in the graduated cylinder, allowing greater accuracy in matching meniscus position with the etched line on the flask.

PART 4   Precision and Accuracy of Measuring Devices

In recording measurement data it is always important for a biologist to be both accurate and precise, but exactly what do we mean by these terms? **Accuracy** is the degree to which an observed value corresponds to the *true* value. **Precision** is the degree to which measurements are reproducible when repeated. Lack of accuracy or precision may be a function of the measuring device or the technique of measurement. Lack of precision may also be a function of experimental design. A measurement may be accurate but not precise, precise but not accurate, both, or neither.

As a rule, only containers designed for measuring should be used to measure. Markings on Erlenmeyer flasks and beakers are only approximate. Always select a measuring device closest in capacity to the volume to be measured. For example, small measuring devices are most accurate for small volumes. However, you should not use a device so small that the measurement must be repeated, because there is always some error in each measurement (for example, using a 1-ml pipette three times to obtain 3 ml would decrease both the accuracy and precision of the measurement).

The following experiment illustrates the difference between precision and accuracy and the degree of precision and accuracy that can be attained with different kinds of measuring devices. Recall that density equals mass per unit volume. By definition, the density of water is 1 g per cubic centimeter (cc) at 4°C. Recall that 1 cc = 1 ml. Therefore, 100 ml of distilled water at 4°C should have a mass of 100 g. You can therefore determine both the accuracy and precision of volumetric flasks and graduated cylinders by determining the mass of the water they contain.

Procedure

1. Place a volumetric flask on a balance. Tare the balance to read 0.000 g.

2. Fill the flask to a point just below the line marking 100 ml. Adjust the water level to the line by adding the remaining milliliter or so with a Pasteur pipette. If you spill any water on the flask, wipe it dry.

   a. Why is it important that the flask be dry? _____

   _____

3. Determine the mass of the water and record it below. Repeat the procedure using two other volumetric flasks of the same size.

   Flask 1 _____ g

   Flask 2 _____ g

   Flask 3 _____ g

   b. For your measurements, what was the range of deviations (smallest to largest) from the true value of 100 g?

   _____

   c. Accuracy *refers to the freedom from error or mistake of a measuring device. How accurate is a volumetric flask as a measuring instrument?* _____

   _____

   d. *The mass of the water in the filled volumetric flask was probably less than 100 g. Why? (Hint: Consider the definition for the density of water as given at the beginning of this part of the exercise.)*

   _____

   _____

   e. *What other factors could affect the accuracy of your measurements?*

   _____

4. Now fill a Nalgene graduated cylinder to the 100-ml mark.

5. Place a 150-ml beaker on the balance and tare the balance to 0.000 g.

6. Pour the water from your graduated cylinder into the beaker and record the mass below. Repeat this procedure four more times. Be sure to dry the beaker thoroughly between measurements and tare the balance before each measurement.

Measurement 1 ———— g

Measurement 2 ———— g

Measurement 3 ———— g

Measurement 4 ———— g

Measurement 5 ———— g

f. *For your measurements, what was the range of deviations (smallest to largest) from the true value of 100 g?*

_____

g. *How accurate is the graduated cylinder compared with the volumetric flask?*

_____

h. *It is possible for a measuring device to be inaccurate but* precise. *How could this be?*

_____

_____

i. *How precise is a graduated cylinder as a measuring device?*

_____

j. *What factors, other than the measuring instrument, could affect the precision of your measurement?*

_____

_____

EXERCISE D   Preparing Solutions

In the biology laboratory, you will often have to prepare **solutions** to use in experiments. Solutions are homogeneous mixtures of **solute,** the dissolved substance, and **solvent,** the dissolving medium. In this exercise, you will learn to prepare the most common type of solution, a molar solution. (Molal, normal, and percent solutions may also be used in the laboratory and can be prepared according to directions in Appendix III, Preparing Solutions.) By making **dilutions** you can use a "stock" solution to prepare less concentrated solutions without wasting chemicals.

||||| Objectives |||||||||||||||||||||||||||||||||||||||||

☐ Prepare a stock NaCl solution of given molarity.

☐ Accurately prepare serial dilutions of a stock solution.

||||| Procedure |||||||||||||||||||||||||||||||||||||||||

A **molar** (1 molar, or 1 M) solution contains 1 mole of a solute in a liter of **solution.** A **mole** of a substance contains Avogadro's number (6.02×10^{23}) of molecules (or atoms or ions) of that substance. For a compound, Avogadro's number is the number of molecules contained in 1 gram molecular weight of the compound. (**Gram molecular weight** is simply molecular weight expressed in grams.) Thus for any compound, 1 mole has a mass equal to its gram molecular weight. The gram molecular weight of NaCl is

58.5 g. Thus a mole of NaCl has a mass of 58.5 g and Avogadro's number of NaCl molecules is contained in that mole (58.5 g of NaCl). (Since this ionic compound dissociates completely in solution, 1 mole *each* of Na^+ and Cl^- are produced when NaCl dissolves in water.)

A mole of $CaCl_2$ has a mass of 111 g (the molecular weight of $CaCl_2$).

To prepare a liter of a 1 M solution of NaCl you would add 58.5 g of NaCl to a 1,000-ml volumetric flask and add water to the 1,000-ml mark. Remember that a molar solution is defined as 1 mole of material per liter of **solution,** not per liter of solvent. The total volume of water and solute that make up the solution must equal 1,000 ml. If you added 1,000 ml of water to 58.5 g of NaCl, you would have more than a liter of solution (this is called a **molal** solution).

a. *Define molar solution.* _____

b. *How would you make 1 l of a 1 M solution of $CaCl_2$ (molecular weight = 111)?* _____

_____

c. *How does a molar (M) solution differ from a molal (m) solution?* _____

_____

To calculate the amount of solute required to make a liter of 0.2 M solution, simply multiply the gram molecular weight by the molarity (0.2 M).

d. *A 0.2 M solution of NaCl contains* _____ *g of NaCl in 1 l of* _____.

Often you need less than a liter of solution of a specific molarity. An easy way to determine the number of grams of solute needed is to use the following formula:

Grams of solute = molecular weight (in grams) × volume (in liters) × molarity

e. *How would you prepare 500 ml of a 0.4 M solution of NaCl?* _____

_____

f. *How would you prepare 250 ml of a 2 M solution of NaCl?* _____

_____

1. Your laboratory instructor will ask you to prepare 100 ml of a NaCl solution. The molarity assigned to you or your group will be selected from a range of 0.2 to 1.2 M. Molarity of NaCl solution _____. NaCl required (grams) _____.

2. Prepare your solution using a 100-ml volumetric flask.

3. After mixing your solution, pour it into a beaker. With your laboratory instructor, measure the specific gravity* of the NaCl solution using a hydrometer. Specific gravity _____.

4. Compare the specific gravity of your solution to known values of NaCl solutions using the standard curve (Figure 2D-1). Specific gravity is plotted on the Y-axis and molarity on the X-axis. Locate the specific gravity of your solution on the Y-axis and draw a horizontal line to the standard curve (a straight line in this case) on the graph. Now drop a vertical line to the X-axis and determine molarity. Assigned molarity _____. Measured molarity _____.

g. *How accurately did you mix your solution?* _____

h. *If your measured value differs from your assigned molarity, what do you think might be the cause?*

_____

---

*The specific gravity of a substance is the ratio of the density of the substance to the density of water.

Figure 2D-1 *Standard curve of specific gravity for NaCl solutions of different molarities.*

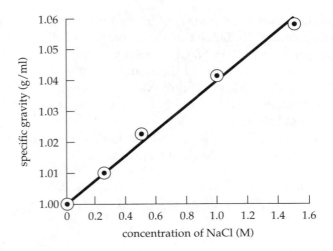

Suppose you were asked to make 100 ml of a 0.001 M solution of NaCl. This would require 0.00585 g of NaCl in 100 ml of solution. You would need to accurately measure 0.00585 g of NaCl. This would be very difficult. To obtain this amount in solution, you could start with 5.85 g of NaCl in a 10-ml solution. Taking 1 ml of this solution (1 ml contains 0.585 g NaCl) and adding it to 9 ml of water would produce a solution of 0.0585 g/ml. Adding 1 ml of this solution (it contains 0.0585 g) to 9 ml of water would produce a solution of 0.00585 g/ml. Then adding 10 ml of this solution to 90 ml of water would produce a solution of 0.00585 g/100 ml—or 100 ml of a 0.001 M solution. This technique is one of dilution. It is called **serial dilution** because it is accomplished in a series of steps (Figure 2D-2). The original solution from which the first dilution was made is called a **stock solution**.

Figure 2D-2 *Serial dilution.*

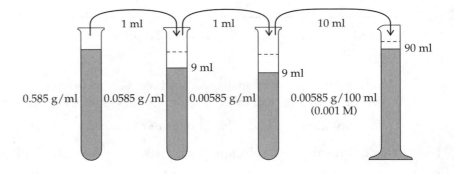

i. How would you prepare 50 ml of a 0.002 M solution of NaCl? (Hint: First determine the total amount (in grams) of NaCl you need in your solution.)

_____

_____

5. Your laboratory instructor has prepared a 4 M solution of NaCl. Use this to prepare 100 ml of a 0.4 M solution. Use the space below for your calculations.

6. Check the accuracy of your work by measuring the specific gravity of your solution and comparing it with those on the standard curve (Figure 2D-1).

j. How accurate was your dilution? _____

Laboratory Review Questions and Problems

1. Using your measurement data from Exercise A (page 2-4), construct a frequency distribution or *histogram*. The abscissa (*X*-axis) should be labeled in centimeters. You will first need to determine the intervals to be used in your frequency distribution. This choice will depend on your data. Refer to Appendix I for assistance.

measure something, convert

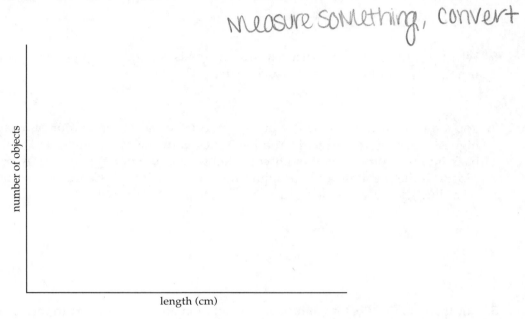

2. What is meant by "taring" the mass of a measuring vessel when using a balance?

3. You place a piece of paper on a balance and find that its mass is 0.63 g. You add a lump of NaOH to the paper and get a mass of 22.21 g. You remember that you forgot to tare the balance before taking this measurement. You tare the balance, add another lump of NaOH, and read 20.64 g. You add another lump of NaOH and read 38.98 g. How much NaOH do you have on the balance, in grams:

_____; milligrams: _____; micrograms: _____; kilograms: _____?

4. On the drawings below, label each type of pipette and indicate how you would measure out the amount of liquid specified for each type of pipette.

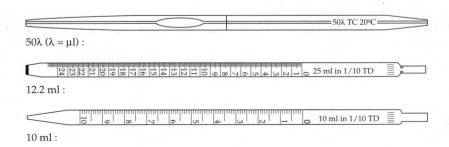

50λ (λ = μl) :

12.2 ml :

10 ml :

5. Draw a representation of the graduations you expect to see on pipettes labeled as follows (to-deliver scales):

5 ml in 1/10

1/10 ml in 1/100

6. When pipetting acids or other harmful materials, should you pipette by mouth? NO
If not, how should you pipette such materials?
pipette pump

7. You are using a 10-ml blow-out pipette graduated in 1/10 ml. (Your pipette reads 0 at the highest marking on the stem.) You have sucked up 8.6 ml of liquid and you cannot put any back into the bottle even though you need to measure out only 6.5 ml. At what mark on the pipette is the meniscus of your liquid now located? If you release liquid from your pipette, at what line should you stop? Draw the pipette and, using two arrows, indicate the start and stop points.

8. Put the quantities below into scientific notation and then convert to the metric units indicated in the second column.

	Scientific Notation	Convert
16 ml	1.6×10^{1} ml	_____ μl
0.0005 g	5.0×10^{-4} g	_____ μg
150 μg	1.5×10^{2} μg	_____ ng
12 nm	1.2×10^{1} nm	_____ mm

9. Given a 50-ml volumetric flask, 50-ml graduated cylinder, 10-ml pipette, 100-ml beaker, and 75-ml Erlenmeyer flask, which measuring instrument would you choose to measure 25 ml of water?

10. Why is a volumetric flask more accurate than a graduated cylinder for measuring volume?

11. What is meant by precision and accuracy in making measurements?

precision—

accuracy— how close you are to the actual value

    a. What determines how precise a measurement is?

        % Rel dev

    b. What determines how accurate a measurement is?

        % error

12. You measure 5 ml of a solution and determine its mass on a balance. You repeat this procedure four times. You try the experiment again. This time you have four new values for mass, as shown below. Choose the statement that best describes your measurements.

Experiment 1 5.7 g, 5.8 g, 5.7 g, 5.8 g

Experiment 2 4.9 g, 5.0 g, 4.8 g, 5.3 g

    a. The measurements in both experiments are accurate and precise.

    b. The measurements in both experiments are not accurate but are precise.

    c. The measurements in experiment 1 are not accurate but are precise; the measurements in experiment 2 are accurate but not precise.

    d. The measurements in experiment 1 are accurate and not precise; the measurements in experiment 2 are not accurate and not precise.

pH and Buffers

OVERVIEW

The concentration of hydrogen ions in solution, expressed as pH, is of great importance to living systems. Because both cell structure and function can be affected by even small changes in pH, maintaining pH within a narrow range is a major goal of cellular homeostasis.

During this laboratory, you will learn how the concentration of hydrogen ions in a solution can be changed or maintained. You will also learn how to prepare various types of solutions and how to use several methods of determining pH.

STUDENT PREPARATION

Prepare for this laboratory by reading the text pages indicated by your instructor. Familiarizing yourself in advance with the information and procedures covered in this laboratory will give you a better understanding of the material and improve your efficiency. After reading Exercise A, Understanding pH, you should complete the problems.

✔ EXERCISE A Understanding pH

Molecules that are dissolved in water may separate (dissociate or ionize) into charged fragments or ions. Often one of these fragments is a hydrogen ion (H^+). The pH of a solution is a measure of the concentration of hydrogen ions (written as $[H^+]$, where [ ] means "concentration of") in that solution, and pH is a measure of the **alkalinity** (*basicity*) or **acidity** of a solution.

Objectives

☐ Define pH.

☐ Define acid and base.

☐ Define neutralization.

☐ Given the molarity of a solution of an acid or base, calculate its pH.

☐ Given the pH of H^+ concentration of a solution, calculate its pOH or OH^- concentration.

Procedure

Water ionizes when a hydrogen atom that is covalently bound to the oxygen of one water molecule leaves its electron behind and, as a hydrogen ion (H^+), joins a different water molecule. Two ions are produced by this reaction: a hydroxide ion (OH^-) and a hydronium ion (H_3O^+). We can express this reaction as follows:

$$2H_2O \rightleftharpoons H_3O^+ + OH^-$$

Convention, however, allows us to express the ionization of water more simply as

$$H_2O \rightleftharpoons H^+ + OH^-$$

In any given volume of pure water, or in any solution, a small but constant number of water molecules are ionized. In pure water, the number of H^+ ions exactly equals the number of OH^- ions, since one cannot be formed without the other being formed. In pure water, $[H^+] = 1 \times 10^{-7}$ M and $[OH^-] = 1 \times 10^{-7}$ M (where M stands for molar concentration or moles per liter).* The product of the molar concentrations of the two ions, $[H^+][OH^-]$, in pure water is always 1×10^{-14}, a number known as the **ion product of water,** and this represents a dissociation constant (K_w) for pure water. (Note that the brackets indicate molar concentration for the substance they enclose.) So, for pure water,

$$[H^+][OH^-] = 1 \times 10^{-14}$$

The concentrations of OH^- and H^+ ions can be written in terms of the number 10 with an exponent. Recall that in a number such as 1×10^{-7}, 10 is the base and $^{-7}$ is the exponent (the power to which 10 is raised). Remember that to find the product of two numbers with exponents, you add the exponents. So, again, for pure water,

$$\mathbf{[H^+][OH^-] = 1 \times 10^{-14}} \qquad \text{or}$$
$$[1 \times 10^{-7}][1 \times 10^{-7}] = 1 \times 10^{-14}$$

Numbers such as these can more easily be expressed as logarithms (base 10). A logarithm is the power to which a base, in this case 10, must be raised to give the desired number. Thus the log of 1×10^{-7} is -7, since this is the power to which 10 must be raised to give the number 0.0000001. pH is defined as the negative logarithm ($\log_{10}$) of the molar hydrogen ion concentration in a solution:

$$\mathbf{pH = -log_{10}\ [H^+]} \quad \text{or} \quad pH = \log_{10}\frac{1}{[H^+]}$$

We use the negative logarithm in working with pH so that the numbers, and the pH scale, are positive—note that $-\log_{10}(1 \times 10^{-7}) = +7$, or simply, 7.

For pure water, we can also determine the pOH, based on the concentration of OH^- ions in a solution:

$$\mathbf{pOH = -log_{10}\ [OH^-]} = \log_{10}\frac{1}{[OH^-]}$$

We can establish a relationship between the pH and the pOH of a solution by using the expression for the ion product of water. In logarithmic form, this can be written as

$$\log_{10}[H^+] + \log_{10}[OH^-] = \log_{10}(1 \times 10^{-14})$$

Since pH is expressed in terms of the *negative* logarithm of the H^+ concentration in solution, we will express the above equation in *negative* logarithmic form:

$$-\log_{10}[H^+] + (-\log_{10}[OH^-]) = -\log_{10}(1 \times 10^{-14})$$

Since the letter "p" in pH stands for "negative logarithm of," we can write this equation as

$$p[H^+] + p[OH^-] = -\log(1 \times 10^{-14}) \qquad \text{or}$$
$$\mathbf{pH + pOH = 14}$$

---

*In a liter of water, there are 3.34×10^{25} water molecules—55.5 moles $\times$ Avogadro's number (6.02×10^{23}). Of these, 1×10^{-7} mole per liter is ionized, so a liter of water contains 6.02×10^{16} H^+ ions. If you divide the number of H^+ ions in a liter by the total number of water molecules in a liter ($6.02 \times 10^{16}/3.34 \times 10^{25}$), you will discover that only 2×10^{-9} or 2 ten-millionths of a percent of all water molecules present in 1 l of water are ionized. That is a very small amount!

Since for pure water, $[H^+] = 1 \times 10^{-7}$ M and $[OH^-] = 1 \times 10^{-7}$ M, then pH = 7 and pOH = 7, so pH + pOH = 14.

It is important to realize that in any solution, as is the case for pure water, the *product* of $[H^+]$ and $[OH^-]$ is *constant*. For pure water, the molar amounts of H^+ and OH^- are equal. If an ionic or polar substance is dissolved in water, it may change the relative amounts of H^+ and OH^- but the product of the two concentrations is always 1×10^{-14} because water is a source of H^+ and OH^- ions and the ion product (which reflects the tendency of water to ionize) remains constant. For instance, if a substance added to a volume of water ionizes to produce sufficient H^+ ions that the H^+ concentration of the solution increases to 1×10^{-5} M (expressed as $[H^+] = 1 \times 10^{-5}$ M), then the OH^- concentration decreases to $[OH^-] = 1 \times 10^{-9}$ M. The ion product of water for the solution will always equal 1×10^{-14} (*remember*: to find the product of two numbers written in logarithmic form, add the logarithms according to the rules for exponents):

$$[H^+][OH^-] = 1 \times 10^{-14}$$
$$[1 \times 10^{-5}][1 \times 10^{-9}] = 1 \times 10^{-14}$$

Likewise, if a substance added to a volume of water increases the $[OH^-]$, the $[H^+]$ will necessarily decrease. It is important to note, however, that adding H^+ to a solution not only increases the number of H^+ ions but *also* decreases the number of OH^- ions, since some of the added H^+ will combine with OH^- to make water. Adding OH^- to a solution can also decrease the H^+ concentration for the same reason— H^+ and OH^- combine to form H_2O! But, no matter what is added, the product $[H^+][OH^-]$ must always be 1×10^{-14}, so if we know the concentration of one ion, we always know the concentration of the other.

We now can use the expression pH + pOH = 14 to generate the pH scale. If we know the pH we can calculate the pOH (pOH = 14 − pH), and if we know the pOH we can calculate the pH (pH = 14 − pOH). The sum of the negative logarithms of $[H^+]$ and $[OH^-]$ is always 14 and the product of the molar ion concentrations, $[H^+][OH^-]$, is always 1×10^{-14}. To be sure that you understand this relationship, examine Table 3A-1 carefully.

A solution of pH = 7 is **neutral,** at the midpoint of the pH scale. Solutions with a pH value lower than 7 are said to be **acidic.** In an acidic solution, the number of H^+ ions exceeds the number of OH^- ions. Solutions with a pH above 7 are **basic** or alkaline: the number of OH^- ions exceeds the number of H^+ ions. Expressed another way, the more acidic a solution, the higher the number of H^+ ions and the lower the number of OH^- ions. The more basic a solution, the lower the number of H^+ ions and the higher the number of OH^- ions. Check Table 3A-1 to make sure this is true. Also, remember that the pH scale is logarithmic, not arithmetic: if two solutions differ by 1 pH unit, then one solution has *ten* times the hydrogen ion concentration of the other.

An **acid** is a substance that causes an increase in the number of H^+ ions and a decrease in the number of OH^- ions in solution. This increase is most often the result of an ionization that produces H^+. Some common acids and their ionization products are

Hydrochloric acid	$HCl \longrightarrow H^+ + Cl^-$
Acetic acid	$CH_3COOH \rightleftharpoons H^+ + CH_3COO^-$
Carbonic acid	$H_2CO_3 \rightleftharpoons H^+ + HCO_3^-$
Phosphoric acid	$H_3PO_4 \longrightarrow H^+ + H_2PO_4^- \longrightarrow H^+ + HPO_4^{2-}$

The more completely the acid ionizes, the more H^+ is released, and the stronger the acid is. For example, if the concentration of an HCl solution is 0.1 mole/l and the HCl ionizes completely, we have 0.1 mole/l of H^+. This could also be expressed as $[H^+] = 1 \times 10^{-1}$ M.

Of the acids listed above, HCl and H_3PO_4 are considered *strong* acids since they ionize completely. CH_3COOH, H_2CO_3, and $H_2PO_4^-$ are relatively *weak* acids. A weak acid might, for example, ionize only 10%, so a 1×10^{-2} M solution would contain only 1×10^{-3} mole/l of H^+. Once $[H^+]$ is known, pH can be calculated.

Table 3A-1   The pH Scale

$[H^+]$ M	pH	$[OH^-]$	MpOH	
10^0 (1.0)	0	10^{-14}	14	
10^{-1} (0.1)	1	10^{-13}	13	
10^{-2} (0.01)	2	10^{-12}	12	
10^{-3} (0.001)	3	10^{-11}	11	**acidic**
10^{-4} (0.0001)	4	10^{-10}	10	
10^{-5} (0.00001)	5	10^{-9}	9	
10^{-6} (0.000001)	6	10^{-8}	8	
10^{-7}	7	10^{-7}	7	**neutral**
10^{-8}	8	10^{-6} (0.000001)	6	
10^{-9}	9	10^{-5} (0.00001)	5	
10^{-10}	10	10^{-4} (0.0001)	4	
10^{-11}	11	10^{-3} (0.001)	3	**basic**
10^{-12}	12	10^{-2} (0.01)	2	
10^{-13}	13	10^{-1} (0.1)	1	
10^{-14}	14	10^0 (1.0)	0	

Example   A solution with $[H^+] = 1 \times 10^{-2}$ M ionizes 10%.

$$[H^+] = 1 \times 10^{-2} \text{ M}$$

$$10\% = 0.10 = 1 \times 10^{-1}$$

$$(1 \times 10^{-2}) \times 10\% = (1 \times 10^{-2})(1 \times 10^{-1}) = 1 \times 10^{-3}$$

$$[H^+] = 1 \times 10^{-3} \text{ M}$$

$$pH = \log_{10} \frac{1}{[H^+]} = \log_{10} \frac{1}{(1 \times 10^{-3})} = 3$$

1. Fill in Table 3A-2 by calculating the pH for each acid.

Table 3A-2   Calculating pH for Acids

Molarity of Acid	Degree of Ionization	$[H^+]$ M	pH
1×10^{-3}	100%		
1×10^{-3}	10%		
1×10^{-3}	1%		
1×10^{-4}	100%		
1×10^{-2}	100%		
1×10^{-1}	100%		

A substance need not give up hydrogen ions itself to cause an increase in the $[H^+]$ in a solution. For instance, an important molecule in biological systems is CO_2 (carbon dioxide) which combines with water to form carbonic acid (H_2CO_3), which ionizes to produce bicarbonate ion (HCO_3^-):

$$CO_2 + H_2O \longleftrightarrow H_2CO_3 \longleftrightarrow H^+ + HCO_3^-$$

The reaction of SO_2 (sulfur dioxide) with atmospheric water is, in part, responsible for acid rain. It dissolves in water to form sulfurous acid (H_2SO_3):

$$SO_2 + H_2O \longleftrightarrow H_2SO_3 \longleftrightarrow H^+ + HSO_3^-$$

A **base** is a substance that causes a decrease in the number of H^+ in solution and an increase in OH^-. In many cases this is achieved by the ionization of the molecule to produce OH^- (hydroxyl ion), which not only adds to the OH^- in solution but also *removes* H^+ from solution by combining with it to form water, thus raising the pH.

$$KOH + HCl \longleftrightarrow K^+ + OH^- + H^+ + Cl^- \longleftrightarrow KCl + H_2O$$

Thus, **neutralization** of an acid by a base produces a **salt** (an ionic compound composed of a negative ion from an acid and a positive ion from a base, such as KCl) and **water.**

Some common bases that ionize to produce OH^- are

Sodium hydroxide $\quad\quad NaOH \longrightarrow Na^+ + OH^-$

Magnesium hydroxide $\quad Mg(OH)_2 \longrightarrow Mg^{2+} + 2\,OH^-$

Potassium hydroxide $\quad\; KOH \longrightarrow K^+ + OH^-$

These are all *strong bases* because they ionize completely in solution.

Ammonia (NH_3), dissolved in water, is also basic. It does not produce OH^- but it can remove H^+ from solution:

$$NH_3 + H_2O \longrightarrow NH_4^+ + OH^-$$

The bicarbonate ion (HCO_3^-) is also basic. It, too, can accept H^+:

$$H^+ + HCO_3^- \longleftrightarrow H_2CO_3$$

2. Fill in Table 3A-3. Remember the relationship between pH and pOH discussed earlier.

Table 3A-3   Calculating pH for Bases

Molarity of Base	Degree of Ionization	$[OH^-]$ M	$[H^+]$ M	pH
1×10^{-3}	100%			
1×10^{-3}	10%			
1×10^{-2}	100%			
1×10^{-5}	100%			
1	100%			

3. Keeping in mind that $[H^+][OH^-] = 1 \times 10^{-14}$, give the following for a *neutral* solution:

$[H^+] = $ _____

$[OH^-] = $ _____

pH = _____

✔ **EXERCISE B** | **Using Indicators to Measure pH**

Indicators are chemicals that change color depending on the pH of the solution and are often used to determine the pH of a solution colorimetrically.

ⅢⅢ Objectives ⅢⅢⅢⅢⅢⅢⅢⅢⅢⅢⅢⅢⅢⅢⅢⅢⅢⅢⅢⅢⅢⅢⅢⅢⅢⅢ

☐ Use a colorimetric pH indicator.

☐ Use an indicator solution or alkacid test paper to determine the pH of some common solutions.

☐ Prepare and use a series of standards for qualitative measurements.

✔ **PART I** **Making a pH Indicator**

A solution of certain **anthocyanins,** the plant pigments responsible for red, blue, and purple colors in flowers, fruits, and autumn leaves, can be used as a pH indicator. At low pH, solutions of anthocyanins turn red; at high pH, they turn blue. In some flowers, however, soil pH affects the uptake of certain metals that can complex with the anthocyanin pigment and prevent its normal color expression. For instance, in hydrangeas, low soil pH promotes the uptake of aluminum, which then complexes with the anthocyanins and prevents the production of "normal" pink color. Instead, the flowers appear blue. Raising the pH of the soil prevents aluminum uptake and the pink color can be expressed. At more neutral pH values, the flowers appear intermediate in color.

A solution of anthocyanins extracted from red cabbage will be used as a pH indicator in this laboratory. At the beginning of class, your instructor will make an extract of anthocyanins by boiling red cabbage in water for 3 minutes and then filtering the cabbage extract through cheesecloth.

To use an indicator to determine the pH of solutions of unknown pH, you must first determine the color changes that occur when the indicator is used with substances whose pH is known. The color changes that occur in the cabbage extract indicator when mixed with substances of pH 2, 4, 6, 7, 8, 10, 12, and 14 will be used as a set of **standards.** You can then determine the pH of various substances by mixing each with cabbage extract and comparing the resulting colors with the colors of the standards.

ⅢⅢ Procedure ⅢⅢⅢⅢⅢⅢⅢⅢⅢⅢⅢⅢⅢⅢⅢⅢⅢⅢⅢⅢⅢⅢⅢⅢⅢⅢⅢⅢⅢⅢⅢ

1. Work in pairs. Put eight clean test tubes in a rack and label them 2, 4, 6, 7, 8, 10, 12, and 14. Use a pipette to measure 5 ml of the appropriate buffer into each test tube (i.e., place the pH 2 buffer in the tube labeled "2," and so on).

2. Use a pipette to measure 3 ml of cabbage extract and add it to each test tube. Cover with Parafilm and invert to mix well.

3. Record the color of each tube below. *Note:* Record both initial and final colors at pH 12 and 14. The pigments are not stable at these pH values.

pH	Color
2	
4	
6	
7	
8	
10	
12	
14	

✔ **PART 2**   **Measuring pH with Cabbage Indicator**

‖‖‖ **Procedure** ‖‖‖‖‖‖‖‖‖‖‖‖‖‖‖‖‖‖‖‖‖‖‖‖‖‖‖‖‖‖‖‖‖‖‖

   1. Obtain two clean test tubes and label them A and B.

   2. In tube A, mix 2 ml of solution A with 1 ml of cabbage extract.

   3. Compare the color in tube A with the colors of the standards. The approximate pH of the unknown solution is the pH of the standard whose color most closely matches the color in tube A.

      pH of A _____ $[H^+]$ _____

   4. Use the same method (steps 2 and 3) to measure the pH of solution B.

      pH of B _____ $[H^+]$ _____

      a. *Which solution is more acidic?* _____

✔ **PART 3**   **Using Alkacid Test Paper**

The cabbage indicator method can be used only with colorless or white solutions. Alkacid test papers, which are impregnated with indicators, are another means of estimating pH and can be used with colored solutions.

‖‖‖ **Procedure** ‖‖‖‖‖‖‖‖‖‖‖‖‖‖‖‖‖‖‖‖‖‖‖‖‖‖‖‖‖‖‖‖‖‖‖‖‖‖

   1. Hold a test paper with forceps. Use a clean stirring rod to apply a drop of solution A to the test paper.

   2. While the paper is still wet, compare its color with the standard pH color scale on the label of the alkacid paper's container.

      pH of A _____

   3. Measure and record the pH values of solutions B, C, and D.

      pH of B _____   pH of C _____   pH of D _____

      a. *Do the results obtained with the alkacid papers match those obtained using the cabbage indicator*
         method? _____

      b. *Explain any discrepancies (alkacid paper doesn't have a standard for pH 7, so a neutral solution will*
         be closer to either 6 or 8). _____

      _____

👁 **EXERCISE C**   **Determining the pH of Some Common Solutions (*Optional*)**

Cabbage indicator and alkacid test paper can be used to determine the pH of common beverages, medicines, and cleaning solutions.

‖‖‖ **Objectives** ‖‖‖‖‖‖‖‖‖‖‖‖‖‖‖‖‖‖‖‖‖‖‖‖‖‖‖‖‖‖‖‖‖‖‖

☐ Understand the relationship of pH to the taste of common beverages and to the activity of medicines and cleaning solutions.

👁 **PART I**   **pH of Beverages**

Many beverages differ in their acidity or alkalinity, often due to CO_2 content (if carbonated) or organic acids.

Formulate a hypothesis that predicts the relative differences in pH values for the beverages listed in the Procedure below.

HYPOTHESIS:

NULL HYPOTHESIS:

Which beverage do you **predict** will have the lowest pH?

What is the **independent variable**?

What is the **dependent variable**?

||||| **Procedure** |||||||||||||||||||||||||||||||||||||||||||

1. Work in pairs. Use the cabbage indicator method (see Exercise B, Part 2) to determine pH values for 7-Up and wine. Use alkacid paper (see Exercise B, Part 3) for determining the pH of coffee and apple juice. Record the values below.

Beverage	pH
Apple juice	
Coffee (black)	
7-Up	
White wine	

2. Arrange these solutions in order of increasing $[H^+]$:

   1. _____

   2. _____

   3. _____

   4. _____

Do your results support your hypothesis? _____ *Your null hypothesis?* _____

Was your prediction correct? _____

a. Did you discover anything about the beverages that surprised you? _____

b. Based on your results, can you think of reasons to drink or not drink any of these beverages in excess?

_____

👁 PART 2   **pH and Activity of Some Common Medicines**

Many medicines, especially those used to guard against acidity in the digestive tract, are fairly alkaline, while other medicines tend to be acidic.

Formulate a hypothesis that predicts the relative differences in pH values for the medicines listed in the Procedure below.

HYPOTHESIS:

NULL HYPOTHESIS:

Which medicine do you **predict** will have the highest pH?

What is the **independent variable**?

What is the **dependent variable**?

IIIII Procedure IIIIIIIIIIIIIIIIIIIIIIIIIIIIIIIIIIIIIIIIIIIII

Work in pairs. Use the cabbage indicator method to determine pH values for the following medicines.

Medicine	pH
Aspirin	
Milk of Magnesia [$Mg(OH)_2$]	
Sodium bicarbonate (Alka-Seltzer) ($NaHCO_3$)	
Maalox	

Do your results support your hypothesis? _____ *Your null hypothesis?* _____

Was your prediction correct? _____

a. *It is often recommended that aspirin be taken with a large glass of milk or water. Based on your results above, why might this be important?* _____

_____

b. *Would apple juice or citrus juice be a good accompaniment to aspirin?* _____ *Explain why or why not.*

_____

_____

Enzymes function best at particular pH values. In the normal human stomach, a pH of 2.0 to 3.0 provides the environment required for the proper functioning of the digestive enzymes found there. The last three medicines in the list above are often used for treatment of "acid indigestion" of the stomach, a condition in which a reduction in pH interferes with efficient enzyme action and thus with digestion.

c. *Based on your results, how would you explain the action of these medicines?* _____

_____

d. *What might happen if an excess of any of these medicines were used?* _____

_____

PART 3   pH and the Action of Some Cleaning Solutions

Cleaning solutions vary in their chemical composition. This, in turn, determines what materials they mix with for cleaning purposes.

Based on your experience with cleaning your clothes, dishes, body, and even your sink, formulate a hypothesis that predicts the relative differences in pH values for the cleaning solutions listed in the following Procedure.

HYPOTHESIS:

NULL HYPOTHESIS:

Which cleaning solution do you **predict** will have the highest pH?

What is the **independent variable**?

What is the **dependent variable**?

||||| **Procedure** |||||||||||||||||||||||||||||||||||||||

Work in pairs. Use the cabbage indicator method to determine the pH values of the cleaning solutions listed below.

Solution	pH
Drano	
Ivory Liquid	
Cascade	
Tide	

Do your results support your hypothesis? _____ *Your null hypothesis?* _____

Was your prediction correct? _____

a. What do all these solutions have in common? _____

_____

👁 **EXERCISE D** | **Soil pH and Plant Growth (*Optional*)**

Since plants obtain most of their nutrients in the form of ions, the kinds and amounts of nutrient ions available from the soil affect plant growth and health. They also have a significant effect on the types of plants that can be grown in soils of various compositions. The chemical composition of soil varies greatly from region to region, due to differences in geology, topography, climate, and plant and animal life. In any soil, the complex equilibria among soil particles, water, and nutrient ions are influenced by soil pH. For example, phosphorus (P), an element essential for plant growth, is taken up in the form of phosphate ions, either as $H_2PO_4^-$ or HPO_4^{2-}. If the soil is acidic, the $H_2PO_4^-$ form is most abundant. HPO_4^{2-} is prevalent under neutral conditions, and PO_4^{3-} exists in alkaline soils. Plants absorb the $H_2PO_4^-$ form most readily, so phosphorus is most available in acidic soils. If the pH is too low, $H_2PO_4^-$ combines with other ions and is precipitated out of the soil. Thus phosphorus is most available at pH 5.5 to 6.5. A similar effect is seen with aluminum, as described in Exercise B, Part 1.

There are at least 13 elements plants must obtain from the soil, and each type of plant thrives within a more or less narrow range of chemical balance. There is no single soil pH that provides the perfect environment for all plants. Instead, plants have adapted to various soil chemistries, and each species has its own range of pH tolerance (Table 3D-1).

Table 3D-1   pH Preference of Selected Species

Plant	pH Range
Eastern hemlock, azalea, rhododendron, gardenia, cranberry, blueberry	4.5 to 6.0
Coleus, iris, tomato, squash, strawberry, tobacco	5 to 7
Gladiolus, cherry, pear, sugar maple, alfalfa, asparagus, yellow poplar	6 to 8

ııııı Objectives ıııııııııııııııııııııııııııııııııııı

☐ Relate soil pH to plant nutrition.

☐ Measure the pH of soil samples.

ııııı Procedure ıııııııııııııııııııııııııııııııııııııı

In this exercise you will determine the pH of several soil samples. Water has been added to these samples to help release some of the ions into solution so the pH can easily be determined.

Based on your experience with gardening and your knowledge of soil and plants common to different environments, formulate a hypothesis that predicts the relative differences in pH values for the soils listed below.

HYPOTHESIS:

NULL HYPOTHESIS:

What type of soil do you **predict** would have the lowest pH?

What is the **independent variable**?

What is the **dependent variable**?

Work in pairs. Use alkacid test paper to measure the pH of each of the following soil samples.

Sample	pH
Potting mix	
Clay	
Sand	
Lime	
Peat moss	

Do your results support your hypothesis? _____ *Your null hypothesis?* _____

Was your prediction correct? _____

a. *Suppose that a soil sample from your garden has a pH of 4. You want to grow* Coleus, *which requires a pH range of 5 to 7. Which of the substances tested could be used to adjust the pH to one suitable for growing* Coleus *plants?* —————————— *Why?* ——————————————————

b. *Suppose that you want to grow azaleas, which require a pH range of 4.5 to 6.0. The soil in your garden has a pH of 7. Which of the substances tested could be used to adjust the pH to one suitable for growing azaleas?* ——————————— *Why?* ——————————————————

c. *Your hydrangea bush produces beautiful pink flowers year after year. One spring you decide you would like to have blue hydrangea flowers instead. [Recall that certain pigments called anthocyanins are often complexed with metals abundant in the soil, and are responsible for pink, purple, and blue color in many flowers (see Exercise B, Part 1).] What should you do?* ——————————————————

✔ **EXERCISE E**   **The pH Meter**

In Exercises B, C, and D, you learned how the pH of a solution can be estimated by comparing experimental indicator colors with known standards. In many cases, however, higher accuracy and greater reliability are necessary. Electrometric methods of pH determination, using pH meters, give more precise results.

The standard laboratory pH meter has a **glass electrode** that is sensitive to hydrogen ion activity ("activity" is defined as the effective concentration of an ionic species in solution; it is usually expressed in moles per liter) and a **reference electrode** which completes the electrical circuit. On some pH meters, a combination electrode performs both functions. Other parts of the pH meter include the following:

Readout meter An analog meter with a pH scale for pH determinations and a millivolt (mV) scale for millivolt measurements.

Mechanical meter zero An adjustment that mechanically zeros the meter pointer.

Function selector A switch that maintains the meter on standby when measurements are not being taken and that selects the measuring mode, either pH or millivolts.

Standardize control A control that allows the meter to be set to the pH of the buffer solution used to standardize the instrument.

Temperature control A control that compensates for the temperature of the solution being measured (it is active only when the function selector is in the pH mode).

⁙⁙⁙ **Objectives** ⁙⁙⁙⁙⁙⁙⁙⁙⁙⁙⁙⁙⁙⁙⁙⁙⁙⁙⁙⁙⁙⁙⁙⁙⁙⁙⁙⁙⁙⁙⁙⁙

☐ Use the pH meter to measure the pH of a solution.

⁙⁙⁙ **Procedure** ⁙⁙⁙⁙⁙⁙⁙⁙⁙⁙⁙⁙⁙⁙⁙⁙⁙⁙⁙⁙⁙⁙⁙⁙⁙⁙⁙⁙⁙⁙⁙⁙

1. Work in pairs. Locate each of the parts (listed above) on the pH meter available in your laboratory. Your instructor will calibrate and standardize the pH meters in your lab by adjusting the pH reading on the analog scale of the meter when its electrodes are placed in a buffer solution of known value (the standard). To obtain valid results when measuring the pH of a solution, the standardization buffer should have a pH close to that of the sample to be tested. In other words, if you are working with acidic solutions, you should standardize the pH meter with a known buffer solution of pH 4, *not* pH 13.

   Because of the range in pH values for the solutions being tested in this exercise, you will need to select a pH meter that has been standardized with a buffer that has a pH close to the estimated or expected pH of the solution you are testing (the cabbage extract indicator or alkacid paper has given you an estimate of pH for each solution). Each pH meter in your laboratory is labeled to indicate the pH of the buffer with which it has been standardized.

2. At one of the pH meters, you will find a 50-ml beaker containing 20 ml of apple juice and a small magnetic stirring bar. Check the approximate pH from your previous results in Exercise C, Part 1, and select the proper pH meter. Record your estimated pH values in Table 3E-1.

3. Set the beaker on the magnetic stirring plate next to the pH meter and turn on the stirrer so that the magnetic bar revolves slowly in the solution.

4. Check to be sure that the function selector is on "standby." Raise the pH electrode out of its storage beaker. Over a waste beaker, rinse the electrode with distilled water from a wash bottle.

5. While taking care that the stirring bar does *not* hit the electrode, immerse the electrode in the beaker of apple juice.

6. Change the function selector from "standby" to "pH."

7. Record the pH in Table 3E-1.

8. Turn the function switch back to "standby."

9. Raise the electrode out of the solution. Over a waste beaker, rinse the electrode with distilled water and place it into its own storage beaker.

10. Repeat this procedure for the remaining solutions listed in Table 3E-1.

lemon juice   2   2.3

Table 3E-1   pH of Common Solutions

Solution	Estimated pH	Measured pH	$[H^+]$ M	$[OH^-]$ M
Apple juice	6	3.36		
7-Up	5	2.83		
Maalox				
Tide	7	7.5		
Ivory Liquid				

NaOH   10   12.30
CaOH   11   11.65

11. Now that you can obtain a more accurate measure of pH (to tenths of pH units), you can calculate $[H^+]$ or $[OH^-]$ by using your calculator. Since pH is expressed as a negative logarithm of $[H^+]$, you need to determine the antilogarithm ("antilog") of the pH expressed as a negative number.

Example   A solution has a pH of 3.2. What is its hydrogen ion concentration? Using a Texas Instruments calculator or other scientific calculator (see your calculator's directions if it uses reverse Polish notation—RPN),

ENTER 3.2

Change it to -3.2 (use the $+/-$ key)

Press (INV)(LOG) = 0.000631, or 6.31×10^{-4}

The hydrogen ion concentration is 6.31×10^{-4} M.

a. What is the hydroxide ion $[OH^-]$ concentration of this solution? _____

12. Given [H⁺], you can also determine the pH.

> **Example**   A solution has an $[H^+]$ of 1.95×10^{-7} M. What is its pH? Using a scientific calculator, enter the number 1.95×10^{-7} into your calculator in scientific notation form:
>
> ENTER 1.95
>
> Press the exponent entry key (EE) to indicate to the calculator that the next numbers entered represent the exponent.
>
> Enter 7 and press the $+/-$ key to convert to -7
>
> Press LOG and the $+/-$ key: $-\log 1.95 \times 10^{-7} = 6.71$
>
> The pH is 6.71.

13. Practice on the following.

Solution	$[H^+]$ M	pH
Urine	5.89×10^{-5}	
Pancreatic juice	7.95×10^{-8}	

Solution	pH	$[H^+]$ M
Lemon juice	2.8	
Milk	6.4	

14. Refer to Appendix III, Preparing Solutions, and then prepare 100 ml of the solutions listed in the table below. Work in pairs.

Solution	Estimated pH	Measured pH	$[H^+]$ M	$[OH^-]$ M
0.1 M NaOH				
0.1 N Ca(OH)₂				
0.2 M KH₂PO₄				
5% NaCl				

15. Use alkacid test paper to estimate the pH of each solution. Then use the pH meter to measure the pH of each solution. Record your results in the table.

16. Now that you know that pH + pOH = 14 and that $[H^+][OH^-] = 10^{-14}$, and you know how to use logarithms to convert from ion concentration to pH or antilogarithms to convert from $(-)$pH to ion concentration, you can convert from any of the following quantities to the other (see next page):

> **Example**   If the pH of a solution is 6, what is its OH^- concentration? First, find the antilog of -6. This will give you the H^+ concentration of the solution. Divide 1×10^{-14} by $[H^+]$ to give $[OH^-]$. Calculate this value. _____
>
> *b. Given that a solution has on OH^- concentration of 1×10^{-4} M, what is its pH?* _____
>
> *c. Given that a solution has $[H^+] = 1 \times 10^{-6}$ M, what is its $[OH^-]$?* _____

$$pH \underset{-\log_{10}[H^+]}{\overset{antilog\ (-)pH}{\rightleftharpoons}} [H^+]$$

$$pOH = 14 - pH \quad\quad pH = 14 - pOH \quad\quad [OH] = \frac{1 + 10^{-14}}{[H^+]} \quad\quad [H^+] = \frac{1 + 10^{-14}}{[OH^+]}$$

$$pOH \underset{-\log_{10}[OH^+]}{\overset{antilog\ (-)pOH}{\rightleftharpoons}} [OH^+]$$

👁 EXERCISE F   Buffers

Physiological processes require that pH remain relatively constant. The pH of blood in our bodies, for example, is usually maintained between 7.3 and 7.5. However, blood returning to the heart contains CO_2 picked up from the tissues (recall that CO_2 combines with water to form carbonic acid), and our diets, as well as the normal metabolic reactions in cells, may contribute an excess of hydrogen ions. The pH must be kept constant by several *buffer systems*.

A **buffer** is defined as a solution that *resists change* in pH when small amounts of acid or base are added. Bicarbonate, phosphate, and protein buffer systems maintain our blood pH. We will use the phosphate buffer system as an illustration.

A buffer is made by mixing a weak acid with its salt in order to have in solution something that can act as an acid (give up hydrogen ions) *and* something that can act as a base (accept hydrogen ions). In a potassium phosphate buffer, the weak acid, $H_2PO_4^-$, is supplied as KH_2PO_4 (monobasic potassium phosphate) and its salt as K_2HPO_4 (dibasic potassium phosphate).

At equilibrium, these substances are ionized to some degree:

$$KH_2PO_4 \rightleftharpoons K^+ + H^+ + HPO_4^{2-}$$
$$K_2HPO_4 \rightleftharpoons 2K^+ + HPO_4^{2-}$$

If hydrogen ions are added to the solution, they can be picked up by HPO_4^{2-}, which acts as a base:

$$H^+ + HPO_4^{2-} \longleftrightarrow H_2PO_4^-$$

If hydroxyl (OH^-) ions are added to the solution, they can be picked up by H^+:

$$OH^- + H^+ \longleftrightarrow H_2O$$

In summary,

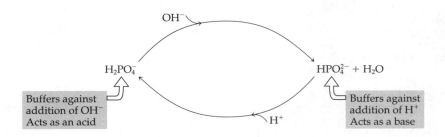

Objectives

- [ ] Define buffer and discuss why buffers are important to organisms.
- [ ] Use a buffer system to show how buffers work.
- [ ] Explain why some substances have buffering capacity and others do not.

Procedure

In this exercise, you will use colorimetric indicators to determine which of four unknown solutions act as buffers. A change in one of the indicators, congo red or thymolphthalein, indicates a change in pH.

The four solutions, A, B, C, and D, were prepared using K_2HPO_4 or KH_2PO_4, or a combination of these, dissolved in distilled water.

1. Locate the four solutions marked A, B, C, and D on your laboratory table.
2. Pour approximately 40 ml of one of the unknown solutions into a 100-ml beaker. Insert a stirring bar and place the beaker on a stirring plate.
3. Add 1 dropperful of congo red. Record the color of the solution in Table 3F-1.
4. Measure the pH using a pH meter and record this value.
5. Using a pipette, add 2 ml of 0.1 N HCl. Record the color of the solution and its pH.
6. Dispose of this solution, and wash and dry the beaker and stirring bar thoroughly.
7. Pour another 40 ml of the same unknown solution into the beaker, insert the stirring bar, and return it to the stirring plate.
8. Add 1 dropperful of thymolphthalein and again measure the color and pH and record in Table 3F-1.
9. Now use a pipette to add 2 ml of 0.1 N NaOH. Record the color and pH.
10. Dispose of the solution and wash and dry the beaker and stirring bar thoroughly.
11. Repeat these steps for all three of the other unknown solutions, to complete Table 3F-1.

Table 3F-1   Determination of Changes in pH and Color

Unknown Solution	Initial Color	Initial pH	Solution Added	Final Color	Final pH
A Congo red			0.1 N HCl		
Thymolphthalein			0.1 N NaOH		
B Congo red			0.1 N HCl		
Thymolphthalein			0.1 N NaOH		
C Congo red			0.1 N HCl		
Thymolphthalein			0.1 N NaOH		
D Congo red			0.1 N HCl		
Thymolphthalein			0.1 N NaOH		

12. Answer the following questions based on your observations of color changes that occurred when either acid or base was added.

    a. Which unknown solution buffered against both acids and bases? _____ *What did this solution contain?* _____

    b. Which unknown solution buffered only against the addition of acid? _____ *What did this solution contain?* _____

    Did the solution act as an acid or base? _____ *Explain.*

    _____

    c. Which unknown solution buffered only against the addition of base? _____ *What did this solution contain?* _____

    Did the solution act as an acid or base? _____ *Explain.*

    _____

    d. Which solution did not buffer against addition of either acid or base? _____ *What did this solution contain?* _____

    e. Which indicator would be the best to use to determine the presence of an acid?

    _____ *What color would it appear at pH = 2?* _____

    pH = 7? _____ *pH = 11?* _____

    f. Which indicator would be the best to use to determine the presence of base?

    _____ *What color would it appear at pH = 2?* _____

    pH = 7? _____ *pH = 11?* _____

Use of indicators gives qualitative evidence of pH change. Qualitative observations can be further supported quantitatively by considering the actual H^+ and OH^- concentration changes.

When adding 2 ml of a 0.1 N acid or base, you are adding the equivalent of 5×10^{-3} M H^+ ions or OH^- ions.

$$\text{Increase in } H^+ \text{ or } OH^- = \frac{\text{volume of acid or base added}}{\text{volume of solution added into}} \times \text{normality of solution added}$$

Note: When dealing with acids and bases it is convenient to use equivalents to indicate the concentration of reacting particles. For example, a solution is 1 N when it contains one gram equivalent weight of reacting particles. (See Appendix III, Normal Solutions.)

In our example,

$$\text{Increase in } H^+ = \frac{0.002 \text{ l}}{0.04 \text{ l}} \times 0.1 \text{ equivalents/liter}$$

where 0.002 l = 2 ml of HCl or NaOH added to 0.04 l or 40 ml of solution. This gives

$$\text{Increase in } H^+ = 0.005 \text{ M (or } 5 \times 10^{-3} \text{ M)}$$

Therefore, if the solution you are testing does not buffer against an acid or a base, you will expect a 5×10^{-3} M increase in its H^+ or OH^- concentration. However, if your solution does buffer, the change in H^+ or OH^- concentrations will not be as dramatic.

13. Calculate the changes in H^+ and OH^- that you observed in your four unknown solutions. To calculate the change in H^+, subtract the initial H^+ concentration from the final H^+ concentration. To calculate the change in OH^-, first convert the pH into pOH. Then, subtract the initial OH^- concentration from the final OH^- concentration. Record these values in Table 3F-2.

Table 3F-2   Calculated Changes in H^+ and OH^- Concentrations

Unknown Solution	Initial $[H^+]$	Final $[H^+]$	Change in $[H^+]$	Initial $[OH^-]$	Final $[OH^-]$	Change in $[OH^-]$
A						
B						
C						
D						

g. *From these results, did the indicators identify the buffering abilities of the unknown solutions accurately?* _____

_____

Laboratory Review Questions and Problems

1. HCl ionizes completely in water. What is the $[H^+]$ of a 0.01 M solution of HCl? What is the pH?

2. Is the hydrogen ion concentration of a pH 3.8 solution higher or lower than that of a solution with a pH of 6.2?

3. If one solution has 100 times as many hydrogen ions as another solution, what is the difference, in pH units, between the two solutions?

4. If solution A contains 1×10^{-6} M H^+ ions and solution B contains 1×10^{-8} M H^+ ions, which solution contains *more* H^+ ions?

5. Make a statement relating hydrogen ion concentration to the acidity and basicity of solutions.

6. HA is an acid that ionizes 10% in solution. What is the $[H^+]$ of a 0.01 M solution of HA? What is its pH?

7. Write the equation for the neutralization of NaOH by HCl.

8. What is the $[H^+]$ of a solution whose pH is 8? What is the $[OH^-]$?

9. Complete the following table.

$[H^+]$	$[OH^-]$	pH
	1×10^{-6}	
		4
1×10^{-3}		

10. You have made the following solutions:

Solution A   Hydrogen ion concentration $= 1 \times 10^{-6}$ M

Solution B   pH $= 5$

State whether each of the following is true or false. If false, explain why.

Solution A contains more H^+ ions than solution B.

Solution A contains 1×10^{-8} M OH^- ions.

Solution B contains 1×10^5 M H^+ ions.

Solution B is acidic and solution A is basic when measured using the pH scale of 0–14.

Solution A is less acidic than solution B.

If the pH of solution A is to be raised by one pH unit, you would want to increase the OH^- concentration to 10^{-7} by adding base.

The two solutions differ in hydrogen ion concentration by a factor of 10 (i.e., one solution has 10 times the hydrogen ion concentration of the other).

11. Assuming complete ionization, what are the $[OH^-]$, $[H^+]$, and pH of a 0.01 M solution of NaOH?

12. Assuming complete ionization, what are the $[OH^-]$, $[H^+]$, and pH of a 0.1 M solution of KOH?

13. Calculate the pH of the listed solutions.

Solution		pH
Maalox	$[H^+] = 3.1 \times 10^{-9}$ M	
Saliva	$[H^+] = 1.95 \times 10^{-7}$ M	
Vinegar	$[OH^-] = 2.4 \times 10^{-12}$ M	

14. Calculate the H^+ and OH^- concentrations of the listed solutions.

Solution	pH	$[H^+]$ M	$[OH^-]$ M
Tomato juice	4.2		
Blood plasma	7.4		
Seawater	8.2		

15. State two reasons why pH maintenance is important to biological systems, and give examples of how pH may affect biological reactions.

16. You have four 1,000-ml beakers filled with four different clear solutions:

   0.1 M NaH_2PO_4

   0.1 M Na_2HPO_4

   0.1 M phosphate buffer, pH 7.2

   Distilled water

Oops! You forgot to label them and they all look alike. You get the congo red and thymolphthalein from the lab and test a sample of each solution, labeling the beakers randomly as A, B, C, and D. You use congo red when HCl is added to the sample, and thymolphthalein when NaOH is added. You get the following results.

	Color Before Addition	Add	Color After Addition
A	Red	HCl	Red
A	Colorless	NaOH	Blue
B	Red	HCl	Blue
B	Colorless	NaOH	Blue
C	Red	HCl	Red
C	Colorless	NaOH	Colorless
D	Red	HCl	Blue
D	Colorless	NaOH	Colorless

What is the identity of solutions A, B, C, and D?

   A

   B

   C

   D

17. You have buffers of pH 2, 4, 6, 8, and 10, but you need a pH 7 buffer for your experiment. Describe how you will make the pH 7 buffer.

18. A gardener planted pink hydrangeas in his yard last year, but this year when the flowers bloomed they were blue. He doesn't understand what happened. He took good care of the plants and mulched them with pine straw, just like the gardening encyclopedia said. What happened to his plants? How could he make them produce pink blooms again?

Using the Spectrophotometer

OVERVIEW

Color provides us with both beauty and useful information. Color is a source of our pleasure in a sunset, in the autumn leaves, or in a beautiful bouquet of flowers. Color can also be an indicator of when vegetables or fruits are ripe, when our coffee is strong enough, or when a storm is coming.

In this laboratory, you will explore how a **spectrophotometer** uses the colors of the light spectrum to determine the concentration of light-absorbing molecules in a solution.

The visible light spectrum, like X-rays, radio waves, and infrared waves, is part of the spectrum of electromagnetic radiation. Types of electromagnetic radiation differ in both wavelength and energy level, but all types travel through space in waves. The height of a wave at its crest is called its **amplitude;** the intensity or the brightness of visible light is proportional to its amplitude. The distance from the crest of one wave to the crest of the next wave is called the **wavelength** (λ); in the visible spectrum, the color of the light we see depends on its wavelength. Wavelength is measured in units called nanometers (1×10^{-9} m). Wavelengths of 400 to 700 nm comprise the "visible light spectrum"— the part of the electromagnetic spectrum that can excite photoreceptors within the human eye.

STUDENT PREPARATION

Prepare for this laboratory by reading Exercise A, Part 1.

✔ **EXERCISE A** | **How the Spectrophotometer Works**

✔ **PART 1** **Principles of Spectrophotometry**

Molecules either absorb or transmit energy in the form of electromagnetic radiation. White light (normal daylight) is made up of all the wavelengths of electromagnetic radiation in the visible spectrum. How objects or chemical substances absorb and transmit the light that strikes them determines their color.

What we see as the color of an object, or a solution, is determined by what wavelengths of light are "left over" to be transmitted or reflected by the object after certain wavelengths are absorbed by its constituent molecules. For example, the pigment chlorophyll, present in the leaves of plants, absorbs a high percentage of the wavelengths of light in the red and violet to blue ranges (Figure 4A-1). Green light, not absorbed by chlorophyll molecules, is reflected from the surface of the leaf—thus most plants appear to be green. A solution of chlorophyll extracted from a leaf would also be green.

The spectrophotometer can be used to measure the amount of light absorbed or transmitted by molecules in a solution. The spectrophotometer operates on the following principle. When a specific

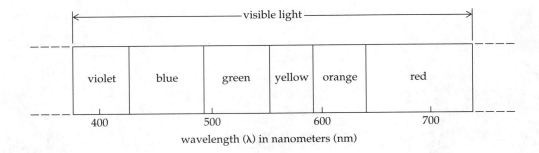

Figure 4A-1   *Electromagnetic wavelengths in the visible light spectrum. For a diagram illustrating the full range of electromagnetic radiation, see Figure 12A-2, page 12-5.*

wavelength of light is transmitted through a solution, the radiant light energy absorbed, **absorbance** (*A*), is directly proportional to (1) the absorptivity of the solution—the ability of the solute molecules to absorb light of that wavelength; (2) the concentration of the solute; and (3) the length of the path of light (usually 1 cm) from its source, through the solution, to the point where the percentage of light energy transmitted or absorbed is measured by a phototube.

Spectrophotometers that employ ultraviolet or visible light are the types most often used to study biological structures and reactions. The investigator selects a wavelength of light that will be maximally absorbed by a particular solute in solution. (If visible light is used and the molecule of interest does not absorb light, it is often possible to set up a chemical reaction that will yield a colored product.) After passing through the solution, the light energy received at the phototube is expressed as the ratio of transmitted light I_T (the light that passes through the sample) to incident light I_0 (the intensity of light at the source before it enters the sample). The light received at the phototube is measured as percent transmittance (*T*), or as the log of its inverse, absorbance (*A*):

$$\%T \text{ (percent transmittance)} = \frac{I_T}{I_0} \times 100$$
$$A \text{ (absorbance)} = \log \frac{I_0}{I_T}$$

By measuring the absorbance (or transmittance) it is possible to determine the concentration of the absorber (molecule) in solution. Concentration can be calculated directly if the molar absorptivity of the molecule (the amount of light at a specific wavelength absorbed by a specified concentration of solute in moles per liter) is known. Usually, however, molar absorptivity is not known and absorbance readings indicate only relative concentrations—a higher absorbance (*A*) resulting from a higher concentration. In such cases, the concentration can be found by locating the absorbance reading of the unknown concentration on a graph of the absorbances of known concentrations (standard curve).

✔   **PART 2**   **Using the Spectrophotometer**

The Bausch & Lomb Spectronic 20 Colorimeter (Figure 4A-2) is an extremely versatile instrument that is useful for the spectrophotometric, or colorimetric, determinations of solutions.

Within the optical system of the spectrophotometer, rotation of a prism (diffraction grating) allows the investigator to select specific wavelengths of light in a range from 375 to 625 nm. Light of a selected wavelength is passed through the sample and is picked up by a measuring phototube, where the light energy is converted to a reading on the meter of the spectrophotometer (Figure 4A-3).

Most spectrophotometers have two scales—one is a linear scale (the *transmittance scale*) given as percent transmittance, the other is a logarithmic scale with the same gradations as the percent transmittance scale (the *absorbance scale*) (Figure 4A-2). (Since transmittance is related to absorbance as the log of the inverse, values of 0.0 absorbance occur at 100% transmittance and values of infinite absorbance

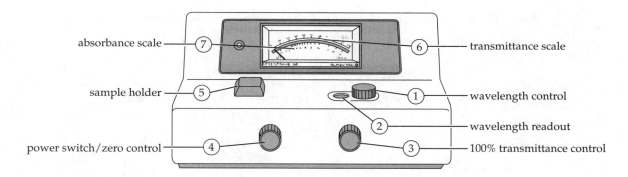

Figure 4A-2 *Features of the Bausch & Lomb Spectronic 20 Colorimeter. The transmittance and absorbance scales run in opposite directions. Why?*

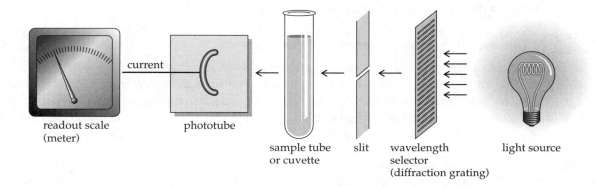

Figure 4A-3 *Operation of the spectrophotometer. Light of 340 to 750 nm is emitted from a light source and passes through a diffraction grating (wavelength selector) that generates light of a particular wavelength. Light intensity is adjusted as the light passes through a slit before reaching the sample. Some light is absorbed by the sample, and the rest is transmitted and detected by a phototube on the other side of the sample. The current generated by the phototube is proportional to the intensity of light hitting it and is registered as transmittance (or its inverse, absorbance) on the readout scale (meter) of the spectrophotometer.*

occur at 0% transmittance.) Examine the readout meter on the spectrophotometer you will be using. Note that the scales run in opposite directions.

IIIII Objectives IIIIIIIIIIIIIIIIIIIIIIIIIIIIIIIIIIIIIIIIIIII

- [ ] Describe the relationship between absorbance and transmittance.
- [ ] Describe how absorbance and transmittance of light are related to the color of an object.
- [ ] Explain how the optical system of a spectrophotometer works.
- [ ] List the steps in the proper use of the spectrophotometer.

IIIII Procedure IIIIIIIIIIIIIIIIIIIIIIIIIIIIIIIIIIIIIIIIIIII

1. Familiarize yourself with the parts of the spectrophotometer shown in Figure 4A-2.

2. Work in pairs. Cut a strip of white paper to just fit the diameter and length of a Spectronic 20 tube or cuvette. Slide the paper into the tube and insert the tube into the sample holder of the spectrophotometer.

3. Leave the sample holder open and place a cylinder of black paper around the opening.

4. Set the wavelength control to 620 nm and adjust the position of the tube containing the white paper until you see the maximum amount of red light on the right side of the paper.

5. Have your partner turn the wavelength dial in both directions and record the range of each wavelength at which you see a particular color. Range of wavelengths:

Red 620-603   Orange 602-585

Yellow 584-576

Green 575-502

Blue 501-460

Violet 459-415

6. Trade positions and let your partner check your observations by repeating steps 3–5. Compare your results with the wavelengths given in Figure 4A-1.

✔ **PART 3**   **Determining Transmittance and Absorbance**

To assure that spectrophotometer readings indicate only the concentration of the solute we wish to measure, a reading must first be obtained using a **blank,** a sample that contains all the components of the solution except the absorbing molecule. For instance, if you are using a reagent that changes color when mixed with a certain solute molecule, a blank should contain *all* the components of the test solution, including the colorimetric reagent, *except* the substance (solute) to be measured.

Example A 1-ml sample of substance X is mixed with 5 ml of water and 1 ml of colorimetric reagent to give a volume of 7 ml in a sample tube. A blank is prepared by mixing 5 ml of water with an additional 1 ml of water (as a substitute for substance X) and 1 ml of colorimetric reagent. The volume in the blank tube is 7 ml. Note that the volume of the blank should always be the same as the volume of the sample.

With the blank inserted into the spectrophotometer, the instrument is adjusted to 100% transmittance (zero absorbance). This step is similar to taring a balance: the transmittance of light through the blank will be less than 100% because of substances (including the colorimetric reagent) present in the blank. However, the instrument can be adjusted to accept this reading as 100% transmittance, so that when the blank tube is replaced by the sample tube to measure absorbance of the sample, the only thing absorbing light will be the sample molecule of interest (the solute that reacts with the colorimetric reagent).

||||| **Procedure** |||||||||||||||||||||||||||||||||||||||||||

1. Prepare a sample tube. Place 10 drops of albumin solution into a spectrophotometer tube and add 1 ml of distilled water. Add 5 ml of Coomassie brilliant blue (a colorimetric reagent used to identify protein) and allow the tube to stand until a blue color develops.

2. Prepare a blank: place 10 drops of water into a spectrophotometer tube and add 1 ml of distilled water and 5 ml of Coomassie blue.

3. Turn the power switch on, and allow a 5-minute warm-up period. The on/off switch is operated by the zero control knob on the left.

4. Use the wavelength control knob to adjust the spectrophotometer to any wavelength between 550 and 600 nm. The selected wavelength is indicated on the wavelength readout in the window next to the knob.

5. When using the Spectronic 20, the meter must be adjusted to read across its full scale—0% transmittance to 100% transmittance. With *no* sample tube in the machine, use the zero (left-hand) control knob to set the scale to 0% transmittance (infinite absorbance). (With no sample tube, the light path is automatically blocked, and no light reaches the phototube; thus, 0% transmittance and infinite absorbance are simulated.) Be sure the cover on the sample holder is closed when you perform this step.

$$\%T = \frac{I_T}{I_o} \times 100 \quad \text{initial light}$$
$$\text{transmitted}$$

$$Abs = \log \frac{I_o}{I_T}$$

6. Insert your blank (be sure it is clean and dry on the outside) into the sample holder, and turn the right-hand control knob to set the meter scale to 100% transmittance, zero absorbance. This adjustment regulates the amount of light reaching the phototube in the absence of the absorber. *Whenever the wavelength is changed, the 100% transmittance adjustment must be reset.* Also, when operating at a fixed wavelength for an extended period of time, periodically check the 100% and 0% transmittance readouts and adjust if necessary.

7. If you are beginning an experiment, repeat steps 5 and 6 to make sure the machine is stable.

8. Insert the sample tube into the chamber; read absorbance directly on the absorbance scale (lower scale). The reading on the absorbance scale is proportional to the concentration of your sample substrate. *Note:* The absorbance scale reads from *right to left*, opposite to the direction of the transmittance scale. Record your data: wavelength __555__ nm, absorbance __.476__, transmittance __33.5__ .

   Do not **discard your sample and blank tubes!**

9. The steps that follow provide a brief checklist for using the spectrophotometer during this laboratory and in later lab work.

QUICK CHECKLIST FOR USING THE SPECTROPHOTOMETER

1. Turn the power on and allow a 5-minute warm-up period before taking sample readings.

2. Select the wavelength.

3. Check that the sample holder is empty and the cover is closed.

4. Use the zero control knob to set the meter to 0% transmittance.

5. Wipe off fingerprints from the reference blank, insert it into the sample holder, and set transmittance to 100%.

6. Wipe off fingerprints from the unknown sample tube, insert it into the sample holder, and read the meter display in percent transmittance or absorbance.

For best results when using the spectrophotometer, always remember the following:

- All solutions *must* be free of bubbles.

- All sample holders *must* be at least one-half full.

- For best performance with test tube holders, be sure that the index mark on the tube or cuvette aligns with the mark on the adapter (if tubes and cuvettes are marked).

- All sample tubes *must* be clean and free of scratches. Use lens paper to remove all fingerprints from the sample tubes and cuvettes.

- During extended operation at a fixed wavelength, make occasional checks for meter drift: use the blank to check for 100% transmittance.

👁 **EXERCISE B**   **Determining the Maximum Absorption Wavelength**

Molecules in solution (solute) will absorb light maximally within a narrow range of wavelengths. When deciding upon the wavelength to use in measuring concentration, an "absorbance spectrum" is generated in which the absorbance of a particular solute, or "absorber," is measured at a continuous selection of wavelengths. A curve such as that shown for bromphenol blue (Figure 4B-1) is generated, and *maximum absorbance* can be determined. The most accurate measurements of absorbance are obtained by selecting a wavelength of light that is maximally absorbed by the solute of interest.

Figure 4B-1 *Absorption spectrum for bromphenol blue.*

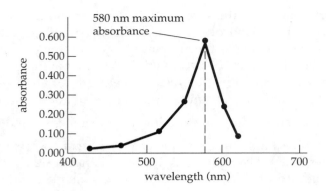

ⅠⅠⅠⅠⅠ Objectives ⅠⅠⅠⅠⅠⅠⅠⅠⅠⅠⅠⅠⅠⅠⅠⅠⅠⅠⅠⅠⅠⅠⅠⅠⅠⅠⅠⅠⅠⅠⅠⅠⅠⅠ

☐ Determine the maximum absorbance wavelength for a light-absorbing substance in solution.

☐ Relate the shape of an absorbance curve to the absorption of light at different wavelengths.

ⅠⅠⅠⅠⅠ Procedure ⅠⅠⅠⅠⅠⅠⅠⅠⅠⅠⅠⅠⅠⅠⅠⅠⅠⅠⅠⅠⅠⅠⅠⅠⅠⅠⅠⅠⅠⅠⅠⅠⅠ

1. Set the wavelength on your spectrophotometer to 540 nm.

2. Use the zero control knob to set the spectrophotometer to 0% transmittance.

3. Adjust the spectrophotometer to 100% transmittance using the same blank as you used in Exercise A.

4. Remove the blank tube and place the albumin sample tube (used in Exercise A, Part 3) into the chamber. Record the absorbance and transmittance readings at 540 nm in Table 4B-1.

5. Readjust the wavelength to 560 nm. Repeat steps 2–4.

   a. *Why do you need to use the blank to adjust the spectrophotometer at each wavelength used for a reading?* _____

   _____

6. Continue to increase the wavelength until you reach 640 nm, repeating steps 2–4 at each wavelength and recording the data in Table 4B-1.

7. Verify the relationship between transmittance and absorbance by calculating absorbance from the transmittance data ($T = \%T/100$). Show your calculations in the last column of Table 4B-1.

Table 4B-1   Absorbance and Transmittance at Various Wavelengths

Wavelength (nm)	Absorbance	% Transmittance	Calculations $A = \log 1/T$ (or $-\log T$)
540	.181	65.9%	-1.82
560	.590	25.7%	-1.41
580	.806	5.7%	-1.20
600	.802	15.9%	-1.20
620	.679	21.4%	-1.33
640	.432	33.0%	-1.52

8. Graph the absorbance data on graph paper.

    b. At what range of wavelengths is absorbance at a maximum for Coomassie blue? _____

    c. Within this range of wavelengths, determine the maximum wavelength for absorption,

      A_{max} = _____

    d. Do you think the wavelength at which absorbance is maximum would change if the concentration of albumin in the sample tube were doubled? _____ *Why or why not?*

_____

EXTENDING YOUR INVESTIGATION: ABSORBANCE AND TRANSMITTANCE

Your instructor will provide you with a tube of solution that appears red/red-orange. Consider the way in which light waves are absorbed or transmitted. What wavelengths of light would you expect to be absorbed by this solution? _590-750_ _____ _abs anything <590_ _____

_____

Formulate a hypothesis that predicts what the absorption spectrum would look like for this solution.

HYPOTHESIS: The Solution absorbs anything less than 590

NULL HYPOTHESIS: The Solution absorbs anything greater than 590.

What do you **predict** will be the outcome of your experiment?

What is the **independent variable?** Solution

What is the **dependent variable?** wavelength

Follow steps 1–8 to test your hypothesis, using the appropriate range of wavelengths for your sample. Collect absorbance data in the table below.

Wavelength	Absorbance
420	.544
470	1.274
520	2.110
570	0.234
620	0.003
670	0.001

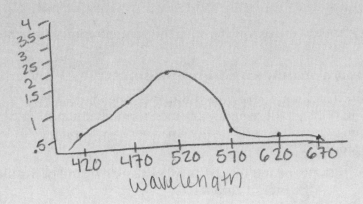

Graph the absorbance data on graph paper. From your results, describe the absorbance properties of the red solution. ___OK_____

_____

Do your results support your hypothesis? Your null hypothesis? yes

Was your prediction correct?

👁 **EXERCISE C** | **Exploring the Relationship between Absorbance and Concentration**

Coomassie brilliant blue forms a colored complex with proteins. In Exercise B, you determined the wavelength at which this complex maximally absorbs. Now, using this wavelength, you can determine the concentration of protein in a solution by measuring the intensity of color developed when Coomassie blue is mixed with the solution.

Color intensity can be accurately measured by determining the amount of light absorbed by the solution. Absorbance is a function of concentration and, by comparing the absorbance of a solution containing an unknown amount of protein to a **standard curve,** a graph of absorbances plotted from a series of samples of known concentrations of the same material, the concentration of protein in an "unknown" can be determined (Figure 4C-1).

Figure 4C-1   *Absorbance measurements of known concentrations of a protein molecule are used to generate the standard curve (in this case the "curve" is linear). An absorbance reading (a) is obtained for a solution containing an unknown amount of protein. By drawing a straight line from (a) to the curve at (b) and dropping a line from (b) to (c), the concentration of the protein can be determined. A_{500} (Y-axis) indicates absorbance at 500 nm.*

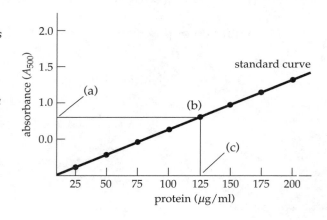

IIIII **Objectives** IIIIIIIIIIIIIIIIIIIIIIIIIIIIIIIIIIIIIIIIII

☐ Describe the relationship between absorbance and the concentration of a light-absorbing substance in solution.

☐ Use a standard curve to determine the concentration of a light-absorbing substance in solution.

IIIII **Procedure** IIIIIIIIIIIIIIIIIIIIIIIIIIIIIIIIIIIIIIIIIIIIII

On your laboratory table you will find a tube marked BSA. It contains 240 μg/ml of bovine serum albumin (BSA), a more highly purified form of albumin. To develop a standard curve for bovine serum albumin, you and your laboratory partner should prepare at least five dilutions (see Appendix III or Laboratory 2, Exercise D).

1. Prepare dilutions as follows. Add 0.5 ml of distilled water to each of five test tubes and label the tubes 1 through 5.

2. To the first tube, add 0.5 ml of BSA stock solution and *mix well*. This will give a protein concentration of 120 μg/ml.

3. Take 0.5 ml of solution from test tube 1 and add it to test tube 2. Mix well.

4. Take 0.5 ml from test tube 2 and add it to test tube 3. Mix well. Repeat this procedure until you have added protein solution to each test tube.

5. Discard 0.5 ml from the last test tube. *a. Why is this necessary?* Because it has 1mL compared to the other .5

6. To prepare a blank, add 0.5 ml of distilled water to another test tube and label it "blank."

b. Why is a blank necessary? Because

c. What type of serial dilution have you performed—1:2, 1:5, 1:10, and so on, or some other?
1:2

7. Tube C on your laboratory bench contains an unknown amount of BSA. Add 0.5 ml of this solution to a test tube, label it "unknown," and set it in the test tube rack with the standards and the blank.

You should now have a series of five test tubes containing protein concentrations of 120, 60, 30, 15, and 7.5 μg/ml; a blank containing only distilled water; and a tube containing an unknown concentration of protein.

8. Add 5 ml of Coomassie blue to each of tubes 1 through 5 and to the blank. Wait at least 3 minutes, but no longer than an hour, then read absorbances at 595 nm. Follow the Spectronic 20 procedure steps outlined in Exercise A, Part 3, or refer to the quick checklist. Record your data in Table 4C-1.

Table 4C-1   Data for Determining Concentration from Absorbance

	Protein Concentration (μg/ml)	Absorbance (595 nm)
Tube 1	120	1.463
Tube 2	60	1.169
Tube 3	30	.745
Tube 4	15	.611
Tube 5	7.5	.573
Unknown		.713

9. Obtain a sheet of graph paper. Label the abscissa (*X*-axis) "Concentration" and the ordinate (*Y*-axis) "Absorbance." Make a graph as large as possible on the paper; plot the absorbance data for the solutions of known concentration. This is your standard curve. (Keep in mind that a "standard curve" can be used as a standard only when known and unknown concentrations have been prepared according to the same procedure. For example, for this standard curve to be useful, knowns and unknowns must be prepared by using 0.5 ml of the sample and 5 ml of reagent.)

10. Add 5 ml of Coomassie blue to your unknown. Wait 5 minutes or the same amount of time as in step 8, and read its absorbance at 595 nm using the Spectronic 20. Record absorbance in Table 4C-1. From the standard curve you have prepared, determine the concentration of your unknown:

_____ μg/ml BSA   (Record this value in Table 4C-1.)

Laboratory Review Questions and Problems

1. You have a solution that appears green. What color light is being transmitted? What color(s) of light are most strongly absorbed?

2. How is absorbance related to transmittance of light through a solution?

3. List the steps in using a spectrophotometer. What is meant by "zeroing" the spectrophotometer?

4. You want to use the Spectronic 20 to find the wavelength of light you should use for determining the concentration of protein in an unknown solution. Using the following diagram, indicate the number of the dial or scale that you would use to accomplish each of the steps listed below.

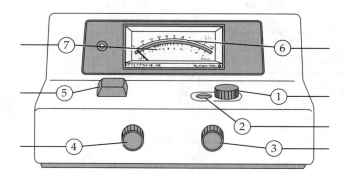

_____ Insert protein sample.

_____ Adjust to 100% transmittance and 0 absorbance using a blank.

_____ Change wavelength.

_____ Adjust to maximum absorbance and 0% transmittance.

_____ Read wavelength.

_____ Read absorbance of sample.

_____ Read transmittance of sample.

5. What is the purpose of a "blank" or reference tube?

6. You are making a standard to be used to measure protein concentration. You have added 2 ml of protein, 3 ml of water, and 4 ml of Coomassie blue. How do you make a blank for this experiment?

7. You generate an absorbance spectrum for a red solution as shown below. Label the axes of this graph (include units where necessary). What wavelength (or color) of light would you expect to show the maximum reading? _____

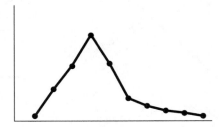

8. a. You are conducting an experiment in which an extract of cytoplasm is mixed with a red dye. When the mitochondria in the cytoplasmic extract are functioning, the red dye fades. You wish to measure the rate of the reaction that causes the decolorization. What wavelength of light might you use to make your measurements if you wanted to measure the decrease in absorbance by the red solution? _____

   b. What would you use as a reference blank?

   c. What would you use as a control for this experiment?

   d. How might you determine the rate (decrease in concentration/time) of the reaction? (*Hint:* How would you measure concentration?)

9. In order to determine the concentration in an unknown solution, you need to prepare a standard curve. You are given a tube containing 5 ml of a 400-μg/ml solution of protein. You are told to make *four* dilutions and record final concentrations for each of the dilutions. Use the tubes below to indicate how you would make these dilutions, and record the final concentration in each tube.

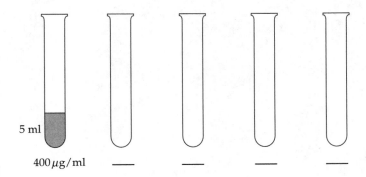

5 ml

400 μg/ml   ____   ____   ____   ____

10. You have developed a standard curve for several dilutions of tyrosine, as shown below (a). Tyrosine can be oxidized by an enzyme, tyrosine oxidase, found in liver. Tyrosine reacts with nitrosonaphthol. You begin with 100 μg of tyrosine. This is mixed with tyrosine oxidase and absorbance readings are obtained (b). Use the graph to complete the table.

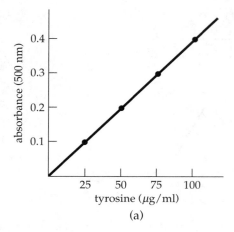

(a)

Time (minutes)	Absorbance (500 nm)	Tyrosine (μg/ml)
0	0.40	
0.5	0.35	
2	0.2	
5	0.1	

a. How much tyrosine has been oxidized during a 5-minute period? _____

b. At what rate does the tyrosine disappear during the first 2 minutes of the reaction? _____

c. During the last 3 minutes? _____

Organic Molecules

OVERVIEW

A cell is a living chemistry laboratory in which most functions take the form of interactions between organic (carbon-containing) molecules. Most organic molecules found in living systems can be classified as carbohydrates, fats, proteins, or nucleotides. Each of these classes of molecules has specific properties that can be identified by simple chemical tests.

In this laboratory you will learn to identify three of the four major types of organic molecules: carbohydrates, fats, and proteins. The fourth major type, the nucleotide, is the basic structural unit of nucleic acids and will be studied during the isolation of the genetic material, DNA, in Laboratory 16.

STUDENT PREPARATION

Prepare for this laboratory by reading the text pages indicated by your instructor. Familiarizing yourself in advance with the information and procedures covered in this laboratory will give you a better understanding of the material and improve your efficiency.

EXERCISE A Testing for Carbohydrates

The basic structural unit of carbohydrates is the **monosaccharide** (or single sugar). Monosaccharides are classified by the number of carbons they contain: for example, **trioses** have three carbons, **pentoses** have five carbons, and **hexoses** have six carbons (Figure 5A-1). They may contain as few as three or as many as 10 carbons.

Monosaccharides are also characterized by the presence of a terminal aldehyde group (Figure 5A-1a, b) or an internal ketone group (Figure 5A-1c). Both of these groups contain a double-bonded oxygen atom that reacts with Benedict's reagent to form a colored precipitate.

When two monosaccharides are joined together, they form a **disaccharide.** If the reactive aldehyde or ketone groups are involved in the bond between the monosaccharide units (as in sucrose, Figure 5A-2a), the disaccharide will not react with Benedict's reagent. If only one group is involved in the bond (as in maltose, Figure 5A-2b), the other is free to react with the reagent. Sugars with free aldehyde or ketone groups, whether monosaccharides or disaccharides, are called **reducing sugars.*** In this exercise, you will use Benedict's reagent to test for the presence of reducing sugars.

Ban

light = energy   wavelength of color is the same to the observer

*These sugars are oxidized (lose electrons to) by the Cu^{2+} in Benedict's reagent, which then becomes reduced (gains electrons); hence the name reducing sugar.

Figure 5A-1 *(a) Glyceraldehyde, a representative triose with an aldehyde reactive group. (b) Ribose, a representative pentose with an aldehyde reactive group. (c) Fructose, a representative hexose with a ketone reactive group. In solution, ribose and fructose occur predominantly in ring forms (d, e). When heated, sugars in the ring form interconvert to the chain form.*

Usually sugars need to be in the chain form to interact with colorimetric substances.

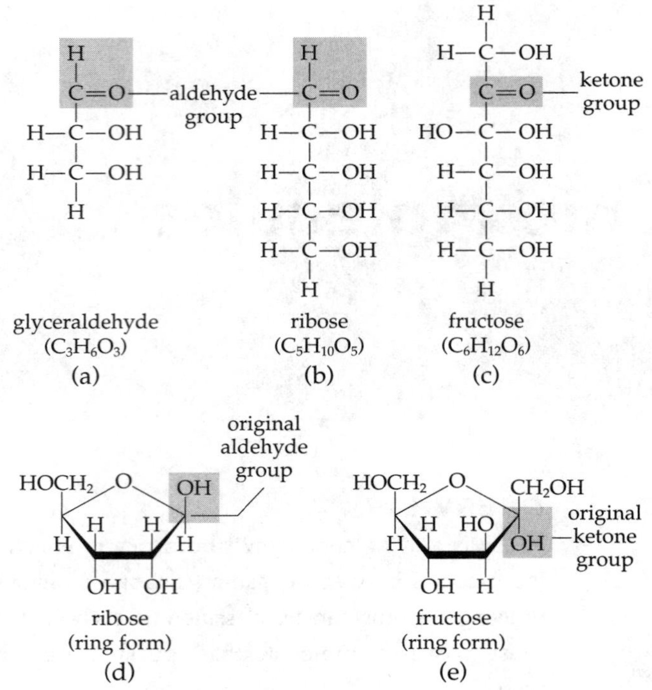

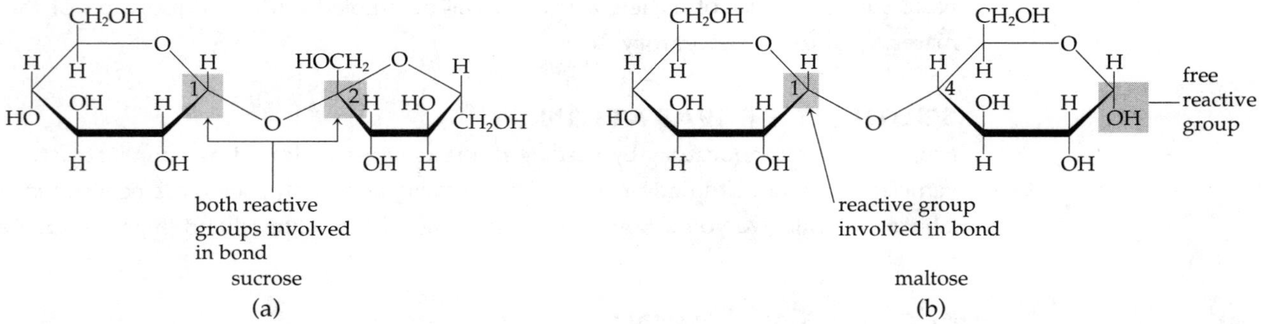

glyceraldehyde
($C_3H_6O_3$)
(a)

ribose
($C_5H_{10}O_5$)
(b)

fructose
($C_6H_{12}O_6$)
(c)

ribose
(ring form)
(d)

fructose
(ring form)
(e)

sucrose
(a)

maltose
(b)

Figure 5A-2 *(a) In sucrose, both reactive groups are involved in the bond between monosaccharides, preventing a reaction with Benedict's reagent. (b) In maltose, only one reactive group is involved in the bond between monosaccharides, leaving the other group free to react with Benedict's reagent. Here the sugars are shown in ring form. The reactive group of maltose is highlighted.*

Monosaccharides may join together to form long chains (**polysaccharides**) that may be either straight or branched. Starch is an example of a polysaccharide formed entirely of glucose units. Starch does not show a reaction with Benedict's reagent because the number of free aldehyde groups (found only at the end of each chain) is small in proportion to the rest of the molecule. Therefore, you will test for the presence of starch with Lugol's reagent (iodine/potassium iodide, I_2KI).

ııııı Objectives ııııııııııııııııııııııııııııııııııııııı

☐ Identify reducing sugars (both monosaccharides and disaccharides) using Benedict's reagent.

☐ Identify polysaccharides using Lugol's reagent.

white-reflect
Black-absorb

green-absorbs
all colors,
reflects green

 **PART I**   **Benedict's Test for Reducing Sugars**

When Benedict's reagent is heated with a reactive sugar, such as glucose or maltose, the color of the reagent changes from blue to green to yellow to reddish-orange, depending on the amount of reactive sugar present. Orange and red indicate the highest proportion of these sugars. (Benedict's test will show a positive reaction for starch only if the starch has been broken down into maltose or glucose units by excessive heating.)

Formulate a **hypothesis** and **predict** what you might expect to find for each of the sugars to be tested with Benedict's reagent (Table 5A-1).

	Hypothesis	Prediction
1. Water		
2. Starch		
3. Glucose		
4. Maltose		
5. Sucrose		
6. Onion juice		
7. Potato slice		
8. Milk		

What is your **null hypothesis** for this group of tests?

What is the **independent variable**?

What is the **dependent variable**?

Procedure

1. Set up a row of eight test tubes. Use a marker to number them 1 through 8.

2. To the test tubes, add 2 ml of the solutions listed in Table 5A-1, matching each number to the number on the tube.

3. Add one dropperful (approximately 2 ml) of Benedict's reagent to each tube.

4. Mix the reagent and the sample by agitating the solution in each tube from side to side. Record the original color of each tube's contents in Table 5A-1, under "Benedict's Test."

5. Heat the test tubes in a boiling water bath for 3 minutes. Record any color changes in Table 5A-1 under "Benedict's Test."

   a. Why did you test water with Benedict's reagent? _____

   b. Which sugars reacted with Benedict's reagent? _____

      Why? _____

   c. Which sugars did not react with Benedict's reagent? _____

      Why not? _____

   d. Explain your results for each of the following:

      Onion juice _____

      Potato slice _____

      Milk _____

Table 5A-1   Data Table for Benedict's and Lugol's Tests

Tube	Benedict's Test		Lugol's Test	
	Original Color Before Boiling	Final Color After Boiling	Original Color Before Adding I_2KI	Final Color After Adding I_2KI
1. Water				
2. Starch				
3. Glucose				
4. Maltose				
5. Sucrose				
6. Onion juice				
7. Potato slice				
8. Milk				

Did the results for each test support your hypotheses? _____

Did your results support your null hypothesis? _____

Did the results for each test agree with your predictions? _____

Explain any discrepancies. _____

PART 2   **Lugol's Test for Starch**

Lugol's reagent changes from a brownish or yellowish color to blue-black when starch is present, but there is no color change in the presence of monosaccharides or disaccharides.

Formulate a **hypothesis** and **predict** what you might expect to find for each of the sugars to be tested with Lugol's reagent (Table 5A-1).

	Hypothesis	**Prediction**
1. Water	_____	_____
2. Starch	_____	_____
3. Glucose	_____	_____
4. Maltose	_____	_____
5. Sucrose	_____	_____
6. Onion juice	_____	_____
7. Potato slice	_____	_____
8. Milk	_____	_____

What is your **null hypothesis** for this group of tests?

What is the **independent variable**?

What is the **dependent variable**?

ΙΙΙΙΙ **Procedure** ΙΙΙΙΙΙΙΙΙΙΙΙΙΙΙΙΙΙΙΙΙΙΙΙΙΙΙΙΙΙΙΙΙΙΙΙ

1. Prepare another eight test tubes as indicated in step 1 above.

2. To the test tubes, add 1 ml of the solutions listed in Table 5A-1, matching each number to the number on the tube.

3. Record the original color of each tube's contents in Table 5A-1.

4. Add several drops of Lugol's reagent (I₂KI) to each tube, mix, and immediately record in Table 5A-1 any color changes that take place. Do *not* heat the test tubes in the Lugol's test.

   a. *Which sugars reacted with Lugol's reagent?* _____

      Why? _____

   b. *Which sugars did not react with Lugol's reagent?* _____

      Why not? _____

   c. *Explain your results for each of the following:*

      Onion juice _____

      Potato slice _____

      Milk _____

Did the results for each test support your hypotheses? _____

Did your results support your null hypothesis? _____

Did the results for each test agree with your predictions? _____

Explain any discrepancies. _____

   d. *From the results of the Benedict's and Lugol's tests, would you conclude that a potato stores its carbohydrates as*

      sugars or as starch? _____ *How do you know?* _____

   e. *How does an onion store its carbohydrates?* _____ *How do you know?*

_____

👁 **EXERCISE B**   Testing for Lipids

The word **lipid** refers to any of the members of a rather heterogeneous group of organic molecules that are soluble in nonpolar solvents such as chloroform ($CHCl_3$) but insoluble in water. Although lipids include fats, steroids, and phospholipids, this exercise will focus primarily on fats.

Triglycerides, a popular topic in discussions of diet and nutrition, are the most common form of fat. They consist of three fatty acids attached to a glycerol molecule (Figure 5B-1). Triglycerides are found predominantly in adipose tissue and store more energy per gram than any other types of compounds.

At room temperature, some lipids are solid (generally those found in animals) and are referred to as **fats,** while others are liquid (generally those found in plants) and are referred to as **oils.** Vegetable oil, a liquid fat, is a mixture of triglycerides.

Since both solid and liquid fats are nonpolar, you will test for their presence by using Sudan IV, a nonpolar dye that dissolves in nonpolar substances such as fats and oils but not in polar substances such as water.

ΙΙΙΙΙ **Objectives** ΙΙΙΙΙΙΙΙΙΙΙΙΙΙΙΙΙΙΙΙΙΙΙΙΙΙΙΙΙΙΙΙΙΙΙΙ

☐ Distinguish between lipid and nonlipid substances using the Sudan IV test.

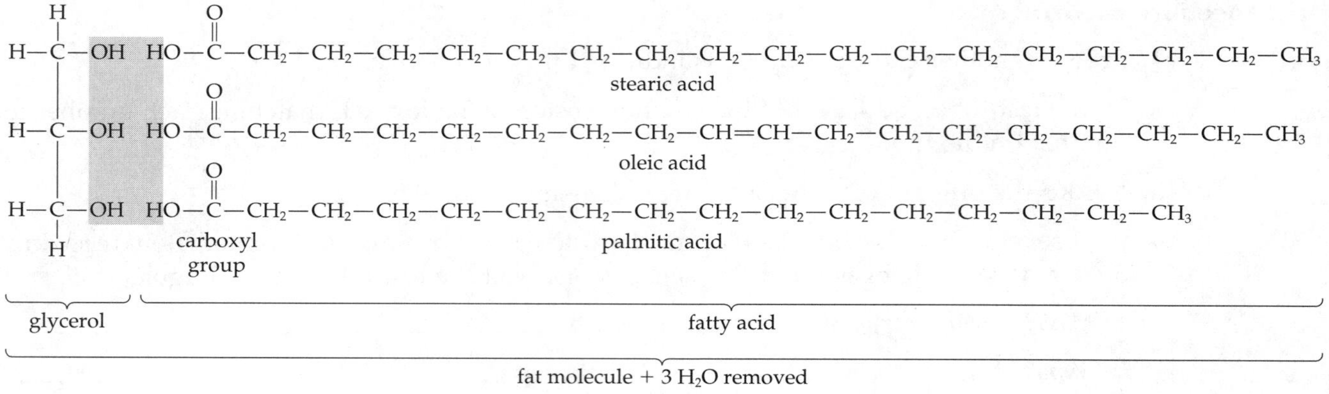

Figure 5B-1   *A triglyceride is composed of three fatty acids and a glycerol molecule. Each bond is formed when a molecule of water is removed by condensation (boxed).*

⁞⁞⁞⁞ Procedure ⁞⁞⁞⁞⁞⁞⁞⁞⁞⁞⁞⁞⁞⁞⁞⁞⁞⁞⁞⁞⁞⁞⁞⁞⁞⁞⁞⁞⁞⁞⁞⁞⁞⁞⁞

1. The familiar "grease spot" is the basis of a very simple test for fats. On a piece of unglazed paper, such as brown wrapping paper, place one drop of oil and one drop of water. Allow the drops to dry.

   a. *Describe the difference between the oil spot and the water spot after a period of drying.*

   _____

The Sudan IV test is a more useful laboratory test for fats. (Since the Sudan IV test is messy, your instructor may demonstrate this method.)

Formulate a **hypothesis** and **predict** what you might expect to find for each of the substances to be tested with Sudan IV (your instructor will tell you which five substances are to be tested; list them here and in Table 5B-1).

	Hypothesis	**Prediction**
1.	_____	_____
2.	_____	_____
3.	_____	_____
4.	_____	_____
5.	_____	_____

What is your **null hypothesis** for this group of tests?

What is the **independent variable**?

What is the **dependent variable**?

2. Label five tubes in sequence, 1 through 5. Add 1 dropperful (1 ml) of each substance you have listed in Table 5B-1 to the appropriate tube. Add 3 drops of Sudan IV to each tube. Mix and then add 2 ml of water to each tube. If fats or oils are present, these will appear as floating red droplets or as a floating red layer colored by Sudan IV. In Table 5B-1, record the reactions that occur in each of the test tubes.

   b. *Why do the droplets float, rather than mix with water?* _____

   _____

Table 5B-1   Data Table for the Sudan IV Solubility Test

Substance	Sudan IV Solubility Reaction
1.	
2.	
3.	
4.	
5.	

c. *Which substances reacted with Sudan IV?* _____

_____

d. *Which substances did not react with Sudan IV?* _____

Why not? _____

Did the results for each test support your hypotheses? _____

Did your results support your null hypothesis? _____

Did the results for each test agree with your predictions? _____

Explain any discrepancies. _____

EXERCISE C   Testing for Proteins and Amino Acids

Proteins are made up of one or more polypeptides, which are linear polymers of smaller molecules called **amino acids** (Figure 5C-1a). Amino acids derive their name from the amino group and the carboxyl group (acidic) that each possesses. Polypeptides are formed when amino acids are joined together by **peptide bonds** between the amino group of one amino acid and the carboxyl group of a second amino acid (Figure 5C-1b).

The biuret reagent reacts with peptide bonds and therefore reacts with proteins, such as egg albumin, but not with free amino acids, such as glycine and alanine. On the other hand, the reagent ninhydrin reacts with the amino group of free amino acids, but not with polypeptides.

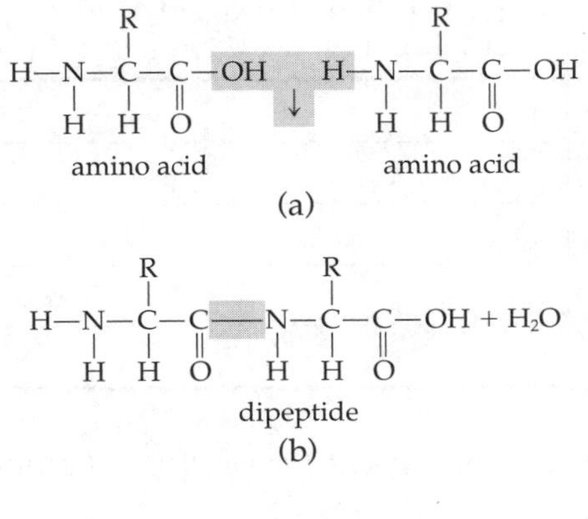

Figure 5C-1   *(a) Structure of an amino acid. Note the presence of an amino (—NH$_2$) group and a carboxyl (—COOH) group. (b) Two amino acids are joined by a peptide bond when a molecule of water is "split out" from their amino and carboxyl groups. A polypeptide is made up of many amino acids joined in this way. "R" represents a side group that is characteristic for each amino acid. (Here and in Figure 5C-2, amino acids are shown in their nonionized form.)*

||||| Objectives ||||||||||||||||||||||||||||||||||||||

☐ Distinguish between free amino acids and proteins (polypeptides) on the basis of their ability to react with either biuret reagent or ninhydrin.

👁 PART I   Testing for Protein with Biuret Reagent

The biuret reagent is light blue, but in the presence of proteins it turns violet. Other types of molecules may cause other color changes, but only the violet color indicates the presence of polypeptides.

Formulate a **hypothesis** and **predict** what you might expect to find for each of the substances to be tested with biuret reagent (Table 5C-1).

	Hypothesis	Prediction
1. Distilled water		
2. Egg albumin		
3. Potato starch		
4. Glucose		
5. Amino acid		

What is your **null hypothesis** for this group of tests?

What is the **independent variable**?

What is the **dependent variable**?

||||| Procedure |||||||||||||||||||||||||||||||||||||||||

1. Obtain five clean test tubes and use a wax pencil to number them from 1 through 5.

2. To the test tubes, add 2 ml of the solutions listed in Table 5C-1, matching each number to the number on the tube.

3. Add one dropperful (approximately 2 ml) of biuret reagent to each tube.

4. After an incubation period of 2 minutes, record your results in Table 5C-1 and determine whether the solution treated contains protein. Base your conclusions only on the presence or absence of the violet color.

Table 5C-1   Data Table for the Biuret Test

Substance	Color with Biuret Reagent after 2 Minutes	Protein Present (+) or Absent (−)
1. Distilled water		
2. Egg albumin		
3. Potato starch		
4. Glucose		
5. Amino acid		

a. *What does this test tell you about the biochemical composition of starch or glucose?*

b. *Why has water been included as one of the test substances?*

c. Which substances reacted with biuret reagent? _____

   Why? _____

d. Which substances did not react with biuret reagent? _____

   Why not? _____

   _____

Did the results for each test support your hypotheses? _____

Did the results support your null hypothesis? _____

Did the results for each test agree with your predictions? _____

Explain any discrepancies. _____

✔ **PART 2** **Testing for Amino Acids with Ninhydrin**

Ninhydrin reagent turns purple or violet in the presence of the free amino groups in amino acids. In the presence of proline, however, it turns yellow. Proline reacts differently because its amino group is not free but is, instead, part of the ring structure of the molecule (Figure 5C-2a, b).

Figure 5C-2   *(a) The amino acid leucine, with free amino group. (b) Proline, with amino group incorporated into a ring.*

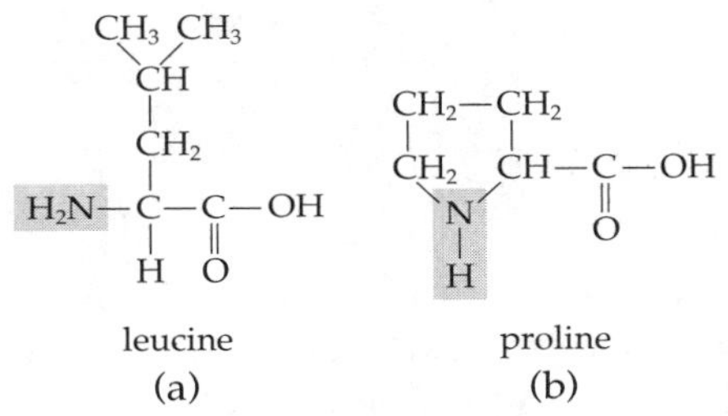

leucine
(a)

proline
(b)

⁞⁞⁞⁞ Procedure ⁞⁞⁞

1. Obtain a piece of filter paper and divide it into four quadrants with a pencil. Letter the quadrants A, B, C, and D.

2. Place one drop of each solution (labeled A, B, C, D) onto the filter paper in the quadrant with the corresponding letter. Allow the spots to dry.

3. Apply one drop of ninhydrin to each spot. Caution: Ninhydrin is poisonous; avoid contact with your skin. Allow the paper to dry at room temperature for 20 to 30 minutes. (The reaction will occur more quickly if the paper is lightly passed over a warm hotplate.)

4. One of the solutions contains proline, two of the solutions contain amino acids other than proline, and one is distilled water. In Table 5C-2, indicate the content of each solution.

Table 5C-2   Data Table for the Ninhydrin Test

Solution	Final Color with Ninhydrin	Type of Molecule in Solution
A		
B		
C		
D		

👁 EXERCISE D | Chromatography of Amino Acids

Amino acids can be classified as **nonpolar, polar uncharged,** or **polar charged** (acidic or basic) molecules, based on the structure of their R groups (Figure 5D-1). Depending on the nature of these R groups, different amino acids will be more or less soluble in different types of solvents. For instance, nonpolar amino acids will be more soluble in nonpolar solvents such as chloroform, whereas polar amino acids will be more soluble in polar solvents such as water. These differences in solubility allow mixtures of amino acids to be separated by **chromatography.**

Figure 5D-1 *Three types of amino acids, classified by R groups. (a) Nonpolar amino acids: R group composed of an unsubstituted hydrocarbon chain, often with branches or rings. (b) Polar uncharged amino acids: R group composed of substituted group, usually nonionized. (c) Polar charged amino acids: R group containing acidic (—COOH) or basic (—NH₂) groups. Here, all amino acids are shown in their fully ionized form at pH 7.*

Chromatography, a method for separating and identifying biological molecules, makes use of the tendency for molecules to show selective attraction to various substances, depending on their polarity (having charged or uncharged polar groups). Since molecules show characteristic differences in attraction, these differences can be used to identify unknown molecules and to separate and identify mixtures of molecules in solution. There are many chromatographic techniques, but most use the same basic method. The mixture to be tested is applied to a solid, stationary matrix material (called the *stationary phase*), which selectively absorbs the molecular mixture. It is then exposed to a motile substance (the *mobile phase*). The solid matrix can be a powder of fine "beads" packed in a vertical column (column chromatography) or spread in a thin layer on a glass plate (thin-layer chromatography). High-quality filter paper can also serve as a solid matrix (paper chromatography). The mobile phase can be a liquid (or a combination of several different liquids) or a gas. As the mobile phase moves through the stationary phase, different molecules within the mixture are separated or *partitioned* according to their relative attraction for each of the two phases. The tendency for different types of molecules in a mixture to bind to the stationary phase or to dissolve in the mobile phase will determine how far they will be moved by the mobile phase.

In a column, changing the pH or type of solvent (polar vs. nonpolar) causes different molecules to move down and finally out of the column, where they can be collected in tubes. In paper or thin-layer chromatography, after the mobile phase has moved across the stationary phase, each type of separated molecule appears as a distinct spot on the stationary phase (the plate or the paper). The stationary phase on which the molecules have partitioned themselves is called a **chromatogram.**

Paper chromatography can be carried out in either an ascending or a descending mode. In ascending chromatography, one end of the paper matrix is placed in the bottom of a chamber containing a solvent. As the solvent moves up the paper, its progress can be noted by observing the "solvent front." In descending chromatography, the solvent is contained in a trough at the top of the chamber.

It is possible to identify separated compounds by the *ratio* of the distance traveled by a compound in a particular solvent to the distance traveled by the solvent. This ratio is known as a compound's R_f value.

$$R_f = \frac{\text{distance of spot from origin}}{\text{distance of solvent (solvent front) from origin}}$$

For precise analysis, it is best to chromatograph unknown compounds together with known compounds, then compare the distances traveled.

Locating substances on chromatograms can be accomplished in various ways—by using fluorescent or other dyes (usually applied to the chromatogram in spray form) or by autoradiography (visualization of spots containing radioactivity).

▥▥▥▥ Objectives ▥▥▥▥▥▥▥▥▥▥▥▥▥▥▥▥▥▥▥▥▥▥▥▥▥▥▥

☐ Distinguish among nonpolar, polar uncharged, and polar charged amino acids based on their solubility properties.

☐ Describe the process of chromatography and how it can be used to separate different types of molecules.

▥▥▥▥ Procedure ▥▥▥▥▥▥▥▥▥▥▥▥▥▥▥▥▥▥▥▥▥▥▥▥▥▥▥▥▥▥▥

1. Place a clean paper towel on a clean counter. Cut a piece of Whatman filter paper into an 8-inch × 11-inch rectangle and place it on the paper towel (the paper towel helps to keep the filter paper clean.)

2. Orient the paper so that the 11-inch-long side is facing you.

3. Using a *pencil* and a ruler, draw a light line 1 inch from the bottom across the full width of the paper. Beginning 1½ inches from the left end of the line, make a series of six small cross marks on this line every 1½ inches (Figure 5D-2a).

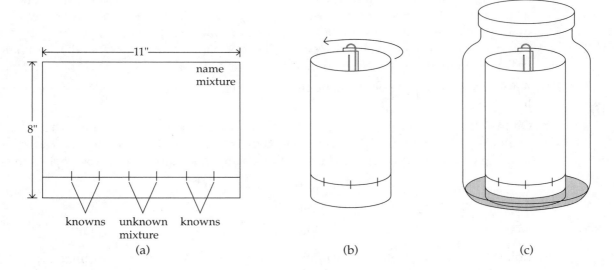

Figure 5D-2 *Preparation of a paper chromatogram. (a) Use a pencil to lightly mark the placement of amino acid samples. Use duplicate spots for the unknown mixture. (b) Roll the paper into a cylinder and hold it together at the top and bottom with plastic paper clips. (c) Place the paper cylinder into a gallon jar containing ¼ to ½ inch of chromatography solvent.*

4. Your instructor will provide you with an "unknown" sample containing a mixture of three amino acids. If several unknown solutions are being used in your laboratory, be sure to record the letter or number that your instructor has used to identify the unknown you selected. Code for unknown solution _____. You will also receive four amino acid "knowns." Record the names of the known samples:

Amino acid 1 _____

Amino acid 2 _____

Amino acid 3 _____

Amino acid 4 _____

Based on the structures of the known amino acids, formulate a hypothesis about the differences in their migration tendencies as measured by chromatography using a nonpolar solvent.

HYPOTHESIS:

NULL HYPOTHESIS:

What do you **predict** about the migration patterns of amino acids 1, 2, and 3?

What is the **independent variable**?

What is the **dependent variable**?

5. Using a wooden applicator stick, apply a small spot of unknown solution (2 mm in diameter) to the filter paper on the line at one of the cross marks located at the center of the paper. Allow the spot to dry (you may use a hair dryer if available). Repeat this process at the same mark two additional times. *Do not* press the applicator stick to the paper hard enough to make a dent in the paper. On a separate sheet of paper, draw and label the chromatogram to indicate the mixture you have applied to each cross mark.

6. Apply another sample of the same unknown to a second mark using the same procedure as in step 5.

7. On the other cross marks, apply spots of the known amino acids—one amino acid per cross mark. Apply each amino acid three times, allowing spots to dry between applications. On the drawing of your chromatogram, be sure to indicate which amino acid was applied to each mark.

Mark 1 _____   Mark 4 _____

Mark 2 _____   Mark 5 _____

Mark 3 _____   Mark 6 _____

8. Roll the chromatography paper into a cylinder and hold it together at the top and bottom with *plastic* paper clips (Figure 5D-2b).

9. Cover the bottom of a gallon glass jar with chromatography solvent (0.25 to 0.5 inch deep).

10. Place the cylinder of chromatography paper into the jar (Figure 5D-2c) and allow it to run for at least 4, preferably 5 to 6, hours. The instructor will remove your chromatograms and mark the solvent front. The chromatograms will then be allowed to air dry.

11. **Next Day:** Spray your chromatogram with ninhydrin or, alternatively, dip the chromatogram into a solution of ninhydrin. Caution: Wear gloves; if you use the spray, work under a hood. Heat the chromatogram according to directions from you instructor. Amino acids will turn purple in the presence of ninhydrin, except for proline, which will turn yellow.

12. Circle spots and record their colors in the table that follows. Determine the R_f values for each spot of unknown amino acid on your chromatogram, then determine the R_f value for each spot of known amino acid. To find the R_f value, estimate the center of each amino acid spot,

Unknown (Code _____)	Color	R_f	Knowns	Color	R_f
Sample 1: Spot 1	_____	_____	Amino acid 1	_____	_____
Spot 2	_____	_____	Amino acid 2	_____	_____
Spot 3	_____	_____	Amino acid 3	_____	_____
			Amino acid 4	_____	_____
Sample 2: Spot 1	_____	_____			
Spot 2	_____	_____			
Spot 3	_____	_____			

measure the distance it has traveled from the origin (cross mark), then measure the distance the solvent front has traveled from the origin. Find the ratio of these distances and record this information in the table.

13. Compare R_f values for the knowns and your unknown mixture. Identify and record the amino acids in your mixture: _____, _____,

_____

Do the R_f values for the known amino acids support your hypothesis? _____ Your null hypothesis? _____
Was your prediction correct? _____
If there are discrepancies between your hypothesis and your results, how might these be explained?

_____

_____

| EXERCISE E | Analyzing Unknowns Qualitatively |

||||| Objectives |||||||||||||||||||||||||||||||||||||

☐ Use knowledge gained in the previous exercises to identify the organic molecules present in unknown solutions.

||||| Procedure ||||||||||||||||||||||||||||||||||||||

A number of commercial products and food samples containing a variety of compounds are available for testing. Using reagents and methods from the previous exercises, identify the kinds of molecules contained in these solutions.

Formulate a **hypothesis** and **predict** what you might expect to find for each test (positive or negative result) based on your knowledge of the substance and the tests performed during this laboratory. (Your instructor will tell you which substances are to be tested).

Substance	Hypothesis	Prediction
_____	_____	_____
_____	_____	_____
_____	_____	_____
_____	_____	_____

_____   _____
_____   _____
_____   _____
_____   _____
_____   _____
_____   _____

Conduct all tests according to directions in Exercises A through C. Record the results in Table 5E-1. Report your results according to your instructor's directions and identify each numbered unknown.

The following hints will help you conduct successful analyses of the unknowns:

- *Never* insert your pipette into the stock bottle of solutions. Instead, pour a small amount into a beaker and pipette from this. Empty and rinse out the beakers at the end of the period.

- The success of this lab relies on using clean glassware and avoiding cross-contamination between solutions. Therefore, wash all glassware with warm soapy water; rinse thoroughly several times with tap water and once with distilled water before using in the next set of tests.

- Make sure you are able to differentiate between a positive reaction (the color change specific for the test being done) and a negative reaction (no color change or a change to a color other than the one specific to the test).

Table 5E-1   Data Table for Analyzing Unknowns Qualitatively

Unknown Tested	Positive (+) or Negative (−) Results				
	Benedict's Test	Lugol's Test	Biuret Test	Ninhydrin Test	Sudan IV Test
1.					
2.					
3.					
4.					
5.					
6.					
7.					
8.					
9.					
10.					

Did you predict the contents of each unknown accurately? (Your instructor will provide a list of expected test results for the unknowns.) _____

If not, investigate the contents of the unknowns by examining the product labels, and explain your results.

_____

_____

Laboratory Review Questions and Problems

1. Explain the limitations of Benedict's test in determining whether or not sugar is present in a certain food product. Why do all monosaccharides, but only some disaccharides, react with Benedict's reagent?

2. What did you learn about the specificity of the biuret reagent?

3. In this lab, you used a biuret reagent to determine the presence of albumin (egg white) in solution. Why didn't you use ninhydrin as the colorimetric reagent?

4. The leaves of many plants are coated with a waxy substance that causes them to shed water. How would you expect this substance to react in the Sudan IV test?

5. Ninhydrin reacts with a mixture of amino acids and turns purple. Could proline be one of the amino acids? _____ How could you find out if the mixture contained proline?

6. Several individual unknowns are tested to determine the type of molecule present. Given the completed table below, indicate whether unknowns 1 through 5 are protein, reducing sugar, starch, lipid, or free amino acids (+ = positive result).

Unknown	Benedict's Test	Lugol's Test	Biuret Test	Ninhydrin Test	Sudan IV Test	Answer
1	−	−	+	−	−	_____
2	+	−	−	−	−	_____
3	−	+	−	−	−	_____
4	−	−	−	+	−	_____
5	−	−	−	−	+	_____

7. Mixtures of unknowns are tested with several colorimetric reagents. Given the results in the following table, determine which of the four choices below best describes the contents of each tube (+ = positive result).

Test Tube	Benedict's Test	Lugol's Test	Biuret Test	Ninhydrin Test	Sudan IV Test
1	−	−	+	+	−
2	+	−	+	−	+
3	+	+	−	−	+

a. Tube 1: reducing sugar and protein

   Tube 2: lipid, free amino acids, and protein

   Tube 3: starch, reducing sugar, and lipid

b. Tube 1: protein and free amino acids

   Tube 2: starch, protein, and lipid

   Tube 3: free amino acids, starch, and protein

c. Tube 1: protein and free amino acids

   Tube 2: lipid, reducing sugar, and protein

   Tube 3: lipid, reducing sugar, and starch

d. Tube 1: free amino acids and lipids

   Tube 2: lipid, starch, and free amino acids

   Tube 3: starch, free amino acids, and reducing sugar

8. You test several unknowns and obtain the following results:

Solution	Results of Lugol's Test	Results of Benedict's Test	Results of Ninhydrin Test
I	Yellow	Blue	Violet
II	Yellow	Orange	Clear
III	Black	Blue	Clear
IV	Brown	Cloudy blue	Yellow
V	Yellow	Blue	Clear

a. Which solution contains starch?

b. Which solution is most likely glucose?

c. Which solution contains an amino acid other than proline?

9. What are some good sources of carbohydrates, proteins, and fats in a typical breakfast, lunch, and dinner?

10. What nutrients (biological molecules) would you obtain from red meat? Do nutritionists recommend that some fats be present in your diet? What would you use these for?

11. Some vitamins should not be taken in excess. Which ones? Why?

12. Four known amino acids (*AA*) produce a chromatograph with R_f valves as follows: $AA_1 = 0.25$, $AA_2 = 0.66$, $AA_3 = 0.50$, $AA_4 = 0.80$. A mixture of three amino acids yields spots that are 2 cm, 5.2 cm, and 6.4 cm from the origin. The solvent front is 8.0 cm from the origin. Which three of the four amino acids are contained in the mixture?

13. Some amino acids are called essential amino acids. What does this mean? Fatty acids with more than one double bond are considered as essential fatty acids. Animals cannot make fatty acids with more than one double bond. What are the sources of essential fatty acids?

14. In winter, plants exchange the saturated lipids in their membranes for unsaturated lipids. Unsaturated lipids are "bent" and keep the membranes more fluid because they cannot be stacked closely together. Of what advantage would this be for herbaceous plants that live through the winter? (Hint: what happens to bacon grease or the grease on top of soup when you put it in the refrigerator?)

Prokaryotic Cells

OVERVIEW

Understanding the nature of cell structure and function is important to an understanding of organisms. All organisms are composed of cells, whether they exist as single cells, colonies of cells, or in multicellular form. Cells are usually very small, and for this reason, a thorough understanding of subcellular structure and function has been possible only through advances in electron microscopy and molecular biology.

There are two general types of cells: **prokaryotic** and **eukaryotic.** These two words have their root in the Greek word *karyon* (nut), which refers to a cell's nucleus. The prefix *pro-* means "before" or "prior to." Thus *prokaryotic* means "before having a nucleus." Prokaryotic cells do not have a membrane-bound nucleus and their genetic material (DNA) is only loosely confined to a nuclear area within the cell. Bacteria, including the cyanobacteria (formerly known as blue-green algae), are prokaryotes. All other organisms are eukaryotes. The prefix *eu-* means "true." The cells of eukaryotes have true, membrane-bound nuclei containing their genetic material.

Prokaryotic and eukaryotic cells also differ in several other ways. Eukaryotic cells are generally larger and contain additional specialized compartments (**membrane-bound organelles**) in which cell functions such as energy production may occur. Prokaryotic cells lack membrane-bound organelles; their cell functions are carried out in the cytoplasm.

During this laboratory you will investigate some of the structural and biochemical properties of prokaryotic cells. During Laboratory 7 you will study eukaryotic cells.

STUDENT PREPARATION

Prepare for this laboratory by reading the text pages indicated by your instructor. Familiarizing yourself in advance with the information and procedures covered in this laboratory will give you a better understanding of the material and improve your efficiency.

EXERCISE A   Producing Protobionts

Evidence suggests that the first cells were prokaryotic. They were probably anaerobic, living in an atmosphere with little or no oxygen, and heterotrophic, consuming organic materials as a carbon source. But how did these first cells come to be? In the 1930s, the Russian scientist A. I. Oparin suggested that molecular aggregates, called **coacervates,** which can be prepared in the laboratory, may resemble the precursors to the first cells. These aggregates are composed of high-molecular-weight substances that, when dissolved in water, aggregate to form viscous droplets.

Although these droplets are not living cells, coacervates and cells have several characteristics in common. They are able to maintain an internal environment that differs from their surroundings and they can selectively absorb substances from their surroundings. Enzymes are included among the high-molecular-weight molecules—usually a combination of lipids, polypeptides, polysaccharides, and nucleic acids—used to form coacervates. Enzymes can catalyze reactions, including syntheses, hydrolyses, and electron transfer.

Coacervates are not alive, hence they are called **protobionts** (*proto-*, "before", plus *bionts*, "life"). However, the formation of coacervates may have allowed organic molecules to interact with one another to form new structural entities—the forerunners of cellular organelles.

ⅠⅠⅠⅠⅠ Objectives ⅠⅠⅠⅠⅠⅠⅠⅠⅠⅠⅠⅠⅠⅠⅠⅠⅠⅠⅠⅠⅠⅠⅠⅠⅠⅠⅠⅠⅠⅠⅠⅠⅠⅠⅠⅠⅠⅠ

☐ Demonstrate how molecules of various molecular weights can be combined to form coacervates.

☐ Relate the biochemical properties of molecules to their role in coacervate formation.

ⅠⅠⅠⅠⅠ Procedure ⅠⅠⅠⅠⅠⅠⅠⅠⅠⅠⅠⅠⅠⅠⅠⅠⅠⅠⅠⅠⅠⅠⅠⅠⅠⅠⅠⅠⅠⅠⅠⅠⅠⅠⅠⅠⅠⅠⅠ

1. In a test tube, mix together 5 ml of 1% gelatin (protein) and 3 ml of 1% gum arabic (carbohydrate). Cover the tube with a piece of Parafilm and shake well.

2. Measure the pH by transferring a drop of the solution to a small section of pH paper using a glass stirring rod. Rinse the stirring rod. pH = _____

3. Prepare a wet mount of the mixture and observe it under the low power of the microscope (10✕). Save the slide.

4. Add acid (1% HCl) to the test tube containing the gelatin/gum arabic mixture, one drop at a time, shaking the mixture after each addition. After each addition wait to see whether the solution becomes cloudy. If it remains clear, add another drop. When the solution turns cloudy, shake well. If it turns clear, add another drop. *Caution:* Excessive acid will cause the cloudiness to disappear. At that point, the acidity required for the formation of coacervates has been exceeded and the procedure must be repeated.

5. Once the solution is permanently cloudy, make another pH reading. pH = _____ Prepare another wet-mount slide of the solution. At this time, coacervates should be visible. If you have difficulty observing the coacervate spheres, adjust the light or use a higher power.

6. Observe the original slide to assure yourself that its solution has no coacervates.

   a. How do you think the pH may be affecting the protein or carbohydrate molecules?

   _____

7. Use congo red, neutral red, or methylene blue to stain the coacervates. Place one drop of coacervate solution on a glass slide and mix with one drop of a stain. Place a coverslip over the preparation and observe under the compound microscope at high power. Use a fresh drop of coacervate solution for each stain.

   b. Do all of the dyes stain the coacervates? _____ *Congo red?* _____ *Neutral red?* _____
   Methylene blue? _____

   Congo red, in the presence of weak acids, changes color from red to blue. It is often used as an indicator but can also be used to stain carbohydrates. *Neutral red* is yellowish at higher pH values but turns red to pink at a slightly acidic pH and blue in strong acids. Often used as an indicator, neutral red stains cell cytoplasm containing carbohydrates and proteins. *Methylene blue* is also used as a general stain for cytoplasm and nuclei.

   c. How would you explain your results from staining the coacervates based on the properties of the stains described here? _____

   _____

   _____

8. Add a drop of 5% NaCl to the edge of the coverslip of each of your stained slides. Observe the coacervates as the NaCl diffuses into the solution.

   d. What happens to the coacervates? _____

   Why? _____

   _____

9. When you have a finished observing the coacervates, add more acid to the mixture in the test tube, a drop at a time. Measure the pH at three-drop intervals. Continue to add acid until the solution clears.

   e. How do changes in pH affect the production of coacervates? _____

   f. Which molecules do you think are responsible for forming the membrane-like barrier between the

   coacervate and its environment? _____

   Why? _____

   _____

EXERCISE B   Examining Bacterial Cells

Present-day bacteria are extremely small (approximately 1 to 2 μm in diameter). Most are heterotrophic, depending on preformed food, but some are photosynthetic and make their own food. Morphologically, bacteria are either round (**cocci**), rod-shaped (**bacilli**), or spiral-shaped (**spirilla**). To view them with the light microscope, you must use an oil-immersion lens (100×). Even then, not much more than their basic shapes will be visible. With the aid of the electron microscope, however, you can study these prokaryotic cells more closely. You can even use special staining techniques to learn about their structure.

ııııı Objectives ııııııııııııııııııııııııııııııııııııı

☐ Give two general characteristics of prokaryotic cells.

☐ Distinguish among the three morphological types of bacteria.

☐ Describe the subcellular structure of a typical bacterium.

☐ Compare the structures of gram-positive and gram-negative bacteria.

✓ PART I   Observing Bacteria Using the Light Microscope

You can use the light microscope to study bacteria, but only their external features will be distinguishable. It is possible to identify the three morphological types of bacteria (coccus, spirillum, and bacillus) by observing their shape (Figure 6B-1). You will also note that bacteria are often found in clusters or in chains. Some have one or more flagella (Figure 6B-2).

ııııı Procedure ııııııııııııııııııııııııııııııııııııı

1. Observe a prepared slide of *Escherichia coli* on demonstration. Describe its shape.

   _____

2. Observe a prepared slide of *Staphylococcus aureus* on demonstration. Describe its shape.

   _____

3. Observe a prepared slide of *Spirillum volutans* on demonstration. Describe its shape.

   _____

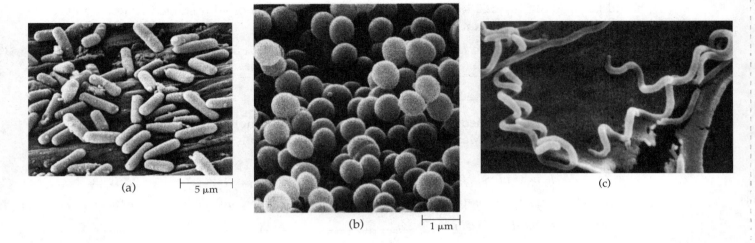

Figure 6B-1   *The cells of many familiar genera of bacteria include the*
(a) rod-shaped bacillus, (b) spherical coccus, and (c) helical spirillum.

Figure 6B-2   *A bacterial cell,*
Proteus mirabilis, *showing*
numerous flagella.

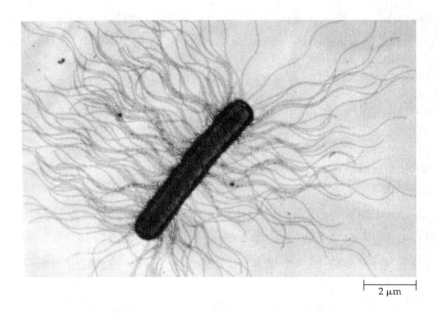

4. Some bacteria have **flagella,** threadlike organelles used in locomotion. Bacterial flagella are composed of the protein flagellin. Observe the flagellated bacteria on demonstration. Do they have one or more than one flagellum? _____

✔ PART 2   Observing Bacteria Using the Transmission Electron Microscope (TEM)

The resolving power of an electron microscope is approximately 1,000 times that of a light microscope. This allows us to observe the subcellular structure of bacteria. The most prominent feature within the bacterial cell is the **nucleoid** region, which may appear as a lighter, fibrous central area. The genetic material of the cell is dispersed throughout this region. The cytoplasm is filled with granular-looking bodies called **ribosomes**. Both a **cell membrane** and a thicker **cell wall** (composed of peptidoglycan; a polymer of amino sugars, and other polymers in some species) surround the bacterium. Outside the cell wall there is often a layer of "slime" that forms a polysaccharide **capsule**. The capsule may help to protect bacteria against attack by the immune system of a host organism and from dehydration. Many bacteria do not produce capsules, and even those with capsules will not die if the capsule is destroyed (Figure 6B-3).

Figure 6B-3 *Convoluted membranes of a large mesosome (arrow) extend throughout this bacterium. The mesosome is continuous with the cell membrane. A thick, slimy capsule surrounds the cell.*

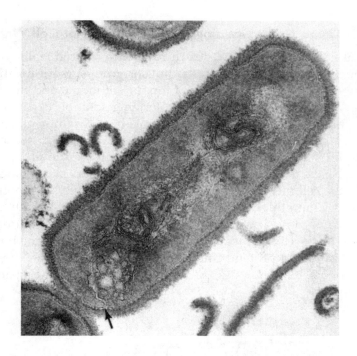

Some bacteria possess **mesosomes,** whorls of membranous material that extend inward from the cell membrane. The function of mesosomes is not yet certain, but they may play a role in cell division or in various energy-releasing reactions. In photosynthetic bacteria (see Part 4 of this exercise), the cell membrane often folds extensively to form an internal membrane system containing chlorophyll and other pigments.

ııııı Procedure ıı

Figure 6B-4 is a transmission electron micrograph of an *Escherichia coli* cell that is in the process of dividing. The two new cells have not yet separated completely. The cell wall of *E. coli* is defined on its inner surface by the cell membrane and on its outer surface by a second, outer membrane. The cell wall exists in the light-colored area between the two membranes; this area is known as the **periplasmic space.** Identify the labeled structures in Figure 6B-4.

Figure 6B-4 Escherichia coli *cell in the process of dividing.*

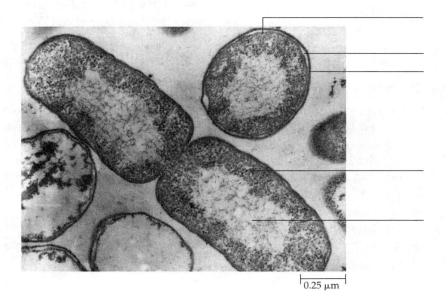

0.25 μm

👁 **PART 3**   **Using Gram Staining to Study Bacterial Cell Walls**

The chemical nature and physical structure of bacterial cell walls differ among species. This difference is the basis for classifying bacteria into two major groups, **gram-positive**, and **gram-negative,** based on their staining properties with **Gram's stain,** a mixture of iodine and the dye crystal violet. When stained with Gram's stain and then treated with a decolorizing solution (alcohol or acetone), gram-positive cells retain the crystal violet color and appear deep purple-blue. Gram-negative cells lose the crystal violet color when washed with alcohol, but are stained red when treated with a second stain, safranin.

The key differences in the nature of the cell wall are shown in Figure 6B-5. Gram-positive bacteria (Figure 6B-5a) have a multilayered, cross-linked network of amino sugars (peptidoglycan) forming a cell wall outside the cell membrane. As many as 20 to 40 layers of peptidoglycan may be present, making the cell wall about 50 nm thick. Polymers of glycerol or ribitol usually extend from the cell membrane into the cell wall. Gram-negative bacteria (Figure 6B-5b) have a much thinner wall of peptidoglycan, usually only several layers thick. The cell wall lies outside the cell membrane but is covered by an additional outer membrane of lipid and protein. Under the electron microscope, the cell wall appears as the lighter layer between the two cell membranes, partly filling the periplasmic space.

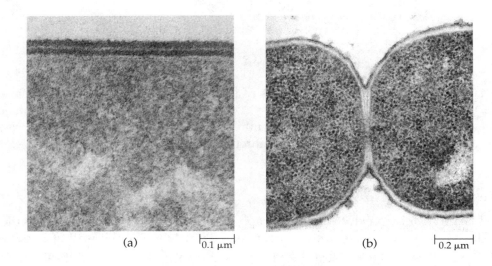

(a)   0.1 μm   (b)   0.2 μm

Figure 6B-5   *Electron micrographs of sections through the cell walls of (a) gram-positive* Bacillus polymyxa *and (b) gram-negative* Escherichia coli. *The wall of a gram-positive bacteria consists of a homogeneous layer of peptidoglycans and polysaccharides seen here as the lower dark band. The upper dark band is a layer of surface proteins. In gram-negative bacteria, a layer of peptidoglycan (cell wall) is sandwiched between inner and outer cell membranes.*

⁙ **Procedure** ⁙⁙⁙⁙⁙⁙⁙⁙⁙⁙⁙⁙⁙⁙⁙⁙⁙⁙⁙⁙⁙⁙⁙⁙⁙⁙⁙⁙

When you go to the doctor's office and a bacterial infection is suspected, a Gram stain test might be performed to determine whether you are infected with gram-positive or gram-negative bacteria. For example, strep infections (from *Streptococcus* species) can readily be identified in this manner.

On the laboratory bench you will find two agar-slant cultures of bacteria. The bacteria are growing on the surface of the nutrient agar; if you look carefully, you can see a "slimy" coating of bacteria growing on the agar. Your challenge is to determine whether the two strains of bacteria, *Escherichia coli* and *Staphylococcus aureus*, are gram-positive or gram-negative.

Using what you know from your reading about the cell wall structure of the two genera of bacteria, form a hypothesis about the Gram staining properties of these two species.

HYPOTHESIS:

NULL HYPOTHESIS: _____

a. What do you **predict** will be the outcome of the Gram stain test? _____

b. What is the **independent variable** in this investigation? _____

c. What is the **dependent variable** in this investigation? _____

1. Work in pairs. Use a clean glass slide. Do not touch the surface of the slide with your fingers. Label one end of the slide "E" and the other "S." One partner should carry out the following steps.

2. Sterilize a wire inoculating loop by flaming it until it is red hot.

3. Using the sterile loop, place two small drops of tap water on the slide approximately 2 cm apart between the "E" and "S" labels.

4. Resterilize the loop. Cool it for a few seconds. With the tip of the loop, remove a small amount of growth from the surface of the *E. coli* agar-slant culture. To maintain sterile conditions, be sure to flame the neck of the test tube after opening it and before closing it.

5. Mix the bacterial cells on the tip of the inoculating loop with the drop of water nearest the "E" label. With a circular motion, spread the spot out to the size of a dime or until it has a pale, milky appearance.

6. Immediately after making the smear, hold the loop *above* the flame to dry the inoculum, *then* flame the loop. (If you plunge the loop into the flame before it is dry, you will splatter bacteria all over yourself and your lab bench.)

7. The second partner should now repeat the above steps using the agar-slant culture of *S. aureus*. Place the *S. aureus* bacterial cells in the drop of water nearest the "S" label.

8. Allow the smears to *air dry*. Hold the slide by the end using forceps and "heat fix" it by passing the slide through the Bunsen burner flame (*right side up*) two or three times. (After heat fixing, the slide should not be too hot to handle. If it is, you have heated it too much.)

9. Place your slide across the edges of a small Petri dish. Place this dish inside a larger Petri dish.

10. Flood the slide drop by drop with **Gram's crystal violet** solution using a Pasteur pipette. Add two drops of 5% sodium bicarbonate buffer to the slide. Stain for 1 minute.

11. Pour off excess stain into the dish provided and *gently* wash the slide with water using a dropper or squirt bottle. Do not allow the dropper to touch the smear.

12. Flood the smears with **Gram's iodine** solution and allow it to react for 1 minute. Pour off the excess and let stand for 1 minute.

13. Wash off the iodine solution as in step 11 and gently blot the slide with bibulous paper or a paper towel.

14. Hold the slide at an angle and apply **Gram's alcohol** drop by drop until the violet color no longer appears in the washes from the smears. This should take only 10 to 15 seconds. *Do not overdo this step.*

15. Quickly rinse off the alcohol with water and blot.

16. Flood the slide with **safranin** and allow it to stain for 1 minute.

17. Wash the slide gently with water. Drain the excess water onto a paper towel and allow the slide to air dry.

18. When the slide is dry, observe it using an oil-immersion objective. Ask your instructor for assistance. You will use one of the oil-immersion microscopes on the demonstration table.*

RESULTS

   d. *Which bacterial species showed a positive test with Gram's stain?* _____

   e. *Which bacterial species showed a negative test with Gram's stain?* _____

CONCLUSION

19. Determine whether *Escherichia coli* and *Staphylococcus aureus* are gram-negative or gram-positive bacteria. *E. coli* _____ *S. aureus* _____

   f. *Do your results support your hypothesis?* _____

   g. *Explain any discrepancies.* _____

👁 **PART 4   Examining Cyanobacteria**

Cyanobacteria are photosynthetic prokaryotes. However, unlike other photosynthetic bacteria, which contain bacteriochlorophyll, cyanobacteria contain **chlorophyll *a*** —the same type of chlorophyll found in eukaryotic green algae and plants. The chlorophyll molecules are not located within chloroplasts, as in higher plants, but are found, instead, within photosynthetic **thylakoid membranes** dispersed throughout the cytoplasm (Figure 6B-6). In addition to chlorophyll *a*, cyanobacteria contain other accessory pigments including the yellow and orange carotenoids and the phycobilins (reddish phycoerythrins and bluish allophycocyanins).

Figure 6B-6 *Photosynthetic membranes fold into stacks inside a bacterial cell.*

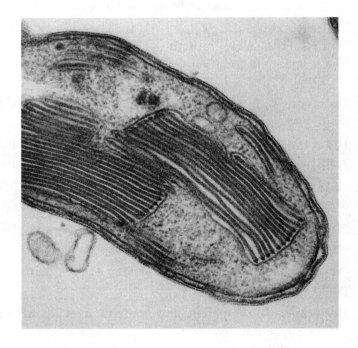

---

*Oil has a greater refractive index than air and thus better resolution can be obtained by replacing the layer of air between the end of the objective and the specimen with a drop of oil into which the objective is carefully immersed. This eliminates much of the refraction and loss of light that occurs when light passes through air before reaching the dry objective. The working distance of an oil-immersion objective (90× to 100×) is less than 1 mm and the objective must almost touch the slide to be in focus. The field of view is much smaller. Light intensity usually needs to be increased when working with oil. Review these principles in Laboratory 1.

ııııı **Procedure** ıı

1. Make a wet mount of one of the cyanobacterial species available in the laboratory. Observe the specimen using the 40× objective.

2. Draw a representative cell in the space below Figure 6B-7. Use this electron micrograph to assist you in labeling the parts of the cell you have drawn.

Figure 6B-7 *Electron micrograph of the cyanobacterium* Anabaena *sp. The large, dark bodies are composed of cyanophycean starch, the chief carbohydrate storage product of cyanobacteria.*

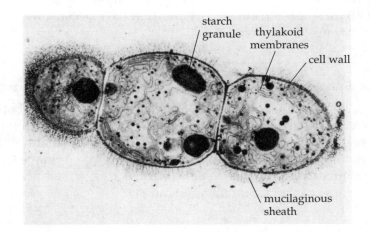

EXERCISE C   **Working with Bacteria**

Microbiologists maintain specific types of bacteria in the laboratory as **cultures;** a culture containing only a single species of bacteria is called a **pure culture.** Since bacteria are found everywhere—in the air, water, soil, and food, on the body surface, and inside other organisms—it is necessary to follow strict procedures to protect laboratory cultures from contamination by unwanted strains of bacteria.

In the laboratory, bacteria are grown on culture media containing the nutrients necessary for their growth. The medium can be liquid (broth medium) or solid if **agar** (a complex carbohydrate product of seaweed) is added to the broth.

Bacteria are transferred from one medium to another or to a new stock of the same medium by **subculturing.** Pure cultures of bacteria, containing only one type of organism, are produced by preparing **streak plates** or **spread plates** that allow discrete **colonies** (small masses of bacteria resulting from multiplication of a single bacterium) to be isolated. These techniques are essential tools for the study of bacterial cells.

ııııı **Objectives** ıı

☐ Subculture bacteria from a broth culture to an agar slant or to a second broth culture.

☐ Prepare a streak-plate culture from a mixture of bacteria to isolate single colonies.

☐ Prepare a spread-plate culture for isolating and culturing single colonies of bacteria.

IIIII **Procedure** IIIIIIIIIIIIIIIIIIIIIIIIIIIIIIIIIIIIIII

All steps should be carried out aseptically (under *sterile* conditions). A few precautions should also be taken: be careful never to lay contaminated or used equipment on laboratory benchtops; if a spill occurs, let your instructor know; dispose of all waste materials in special containers.

✔ **PART 1**   **Techniques for Transferring Cultures**

It is often necessary to begin new cultures by transferring bacterial cells from an existing culture to a new growth medium so as to maintain cultures that have become overgrown and have used up all the available nutrients.

IIIII **Steps for Subculturing Bacterial Cells** IIIIIIIIIIIIIIIIIIIIIIIIIIIIIIIIIIIIIIIIIIIII

1. Obtain a 24-hour nutrient-broth stock culture of *Escherichia coli* (a gram-negative bacillus) or *Serratia marcescens* (a gram-negative, motile bacillus with red pigment) and an agar-slant tube. Label the tube to be inoculated with the name of the bacterium and your initials.

2. Hold both the stock tube and agar-slant tube in one hand using your thumb to separate them into a **V.**

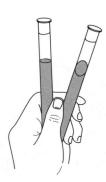

3. Obtain a wire inoculating loop from your instructor. Sterilize the loop by holding it in the hottest (blue) part of a Bunsen burner flame until the loop glows red. Then, continue to pass the lower two-thirds of the shaft through the flame. Once you have "flamed" the loop, never put it down; hold it for 10 to 20 seconds to allow it to cool. (Alternatively, a sterile, disposable plastic inoculating loop can be used— *do not* place it in the flame.)

4. Uncap both tubes by grasping the caps with the second, third, and fourth fingers of the hand holding the inoculating loop. Once removed, *do not* put the caps down or sterility will not be maintained.

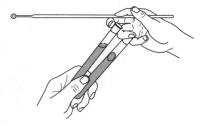

5. Flame the necks of the two tubes briefly by passing them through the Bunsen burner flame.

6. Cool the transfer loop by touching it to the sterile inside surface of the culture tube and then (a) insert the loop into the broth culture and shake slightly. (b) Remove the bacteria-laden loop and insert it into the agar slant until you lightly touch the surface of the agar. Drag the loop across the surface of the agar in a zigzag line and remove. (To transfer a culture from a broth to a broth, see steps 8 to 10; to transfer from a broth to an agar plate, see Part 2 of this exercise.)

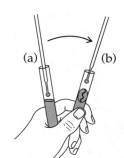

(a)   (b)

(continued next page)

(continued from previous page)

7. After inoculating, (a) reflame the necks of the tubes and (b) replace the caps on the correct culture tubes. (c) Always reflame the inoculating loop after use to destroy any remaining bacteria.

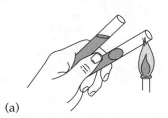

(a)

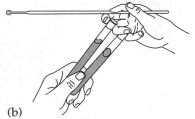

(b)

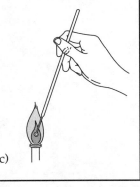

(c)

8. Resterilize the inoculating loop as in step 3.

9. Obtain a tube of sterile nutrient broth and, using the same stock culture tube as before, label the tube to be inoculated, uncap the two tubes, flame the necks of the tubes, and transfer a sample of bacteria from the culture tube to the nutrient broth with the inoculating loop. After inserting the loop into the sterile nutrient broth, shake the loop slightly to dislodge the bacteria.

10. After inoculating, reflame the necks of the tubes, replace the caps, and reflame the inoculating loop.

11. Incubate the tubes containing the inoculated broth and agar at 25°C for 24 to 48 hours. Examine your cultures for the presence of bacterial growth. You can recognize growth as turbidity in the broth culture and the appearance of a whitish (*E. coli*) or orange-red (*S. marcescens*) growth on the surface of the agar slant.

a. Why is the inoculating loop flamed before and after each transfer? _____

_____

b. Why is it important NOT to place the caps to the tubes on your laboratory table during the transfer

procedure? _____

_____

c. Why is it important to cool the inoculating loop before transferring cells? _____

_____

👁 PART 2 Isolating Pure Cultures

The isolation of discrete colonies of bacteria is important in the study of the morphological and biochemical characteristics of bacteria and in recombinant DNA studies.

1. Obtain a nutrient broth culture of a mixture of *S. marcescens* and *E. coli* and a Petri dish of nutrient agar. Label the Petri dish with the names of the bacteria you are using and with your initials.

2. To prepare a **streak plate,** obtain a loopful of culture (follow sterile procedures as in Part 1) and, after carefully lifting one side of the lid of the Petri dish, deposit the bacteria (Figure 6C-1) onto the surface of the agar toward one side of the dish. When performing this type of transfer, always keep the lid of the Petri dish above the agar surface—*never* remove the lid completely and lay it down on the laboratory bench and *never* completely uncover the agar surface. By lifting only one side of the lid, you will prevent dust and other contaminants

Figure 6C-1 *Preparation of a streak plate.*

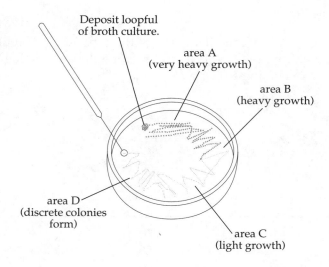

Deposit loopful
of broth culture.

area A
(very heavy growth)

area B
(heavy growth)

area D
(discrete colonies
form)

area C
(light growth)

from coming into contact with the surface of the agar and the inner surface of the lid of the Petri dish.

3. Reflame and cool the loop (you may stab it into the agar at the side of the plate several times or touch it to the surface of the agar until it no longer sizzles). Drag the loop through the bacteria rapidly, moving back and forth in a zigzag pattern (Figure 6C-1, area A).

4. Reflame and cool the loop. (In preparation of a streak plate, the loop is reflamed to help obtain a more dilute solution of bacteria.) Turn the Petri dish 90° and draw the loop through the first streak and continue streaking in a zigzag manner (Figure 6C-1, area B).

5. Again turn the agar plate 90° and repeat step 4 (Figure 6C-1, area C).

6. Once again, turn the plate 90°, and touching the loop to a corner of the streaks in area C, drag the loop across the surface in a zigzag motion. Be careful not to touch any previously streaked areas. Reflame the loop at the end of this first inoculation procedure.

7. Incubate plates for 48 to 72 hours at 25°C in an *inverted* position. (This precaution prevents the condensation that forms on the lid of the Petri dish from dropping onto the culture.)

8. To prepare a **spread plate,** obtain a nutrient broth culture of a mixture of *E. coli* and *Micrococcus luteus* (a gram-positive coccus bacterium with yellow pigment) and a sterile Petri dish of nutrient agar. Label the agar plate with the names of the organisms you are using and with your initials.

9. Obtain a glass spreading rod (this may be a glass rod bent into the shape of an **L** or a rod with a triangular bend on the end). Place the spreading rod into a beaker of 95% ethyl alcohol. Be sure that the beaker contains enough alcohol to cover the lower bent portion of the rod.

10. With a sterile loop, place a sample of the bacterial culture from the broth tube onto the center of the agar. (Remember to hold the lid face down above the culture plate to help prevent contamination.) Replace the cover and reflame the inoculating loop before putting it aside.

11. Remove the spreading rod from the alcohol. Keep the bent portion pointing downward (to prevent alcohol from running up the handle and onto your fingers). Pass the rod through the flame of the Bunsen burner and allow the alcohol to burn off completely. Allow the rod to cool for 10 to 15 seconds.

12. Lift the lid of the Petri dish and touch the rod to the clean surface of the agar to make sure it is cool before touching it to the bacteria in the middle of the plate.

13. Lightly move the spreading rod back and forth across the surface of the agar while spinning the plate around. (You may use a turntable to spin the plate if available, or you can do this by hand using the fingers of the hand holding the lid—you may wish to practice this technique on an empty dish first.)

14. Recover the plate and incubate in an *inverted* position at 25°C for 48 to 72 hours.

a. *Observe the growth on both the streak plate and spread plate. Describe and make a drawing of each.*

Streak plate

Spread plate

b. *On each of the plates, locate two colonies that are morphologically different (I and II) and record your observations in Table 6C-1. Colonies may differ in the following ways: (1) color; (2) form: circular, irregular, or spreading; (3) elevation: flat or raised; and (4) size.*

c. *Which organism do you think is present in each of the colonies you observe? Record this information in Table 6C-1.*

Table 6C-1   Isolating Pure Colonies of Bacteria

	Description	Organism
Streak plate Colony I		
Colony II		
Spread plate Colony I		
Colony II		

d. *How might you prepare a pure culture of* E. coli, S. marcescens, *or* M. luteus *from the streak or spread plates?* _____

_____

_____

15. Compare the bacterial colonies you have observed with the viral plaques on demonstration. Viral plaques are formed by bacteriophages (viruses that attack bacteria). They are prepared as follows: A layer of harder agar is overlaid with a layer of soft agar containing a mixture of bacterial cells and bacteriophages. The bacteria reproduce and spread throughout the agar to

form a confluent layer, or "lawn," across the surface of the plate, except where a bacterio-phage infection occurs. When the bacteriophage replicates inside the host cell, it causes the bacterial cell to *lyse* (dissolve); new virus particles then infect nearby cells. The process of viral replication and cell lysis continues, producing a clear, circular area called a **plaque** in the bacterial lawn where the bacterial cells were destroyed.

Laboratory Review Questions and Problems

1. Label the parts of the typical gram-negative bacterial cell shown below.

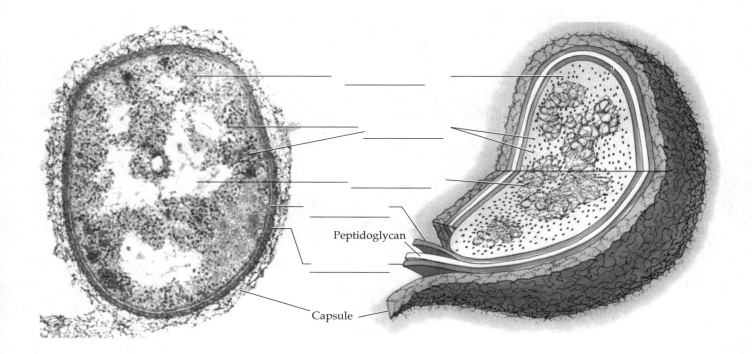

Peptidoglycan

Capsule

2. Coacervates are called protobionts. Why are they not called cells? What characteristics do they share with cells?

3. All cells are surrounded by a membrane that separates the external environment from the internal environment. A typical cell membrane is formed by a lipid bilayer in association with proteins. The protobionts made during this laboratory were composed of carbohydrate and protein. How is a barrier formed in these protobionts? Which molecules could form a membrane-like barrier?

4. Describe the structure of a bacterial chromosome. How does it differ from a eukaryotic chromosome?

5. Bacterial flagella are different in composition and structure from the flagella of eukaryotes. What types of differences exist? Do the flagella of prokaryotes move in the same way as the flagella of eukaryotes?

6. Cyanobacteria can carry out photosynthesis, yet they do not contain chloroplasts. Where is chlorophyll found in these organisms?

7. Your doctor tells you that you have a bacterial infection in your salivary glands. You are given the opportunity to view the slide made from a sample of your saliva. It is labeled "gram-negative, coccus." Describe what you will see on the slide.

8. Explain how the structure of the bacterial cell wall affects the results of the Gram staining procedure.

9. Your laboratory instructor is going to observe and grade your technique while you inoculate a broth culture and make a streak plate from an agar slant of *E. coli*. List four things that you should be careful to do (or not do) to receive an "A" for your aseptic technique.

10. You are given a broth culture of *E. coli* mixed with several other species of bacteria. Explain how you would establish a pure culture of *E. coli* for further laboratory investigations.

11. You are in the microbiology laboratory and are given the task of cleaning up after the last three lab sessions. You decide to sort the plates into "like" types. You find several (A) with small white circular areas of growth and others (B) that appears fairly "slimy" all over but with small circular clear areas dotted at various locations around the plate. You have others (C) that have irregular areas of red-colored growth and several (D) with a yellow-colored growth. You tentatively try to identify what is in the plates so you will know how to handle them. What do you decide about each?

Eukaryotic Cells

OVERVIEW

All eukaryotic organisms are composed of cells, whether they exist as single cells, colonies of cells, or in multicellular form. Your body is composed of 50 to 100 trillion cells, most of which are very small, with specialized structures that allow for a diversity of functions.

All eukaryotic cells have their genetic material enclosed by a nuclear membrane, the nuclear envelope. In addition, a variety of subcellular membrane-bound organelles are present. These include plastids, mitochondria, lysosomes, microbodies, and Golgi complexes. Internal membrane systems, or endoplasmic reticula, divide the cell into specialized compartments. Non-membrane-bound organelles, such as ribosomes, centrioles, microtubules, and microfilaments, are also present in eukaryotic cells.

During this laboratory you will investigate the structure of plant and animal cells and will learn how biochemical analysis, cell fractionation, and transmission electron microscopy have allowed scientists to unlock the secrets of eukaryotic cells.

STUDENT PREPARATION

Prepare for this laboratory by reading the pages indicated by your instructor. Familiarizing yourself in advance with the information and procedures covered in this laboratory will give you a better understanding of the material and improve your efficiency.

EXERCISE A | Examining Plant Cells

The cells of plants are eukaryotic, containing both a membrane-bound nucleus and membrane-bound organelles. A large central **vacuole** surrounded by a membrane (the **tonoplast**) is used for storing water, pigments, and wastes. Within the cytoplasm are various types of **plastids** responsible for photosynthesis and for storing starch or pigments. A **cell wall** composed of cellulose surrounds the plant cell.

Objectives

☐ Identify the structures of a typical plant cell.

Procedure

1. Prepare a wet-mount slide of an *Elodea* leaf. Observe the thick cell wall, thinner cell membrane, cytoplasm, nucleus, and chloroplasts. A large central vacuole should be apparent. These structures characterize a generalized plant cell.

2. On the top half of a separate sheet of paper, draw a representative *Elodea* cell and label its parts.

3. Prepare a fresh wet-mount slide of *Elodea*. Observe your slide using high power (40×).

   Do the chloroplasts appear to move? _____ Describe their movement.

4. Prepare a wet-mount slide of onion tissue. Onions (*Allium*) have layers of modified leaves (scales) that can easily be separated from one another. Peel off a portion of one layer and examine the concave side of the piece you have obtained. The surface is covered by a thin layer of cells, the epidermis.

5. Remove a small piece of the epidermis (approximately 3 × 8 mm) by breaking the scale gently, leaving the epidermis intact. Peel the epidermis from one of the halves of the scale. Prepare a wet-mount slide of the isolated epidermis.

6. Observe the onion cells using low power (10× objective) and then high power (40× objective).

7. If it is difficult to see the cells, add a drop of Lugol's solution (I_2KI) at the edge of the coverslip. Does this solution stain the cells as it reaches them? _____

8. Draw a representative onion cell on the lower half of the sheet of paper you used for your drawing of *Elodea*. Compare the onion cell with the *Elodea* cell. Since they are both plant cells, they should be similar. You will note that onion cells lack one structure that is very conspicuous in *Elodea* cells. What is this structure? _____

   a. List the similarities and differences between *Elodea* cells and onion cells.

Similarities	Differences
_____	_____
_____	_____
_____	_____
_____	_____

9. Use a razor blade to slide a piece of tissue, as thin as possible, from a potato. Be careful not to cut your fingers. Prepare a wet-mount slide; use a drop of water.

10. Study the slide at low power (10× objective) and then at high power (40× objective). Add a drop of Lugol's solution (I_2KI) to the side of the coverslip and observe the cells as the iodine solution makes contact with them.

    b. *How does the reaction of iodine with the potato cells compare with what you observed in your onion epidermis preparation (step 7)?* _____

    _____

    c. *What does this tell you about the differences between the storage products in onions and potatoes?*

    _____

    _____

    d. *Do you see any chloroplasts?* _____ *Why or why not?* _____

    e. *You will probably see some small oval-shaped blue-black structures. These* **leucoplasts** *store starch. Why did they turn blue?* _____

    _____

11. If a banana is available, scrape a small amount of tissue from its surface and spread it onto a slide. Add a drop of Lugol's solution and a coverslip. Observe the preparation using high power (40× objective).

f. How do these observations compare with what you saw in the potato?

_____

Plastids are membrane-bound organelles unique to plants. You have already been introduced to two types—chloroplasts (containing chlorophyll) and leucoplasts (containing starch). **Chromoplasts** contain several types of pigment including carotenoids, which give plants an orange or yellow color.

> **12.** Use a razor blade to slice a piece of tissue, as thin as possible, from the outer portion of a peeled carrot. Prepare a wet-mount slide.
>
> *g. Can you see the chromoplasts? Describe them.* _____

_____

EXTENDING YOUR INVESTIGATION: CYTOPLASMIC STREAMING

Many environmental conditions can affect the natural functions of cells, including the movement of chloroplasts. Microfilaments, composed of the protein actin, direct the movement of chloroplasts within the cell cytoplasm. What changes in the cellular environment do you think might disrupt, increase, or decrease the rate of chloroplast movement? Formulate a hypothesis that predicts how alterations in the environment might affect chloroplast movement in *Elodea*.

HYPOTHESIS:

NULL HYPOTHESIS:

Can you **predict** how changing the environment might affect chloroplast movement?

What is the **independent variable?** _____

What is the **dependent variable?** _____

Design an experimental procedure to test your hypothesis.

PROCEDURE:

Describe what you observed during your experiment.

RESULTS:

Do your results support your hypothesis?

Your null hypothesis?

What do you **conclude** about how the environment affects chloroplast movement?

✔ **EXERCISE B | Examining Animal Cells**

The cells of animals, like those of plants, are eukaryotic with a membrane-bound nucleus and membrane-bound organelles. Unlike plant cells, however, they have no central vacuole (although they may have small vacuoles) and no plastids. A plasma membrane surrounds the cell, but there is no cell wall.

✔ **PART 1   Studying Animal Cells Using Light Microscopy**

Animal cells can be studied using the light microscope, but most of the cellular organelles within the cytoplasm are not visible without the use of special staining techniques. The nucleus and **nucleolus,** where ribosomes are manufactured, are usually apparent in most cells.

To study the structure of animal cells you will use prepared slides of animal **tissues.** These are collections of cells that have a similar function. The cells are usually organized into sheets.

IIIII **Objectives** IIIIIIIIIIIIIIIIIIIIIIIIIIIIIIIIIIIIIIII

☐ Identify the structure of a typical animal cell.

IIIII **Procedure** IIIIIIIIIIIIIIIIIIIIIIIIIIIIIIIIIIIIIIII

1. Observe a prepared slide of frog epithelium (simple squamous or cuboidal). Examine the slide using the 10× and the 40× objectives.

2. Sketch one or more of the cells in the space below. Label the nucleus, cytoplasm, and plasma membrane.

3. If available, observe a slide of salamander epithelium that has been stained to demonstrate mitochondria—membrane-bound organelles used for cellular energy production. Use the 40× objective to examine the cells.

   a. Do the cells you observe have a cell wall? _____ Plastids? _____ Chlorophyll? _____

   b. List the similarities and differences between the plant cells and the animal cells you have observed.

   Similarities   Differences

   _____   _____

   _____   _____

   _____   _____

   _____   _____

✔ **PART 2   Studying Animal Cells Using Cytochemical Stains**

Staining very thin sections of tissue highlights the internal structures of cells, enabling us to examine them in more detail. Stains may be general, staining many parts of the cell, or specific, reacting only with particular biochemical macromolecules. If a certain type of macromolecule happens to be located within a certain organelle, specific cytochemical stains allow us to determine the chemical composition of that organelle.

ııııı Objectives ıııııııııııııııııııııııııııııııııııııı

☐ Describe how scientists can use the microscope to study the biochemistry of cells.

☐ Describe how the biochemical composition of cellular organelles can be determined and give specific examples.

ııııı Procedure ıııııııııııııııııııııııııııııııııııııı

On demonstration you will find several examples of slides stained with special cytochemical stains specific for biochemical compounds such as DNA, RNA, protein, starch, and lipid. In each pair of slides you will study, one tissue sample has been treated with enzymes to remove the substance being studied. Examine the pairs of demonstration slides listed below and record your observations.

1. DNA in mammalian liver (Feulgen stain):

   Color of nuclei _____

   Color of cytoplasm _____

   Other characteristics _____

   _____

   Mammalian liver treated with DNase to remove DNA (Feulgen stain):

   Color of nuclei _____

   Color of cytoplasm _____

   Other characteristics _____

   _____

   a. Based on your observations, what does Feulgen reagent stain? _____

2. RNA in mammalian liver (methyl-green-pyronin stain):

   Color of nuclei _____

   Color of cytoplasm _____

   Other characteristics _____

   _____

   Mammalian liver treated with RNase to remove RNA (methyl-green-pyronin stain):

   Color of nuclei _____

   Color of cytoplasm _____

   Other characteristics _____

   _____

   b. Based on your observations, what does methyl-green-pyronin stain? _____

3. Glycogen in mammalian liver (carmine stain):

   Color of nuclei _____

   Color of cytoplasm _____

   Other characteristics _____

   Mammalian liver treated with amylase to remove glycogen (carmine stain):

   Color of nuclei _____

   Color of cytoplasm _____

   Other characteristics _____

   c. Based on your observations, what does carmine stain indicate is present in liver cells?

   _____

4. Fat tissue (osmium tetroxide):

   Identifying characteristics _____

   _____

   Fat tissue treated with lipase to remove lipid (osmium tetroxide):

   Identifying characteristics _____

   _____

   d. Based on your observations, how does osmium tetroxide indicate the presence of fat?

   _____

5. Based on the observations made above, describe what you would observe in the following slides.

   e. Carmine stain used on a liver sample from a starved rat (with no stored glycogen).

   _____

f. Feulgen stain used on a chromosome preparation from Drosophila.

_____

g. Osmium tetroxide used to stain a section of adipose (fat) tissue from an obese mouse.

_____

✔ **EXERCISE C** ▌ **The Strange Shapes of Cells**

You have just studied some representative plant and animal cells, all of which have a fairly simple shape and are of an approximately similar size. You should realize, however, that cells come in all sizes and shapes with a multitude of specializations for function or for interaction with other cells.

The smallest bacterial cells (mycoplasmas) are about 0.1 μm in diameter. Most other bacteria are on the order of 1 to 2 μm in diameter (Laboratory 6). A human liver cell has a diameter of 20 μm. A human ovum is approximately 100 μm in diameter—1,000 times larger than the smallest cells. Then, of course, there are the unusual cases to consider: for example, the ostrich's ovum, a single cell approximately 1,000 times the size of a human ovum.

▐▐▐▐▐ **Objectives** ▐▐▐▐▐▐▐▐▐▐▐▐▐▐▐▐▐▐▐▐▐▐▐▐▐▐▐▐▐▐▐▐▐▐▐▐▐

☐ Recognize diverse forms of specialization in cell structure.

▐▐▐▐▐ **Procedure** ▐▐▐▐▐▐▐▐▐▐▐▐▐▐▐▐▐▐▐▐▐▐▐▐▐▐▐▐▐▐▐▐▐▐▐▐▐▐

Observe examples of some of the diverse cell types that may be available in the laboratory. Sketch and label these on a separate of paper and insert your drawings into your laboratory manual.

Nerve cell A highly specialized cell consisting of a cell body (containing the nucleus) and long cytoplasmic extensions for transmission of nerve impulses.

Stentor A ciliated, single-celled eukaryote belonging to the kingdom Protista.

Acetabularia An unusually large, single-celled green alga (2 to 5 cm) possessing a foot (containing the nucleus), a stalk, and a cap.

Volvox A colony of biflagellated green algal cells.

Human sperm cell A highly differentiated sex cell consisting of three different regions: the head (containing the nucleus), the midpiece (with a large mitochondrion), and the tail composed of a flagellum.

Starfish egg A large, highly differentiated sex cell containing an extremely large nucleus called a germinal vesicle. The **nucleolus** (site of ribosome synthesis) is visible within the nucleus as a dark dot.

👁 **EXERCISE D** ▌ **Cell Fractionation: A Study of Eukaryotic Cells**

Cells can be broken apart by several means such as electric shock, sonication (vibration), or grinding. The cell membranes and endoplasmic reticulum are broken open, but the membrane fragments quickly "reseal" to form smaller vesicles. If gentle techniques are used, the cell organelles, including nuclei, mitochondria, and chloroplasts, remain intact and retain most of their original biochemical function.

A suspension of cell parts and organelles is called a **homogenate.** The many components of a cell homogenate can be separated by **differential centrifugation:** the use of different amounts of centrifugal force (generated at different speeds) to sediment cell parts of various sizes and densities. At low speeds, large nuclei and unbroken cells sediment to form a **pellet** at the bottom of the centrifuge tube. The remaining cellular organelles are found in the **supernatant** above the pellet. At slightly higher speeds, chloroplasts and mitochondria can be separated in the pellet. The material still remaining in the supernatant is the **microsomal fraction** containing small cellular organelles (lysosomes, microbodies, ribosomes) and membrane vesicles.

If this exercise, you will separate the organelles from the cells of pea seeds using differential centrifugation. You will identify the types of organelles in the many cell fractions by using specific colorimetric staining reactions.

␣␣␣␣␣ Objectives ␣␣␣␣␣␣␣␣␣␣␣␣␣␣␣␣␣␣␣␣␣␣␣␣␣␣␣␣␣␣␣␣␣␣␣␣␣

☐ Describe how cell structure can be studied by fractionating cells.

☐ Determine the presence of cellular organelles within a cell fraction.

␣␣␣␣␣ Procedure ␣␣␣␣␣␣␣␣␣␣␣␣␣␣␣␣␣␣␣␣␣␣␣␣␣␣␣␣␣␣␣␣␣␣␣␣␣

Your instructor will homogenize a sample of pea seeds that have been soaked overnight to soften. Stir the homogenate before using. See Figure 7D-1 for a flow chart of this experiment.

1. Work in pairs. Fill a 15-ml centrifuge tube with homogenate until it is three-fourths full. Have your laboratory partner do the same so that there is an identical amount of material in the two tubes. Write your initials on the side of the tube.

2. Place your tube in a clinical centrifuge *opposite* your lab partner's tube. (The rotor must always be balanced with equal weights or volumes opposite each other.)

3. Close the top of the centrifuge and slowly increase the speed to 200 × gravity (200 × g; setting 4 on a clinical, table-top centrifuge). Centrifuge for 3 minutes.

4. While your material is in the centrifuge, obtain some of the cellular debris left on the cheesecloth when your instructor prepared the "pea soup." Place a small amount of this material on a slide and add a drop of water. Cover with a coverslip and observe the material using the light microscope (magnification 40×)

   a. *Describe what you observe.* _____

   _____

   b. *Do you see cell wall fragments?* _____ *If yes, describe them.*

   _____

   c. *Do you see intact cells?* _____

   d. *Add a drop of I₂KI solution at the edge of the coverslip. What does this stain?*

   _____

   e. *Describe what you see.* _____

5. After centrifugation is complete, use a Pasteur pipette to remove a small amount of sediment from the bottom of each tube and save it in a clean test tube. Return the original tubes to the centrifuge (a balanced pair) and slowly increase the speed to 1,300 × g (setting 7 on the clinical centrifuge). Centrifuge for 10 minutes.

6. While your material is in the centrifuge, investigate the contents of the sediment. Place a drop of the sediment on a slide and mix it with a drop of I₂KI. Cover with a coverslip and observe at 40×. Pea seedlings store food reserves as starch grains. Starch grains have characteristic shapes and are sometimes used for classifying plants.

   f. *Describe the structure of the starch grains you observe.* _____

   _____

   g. *Why are they stained blue?* _____

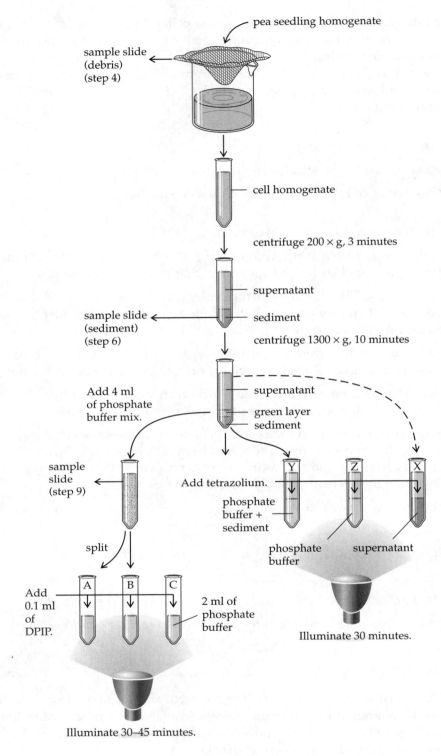

Figure 7D-I *Cell fractionation procedure.*

h. Why are there so many starch grains in pea seeds? _____

_____

7. After centrifugation is complete, carefully remove your sample tubes from the centrifuge.

   i. Describe what the contents of the tube look like. _____

8. Nuclei and chloroplasts will be in a green layer above the sediment. Do you see this layer? Hold the tube in front of a light source and, using a Pasteur pipette, *carefully* remove material from this layer. (You will have to place the tip of the Pasteur pipette next to the side of the tube. You may wish to mark the pipette tip with a waterproof black marker to make it easier to see.) After removing this material, store the centrifuge tube on ice.

9. Place the material you remove into a small test tube on crushed ice. Add 4 ml of phosphate buffer and resuspend the contents. Use one drop of the material to make a wet-mount slide. Observe using high power (greater than 40× if possible).

   j. Describe what you see. _____

10. Mark two test tubes "A" and "B," and put 2 ml of the sample into each tube. Cover tube with aluminum foil. Keep both tubes on ice. Place 2 ml of phosphate buffer into a third tube and label this tube "C." Add 0.1 ml of DPIP to each tube. Cover with parafilm and mix by inverting. Use the spectrophotometer to read the absorbance at 600 nm for each of the three tubes and record your data in the "before" column of Table 7D-1. If a spectrophotometer is not available, compare the colors of the three tubes and record this information as qualitative observational data. Tube C is your blank.

Table 7D-1

Tube	Color		Absorbance (600 nm)		
	Before	After	Before	After	Difference
A					
B (dark)					
C					

   DPIP (2,6-dichlorophenol-indophenol) is a blue dye that can act as a hydrogen and electron acceptor. During the light-dependent reactions of photosynthesis, DPIP can substitute for $NADP^+$ and is reduced by the addition of hydrogen and electrons. When DPIP is reduced it becomes colorless.

   k. If DPIP, when mixed with the chloroplast sample, loses its blue color, what is probably happening

   in the sample? _____

11. Form a hypothesis that predicts what changes you might expect in the three tubes and why this is so.

   HYPOTHESIS:

   NULL HYPOTHESIS:

12. Illuminate the three tubes with a 100-watt light. After 30 to 45 minutes, record your results in Table 7D-1 by observing the color in the three tubes or, if you are using a spectrophotometer, by reading the absorbance at 600 nm.

  l. Do your results support your hypothesis? _____ How do tubes A and B compare?

  m. What do you conclude about the contents of the green layer formed during the cell fractionation?

  _____

13. Locate the centrifuge tube on ice (end of step 8). Into a clean test tube add enough of the yellow-green supernatant layer to fill the tube approximately one-third to one-half full, and label this tube "X." Carefully pour off the remaining supernatant from the centrifuge tube. Suspend a sample of the sediment in an amount of phosphate buffer equal to the contents of tube X. This should produce a slightly cloudy, translucent mixture. Label the tube "Y." Into a third tube add an amount of phosphate buffer equal to that in tubes X or Y, and label this "Z."

14. You will use the dye **tetrazolium*** to test for mitochondrial activity. Tetrazolium will indicate whether the mitochondrial electron transport system is present and working. Tetrazolium can act as an electron acceptor (in place of cytochrome in the electron transport system) and will turn red when reduced by addition of electrons.

  n. The yellow-green supernatant contains chlorophyll pigments from broken chloroplasts, but what

  else might it contain? _____

  o. How does the size of mitochondria compare with that of chloroplasts? _____

15. Form a hypothesis that predicts which cellular fraction (X or Y) contains mitochondria and why this is so.

  HYPOTHESIS:

  NULL HYPOTHESIS:

16. Add tetrazolium solution to tubes X, Y, and Z to fill each tube approximately two-thirds full (the amount of tetrazolium should be equal to the amount of homogenate in the tube).

17. Place all three tubes in a beaker of warm water (35–40°C) in front of a 100-watt light for 30 minutes (check the water temperature after 15 minutes and replace with warm water if necessary).

18. RESULTS: After 30 minutes, record your observations in the following table.

  p. Do your results support your hypothesis? _____ Your null hypothesis? _____

  q. What can you conclude about the size of mitochondria relative to the size of chloroplasts in pea

  seeds? _____

Tube	Color
X	
Y	
Z	

✔ **EXERCISE E   A Closer Look at Eukaryotic Cells**

Figure 7E-1 shows two pictures, or *micrographs,* of an onion cell nucleus. (A micrograph is a photograph taken using a microscope.) One is a photomicrograph taken under the light microscope. The other is an

_____

**Tetrazolium is a poison. If desired, see alternate directions in Preparator's Guide.*

electron micrograph. Note that the overall magnification of the two onion cells is approximately equal. Why, then, do the cells look so different? The reason is that the light microscope and electron microscope differ in their ability to resolve, or distinguish, fine detail. An electron microscope is able to resolve two structures as separate when they are only 0.0001 μm apart, whereas the light microscope can resolve objects only when they are approximately 0.5 μm apart (Laboratory 1).

Figure 7E-1 *Photograph of an onion cell nucleus using (a) a light microscope and (b) an electron microscope.*

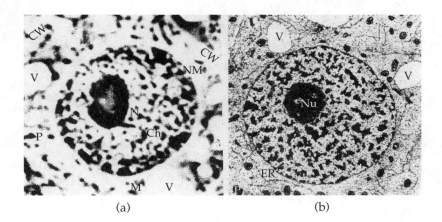

(a)   (b)

||||| **Objectives** |||||||||||||||||||||||||||||||||||||||||||

☐   Recognize membrane-bound and non-membrane-bound organelles in transmission electron micrographs.

||||| **Procedure** |||||||||||||||||||||||||||||||||||||||||||

Eukaryotic cells are compartmentalized into many subcellular membrane-bound organelles. Non-membrane-bound particles and organelles are also present.

1. In the following transmission and scanning electron micrographs, indicate the type of organelle that is the main subject of the micrograph and describe its function. (The organelle in micrograph 6 is named for you.)

2. Use your knowledge gained from lectures and from reading your textbook to identify the lettered structures and list their functions (where indicated). If you are not sure of the identity of a particular structure, look for examples in labeled micrographs available in the laboratory or in your text.

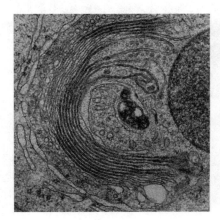 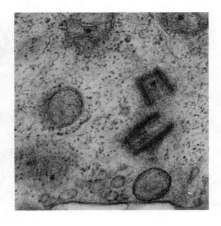 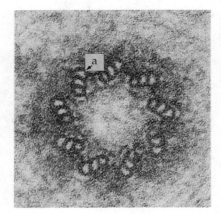

1. Organelle _____   2. Organelle _____   3. Organelle _____

   Function _____   Function _____   Function _____

   **a** _____

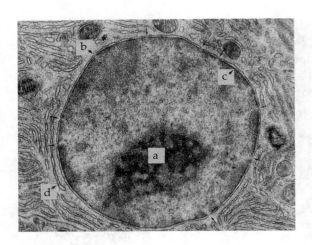

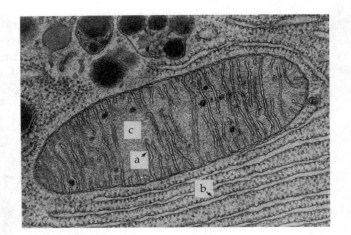

4. Organelle _____

Function _____

Part   Function

a _____   _____

b _____   _____

c _____   _____

d _____   _____

5. Organelle _____

Function _____

Part   Function

a _____   _____

b _____   _____

c _____   _____

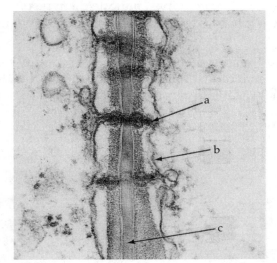

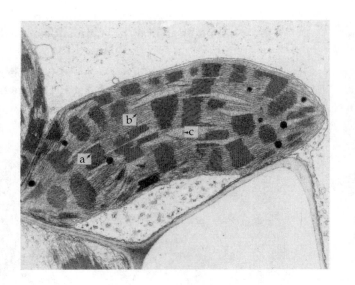

6. Cell junction

Function _____

Part

a _____

b _____

c _____

7. Organelle _____

Function _____

Part   Function

a _____   _____

b _____   _____

c _____   _____

Laboratory Review Questions and Problems

1. Label the parts of the generalized animal cell and plant cell shown below.

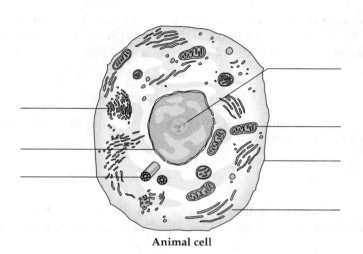

Animal cell

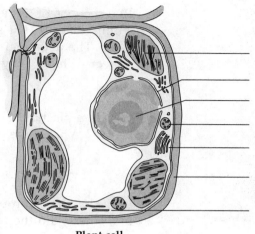

Plant cell

2. Complete the following table. If possible, add extra information on the structure of each organelle.

Organelle(s) or Other Structure(s)	In Prokaryotes?	In Eukaryotes?	Function
Nucleus with nuclear membrane			
Nucleolus			
Chromosomes			
Cytoplasm			
Mitochondria			
Golgi complexes			
Endoplasmic reticulum			
Ribosomes			
Centrioles			
Chloroplasts			
Cell wall			
Flagellum			

3. A pathologist is studying a set of slides made from a patient's tissues, which have been stained with a series of cytochemical stains. The pathologist observes that the cell nuclei on all of the slides are red. The physician has informed the pathologist that all of the slides have been stained with Feulgen stain. Why are the nuclei red?

Two sections of tissue treated with osmium tetroxide appear to contain large dark bodies within the cells. The slides were not labeled to indicate the organ from which the tissue was taken. The pathologist's apprentice wants to label them as liver. Is this label likely to be correct? Why or why not?

4. What are plastids? How many types of plastids did you observe during this laboratory period? What is the function of each type?

5. In *Elodea*, the chloroplasts appear to move within the cytoplasm. When you observed this movement, did they all move in the same direction? What cellular structures guide this movement?

6. Why are central vacuoles important to plant cells? For growth? Waste disposal? Storage? For improving the surface-to-volume ratio?

7. Describe the structure of a plant cell wall. Distinguish between the primary cell wall, the secondary cell wall, and the middle lamella.

8. What is the cytoskeleton of a cell? What are its parts? Can it be observed using the light microscope? Why or why not?

9. What is the largest cell that you observed during this laboratory period? _____ The smallest? _____ Why are cells (even fairly "large" ones) usually small?

10. To study subcellular organization it may be necessary to homogenize the cells. What is the purpose of homogenization?

11. What physical properties of cell organelles determine their behavior during differential centrifugation? Why are different speeds of centrifugation used? Rank the cell parts that you have studied by their size (smallest to largest).

12. Organelles are often studied by separating them on a *sucrose gradient*. A centrifuge tube is filled with a sucrose solution of increasing concentration from the top of the tube to the bottom (5% at the top to 20% at the bottom). What would you expect to observe if you put a layer of cell homogenate on top of the sucrose and then subjected the sample to high-speed centrifugation?

13. Both 2,6-dichlorophenol-indophenol (DPIP) and tetrazolium are used as colorimetric indicators of metabolic functions in cells. For what functions were they used as tests in this laboratory? What do DPIP and tetrazolium have in common in their mechanisms of action? What is reduction? How can you tell that DPIP and tetrazolium have been reduced? When does this reduction take place during (a) photosynthesis and (b) cellular respiration?

Osmosis and Diffusion

OVERVIEW

Cells consist of highly complex organizations of molecules, whose behavior can be explained by the laws of physics and chemistry. One of these laws, the **law of diffusion,** is of particular importance to our understanding of the movement of molecules into and out of cells. The law of diffusion states that molecules tend to move from areas of higher chemical potential to areas of lower chemical potential.

Chemical potential (μ) is a measure of free energy available to do the work of moving a mole of molecules from one location to another and, in some cases, through a barrier such as a cell membrane. In a solution, the greater the concentration of molecules of a dissolved substance (**solute**), the higher is the chemical potential (free energy per mole) of that substance. Thus, we can also say that molecules tend to move from areas of higher concentration (higher chemical potential) to areas of lower concentration (lower chemical potential) (Figure 8-Ia).

Osmosis, a special case of diffusion with special relevance for cells, is the movement of water molecules from regions of higher water potential to regions of lower water potential across a semipermeable or selectively permeable membrane. **Water potential (ψ)** is a measure of the chemical potential, or free energy per mole, of water molecules. Water potential is affected by the amount of other substances (solutes) dissolved in the

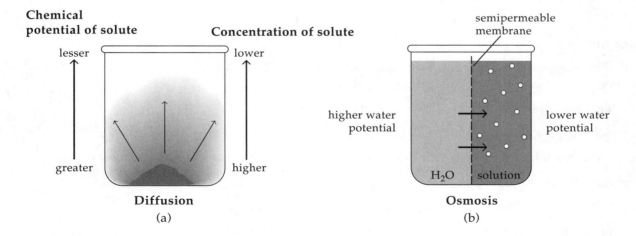

Figure 8-I *Diffusion and osmosis.*

water. The addition of solute to water lowers its water potential (the chemical potential of water molecules in the solution). Pure water, not under pressure (that is, at atmospheric pressure), has a water potential of zero, but as solute is added to pure water, the water potential becomes negative. The more solute is dissolved in water, the lower (more negative) is the water potential of the solution.

If pure water is separated from a solution by a semipermeable membrane, water molecules will move across the membrane from an area of higher water potential to an area of lower water potential. In other words, water tends to move across membranes toward areas with a higher concentration of solutes, where the water potential is lower (Figure 8-1b).

Both diffusion and osmosis are examples of **passive transport** because molecules, whether water or solute, move "down" concentration or free-energy gradients. To move molecules *against* a gradient requires an energy input: this type of transport is called **active transport.**

During this laboratory period, you will examine the principles of solute and water movement in both artificial and living systems.

STUDENT PREPARATION

Prepare for this laboratory by reading the pages indicated by your instructor. Familiarizing yourself in advance with the information and procedures covered in this laboratory will give you a better understanding of the material and improve your efficiency.

✔ | EXERCISE A | Brownian Movement

In order to understand how substances pass through a membrane, it is important to realize that molecules, when at temperatures above absolute zero, are in constant motion. Molecular motion is a form of energy: the translational, vibrational, and rotational kinetic energies of molecules. Although individual molecules are impossible to see, their existence is revealed by the jiggling—called **Brownian movement**— of minute particles suspended in water.

⁞⁞⁞⁞⁞ **Objectives** ⁞⁞⁞

☐ Define Brownian movement.

☐ Describe the effects of increased temperature on the movement of water molecules.

⁞⁞⁞⁞⁞ **Procedure** ⁞⁞

Place a drop of powdered carmine suspension (carmine is not soluble in water) on a slide and cover with a coverslip. Examine the slide at high power (40×).

a. *Do the carmine particles move randomly or in a definite path?* _____

b. *Can you see the water molecules?* _____ *Are the water molecules moving?* _____

c. *Is the movement of a carmine particle due to the movement of its own molecules or to bombardment*

   by water molecules? _____ *Explain.* _____

   _____

d. *Would an increase in temperature increase or decrease the rate of Brownian movement?* _____

   Why? _____

👁 **EXERCISE B**   **Diffusion**

The molecules in solids, liquids, and gases are in constant motion. In liquids and gases, the molecules are free to move and tend to migrate from regions of higher chemical potential (μ) and higher concentration to regions of lower chemical potential and lower concentration. They will move until they are uniformly distributed throughout the medium.

How fast does diffusion occur? The rate is dependent on the particle size and on the difference in chemical potentials of the molecules in the two regions: the larger the difference, the faster the rate of diffusion.

⊪⊪⊪ **Objectives** ⊪⊪⊪⊪⊪⊪⊪⊪⊪⊪⊪⊪⊪⊪⊪⊪⊪⊪⊪⊪⊪⊪⊪⊪⊪⊪⊪⊪⊪⊪⊪⊪⊪

☐ Define diffusion.

☐ Describe how diffusion of a gas in a gas or of a liquid in a liquid can be observed.

☐ Describe the effect of the molecular weight of a substance on its rate of diffusion.

👁 **PART 1**   **Diffusion of a Gas in a Gas**

⊪⊪⊪ **Procedure** ⊪⊪⊪⊪⊪⊪⊪⊪⊪⊪⊪⊪⊪⊪⊪⊪⊪⊪⊪⊪⊪⊪⊪⊪⊪⊪⊪⊪⊪⊪⊪⊪⊪⊪⊪⊪⊪

We can study the process of diffusion of a gas in a gas by observing the movement of ammonium hydroxide vapors in air.

1. Wet a 4- to 6-inch strip of filter or chromatography paper with a solution of phenolphthalein (phenolphthalein is an indicator of pH changes).

2. Add 10 to 20 ml of ammonium hydroxide solution to a 250-ml graduated cylinder. As you pour, avoid wetting the sides of the container.

3. Suspend the filter paper from the hook in the bottom of the cork provided by your instructor. Immediately lower the phenolphthalein-soaked strip into the vessel and secure the cork. The bottom of the strip should *not* touch the ammonium hydroxide.

   a. *Describe the changes that take place in the filter paper over the next several minutes.*

   _____

   b. *Does the filter paper change color all at once?* _____

   c. *Explain the results of your experiment by relating the chemical potential of the ammonium*

      hydroxide molecules to the process of diffusion. _____

   _____

👁 **PART 2**   **Diffusion of a Liquid in a Liquid**

We can make some observations about the rate of diffusion and changes in concentration of a diffusing substance by observing the diffusion of a colored liquid in water. The rate of diffusion is determined by the magnitude of difference between the concentrations of a diffusing substance in two different locations, the area of origin and the area of destination.

⊪⊪⊪ **Procedure** ⊪⊪⊪⊪⊪⊪⊪⊪⊪⊪⊪⊪⊪⊪⊪⊪⊪⊪⊪⊪⊪⊪⊪⊪⊪⊪⊪⊪⊪⊪⊪⊪⊪⊪⊪⊪⊪

1. Place a clear plastic Petri dish on a white piece of paper. Center a thin plastic ruler (white or clear) under the bottom of the Petri dish.

2. Pour 30 ml of distilled water into the Petri dish. Allow it to stand undisturbed for a few minutes to minimize water convection currents.

3. Gently add one drop of blue food coloring or India ink to the center of the dish. Cover the Petri dish with its lid to prevent disturbance from air currents.

4. Find the rate of diffusion of the dye or ink into the water by measuring the diameter (D) in millimeters of the pigmented spot at 1-minute intervals for 15 minutes. Record your results in Table 8B-1.

Table 8B-1   Diffusion of a Liquid in a Liquid

Minute	Diameter (D)	Radius ($D/2$)
1		
2		
3		
4		
5		
6		
7		
8		
9		
10		
11		
12		
13		
14		
15		

5. Graph the radius ($D/2$) of the pigmented spot against time, in 2-minute intervals (use Figure 8B-1).

Figure 8B-I *Graph the radius of the pigmented spot against time.*

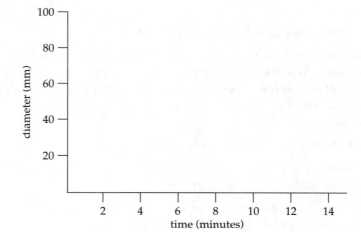

6. Choose three 2-minute intervals and calculate the rate of diffusion from the data in your graph. Use the following equation and record your results in Table 8B-2.

$$\text{Rate of diffusion} = \frac{(\text{radius at start of 2-min interval}) - (\text{radius at end of 2-min interval})}{2 \text{ min}}$$

$$= \underline{\hspace{1cm}} \text{ mm/min}$$

Table 8B-2   Rate of Diffusion

Time Interval	Rate of Diffusion (mm/min)
Minute _____ to _____	_____
Minute _____ to _____	_____
Minute _____ to _____	_____

In diffusion, the random movements of individual molecules produce a net movement from an area of greater concentration to an area of lesser concentration. Eventually, both types of molecules (in this case, water molecules and pigment molecules) will be evenly distributed.

a. Does the net movement of molecules slow down as equilibrium is reached? _____ Why?

_____

b. What can you say about the chemical potential of the pigment molecules at different distances from the center of the spreading pigment spot? _____

c. How is the chemical potential of the pigment molecules related to their rate of movement?

_____

d. Does net diffusion eventually come to an end? _____ Why? _____

_____

EXTENDING YOUR INVESTIGATION: DOES TEMPERATURE AFFECT THE RATE OF DIFFUSION?

Would using hot water or cold water (refrigerated overnight) in the experiment in Exercise B, Part 2, have an effect on diffusion rates? What do you think would happen? Formulate a hypothesis.

HYPOTHESIS:

NULL HYPOTHESIS:

What do you **predict** will happen during your experiment?

What is the **independent variable** in this investigation?

What is the **dependent variable** in its investigation?

PROCEDURE: Repeat the experimental procedure in Exercise B, Part 2, using hot water and cold water (refrigerated overnight).

RESULTS: Record your data in the tables below.

Diffusion of a Liquid in Hot Water

Minute	Diameter (D)	Radius ($D/2$)
1		
2		
3		
4		
5		
6		
7		
8		
9		
10		
11		
12		
13		
14		
15		

Diffusion of a Liquid in Cold Water

Minute	Diameter (D)	Radius ($D/2$)
1		
2		
3		
4		
5		
6		
7		
8		
9		
10		
11		
12		
13		
14		
15		

What did you observe during your experiment?

Determine the rates of diffusion of the colored liquid in both hot and cold water and record them below.

Time Interval	Rate (mm/min) HOT	Rate (mm/min) COLD
Minute _____ to _____	_____	_____
Minute _____ to _____	_____	_____
Minute _____ to _____	_____	_____

Do your results support your hypothesis?

Your null hypothesis?

Was your prediction correct?

What do you **conclude** about the effects of temperature on the rate of diffusion?

PART 3   **Effect of Molecular Weight on the Rate of Diffusion**

In this experiment, two different gases are used. Ammonia (NH_3) and hydrogen chloride (HCl) react chemically to form a white salt, ammonium chloride (NH_4Cl):

$$NH_3 + HC \longrightarrow NH_4Cl$$

The distance that substances (in this case, two gases) travel per unit of time (the rate of diffusion) is related to the molecular weights of the diffusing substances:

IIIII Procedure IIIIIIIIIIIIIIIIIIIIIIIIIIIIIIIIIIIIIIIIIII

1. Your instructor will soak a cotton plug with HCl and another with NH_4OH. Both plugs will be inserted simultaneously into opposite ends of a large glass tube and the ends will be capped with rubber stoppers (Figure 8B-2). *Note the time of insertion.* Time _____

    Ammonium hydroxide (NH_4OH) dissociates into ammonia (NH_3, a gas) and water. Ammonia diffuses toward the opposite end of the tube. At the same time, hydrochloric acid vapors (HCl) diffuse toward the ammonia. The overall reaction is

    $$NH_4OH + HCl \longrightarrow NH_4Cl + H_2O$$

2. When the two gases meet, they react to form a white ring (ammonium chloride) around the glass tubing. Record the time at which the white ring is first seen. Time _____

    The diffusion rate (r) of a molecule is inversely related to the square root of its molecular weight (MW),

    $$r = \frac{1}{\sqrt{MW}}$$

Figure 8B-2 *Experimental apparatus for determining the effect of molecular weight on the rate of diffusion.* **Caution:** *Both HCl and NH_4OH are poisonous. They can cause burns and can be fatal if swallowed. Carry out this experiment in a fume hood.*

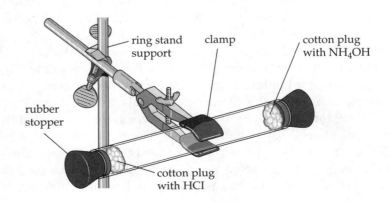

and the ratio of the distances traveled by NH_3 (d_1) and HCl (d_2) should be close to the ratio of the diffusion rates:

$$\frac{r_1 = 1/\sqrt{MW_1}}{r_2 = 1/\sqrt{MW_2}} = \frac{r_1}{r_2} \propto \frac{d_1}{d_2}$$

The molecular weight of ammonia gas (NH_3) is 17; the molecular weight of hydrogen chloride gas (HCl) is 36.

3. Measure the distance to the white ring from the front edge of each of the cotton plugs.

   • Distance of NH_4Cl precipitate from NH_4OH plug, d_1 = _____ mm

   • Distance of NH_4Cl precipitate from HCl plug, d_2 = _____ mm

   • Ratio d_1/d_2 = _____

4. Now calculate the ratio of diffusion rates from the molecular weights.

   • Diffusion rate for NH_3, r_1 = $\dfrac{1}{\sqrt{MW_1}}$ = _____

   • Diffusion rate for HCl, r_2 = $\dfrac{1}{\sqrt{MW_2}}$ = _____

   • Ratio r_1/r_2 = _____

   a. *Is the ratio of the distances approximately equal to the ratio of the rates?* _____ *Why?*

   _____

   b. *Was the white precipitate in the center of the tube?* _____ *Why or why not?*

   _____

   c. *Which gas traveled faster?* _____

   Why was the white ring of NH_4Cl not seen immediately after the plugs were inserted?

   _____

   d. *What is the relationship between molecular weight and the rate of diffusion of a gas?*

   _____

👁 EXERCISE C   Diffusion Across a Selectively Permeable Membrane

The cell membrane is a **selectively permeable** membrane. Small hydrophobic solute molecules, water, and other very small polar, uncharged molecules can move freely through the cell membrane. But larger molecules, and even small charged ions, may pass more slowly or sometimes not at all. All cell membranes are dynamic systems that can alter their lipid and protein contents to change their pore size. By regulating membrane permeability in this way, cells allow specific molecules or ions to move into or out of the cell as needed.

Solutes that diffuse through a selectively permeable membrane always move from the solution that contains more of the solute to the solution containing less solute (Figure 8C-1). The solutes diffuse from an area of higher chemical potential to an area of lower chemical potential. If the two solutions contain equal concentrations of a solute, the chemical potentials are equal and no net movement of particles occurs.

In the laboratory, molecules of different sizes can be separated by **dialysis** using artificial membranes made into tube-shaped bags (dialysis bags). These membranes are **semipermeable**. (Since the membranes are nonliving, the pore sizes cannot be changed to "select" molecules of differing sizes.) The size of the pores in the dialysis tubing determines which substances can pass through. Molecules larger than the pore size remain inside or outside the bag, while smaller molecules and ions diffuse through the membrane. In

Figure 8C-1 *When two solutions are separated by a selectively permeable membrane, solute will move from an area of higher solute concentration and higher chemical potential to an area of lower solute concentration and lower chemical potential.*

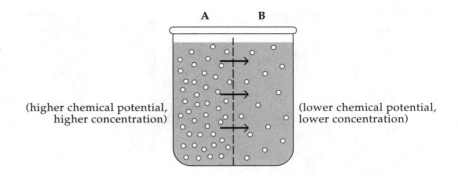

(higher chemical potential, higher concentration)

(lower chemical potential, lower concentration)

all cases, molecules move from an area of higher concentration or higher chemical potential to an area of lower concentration or lower chemical potential.

In this experiment, solutes dissolved in water are separated by a semipermeable membrane. A solution of starch and glucose is placed inside a dialysis bag, and I$_2$KI (Lugol's solution) is added to the water outside the bag. Molecules can move both ways across the dialysis bag membrane *if* they are small enough.

a. *Which molecule(s) do you think will move through the dialysis bag?* _____

   Why? _____

b. *Which molecule is least likely to move through the dialysis bag?* _____

   Why? _____

Formulate a hypothesis that offers a tentative explanation for how molecular size might be related to the movement of glucose, starch, and I$_2$KI molecules in this experiment.

   HYPOTHESIS:

   NULL HYPOTHESIS:

Which way do you **predict** that the solute molecules used in this experiment might flow?

What is the **independent variable** in this investigation?

What is the **dependent variable** in this investigation?

Use the following procedure to test your hypothesis.

IIIII Procedure IIIIIIIIIIIIIIIIIIIIIIIIIIIIIIIIIIIIIIIIIIII

1. Work in pairs. Obtain a piece of dialysis tubing and tie a knot in one end of it.

2. Add 4 Pasteur pipettefuls of 15% glucose solution to the bag, then use a different Pasteur pipette to add 4 pipettefuls of 1% starch solution to the glucose in the bag.

3. Hold the bag closed and mix its contents. Record its color in Table 8C-1. Carefully rinse off the outside of the bag in tap water.

4. Fill a beaker two-thirds full with water. Add about 4 dropperfuls of Lugol's reagent (I$_2$KI) to the water in the beaker. In Table 8C-1, record the color of the solution in the beaker.

Figure 8C-2 *Setup for dialysis experiment.*

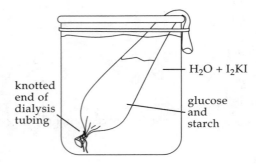

knotted end of dialysis tubing

H$_2$O + I$_2$KI

glucose and starch

Table 8C-1   Data for Dialysis Experiment

	Original Contents	Original Color	Final Color	Color After Benedict's Test
Bag	Glucose and starch			
Beaker	H$_2$O + I$_2$KI			

5. Place the bag in the beaker so that the untied end of the bag hangs over the edge of the beaker. Do *not* allow the liquid to spill over the top of the bag. If the bag is too full, remove some of the liquid and rinse the bag again. Place a rubber band around the beaker so that the top of the bag is held securely in place (Figure 8C-2).

6. Allow the setup to stand until you see a distinct color change in the bag or in the beaker. (You may wish to go on to another exercise and then return to check your setup after 30 minutes or more.) In Table 8C-1, record the final color of the solution in the bag and the solution in the beaker.

7. After you have observed the color change, stir the contents of the beaker, remove 3 ml of the solution from the beaker, and test with Benedict's reagent according to the directions in Laboratory 5. (You may use TesTape if available.) Record your results in Table 8C-1.

8. Now use Benedict's reagent (or TesTape) to test 3 ml of the solution from the bag. Record your results in Table 8C-1.

c. *I$_2$KI reagent is used to test for the presence of which type of molecule?* _____

   What color change occurs in the presence of this molecule? _____

d. *TesTape or Benedict's reagent is used to test for the presence of which type of molecule?* _____

   What color change occurs in the presence of this molecule? _____

e. *Which substance or substances entered the bag? Which one(s) left the bag? Use Table 8C-2 to give evidence for your answer in terms of the color changes that occurred.*

Do your results support your hypothesis? _____ *Your null hypothesis?* _____

What do you **conclude** *about the effects of molecular size on the diffusion of molecules across a semipermeable membrane?* _____

Plasma membranes remain selectively permeable as long as they are alive and normal. A membrane may become temporarily more permeable if the cell is in a state of shock; however, upon cell death, the membrane becomes completely permeable. Anesthetics, certain toxins, and low and high temperatures change the permeability of cell membranes.

Table 8C-2   Evidence: Substances Leaving and Entering

	Outside	Inside
I_2KI		
Starch		
Glucose		

9. Examine the boiled red cabbage and boiled beet cubes on demonstration. A pigment located in the cell vacuoles of these plants is responsible for their color.

   f. *Where do you find the red pigment after boiling?* _____

   g. *What does boiling do to the structure of a membrane?* _____

   h. *How has this affected the differential permeability of the vacuolar membrane?*

   _____

👁 **EXERCISE D   A Look at Osmosis**

The movement of water molecules across a semipermeable or selectively permeable membrane is a special case of diffusion known as **osmosis.** Water molecules move from an area where the chemical potential is higher to an area where the chemical potential is lower. Recall that the chemical potential of water molecules is called water potential. Thus, as a rule, water molecules will always diffuse from an area of higher water potential to an area of lower water potential.

Water potential (ψ) results from the combined actions of **osmotic potential** (ψ_π), which is dependent on solute concentration in a solution, and **pressure potential** (ψ_p), which results from the exertion of positive pressure or negative pressure (tension) on a solution. We can express this relationship as

$$\underset{\substack{\text{water} \\ \text{potential}}}{\psi} = \underset{\substack{\text{osmotic} \\ \text{potential}}}{\psi_\pi} + \underset{\substack{\text{pressure} \\ \text{potential}}}{\psi_p}$$

In general, added positive pressure causes the water potential of a solution to become more positive (higher). Addition of solute causes the water potential of a solution to become more negative (lower) if

Figure 8D-1 *Pure water has a water potential (ψ) of 0 (if the water is not under pressure). The addition of solute (indicated by the solid black arrow) lowers water potential. In (a), solute has been added to pure water and water potential declines. Pressure (indicated by the open arrow) increases water potential. In (b), pressure has been applied to pure water. As you can see in (c) through (e), the net water potential (ψ) depends on the effects of both solute concentration (ψ_π) and pressure (ψ_p).*

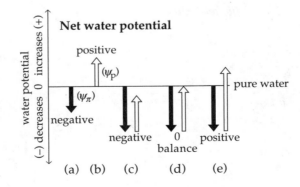

there is no pressure to offset the effects of adding solute (Figure 8D-1). As a result, water potential (ψ) can be negative as in (c) and zero as in (d) or positive as in (e) (Figure 8D-1).

a. *The water potential of solutions not under positive pressure is always* _____.

b. *Can the water potential of a solution ever be positive?* _____ *How?* _____

c. *How would negative pressure (tension) affect water potential?* _____

Differences in water potential result in a tendency of water to leave one area in favor of another. The addition of solute to water *lowers* the **osmotic potential** of a solution (makes ψ_π more negative) and, therefore, lowers the water potential of that solution. Thus, in the absence of other factors (such as pressure; that is, $\psi_p = 0$), osmosis results in the net movement of water across a semipermeable membrane from an area of lower solute concentration (higher water potential) to an area of higher solute concentration (lower water potential).

Note: If two solutions, separated by a semipermeable membrane, have identical osmotic potentials, they are **isotonic** (they contain the same amounts of non-penetrating solutes, even if the chemical composition of the solutes is different). If they are not isotonic, then the solution with the greater concentration of solute (and more negative osmotic potential) is **hypertonic** to the other solution. A solution that is **hypotonic** has fewer solute particles and a less negative osmotic potential than the solution to which it is compared. During osmosis, water will move from a hypotonic solution into a hypertonic solution.

Sucrose is a disaccharide sugar. Suppose pure water is separated from a sucrose solution by a semipermeable membrane that is not permeable to the sugar; sugar is a non-penetrating solute (Figure 8D-2).

Which way will water molecules move? _____

Figure 8D-2 *When non-penetrating solute, in this case sucrose, is added to pure water (left-hand side of beaker), water potential declines from zero to a negative value. Water always moves from an area of higher water potential to an area of lower water potential.*

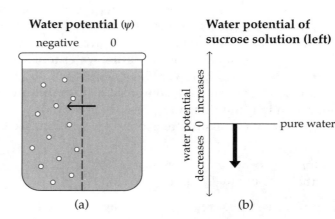

Now, suppose you put the sucrose solution into a dialysis bag, permeable only to water, tie both ends of the bag, and place the bag in a beaker of pure water (Figure 8D-3).

d. *Where is the water potential higher—inside or outside the bag?* _____ *Why?*

Figure 8D-3 *If a sucrose solution is placed in a dialysis bag and the bag is submerged in pure water, osmosis will move water from an area of higher water potential (pure water, $\psi = 0$) to an area of lower water potential (sucrose solution with a negative ψ).*

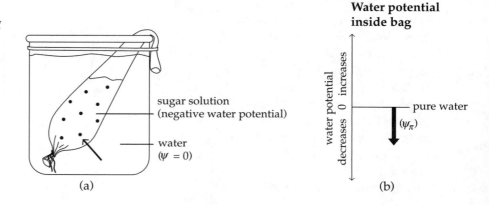

If solute is added to the water outside the bag, water will continue to flow from the beaker into the bag as long as ψ of the solution in the bag is more negative (lower) than ψ of the solution in the beaker (Figure 8D-3). The **rate** of diffusion depends on the size of the difference between the two water potentials (the larger the difference, the faster the flow).

Don't forget that **osmotic potential** (ψ_π) is a measure of the tendency for water to move into a solution. The more solute is added, the greater the osmotic potential (ψ_π) becomes (a *larger* negative value) and this large negative value lowers the overall water potential. Look at the size of the solid arrows (ψ_π) in Figure 8D-4. The arrow is smaller (a *smaller* negative value) for the solution in the beaker than for the solution in the bag (a *greater* negative value). You might think of osmotic potential as the "potential" for one solution (the one with the greater negative value of ψ_π) to "suck" water in from the other solution.

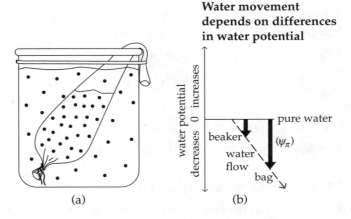

Figure 8D-4 *The more solute in a solution, the greater its osmotic potential (ψ_π) and the lower its water potential. Water tends to move toward regions of more negative water potential, so water moves from the beaker into the bag (dashed line).*

e. In Figure 8D-4, where is osmotic potential greater—inside or outside the bag? _____ Why?

f. Which way will water move? _____ Why? _____

If water potentials on the two sides of a selectively permeable membrane (such as a dialysis bag or cell membrane) eventually become equal, the *net* water flow will stop. The solutions will be at equilibrium. Individual water molecules, however, will continue to move back and forth across the membrane while the solution is at equilibrium (water movement does not stop; just *net* water movement ceases).

But, how can water potentials become equal if the bag always contains sucrose and the liquid outside the bag is pure? As water moves into the dialysis bag, the bag will swell. The same thing would happen to a cell (which contains solute in its cytoplasm or, in plant cells, in its vacuole). But, if a barrier (such as the wall of a tied dialysis bag or a cell wall) prevents indefinite expansion, then pressure will build up inside the compartment (bag or cell) into which the water is moving. An increase in positive pressure raises the pressure potential (ψ_p) and, because of this, the water potential inside the compartment becomes more positive (less negative). The pressure will "push" on the walls of the bag. The pressure that builds up (due to osmosis) is hydrostatic pressure, and it will build up until it equals the **osmotic pressure** (the amount of pressure necessary to stop water movement into the cell). See Figure 8D-5. When osmotic pressure is exerted against the walls of a cell, it is called **turgor pressure.** Turgor pressure is what makes the celery in your salad firm and keeps green plants standing up straight. Loss of turgor results in wilting.

As pressure builds inside a dialysis bag (or cell), the water potential gradient (difference) between the solution outside the bag and the solution inside the bag decreases. Water continues to flow into the bag until enough pressure builds up within the bag to offset the tendency for water to be drawn inward by the solute. Net water movement into the bag stops. At this point, the water potential inside the bag is equal to the water potential outside the bag (Figure 8D-5).

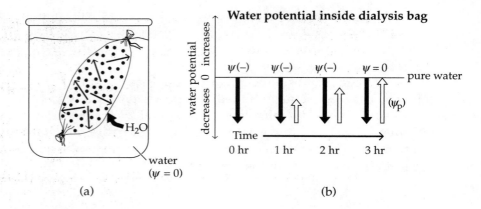

Figure 8D-5 *(a) As water moves into a dialysis bag containing a sucrose solution (assume the bag is not permeable to sucrose), pressure builds up over time (b) until the positive pressure exerted on the walls of the bag (ψ_p) balances the inflow of water.*

IIIII Objectives IIIIIIIIIIIIIIIIIIIIIIIIIIIIIIIIIIIIIIIII

☐ Define osmotic potential and osmotic pressure.

☐ Describe how to measure osmotic potential.

☐ Define turgor.

☐ Describe what causes plasmolysis in living plants.

☐ Relate osmotic potential to solute concentration and water potential.

PART I   Measuring Osmotic Potential

In this experiment, you will investigate the relationship between osmotic potential and the movement of water through a selectively permeable membrane by the process of osmosis. You will place several dialysis bags containing solutions of different sucrose concentrations into beakers containing distilled water. The direction of water movement can be determined by finding the mass of (weighing) the bags before and after placing them in distilled water: the dialysis bags are permeable to water, but not to sucrose, and can gain or lose water.

At atmospheric pressure, the water potential of the pure water in the beakers can be assumed to be zero ($\psi = 0$); no solute is present, thus osmotic potential is zero ($\psi_\pi = 0$) and pressure potential is also zero ($\psi_p = 0$).

The water potential of the sucrose solutions in the dialysis bags will be negative (recall that in the absence of positive pressure, the addition of solute to pure water decreases osmotic potential), thus the water potential of the solutions inside the bags will be lower (ψ is negative) than the water potential of the water outside the bags, where ψ is zero. Can you predict which way water will move?

IIIII Procedure IIIIIIIIIIIIIIIIIIIIIIIIIIIIIIIIIIIIIIIII

1. Obtain six 25-cm strips of presoaked dialysis tubing and keep them in fresh distilled water.

2. Tie a knot at one end of each piece of dialysis tubing to form six bags. Use a pipette to put 15 ml of each of the solutions listed in Table 8D-1 into separate bags. After adding the solution, remove most of the air from the bag by drawing the unfilled portion between two fingers. Tie a knot near the open end to seal the solution within the tube. You should have 1.5 to 2 times as much empty space in the tube as that taken up by the volume of the solution. This will leave enough unfilled space within the tube to accommodate the possible accumulation of water. Be sure to keep track of which tube contains which solution.

3. Carefully blot the outside of each bag. Determine the initial mass of each bag and record it in Table 8D-1.

4. Fill six 250-ml beakers (or plastic cups) three-quarters full with distilled water.

Table 8D-1   Data for Measuring Osmotic Potential

Contents of Dialysis Tube	Initial Mass	Final Mass	Percent Change in Mass
Distilled water			
0.2 M sucrose			
0.4 M sucrose			
0.6 M sucrose			
0.8 M sucrose			
1.0 M sucrose			

Figure 8D-6 *Percent change in mass of dialysis bags containing different molarities of sucrose.*

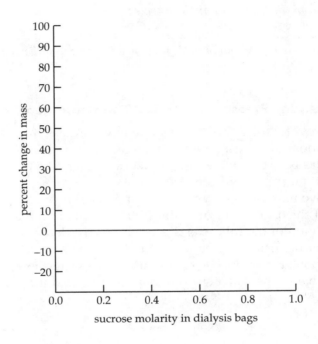

5. Place each bag in one of the beakers of distilled water and label the container to indicate the molarity of the solution in the dialysis bag. Make sure that all parts of the bag are completely covered by water.

6. Let stand for 1 hour.

7. At the end of the required time, remove the bags from the water and carefully blot them. Determine the mass of each bag and record your data in Table 8D-1.

8. Calculate the percent change in mass for each bag.

$$\% \text{ Change} = \frac{\text{final mass} - \text{initial mass}}{\text{initial mass}} \times 100$$

Graph your data in Figure 8D-6.

a. *What is the relationship between the increase in mass and the molarity of sucrose in the dialysis bags?* _____

_____

b. *How does the increase in sucrose concentration affect the water potentials of the various solutions inside the dialysis bags?* _____

c. *Does it affect the water potential of the solution outside the bags?* _____

d. *A solution (A) that contains more non-penetrating solute than another solution (B) is often said to be hypertonic (hyper- = more than; solution A is hypertonic to solution B). Likewise, solution B can be described as being hypotonic (hypo- = less than) to solution A. Use the terms* hypertonic *and* hypotonic *to describe the difference in water potential between solutions A and B.* _____

e. *If two solutions are* isotonic *to one another, how do their solute concentrations compare?*

_____

f. *Predict what would happen in your experiment if all the bags were placed in a 0.4-M sucrose solution instead of in distilled water.* _____

_____

g. *What would a graph of percent change in mass look like for this experiment? Draw it on Figure 8D-6.*

PART 2   Measuring Pressure Potential: The Osmometer

The osmometer provides a way to demonstrate how changes in pressure potential can affect the water potential of a solution. In the osmometer on demonstration (Figure 8D-7), a dialysis bag containing a 40% sucrose solution has been suspended in a beaker of water. The bag is semipermeable—allowing passage of solvent (water) but no passage of solute (sucrose).

Water will move into the bag (where water potential is negative) and the sucrose solution will rise in the glass tubing of the osmometer until the water potential on both sides of the membrane is equal. At this point, the solution in the tube exerts enough positive hydrostatic pressure (pressure resulting from the movement of water or "osmotic pressure") on the contents of the dialysis bag to equal, or offset, the negative osmotic potential of the sucrose solution. The water potential of the solution in the bag will be zero ($\psi = 0$), like that of the water surrounding the bag, and *net* movement of water into the bag stops.

Figure 8D-7 *A simple osmometer.*

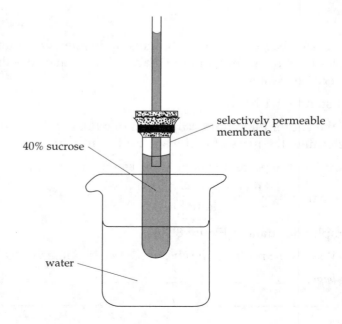

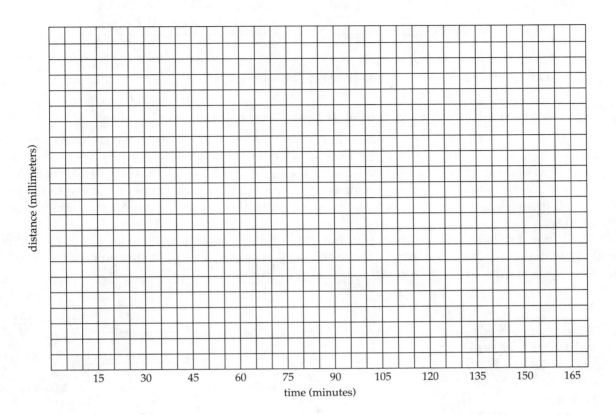

Figure 8D-8   *Graph osmotic pressure in millimeters of solution traveled up the osmometer tube.*

IIIII **Procedure** IIIIIIIIIIIIIIIIIIIIIIIIIIIIIIIIIIIIIIIIIIIII

1. At 15-minute intervals during the laboratory period, observe the changes in the osmometer on demonstration. On a separate sheet of paper, record the time and the height, in millimeters, of the solution in the tube.

2. Plot your data in Figure 8D-8.

   a. *Why does water move into the bag?* _____

   _____

   b. *In which solution (inside the bag or outside) is the water potential higher?* _____

   In which is solute concentration higher? _____

   c. *Which solution (inside the bag or outside) is hypertonic?* _____ *Hypotonic?*

   _____

   d. *How is osmotic pressure (the tendency to resist further net water movement) exerted within the*

   dialysis bag? _____

   e. *What does this pressure do to the water potential of the solution in the osmometer?*

   _____

   f. *What is the water potential inside the bag when the fluid stops rising in the glass tube?*

   _____

How do you know? _____

_____

 g. *Does osmosis stop after a certain period of time? (Be specific.)* _____

PART 3   Measuring the Water Potential of Living Plant Cells

When a solution, such as that inside a potato cell, is separated from pure water by a selectively permeable membrane, water will move (by osmosis) from the surrounding area where water potential is higher ($\psi = 0$) into the cell where water potential is lower due to the presence of solutes (ψ is negative). (We will assume, for purposes of explanation, that solute is not diffusing.) The movement of water into the cell causes the cell to swell, and the cell membrane pushes against the cell wall to produce an increase in pressure (**turgor**).

 Positive turgor pressure inside the cell continues to build up until the water potential of the cell equals the water potential of the pure water outside the cell ($\psi_{cell} = \psi_{outside} = 0$). At this point, a dynamic equilibrium exists and net water movement ceases (Figure 8D-9).

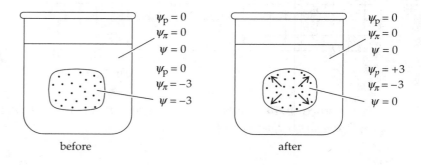

Figure 8D-9 *Water moves into a cell from an area of higher water potential outside the cell ($\psi = 0$) to an area of lower water potential inside the cell ($\psi = -3$). Net water movement ceases when enough pressure builds up in the cell so that the water potential inside the cell equals the water potential outside the cell ($\psi = 0$ for both). Recall that $\psi = \psi_{\pi} + \psi_{p}$.*

before   after

(Figure 8D-9 labels, before:) $\psi_p = 0$, $\psi_\pi = 0$, $\psi = 0$; $\psi_p = 0$, $\psi_\pi = -3$, $\psi = -3$

(Figure 8D-9 labels, after:) $\psi_p = 0$, $\psi_\pi = 0$, $\psi = 0$; $\psi_p = +3$, $\psi_\pi = -3$, $\psi = 0$

 If solute is added to the water outside the potato cells, the water potential outside the cells is decreased, and, eventually, if enough non-penetrating solute is added to the water, the water potential outside the cell will end up being the same as the water potential inside the cell; then no net movement of water will occur. This does not mean, however, that the solute concentrations inside and outside the cell are equal, because water potential inside the cell results from a combination of both pressure potential (ψ_p) and osmotic potential (ψ_π) (Figure 8D-10).

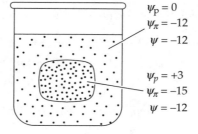

Figure 8D-10 *No net water movement occurs when the water potentials of the cell and its surroundings are equal ($\psi = -12$ for both), but ψ_π for the cell and solution are not the same. It is ψ_p inside the cell that makes it possible for* $\psi_{cell} = \psi_{solution}$.

(Figure 8D-10 labels:) $\psi_p = 0$, $\psi_\pi = -12$, $\psi = -12$; $\psi_p = +3$, $\psi_\pi = -15$, $\psi = -12$

 In this experiment, you will calculate the water potential of potato-tissue cells soaked in different molarities of sucrose. The original turgor pressure of the cells cannot be measured, and thus the osmotic potential of the cells cannot be determined. However, you will be able to measure the water potential of the cells by determining at what molarity of sucrose net movement of water into the potato tissue stops.

Your instructor may ask you to compare different types of potatoes or even different types of vegetables. Does one type of potato or vegetable have a higher water potential than another? Does a potato stored in the refrigerator have a higher water potential than one stored on the shelf in the pantry? Consider the available materials and formulate a hypothesis.

HYPOTHESIS:

NULL HYPOTHESIS:

What do you **predict** would occur in an experiment to test your hypothesis?

What is the **independent variable** in your investigation?

What is the **dependent variable** in your investigation?

Now use the following procedure to test your hypothesis.

Procedure

1. Work in pairs within groups of 4. Each pair of students in the group should obtain 100 ml of each of three sucrose solutions or distilled water, as listed in Table 8D-2. Pour each solution into a separate 250-ml beaker or plastic cup and label each container to indicate the molarity of sucrose it contains.

2. Each pair of students should use a cork borer (approximately 5-mm inner diameter) to cut 12 potato cylinders. Cut each cylinder to 3 cm in length. Do not include any skin on the cylinders. For the group's experiment, all cylinders should be cut from the same potato.

3. Keep your potato cylinders in a covered beaker or Petri dish until it is your turn to use the balance for finding the mass of the cylinders.

4. Quickly determine the mass of four cylinders together, and record it in the table below. Put all four cylinders into one of the three sucrose solutions.

5. Determine the masses of the other groups of four cylinders, and record them in the table. Put these cylinders into the other two solutions.

6. Cover all beakers to prevent evaporation.

7. Let the beakers stand for 2 hours or overnight, according to directions from your instructor.

8. Record the temperature of each solution after standing.

9. Remove the cores from one of the beakers, blot them gently by rolling lightly on a paper towel, and find their mass. Then do the same for the other groups of cylinders.

10. Record the final masses of your other two sets of cylinders.

Contents of Beaker	Initial Mass	Final Mass
1.		
2.		
3.		

11. Obtain the data for the remaining sucrose solutions from other members of your group. Record all data in Table 8D-2.

Table 8D-2   Group Data

Contents of Beaker	Initial Mass	Final Mass	Percent Change in Mass
Distilled water			
0.2 M sucrose			
0.4 M sucrose			
0.6 M sucrose			
0.8 M sucrose			
1.0 M sucrose			

12. Calculate percent change in mass as in Exercise D, Part 1, and graph your data in Figure 8D-11.

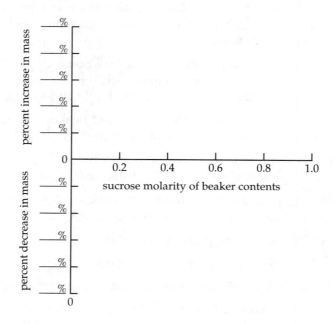

Figure 8D-11   *Percent change in weight of potato cores at different molarities of sucrose. Label the vertical axis in appropriate percent intervals.*

13. Determine the osmolarity of the sucrose solution in which the mass of the potato cores does not change. To do this, draw the straight line that best fits your data points (Figure 8D-11). The point at which this line crosses the *X*-axis represents the molar concentration of sucrose with a water potential equal to that of the potato tissue. At this concentration, there is no net gain or loss of water from the potato-tissue cells.

Molar concentration = _____ *M*

14. The osmotic potential of a sucrose solution can be calculated from the molarity of the solution by using the formula

$$\psi_\pi = -iCRT$$

where

i = ionization constant (for sucrose, this is 1 because sucrose does not ionize in water; for NaCl, $i = 2$)

C = osmotic molar concentration

R = pressure constant (handbook value: $R = 0.0831$ liter bar/mol K)

T = temperature in degrees Kelvin (273 + °C of solution)

The units of measure will cancel as in the following example. For a 1.0-M sugar solution at 30°C under standard atmospheric conditions:

$$\psi_\pi = -1 \left(1.0 \; \frac{\text{mol}}{\text{liter } H_2O}\right) \left(0.0831 \; \frac{\text{liter bar}}{\text{mol K}}\right) (303 \; \text{K}) = -25.18 \text{ bars*}$$

$$\psi_\pi = -i \quad \times \quad C \quad \times \quad R \quad \times \quad T$$

15. Knowing the osmotic potential of the sucrose solution (ψ_π) and knowing that the pressure potential of the solution is 0 ($\psi_p = 0$), since the solution is open to the atmosphere, allows you to calculate the water potential of the solution. It will be equal to the osmotic potential.

$$\psi = \psi_p + \psi_\pi$$
$$= 0 + \psi_\pi$$
$$= \psi_\pi$$

The water potential of the solution at equilibrium will be equal to the water potential of your potato cells. *a. Why?* _____

16. Determine the water potential of your potato cells.

$\psi_{\text{potato cells}} = $ _____

b. Is the osmotic potential of the potato cells the same as that of the sucrose solution when

$\psi_{\text{potato}} = \psi_{\text{solution}}$? _____ *Why or why not?* _____

_____

Do your results support your hypothesis? _____ *Your null hypothesis?* _____

Was your prediction correct? _____

*What do you **conclude** about the water potential of your experimental material(s)?* _____

_____

c. If a potato is allowed to dehydrate by sitting in the open air, would the water potential of the potato cells become

higher or lower? _____ *Why?* _____

_____

A pressure of 1 bar is just slightly less than 1 atmosphere. [An atmosphere (atm) is the pressure exerted at sea level by an imaginary column of air—approximately 1 kg of pressure per cm^2.] Plant biologists usually measure water potential (ψ) in a unit of pressure called the megapascal (MPa): 1 atm = 0.1 MPa. We can also say that

$1 \text{ bar} = 0.1 \text{ MPa}$

To express the above answer in MPa, simply move the decimal point one place to the left:

$-25.18 \text{ bars} = -2.518 \text{ MPa}$

EXTENDING YOUR INVESTIGATION: WATER POTENTIAL OF DIFFERENT TUBER TYPES

A study of various types of tubers was conducted to determine the percent carbohydrate and percent water content. The data are given below.*

Tuber	Percent Carbohydrate	Percent Water
Beets	9.6–9.9	87.6
Carrots	9.3	88.2
Parsnips	13.5–18.2	78.6
Sweet potato	27.9	68.5
Turnip	7.1–8.1	90.9
White potato	19.1	77.8

When students in an introductory biology class used some of these tubers for their water potential study (as in Exercise D, Part 3), they recorded the following results:

Sweet potato $\psi = -10.93$ bars

Turnip $\psi = -18.39$ bars

The students had hypothesized that sweet potatoes would have a lower water potential than turnips on the basis of carbohydrate and water content. Their hypothesis was *not* supported by their data. How might you explain this?

---

*Data from A. E. Leach and A. L. Winston, Food Inspection and Analysis, *John Wiley & Sons, New York, 1920; and Morris J. Jacobs (ed.),* The Chemistry and Technology of Food and Food Products, *Interscience Publishers, New York, 1944.*

PART 4   Observing Osmosis in a Living System

If a plant cell is immersed in a solution that has a higher solute concentration than that of the cell (Figure 8D-12), water will leave the cell, moving from an area of higher water potential to an area of lower water potential. The loss of water from the cell will cause the cell to lose turgor. Macroscopically, you can see the effects of loss of turgor in wilted house plants or limp lettuce. Microscopically, increased loss of water and loss of turgor become visible as a withdrawal of the protoplast from the cell wall (**plasmolysis**) and as a decrease in the size of the vacuole (Figure 8D-13).

Procedure

1. The various plant materials on display were placed into either pure water or a saltwater solution.

Figure 8D-12 *When a plant cell is placed in a solution of (a) equal solute concentration ($\psi_{\pi(cell)}$ = $\psi_{\pi(surrounding)}$) or (b) greater solute concentration, water may leave the cell (why?), resulting in loss of turgor and, eventually, plasmolysis.*

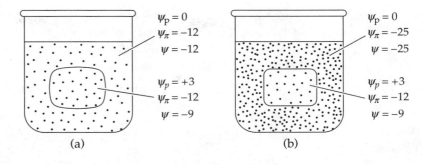

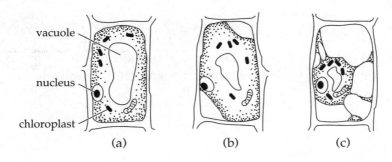

Figure 8D-13 *Plasmolysis in an epidermal cell of a leaf. (a) Under normal conditions, the plasma membrane is pressed against the cell walls. A large vacuole occupies the center of the cell, pushing the cytoplasm and nucleus to the periphery. (b) When the cell is placed in a solution with a higher concentration of solutes than that of the cell, water passes out of the cell, and the cell contents contract. (c) In an even more concentrated solution, the cell contents contract still further.*

a. *Compare the effects of water and salt solutions. What has happened?*

_____

b. *Does cell turgor control the overall turgor of tissues?* _____ *The overall turgor of the*

   plant part? _____

c. *Can plant cells burst?* _____ *Explain.* _____

_____

2. Obtain a filament of the green alga *Spirogyra* or a leaf from the tip of an *Elodea* plant. Place it in a drop of water on a slide, cover it with a coverslip, and examine the material first at low power (10×) and then at high power (40×). Locate a region of healthy cells and sketch the location of the chloroplasts in the left side of the space below.

3. While touching one corner of the coverslip with a torn piece of paper towel to draw off the water, add a drop of concentrated salt solution to the opposite corner of the coverslip. Be sure that the salt solution moves under the coverslip. Wait about 5 minutes, then examine as before. Record your observations in the space beside your drawing in step 2.

d. *What happened when the water in which the cells were mounted was replaced by the salt solution?*

_____

e. *Assuming that the cells have not been killed, what should happen if the salt solution were to be*

replaced by water? _____

Laboratory Review Questions and Problems

1. Dialysis bags containing a 0.5% sucrose solution are placed in beakers containing the sucrose solutions indicated below. Dialysis bags are not permeable to sucrose.

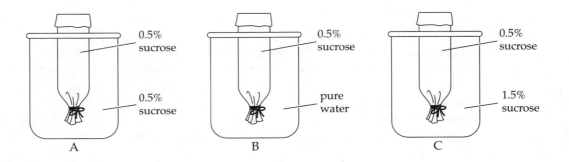

Indicate the letter of the beaker in which the following events or situations occur.

a. The dialysis bag remains the same size. _____

b. The dialysis bag swells. _____

c. Water moves from the dialysis bag into the beaker. _____

d. The solution inside the bag is hypotonic to its surroundings. _____

e. Water potential inside the bag is higher than water potential outside the bag. _____

f. The solution inside the bag is isotonic to the solution outside the bag. _____

2. How does the method of salt or sugar curing (a process in which meat is packed in salt or sugar) help to preserve meat?

3. In some restaurants, sliced potatoes are soaked in water before they are fried. Can you provide an explanation for this practice?

4. Why is it important that solutions administered intravenously be isotonic to the recipient's blood? What would happen if an injected solution were hypertonic to your blood? What if it were hypotonic?

5. A gardener's favorite bush died several days after she applied twice the recommended amount of fertilizer. What probably happened?

6. A plant cell is placed in a beaker of pure water. What will happen to the cell?

7. An animal cell is placed in a beaker of pure water. What will happen to the cell? What is the reason for the difference between what happens to plant and animal cells placed in pure water?

8. A dialysis bag is attached to the end of a glass tube to make a simple osmometer. The dialysis bag contains a 1 M sucrose solution with a water potential of $\psi = -35$ bars. Since the osmometer is open to the air, ψ_p is initially 0. What is ψ_π of the sucrose solution in the bag? _____ If the osmometer is placed in a beaker of water and osmosis occurs, what will ψ, ψ_p, and ψ_π be at equilibrium for the solution in the bag and the water in the beaker?

	ψ	ψ_p	ψ_π
Bag			
Beaker			

9. A plant cell, when initially placed in pure water, has an osmotic potential of -4 bars and a pressure potential of $+2$ bars.

   a. Which way will water diffuse?

   b. When will net diffusion stop?

   c. When equilibrium is reached, what has happened to the cell's osmotic potential and pressure potential values? $\psi_\pi = $ _____ $\psi_p = $ _____

10. A protozoan cell is placed in a 0.5 M sucrose solution at 27°C. Assume the cell has an initial osmotic potential of -2 bars. Because it lacks a cell wall, it cannot generate a turgor pressure and will always have a pressure potential of 0.

    a. When the cell is placed in a sucrose solution, which way will water diffuse?

    b. When will net diffusion stop?

    c. What will be the appearance of the cell when equilibrium is reached?

d. What will be the cell's osmotic potential and pressure potential values at equilibrium? (Assume that the quantity of water lost by the cell will not appreciably change $\psi_{outside}$.)

ψ_π = _____ ψ_p = _____

11. A plant cell with a rigid cell wall is placed in a 0.2 M solution of NaCl at 27°C and is allowed to equilibrate. Assume the cell has an initial osmotic potential of −8 bars and an initial pressure potential of +2 bars. Based on this information:

a. In which direction will water diffusion occur?

b. What will be the cell's water potential at equilibrium?

c. As the cell becomes less turgid, what happens to ψ_p?

d. What will be the cell's osmotic potential and pressure potential at equilibrium? (Assume that the quantity of water lost by the cell will not appreciably change $\psi_{outside}$.)

ψ_π = _____ ψ_p = _____

e. Will the cell be more turgid or less turgid at equilibrium than when it was first introduced into the solution?

12. Assume an animal cell with a volume of 1 ml and an osmotic potential of −5 bars is placed into a 0.8 M solution of sucrose at 27°C and is allowed to equilibrate. (Remember that animal cells cannot build up pressures in excess of atmospheric pressure, therefore the pressure potential of animal cells is always 0.)

a. Which way will diffusion occur?

b. What is the water potential of the cell at equilibrium? (Assume $\psi_{outside}$ does not change.)

c. What is the volume of the cell at equilibrium? *Hint:* We can obtain a good estimate of volume changes by recognizing two facts: (1) as the cell shrinks, the cytoplasmic solute concentration increases; (2) the solute concentration is directly related to osmotic potential. Thus we can use van't Hoff's law, which expresses the classical chemical relationship between concentration (M) and volume (V) to determine volume changes: $M_1 V_1 = M_2 V_2$; however, here we substitute ψ_π for M:

$M_1 \ (\psi_{\pi_1})$ = −5 bars

V_1 = 100% initial volume

$M_2 \ (\psi_{\pi_2})$ = osmotic potential of cell at equilibrium

$V_2 = \dfrac{M_1 V_1}{M_2}$

The answer will be expressed as a percent of the cell's initial volume.

Mitosis

OVERVIEW

According to the cell theory, all cells come from preexisting cells. New cells are formed by the process of cell division, which involves both division of the cell's nucleus (**karyokinesis**) and division, of the cytoplasm (**cytokinesis**).

There are two types of nuclear division: **mitosis** and **meiosis.** Mitosis typically results in the production of two daughter nuclei that are genetically identical to each other and to the parent nucleus. Formation of an adult organism from a fertilized egg, asexual reproduction, regeneration, and maintenance or repair of body parts are all accomplished by mitotic cell division. Meiosis, on the other hand, reduces the chromosome number in daughter nuclei to half that of the parent cell. Gametes (sex cells) in animals and spores in plants are both produced by meiotic division.

During this laboratory period you will focus on mitosis; you will investigate meiosis in Laboratory 13.

STUDENT PREPARATION

Prepare for this laboratory by reviewing the text pages indicated by your instructor. Do Exercise A before coming to the laboratory and familiarize yourself with the contents of Exercises B through D.

✔ **EXERCISE A** | **The Cell Cycle—Interphase and "Getting Ready"**

▐▐▐▐ **Objectives** ▐▐▐▐▐▐▐▐▐▐▐▐▐▐▐▐▐▐▐▐▐▐▐▐▐▐▐▐▐▐▐▐▐▐▐▐▐▐▐

☐ Name, identify, and describe the events occurring during the four phases of the cell cycle.

▐▐▐▐ **Procedure** ▐▐▐

The term **cell cycle** is used to describe the life history of actively dividing cells. Repeated cellular divisions are separated by periods of growth and preparation for the next division.

Even though it is a continuous process, the cell cycle can be separated into several stages (Figure 9A-1). After a cell divides, it enters **interphase,** which consists of three stages: G_1, S, and G_2. Interphase is often referred to as a "resting" stage, but cells in interphase are not really at rest. Indeed, they are actively preparing for the next division.

Figure 9A-1 *Stages of the cell cycle. G_1 and G_2 stand for the first and second "gaps" in the cell cycle. S stands for synthesis of DNA. Mitosis is also called "M phase."*

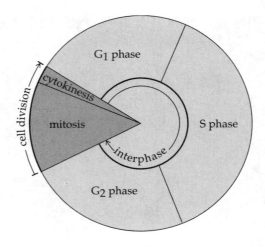

Figure 9A-2 *Electron micrograph of an isolated chief cell from mouse gastric mucosa. Note the darkly stained chromatin around the inner surface of the nuclear membrane.*

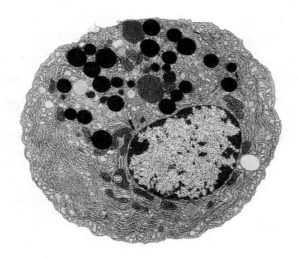

During interphase, DNA, with its chromosomal proteins, exists in a highly uncoiled state. Thus, when cell contents are stained, distinct chromosomal structures are not visible at this stage. Chromosomes appear, instead, as a granular material called **chromatin** within the nucleus (Figure 9A-2).

Events of G_1 During G_1 (first "gap"), the cell approximately doubles in size, and its enzymes and organelles double in number.

a. Why are the activities of G_1 important for a cell that is preparing to divide? _____

_____

_____

Late in the G_1 period, a critical **restriction point** is reached that determines whether or not the cell will continue to divide. If the cell does not proceed to the S phase, it exits the cell cycle and becomes "non-dividing." Nondividing cells are considered to be in the **G_0 phase.** Most cells of the body share this fate. Some, such as nerve cells, may never divide again; others can resume dividing if necessary. An example of the latter is the remarkable ability of the liver to regenerate after damage or transplant.

In order to pass the restriction point, a cell must reach a certain size. The ratio **volume of cytoplasm/ genome size** appears to be of most importance. Other environmental factors such as cell density or adherence to a substrate may also play a part in regulation at this point in the cell cycle.

In addition, **protein kinase** enzymes fluctuate in concentration throughout the cell cycle. Kinase enzymes control cell cycle activity by phosphorylating specific proteins. Regulatory proteins, called **cyclins,** can in turn regulate protein kinases. Protein kinases controlled by cyclins are called **Cdk** proteins (cyclin-dependent kinases). A specific cyclin-Cdk enzyme complex is required for cells to pass the restriction point.

Events of S Before the S phase, each chromosome consists of a double-stranded helix of DNA. During the S phase, the two strands of the DNA helix unwind and separate, and each duplicates by a process called **replication.** By the end of the S phase, each chromosome is composed of two helices of DNA, called **chromatids,** joined at a region of the chromosome known as the **centromere** (Figure 9A-3).

Figure 9A-3 *Replication of chromosomal DNA. The two strands of a helix of DNA separate, and each directs the synthesis of its other half so that each chromatid will be composed of a double-stranded helical DNA molecule. The small oval shape located at the centromere region is a kinetochore—a small proteinaceous body to which spindle fibers attach during mitosis.*

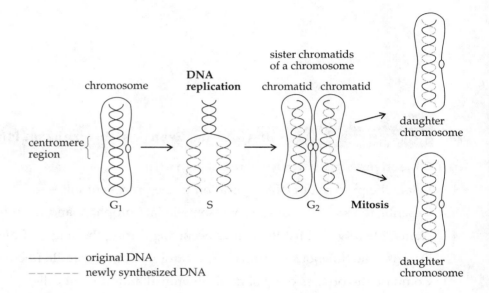

Distinct chromosomes are not yet visible during this phase. The DNA molecules are structurally intact, but are largely uncoiled and dispersed as chromatin. Chromosomal proteins, also synthesized during the S phase, will eventually associate with DNA to help coil it into tightly packed chromosome structures prior to the onset of mitosis.

b. Why is it necessary for DNA to duplicate before cells divide? _____

_____

Events of G_2 During the G_2 phase, structures involved directly in cell division are synthesized. **Spindle fibers** begin to assemble. These will become attached to chromosomes to guide their movement during cell division. In animal cells, a pair of **centrioles** completes division to form two pairs of centrioles. These will also play a role in the movement of chromosomes during the mitotic process. Cells of higher plants have spindle fibers but usually lack centrioles.

Regulatory proteins also serve an important role in the transition from the G_2 phase to the **M phase** (mitosis). This transition is regulated by a buildup of cyclins at the end of the G_2 phase and their regulation of a Cdk protein kinase as part of an enzyme complex known as **MPF** (M-phase promoting factor). At the end of the M phase, MPF activates an additional enzyme that destroys cyclin and turns off MPF activity.

c. Summarize the events of interphase in the following figure.

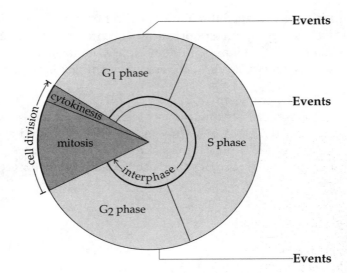

EXERCISE B   Simulating the Events of Interphase, Mitosis, and Cytokinesis

IIIIII Objectives IIIIIIIIIIIIIIIIIIIIIIIIIIIIIIIIIIIIIIIII

- ☐ Relate the process of DNA replication to the process of mitosis.
- ☐ Describe the structure of a chromosome in late prophase and identify its parts.
- ☐ Name, identify, and list the events occurring during the stages of mitosis.
- ☐ Describe the functions of centrioles, centromeres, and spindle fibers in mitosis.
- ☐ Compare the process of cytokinesis in animal and in plant cells.

IIIIII Procedure IIIIIIIIIIIIIIIIIIIIIIIIIIIIIIIIIIIIIIIIIIII

The purpose of this exercise is to simulate major features of mitotic cell division, *not* to duplicate it. The materials used will limit how closely you can approximate the actual process.

On your laboratory bench, you will find a chromosome simulation kit consisting of two strands of red beads and two strands of yellow beads, four hollow plastic cylinders, and four pieces of string. The strands of plastic beads represent chromosomes. The magnetic piece in each "chromosome" represents the centromere region. The four hollow plastic cylinders represent centrioles. The string will be used to simulate spindle fibers.

Figure 9B-1 shows a simulation of a chromosome as it would appear before the S phase of interphase, when DNA duplication occurs. It represents one double-stranded helical DNA molecule. Figure 9B-2 shows a simulated chromosome as it would appear after DNA replication. Each half of this duplicated chromosome is called a chromatid and contains its own helical DNA molecule. The two chromatids attached at the centromere region are identical and are called **sister chromatids.**

Figure 9B-1 *Early interphase. A chromosome, as it appears before DNA duplication in interphase, is composed of one double-stranded DNA molecule (see Figure 9A-3, G_1 phase).*

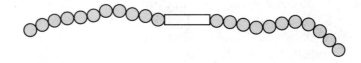

Figure 9B-2 *Mid-interphase. A chromosome as it appears after DNA duplication interphase. Each sister chromatid contains one double-stranded DNA molecule (see Figure 9A-3, G$_2$ phase).*

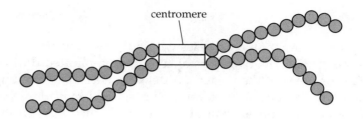

centromere

a. Next to each diagram (Figures 9B-1 and 9B-2), draw its representative DNA structure. Is the helix double-stranded or single-stranded in each chromatid? _____

Interphase

Place one strand of red beads and one strand of yellow beads near the center of your work area. Together these two strands of beads represent a **homologous pair** of chromosomes.*

b. What is the significance of using two different colors for the two homologous chromosomes?

_____

Could you have used two red strands, one short and the other long? _____ *Why or why not?*

_____

Keep in mind that chromosomal DNA would not be seen in the form of chromosomes at this stage, but would exist as part of the diffuse chromatin contained within the nucleus. Position two of the hollow plastic cylinders "centrioles") at right angles to each other outside an imaginary nuclear membrane or envelope surrounding the chromosomes.

DNA synthesis occurs during interphase to produce chromosomes consisting of two chromatids. Simulate this process by bringing the magnetic centromere of the second red strand on the table. Do the same with the yellow strands. Two chromosomes, each consisting of two sister chromatids, are now before you (Figure 9B-3). Simulate centriole replication by placing two additional plastic cylinders next to the original centriolar bodies.

c. Why do you need to replicate the centrioles? _____

_____

Figure 9B-3 *Late interphase. Chromosomal DNA has already duplicated and the centrioles have replicated by the end of interphase. The dashed line in (a) represents the nuclear envelope.*

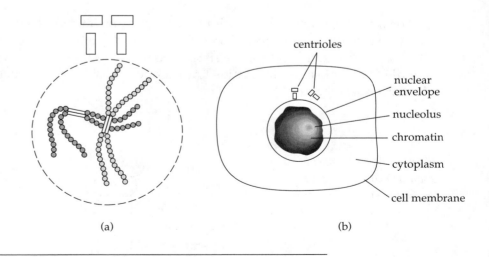

centrioles

nuclear envelope

nucleolus

chromatin

cytoplasm

cell membrane

(a)   (b)

In diploid organisms, chromosomes of somatic cells occur in matched pairs that resemble each other in size, shape, and location of their centromeres. Matched pairs are called homologous chromosomes. One member of each pair is paternal and the other, maternal, in origin.

Mitosis
Prophase

During prophase, chromatin condenses within the nucleus until the chromosomes and their individual chromatids are visible. The two pairs of centrioles, replicated during interphase, separate and migrate to opposite sides or **poles** of the nucleus. A dense material surrounding the centrioles is associated with microtubule formation and is often referred to as a *microtubule organizing center*. This material, in combination with the centriole pair at its center, is called a **centrosome.** Spindle fibers composed of microtubules begin to form. The assembly of microtubules is initiated by the centrosomes and requires the polymerization of globular **tubulin** protein molecules into fibers (similar to lengthening a string of beads). As the spindle fibers lengthen, the nuclear membrane and nucleoli disappear. The spindle fibers attach to a protein structure, the **kinetochore,** located at the centromere region of two sister chromatids. **Kinetochore microtubules** extend from the centrosomes to the kinetochores of chromosomes. Some fibers reach all the way across the cell from centrosome to centrosome, while others seem to stop short of attachment. Kinetochore microtubules shorten during anaphase by depolymerization of tubulin fibers at their kinetochore ends.

Begin your simulation of the events of prophase by separating the two pairs of cylindrical beads, moving one pair to the opposite side of the imaginary nuclear membrane that surrounds the chromosome bundle. Approximately 50 cm from each side of the chromosomes, tape down one centriole of each pair so that it is pointing toward the nucleus (Figure 9B-4).

Figure 9B-4   *Early prophase. Centrioles have migrated to opposite sides of the nucleus and spindle fibers will soon become attached to the centromere regions of the chromosomes.*

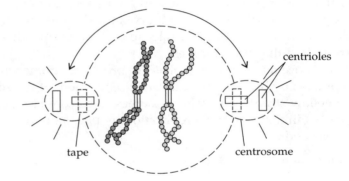

centrioles

tape   centrosome

Form a loop at one end of each piece of string in your kit. Draw the loop of one piece of string tightly around the centromere region of one of the sister chromatids (Figure 9B-5); repeat this process for the other sister chromatid and for the sister chromatids of the homologous chromosome. Extend the strings (spindle fibers) of sister chromatids toward opposite sides of the nucleus; thread each string through one of the cylinders (centrioles).

Figure 9B-5   *Attachment of spindle fibers to the centromere regions of sister chromatids.*

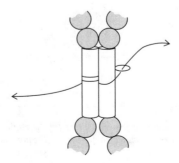

Metaphase

During metaphase, the centromere regions of sister chromatids are attached by spindle fibers to opposite centrioles and the chromosomes line up in single file at the middle, or equator, of the spindle apparatus (Figure 9B-6).

d. Why is it important that the chromosomes line up in single file during metaphase?

_____

Figure 9B-6 *(a) The mitotic spindle apparatus of animal cells consists of astral fibers (the short fibers that surround the centrioles), spindle fibers that attach to chromosomes, and spindle fibers that extend between centrioles. (b) In cells of higher plants, the mitotic spindle apparatus has spindle fibers but lacks centrioles and astral fibers.*

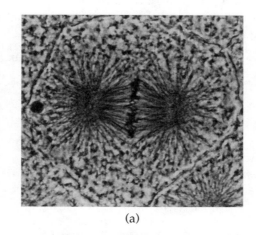

(a)

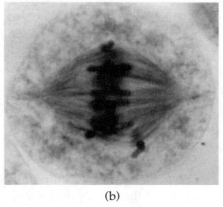

(b)

To simulate metaphase, center the chromosomes between the centrosomes (Figure 9B-7).

Figure 9B-7 *Metaphase. Chromosomes line up in single file on the mitotic spindle apparatus. Chromosomes appear to be perpendicular to the long axis of the spindle apparatus and are equidistant from each pole of the cell.*

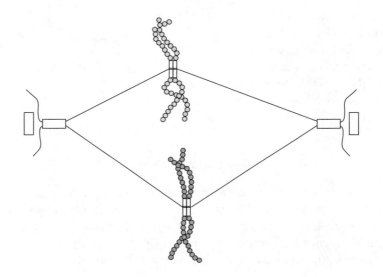

Anaphase

During anaphase, the sister chromatids of each chromosome separate at the region of their centromeres and move toward opposite poles. Recall that the spindle fibers shorten at their kinetochore ends and the chromosomes follow. Thus they appear to be pulled toward opposite sides of the cell. The separated chromatids are now called **daughter chromosomes.** The name of each of the two DNA strands, formerly held together at the kinetochore, has changed from chromatid to chromosome. When a strand of DNA has its own centromere region, it is called a chromosome.

To simulate anaphase, pull on the strings until the magnetic centromeres are separated. Continue pulling the daughter chromosomes toward the centrioles. Note that the arms of each chromosome follow the centromere toward the pole (Figure 9B-8).

e. During mitosis, when would you change the description of DNA strands from chromatids to

chromosomes? _____

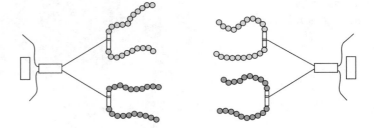

Figure 9B-8 *Anaphase. Centromere regions of sister chromatids separate, and daughter chromosomes move to opposite ends of the cell.*

Telophase

The spindle apparatus disappears during telophase and nuclei are re-formed as the nuclear membrane appears. Nucleoli also reappear. There are now two nuclei, one for each of the daughter cells. Chromosomes decondense and re-form diffuse chromatin.

f. How many helices of DNA are in each of the daughter chromosomes at this time? _____

To simulate telophase, remove the strings. Near each pair of centrioles, pile up the one red and one yellow chromosome drawn to that side of the cell during anaphase (Figure 9B-9). These represent the daughter nuclei and daughter cells. Note that you began cell division with two chromosomes: a red one and a yellow one.

Figure 9B-9 *Telophase. Sister chromatids have separated, forming daughter chromosomes, and new nuclear membranes form around the separated chromosomal bundles.*

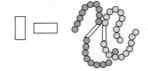 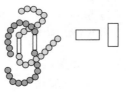

g. How many chromosomes do you have in each of your daughter nuclei? _____

Note: The number of chromosomes in a cell is determined by counting the number of visibly separate centromeres, regardless of whether the centromere region joins two sister chromatids in a chromosome or appears in an undivided chromosome composed of only one chromatid.

Repeat the procedure outlined above until you are thoroughly familiar with the major events of mitosis. Have your instructor check your work if you experience difficulty.

After you are comfortable with this simulation of the mitotic process, combine your chromosome kit with that of another student. Shorten the two yellow and two red strands from one of the kits so that there are only five beads on each side of the centromere region.

Now simulate mitosis in a nucleus with four chromosomes (two homologous pairs, one pair consisting of one short red and one short yellow chromosome, and the other of one long red and one long yellow chromosome). Draw each stage (prophase, metaphase, anaphase, and telophase) on a separate piece of paper and insert this into your laboratory manual. You should end up with two nuclei, each containing four chromosomes—two red and two yellow. Have your instructor check your work.

When you are finished, reattach the beads to the original chromosomes and separate the two kits in preparation for the next class.

Cytokinesis

Division of the cytoplasm (cytokinesis) usually occurs as telophase of mitosis progresses. In animal cells, microtubules play a role in the formation of a furrow that constricts the cytoplasmic mass into two daughter cells (Figure 9B-10a).

In plant cells, a cell plate is formed along the midplane of the dividing cell as many vesicles (produced by Golgi bodies) become joined together. This creates a membrane-bound space, the cell plate. Membranes of either end of the cell plate fuse with the cell membrane to produce two separate daughter cells. The area between the cells becomes impregnated with pectin to form the middle lamella, and the two daughter cells synthesize cell walls (Figure 9B-10b).

h. *How does the number of chromosomes in the two daughter cells that you formed in your simulations compare with the number of chromosomes in the parent cell?* _____

i. *How do the two cells that you formed in your simulation compare with the parent cell with regard to which chromosomes are present?* _____

_____

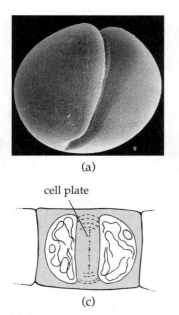

(a)

cell plate

(c)

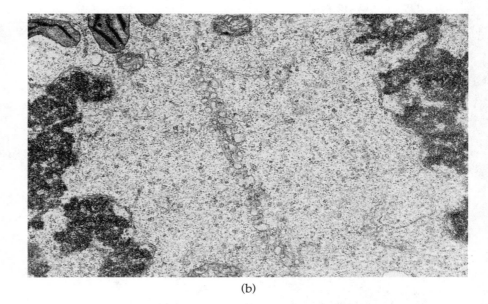

(b)

Figure 9B-10   *(a) Cytokinesis in an animal cell. (b) and (c) Cytokinesis in a plant cell.*

j. Would the constitution of the daughter cells be any different if you had lined up the metaphase chromosomes in a different order? _____

k. If the red chromosomes represent paternal chromosomes and the yellow chromosomes represent maternal chromosomes, what can be said about the genetic constitution of the daughter cells?

_____

l. Keeping your answer to the previous question in mind, briefly describe how mitosis maintains a constant chromosome number and why daughter nuclei are always genetically identical. _____

_____

✔ **EXERCISE C**   **Mitosis in Living Tissues—Onion Root Tips**

║║║ **Objectives** ║║║║║║║║║║║║║║║║║║║║║║║║║║║║║║║║║║║║║║║

☐ Identify the stages of mitosis in living tissues from the onion root tip.

☐ Relate the apparent frequency of mitotic figures to the growth pattern of root-tip tissue.

║║║ **Procedure** ║║║║║║║║║║║║║║║║║║║║║║║║║║║║║║║║║║║║║║║

Mitosis is easily studied using the root tips of actively growing plants. Roots of the common garden onion (*Allium cepa*) provide good material for such a study (Figure 9C-1). (*Allium cepa* has 16 chromosomes.)

1. Obtain a stained root tip for microscopic examination and place it on a slide in a drop of water. Add a coverslip, gently pressing downward with the eraser end of a pencil to spread the tissues apart.

2. Observe the preparation using a microscope (10×) and squash it further if necessary; do not be too heavy-handed initially.

Figure 9C-1   *Median longitudinal section of an onion root tip.*

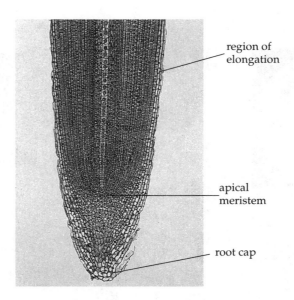

region of elongation

apical meristem

root cap

3. Use the microscope to examine the *Allium* root-tip slide. Scan the entire length of the slide to observe cells far from the tip and cells right at the tip. Are the cells in these areas dividing? Locate the area of cell division, the region where the greatest number of mitotic figures are represented, and draw a box around this area in Figure 9C-1. Label this area "region of cell division."

At the apex of the region of cell division, certain cells called "initials" give rise to new cells for root growth. When an initial cell divides, one daughter cell goes on to divide a few more times (in the region of cell division) before differentiating into specialized cells. The other daughter cell continues as an initial cell and remains in the apical region (called the *apical meristem*; **meristem** refers to undifferentiated plant tissue from which new cells are formed). Covering the apical meristem is the *root cap*. Behind the region of cell division, cells elongate and later differentiate as the root lengthens.

a. What might be the purpose of the root cap? _____

4. Switch to high power (40×). Use Figure 9C-2 to identify cells in interphase, prophase, metaphase, anaphase, and telophase. Draw and label these cells on a separate sheet of paper. Insert the drawings into your notebook.

Figure 9C-2 *Stages of mitosis in the onion root tip.*

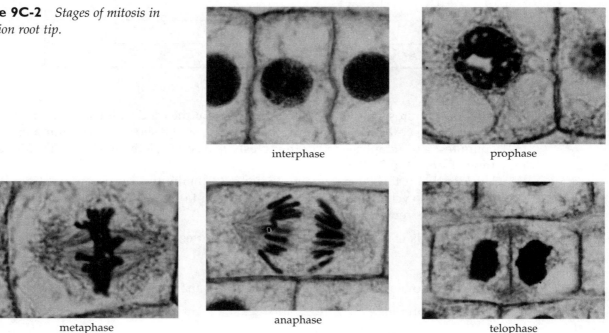

interphase

prophase

metaphase

anaphase

telophase

👁 **EXERCISE D**   **Phases of the Cell Cycle in the Onion Root Tip**

Now that you are familiar with the events of mitosis and the stages of the cell cycle, you can estimate the relative duration of each phase of the cell cycle by recording the frequency with which you find each phase in regions where cell division is actively taking place. The frequency of a phase is an indication of the relative length of that phase.

ııııı **Objectives** ııııııııııııııııııııııııııııııııııııı

☐ Use observations of cells in an onion root tip to describe the relative duration of the phases of the cell cycle.

ııııı **Procedure** ıııııııııııııııııııııııııııııııııııııı

1. Obtain a prepared slide of onion root tip or use your fresh onion root-tip squash if the preparation is good and many mitotic figures are visible.

2. With the 40× objective in place, examine a single field of view (the area visible in the microscope with the slide stationary) in the apical meristem region and count the number of cells in the phases of the cell cycle listed in Table 9D-1. Make sure you are surveying the actively dividing area of the onion root tip. Repeat this count in at least two more nonoverlapping fields. Use Table 9D-1 to collect and calculate your results.

Table 9D-1   Percentage of Cells in Each Phase of the Cell Cycle

	Number of Cells				Percent of Grand Total (Total/Grand Total × 100)
	Field 1	Field 2	Field 3	Total	
Interphase					
Prophase					
Anaphase					
Telophase					
Grand total					

The duration of mitosis varies for different tissues in the onion. However, prophase is always the longest phase (1–2 hours), and anaphase is always the shortest (2–10 minutes). Metaphase (5–15 minutes) and telophase (10–30 minutes) are also of relatively short duration. Interphase may range from 12 to 30 hours.

Consider that it takes, on average, 16 hours (960 minutes) for onion root-tip cells to complete the cell cycle. You can calculate the amount of time spent in each phase of the cell cycle from the percentage of cells in that stage:

Percentage of cells in stage × 960 minutes = minutes of cell cycle spent in stage
                                     (16 hours)

3. Calculate the following (convert the times to hours and minutes):

Time spent in: prophase   _____ hr   _____ min

metaphase   _____ hr   _____ min

anaphase   _____ hr   _____ min

telophase   _____ hr   _____ min

Total time spent in mitosis   _____ hr   _____ min

Time spent in interphase   _____ hr   _____ min

a. What percentage of the cell cycle is spent in mitosis? _____ *In interphase?* _____

b. How do your results compare with what is generally known about the Allium cepa *cell cycle, as described above?* _____

_____

_____

c. If your data show little agreement with the information given on the Allium cepa *cell cycle, would the cell cycle for the root-tip cells you observed have to be longer or shorter than the average of 16 hours in order to explain your results?* _____

_____

4. If you are using a prepared slide of an onion root tip, you will be observing a longitudinal section through the root tip.

    d. *How does a longitudinal section compare with a cross section?* _____

    e. *Why do some of the cells appear to be empty?* _____

Laboratory Review Questions and Problems

1. The diagram below shows the relative amount of DNA present in a cell progressing through the stages of the cell cycle. Label the line segments between the hatch marks to show G₁, S, G₂, and mitosis.

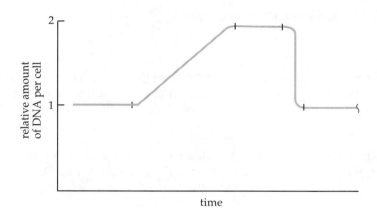

2. Complete the following table.

	Number of Chromatids per Chromosome
G₁	
S	
G₂	
Prophase	
Metaphase	
Anaphase	
Telophase	

3. How do protein kinases and cyclins regulate the cell cycle?

4. What role does cell division play in the growth of an organism?

5. What is the function of the centromere region of a chromosome?

6. What is a centrosome? How are centrioles and centrosomes related?

7. What is the function of spindle fibers? How do they lengthen and shorten?

8. When does a chromatid become a chromosome?

9. You have a diploid cell containing eight chromosomes ($2n = 8$). What would this cell look like at metaphase? How many chromosomes would each of the daughter cells have?

10. You are given two dishes of beads, one containing red beads and the other yellow. Describe how you would simulate the chromosomes of a diploid cell for which $2n = 4$.

11. Can a haploid cell undergo mitosis? Why or why not? If you had a haploid cell with four chromosomes ($n = 4$), what would it look like during metaphase?

12. Indicate the stage of mitosis (interphase, prophase, metaphase, anaphase, telophase) during which each of the following events occurs.

_____ Nuclear membrane disappears.

_____ Centrioles replicate.

_____ Chromosomes are arranged in single file between the poles of the cell.

_____ Chromatids separate.

_____ Spindle fibers form.

_____ Cell plate forms.

_____ DNA replication occurs.

_____ Chromosomes first become visible as long thin strands.

_____ Chromosomes move to opposite poles of the cell.

Enzymes

OVERVIEW

Without enzymes, most biochemical reactions would take place at a rate far too slow to keep pace with the metabolic needs and other life functions of organisms. **Enzymes** are catalysts that speed up chemical reactions but are not themselves consumed or changed by the reactions.

The cell's biological catalysts are proteins. These enzymes have a very complex three-dimensional structure consisting of one or more polypeptide chains folded to form an **active site**—a special area into which the **substrate** (material to be acted on by the enzyme) will fit.

Changes in temperature, alterations in pH, the addition of certain ions or molecules, and the presence of inhibitors all may affect the structure of an enzyme's active site and thus the activity of the enzyme and the rate of the reaction in which it participates. The rate of an enzymatic reaction can also be affected by the relative concentrations of enzyme and substrate in the reaction mixture.

During this laboratory period, you will investigate how changes in pH, temperature, substrate concentration, and enzyme concentration affect the enzymatic activity of catecholase. We will also observe the action of rennin during the process of cheesemaking.

STUDENT PREPARATION

Prepare for this laboratory by reviewing the text pages indicated by your instructor. Familiarizing yourself in advance with the information and procedures covered in this laboratory will give you a better understanding of the material and improve your efficiency.

EXERCISE A   Investigating the Enzymatic Activity of Catecholase

During this exercise you will study the activity of the enzyme **catecholase** contained in some fruits and vegetables. Peeled potatoes and bruised fruits turn brown when exposed to air because catecholase facilitates a reaction between catechol and oxygen. In the presence of oxygen, the compound **catechol** is oxidized by the removal of two hydrogen atoms. Catechol is thus converted to benzoquinone, and oxygen is reduced by the addition of two hydrogen atoms to form water. Benzoquinone molecules then link together to form long, branched chains. These chains are the structural backbones of the red and brown melanoid pigments that cause darkening.

Throughout this exercise, work in pairs. Each pair of students should perform all four parts of the exercise. If you will be using the spectrophotometer to collect results, use the procedures labeled "quantitative." If you will be determining enzyme activity by observing color changes, use the procedures "qualitative."

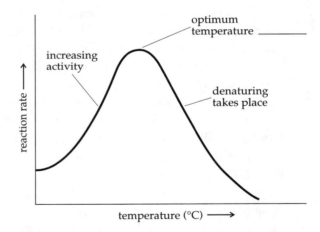

catechol   enzyme catecholase   benzoquinone

Note: Your instructor will dispense an extract of potato juice containing the enzyme catecholase to each group. The dropper bottle should be kept tightly covered at all times. Keep in mind that whenever you use potato juice in the following experiments, you are using an enzyme—catecholase—preparation.

ⅢⅢⅢ Objectives ⅢⅢⅢⅢⅢⅢⅢⅢⅢⅢⅢⅢⅢⅢⅢⅢⅢⅢⅢ

☐ Determine the effects of temperature on the enzymatic activity of catecholase.

☐ Describe the effects of changes in enzyme concentration on the rate of an enzyme-catalyzed reaction.

☐ Describe the relationship between substrate concentration and the maximum velocity of an enzyme-catalyzed reaction.

👁 PART I   The Effect of Temperature on Enzyme Activity

Like most chemical reactions, the rate of an enzyme-catalyzed reaction increases as temperature increases, up to a point at which the rate is maximum. The rate then abruptly declines with a further increase in temperature (Figure 10A-1). Above 40°C, most enzymes active in living tissue become **denatured**—their secondary or tertiary protein structure breaks down.

Figure 10A-1 *Effect of temperature on reaction rate.*

optimum temperature _____

increasing activity

denaturing takes place

reaction rate ⟶

temperature (°C) ⟶

a. *From what you have learned about proteins from reading your text or from lecture discussions, describe the significance of the secondary and tertiary structures of a protein.*

_____

_____

b. *Why would changing the secondary or tertiary structure of a protein affect its enzymatic activity?*

_____

_____

ⅢⅢⅢ Procedure (Quantitative) ⅢⅢⅢⅢⅢⅢⅢⅢⅢⅢⅢⅢⅢⅢⅢⅢⅢⅢⅢⅢⅢⅢⅢ

1. Refer to Laboratory 4, Exercise A, for directions on how to use the spectrophotometer (Spectronic 20) and for an explanation of the theory behind its operation. Set the wavelength to 420 nm and adjust the machine to 100% transmittance and 0% absorbance. Prepare three

tubes, each containing 3 ml of pH 7 phosphate buffer, 10 drops of potato juice, and 10 drops of water, as blanks for the three temperatures to be used. Label the tubes "10B," "24B," and "50B."

2. Fill three additional test tubes with 3 ml of pH 7 phosphate buffer. Label the tubes "10," "24," and "50." These are your "experimental" tubes.

3. Place the first experimental tube and its corresponding blank in an ice-water bath, leave the second tube and its corresponding blank at room temperature, and place the third tube with its blank in a beaker of warm water (approximately 50°C). Allow 10 minutes for the buffer in the first pair of tubes to reach 10°C (or less), the buffer in the second pair of tubes to reach 24°C (room temperature), and the buffer in the third pair of tubes to reach 50°C. Meanwhile, continue to step 4.

4. Take two test tubes containing potato juice and put one on ice (10°C) and the other in a warm-water bath (50°C). Take two test tubes containing catechol and put one on ice and the other in the warm-water bath.

5. After 10 minutes, add 10 drops of catechol (from a tube at the same temperature, or for the 24°C sample, from the room temperature stock) to each experimental tube. Cover each tube with Parafilm and invert several times to mix.

6. Use the blank for 10°C (tube 10B) to adjust the Spectronic 20 to 0% absorbance. Wipe condensation off the tube before reading. Add 10 drops of potato juice (from the tube at 10°C) to your 10°C experimental tube (tube 10). (Water has been used to replace this in your blank.) Cover the tube with Parafilm and invert it several times to mix the contents. Quickly read absorbance and record your reading in Table 10A-1 as the "0 minutes" reading. Return both the experimental tube and the blank to the 10°C beaker.

7. Repeat the procedure in step 6 for your room temperature (24°C) and 50°C experimental tubes. Remember to use appropriate blanks at each temperature. (Note: You may want to complete all readings at the first temperature and then follow with the other two sets of tubes, but this will take more time.)

8. At 2-minute intervals, determine the absorbance for each of your three experimental tubes (always adjusting the Spectronic 20 with the appropriate blank). Record your data in Table 10A-1. Allow the experiment to run for 10 minutes.

Table 10A-1   Effect of Temperature on Enzyme Activity

Minutes	Absorbance (420 nm)		
	At 10°C	At 24°C	At 50°C
0			
2			
4			
6			
8			
10			

9. Graph your data for the 6-minute time (use graph paper). Label the X-axis "temperature" and the Y-axis "absorbance (420 nm)."

   c. At what temperature was the amount of product produced by the enzyme-catalyzed reaction greatest? _____

 d. How would product produced per minute relate to the rate of an enzyme-catalyzed reaction?

_____

10. To determine rate, graph the data for each of the three tubes. Label the X-axis "time" and the Y-axis "absorbance (420 nm)." Find the slope of the straightest portion of each curve and compare these rates (absorbance/min).

 e. At what temperature was the rate of the enzyme-catalyzed reaction greatest? _____ Write this value in the space provided in Figure 10A-1.

 f. Do your results support the explanation given at the beginning of this exercise? _____ If not, what might be the reason for this? _____

_____

11. Skip to "Final Step" on p. 10-5.

▮▮▮▮▮ Procedure (Qualitative) ▮▮▮▮▮▮▮▮▮▮▮▮▮▮▮▮▮▮▮▮▮▮▮▮▮▮▮▮▮▮▮▮▮▮▮

1. Fill three test tubes with 3 ml of pH 7 phosphate buffer. Label the tubes "10," "24," and "50."

2. Place the first test tube in an ice-water bath, leave the second tube at room temperature, and place the third tube in a beaker of warm water (approximately 50°C). Allow 10 minutes for the buffer in the first tube to reach 10°C (or less), the buffer in the second tube to reach 24°C (room temperature), and the buffer in the third tube to reach 50°C. Meanwhile, continue to step 3.

3. Take two test tubes containing potato juice and put one on ice (10°C) and the other in a warm-water bath (50°C). Take two test tubes containing catechol and put one on ice and the other in the warm-water bath.

4. After 10 minutes, add 10 drops of catechol (from a tube at the matching temperature, or for the 24°C sample, from the room temperature stock) to each tube of buffer.

5. Add 10 drops of potato juice (again, at the matching temperature) to each tube.

6. Allow the tubes to stand for 5 minutes (shake the mixture several times during the 5 minutes). Record your results in Table 10A-2 by indicating the intensity of color in each tube after 5 minutes (0 indicates no color; +, ++, and +++ indicate increasing intensities of color).

Table 10A-2   Effect of Temperature on Enzyme Activity

Minutes	Intensity of Color		
	10°C	24°C	50°C
5			

 c. At what temperature was the amount of product produced by the enzyme-catalyzed reaction greatest? _____

 d. How would product produced per minute relate to the rate of an enzyme-catalyzed reaction?

_____

 e. What was the purpose of shaking the tubes? _____

_____

f. *At what temperature was the rate of the enzyme-catalyzed reaction greatest?* _____ *Write this value on the blank provided in Figure 10A-1.*

Final Step Indicate the relationship between temperature and enzymatic activity by writing the temperatures used in this experiment at appropriate positions along the horizontal axis of Figure 10A-1. To assure accuracy, it may be helpful first to mark highest and lowest temperatures on the graph itself.

Cooking hint: Placing peeled potatoes into ice water before cooking keeps them from darkening in color. Why? _____

👁 **PART 2**   **The Effect of pH on Enzyme Action**

The presence of various ions can interfere with the pattern of positive and negative charges within a protein molecule, thus changing the way the protein folds. In enzymes, the shape of the active site may be changed. We should expect, then, that changes in pH (reflecting the concentrations of hydrogen and hydroxide ions) would affect the action of enzymes.

The most favorable pH value—the point at which the enzyme is most active—is known as the optimum pH (Figure 10A-2). Extremely high or low pH values usually result in a complete loss of enzyme activity due to **denaturation,** the breakdown of the secondary and tertiary structure of a protein. Denaturation can also be caused by extremes in temperature, as explained in Part 1 of this exercise.

Figure 10A-2   *Effect of pH on reaction rate.*

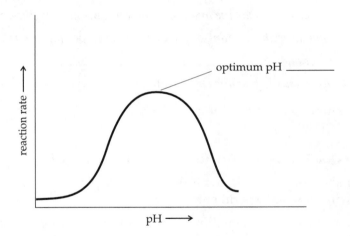

a. *How do you think extremes of pH might cause denaturation and loss of enzyme activity?*

_____

_____

‖‖‖ **Procedure (Quantitative)** ‖‖‖‖‖‖‖‖‖‖‖‖‖‖‖‖‖‖‖‖‖‖‖‖‖‖‖‖‖‖‖‖‖‖‖‖‖‖

1. Prepare five tubes for blanks, each containing 3 ml of a different phosphate buffer (pH 4, pH 6, pH 7, pH 8, pH 10), 10 drops of potato juice, and 10 drops of water. Be sure to label blank tubes with a "B" and the pH value. These will also serve as controls.

2. Set the spectrophotometer wavelength to 420 nm and adjust the machine to 100% transmittance and 0% absorbance.

3. Label five additional test tubes with the following pH values: pH 4, pH 6, pH 7, pH 8, and pH 10. These will serve as experimentals.

4. Fill each experimental tube with 3 ml of the matching phosphate buffer.

5. Add 10 drops of potato juice to each experimental tube. Cover each tube with Parafilm and invert the tube several times to mix the contents.

6. Add 10 drops of catechol to each tube (in the blanks, water substitutes for catechol). Cover the tubes with Parafilm and invert each several times to mix the contents.

7. Allow the tubes to stand for 5 minutes (mix at 1-minute intervals). At the end of this time, record absorbance readings (at 420 nm). Be sure to use the proper blank to adjust the spectrophotometer before each reading. Record your data in Table 10A-3.

Table 10A-3   Effect of pH on Enzyme Activity

Absorbance (420 nm)				
pH 4	pH 6	pH 7	pH 8	pH 10

8. Graph your data on graph paper. Label the X-axis "pH" and the Y-axis "absorbance (420 nm)."

   b. *At what pH was the amount of product in this enzyme-catalyzed reaction greatest?* _____

   c. *How does the amount of product produced per minute relate to the rate of the enzyme-catalyzed reaction?* _____

   d. *At what pH was the rate greatest?* _____

9. Skip to "Final Step" on p. 10-7.

ⅢⅢⅢ Procedure (Qualitative) ⅢⅢⅢⅢⅢⅢⅢⅢⅢⅢⅢⅢⅢⅢⅢⅢⅢⅢⅢⅢⅢⅢⅢ

1. Label five test tubes with the following pH values: pH 4, pH 6, pH 7, pH 8, and pH 10. These will serve as experimental tubes.

2. Fill each experimental tube with 3 ml of the matching phosphate buffer.

3. Add 10 drops of catechol to each tube. Cover the tube with Parafilm and invert several times to mix the contents.

4. Be prepared to observe color changes. Allow the tubes to stand for 3 to 5 minutes. Invert and mix each tube at 1-minute intervals. The rate of the enzyme-catalyzed reaction (measured by the amount of product produced during a given period of time) will be qualitatively proportional to the intensity of color developed in each reaction mixture. Record your results by indicating the intensity of color in each tube (0 = no color; +, ++, and +++ = increasing intensities of color). Record your data in Table 10A-4.

Table 10A-4   Effect of pH on Enzyme Activity

Intensity of Color				
pH 4	pH 6	pH 7	pH 8	pH 10

b. At what pH was the amount of product in this enzyme-catalyzed reaction greatest? _____

c. How does the amount of product produced per minute relate to the rate of the enzyme-catalyzed reaction? _____

d. At what pH was the rate greatest? _____

Final Step Indicate the relationship between pH and enzymatic activity by writing the pH values used in this experiment at appropriate positions along the horizontal axis of Figure 10A-2. To assure accuracy, it may be helpful to first mark pH values on the graph itself.

Cooking hint: Sprinkling lemon or orange juice over peeled fruit keeps the fruit from darkening. Citrus juice contains citric and ascorbic acids. *Why does citrus juice keep fruit from darkening?*

PART 3   **The Effect of Enzyme Concentration on Enzyme Activity**

When a substrate fits into the active site of an enzyme, an enzyme–substrate complex is formed. The enzyme assists with the chemical reaction and a product is formed. The reaction rate is usually directly proportional to the enzyme concentration (Figure 10A-3). The reaction rate will increase in proportion to an increasing enzyme concentration only if substrate is present in excess amounts so that the reaction is not limited by substrate availability.

Figure 10A-3 *Effect of enzyme concentration on reaction rate.*

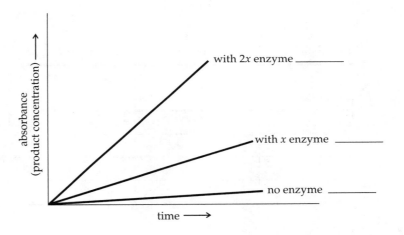

a. If an adequate amount of an enzyme is present, how might availability of a substrate affect the production of a biochemical product in the cells of your body? _____

Procedure (Quantitative) ∎∎∎

1. Label four test tubes as noted in the table below and add the required amounts of phosphate buffer, potato juice, and water to each tube. Cover each tube with Parafilm and invert the tube several times to mix. These tubes will serve as blanks.

Tube	pH 7 Phosphate Buffer	Potato Juice	Water
Blank A	3 ml + 20 drops	—	10 drops
Blank B	3 ml + 15 drops	5 drops	10 drops
Blank C	3 ml + 10 drops	10 drops	10 drops
Blank D	3 ml + 0 drops	20 drops	10 drop

2. Label four additional tubes as noted below. These tubes will serve as experimentals. Add the required amounts of phosphate buffer and potato juice. DO NOT ADD CATECHOL YET! Cover each tube with Parafilm and invert it several times to mix the contents.

Tube	pH 7 Phosphate Buffer	Potato Juice	Catechol (ADD LAST)
A	3 ml + 20 drops	—	10 drops
B	3 ml + 15 drops	5 drops	10 drops
C	3 ml + 10 drops	10 drops	10 drops
D	3 ml + 0 drops	20 drops	10 drops

3. Set the spectrophotometer wavelength to 420 nm and adjust the machine to 100% transmittance and 0% absorbance.

4. Adjust the spectrophotometer using blank A. Add catechol to experimental tube A. Cover the tube with Parafilm and invert it several times to mix. Immediately read absorbance and continue to take readings every 2 minutes for 6 minutes. Record your data in Table 10A-5.

Table 10A-5   Effect of Enzyme Concentration on Enzyme-Catalyzed Reactions

Minutes	Absorbance (420 nm)			
	A	B	C	D
0				
2				
4				
6				

5. Now repeat the procedure in step 4 using tubes B, C, and D, and the appropriate blanks. Record your data.

6. Graph your data on graph paper. Label the X-axis "time (minutes)" and the Y-axis "absorbance (420 nm)."

7. The absorbance is proportional to the amount of product produced by this enzyme-catalyzed reaction. The slope of the plotted line, expressed in absorbance units per unit of time, is an indication of the rate of the reaction. The slope of the line can be determined by taking any two points on a straight line plotted in step 6 and performing the following operation:

$$\text{Slope} = \frac{\text{absorbance}_2 - \text{absorbance}_1}{\text{time}_2 - \text{time}_1}$$

b. *How do the rates of the reactions in tubes A, B, C, and D differ?*

_____

_____

c. How does increasing enzyme concentration in the presence of unlimited substrate affect the rate of the enzyme-catalyzed reaction? _____

_____

8. Skip to "Final Step" on p. 10-10.

▌▌▌▌▌ Procedure (Qualitative) ▌▌▌▌▌▌▌▌▌▌▌▌▌▌▌▌▌▌▌▌▌▌▌▌▌▌▌▌▌▌▌▌▌▌▌▌▌▌▌

1. Label four test tubes as noted in the table below. Add the indicated amounts of phosphate buffer and potato juice to each tube. DO NOT ADD CATECHOL YET! Cover each tube with Parafilm and invert it several times to mix the contents.

2. Add 10 drops of catechol to each tube. Cover each tube with Parafilm and invert it several times to mix the contents. (Do not use the same piece of Parafilm on all tubes!)

Tube	pH 7 Phosphate Buffer	Potato Juice	Catechol (ADD LAST)
A	3 ml + 20 drops	—	10 drops
B	3 ml + 15 drops	5 drops	10 drops
C	3 ml + 10 drops	10 drops	10 drops
D	3 ml + 0 drops	20 drops	10 drops

3. Allow the tubes to stand for 3 to 4 minutes; mix the contents of each tube at 1-minute intervals.

4. The rate of the enzyme-catalyzed reaction (measured as the amount of product produced during a given period of time) will be proportional to the intensity of color developed in each reaction mixture. In Table 10A-6, record your results by indicating the intensity of color in each tube after 3 or 4 minutes (0 = no color; +, ++, and +++ = increasing intensities of color).

b. What changes did you observe throughout the period during which the reaction was occurring?

_____

_____

Table 10A-6   Effect of Enzyme Concentration on Enzyme-Catalyzed Reactions

Intensity of Color			
A	B	C	D

c. At what enzyme concentration was the amount of product in this enzyme-catalyzed reaction greatest? _____

d. At what enzyme concentration was the rate greatest? _____

Final Step Record your observations of the relationship between enzyme concentration and the rate of reaction by labeling the lines in Figure 10A-3 with A, B, or C, representing the activity of catecholase in tubes A, B, and C. Add a line to represent the rate of the reaction observed in tube D.

e. *Do your results support the concept that reaction rate is proportional to enzyme concentration?* _____

   Explain. _____

_____

PART 4   The Effect of Substrate Concentration on Enzyme Activity

If the amount of enzyme is kept constant but the amount of substrate is gradually increased, then the **velocity,** the rate of speed, at which the enzyme works (converts substrate to product) will increase until it reaches a maximum. At this point, increases in substrate concentration will not increase the velocity of the reaction because all of the available enzyme is participating in the enzyme–substrate complex. Thus, every enzyme has a "maximum" velocity at which it will work (Figure 10A-4).

Figure 10A-4   *Effect of substrate concentration on reaction rate (velocity).*

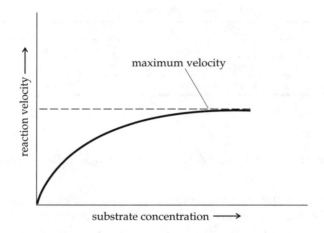

a. *What consequences would such maximum velocities have for cell metabolism?*

_____

ⅠⅠⅠⅠⅠ Procedure (Quantitative) ⅠⅠⅠ

1. Work in groups of four (two student pairs). Dilute the stock potato extract 1:1.

2. Label 10 tubes with the numbers indicated in the following table—make *two* blanks. One pair of students will work with the first four tubes and the other pair with the remaining four. Each pair of students will need a blank.

   Note: If there are enough students in your laboratory section, your instructor will assign two substrate concentrations to each pair (of course, you must make a blank also—therefore every student pair will need three tubes). Work with three other pairs of students in your laboratory to construct a team.

3. Add phosphate buffer and catechol to all tubes, as indicated.

4. Set the spectrophotometer to 420 nm and adjust it to 100% transmittance and 0% absorbance.

5. Add 30 drops of diluted potato juice to your blank. Cover the tube with Parafilm, and invert it several times to mix the contents. Use the blank to readjust the spectrophotometer to 0% absorbance.

Tube	pH 7 Phosphate Buffer	Catechol
Blank	5 ml + 48 drops	0 drops
1	5 ml + 47 drops	1 drop
2	5 ml + 46 drops	2 drops
4	5 ml + 44 drops	4 drops
8	5 ml + 40 drops	8 drops
16	5 ml + 32 drops	16 drops
24	5 ml + 24 drops	24 drops
32	5 ml + 16 drops	32 drops
48	5 ml + 0 drops	48 drops

6. Make sure that you know who is processing (reading absorbance for) which experimental tubes before you proceed with the experiment.

7. Now add 30 drops of diluted potato juice to one of your experimental tubes. Mix and immediately record the absorbance in the appropriate "0 minutes" column in Table 10A-7. Record the absorbance reading every 2 minutes for 6 minutes.

   Caution: If two student pairs are using one spectrophotometer, be sure that one pair *follows* the other and uses the appropriate blank. "Zero time" is critical and must be read immediately!

Table 10A-7   Effect of Substrate Concentration on Enzyme-Catalyzed Reactions

Tube	Absorbance (420 nm)			
	0 minutes	2 minutes	4 minutes	6 minutes
1				
2				
4				
8				
16				
24				
32				
48				

8. Repeat step 7 with your other experimental tube(s). Remember to use your blank to adjust the spectrophotometer if others have used the machine.

9. Using graph paper, plot absorbance on the Y-axis and time on the X-axis. Your graph should contain a line for each experimental tube.

   b. *Does the amount of enzyme (30 drops of diluted potato juice) ever limit the reaction?* _____

      If so, when? _____

   c. *If you increased the amount of enzyme as you increased substrate so that neither enzyme nor substrate was limiting, what would happen to the reaction?* _____

      _____

d. *What would the curve look like?* _____

10. Now, determine V_0 (initial velocity) for each of your enzyme concentrations. V_0 is the slope of the line tangent to the initial portion of your curve. Record V_0 in the appropriate space in the following table. (Your classmates will provide the missing values for this table.)

Tube	Substrate Concentration	V_0
1		
2		
4		
8		
16		
24		
32		
48		

11. Using graph paper, plot V_0 vs. substrate concentration for all tubes. This will plot as a single reaction curve. Place V_0 on the Y-axis and substrate concentration on the X-axis.

    e. *Do your observed results indicate that the enzyme has a maximum velocity, V_{max}, at which it will*

    work? _____ *Explain.* _____

    _____

12. Every enzyme has a K_m value that is a measure of its activity. K_m is defined as the substrate concentration at which the reaction velocity is $\frac{1}{2}V_{max}$. You can find the K_m for catecholase by drawing a horizontal line on your graph at $\frac{1}{2}V_{max}$. Where it intersects your curve, drop a vertical line to the X-axis. Where this line crosses the X-axis, you can determine K_m (Figure 10A-5).

    For catecholase $V_{max} =$ _____ $K_m =$ _____

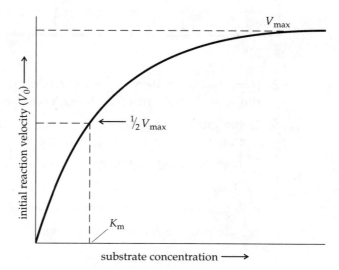

Figure 10A-5 *Determining K_m for an enzyme.*

Figure 10A-6

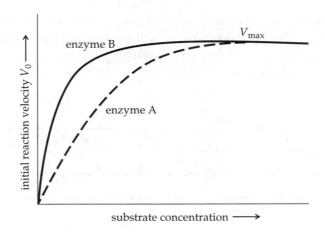

f. *In Figure 10A-6, which enzyme (A or B) has the higher K_m?* _____

g. *Which enzyme has the fastest rate (steeper slope and higher V_0 values at a given substrate concentration)?* _____ *Remember, K_m is a measure of affinity; does the enzyme with the fastest rate have the higher or lower K_m?* _____

h. *From your answers, describe the relationship between the size of K_m and the affinity of an enzyme for its substrate.* _____

13. Indicate the effects of substrate concentration on the rate of the catecholase reaction by writing the substrate values (1, 2, 4, 8, 16, and 24) at appropriate positions along the horizontal axis of Figure 10A-4.

||||| **Procedure (Qualitative)** ||||||||||||||||||||||||||||||||||||||||||

1. Work in groups of four (two student pairs). Dilute the stock potato extract 1:1.

2. Label eight test tubes with the numbers indicated in the table below. Each student pair should fill four of the tubes with phosphate buffer and catechol, as indicated. Cover each tube with Parafilm and mix by inverting the tube several times.

Tube	pH 7 Phosphate Buffer	Catechol
1	5 ml + 47 drops	1 drop
2	5 ml + 46 drops	2 drops
4	5 ml + 44 drops	4 drops
8	5 ml + 40 drops	8 drops
16	5 ml + 32 drops	16 drops
24	5 ml + 24 drops	24 drops
32	5 ml + 16 drops	32 drops
48	5 ml + 0 drops	48 drops

3. Add 30 drops of diluted potato juice extract to each of the eight tubes. Cover each tube with Parafilm and mix by inverting the tube several times.

4. Incubate the mixtures for 5 minutes; mix each tube at 1-minute intervals.

5. The rate of the enzyme-catalyzed reaction (measured as the amount of product produced during a given period of time) will be proportional to the intensity of color developed in each reaction mixture. Record your results by indicating the intensity of color in each tube after 5 minutes (0 = no color; +, ++, and +++ = increasing intensity of color).

Intensity of Color							
1	2	4	9	16	24	32	48

b. *Do your observed results support the idea that an enzyme has a maximum velocity at which it would work?* _____ *Explain.* _____

_____

6. Indicate the effects of substrate concentration on the rate of the catecholase reaction by writing the substrate values (1, 2, 4, 8, 16, and 24) at appropriate positions along the horizontal axis of Figure 10A-4.

EXTENDING YOUR INVESTIGATION: MAKING JUICES JUICIER

Pectins are large polysaccharide molecules located primarily in the cell walls of plants. These molecules contribute to the sturdiness of plant cells. Pectin compounds are especially prevalent in fruits. When fruits are squeezed to make juices, pectins that remain in the cell walls help to keep particles of the fruit intact, as you may notice in the "lumpy" texture of orange or tomato juice. When pectin molecules are released into the juice during the process of squeezing, they help to hold the remaining particles in suspension and increase the juice's viscosity, a characteristic we recognize as the "thickness" or "body" of, for example, prune juice or fruit nectars. Pectins are added to fruit juices to make jellies and jams.

This thickening effect and the ability to keep particles in suspension are due, in part, to the size of the pectin molecules. They are large enough not to dissolve completely in water, but not so large and heavy that they exist as separate particles that separate easily from a liquid and sink, as would, say, particles of clay stirred into water.

Pectinases are enzymes that break down pectin molecules. In the commercial preparation of "clear" (as opposed to "homestyle") fruit juices, pectinases are often added to processed fruit to increase the amount of juice released and to allow particles of fruit to settle out of the liquid.

You will be given some applesauce and some pectinase. Formulate a hypothesis about the effect of pectinase on the applesauce.

HYPOTHESIS:

NULL HYPOTHESIS:

What do you **predict** will be the effect of pectinase on the applesauce?

Identify the **independent variable** in this investigation.

Identify the **dependent variable** in this investigation.

Now investigate the effects of pectinase on applesauce using the following procedure.

PROCEDURE:

1. Place approximately 25 ml of applesauce into each of two beakers, one labeled "no enzyme" and the other labeled "pectinase."

2. Add 0.5 ml of distilled water to the "no enzyme" beaker and 0.5 ml of pectinase to the "pectinase" beaker.

3. Use separate spatulas to stir the material in each beaker.

4. Let stand 10 minutes.

5. Place cheesecloth in a funnel and place the funnel into a graduated cylinder.

6. With the aid of the spatula, pour the contents of each beaker into a separate funnel and collect the filtrate.

7. Record the amount of juice collected in each cylinder after 5 minutes.

RESULTS:

Amount of juice collected without pectinase _____ ml

Amount of juice collected with pectinase _____ ml

Do your results support your hypothesis?

Your null hypothesis?

Was your prediction correct?

What do you **conclude** about the role of pectinase in making juices juicier?

EXERCISE B   **The Essentials of Cheesemaking**

The enzyme **rennin** is used to coagulate the casein of milk during cheese production and serves as a good example of one of the many uses of enzymes. Although there is great variety among cheeses, the methods of production are basically very similar. Cheese is the solid portion (curd) of milk which has been separated from the liquid portion (whey). The sequence of procedures usually followed in cheesemaking is described below.

The Process of Cheesemaking

- Lactic acid-forming bacteria added to milk convert lactose to lactic acid, which sours the milk (25–30°C, 10–75 minutes). Amateur cheesemakers may use buttermilk as a starter (ripen for 4–12 hours).

- Coloring may be added for eye appeal. It has no effect on the quality of the cheese.

- Rennin (pH 5.5–6) coagulates ripened milk to form a precipitate (curd); liquid (whey) remains.

- Cut and stir—the smaller the curd, the harder and drier the cheese.

- Cook (38–39°C)

(continued on next page)

The Process of Cheesemaking (continued)

- Drain and salt.
- Press in cheesecloth-lined molds. Cheese may be inoculated with bacteria or mold to produce special cheeses.
- Seal in wax or treat periodically with brine or vegetable oil to keep contaminating organisms from growing on the surface.
- Cure and age to allow characteristic cheese flavor to develop.

ıııı **Objectives** ıııııııııııııııııııııııııııııııııııııı

☐ Apply the principles of enzyme activity to cheesemaking.

ıııı **Procedure** ıııııııııııııııııııııııııııııııııııııı

1. Work in pairs. Measure 250 ml of preripened milk (pH 5 to 5.6) into a beaker and *gently* warm it to 32°C.

2. Add 0.5 ml (10 drops) of Rennilase* to the milk while stirring continuously. Remove from the heat and leave *undisturbed* (without jarring) for 15 minutes or until the milk coagulates.

3. Using a glass rod, break up the curds and separate these from the whey. Filter the mixture through several layers of cheesecloth. Put the curds in a sterile Petri dish; add salt, pepper, or other seasoning to taste and mix. *Keep this preparation clean* if you wish to eat the cheese.

   a. *At what pH and temperature is the action of rennin most effective in making cheese?*

   _____

   b. *Name two of your favorite types of cheese. What changes in the cheesemaking process do you think were required to produce each of these?* _____

   _____

   _____

   _____

   c. *Dehydrated whey powder is a highly nutritious by-product of the cheese industry. Can you think of some uses for whey powder?* _____

   _____

Laboratory Review Questions and Problems

1. After discussing the results of all four experiments in Exercise A with your classmates, briefly summarize *your* conclusions on the effects of temperature, pH, enzyme concentration, and substrate concentration on the action of enzymes.

---

*Rennin is the enzyme that coagulates milk. It can be extracted in its crude form (rennet) from the lining of a portion of the cow's stomach (abomasum). Advances in genetic engineering, however, have allowed microbiologists to produce a bacterium having milk-coagulating properties. These bacteria can be grown in large numbers to produce greater amounts of the enzyme rennin. Rennilase is a trademark name for this product.

2. What important consequences do these properties of enzymes (from question 1) have for the metabolism of cells and organisms?

3. Define enzyme denaturation in terms of protein structure. What environmental factors can denature enzymes?

4. What does it mean to say that an enzyme-catalyzed reaction is either enzyme-limited or substrate-limited?

5. The K_m of an enzyme for a particular substrate is defined as the substrate concentration at which the reaction occurs at half its maximum rate. From Exercise A, Part 4, what would be the K_m (expressed here in drops) of catecholase? _____

6. The K_m usually, but not always, is inversely related to the strength of binding of a substrate to an enzyme. (In such cases $1/K_m$ is a measure of binding affinity.) Would an enzyme have lower or higher K_m values for more tightly bound substrates? _____

7. If a competitive inhibitor is introduced to a series of tubes in which enzyme and increasing amounts of substrate are mixed, the kinetics of the reaction would be as follows:

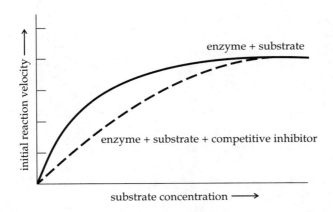

Competitive inhibitors bind to the active site of enzymes and compete with the substrate's ability to bind.

a. Is the K_m lower, higher, or the same for this reaction when a competitive inhibitor is added? _____

b. Explain why this is so. (*Hint:* Why is the maximum velocity the same for both reactions?)

c. How do the rates of the reactions compare?

8. If a noncompetitive inhibitor is added to tubes containing enzyme and substrate as above, the kinetics of the reaction would be as follows:

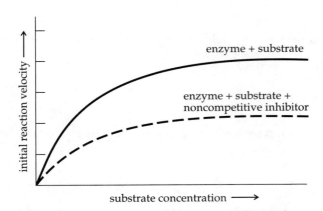

Noncompetitive inhibitors bind to areas on an enzyme other than the active site. They do not compete with the substrate, but affect reaction rate by changing the structure of the enzyme.

a. Is the K_m higher, lower, or the same for this reaction when a noncompetitive inhibitor is added? _____

b. Is the maximum velocity higher, lower, or the same? _____

c. Explain why this is so, based on what you know from your reading about the action of noncompetitive inhibitors.

OVERVIEW

Many metabolic reactions within the cell do not occur spontaneously but require a source of chemical energy in the form of ATP (adenosine triphosphate). The major source of ATP for most cells is the oxidation of glucose, a series of enzymatic reactions that results in the breakdown of carbon compounds into carbon dioxide, water, and energy.

The oxidation of glucose takes place in two major stages (Figure 11-1). The first is **glycolysis,** an **anaerobic** process (one that can proceed in the absence of oxygen).

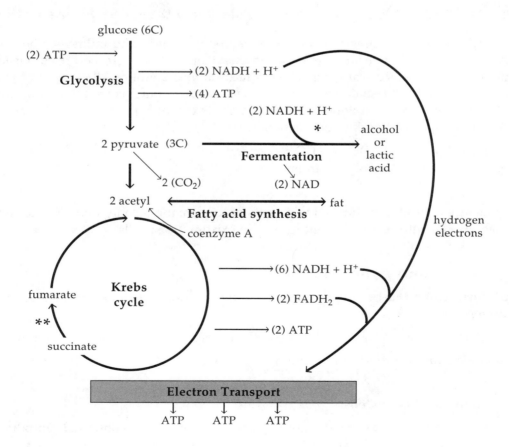

Figure 11-1 *Overview of cellular respiration.*

Glycolysis occurs in the cytoplasm of both **aerobic** (oxygen-requiring) and anaerobic organisms. The end product of glycolysis is pyruvic acid.

When oxygen is unavailable, the conversion of glucose to pyruvic acid is the major source of energy. In some organisms, such as yeasts, pyruvic acid can be further metabolized by a second series of reactions, the anaerobic process of **fermentation,** which results in the production of alcohol and carbon dioxide. Lactic acid can also be formed by the anaerobic metabolism of pyruvic acid. For example, when the oxygen supply available to muscle cells is depleted during strenuous exercise, lactic acid is produced. Eventually, as oxygen becomes available, lactic acid is used to resynthesize pyruvate.

When oxygen is available, the second stage in the oxidation of glucose is aerobic cellular respiration, which consists of the **Krebs cycle** (or *citric acid cycle*) and **electron transport.** These reactions, which take place in the mitochondria, greatly increase the energy harvest from the oxidation of glucose.

During this laboratory, you will study the processes of fermentation and respiration by making observations about the products of cellular reactions.

STUDENT PREPARATION

Prepare for this laboratory by reading the text pages indicated by your instructor. Familiarizing yourself in advance with the information and procedures covered in this laboratory will give you a better understanding of the material and improve your efficiency.

✔ 👁 **EXERCISE A**   Production of Carbon Dioxide and Ethanol by Fermentation

Yeasts are simple unicellular organisms related to mushrooms, molds, and mildews. They are called **heterotrophs** because they do not carry on photosynthesis, but obtain their food from outside sources such as grapes or grain. Yeasts are also classified as **facultative anaerobes**—they can live in aerobic or anaerobic environments. Under anaerobic conditions, yeasts carry out fermentation to produce alcohol and carbon dioxide. The alcohol in wine, beer, and other beverages is produced by the metabolic reactions of yeasts grown on grapes and grains such as barley.

$$C_6H_{12}O_6 \longrightarrow 2C_2H_5OH + 2CO_2 + energy$$

glucose   ethanol   carbon
                              dioxide

Very little net energy is produced during the process of fermentation—only two ATPs for every glucose molecule metabolized—but this is sufficient to sustain existing yeast cells.

||||| **Objectives** ||||||||||||||||||||||||||||||||||||||||||||

☐ Describe the structure of a yeast cell with reference to the location of metabolic reactions involved in glycolysis and fermentation.

☐ Describe how to demonstrate carbon dioxide and ethanol production during fermentation.

☐ List the requirements for anaerobic fermentation

✔ **PART I**   Examining Yeast Cells

Yeast cells are single-celled eukaryotes (Figure 11A-1). They can often be observed reproducing by budding.

Figure 11A-1 *Yeast cells. The cell on the right has almost completed the process of budding from the one on the left, which is in the process of producing three new yeast cells.*

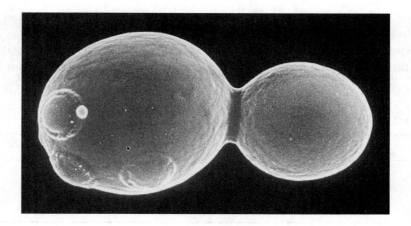

⸀⸀⸀⸀⸀ **Procedure** ⸀⸀⸀

1. Using a Pasteur pipette, place several drops of the stock yeast solution onto the center of a microscope slide.

2. Add a drop of neutral red and apply a coverslip.

3. Use high power (40× objective) to observe the yeast.

   a. Why was the neutral red added to the yeast sample?

   _____

4. Under optimal conditions, yeast multiply by budding, a process involving an outpocketing of the cytoplasm and the subsequent pinching off of a new cell. Scan your slide slowly to see if you can detect cells with buds.

   b. How do you know that yeasts are eukaryotic, not prokaryotic? _____

   _____

   c. In what parts of the yeast cell does glycolysis take place? _____

👁 **PART 2**   **Production of Carbon Dioxide by Fermentation**

Under anaerobic conditions, yeast cells break down sugars, releasing carbon dioxide gas. Evidence that fermentation is taking place in a yeast culture can be provided by bubbling the gas into an indicator solution.

⸀⸀⸀⸀⸀ **Procedure** ⸀⸀⸀

There are two flasks on demonstration: one contains a sugar and yeast solution and the other contains a 10% sucrose solution.

Each flask has a bent **U**-tube extending from the flask into a cylinder filled with a bromothymol blue indicator solution. If carbon dioxide passes from the **U**-tube into the cylinder, the solution in the cylinder will turn yellow.

a. In which flask is carbon dioxide being produced? _____

b. What is the source of the carbon dioxide? _____

c. As mixtures of grapes and yeast ferment, bubbles appear. What causes the formation of these bubbles?

_____

👁 PART 3   Requirements for Fermentation in Yeast

In this experiment, the class will study the role of yeasts as well as the food source requirements for fermentation. You will be assigned to one of four treatment groups by your laboratory instructor (Table 11A-1).

Table 11A-1

Group	Yeast	Boiled Yeast	Sugar	Water
I	10 ml	—	5% glucose	—
II	10 ml	—	5% sucrose	—
III	—	10 ml	5% sucrose	—
IV	10 ml	—	—	Distilled water

During the processes of glycolysis and fermentation, yeasts use sugars, but not all sugars are used at the same rate. Why do you think this might happen? Can you formulate a hypothesis about the requirements that must be met in order to maximize the rate of fermentation?

HYPOTHESIS:

NULL HYPOTHESIS:

What do you **predict** will happen in the experiment (consider all treatments)?

What do you predict will happen in your treatment compared with other treatments?

What is the **independent variable** in this experiment?

What is the **dependent variable** in this experiment?

Now measure the fermentation rates for the four treatments by using the following procedure.

ⅠⅠⅠⅠⅠ Procedure ⅠⅠⅠⅠⅠⅠⅠⅠⅠⅠⅠⅠⅠⅠⅠⅠⅠⅠⅠⅠⅠⅠⅠⅠⅠⅠⅠⅠⅠⅠⅠⅠⅠⅠⅠⅠⅠⅠⅠ

1. In each assigned laboratory group, work in pairs. Obtain two plastic fermenting vials, one large and one small.

2. Add 10 ml of yeast suspension (boiled yeast suspension if you are assigned to Group III) to the smaller vial.

3. Finish filling the small vial to its brim with the sugar solution or with distilled water, as designated for your group.

4. Hold the small vial upright. Invert the large vial so that the bottom side is upward and lower it to cover the small vial.

5. Hold the two vials tightly together at their ends and invert the apparatus so that the small vial is now upside down within the larger vial (Figure 11A-2). As CO_2 is produced by the

Figure 11A-2 *Fermentation apparatus.*

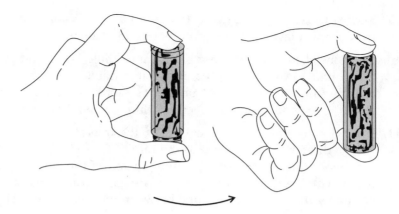

process of fermentation, it will collect in the upper portion of the small vial and the yeast mixture will be pushed downward and out into the larger vial.

6. If there is an air space at the top of the small vial, measure its length using a millimeter ruler and use this measurement as the zero point.

7. Measure the length of the gas column at 5-minute intervals for as long as possible (up to 45 minutes). Foaming within the large vial may obscure the length of the gas column. If you find it difficult to read, move the small vial until its side is pressed against the side of the outer vial.

8. Record your readings in Table 11A-2, and remember to subtract the value for the zero point from each reading.

Table 11A-2   Class Data for Yeast Fermentation Experiment, Recorded as Length of Gas Column (in millimeters)

	Group I					Group II					Group III					Group IV				
	Student Pair:					Student Pair:					Student Pair:					Student Pair:				
Minutes	1	2	3	4	Average	1	2	3	4	Average	1	2	3	4	Average	1	2	3	4	Average
0																				
5																				
10																				
15																				
20																				
25																				
30																				
35																				
40																				
45																				

9. Record the data for all teams (student pairs) in your group. Average the data for each time period and place all group data on the blackboard.

10. Copy the class data for other groups into Table 11A-2 and plot the data (using the averages) on graph paper. Always graph dependent variables along the vertical (Y) axis; in this case, "gas column (mm)"; "time (min)" should be plotted on the horizontal (X) axis. Be sure to label your graph. Use either colored pencils or different symbols to designate the four different groups.

11. Since the vials are uniform cylinders, the rate of increase in the length of the space filled by CO_2 is directly proportional to the rate of CO_2 production. To determine the actual rate, pick any two points on the straight-line portion of the curve (or a straight line approximated to the curve). Divide the difference in the gas column height (in millimeters) between the two points by the difference in time between the two points. The result will be the rate of CO_2 production in millimeters per minute.

   Millimeters can be converted to volume (milliliters) by using the formula for the volume (V) of a cylinder, $\pi r^2 h = V$, where r is the radius of the cylinder (in millimeters) and h is the height of the gas column (in millimeters). This will allow fermentation rates to be expressed as milliliters per minute. Record fermentation rates in Table 11A-3.

Table 11A-3

Treatment Group	Fermentation Rate (ml/min)
I	
II	
III	
IV	

a. What do the data from each group tell you about the process of fermentation?

Group I _____

Group II _____

Group III _____

Group IV _____

b. Why was Group IV included in this experiment? _____

c. What is the effect of boiling the yeast (Group III)? _____

d. Group III serves as a control for this experiment. What other setup could be included as a control for this experiment? _____

Do your results support your hypothesis? _____ Your null hypothesis? _____

What do you **conclude** about the role of yeasts in the process of fermentation? _____

_____

_____

What do you conclude about the importance of the type of sugar used as a food source for yeasts in the process of fermentation? _____

_____

✔ **PART 4** **Production of Ethanol by Yeast**

Some yeasts are more metabolically active than others and can produce more alcohol. Lugol's solution (I_2KI) can be used to test for metabolic activity in yeast. In this experiment, you will be looking at the **fermentation** reaction; see* in Figure 11-I.

||||| **Procedure** |||||||||||||||||||||||||||||||||||||||||||

1. Obtain a yeast culture that has been incubating for 24 hours.

2. Using a 10-ml pipette, transfer a 5-ml sample of the clear solution from the yeast culture into a test tube.

3. Add 2 ml of 10% NaOH down the side of the tube and mix by gently shaking.

4. Slowly add 3 ml of Lugol's reagent (I_2KI), one drop at a time, while gently shaking the solution.

5. Allow the mixture to sit at room temperature for about 5 minutes.

6. Check for the presence of a yellow, flaky precipitate at the bottom of the vial. This layer contains iodoform, a compound produced when alcohol reacts with iodine.

   a. *What is the source of the alcohol that reacts to form the iodoform layer?*

👁 **EXERCISE B**   **The Krebs Cycle Reactions in Bean Seeds**

Within the mitochondria of aerobic organisms, pyruvic acid, the product of glycolysis, is further metabolized by a cyclic series of reactions called the **Krebs cycle** (or citric acid cycle). During the reactions of the Krebs cycle, special electron carriers (coenzymes FAD and NAD^+) are reduced (gain electrons) and a small amount of ATP is produced. Carbon dioxide is also produced.

It is the hydrogen electrons stripped from glucose and its breakdown products that are used to reduce the electron carriers FAD and NAD^+ to $FADH_2$ and $NADH + H^+$. When these reduced electron carriers are later oxidized by the electron transport chain, the released energy is used to phosphorylate ADP, thus regenerating ATP. The final electron acceptor of the electron transport chain is oxygen.

The Krebs cycle and electron transport reactions will occur only if oxygen is available to accept electrons. These pathways are part of the aerobic respiration process. The final products of respiration are carbon dioxide (CO_2), water (H_2O), and energy; all carbon-containing substrate molecules have been completely oxidized.

$$C_6H_{12}O_6 \; + \; 6O_2 \; \longrightarrow \; 6CO_2 \; + \; 6H_2O + energy$$
$$\text{glucose} \quad \text{oxygen} \qquad \text{carbon} \quad \text{water}$$
$$\text{dioxide}$$

In this exercise, you will examine one of the reactions of the Krebs cycle—the conversion of **succinate** to **fumarate,** catalyzed by the enzyme succinic dehydrogenase, using the coenzyme FAD (flavin adenine dinucleotide) (see ** in Figure 11-I).

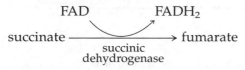

During this reaction, two hydrogen atoms are removed from succinate and are transferred to FAD, reducing it to $FADH_2$. For test purposes, certain dyes such as 2,6-dichloro-phenol-indophenol (DPIP) can function as hydrogen acceptors in place of FAD. DPIP is decolorized as it accepts hydrogen from succinate. The amount or rate of decolorization provides a means for observing and measuring the reaction.

IIIII **Objectives** IIIIIIIIIIIIIIIIIIIIIIIIIIIIIIIIIIIIIIIIII

☐ List the requirements for the Krebs cycle reaction in which succinate is transformed to fumarate.

☐ Describe how DPIP can be used to study a reaction in which hydrogen is transferred from one compound to another.

☐ Describe how FAD functions as a coenzyme with succinic dehydrogenase.

IIIII **Procedure** IIIIIIIIIIIIIIIIIIIIIIIIIIIIIIIIIIIIIIIIIIIIIII

1. Work in pairs within groups of four. In this exercise, you will observe the decolorization of DPIP as it accepts hydrogen electrons from succinate (usually the function of FAD). As the DPIP becomes less blue, it will absorb less light (and consequently transmit more light). To observe the changes in absorbance and transmittance, you will use a spectrophotometer (the Spectronic 20).

    Note: For a complete discussion of the operation of a spectrophotometer, review Laboratory 4. In this instrument, light is passed through a grating that, like a prism, separates it into bands of certain wavelengths. Colored indicator dyes such as DPIP absorb a maximum amount of light in the 600-nanometer (nm) wavelength range. If a sample is placed in the path of this light, a photocell on the other side of the sample will measure how much light has passed through (been transmitted). On separate scales, the instrument shows the percent of light transmitted and the percent of light absorbed.

2. Obtain 8 spectrophotometer tubes per group. Each pair of students will use 4 tubes.

3. Since you want to measure only decolorization of DPIP and not absorbance by other materials in your solution, a "blank" must be prepared. A blank contains all materials present in the sample except the one to be measured (DPIP). Each pair of students should prepare a blank as indicated in Table 11B-1. Make sure the contents are mixed by covering the tube with Parafilm and inverting it several times. With the machine empty, use the left-hand knob to set the spectrophotometer to 100% absorbance and 0% transmittance. Now insert the blank and use the right-hand knob to set the machine to 100% transmittance and 0% absorbance. Absorbance by DPIP alone can now be measured in the experimental tubes.

Table 11B-1

Tube	Lima Bean Juice	DPIP	Phosphate Buffer	Malonate	Do not add until step 6: Succinate
1	0.3 ml	0.3 ml	4.3 ml	—	0.1 ml
2	0.3 ml	0.3 ml	4.4 ml	—	—
3	0.3 ml (boiled)	0.3 ml	4.3 ml	—	0.1 ml
4	0.3 ml	0.3 ml	4.0 ml	0.3 ml	0.1 ml
5	0.3 ml	0.3 ml	4.1 ml	—	0.3 ml
6	0.3 ml	0.3 ml	3.8 ml	0.3 ml	0.3 ml
Blank	0.3 ml	—	4.6 ml	—	0.1 ml

4. Your instructor will homogenize and then centrifuge a solution containing 50 g of soaked lima bean seeds. The supernatant (the clear liquid remaining above the sediment) will contain isolated mitochondria. (One sample will be boiled for 5 minutes.)

5. To test tubes 1 through 6, add the reagents in the amounts indicated in Table 11B-1. *Do not* add succinate yet. Mix the contents of each tube by covering the tube with Parafilm and

inverting it several times. Use a *different* piece of Parafilm for each tube to avoid contamination. (After all ingredients have been added, all tubes should contain 5 ml of solution.)

Caution: If another group of students has used the Spectronic 20, use your blank to readjust the machine before making experimental readings. Remember, the blank tube contains all reagents except the dye (DPIP) and will allow you to zero the spectrophotometer to measure only those changes in color that result from changes in DPIP in the sample tubes.

6. Add succinate (see Table 11B-1) to one tube at a time (except tube 2). After adding the succinate, place a piece of Parafilm over the tube and invert the tube to mix in the succinate. Immediately read the absorbance at 600 nm. Do this for each of the three experimental tubes assigned to your student pair.

7. Read absorbances at 2-minute intervals and record your data in Table 11B-2. Take turns making absorbance readings. Remember to readjust the Spectronic 20 with your blank if necessary.

Table 11B-2   Absorbances of Bean Seedling Solutions

Minutes	Tube 1	Tube 2	Tube 3	Tube 4	Tube 5	Tube 6
0						
2						
4						
6						
8						
10						
12						
14						
16						

8. Graph your data on a sheet of graph paper. Place absorbance on the *Y*-axis and time (in minutes) on the *X*-axis. Use a different symbol or colored pencil to indicate each tube.

9. Discuss the results of your experiment. Describe what the data from each group show about the reaction succinate ⟶ fumarate.

Tube 1 _____

Tube 2 _____

Tube 3 _____

Tube 4 _____

Tube 5 _____

Tube 6 _____

a. How does this reaction indicate that the bean seedlings used in the test were alive?

_____

b. *How do the results in tube 3 support your answer to question a?* _____

_____

c. *Which tube is used as a control?* _____

The malonate molecule is sufficiently similar in shape to the succinate molecule that it binds to the active site of the enzyme succinic dehydrogenase (Figure 11B-1). Succinic dehydrogenase cannot act on malonate. When malonate is bound to the enzyme, no $FADH_2$ or fumarate is produced. Thus, malonate is a competitive inhibitor of succinic dehydrogenase (Figure 11B-2).

Figure 11B-1 *Succinate and malonate are similar in shape, and both can bind to the enzyme succinic dehydrogenase.*

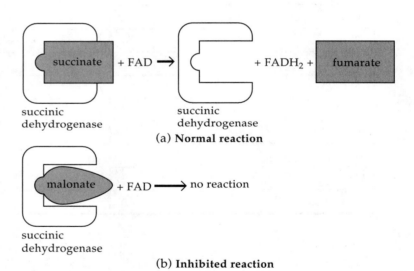

Figure 11B-2 *(a) The reaction of succinic dehydrogenase and its substrate, succinate. (b) When malonate occupies the active site for succinate in the succinic dehydrogenase enzyme, no reaction occurs.*

10. Now, describe the effects of malonate in your experiment. If succinate and malonate are both present in a mixture, they will compete for the enzyme's active site.

    d. *What did you observe in tube 4?* _____

    _____

    e. *Based on what you know about competition, how do you explain your observations?*

    _____

    _____

11. Compare your results for tube 4 with those for tube 6.

    f. *In which tube is more fumarate produced?* _____ *How do you know?*

    _____

g. What is different about the contents of tubes 4 and 6? _____

_____

h. How does increasing the amount of succinate affect malonate's ability to inhibit this reaction?

_____

12. From your textbook reading, explain the differences between the actions of competitive and noncompetitive inhibitors. Mercury (Hg) is a noncompetitive inhibitor of succinic dehydrogenase.

   i. How would Hg have affected the enzymatic breakdown of succinate by succinic dehydrogenase?

   _____

   _____

   j. If you had included Hg (mercuric chloride, or $HgCl_2$) in any of the reaction mixtures, would you have expected to see DPIP decolorize? _____ Why or why not?_____

   _____

   k. If you increased the amount of succinate 10-fold in the presence of $HgCl_2$, would you expect to see DPIP decolorize? _____ Why or why not? _____

   _____

EXTENDING YOUR INVESTIGATION: STUDYING INHIBITION

How would changing the amounts of succinate and malonate change the activity of succinic dehydrogenase? If 10 times more succinate than malonate is available, what is the chance that succinate will combine with the enzyme before malonate? What would be the consequence of this? Do you think it would be possible to overcome competitive inhibition? Formulate a hypothesis to explain the interaction of malonate and succinate when combined in different ratios with succinic dehydrogenase.

   HYPOTHESIS:

   NULL HYPOTHESIS:

What do you **predict** might happen as the succinate/malonate ratio increases?

What is the **independent variable** in this investigation?

What is the **dependent variable** in this investigation?

Design an experiment to test your hypothesis. Follow the same type of procedure as in Exercise B, combining lima bean juice, DPIP, phosphate buffer, succinate, and malonate.

PROCEDURE:

RESULTS: Record your results in the table below. What do the data collected from your experiment tell you about the competitive inhibitor, malonate?

Minutes	Tube 1	Tube 2	Tube 3	Tube 4	Tube 5	Tube 6
0						
2						
4						
6						
8						
10						
12						
14						
16						

Do your results support your hypothesis?

Your null hypothesis?

Was your prediction correct?

What do you **conclude** about the effect of competitive inhibitors on enzymatic reactions?

✓ EXERCISE C   **Heat Production During Respiration in Seedlings**

When respiration occurs, energy is produced. Some of this energy is used to produce ATP and some is given off as heat.

||||| **Objectives** |||||||||||||||||||||||||||||||||||||

☐ Detect respiration in growing tissues by measuring heat generation.

||||| **Procedure (Class Demonstration)** |||||||||||||||||||||||||||||||||||||||||

1. Your instructor has placed some moist cotton into the bottom third of an insulated pint bottle (Thermos) (Figure 11C-1). The remainder of the bottle has been filled with germinating seeds presoaked in water.

Figure 11C-1 *Setup for measuring the production of heat by seedlings in a Thermos bottle.*

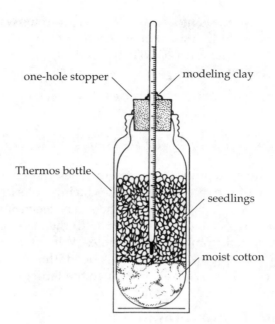

one-hole stopper

modeling clay

Thermos bottle

seedlings

moist cotton

2. A thermometer has been inserted so that its tip extends into the mass of seeds.

3. In the table that follows, record the time, the temperature inside the Thermos, and room temperature. Record these data several times during the laboratory period.

Time	Thermos Temperature	Room Temperature

a. *Do your temperature readings indicate that the seeds are respiring?* _____ *How do you know?*

_____

b. *What source of energy are the germinating seedlings using?* _____

c. *What influence might small increases in temperature have on respiration rates?*
_____ *Large increases in temperature?* _____

d. *Explain your answers to question c in terms of your knowledge of enzyme structure and function.*

_____

e. *How is metabolic heat useful to plants and animals in their natural environments?*

_____

_____

The heat production you have measured shows that respiration is not 100 percent efficient. When 1 mole of glucose (the molecular weight in grams) is oxidized in a respiring cell, approximately 434 kilocalories of useful energy are generated in the form of ATP. In contrast, the complete oxidation of glucose yields 687 kilocalories of energy measured as heat.

> *f. Using these figures, calculate the efficiency of respiration as a percent.* _____

✔ **EXERCISE D**   **Respiration in Plant Embryos**

Several tests are used by the agricultural industry to check the viability of seeds before planting. One of these tests involves the use of the dye tetrazolium, which is colorless when oxidized but becomes reddish when reduced. The test relies on whether or not the seed's electron transport system is working. When tetrazolium is added to a living cell, it will interact with the electron transport system to accept hydrogen electrons as they are transferred from the cytochromes. When tetrazolium accepts these electrons, it is reduced and turns red or a deep pink. If the seed is dead, the electron transport system will not be functioning, no hydrogens will be available to reduce the tetrazolium, and the seed will remain colorless.

⁞⁞⁞⁞ **Objectives** ⁞⁞⁞⁞⁞⁞⁞⁞⁞⁞⁞⁞⁞⁞⁞⁞⁞⁞⁞⁞⁞⁞⁞⁞⁞⁞⁞⁞⁞⁞⁞⁞⁞⁞⁞

☐ Explain how the indicator TTC (2,3,5-triphenyl tetrazolium chloride) can be used to determine whether seeds are viable.

⁞⁞⁞⁞ **Procedure** ⁞⁞⁞⁞⁞⁞⁞⁞⁞⁞⁞⁞⁞⁞⁞⁞⁞⁞⁞⁞⁞⁞⁞⁞⁞⁞⁞⁞⁞⁞⁞⁞⁞⁞⁞

1. On the demonstration table you will find two groups (A and B) of seeds (beans or corn). One lot has been boiled and is dead. The other is alive. Test the two groups of seeds for viability. Cut three seeds from each group in half along their long axis with a sharp razor blade (Figure 11D-1).

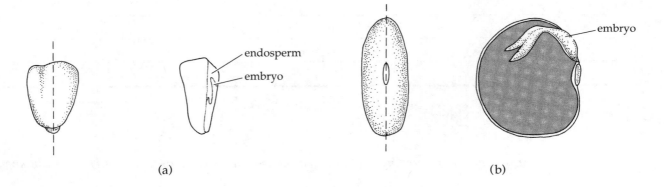

(a)   (b)

Figure 11D-1   *Seeds cut to show embryo. (a) Corn. (b) Bean.*

2. Compare your cross section of the seed with that shown in Figure 11D-1 to be sure that you have exposed the embryo. Locate the embryo, endosperm (food for the growing seed), and seed coat of your seeds.

> *a. Which tissue would you expect to stain red after application of the TTC reagent?*
>
> _____

3. Place several drops of TTC into two Petri dishes. Label one Petri dish for the A group of seeds and another for the B group. Place the seed halves with their cut side down in the Petri dishes; add enough TTC to completely cover the seeds.

   Caution: TTC is a poison; avoid contact with skin. Wash immediately if you come into direct contact with TTC.

4. At the end of the laboratory period, turn the seed halves over and examine them. Use forceps!

   b. *Which group of seeds is alive and respiring?* _____ *How do you know?* _____

   _____

Cyanide affects the electron transport system by binding to components of the system (the cytochromes), thereby inhibiting the transfer of electrons.

   c. *If seeds were treated with cyanide, what results would they show in the tetrazolium test?*

   _____ *Explain.* _____

   _____

   _____

Laboratory Review Questions and Problems

1. Where in the cell do the following processes occur? glycolysis _____

   fermentation _____ Krebs cycle _____ electron transport _____

2. Why is fermentation considered to be an inefficient process?

3. Why is the process of fermentation called anaerobic?

4. Complete the following chart.

	Requirements	Products	Final Electron Acceptor
Alcoholic fermentation			
Lactic acid fermentation			
Aerobic respiration			

5. You have just finished running a 3-minute mile. Your legs are cramping badly and you are breathing very rapidly. How could heavy breathing be beneficial?

6. You have three unlabeled solutions—A, B, and C. You know that one is an enzyme solution, another is a solution of that enzyme's substrate, and the third contains an inhibitor of enzyme activity. You know that when the enzyme and the substrate react, the solution will turn red. You mix various amounts of the three solutions. From the results, determine the identities of solutions A, B, and C. Is the inhibitor competitive or noncompetitive?

Reaction	Buffer (ml)	Solution A (ml)	Solution B (ml)	Solution C (ml)	Color
1	0.4	0.1	0.1	0.1	Medium pink
2	0.5	—	0.1	0.1	Red
3	0.3	0.2	0.1	0.1	Light pink
4	0.4	0.2	—	0.1	Colorless
5	0.3	0.1	0.2	0.1	Dark pink
6	0.5	0.1	0.1	—	Colorless
7	—	0.1	0.4	0.2	Red

7. Ten glucose molecules are broken down during glycolysis. How many pyruvate molecules are produced? _____ Each of the 3-carbon pyruvate molecules can yield a 2-carbon acetyl fragment after CO_2 is removed. How many acetyl compounds can be produced from the 10 glucose molecules? _____ If these are used to produce a fatty acid, how many carbons would be in the fatty acid chain? _____ Draw the fatty acid in the space below (see Laboratory 5).

8. A farmer has a large bag of pea seeds left over from last year's planting. He would like to save some money by planting the seeds, but is not sure that a sufficient percentage of the seeds remains viable. How could he determine what percentage of the seeds would be expected to germinate?

Photosynthesis

OVERVIEW

During the process of **photosynthesis,** plants capture a small fraction of the sun's energy and store it in the chemical bonds of carbohydrates. The carbon source for the organic compounds is the inorganic atmospheric gas carbon dioxide, which is reduced by the addition of electrons from H_2O to form carbohydrate ($C_6H_{12}O_6$). Oxygen (O_2) is released when water is "split" to provide hydrogen electrons ($2H_2O \longrightarrow 4H + O_2$). The general formula for photosynthesis is:

$$6CO_2 + 6H_2O + 686 \text{ kilocalories} \longrightarrow C_6H_{12}O_6 + 6O_2$$

carbon dioxide   water   light energy   glucose   oxygen

Photosynthesis is the source of virtually all energy used by organisms (with the exception of a few chemosynthetic organisms); photosynthesis is the only method by which chemical energy is added to the ecosystem. As we examine the factors involved in energy capture and storage, remember that cells that are photosynthesizing are also carrying on cellular respiration, as are all animal organisms. Cellular respiration uses the O_2 produced by photosynthesis to break down glucose. Energy trapped in the chemical bonds of glucose is released for cellular work, and carbon is returned to the atmosphere in the form of CO_2. This CO_2 serves as a carbon source for further photosynthesis (Figure 12-1).

Figure 12-1 *Note that the product of photosynthesis, glucose, is the starting reactant for respiration. The photosynthetic reactions begin with water and carbon dioxide. Thus, with the addition of energy from sunlight, water and carbon dioxide are constantly recycled into higher-energy glucose molecules. The energy-requiring processes of life often obtain energy by breaking down glucose.*

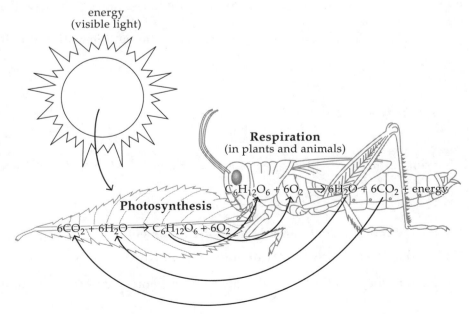

energy (visible light)

Respiration (in plants and animals)

$C_6H_{12}O_6 + 6O_2 \longrightarrow 6H_2O + 6CO_2 + \text{energy}$

Photosynthesis

$6CO_2 + 6H_2O \longrightarrow C_6H_{12}O_6 + 6O_2$

STUDENT PREPARATION

Prepare for this laboratory by reading the text pages indicated by your instructor. Familiarizing yourself in advance with the information and procedures covered in this laboratory will give you a better understanding of the material and improve your efficiency.

✔ 👁 **EXERCISE A**   **The Energy-Capturing Reactions**

Photosynthesis takes place in two stages. The first, the energy-capturing reactions, called **light-dependent reactions,** must take place in light. The light-dependent reactions are summarized by the equation:

$$H_2O + ADP + P_i + NADP^+ \xrightarrow[\text{chloroplasts}]{\text{light}} \tfrac{1}{2}O_2 + ATP + NADPH + H^+$$

Light energy is used to split water and energize the electrons of chlorophyll molecules. Some of this energy is then captured and used to form ATP and NADPH; O_2 is released as a by-product. You can determine the rate of this reaction by measuring the amount of product made, in this case O_2. Both light intensity and light quality (color or wavelength) affect the rate of the energy-capturing reaction.

▥▥▥ Objectives ▥▥▥▥▥▥▥▥▥▥▥▥▥▥▥▥▥▥▥▥▥▥▥▥▥▥▥▥▥▥▥▥

- ☐ Explain how to measure the rate of photosynthesis.
- ☐ Discuss the effect of light intensity on photosynthesis.
- ☐ Discuss the effect of light quality (wavelength) on photosynthesis.

👁 **PART I**   **Determining the Effect of Light Intensity**

The rate of photosynthesis can be affected by the amount of light available to plants. Since oxygen is a product of photosynthesis, measuring its production allows us to measure the rate of photosynthesis.

In this exercise, you will work with small pieces (disks) of spinach leaf tissue. The leaf tissue is riddled with gas-filled intercellular spaces. If leaf disks cut from spinach leaves are placed in a sodium bicarbonate ($NaHCO_3$) solution and subjected to a vacuum, sodium bicarbonate will replace the gases in the intercellular spaces and will serve as a carbon source for photosynthesis. Replacement of gas by liquid will cause the disks to sink to the bottom of the flask. As the light-dependent reactions of photosynthesis proceed, oxygen gas will be evolved and will diffuse into the intercellular spaces. When enough oxygen accumulates in the intercellular spaces, each leaf disk will regain its buoyancy and turn on edge or float to the surface. Thus, by observing the flotation of spinach leaf disks, you can indirectly measure oxygen production as an indicator of photosynthetic activity. An increase in the rate of photosynthesis should result in an increase in the number of floating disks.

Formulate a hypothesis about the effects of different intensities of light on photosynthesis.

HYPOTHESIS:

NULL HYPOTHESIS:

What do you **predict** will happen in an experiment used to test your hypothesis?

What is the **independent variable** in your experiment?

What is the **dependent variable in** your experiment?

Now, determine the effects of light intensity on photosynthesis by using the following procedure.

⁞⁞⁞⁞⁞ Procedure ⁞⁞⁞⁞⁞⁞⁞⁞⁞⁞⁞⁞⁞⁞⁞⁞⁞⁞⁞⁞⁞⁞⁞⁞⁞⁞⁞⁞⁞⁞⁞⁞⁞⁞⁞⁞⁞

1. Work in pairs. Each pair will be assigned a light bulb of a different wattage: 40, 60, 100, or 150 W. Several student pairs will use no light.

2. Place the bulb in the flood light assembly and adjust the light so that it is 27.5 cm above the base of the ring stand. The light intensities delivered from this height (W/m^2) will differ with the wattage of the bulb being used (see Table 12A-1).

Table 12A-1

Bulb (W)	Intensity (W/m^2)
None	0.000
40	0.036
60	0.064
100	0.100
150	0.154

3. Using a cork borer, cut 50 disks from spinach leaves. Avoid cutting through areas with large veins. Cut leaves on a paper towel or Styrofoam board. Immediately put the disks into your instructor's 500-ml flask (the flask contains $NaHCO_3$). All groups should place their disks in the same flask.

4. Your instructor will use a water aspirator to sink the disks in a vacuum, using the following procedure:

   • Attach a piece of vacuum tubing to the sidearm. Put a rubber stopper firmly in the mouth of the flask, and press a piece of tape securely over the hole of the stopper.

   • Turn the water to full force. When the solution begins to bubble, release the vacuum by peeling back the tape.

   • Swirl the flask, and apply the vacuum again.

   • Repeat this procedure three or four times until most of the disks fall to the bottom of the flask when the vacuum is released. As the solution infiltrates the disks, they will become a darker green.

Figure 12A-1 *Setup for experiment demonstrating the effects of light intensity on photosynthesis. Adjust the space between the lamp and the Petri dish to 27.5 cm. (This setup, using different lamp-to-Petri-dish distances, will also be used for Exercise A, Part 3.)*

light source

7.5 cm

beaker of water (heat sink)

Petri dish containing spinach leaf disks

5. During this time, obtain a Petri dish with a clear lid (the dish is wrapped with black tape on the bottom to exclude excess light). If you are assigned to the "no light" treatment, you will use a Petri dish with both top and bottom covered with black tape. Fill the Petri dish approximately two-thirds full with fresh 0.2% $NaHCO_3$ solution.

6. Turn off the room lights and, using forceps, gently transfer 20 sunken spinach disks to your Petri dish and cover the dish—replace any floating disks until all 20 are sunken disks.

7. Center the Petri dish on the base of the ring stand under the light and cover the dish. If you are using no light, leave the dish on the laboratory table or place it in a dark drawer.

8. Place a 2-liter beaker, filled to the 1,600-ml mark with tap water, on top of the Petri dish to act as a heat filter (see Figure 12A-1).

9. Turn on the light and record the time. After 10 minutes, count the number of spinach disks that are floating. Record your results below.

Treatment	Number of Disks Floating (10 minutes)

10. When all groups have finished, record the class data and compute and record the average (arithmetic mean) for each light intensity in Table 12A-2.

Table 12A-2   Effects of Varying Light Intensities on Photosynthesis, Measured as Number of Floating Disks

Intensity (W/m²)	Group						Mean
	1	2	3	4	5	6	
0.000							
0.036							
0.064							
0.100							
0.154							

11. Graph your data as percent of floating disks: (average/20) × 100. Make a bar graph using the grid in Figure 12A-2.

   a. *Why is it important to use mean values when making the graph:?*

   _____

   _____

Do your results support your hypothesis? _____ *Your null hypothesis?* _____

*From your graph, what do you **conclude** about the effects of light intensity on the rate of photosynthesis?*

_____

_____

Is the effect of increasing light intensity significant? To determine this, your instructor will collect data from all laboratory classes and will use these data to perform a chi-square median test. The results will assist you in determining whether to accept or reject your null hypothesis.

Figure 12A-2

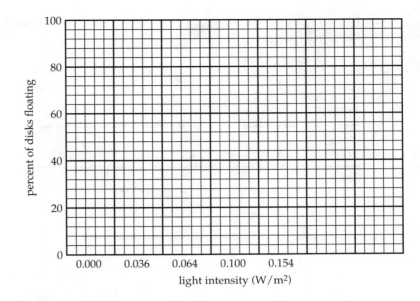

PART 2   The Spectrum of Visible Light

White light is composed of all the wavelengths (colors) in the spectrum of visible light (Figure 12A-3). These wavelengths are measured in nanometers. We see colors because objects contain **pigments** that selectively absorb some wavelengths of visible light and reflect or transmit others. What we recognize as an object's color is composed only of those wavelengths of light that are transmitted or reflected.

The various pigments found in chloroplasts—including chlorophylls *a* and *b*, xanthophylls, and carotenes—absorb different wavelengths of light, thus making use of a wide range of light energy for photosynthesis.

ⅠⅠⅠⅠⅠ Procedure ⅠⅠ

1. Use the prism on the demonstration table to project a spectrum onto a piece of white paper.

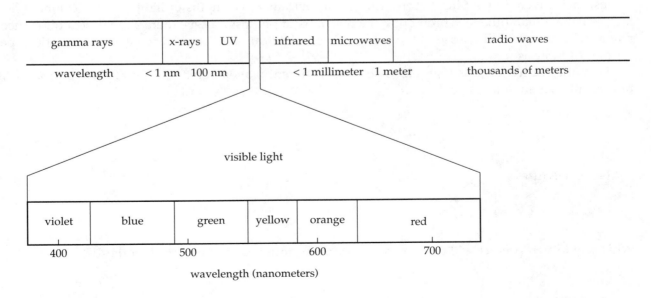

Figure 12A-3   *The spectrum of white light.*

2. Hold various colored filters between the light source and the prism. What do you predict will happen? Complete Table 12A-3.

Table 12A-3   Absorption and Transmission Properties of Colored Filters

Filter Pigment	Colors Absorbed	Colors Transmitted
Blue		
Green		
Red		

a. *Why do green plants appear green? (What colors of light would you expect plants to absorb or reflect?)* _____

_____

b. *Why does a black piece of material appear black?* _____

_____

👁 **PART 3**   **Determining the Effects of Light Quality—The Action Spectrum**

An **action spectrum** defines the relative effectiveness of different wavelengths of light (colors) for light-dependent processes such as photosynthesis. In this experiment, you will test the effects of colors of light on photosynthesis in sections of spinach leaves.

As in Exercise A, Part 1, spinach leaf disks placed in sodium bicarbonate ($NaHCO_3$) and subjected to a vacuum will sink as the solution replaces the gases trapped in the intercellular spaces. In the presence of different colors of light, photosynthesis will produce varying amounts of oxygen. Accumulation of oxygen in the leaf disks will cause them to float.

You will test the effects of different wavelengths of light by placing the spinach leaf disks into Petri dishes, each covered by a filter that transmits only certain wavelengths of light. If some colors of light are more effective than others in promoting the reactions of photosynthesis, there should be a difference in the rate at which the leaf disks rise to the surface. These data can then be used to generate an action spectrum for photosynthesis.

Formulate a hypothesis that addresses how different wavelengths of light affect the process of photosynthesis in spinach leaf disks.

HYPOTHESIS:

NULL HYPOTHESIS:

What would you **predict** will occur in an experiment designed to test your hypothesis?

What is the **independent variable** in your experiment?

What is the **dependent variable** in your experiment?

Use the following to carry out your investigation.

｜｜｜｜｜ **Procedure** ｜｜｜｜｜｜｜｜｜｜｜｜｜｜｜｜｜｜｜｜｜｜｜｜｜｜｜｜｜｜｜｜｜｜｜｜｜｜｜

1. Work in pairs. Each pair should obtain one of five specially prepared Petri dishes. The bottoms of the Petri dishes are wrapped in black tape to eliminate extra light. The tops are covered with colored filters (clear or none, blue, green, or red), or with black tape if you are assigned to the dark treatment). Fill your dish two-thirds full with the 0.2% $NaHCO_3$ solution. Place your dish on the stand portion of a ring stand. (Students assigned to a "dark" treatment will place dishes in the dark in a drawer.)

2. Attach a 150-W photoflood lamp to your ring stand. Because filters absorb different amounts of light, adjust the lamp-to-Petri-dish distance to make all light intensities equal as follows: no filter (white light), 27.5 cm; blue, 23.5 cm; green, 21.0 cm; red, 26.5 cm (dark, in box or drawer) (see Figure 12A-1).

3–8. Follow steps 3–8 in Exercise A, Part 1, substituting (in step 5) Petri dish lids with colored filters for clear Petri dish lids, when assigned.

9. Turn on the light and record the time. After 10 minutes, count the number of spinach disks that are floating and record your results below.

Treatment	Number of Disks Floating (10 minutes)

10. When all groups have finished, record the class data and compute and record the average (mean) for each color (wavelength) of light in Table 12A-4.

Table 12A-4   Effects of Different Wavelengths of Light on Photosynthesis, Measured as Number of Floating Disks

Wavelength (λ)	Group						Mean
	1	2	3	4	5	6	
White							
Blue							
Green							
Red							
Dark (no light)							

11. Graph your data as percent of floating disks: (average/20) × 100. Make a bar graph using the grid in Figure 12A-4, then superimpose an action spectrum curve along the tops of all the bars in your graph.

   a. Why is this called an action *spectrum?* _____

Figure 12A-4

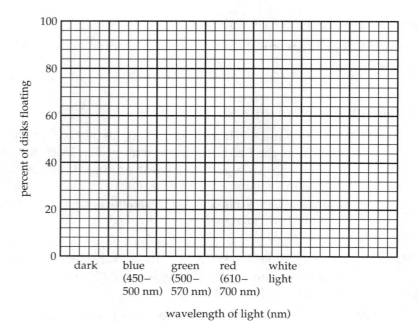

Do your results support your hypothesis? _____ *Your null hypothesis?* _____

b. Which wavelengths of light are most effective in promoting photosynthesis? _____

c. Which are least effective? _____

From your graph, what do you **conclude** *about the effects of different colors of light on the rate of photosynthesis?*

_____

Is the difference in photosynthetic rate at specific wavelengths significant? To determine this, your instructor will collect data from all laboratory classes and will use these data to perform a chi-square median test. The results will assist you in determining whether to accept or reject your null hypothesis.

 EXERCISE B   The Pigments of Chloroplasts

Chlorophyll is the pigment responsible for the green color of plants. However, plants also contain pigments that absorb wavelengths of light other than those used by chlorophyll. *From your results in Exercise A, Part 3, why would this be necessary?* _____

▮▮▮▮▮ **Objectives** ▮▮▮▮▮▮▮▮▮▮▮▮▮▮▮▮▮▮▮▮▮▮▮▮▮▮▮▮▮▮▮▮▮▮▮▮▮▮▮

☐ Define "pigment." Explain the meaning of "absorption" and "transmission" of light by pigments.

☐ Describe the purpose and technique of chromatography.

☐ Name the pigments present in green plants.

☐ Explain how a spectrophotometer (Spectronic 20) works (*optional*).

☐ Using plant pigments as an example, explain what an absorption spectrum shows.

☐ Using photosynthesis and plant pigments as an example, discuss the relationship between absorption and action spectra.

☐ Give evidence that light is necessary for chlorophyll formation.

PART I   How Plant Pigments Use the Light Spectrum

The process of **chromatography,** which separates complex mixtures into their component parts based on their solubility in different kinds of solvents, can help identify some of the pigments used in photosynthesis. Acetone has been used to extract pigments from spinach leaves.

ⅢⅢⅢ Procedure ⅢⅢⅢⅢⅢⅢⅢⅢⅢⅢⅢⅢⅢⅢⅢⅢⅢⅢⅢⅢⅢⅢⅢⅢⅢⅢⅢ

1. Obtain a piece of chromatograph paper precut to fit a large test tube and, using a *pencil*, draw a baseline 1.5 cm from the bottom or pointed end (Figure 12B-1a). (Try not to touch the paper because oil from your skin will interfere with the chromatography process.)

2. Use a capillary pipette to streak chlorophyll extract along the pencil line. The streak should be as thin as possible. (Dry with a hair dryer if available.) Reapply the chlorophyll extract at least 5 more times, keeping the pigment line as narrow as possible. (Alternatively, obtain a large spinach leaf or ivy leaf. Place the leaf with its lower surface down across the chromatography paper. Roll a dime or quarter across the leaf, pressing hard, to make a green pigment line across the paper as close to your pencil line as possible. Roll the coin back and forth to squeeze out as much pigment as possible.)

Figure 12B-1 *Setup for chromatography of plant pigments.*

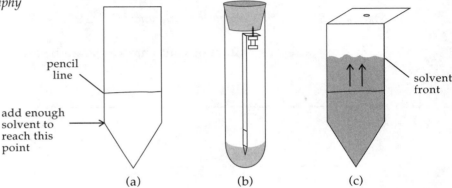

pencil line

add enough solvent to reach this point

solvent front

(a)   (b)   (c)

3. Obtain a stoppered test tube containing chromatography solvent. Fold the top of the chromatography paper about 0.5 cm from the end (Figure 12B-1b). Remove the rubber stopper and tack the strip of chromatography paper to the center of the stopper. Carefully place the paper strip into the tube, pointed end first, and replace the stopper. The chromatography solvent should extend up the paper, covering the point, but should be below the pigment line.

   Warning: Do not inhale the solvent fumes. Make sure the room is well ventilated or a hood is used. Petroleum ether and acetone are very flammable. Keep away from hot plates, open flames, or other sources of high heat.

4. Place the tube into a test tube rack and do not disturb it while the chromatography procedure continues.

5. Allow the solvent to rise up the paper until it is approximately 1 cm from the top of the paper. Remove the paper and mark the location of the solvent front (Figure 12B-1c) using a pencil.

6. After the chromatogram has dried, use a pencil to circle each pigment spot. Record the color of each pigment spot and the distance the pigment has traveled from the origin. Use the center of the spot as your point of reference. The colors will fade quickly, so make your observations immediately.

Two kinds of green chlorophylls can be observed. Chlorophyll *a* is a bluer-green than chlorophyll *b*, which appears yellow-green. Chlorophyll *b* is found closer to the origin. Also present are two types of yellow-to-orange pigments called carotenoids; carotenes are orange-yellow and xanthophylls are yellow.

All four of the pigments you observe are found in the thylakoid membranes of chloroplasts. Both chlorophylls *a* and *b* consist of a large ring component (a tetrapyrrole ring, like that of hemoglobin), associated with a magnesium ion (Mg^{2+}). A long, nonpolar hydrocarbon chain attached to the ring anchors the charged ring portion of the pigment molecule to the nonpolar lipid layers of the thylakoid membranes. The carotenes and xanthophylls are pure hydrocarbon chains, with alternating single and double bonds between carbons. Both xanthophylls and carotenes are very nonpolar and are embedded in the thylakoid membranes.

Recall that the chromatography solvent is nonpolar. Only nonpolar materials dissolve in nonpolar solvents. The more nonpolar a pigment is, the further it will travel with the solvent front (leading edge of the solvent traveling up the paper). The more polar a material is, the greater will be its tendency to stay in place, tightly bound to the water in the paper. In this way, the photosynthetic pigments become partitioned on the chromatography paper.

a. *Which pigment appears to be the most nonpolar?* _____

b. *Which pigment is the most polar?* _____

c. *What general structural feature of chlorophylls* a *and* b *causes them to remain close to the origin? (Hint: Charged ions are polar and hydrophilic.)* _____

d. *How many of the pigments present are chlorophyll?* _____

e. *How many of the pigments present are carotenoids?* _____

   7. Draw a representation of your chromatogram below. Label the pigments.

  PART 2   **Absorption Spectra of Chloroplast Pigments**

Having isolated and identified the pigments found in chloroplasts, you can now determine the wavelengths of light transmitted by each pigment and then by all the pigments in the chloroplast extract.

Using a spectrophotometer (Spectronic 20), white light will be passed through a grating, which, like a prism, separates light into bands of specific wavelengths. You select a wavelength—for example, in the green range—to pass through the chlorophyll extract. Green light shines on the sample, and a photocell on the other side of the sample measures how much light has passed through the sample (has been transmitted). The spectrophotometer then records, on separate scales, the percentage of light transmitted and the percentage of light absorbed (**absorbance value**). You can determine the absorbance for each pigment at different wavelengths and generate the absorption spectrum for each pigment. (See Laboratory 4, Exercise A.)

⦚⦚⦚⦚⦚ Procedure ⦚⦚⦚

   1. Work in groups of four. Your instructor will assign each group a pigment isolated from chloroplasts.

   If you are not already familiar with the use of the spectrophotometer from Laboratory 4, follow these instructions. Insert a spectrophotometer tube containing a white strip of paper into the sample holder. Rotate the tube until you see colored light reflected from the paper. Turn the wavelength knob. As you turn the knob, you will see the different colors of light in the spectrum. What color do you see at 450 nm? _____ 550 nm? _____

650 nm? _____ As you vary the wavelength of light, you vary the color of light that can be absorbed by a sample (in this case, a chlorophyll extract in the sample holder). The instructions for using the Spectronic 20 are beside the instrument on the demonstration table. Ask your instructor for assistance if necessary.

2. The chloroplast pigments (chlorophyll *a*, chlorophyll *b*, xanthophylls, carotenes) are dissolved in different solvents—methanol, petroleum ether, and ethyl ether. Since the solvents themselves absorb a small amount of light, this amount must be subtracted from the total absorption by the pigment (analogous to taring a balance). This is done by using a tube containing just solvent (a solvent *blank*). First, with no tube in the machine, use the left-hand knob to adjust to 100% absorbance and 0% transmittance. Select a pigment and correct for absorbance by the solvent; place a tube containing the solvent (blank) in the machine and use the right-hand knob to adjust to 100% transmittance and 0% absorbance. For each pigment, be sure that the blank is the same solvent that was used to separate the extract.

3. Place the sample of chlorophyll pigment extract into the proper Spectronic 20 tube. Read the absorbance value. Take a reading every 25 nanometers (nm) from 400 to 700 nm. (Since absorbance by the solvent blank changes with different wavelengths, you must rezero the machine, using the blank, every time you change the wavelength.)

4. Record your results in Figure 12B-2.

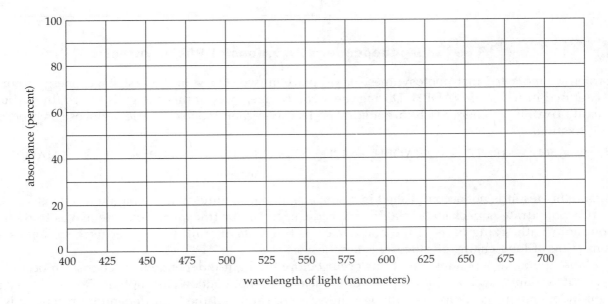

Figure 12B-2   *Graph the absorption spectrum of chloroplast pigments.*

5. Trace your graph onto your instructor's transparency. At the end of the period, examine all the graphs. Roughly sketch the curves for the other pigments on your own graph.

    In Exercise A, you used spinach disks to investigate the wavelengths of light that are most effective in photosynthesis, and you generated an *action spectrum*. Now you have examined the wavelengths of light absorbed by individual chloroplast pigments and have generated an *absorption spectrum*.

    a. In your own words, describe the difference between an action spectrum and an absorption spectrum.

_____

6. Superimpose your results from your action spectrum (Exercise A, Part 3) onto Figure 12B-2 at the appropriate wavelengths.

   b. What can you conclude about which pigments are responsible for photosynthesis?

   _____

✔ **PART 3**   **The Role of Light in Chlorophyll Synthesis**

IIIII **Procedure** IIIIIIIIIIIIIIIIIIIIIIIIIIIIIIIIIIIIIIIIII

Observe the two flats of wheat seedlings on demonstration. Record your observations below.

a. Appearance of light-grown seedlings: _____

b. Appearance of dark-grown seedlings: _____

c. What pigments are present in each set of plants? _____

d. Is light necessary for the synthesis of chlorophyll? _____

e. Is light necessary for photosynthesis? (In Exercise A, what happened to the spinach disks placed in the dark?)

_____

f. If chlorophyll is the pigment responsible for capturing the sun's rays, explain how the dark-grown seedlings

   managed to survive and grow. _____

   _____

👁 **EXERCISE C**   **The Light-Independent Reactions of Photosynthesis**

Earlier, we indicated that photosynthesis takes place in two stages. In the first stage, energy from the sun is captured in ATP and NADPH. During the second stage, most of this energy is stored, by **carbon dioxide fixation,** in energy-rich carbohydrates. The six-carbon monosaccharide, glucose, is produced from CO_2 fixation.

$$6CO_2 + 18ATP + 12NADPH + 12H^+ \longrightarrow \underset{\text{glucose}}{C_6H_{12}O_6} + 18ADP + 18P_i + 12NADP^+ + 6H_2O$$

Again, this result is not accomplished in a single step, but requires many separate reactions.

   Carbohydrate is packaged as a disaccharide, sucrose, for transport from the leaves to the nonphotosynthesizing parts of the plant, such as the roots or fruit. For longer-term storage, starch, consisting of long chains of glucose, is synthesized.

   Must the light-dependent reactions occur before the light-independent reactions can occur? In some variegated plants, such as *Coleus*, parts of the leaves do not contain chlorophyll. Do they carry out light-independent reactions even though they cannot carry out the light-dependent reactions without chlorophyll?

IIIII **Objectives** IIIIIIIIIIIIIIIIIIIIIIIIIIIIIIIIIIIIIIIIII

☐ Define carbon dioxide fixation.

☐ Discuss and give evidence for the role of plant pigments in carbon dioxide fixation.

IIIII **Procedure** IIIIIIIIIIIIIIIIIIIIIIIIIIIIIIIIIIIIIIIIII

   1. Obtain a leaf from a variegated *Coleus* plant.

   2. Place a sheet of white paper over the leaf, hold it up to the window (or use a light table if available), and trace the outline of the pattern of colors. Label each section with its color.

3. Put the leaf into a boiling-water bath for 2 to 3 minutes. Notice that pink and purple pigments (anthocyanins) are removed by this treatment.

4. Transfer the leaf to hot alcohol (heated in a beaker placed in a water bath). <u>Caution:</u> Be very careful heating the alcohol—do not allow it to boil! Leave the leaf in the alcohol until all of the pigment has been leached out.

5. Place the leaf in a Petri dish and pour Lugol's solution (I_2KI) over it. Starch will stain dark blue-black.

6. Indicate the starch-containing areas on your leaf tracing, and complete Table 12C-1.

Table 12C-1 Relation of Plant Pigments to Starch Formation

Pigment	Starch Present?
None	
Anthocyanins	
Carotenes/xanthophylls	

a. *In which parts of the plant did the carbon-fixing reactions result in the synthesis of starch?* _____

b. *Was chlorophyll present in these areas?* _____

c. *Would you conclude that the energy-capturing reactions are a necessary preliminary to the carbon-fixing reactions?* _____

d. *If you placed a plant that contained starch in the dark and tested a week later for starch, what would your results show?* _____

EXTENDING YOUR INVESTIGATION: DO ALL PLANTS STORE STARCH?

Plants store glucose molecules in different ways. Glucose molecules may be synthesized into the large starch molecules, amylose and amylopectin. Some plants use glucose to synthesize other storage compounds such as sucrose (glucose + fructose), and sometimes glucose is simply stored as glucose.

Pieces of potatoes and onions will be available to you. How do you think glucose is stored in each one? Formulate a hypothesis.

HYPOTHESIS:

NULL HYPOTHESIS:

What do you **predict** you would find if you tested onions and potatoes for storage of the products of photosynthesis?

In this investigation you will use I_2KI and Benedict's reagent to test for the presence of starch and glucose in potatoes and onions.

What is the **independent variable** in this investigation?

What is the **dependent variable** in this investigation?

Use the procedure outlined below to conduct your investigation; or, using the materials available, design your own experiment.

PROCEDURE

1. Obtain a piece of potato and a piece of onion and place them in a plastic Petri dish.

2. Add several drops of Lugol's solution (I_2KI) to each.

   a. What do you observe? _____

   Recall that I_2KI turns blue-black in the presence of starch.

   b. Which material contains starch? _____

3. Obtain a new piece of potato and a new piece of onion. Use a pair of forceps or razor blade to crush the material. Add the onion pieces to one test tube and the potato pieces to another test tube. Label the tubes.

4. Add 2 ml of water to each tube. Add 2 ml of water to a third, empty tube.

5. Add 2 ml of Benedict's reagent to each of the three tubes.

6. Place the tubes in a beaker of water containing boiling chips and boil the material in the tubes for 3 to 5 minutes.

   c. What is the purpose of the tube that does not contain plant material?

   _____

7. Record your observations in the table below.

RESULTS:

Material	Color at Start	Color After 3–5 min	Presence of Precipitate
Onion			
Potato			
No material			

What do you think happened?

Do your results support your hypothesis?

Your null hypothesis?

Benedict's reagent is a test for certain sugars (reducing sugars) such as glucose. In the presence of glucose, a red precipitate forms. In the presence of sucrose (not a reducing sugar), no precipitate forms. From your results, what do you **conclude?**

Do potatoes store starch? _____ sugar? _____

Do onions store starch? _____ sugar? _____

What types of sugars are stored in the plant material you have studied?

Plant Material	Types of Storage Sugar

Laboratory Review Questions and Problems

1. In order for photosynthesis to occur in green plants, the following must be present:

a. _____ as the energy source.

b. _____ as the carbon source.

c. _____ for the absorption of light energy.

d. _____ as the electron donor.

2. According to your spectrophotometric data in Exercise B, which colors of light are used in photosynthesis?

3. a. Explain the difference between an action spectrum and an absorption spectrum.

b. For chlorophylls, carotenes, and xanthophylls, what is the relationship of the action spectrum to the absorption spectrum?

c. Why do plants contain so many pigments?

4. When using the I_2KI test in Exercise C, why did you extract the pigments from the leaf before adding I_2KI?

5. Many plants contain water-soluble red pigments called anthocyanins. Why were these not visible in the chromatogram of the chlorophyll extract?

6. The processes of photosynthesis and respiration were studied in separate labs, but, as you know, any cell that is carrying on photosynthesis is also carrying on respiration. On average, if a plant is to grow, the rate of photosynthesis must exceed the rate of respiration by a factor of at least three.

   The following graph shows the effect of temperature on the rates of photosynthesis and respiration of one plant. The temperature at which the two rates are equal is referred to as the *compensation point* and is not the same for all plants.

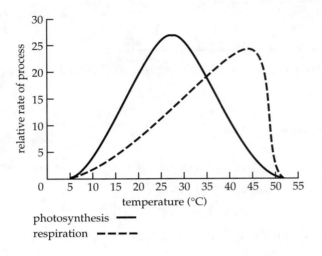

photosynthesis ▬▬

respiration ▬ ▬ ▬

a. At what temperature is the compensation point reached in this example? _____

b. At what temperature(s) would you expect growth to be most rapid? Explain.

c. As temperature rises, what happens to the rate of photosynthesis? Of respiration? Why?

d. Certain fruits, such as apples, are frequently stored under refrigeration in a carbon dioxide–rich atmosphere. Explain the reason for this.

7. Explain how each of the following could limit the rate of photosynthesis.

   CO_2 concentration

   Light quantity

   Temperature

   Water

Meiosis: Independent Assortment and Segregation

OVERVIEW

Sexual reproduction allows the genes of two individuals to combine and provides the variability upon which evolution can work.

In animals that reproduce sexually, the production of sex cells, or gametes, requires that each parent's chromosomes be reduced to half the normal number. This halving of the

Figure 13-1 *Meiosis in (a) animals and (b) plants.*

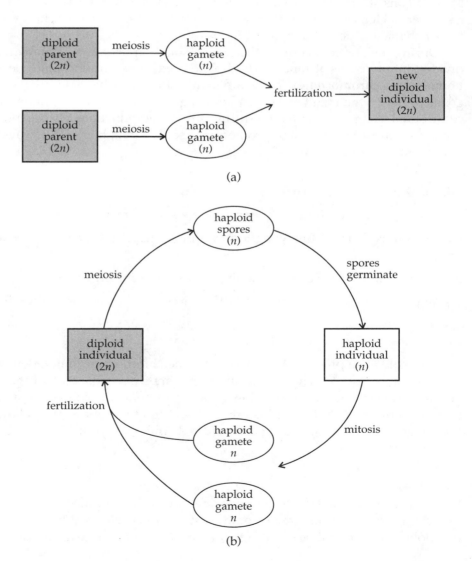

(a)

(b)

parent's chromosome number from the **diploid**, or **2n**, number to the **haploid**, or **n**, number is the result of **meiosis**. Combining two haploid (*n*) gametes during fertilization then restores the chromosome number to the number that is characteristic of the diploid (2*n*) organism (Figure 13-1).

In plants, meiosis results in the production of haploid (*n*) **spores**, which germinate to form haploid individuals. These individuals then give rise to haploid gametes by mitosis. Subsequent fusion of two haploid gametes during fertilization once again results in a diploid (2*n*) individual.

During this laboratory period, you will examine the process of meiosis and discover the genetic consequences of sexual reproduction.

STUDENT PREPARATION

Prepare for this laboratory by reading the text pages indicated by your instructor. Familiarizing yourself in advance with the information and procedures covered in this laboratory will give you a better understanding of the material and improve your efficiency.

✔ | **EXERCISE A** | **Simulation of Chromosomal Events During Meiosis**

Meiosis consists of two nuclear divisions (meiosis I and meiosis II) and results in the production of four daughter nuclei, each of which contains only half the number of chromosomes (and half the amount of DNA) characteristic of the parent (Figure 13A-1).

During meiotic reduction of the chromosome number to half, however, chromosomes are not just divided into two sets at random. In diploid organisms, chromosomes occur in matched pairs called **homologous chromosomes.** These are identical in size, shape, location of their centromeres, and types of genes present. One member of each homologous pair is contributed by the male parent and one is contributed by the female parent during sexual reproduction. Meiosis provides a precise mechanism for separating these homologous chromosomes so that daughter cells always carry one member, or homologue, of each chromosome pair.

▐▐▐▐▐ **Objectives** ▐▐▐▐▐▐▐▐▐▐▐▐▐▐▐▐▐▐▐▐▐▐▐▐▐▐▐▐▐▐▐▐▐▐▐▐▐▐

☐ List and explain the principal events of the stages of meiosis.

☐ Define and explain the following terms: diploid, haploid, homologous chromosomes, alleles, synapsis, and tetrad.

☐ Explain the difference between the first and second meiotic divisions.

☐ List and explain the similarities and differences between meiosis and mitosis.

▐▐▐▐▐ **Procedure** ▐▐

You can study the process of meiosis using the chromosome simulation kit used earlier to demonstrate mitosis (Laboratory 9). The yellow and red strands of beads, both of the same length, represent two homologous chromosomes. A yellow strand represents the contribution of one parent and a red strand represents the contribution of the other parent. In other words, the yellow and red chromosomes are a homologous pair. The second yellow and red strands in your kit are to be used as chromatids for each of these chromosomes.

Interphase

Place one strand of red beads and one strand of yellow beads near the center of your work area. (Recall that chromosomes at this stage would exist as diffuse chromatin and not as visible structures.) Position the two hollow, cylindrical beads at right angles to each other near the chromosomes. These hollow beads

Figure 13A-1   *An overview of meiosis.*

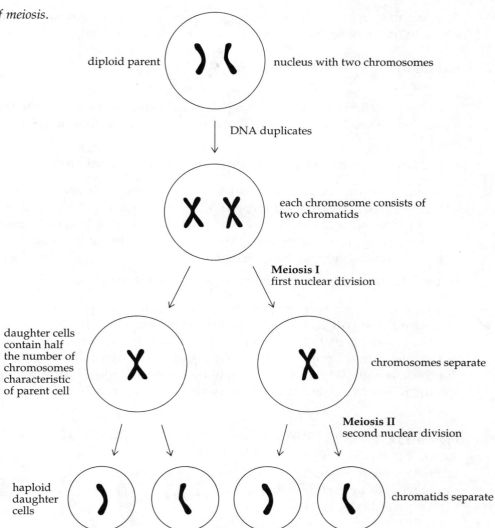

diploid parent   nucleus with two chromosomes

DNA duplicates

each chromosome consists of
two chromatids

Meiosis I
first nuclear division

daughter cells
contain half
the number of
chromosomes
characteristic
of parent cell

chromosomes separate

Meiosis II
second nuclear division

haploid
daughter
cells

chromatids separate

represent a pair of centrioles. DNA synthesis occurs during interphase prior to meiosis, and each chromosome, originally composed of one strand, is now made up of two strands, or **sister chromatids,** joined together at the **centromere** region. (Within the centromere region where two chromatids are most closely associated, a platelike protein structure or **kinetochore** is attached to each chromatid. Spindle fiber microtubules attach to the kinetochores during division.) A chromosome composed of two chromatids is called a **dyad** (or **bivalent**). Simulate DNA replication by bringing the magnetic centromere region of the second red strand into contact with the centromere region of the first red strand. Do the same with its homologue, the yellow strand (Figure 13A-2).

Figure 13A-2   *A homologous pair
of chromosomes, each containing two
chromatids, following DNA synthesis
during interphase.*

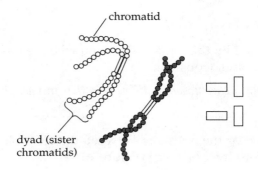

chromatid

dyad (sister
chromatids)

Centriole replication also takes place prior to division. Use two additional cylindrical beads to simulate centriole replication. Place these next to the two original centriolar bodies.

Meiosis I

Prophase I

Homologous chromosomes come together and **synapse** (closely apply themselves to each other), pairing along their entire length. Here, you should recognize the first major difference between mitosis and meiosis.

a. Did homologous chromosomes synapse during prophase of mitosis? _____

A **tetrad,** consisting of four chromatids or two dyads, is formed. Entwine the two chromosomes as shown in Figure 13A-3. During this time, the pair of homologous chromosomes seems to shorten and thicken. Within the tetrad, a ladderlike protein structure, the **synaptonemal complex,** helps to align the tightly paired homologous chromosomes. At this site, segments of two non-sister chromatids (each belonging to a different homologue) may be exchanged by breaking and rejoining. This process, called **crossing-over,**

further increases genetic variability (Figure 13A-3b). *Why?* _____

You will not include crossing-over in this simulation, but you should be aware that it can happen during prophase I. You will examine this process in greater detail in Exercise D.

The centrioles that replicated prior to division begin to move to opposite sides (poles) of the nucleus as the nuclear membrane breaks down. Separate the two pairs of centrioles and move them to each side of the chromosomes (Figure 13A-3a). Spindle fibers also appear during prophase. You will not simulate spindle fibers. Imaginary spindle fibers are shown as dotted lines in all diagrams.

Figure 13A-3 *(a) Tetrad formation during prophase I. Centrioles begin to migrate to opposite sides of the nucleus as the nuclear membrane begins to break down. (b) Crossing-over occurs, and homologous chromosomes begin to untangle forming chiasmata. Note new gene combinations on the inner, non-sister chromatids.*

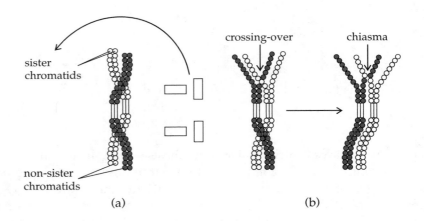

By the end of prophase I, each tetrad can be clearly seen to contain four separate chromatids. Sister chromatids are linked at their centromere, while non-sister chromatids that have crossed over appear to be held together at **X**-shaped locations called **chiasmata** (singular, **chiasma**) (Figure 13A-3b).

Metaphase I

Chromosomes have untwined by this time and can now be seen as dyad chromosomes. They now line up in the center of the cell in homologous pairs.

b. How does this arrangement of chromosomes differ from that in metaphase of mitosis?

_____

Position the chromosomes near the midpoint between the centrioles and at right angles to the imaginary spindle fibers extending from the centrioles (Figure 13A-4).

Figure 13A-4 *Metaphase I—homologous chromosomes line up in pairs.*

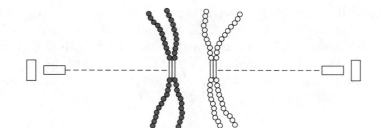

Anaphase I

During anaphase, the homologous chromosomes separate (Figure 13A-5) and are pulled to opposite sides of the cell by **kinetochore microtubules** (see Laboratory 9, Exercise B). This represents a second significant difference between the events of mitosis and meiosis.

Figure 13A-5 *Anaphase I—homologous chromosomes separate and move to opposite poles of the cell.*

c. What is this difference? _____

_____

d. What happens to the chromatids of each chromosome during anaphase of meiosis?

_____

Telophase I

Place each chromosome near its centriole pair (Figure 13A-6). Centriole duplication takes place at the end of telophase in preparation for the next division. Place a second pair of centrioles near the first and at right angles to it.

Formation of a nuclear membrane and division of the cytoplasm, **cytokinesis,** often occur at this time to produce two cells, but this is not always the case. Notice that each chromosome within the two daughter cells still consists of two chromatids.

Figure 13A-6 *Telophase I—homologues are found at opposite ends of the cell. These may be separated into two cells by cytokinesis. In most organisms, centrioles duplicate at this stage.*

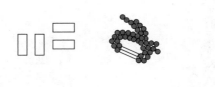 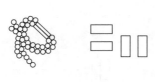

e. Compare the amount and arrangement of genetic material in each cell following telophase I of meiosis and telophase of mitosis. _____

f. How many of each type of chromosome do you see per cell? _____ How many chromatids does each chromosome have? _____

A second division is necessary to separate the chromatids of the chromosomes in the two daughter cells formed by this first division. This will reduce the amount of DNA (number of chromatids) to one double-helical strand per chromosome, typical for a nondividing cell. This second division is called **meiosis II**. It resembles mitosis except that (1) it is part of the continuing process of meiosis and thus is called meiosis II, and (2) only one homologue from each homologous pair of chromosomes is present in each daughter cell undergoing meiosis II.

Meiosis II

The following simulation procedures apply to *both* chromosome groups (daughter cells) produced by meiosis I.

Interphase II (Interkinesis)

The amount of time spent "at rest" following telophase I depends on the type of organism, the formation (or not) of new nuclear membranes, and the degree of chromosomal unwinding. Because interphase II does not necessarily resemble interphase I, it is often given a different name—**interkinesis.** DNA replication does not occur during interkinesis. This represents a third major difference between mitosis and meiosis.

Prophase II

Separate the pairs of duplicated centrioles and tape them down on opposite sides of each chromosome group (Figure 13A-7).

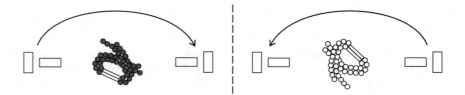

Figure 13A-7 *Prophase II—duplicated centrioles move to opposite poles in the two daughter cells. Chromosomes shorten and thicken.*

g. *Does this action duplicate what you did during prophase I of meiosis?* _____

h. *What is different about prophase I and prophase II of meiosis?* _____

_____

Metaphase II

Orient the chromosome so that it is centered between the centrioles during metaphase II (Figure 13A-8).

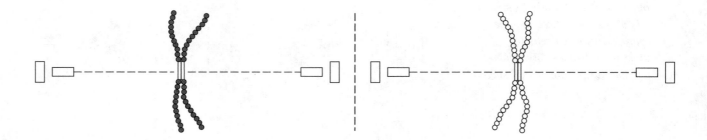

Figure 13A-8 *Metaphase II—chromosomes line up in single file. In this simulation, only one chromosome composed of two chromatids is present in each daughter cell, but in a cell containing many chromosomes, all of the chromosomes would now be lined up in the center of the cell, in single file.*

i. How does metaphase II differ from metaphase I? _____

_____

j. How does metaphase II of meiosis remind you of metaphase of mitosis?

_____

Anaphase II

Sister chromatids now appear to be more loosely associated at the centromere region. When completely separated, each chromatid will have its own centromere region and can be referred to as a chromosome. Separate the sister chromatids of the chromosome and pull the new daughter chromosomes toward the centrioles on opposite sides of each daughter cell (Figure 13A-9).

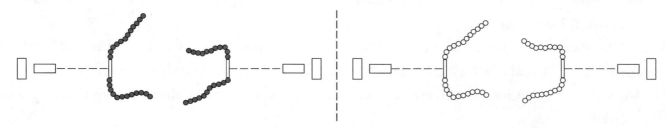

Figure 13A-9 *Anaphase II—the chromatids of each chromosome are separated and move to opposite poles of the cell.*

Telophase II

Pile each chromosome near its centriole. Formation of a nuclear membrane and division of the cytoplasm, cytokinesis, occur at this time (Figure 13A-10).

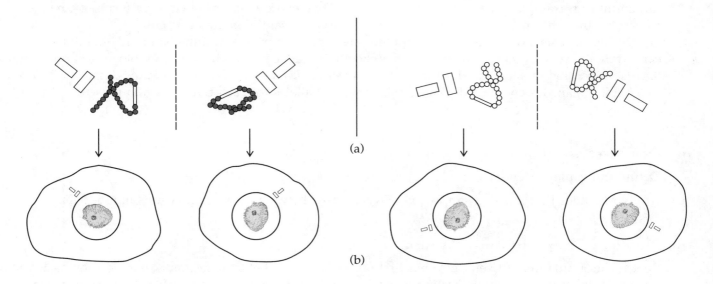

(a)

(b)

Figure 13A-10 *Telophase II—four haploid daughter cells are formed following cytokinesis. New nuclear membranes are formed within each daughter cell; one pair of centrioles is present outside the nuclear membrane. (a) Simulation. (b) Diagram of cells.*

k. *How many cells have you formed during the process of meiosis?* _____

l. *How many cells were formed during the process of mitosis?* _____

m. *Are the cells formed in meiosis haploid (n) or diploid (2n)?* _____

n. *If the same set of chromosomes with which you began this exercise were to undergo mitosis, would the resulting cells be haploid or diploid?* _____

o. *List three major differences between meiosis and mitosis:* _____

_____

Place a single short red strand of beads and a single long yellow strand of beads on your laboratory table. *Could a cell like this exist?* _____ *Would it be haploid or diploid?* _____ *Could this cell carry out mitosis?* _____ *Meiosis?* _____ Demonstrate these processes if they can occur. Use the strands of beads available in your kit.

Place a long red strand and a long yellow strand of beads on your laboratory table. *Could a cell like this exist?* _____ *Would it be haploid or diploid?* _____ *Could this cell carry out mitosis?* _____ *Meiosis?* _____ Demonstrate these processes if they can occur. Use the strands of beads available in your kit.

EXERCISE B | Mendel's First Law: Alleles Segregate During Meiosis

Since there is a pair (the homologues) of each type of chromosome in a diploid organism, there will also be a pair of each type of gene: one gene on one chromosome and the second on its homologue. Genes for a particular trait are found at the same **locus** (physical place or location) on each of the homologous chromosomes. Each of these two genes is called an **allele.**

In some cases, the alleles on the two homologues are identical and the organism's **genotype** (gene content) is said to be **homozygous.** In other cases, the two alleles control alternative expressions of the same trait (for example, green and yellow are alternative expressions, or alternative forms, of the seed-color gene in corn). In this case the organism's genotype is said to be **heterozygous.**

During meiosis, homologous chromosomes are separated from each other, and only one may be carried in a particular gamete or spore. Thus the alleles carried on each of the homologous chromosomes are also separated or **segregated.** *Mendel's first law states that alleles segregate in meiosis* (Figure 13B-1). When two haploid gametes combine during fertilization, two alleles for each trait are again present in the offspring.

Objectives

☐ Define and explain Mendel's first law.

☐ Apply Mendel's first law to a simple monohybrid cross between two heterozygous individuals.

Procedure

Repeat the simulation of Exercise A, but first take a piece of label tape and mark one bead on each yellow strand (chromatid) as *A* (same location on each strand). Mark one bead on each red strand as *a.* Make sure that *A* and *a* appear at the same locus on the two homologues (Figure 13B-2).

a. *Why do the two chromatids of a chromosome have the same alleles (A and a) on one dyad?*

_____

Figure 13B-1 *Alleles segregate during meiosis.*

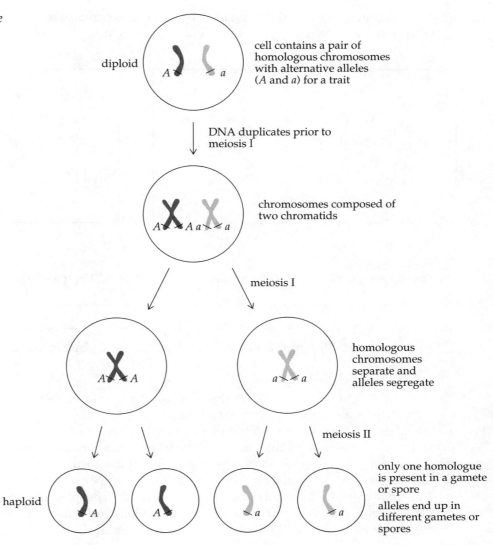

diploid — cell contains a pair of homologous chromosomes with alternative alleles (*A* and *a*) for a trait

A   *a*

DNA duplicates prior to meiosis I

A *A a* *a* — chromosomes composed of two chromatids

meiosis I

A *A* | *a* *a* — homologous chromosomes separate and alleles segregate

meiosis II

haploid — only one homologue is present in a gamete or spore

alleles end up in different gametes or spores

A   *A*   *a*   *a*

Figure 13B-2 *Alleles on homologous chromosomes.*

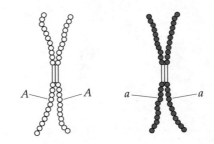

A — *A*   *a* — *a*

Diagram the products of this meiosis below. Use colored pencils to show individual chromosomes. Indicate *A* and *a* as they appear during your simulation.

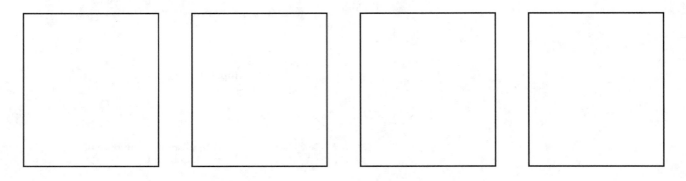

b. What do you conclude about the final distribution of the A and a alleles in the daughter cells?

_____

_____

c. Do your observations support Mendel's first law? _____ *Explain.* _____

_____

You have just demonstrated that during meiosis the two factors (alleles) for any trait segregate so that each ends up in a different gamete or spore.

Mendel's first law can be verified by examining kernel color on an ear of **hybrid** corn (formed from the union of gametes from two parents that differ in one or more inheritable characteristics). Each kernel represents a seed developed from a fertilized egg of the corn plant. Thus kernel color is determined by the combination of alleles contributed by the gametes: the sperm and the egg (Figure 13B-3).

In corn an allele (*R*) produces purple-colored seeds. The allele *R* is **dominant** to the allele *r* for "no purple" or yellow seeds. If two *R* alleles or two *r* alleles are present in the seed, it is said to be **homozygous.** The seed will be purple (*RR*) or yellow (*rr*), respectively. If the two alternative alleles, *R* and *r*, are both

Figure 13B-3 *Corn life cycle.*

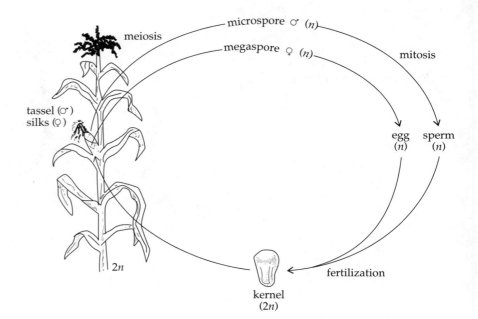

present in the seed, then the **heterozygous** kernel will be purple due to the dominance of the R gene. The **genotype** (genetic constitution of alleles) is Rr, but the **phenotype** (appearance) of the kernel is purple.

A cross is made between a homozygous dominant plant (RR) and a homozygous recessive plant (rr). The cross $RR \times rr$ represents the fertilization event in the parental (P_1) generation.

d. *What color seeds would the RR parent have?* _____

e. *What is the genotype of gametes produced by the RR parent?* _____

f. *What color seeds would the rr parent have?* _____

g. *What is the genotype of gametes produced by the rr parent?* _____

h. *Consider the genotypes of gametes that can be produced by the homozygous parents, RR and rr. All offspring*

   resulting from the union of one gamete from each parent would have the genotype _____

i. *What is the phenotype of this F_1 generation?* _____

When F_1 individuals make gametes, their alleles for seed color will segregate.

j. *What are the genotypes of gametes produced by F_1 individuals?* _____

The consequences of this segregation of alleles will become apparent in the next (F_2) generation (offspring produced by F_1 individuals) when one examines all the possibilities for the genotypes that would be present in the F_2 individuals.

The possible combinations of alleles that may be produced in each parent's gametes, and the results of these combinations in the genotypes of the offspring, can be determined by using a **Punnett square.** All of the possible genotypes of gametes that can be produced by one parent are listed across the top of the square; all genotypes of gametes that can be produced by the other parent are listed along the side. In the Punnett square below, one type of gamete from each F_1 parent has already been listed, and one possible combination is shown. Fill in the blanks for the other gamete genotype for each parent, and then complete the other three combinations in the square to determine the possible genotypes of the offspring. (*Note:* By convention, the dominant allele for each trait is written first: for example, Rr, not rR.) Next to the Punnett square, list the genotypes and phenotypes of the four types of individuals produced in the F_2 generation.

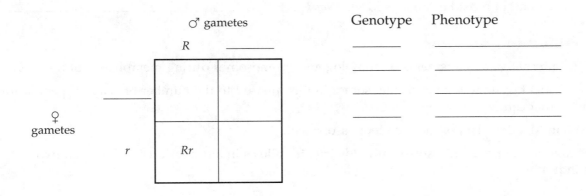

k. *How many different kinds of genotypes are present in the F_2 generation?* _____

l. *Indicate below the proportion (ratio) of individuals with these genotypes:*

genotype (homozygous dominant)	number	genotype (heterozygous)	number	genotype (homozygous) recessive)	number
_____	:	_____	_____	:	_____	_____

m. *How many different kinds of phenotypes are present in the F_2 generation?* _____

n. *What is the proportion (ratio) of individuals showing these phenotypes?*

       number   phenotype   number   phenotype

       _____   purple :   _____   yellow

Your instructor will provide a plastic box containing a series of corn ears that demonstrate crosses. Obtain one of the corn ears labeled "Monohybrid Cross F_2." Count and record the total number of purple kernels from at least four rows. Repeat with the yellow kernels.

       Number of purple kernels _____ Number of yellow kernels _____

       Proportion of purple kernels to yellow kernels _____ : _____ .

The phenotypic ratio of purple to yellow kernels in the F_2 generation should be 3:1.

o. *How do your results compare with this expected phenotypic ratio?* _____

p. *Can you determine the genotypic ratio by examining the ear of corn?* _____ *Why or why not?*

EXERCISE C   Mendel's Second Law: Alleles of Unlinked Genes Assort Independently

Now let us consider meiosis involving two sets of homologous chromosomes. Alleles for trait A (*A* or *a*) are found on one pair of homologues. Alleles for an entirely different trait B (*B* or *b*) are found on the other pair of chromosomes. Assume that two parents are each heterozygous for both genes. Each parent would have the genotype *AaBb*. It is possible for these parents to produce gametes *AB* and *ab* or *aB* and *Ab*, depending on how the pairs of homologous chromosomes are arranged at metaphase I of meiosis. The alleles for the two genes sort themselves out independently. The behavior of *A* is not linked to that of *B* (**unlinked genes**) because the genes are on separate chromosomes. So, for example, the combination *AB* is as likely as the combination *ab*. *Mendel's second law states that alleles of unlinked genes assort independently* (Figure 13C-1).

Since many gametes are produced at one time, a parent can produce gametes of all four genotypes: *Ab, ab, aB,* and *Ab*. When considering the possible genotypes for offspring, all gamete genotype

possibilities for each parent must be considered. *a. Why?* _____

⦚⦚⦚⦚⦚ Objectives ⦚⦚⦚⦚⦚⦚⦚⦚⦚⦚⦚⦚⦚⦚⦚⦚⦚⦚⦚⦚⦚⦚⦚⦚⦚⦚⦚⦚⦚⦚⦚⦚

☐ Demonstrate the alternative arrangements of homologous chromosomes during metaphase I of meiosis.

☐ Relate the arrangement of homologous chromosomes in metaphase I to the number of types of genetically different gametes that can be produced.

☐ Define and explain Mendel's law of independent assortment.

☐ Verify that the law of independent assortment holds true for alleles in a dihybrid cross between two heterozygous individuals.

⦚⦚⦚⦚⦚ Procedure ⦚⦚⦚⦚⦚⦚⦚⦚⦚⦚⦚⦚⦚⦚⦚⦚⦚⦚⦚⦚⦚⦚⦚⦚⦚⦚⦚⦚⦚⦚⦚⦚⦚

Combine your simulation kit with that of the person sitting next to you. Remove five beads from one side of the centromere on all four strands of one kit to form chromosomes of a different length from those in the other kit.

Using two pairs of homologous chromosomes, repeat all the steps of meiosis. Remember that your homologues are alike in length, not color. You have one homologous pair (one red dyad and one yellow dyad) in which the centromere is in the middle and the four chromosome arms are of equal length, and a second homologous pair (one red dyad and one yellow dyad) with short arms on one side of the centromere region and long arms on the other. Mark the chromosomes with tape to indicate the alleles they are carrying, as shown in Figure 13C-2.

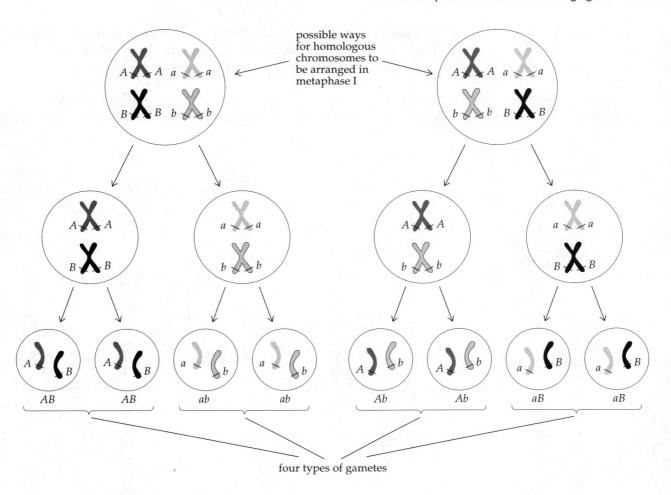

possible ways for homologous chromosomes to be arranged in metaphase I

four types of gametes

Figure 13C-1 *Possibilities for allele combinations in haploid gametes or spores are dependent upon independent assortment of alleles present on different chromosomes.*

Figure 13C-2 *Mark alleles at the same locus on each chromatid and on homologous chromosomes.*

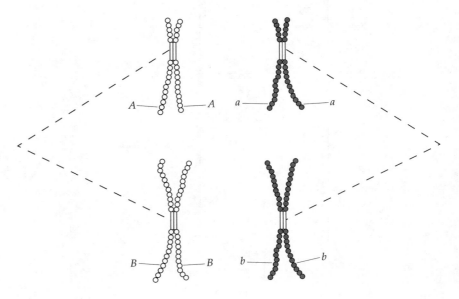

If you begin with two pairs of homologous chromosomes, there are two possibilities for their alignment during metaphase I of meiosis, depending on which side you place the yellow and red homologues. Draw both of these possibilities in Figure 13C-3. Indicate the alleles present on each chromosome.

Follow possibility 1 through meiosis, using the red and yellow beads. As you proceed through your simulation for possibility 1, figure out how the chromosomes and alleles would be distributed in

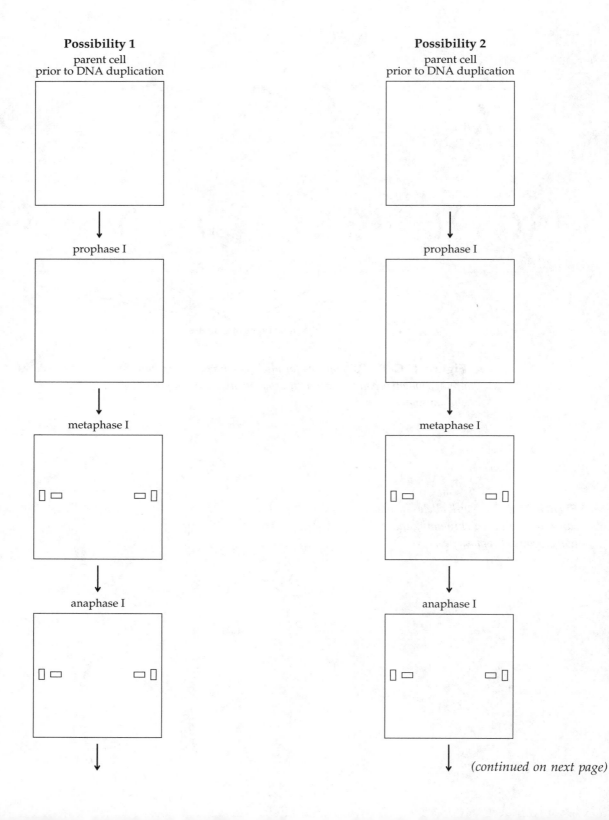

(continued on next page)

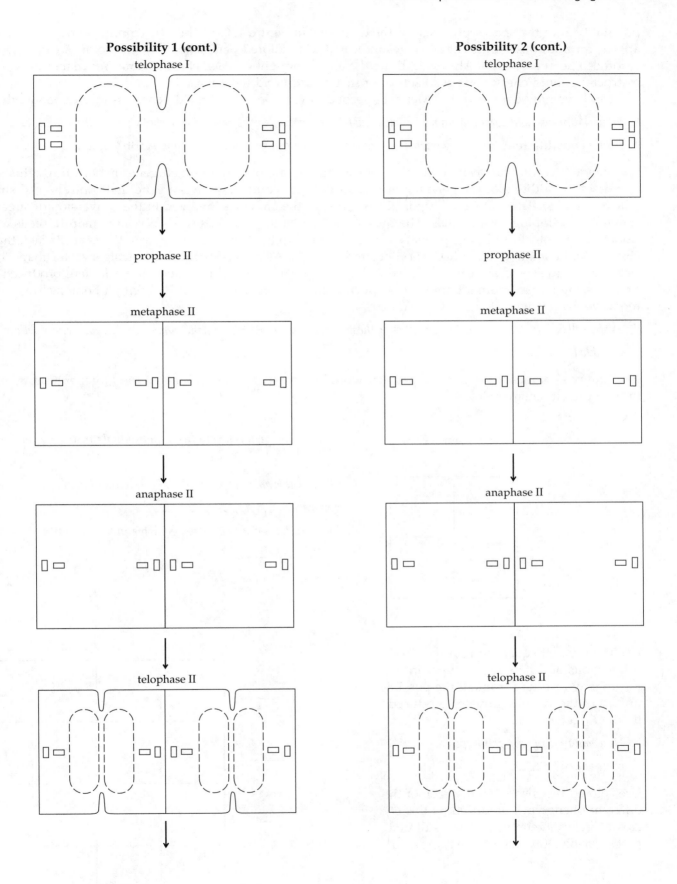

Figure 13C-3   *Meiosis with two pairs of homologous chromosomes.*

possibility 2. In the spaces provided in the diagrams in Figure 13C-3, draw the chromosomes as they appear for both possibilities at all stages indicated. Use colored pencils to show the color of each chromosome at each step. Always indicate the alleles present on each chromosome. Your laboratory instructor should check your bead setup for metaphase I and metaphase II.

    You have now demonstrated Mendel's second law: alleles of unlinked genes assort independently.

b. How many possible combinations of alleles exist if you consider the results from both possibilities? _____

c. How many different types of gametes can be made by an individual of genotype AaBb? _____

    Mendel's second law can be verified by tracing the fate of two unlinked genes in corn through a series of crosses. In addition to the gene for seed color (alleles R and r), there is a gene that controls seed shape (alleles S and s). The S allele is responsible for one of the enzymes involved in the conversion of sugar to starch in developing corn kernels. The presence of starch in SS and Ss seeds gives the mature seeds a full, smooth appearance, and their phenotype is considered to be "smooth." However, the ss seeds lack this enzyme and contain a much higher proportion of sugar. As a consequence, they taste sweeter than "smooth" kernels and appear "wrinkled." The ears of corn with which you will work were produced by first crossing a plant from a homozygous purple, smooth-seeded plant ($RRSS$) with a homozygous recessive yellow, wrinkled-seeded plant ($rrss$).

d. Which alleles are present in the gametes of these parent types: RRSS (purple, smooth) _____ *rrss (yellow,*

    wrinkled)? _____

Find the possibilities for the genotypes that would be present in individuals of the F_1 generation by filling in the Punnett square below.

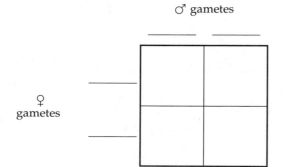

e. What is the genotype of all the individuals in the F_1 generation?

_____

f. What is the phenotype of all the individuals in the F_1 generation? _____

g. Which alleles are present in gametes produced by the F_1 plants?

_____

Use the Punnett square to find the proportions of different genotypes in the F_2 progeny resulting from all the possible unions of the various gametes produced by the F_1 generation.

h. How many different genotypes are represented? _____

Use colored pencils to identify each of the different phenotypes produced by this cross. Circle all genotypes that result in a particular phenotype with the same color.

♂ gametes

♀ gametes

i. Indicate the proportions (ratios) of individuals showing the following phenotypes: purple, smooth _____:

purple, wrinkled _____: *yellow, smooth* _____: *yellow, wrinkled* _____

On demonstration you will find ears of the P, F_1, and F_2 generations of this cross. In the F_2 of a dihybrid cross involving dominant unlinked genes, one expects to find a phenotypic ratio of 9:3:3:1. Select an ear of corn labeled "F_2 of dihybrid cross." There are four types of kernels present on each ear.

Count at least five rows of kernels and record the number of each type found under "Observed" below. To compare the observed data with the numbers expected of each phenotype, divide the total number of all seeds counted by 16 (round to the nearest whole number) and multiply this value by 9, 3, 3, and 1, respectively. Record this information under "Expected."

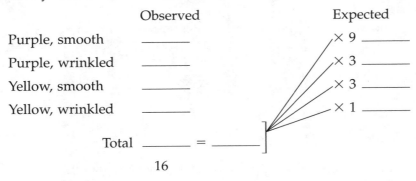

         Observed   Expected

Purple, smooth _____   × 9 _____

Purple, wrinkled _____   × 3 _____

Yellow, smooth _____   × 3 _____

Yellow, wrinkled _____   × 1 _____

     Total _____ = _____

         16

j. Do your results support what you predicted from your Punnett square? _____

k. If the genes were linked, rather than unlinked, would the alleles assort independently? _____

l. Would you expect to see the same genotypic-to-phenotypic ratios in genetic crosses if the genes were linked? _____

Why or why not? _____

_____

EXTENDING YOUR INVESTIGATION: MEIOSIS AND LINKED GENES

Suppose that two genes, A and B, are located on the same chromosome. These are **linked** genes and tend to stay together during meiosis. Would they obey Mendel's second law? If two individuals, $AABB$ and $aabb$, are crossed, all F_1 individuals are $AaBb$. Formulate a hypothesis to predict how the genotypes of offspring in the F_2 generation would be determined.

HYPOTHESIS:

NULL HYPOTHESIS:

What do you **predict** will happen when you cross two $AaBb$ individuals?

Identify the **independent variable** in this experiment.

Identify the **dependent variable** in this experiment.

Design a procedure to test your hypothesis. (Use the bead kits from Exercise A.)

PROCEDURE:

RESULTS: Diagram your results.

From your results, describe how the genotypes in the F_2 generation are determined.

Do your results support your hypothesis?

Your null hypothesis?

Was your prediction correct?

What do you **conclude** about the effects of linked genes on the genotypes of offspring in the F_2 generation?

EXERCISE D   Meiosis and Crossing-Over in *Sordaria (Optional)*

The fungus *Sordaria fimicola* is often used to study the processes of gene segregation and crossing-over during meiosis. This common fungus spends most of its life cycle in the haploid condition (Figure 13D-1). Its body is composed of haploid (n) cells attached end to end to form hairlike filamentous hyphae that intertwine to form a mass (*mycelium*). When hyphal cells of two different mycelia come together, they fuse and the haploid nuclei from one hypha migrate into the other. The haploid nuclei combine within the dikaryotic cells to form a diploid ($2n$) zygote nucleus. The diploid nucleus in the hyphal tip immediately undergoes meiosis to form four nuclei, returning the organism to its haploid state. These haploid nuclei then divide mitotically to yield a total of eight haploid cells. The cells develop thick, resistant cell walls and are called **ascospores.** The ascospores are arranged in a linear array within a sac called an **ascus** (plural, *asci;* see Figure 13D-2). Many such asci grouped together line the inside of a fruiting body (**ascocarp**), formed from tightly fused hyphae. In *Sordaria* the ascocarp, called a **perithecium,** is flask-shaped with a small hole through which mature spores escape when the asci rupture.

Ascospores of *Sordaria* are normally black. However, several different genes can be involved in determining spore color, and each of these genes can have several allelic forms. Black spores can only be produced if both of two genes controlling color are normal or wild-type (g^+ and t^+). A mutation in one such gene can result in gray spores (t^+ and g^-), whereas a mutation in a different gene can result in tan spores (t^- and g^+). If both tan and gray mutant alleles are present (t^- and g^-), the cumulative effect of both mutations is that the ascospores are colorless (Figure 13D-3). Ascospores are haploid, so only one allele is present for each gene. (There is no homologous chromosome carrying an alternative or duplicate allele at the same locus.) As a result, the spore's phenotype (physical characteristic) is equivalent to its genotype—the expression of the allele cannot be "masked" by a different (and perhaps dominant) allele at the same locus on the homologous chromosome. This equivalence is one of the major reasons that fungi such as *Sordaria* (an ascomycete) are used extensively in genetic research. It is also of critical importance in this exercise.

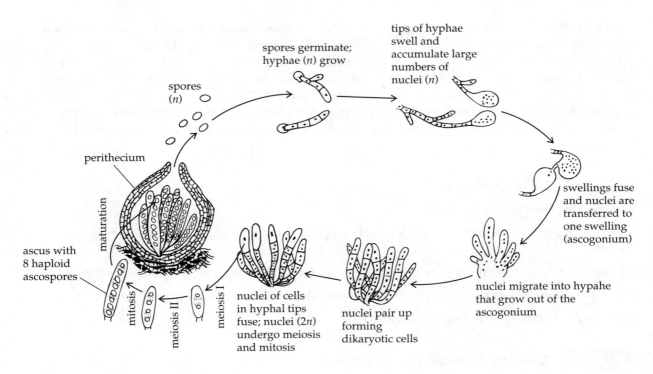

Figure 13D-1 *Sordaria fimicola life cycle.*

Figure 13D-2 *Photomicrograph of* Sordaria fimicola *asci.*

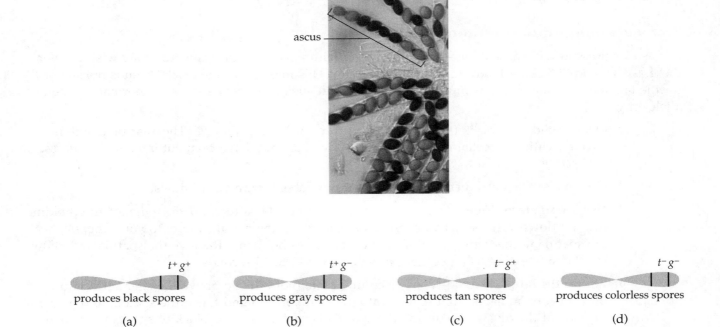

Figure 13D-3 *Two different genes control ascospore color. When both genes are represented by the wild-type allele, the spores are black (a). A mutation in either one of the genes can cause gray color (b) or tan color (c) if the allele at the second locus remains wild type (+). If both genes are mutant, the ascospores are colorless (d).*

The genetics of both the process and the results of crossing-over in *Sordaria* can be studied by observing the phenotype (color) and arrangement of ascospores in asci of hybrids that are produced by mating a wild-type black strain (t^+g^+) and a mutant strain (either t^+g^- or t^-g^+). In this exercise, you will begin your study with the behavior of the gene at the tan locus (remember, this is a mutation, t^-, of the wild-type gene, t^+). Thus, you will cross a black strain (t^+g^+) with a tan strain (t^-g^+). This study will require 7 to 10 days for completion. You will set up crosses between wild-type and tan-type mutant *Sordaria* during this laboratory period and then analyze your results during next week's laboratory period (Laboratory 14).

‖‖‖ Objectives ‖‖‖‖‖‖‖‖‖‖‖‖‖‖‖‖‖‖‖‖‖‖‖‖‖‖‖‖‖‖‖‖‖

☐ Describe the life cycle of *Sordaria fimicola*.

☐ Describe the role of mitosis and meiosis in the formation of asci in an ascomycete fungus such as *Sordaria fimicola*.

☐ Demonstrate how to make a hybrid cross between two different strains of *Sordaria fimicola*.

☐ Explain how meiosis and crossing-over result in different arrangement of ascospores in asci.

☐ Identify nonhybrid and hybrid MI and MII asci.

✔   **PART 1**   Crossing *Sordaria* Strains (Week 1)

If mycelia from two different strains of *Sordaria* are placed on the same agar plate, the mycelia will fuse and hybrid asci will form (however, it is important to understand that a genetic strain may also mate with itself).

Note: Your instructor may have already made crosses for you. If so, the crosses have been allowed to incubate for 7 days and the plate you receive will contain mature perithecia. In this case, proceed to the directions for Part 2. Your instructor may also ask you to proceed with the analysis of crossovers and mapping of the *Sordaria* chromosome. If this is the case, continue with Exercise A in Laboratory 14.

If you are asked to set up your own crosses, proceed as follows.

‖‖‖ Procedure ‖‖‖‖‖‖‖‖‖‖‖‖‖‖‖‖‖‖‖‖‖‖‖‖‖‖‖‖‖‖‖‖‖‖‖‖‖

In this exercise you will cross a wild-type black strain with a mutant tan strain. You may wish to refer to these as "+" and "t," respectively, to simplify notation. (This means that the black strain is normal or t^+ for the tan allele, the tan strain is mutant or t^- for the tan allele, and both strains are normal for the g allele, or g^+.)

1. Obtain wild-type black (+) and tan (t) cultures of *Sordaria fimicola*. The agar plate will be covered with a mycelial mat of hyphae and will already have been cut into small squares. *Do not open the lid yet!*

2. Obtain a sterile Petri dish of crossing agar and label it with your initials.

3. **Use aseptic technique.** Place the end of a small metal spatula into 95% ethanol and, holding the end downward (so that alcohol does not run up the handle and onto your fingers), heat the spatula in the flame of a Bunsen burner or alcohol lamp. Remove the spatula and allow the flame to extinguish. Let the spatula cool for 10 to 15 seconds.

4. Lift up the lid of the wild-type (+) culture plate (hold the lid above the agar surface to prevent contaminants from falling onto the agar surface) and touch the spatula to the agar at the side of the dish to make sure the spatula is cool. Remove a block of agar containing a portion of the mycelium.

5. Transfer the agar block to the Petri dish of crossing agar, turning it upside down so the mycelium contacts the surface of the crossing agar. Place the block at one of the positions marked + in Figure 13D-4. Reflame the spatula and repeat this procedure with a second block of agar from the wild-type (+) black strain and place the block in the second position marked + in Figure 13D-4. Blocks should be approximately 2 cm apart.

Figure 13D-4 *Blocks of agar covered with mycelia from different genetic strains of* Sordaria *are placed on crossing agar. In week 2 of this experiment (Laboratory 14), you will collect perithecia from the areas where the two strains will have fused.*

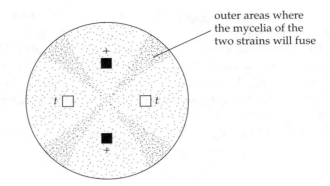

outer areas where the mycelia of the two strains will fuse

6. Reflame the spatula and transfer two blocks from the mutant (*t*) tan strain to the marks indicated in Figure 13D-4.

7. Allow the plates to incubate in the dark at 22 to 24°C for approximately 7 days.

PART 2   Analyzing Hybrids and Crossovers (Week 2)

Your instructor will provide you with a plate of *Sordaria fimicola* that contains a cross of black and tan strains. The plate has been incubated for a week and a mycelial mat covers the surface. You will notice areas covered by black and tan hyphae and a dark **X**-shaped area where hyphae growing out from the original agar blocks have now fused (Figure 13D-4). The small black structures (that look much like poppy seeds) are the perithecia or fruiting bodies of the fungus (see Figure 13D-1).

Recall that when mycelia of the black strain (referred to here as + to simplify notation) fuse with mycelia of the mutant tan strain (referred to simply as *t*), their haploid nuclei fuse to form diploid nuclei at the tips of specialized elongated hyphae that will develop into asci. A diploid nucleus divides by meiosis (I and II) to form four nuclei (tetrad stage). Each of these nuclei then replicates by mitosis to form two new haploid nuclei, for a total of eight haploid nuclei. Each of these serves as the nucleus for a spore called an **ascospore.** There are eight ascospores in each ascus of *Sordaria fimicola* (Figure 13D-5).

A key factor that makes *Sordaria fimicola* so useful for genetic studies is that the eight ascospores within the ascus are ordered; they are always lined up in a pattern that is related to the way in which they

Figure 13D-5 *A diploid* Sordaria *nucleus resulting from the fusion of two haploid nuclei undergoes meiosis I and II to form four haploid nuclei (tetrad stage). Each of these nuclei will, by mitosis, form two haploid nuclei, for a total of eight ascospore nuclei in a single ascus.*

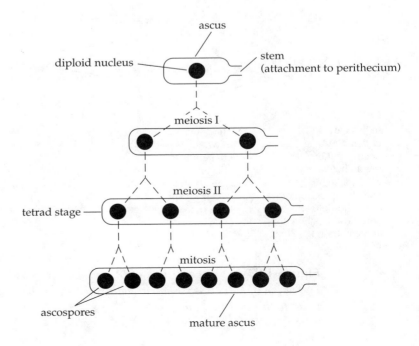

ascus

diploid nucleus

stem (attachment to perithecium)

meiosis I

meiosis II

tetrad stage

mitosis

ascospores

mature ascus

were produced by meiosis. Because of the elongated nature of the ascus and the fact that the spindles of the meiotic and mitotic divisions do not overlap, you can identify the nucleus from which each ascospore was made. (Trace the lineage of the third ascospore from the left in Figure 13D-5.)

If a black strain has hybridized with a tan strain and no crossing-over has occurred during meiosis I, the asci produced will contain four black and four tan ascospores. Segregation of the wild-type black allele and the tan mutant allele to two different nuclei has occurred at meiosis I (see Figures 13D-5 and 13D-6a). Each of these nuclei will give rise to four haploid ascospores as a result of meiosis II and mitosis. To designate the point at which segregation of alleles occurred, these "4+4" asci are called **MI asci.**

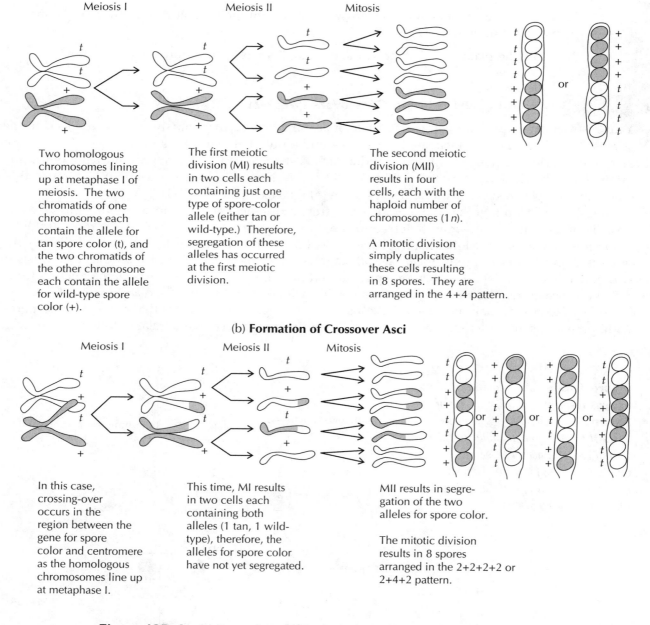

(a) Formation of Non-Crossover Asci

Meiosis I

Two homologous chromosomes lining up at metaphase I of meiosis. The two chromatids of one chromosome each contain the allele for tan spore color (t), and the two chromatids of the other chromosone each contain the allele for wild-type spore color (+).

Meiosis II

The first meiotic division (MI) results in two cells each containing just one type of spore-color allele (either tan or wild-type.) Therefore, segregation of these alleles has occurred at the first meiotic division.

Mitosis

The second meiotic division (MII) results in four cells, each with the haploid number of chromosomes (1n).

A mitotic division simply duplicates these cells resulting in 8 spores. They are arranged in the 4+4 pattern.

(b) Formation of Crossover Asci

Meiosis I

In this case, crossing-over occurs in the region between the gene for spore color and centromere as the homologous chromosomes line up at metaphase I.

Meiosis II

This time, MI results in two cells each containing both alleles (1 tan, 1 wild-type), therefore, the alleles for spore color have not yet segregated.

Mitosis

MII results in segregation of the two alleles for spore color.

The mitotic division results in 8 spores arranged in the 2+2+2+2 or 2+4+2 pattern.

Figure 13D-6 *(a) Formation of MI non-crossover asci in* Sordaria. *(b) Formation of MII crossover asci. The symbol + has been substituted for t⁺ to indicate the wild-type allele.*

a. Which alleles are responsible for producing black spore color in Sordaria? _____

b. Which alleles are responsible for producing tan spore color in Sordaria? _____

c. In the space below, show how the chromosomes must be aligned to produce an ascospore pattern of four black spores at the top of the ascus and four tan spores at the base of the ascus. (See example in Figure 13D-6a.)

If crossing-over occurs at meiosis I, the result is two nuclei in which both alleles t^+ and t^- (or + and t in our notation) are present in each daughter nucleus. These alleles do not segregate until meiosis II, when the chromatids of each chromosome separate. Depending on the arrangement of the chromatids when they separate in meiosis II, four different color patterns of ascospores within the asci are possible (Figure 13D-6b). Asci showing these 2+2+2+2 or 2+4+2 color patterns are designated **MII asci.**

Since in *Sordaria* a genetic strain can mate with itself, asci containing spores of only one color (all black or all tan) are also possible, but these are not referred to as hybrids.

Figure 13D-7 shows meiosis I crossovers between different pairs of non-sister chromatids (these belong to homologous chromosomes present in a diploid fusion nucleus). Complete the information in the figure to show the results of meiosis II and mitosis, then match each ascospore to one of the patterns shown in Figure 13D-6b. (The first possibility has been completed for you.)

||||| Procedure |||||||||||||||||||||||||||||||||||||||

1. Obtain a plate of *Sordaria* that contains a cross of black and tan strains (a cross plate). Identify the locations where mycelia of the two strains have overlapped and fused. Dark lines of tiny perithecia should be visible in these areas. Use a toothpick to gently scrape the surface of the plate to collect perithecia. (It is usually best to collect perithecia toward the outer rim of the dish.) Do *not* scrape up any of the agar onto your toothpick!

2. Place the perithecia in a drop of water on a slide. Cover with a coverslip and gently press on the coverslip (use a small cork or pencil eraser) to rupture the perithecia. Be gentle so that the ascospores remain in the asci (see Figure 13D-2). If you press too hard, you will observe (in step 3) single ascospores dispersed throughout the slide. In this case, discard your slide and try again.

3. View the slide using the 10× objective and locate a group of asci containing at least some hybrid asci (asci that contain both black and tan ascospores). In the space below, draw a diagram of each type of ascus that you observe. Indicate whether the ascus is hybrid and if so, whether it is MI or MII. Try to find an example of each type of ascus shown in Figure 13D-6a, b. You may have to inspect several perithecia and perhaps make several slides.

d. What type of ascus is most common? _____ Why? _____
_____

e. Occasionally you will see asci with a 2+3+1+1+1 pattern or a 2+2+1+2+1 pattern. What might be the explanation for these patterns? _____
_____

f. In your own words, explain the difference between an MI and MII ascus.
_____

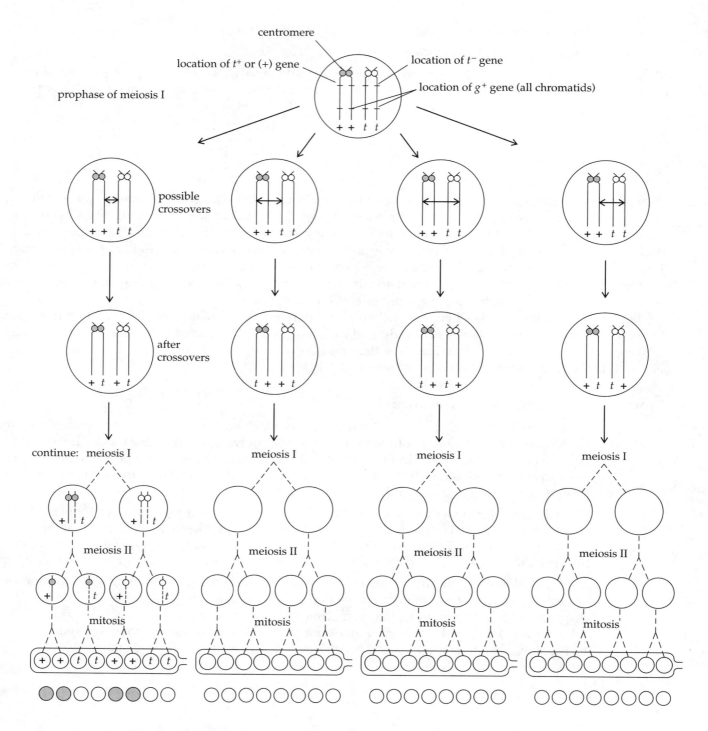

Figure 13D-7   *A diploid fusion nucleus contains a homologous pair of chromosomes carrying alleles for ascospore color. Each chromatid has a gene for tan (normal +, or mutant t) and a normal gray gene (g⁺). (Centromere color indicates future spore color for the combination of alleles present on that homologue.) Crossing-over occurs during prophase I. Non-sister chromatids exchange alleles, with several different patterns of recombination being possible (as shown by arrows ←——→). After crossing-over is complete, meiosis I continues, followed by meiosis II. The results of different patterns of recombination can be seen in the patterns of spore color in resultant asci. (Modified from Richard P. Nickerson,* Genetics, *p. 161, Scott Foresman/Little Brown, 1990.)*

Be sure that you can identify all of the types of MI and MII asci. During the next laboratory period you will calculate the frequency of crossover events by determining the percentage of MI and MII asci on your cross plate. Using the crossover frequencies, it is possible to map the *Sordaria* chromosome (see Laboratory 14, Exercise A).

Laboratory Review Questions and Problems

1. Indicate the stage of meiosis during which each of the following events or situations occurs.

_____ Chromosomes synapse

_____ Chromosomes line up in single file

_____ DNA duplicates

_____ First cell stage to contain only one homologue from each chromosome pair

_____ Homologous chromosomes separate

_____ Chromatids separate

_____ Crossing-over

2. List five ways in which meiosis differs from mitosis.

3. Complete the table below.

	Number of Chromatids per Chromosome	Number of Chrosomes per Cell
G_1 } Interphase S } preceding G_2 } meiosis		
Prophase I		
Metaphase I		
Anaphase I		
Telophase I		
Interphase		
Prophase II		
Metaphase II		
Anaphase II		
Telophase II		

4. During gamete formation, if alleles did not segregate at meiosis I, what might the consequences be for a zygote resulting from the union of such gametes?

5. Because of independent assortment, it is possible for a single human to produce 2^{23} different types of gametes. Explain how this is possible.

6. During gamete formation, genetic variation, in addition to that due to independent assortment, can be increased by crossing-over events during meiosis I. Explain what consequences this could have for possible offspring.

7. How might recombination of genes due to crossing-over be important to the process of evolution?

8. Two pea plants, Rr and RR, are crossed. What are the genotypes and phenotypes of the F_1 offspring? Did segregation of alleles occur? Did independent assortment of alleles occur?

9. Two pea plants, $RrSs$ and $rrSs$, are crossed. What are the genotypes and phenotypes of the offspring? Did segregation of alleles occur? Did independent assortment of alleles occur?

10. What does it mean to say that two genes are "linked"? Can linked genes assort independently? Why or why not?

11. In a certain species of flower, genes for petal color, petal shape, and plant height are linked by their presence on the same chromosome.

Dominant	Recessive
Blue (*B*)	White (*b*)
Frilly petal (*F*)	Smooth petal (*f*)
Tall (*T*)	Short (*t*)

A tall, blue, frilly plant, heterozygous for each trait is crossed with a white, smooth, short plant. Among the offspring are some unexpected blue, smooth, short plants as well as some white, frilly, tall plants. Explain how this could have happened.

12. Two strains of *Sordaria* (black and gray) are crossed, and asci have the following color patterns. (+ represents wild-type, t^+):

Group	Ascopore Pattern
I	++++*gggg*
II	*gggg*+++
III	*gg*++*gg*++
IV	++*gg*++*gg*
V	++*gggg*++
VI	*gg*++++*gg*

a. Which asci result from the absence of crossing-over?

b. Which asci are MI asci?

c. Which asci are MII asci?

d. Why can a single cross produce all of the types of asci shown?

e. Explain why only half of the ascospores in groups III, IV, V, and VI are the result of recombination (*think*, and review Figure 13D-7).

13. From your study in Laboratory 9 and Laboratory 13, you should understand the basic differences between the processes of mitosis and meiosis and the genetic consequences of each. To summarize your understanding, start with a cell having a diploid chromosome number equal to six ($2n = 6$) and draw the processes of mitosis and meiosis. (You may wish to use colored pencils.) Note the n number ($1n$, $2n$, $4n$, etc.), where n represents a unique set of chromosomes (e.g., if diploid, there are two sets of homologous chromosomes, or $2 \times n = 2n$). Also note the c number, which refers to the number of chromatids (or copies of DNA) per chromosome.

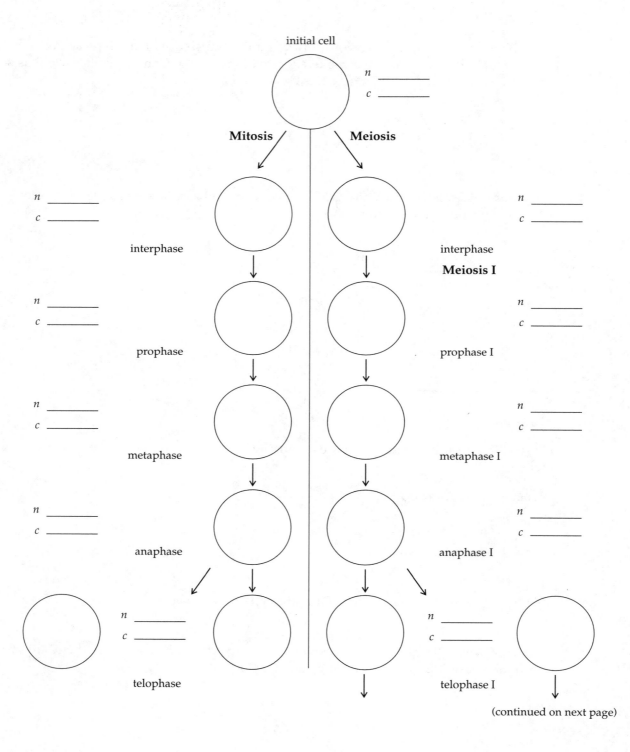

(continued on next page)

(continued from previous page)

Meiosis II

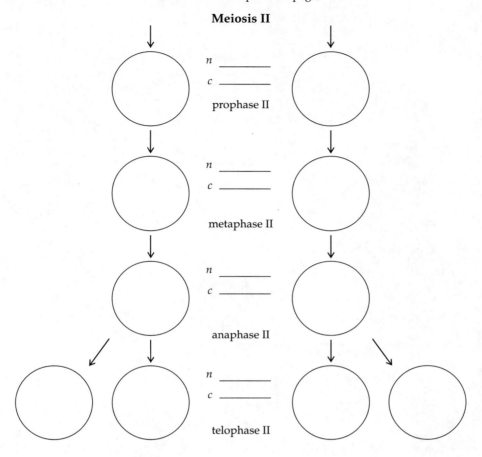

n _____
c _____

prophase II

n _____
c _____

metaphase II

n _____
c _____

anaphase II

n _____
c _____

telophase II

Genes and Chromosomes: Chromosome Mapping

OVERVIEW

Genes are located on chromosomes in particular positions called **loci.** Genes can be assigned to the sex chromosomes (*X* and *Y*) by studying inheritance patterns among family groups, but assigning genes to specific autosomes (non-sex chromosomes) is more difficult. Since segregation of alleles for linked genes is not independent, linkage studies make it possible to assign two genes to the same chromosome, but this information does not usually allow us to identify a particular gene as belonging to a specific chromosome. Several methods, including the use of restriction enzymes and recombinant DNA techniques (Laboratory 18), have made it possible not only to assign a gene to a particular chromosome, but to determine, or **map,** its location with respect to other genes on that chromosome.

In this laboratory, you will map the genes for spore color in the fungus *Sordaria.* Using linkage analysis, you will be able to determine the relative distance between the centromere and known mutant genes. You will also study the giant chromosomes of the fruit fly *Drosophila,* which will allow you to visualize genes on a eukaryotic chromosome. By using bacterial conjugation as a means of genetic recombination, you will be able to map several genes on the circular chromosome of the bacterium *Escherichia coli.* Finally, you will have the opportunity to use restriction enzymes and electrophoresis to compile a restriction map of the phage lambda (λ).

STUDENT PREPARATION

Prepare for this laboratory by reading the text pages indicated by your instructor. Familiarizing yourself in advance with the information and procedures covered in this laboratory will give you a better understanding of the material and improve your efficiency.

EXERCISE A | Mapping the Genes of *Sordaria fimicola*

In *Sordaria fimicola,* ascus formation requires both meiotic and mitotic divisions to produce eight haploid ascospores. During the first nuclear division of meiosis (meiosis I, or MI), crossing-over often occurs between non-sister chromatids of homologous chromosomes, resulting in recombination of the genes on the two homologues (Figure 14A-1). The greater the distance between the two genes on the chromosome, the more likely it is that crossing-over will occur; the genes are "loosely" linked. If the genes are close together, then crossing-over is less likely; the genes are said to be "tightly" linked. Thus the frequency with which two genes recombine is related to how far apart the two genes are on the chromosome. By using recombination frequencies, it is possible to produce a *linkage map* for genes on a chromosome.

Figure 14A-1 *(a) Two homologous chromosomes carrying different alleles of a gene for a specific trait. The alleles are located at the same loci on the homologues. (b) Crossing-over occurs. (c) Alleles are exchanged, and each chromosome carries alleles that are different from those on the original chromosome.*

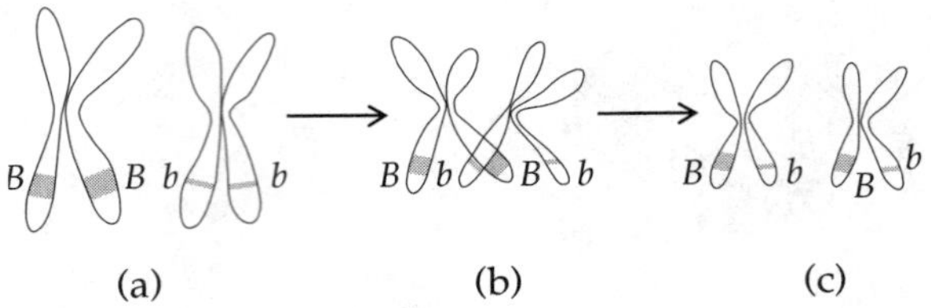

(a)   (b)   (c)

PART I   **Mapping *Sordaria* Chromosomes (Week 2; see Laboratory 13, Exercise D)**

During ascus development, if no crossing-over occurs in meiosis I, a 4+4 pattern of ascospores is produced. These asci have an **MI pattern.** If crossing-over does occur, then the pattern of ascospores is referred to as **MII** and is either 2+2+2+2 or 2+4+2.

From the numbers of asci with MI and MII patterns following a cross, you can "map" the location of the gene for tan color (*t*) on the *Sordaria* chromosome. You will determine the distance of the tan gene locus from the centromere (centromere-to-locus distance). Note that this exercise considers just the gene for tan color, since the presence of a normal allele (t^+) or a mutant allele (t^-) will alter ascospore color. For this reason you will map the distance from the centromere, not the distance from another gene. (In more typical linkage mapping, recombination between two genes is studied to determine map distance between the genes.)

Review basic information about *Sordaria* genetics in Laboratory 13, Exercise D, before continuing. Be sure that you can recognize MI and MII ascus patterns.

IIIII Objectives IIIIIIIIIIIIIIIIIIIIIIIIIIIIIIIIIIIIIIIIIII

☐ Explain how meiosis and crossing-over result in different arrangements of ascospores within asci.

☐ Calculate the map distance between a gene for ascospore color and the centromere of the chromosome on which the gene is found.

IIIII Procedure IIIIIIIIIIIIIIIIIIIIIIIIIIIIIIIIIIIIIIIIIIIIIII

1. Your instructor will provide you with several brown paper bags used to simulate *Sordaria* perithecia from a black × tan cross. Each bag contains 10 asci. Remove the asci and determine whether they are nonhybrid or hybrid, MI or MII asci. Your instructor will ask you to report the number of MI and MII asci in your bag; your response should be verified (or corrected) by your laboratory partner. (You do *not* count the nonhybrid asci containing all like-color ascospores. *Why?*) After totaling the number of MI and MII asci for the entire class, you will be asked to determine the frequency of crossing-over and the map distance from the centromere to the tan gene, using steps 7 and 8 of this procedure. Record class data in Table 14A-1.

Table 14A-1   Class Data

Total asci counted	
Total MI asci	
Total MII asci	

a. *What is the map distance of the tan gene from the centromere on the* Sordaria *chromosome as simulated in this cross?* _____

2. Now, obtain your *Sordaria* cross (from Laboratory 13). Alternatively, your instructor has set up the *Sordaria* crosses and will provide you with a culture.

3. Where the mycelia of the two strains overlap and fuse, dark lines of tiny perithecia will be visible. Use a toothpick or spatula to gently scrape the surface of the agar to collect perithecia (see Figure 13D-4). It is usually best to collect perithecia toward the outer rim of the dish.

4. Place the perithecia in a drop of water on a slide. Cover with a coverslip and gently press on the coverslip (use a small cork) to rupture the perithecia. Be gentle so that the ascospores remain in the asci (see Figure 13D-2).

5. View the slide using the 10× objective and locate a group of hybrid asci (recall that asci produced by fusion of two identical strains, both black or both tan, will result in ascospores that are all of the same color within an ascus—disregard these asci). Hybrid asci contain both black and tan ascospores within each ascus.

6. Count at least 50 hybrid asci and score them as either MI asci (4+4 arrangement) in which alleles segregated in meiosis I, or MII asci (2+2+2+2 or 2+4+2) in which alleles segregated in meiosis II. (Remember, do *not* count the nonhybrid asci.) Record your results in Table 14A-2. Determine the number of M1 and M2 asci counted by all students and record class results in Table 14A-3.

Table 14A-2   Your Data

Number of MI Asci Showing No Crossover (4:4)		Number of MII Asci Showing Crossover (2:2:2:2) or (2:4:2)		Total MI + MII Asci	Percentage of Asci Showing Crossover	$\dfrac{\text{Frequency}}{2}$ (Map Units)
○○○○●●●●		○○●●○○●●				
●●●●○○○○		●●○○●●○○				
		○○●●●●○○				
		●●○○○○●●				

Recall that in this exercise, you are studying only one gene, and you will map its distance from the centromere by determining the frequency of crossover events involving that gene—crossovers that occur somewhere between the centromere and the gene and result in its recombination with the chromatid of a different chromosome.

The frequency of crossing-over between two genes is largely controlled by the distance between genes (or between gene and centromere, as in this case); the probability of a crossover occurring between two particular genes on the same chromosome increases as the distance between those genes increases. The frequency of crossing-over is, therefore, proportional to the distance between genes. An arbitrary unit of measure, the **map unit,** is generally used to describe distances between linked genes. A map unit is equal to a 1% frequency of crossovers. For instance, when there is a 30% frequency of crossing-over between two genes, these genes are said to be 30 map units apart.

7. Determine the frequency of crossing-over (percentage of crossovers) by dividing the number of MII crossover asci by the total number of asci counted, and multiplying by 100:

$$\text{Frequency of crossing-over} = \frac{\text{MII}}{\text{MI} + \text{MII}} \times 100$$

8. In *Sordaria,* since only 4 of the 8 ascospores carry recombinations, the frequency of recombination is one-half of the frequency of crossing-over.

b. Why do only half of the ascospores carry recombinant chromosomal strands?

_____

_____

The relationship between frequency of crossing-over and map distance is expressed as:

$$\text{Number of map units} = \frac{4 \times \text{number of recombinant (MII) asci}}{8 \times \text{total number of asci}} \times 100$$

or

$$\text{Number of map units} = \frac{\text{frequency of crossing-over}}{2}$$

9. Record the map distance of gene t or t^+ from the centromere. _____ Record this map distance in Table 14A-2. Published results indicate that the map distance of the tan spore-color gene from the centromere in *Sordaria fimicola* is 26 map units.

   c. How closely do your data fit this measurement? _____

PART 2   The Chi-Square Test (*Optional*)

To test how closely your class crossover data agree with or "fit" the value for the map distance of the mutant tan gene obtained by others (26 map units), you can perform a chi-square test (see Appendix I). The chi-square test is a statistical test that allows you to determine if differences between your data (what was *observed*) and known or hypothetical values (what was *expected*) are significant.

IIIII Objectives IIIIIIIIIIIIIIIIIIIIIIIIIIIIIIIIIIIIIIIIIIIIIII

☐ Perform a chi-square test to determine the probability that your data "fit" the expected map distance.

IIIII Procedure IIIIIIIIIIIIIIIIIIIIIIIIIIIIIIIIIIIIIIIIIIIIIII

1. As you determine whether or not your observed data "fit" within the limits of what is expected, the chi-square analysis is actually testing the validity of your **null hypothesis** (Laboratory I). Your null hypothesis states: There will be no difference between the predicted (expected) map distance of 26 units and the observed map distance calculated from class data.
   If there *is* a difference (your map distance is not 26 units), you want to know how significant this difference is. Is the difference due just to chance (perhaps sampling errors)? Or is it due to something real (a chromosomal rearrangement or deletion, etc.)? If the results of the chi-square test indicate that the observed data do *not* vary significantly from the expected, then you accept the null hypothesis: there is no difference. If, however, the results indicate that the

Table 14A-3   Class Data

Number of MI Asci		Number of MII Asci		Total MI + MII Asci	Percentage of Asci Showing Crossover	$\frac{\text{Frequency}}{2}$ (Map Units)
○○○○●●●●		○○●●○○●●				
●●●●○○○○		●●○○○●●○				
		○○●●●●○○				
		●●○○○○●●				

observed data vary significantly from the expected, then you *reject* the null hypothesis and accept your **alternative hypothesis** (see Laboratory I).

Your instructor will use the data from the *Sordaria* simulation to demonstrate how to perform the chi-square test following the steps outlined below.

a. How well do the data from the class simulation fit the expected result of 26 map units?

_____

2. Use the class data on *Sordaria* cross plates from Table 14A-3.

Chi-square is calculated as shown in the following example.

Example If 26 map units is the expected locus-to-centromere distance for the tan gene on the *Sordaria* chromosome, you should expect 52% of all asci you observe to be crossover asci. (Recall that you divided the percentage of crossovers by 2 to calculate map distance; 26% $\times$ 2 = 52%.) Suppose 1,000 asci are counted by the class; 52% of 1,000 = 520. Thus, 520 of the asci should be crossovers and 480 should be non-crossovers. This is what is *expected*. What was actually observed by the class was

$$\text{Crossovers} = 510$$

$$\text{Non-crossovers} = 490$$

Using the formula

$$\chi^2 = \sum \frac{(\text{observed} - \text{expected})^2}{\text{expected}}$$

for both crossovers and non-crossovers, the expected (or hypothetical values) are filled in as follows:

$$\chi^2 = \frac{(\text{observed crossovers} - 520)^2}{520} + \frac{(\text{observed non-crossovers} - 480)^2}{480}$$

Since 510 crossovers and 490 non-crossovers were observed in class,

$$\chi^2 = \frac{(510 - 520)^2}{520} + \frac{(490 - 480)^2}{480}$$

$$\chi^2 = \frac{(-10)^2}{520} + \frac{(10)^2}{480} = \frac{100}{520} + \frac{100}{480} = 0.192 + 0.208 = 0.400$$

Do observed and expected results differ in this case, where $\chi^2 = 0.400$? To decide, you must refer to a Critical Values of χ^2 Table (Appendix I). Table 14A-4 is an abbreviated table for use in this laboratory.

Table 14A-4   Critical Values of Chi-Square

Degrees of Freedom (df)	Probability (p)				
	0.9 (9 in 10)	0.5 (1 in 2)	0.2 (1 in 5)	0.05 (1 in 20)	0.001 (1 in 100)
1	0.016	0.46	1.64	**3.84**	6.64
2	0.21	1.39	3.22	5.99	9.21
3	0.58	2.37	4.64	7.82	11.35

First, you must determine the **degrees of freedom** (df) for your experiment. In this example, the degrees of freedom are 1 less ($n - 1$) than the number of attributes being observed ($n = 2$, since crossover and non-crossover are the only attributes being studied). Next, you determine a **probability value** (p value). For most scientific studies, the minimum probability for rejecting a null hypothesis is usually $p = 0.05$. In selecting a probability of $p = 0.05$, you set a "level of rejection" for your null hypothesis. If you reject your null hypothesis, you have a 1 in 20 (or 5%) chance of being wrong in doing so (a fairly low probability of making the wrong decision!). Having determined the df and p values, you next find the **critical value** from the chi-square table, in this case 3.84.

- If the calculated chi-square value is **greater than or equal to** the critical value, then you **reject** the null hypothesis (and accept the alternative hypothesis). You conclude that deviations from the expected are sufficiently large to be meaningful (significant), so there must actually be a difference (you reject the null or no-difference statement).

- If the calculated chi-square value is **less than** the critical value, then you **accept** the null hypothesis. You conclude that deviations from the expected are sufficiently small that there is no difference (you accept the null or no-difference statement).

The chi-square value for this example ($\chi^2 = 0.40$) is much smaller than 3.84. This means that you *accept* the null hypothesis: that class observations agree with (there is no difference from) the published or known value of 26 map units for the distance of the tan gene from the centromere on the *Sordaria* chromosome. [*Note:* If you rejected the null hypothesis based on $\chi^2 = 0.40$, you would have a *greater* chance of being wrong in making this decision—almost a 1 in 2 chance (see $p = 0.5$ for a critical value of 0.46), and this is not an acceptable degree of error.]

3. Now, apply the chi-square test to your class results recorded in Table 14A-3. Perform your calculations in the space below.

b. What is your null hypothesis for this investigation? _____

What is your alternative hypothesis? _____

c. How well do your class data fit the expected value of 26 map units? _____

d. Do you accept or reject your null hypothesis? _____

e. What do you conclude about the distance of the tan gene from the centromere on the Sordaria

chromosome? _____

_____

✔ | **EXERCISE B** | **Examining the Giant Chromosomes of *Drosophila***

During development, cells of the *Drosophila* salivary glands pass through the cell cycle many times, their DNA replicating in preparation for division. However, the cells do not divide and as a result, the number of DNA strands comprising each chromosome continues to increase; the chromosomes become multistranded and are called *polytene* chromosomes. The DNA content of polytene chromosomes is approximately 1,000 times greater than the normal DNA content of chromosomes containing a single helical strand of DNA.

The many DNA strands of a polytene chromosome condense and fold in the same manner as the single helical strand of DNA in other chromosomes. Highly condensed or folded areas stain darkly and give chromosomes a banded appearance. Since all of the DNA strands in the giant polytene chromosome are duplicates of one another, the folded portions are in register with one another and the bands appear to stretch across the entire chromosome, giving the chromosomes a very dramatic "striped" appearance (Figure 14B-1).

During transcription of messenger RNA, some of the banded regions of the polytene chromosomes uncoil and expand to form "puffs." Thus, it is possible to visualize genes on these giant chromosomes. If the protein product synthesized by the gene can be identified, so can the gene. Proof of its location can also be obtained by isolating the mRNA produced by the "puff" region, labeling it in some way, and

Figure 14B-1 *The polytene chromosomes of* Drosophila melanogaster.

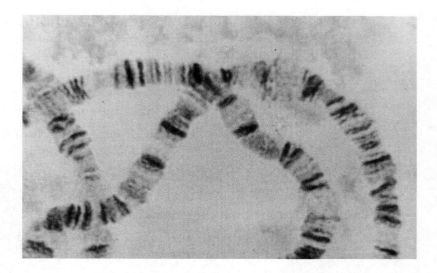

hybridizing this mRNA with the chromosome. Where the bases are complementary to the DNA, they bind, allowing the genes to be identified by autoradiographic or immunofluorescent techniques.

In this exercise you will prepare chromosome "squashes" of *Drosophila* polytene chromosomes and observe their banded nature.

IIIII Objectives IIIIIIIIIIIIIIIIIIIIIIIIIIIIIIIIIIIIIII

☐ Describe how the banded pattern of the giant chromosomes of the *Drosophila* salivary gland is related to the arrangement of genes on the chromosome.

IIIII Procedure IIIIIIIIIIIIIIIIIIIIIIIIIIIIIIIIIIIIIIII

1. Use the dissecting microscope. If your microscope does not have a built-in illuminator, use a black background or a mirror and a transparent glass plate. Obtain a clean slide and place a drop of 0.7% saline toward one end.

2. Place a *Drosophila* larva in the saline, and with two dissecting needles decapitate the larva. Place one needle at the middle of the larva and the other just behind the head. Often you can see the two salivary glands separating from one another in a **V** at this point, and it is best to place the point of the needle in the middle of the **V**. Pull the needle at the head with a quick

Figure 14B-2 *Dissection of the salivary glands of a* Drosophila melanogaster *larva.*

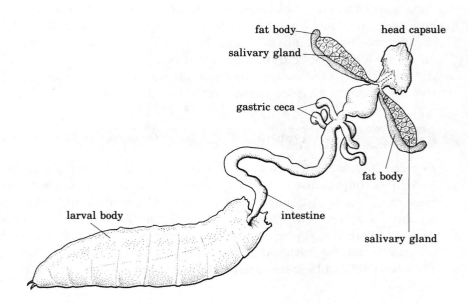

jerk and then relax, and the salivary glands will slowly be pushed out of the body. The glands are elongated and semitransparent—the individual cells are large (grapelike in appearance) and are lined up along the lumen of each gland. The lumen connects to the digestive tract through a duct (Figure 14B-2).

3. Remove the opaque fat body material adhering to the glands. If you are using transmitted light, the fat body material will appear grayish. With reflected light, it will appear white.

4. Place a drop of acetocarmine or aceto-orcein stain next to the drop of saline and, with a dissecting needle, transfer the glands from the saline to the stain. Examine the drop of stain to make sure the glands were transferred.

5. Stain for 10 minutes. Make sure that the drop of stain does not dry up.

6. Place a coverslip on the preparation.

7. Place the slide between the folds of a paper towel and press down on the coverslip firmly. The eraser end of a pencil can also be used for pressing.

8. Examine the slide using low power (10×) to locate the chromosomes. Examine the chromosomes to observe the banded pattern.

   a. *Do you see any bulges along the length of the chromosome?* _____ *What do they represent?*

   _____

   b. *How many chromosomes do you see?* (Note: *Homologous chromosomes are synapsed along their*

   entire length, so what appears as one chromosome is actually two.) _____

Note: The chromosome studies in Exercises A and B have introduced you to some of the basic, and more classical, techniques for locating genes on eukaryotic chromosomes. Rapid advances in biotechnology have made it possible to map eukaryotic genes more quickly and more accurately. To do this, special enzymes, **restriction endonucleases** (see Exercise D), are used to chop DNA into small fragments. These enzymes recognize specific nucleotide sequences in the DNA and always cut the DNA at the same sites. (Mutations in the DNA may alter the nucleotide sequence of a site so that the restriction endonuclease no longer cuts the DNA at the original position: a fragment with a "new length" is produced.)

   Fragments produced by restriction endonucleases are called RFLPs (pronounced "rif-lips"). (See Laboratory 15, Exercise F.) By studying how frequently characteristic RFLPs appear in several generations of families that exhibit a particular genetic trait or disorder, geneticists can determine the approximate location of the gene for that disorder on a particular chromosome. Just as in linkage studies, where two genes are said to be linked if they constantly appear together (they are so close together on the chromosome that no crossing-over can occur), the RFLP and a particular trait are assumed to be "linked" and, thus, in close proximity to one another on the chromosome. If the RFLP is hybridized to the DNA of the chromosome, the physical location of the gene can be determined. In addition, the RFLP can serve as a "marker" for a specific trait. If the DNA of a patient contains the RFLP associated with a particular disorder, it is likely that the person will manifest that disorder. This is especially important as a diagnostic tool for disorders that are characterized by late onset.

👁 **EXERCISE C** | **Mapping the Chromosome of *Escherichia coli***

Reproduction in bacteria is primarily an asexual process involving fission, but a type of sexual reproduction known as **conjugation** can also occur. During this process, genetic material is transferred from one bacterium (the + or donor strain) to another bacterium (the − or recipient strain). Donor cells contain a fertility factor, or *F* factor, carried by a plasmid (a small extrachromosomal, circular piece of DNA), which can be transferred during conjugation. These donors (males) are designated as F^+. If the *F* factor becomes integrated into the bacterial chromosome of the donor, the chromosome (although usually not the entire chromosome) can be transferred to the recipient bacterial cell. If recombination takes place, the recipient may express characteristics originally unique to the donor. Bacteria containing the *F* factor integrated into the chromosome (as an **episome**) are called **Hfr** (high frequency of recombination) **cells.**

The amount of chromosomal DNA that can be transferred from an Hfr cell to a recipient cell is determined by how long the cells remain in contact (during conjugation, the cells are attached by **pili;** DNA transfer occurs through cytoplasmic "bridges"). The farther away a given gene is from the leading point of the chromosome being transferred, the less chance the gene has of being transferred before the bridge is broken. By interrupting mating at specific times, it is possible to construct a circular map of the bacterial chromosome. The map distances are in "minutes," referring to the time it took to transfer certain genes (Figure 14C-1).

In this exercise, you will work with two strains of *E. coli.* The donor, the Hfr strain, is streptomycin-sensitive: it does not carry the gene for streptomycin resistance and thus will be killed by this antibiotic. This "wild type" also carries alleles for synthesizing the amino acids proline, leucine, and threonine, and the vitamin thiamine. Thus this strain is designated *Str^s pro$^+$ leu$^+$ thi$^+$ thr$^+$*. The wild-type strain can live on minimal medium that contains only glucose, ammonia, and inorganic salts, because it can make all of the amino acids needed for growth. (Remember: if streptomycin is present in the medium, these cells will die.)

The mutant strain, which is the recipient, carries the gene for streptomycin resistance, but does not carry the alleles for synthesizing proline, leucine, threonine, and thiamine. This mutant strain is designated *Str^r pro$^-$ leu$^-$ thi$^-$ thr$^-$*. For the recipient to survive (prior to recombination), the minimal medium must be supplemented with the amino acids proline, leucine, and threonine and the vitamin thiamine; streptomycin in the medium will not harm the recipient strain.

Figure 14C-1   *Genetic map of* Escherichia coli *showing several genes.*

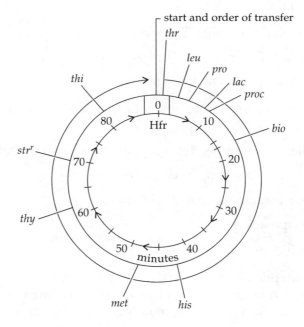

You can determine whether particular genes have been transferred from the donor to the recipient by testing to see if the recipient has acquired the ability to synthesize any of the nutrients that had previously been synthesized only by the donor strain. For example, the transfer of the *thr$^+$* gene can be detected in the recipient by the fact that the recipient, previously *thr$^-$* (unable to synthesie threonine), is now *thr$^+$*, due to recombination. [Since this gene is close to the origin of replication and transfer (Figure 14C-1), it is the most likely gene to be transferred to the recipient.] The recipient can now synthesize its own threonine and no longer requires a medium containing threonine for its survival.

a. *If you had unlabeled cultures of* Str^s pro^+ leu^+ thi^+ thr^+ *and* Str^r pro^- leu^- thi^- thr^-, *what could you do to tell them apart?* _____

_____

_____

b. *What kind of test could you devise to tell whether recombination had occurred between* Str^s pro^+ leu^+ thi^+ thr^+ *and* Str^r pro^- leu^- thi^- thr^-? _____

_____

_____

ıııııı **Procedure** ıı

1. Work in pairs. Review Appendix III, Preparing Serial Dilutions. Obtain a broth culture of *E. coli* strain Str^s pro^+ leu^+ thi^+ thr^+ labeled "D" for "donor strain."

2. Obtain a broth culture of *E. coli* strain Str^r pro^- leu^- thi^- thr^- labeled "R" for "recipient strain."

3. Also obtain from your instructor four agar plates containing minimal medium supplemented with streptomycin and thiamine (STR/THI) and two agar plates containing minimal medium (M) supplemented with the amino acids proline, leucine, threonine, and the vitamin thiamine, but without streptomycin (M/PLTT).

   Notice that both types of plates you are using are supplemented with thiamine. Examine Figure 14C-1.

   c. *Where is the thiamine (thi) gene located?* _____ *What is the chance that it will be transferred from donor to recipient during conjugation?* _____ *Why would you supplement plates with this vitamin?* _____

4. Before proceeding, gather the following equipment and set up your work area on a clean surface:

   1 sterile, *empty* capped test tube (the "conjugation" tube)

   5 tubes containing 9.0 ml of sterile distilled water

   1 beaker containing 95% ethyl alcohol

   1 glass spreading rod

   10 sterile 1-ml pipettes

5. Using aseptic technique, first flaming the mouth of the tube and using a sterile pipette, transfer 1.0 ml of "D" suspension into the conjugation tube. Aseptically transfer 1.0 ml of "R" suspension into the conjugation tube. *Note the time* of addition of culture "R" to culture "D" in the conjugation tube: _____ Gently agitate the mixture by rotating the tube between the palms of your hands. Allow the culture to incubate at 37°C for 30 minutes. Be sure that the mating mixture is *not* disturbed during this time.

6. Aseptically pipette 1.0 ml of "D" culture into a 9.0-ml sterile water blank and mix thoroughly. Note that the donor culture is now diluted to 1:10. Label the tube "D 1:10."

7. Using the diluted donor culture, prepare a spread-plate (see Laboratory 6, Exercise C). Use a sterile 1-ml pipette to remove 0.1 ml of dilute culture and transfer it to the center of the STR/THI plate. Dip a glass spreading rod in 95% ethyl alcohol and flame it. Let it cool for a

few seconds and touch it to the outer edge of the agar plate—if it sizzles, it is still too hot to use. When the rod has cooled, spread the inoculum over the surface of the plate. To avoid contamination, hold the lid above the plate as you are working with the bacteria.

8. Use a sterile 1-ml pipette to remove 0.1 ml of dilute "D" culture and transfer it to the center of a M/PLTT plate. Spread the bacteria as described in step 7.

   d. *Do you expect the donor strain to grow on the STR/THI plate? _____ Why or why not?*

   e. *Do you expect the donor strain to grow on the M/PLTT plate? _____ Why or why not?*

   f. *Why did you prepare both the STR/THI and the M/PLTT plates using only donor cells?*

9. Now repeat the procedure outlined in steps 6 through 8, but this time use the recipient strain and fresh STR/THI and M/PLTT plates. (Label the diluted recipient strain culture "R 1:10.")

   g. *Do you expect the recipient strain to grow on the STR/THI plate? _____ Why or why not?*

   h. *Do you expect the recipient strain to grow on the M/PLTT plate? _____ Why or why not?*

   i. *Why did you prepare both the STR/THI and M/PLTT plates using only recipient cells?*

10. After conjugation has proceeded for 30 minutes (step 5), remove the culture from the 37°C water bath or incubator and vigorously agitate the mixture by rotating the tube between the palms of your hands (create a vortex, if possible).

11. Aseptically transfer 1.0 ml of the conjugation mixture to a sterile water blank (9.0 ml). Mix thoroughly (roll between your palsm) and label the tube "D × R 1:10."

12. Aseptically pipette 1.0 ml of the D × R 1:10 suspension into a second sterile water blank (9 ml). Mix thoroughly (roll between your palms) and label the tube "D × R 1:100."

13. Aseptically transfer 0.1 ml of the D × R 1:10 dilution to the surface of one of the two unused STR/THI plates. Mark the plate "STR/THI 1:10" and label it with your name and the date.

14. Similarly, transfer 0.1 ml of the D × R 1:100 dilution to the surface of the other STR/THI plate. Mark the plate "STR/THI 1:100" and label it with your name and the date.

15. Sterilize a glass spreading rod by dipping it in alcohol and passing it quickly through a flame. Spread the bacteria on the STR/THI 1:10 plate as in step 7. Resterilize the spreading rod and spread the bacteria on the STR/THI 1:100 plate.

16. Be sure that all six of your plates are clearly labeled. Tape them together. Invert them and place them in an area designated by your instructor. They will be incubated for 2 days at 37°C and then refrigerated until the next laboratory period.

17. During the next laboratory period, complete Table 14C-1 using ✔ to indicate growth and 0 to indicate no growth. Interpret your results.

   j. *Did conjugation occur? _____ How do you know?*

Table 14C-1   Record of Growth (✓) and No Growth (0) of *E. coli*

	Donor (1:10)	Recipient (1:10)	D × R 1:10	D × R 1:100
STR/THI				
M/PLTT			✕	✕

    k. *With reference to the circular map of* E. coli, *which genes*$^{(+)}$ *are found in the recombinant cells that were mutant*$^{(-)}$ *in the recipient cells?* _____

_____

    l. *It is unusual for the entire* E. coli *chromosome to be transferred during the process of conjugation. What is the consequence of this fact for genes that are farther away from the origin of replication and transfer?* _____

_____

    m. *Why was streptomycin included in the agar used for plating recombinant cells?*

_____

If the recipient bacterial strain had also been *met*⁻ (lacking the ability to synthesize the amino acid methionine) and recombination with the same wild-type donor cell occurred, what would have happened under the following conditions? (Refer to Figure 14C-1.)

    • Conjugation is interrupted after 10 minutes and recombinants are plated onto minimal medium containing streptomycin. *n. Would recombinants grow?* _____

    o. *Would these recombinants grow on medium containing both streptomycin and methionine?*

_____

    • Conjugation is allowed to continue for 60 minutes and recombinants are plated onto minimal medium containing streptomycin. *p. Would these recombinants grow?* _____

    q. *What does this indicate about the relative positions of the genes for proline, leucine, threonine, methionine, and thiamine?* _____

_____

✔ 👁 **EXERCISE D**   **Restriction Endonucleases: Mapping Bacteriophage Lambda**

Restriction endonucleases are essential tools in recombinant DNA methodology. Several hundred have been isolated from a variety of prokaryotic organisms. In the nomenclature of restriction endonucleases, the letters refer to the organism from which the endonuclease was isolated. The first letter of the name stands for the genus name of the organism. The next two letters represent the initial letters of the second word of the species name. The fourth letter (if there is one) represents the strain of the organism. Roman numerals indicate whether the particular endonuclease was the first isolated, the second, and so on. For example:

    *Eco*RI   *E* = genus *Escherichia*

              co = *coli*

              R = strain RY 13

              I = first endonuclease isolated

HindIII   H = genus *Haemophilus*

           in = *influenza*

           d = strain Rd

          III = third endonuclease isolated

Each restriction endonuclease "recognizes" a specific DNA sequence (usually a 4- to 6-base-pair sequence of nucleotides) in double-stranded DNA and digests phosphodiester bonds at specific sites in the sequence. For example, the restriction endonuclease *Eco*RI cuts double-stranded DNA as follows:

$$\downarrow$$
$$-G-A-A-T-T-C-$$
$$-C-T-T-A-A-G-$$
$$\uparrow$$

Since this endonuclease cuts at specific sites on each strand, "**sticky ends**" are produced. (Sticky ends make it possible to insert DNA pieces, usually containing a gene of interest, into plasmids or viral vectors that have been cut with the same endonuclease, providing complementary sites for attachment—see Laboratory 18.) Other restriction endonucleases cut DNA in the same position on both strands to produce fragments with "**blunt ends**."

    Fragments of DNA produced by restriction enzyme cleavage can be separated by gel electrophoresis. When any molecule enters an electrical field, its mobility, or the speed at which it will move, is influenced by the charge on the molecule, the strength of the electrical field, the size and shape of the molecule, and the density of the medium (gel) through which it is migrating. Consequently, it is possible to separate heterogeneous populations of molecules (such as fragments of DNA). When all molecules are positioned at a uniform starting site on a gel, the gel is placed in a chamber containing a buffer solution and electrodes, and an electric current is applied, the molecules will migrate and form bands (concentrations of homogeneous molecules). Since the phosphates in the DNA backbone are negatively charged at neutral pH, DNA fragments will migrate through the gel toward the positive electrode.

    After electrophoretic separation and staining of the gels with colored or fluorescent dyes, the sizes of DNA fragments can be determined by comparing them with markers of known molecular weight or other fragments of known sizes. Once this information is available, it is possible to construct a **restriction map** of the DNA in question.

✔ **PART I**   **Estimating DNA Fragment Size Using Gel Electrophoresis**

In this exercise, you will use an agarose gel to separate a mixture of bacteriophage lambda (λ) DNA fragments produced using two restriction enzymes, *Eco*RI and *Hin*dIII. (The density of an agarose gel can be varied to improve the resolution of similar-sized molecules.) In agarose, the migration rate of linear fragments of DNA is inversely proportional to their size; the smaller the DNA fragment, the faster it migrates through the gel. The size of the fragments is measured in number of base pairs (bp).

    The size of fragments produced by a specific endonuclease (*Eco*RI in this exercise) can be determined by using *standard* fragments of known size (fragments produced by *Hin*dIII in this case). These fragments are electrophoresed together and, after visualizing the bands by staining with dyes such as methylene blue or ethidium bromide (a fluorescent dye), a standard curve can be plotted for *Hin*dIII bands of known size. The sizes of the standard fragments are plotted on the *Y*-axis and the distances they migrated are plotted on the *X*-axis. By plotting the migration distances of the fragments of unknown size on the *X*-axis, you will be able to determine the size of the *Eco*RI fragments from the standard curve.

‖‖‖‖‖ **Objectives** ‖‖‖

☐ Explain the principles of electrophoresis as they pertain to separating and identifying DNA fragments.

☐ Determine DNA restriction fragment sizes using a standard curve.

ⅠⅠⅠⅠⅠ Procedure ⅠⅠ

Work in groups of four.

A. Casting Agarose Gels

1. Tightly tape the ends of the gel bed with masking tape to make a good seal (Figure 14D-1).

2. Place the gel comb across the gel bed, putting the ends of the comb into the notches on the side of the bed nearest the end of the bed. (The comb creates "wells" in the gel into which you will load various samples for testing.)

3. Obtain a bottle of 1% agarose (50 ml) from your instructor. *Caution:* The bottle will be hot since it has just been heated to melt the agarose. (Add 2 drops of Carolina Blu stain if required by your instructor.)

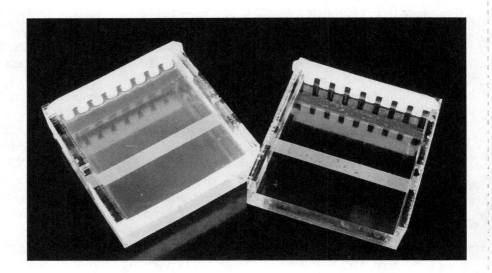

Figure 14D-1 *Preparation of the gel bed. The bed on the left has been taped and the agarose gel has been poured into it.*

4. Use a Pasteur pipette to transfer some agarose solution into the gel bed at the edges of the tape to make sure there are no leaks, but do this quickly and do not let it solidify. Now, carefully pour the remaining agarose into the gel bed. The agarose should cover only about one-third the height of the teeth (about 5 to 7 mm thick; you do not want your gel to be too thick, or too thin). If bubbles appear around the teeth of the comb, remove the comb and reinsert it. If bubbles appear in the gel, use a Pasteur pipette to draw them off to the sides.

5. Keep the gel bed completely immobile while the gel is setting (approximately 15 minutes). The final appearance of the gel will be cloudy.

6. When ready, remove the tape from the ends of the gel bed.

7. Carefully lower the gel bed into the chamber with the comb nearest to the *negative* (black) electrode (Figure 14D-2). Make sure that the gel bed is properly seated and centered.

8. Fill the electrophoresis chamber with running TBE buffer (tris-borate-EDTA) until the buffer covers the surface of the gel. (Make sure the 10× stock buffer has been diluted 1:10.) (Add 12 drops of Carolina Blu stain to 1 liter of buffer, if required by your instructor.)

9. Carefully remove the comb (pull it straight up). Make sure that the sample wells are filled with buffer. Remove any bubbles by using a Pasteur pipette to blow gently into the buffer above the wells. *Do not stick the end of the pipette into the wells—you might puncture the gel!*

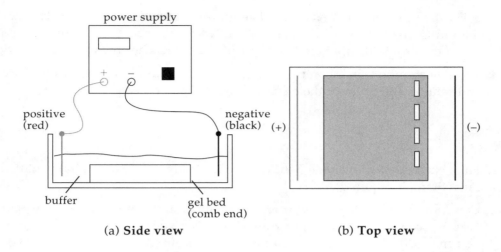

Figure 14D-2 *Setup for the electrophoresis system.*

B. Loading Samples

1. If a practice gel is available, practice loading the samples using the gel-loading solution only.

   a. To load practice samples on the gel, use a small micropipette or plastic transfer pipette. Pull a small amount of the practice gel-loading solution into the end of the pipette. (Do not allow the solution to move up into the body of the pipette or bubbles will be introduced into the well of the agarose gel during loading.)

   b. Dip the pipette into the buffer and hold the tip of the pipette slightly *above* the well in the gel. Gently dispense the solution. The loading dye is denser than the buffer and will move into the well. (Do *not* place the tip of the pipette into the well or you might puncture the gel.)

2. After practicing, you are ready to load the gels. Obtain a microtest tube containing phage lambda DNA digested with *Eco*RI endonuclease.* FIll one well of the electrophoresis apparatus with approximately 20 μl of this solution. The DNA is mixed with a solution containing tracking dye that will make it possible to trace the process of the DNA migration in the agarose gel.

3. Obtain a microtest tube containing phage lambda DNA digested with *Hind*III endonuclease. Follow your instructor's directions and fill a second well with 20 μl of this *Hind*III digest. The DNA fragments from this digest are of known size and will serve as a "standard" for measuring the size of the *Eco*RI fragments from step 2.

4. Load 20 μl of undiluted phage lambda DNA (control) into a third well.

C. Electrophoresis

1. Place the top on the electrophoresis chamber and connect the electrical leads (black to black and red to red). If using an Edvotek chamber, set the voltage to 50 volts. If using a Cabisco apparatus, set to 80 volts and check for a current reading of 50 to 100 milliamperes. When the current is flowing, you should see bubbles on the electrodes.

2. Allow electrophoresis to continue for a minimum of 1½ hours, or until the loading dye has moved at least 5 to 7 cm from the wells. The tracking dye will eventually form two bands of color. A purplish band (bromophenol blue) will be seen farthest from the wells. A slower-moving aqua-colored band (xylene cyanol) will migrate through the gel at a rate

---

*DNA for this procedure has been predigested to save time. However, your instructor may wish to have you perform the digests rather than use predigested material. If this is the case, your instructor will give you separate directions.

equivalent to that of a DNA fragment approximately 2,000 bp long. This aqua band will be migrating just in front of your smallest DNA fragments. Turn off the electrophoresis apparatus when the bromophenol blue band (purple band) reaches the opposite end of the gel.

3. After electrophoresis is completed and the power supply is turned off, disconnect the leads and remove the cover of the electrophoresis chamber.

D. Staining
Wear gloves!

1. Fill a staining tray (or large Petri dish) with methylene blue staining solution (or Carolina Blu Final Stain, if required).

2. Carefully remove the gel bed from the chamber and gently transfer the gel to a staining tray. Use the scooper provided with your kit or keep your hands under the gel during the transfer. You may wish to remove a small piece of gel from the upper right-hand corner to keep track of the gel's orientation. *Do not stain in the electrophoresis apparatus.*

3. Stain for 30 minutes.

4. Carefully transfer the gel to a tray containing approximately 500 ml of distilled water to destain. Rinse several times and then let the gel destain for 1 to 24 hours. *Do not* change the water during this time or the bands will fade.

5. Transfer the gel to a visible-light box or overhead projector for examination.

E. Determining Fragment Size

1. After observing the gel on the light box, carefully wrap the gel in plastic wrap and smooth out all the wrinkles, or overlay with a transparency sheet.

2. Use a permanent marking pen to trace the outlines of the sample wells and the location of the bands.

3. Remove the plastic wrap and flatten it out on a white piece of paper on the laboratory bench. Save the gel in a plastic bag. Add several drops of the water used for destaining. Close the bag tightly and store at 4°C.

4. If the exercise was done as a demonstration, your instructor will transfer the marks onto an overhead transparency and will make copies of the transparency for each student. If you have run your own gel, you can make measurements directly from the plastic wrap.

5. For the *Hind*III fragments, measure the migration distance in centimeters (to the nearest millimeter). Measure from the front edge of the sample well to the front edge of each band on your gel.

Table 14D-1

*Hind*III Fragment (bp)	Distance Traveled (cm)
23,130	
9,416	
6,557	
4,361	
2,322	
2,027	

The distance a fragment migrates is related to its molecular weight. The greater its molecular weight, the shorter the distance the fragment will travel through the gel. For simplicity, we will use base-pair length instead of molecular weight. The known *Hin*dIII fragment lengths are given in Table 14D-1. Indicate the distance in centimeters that each has traveled. You will identify each fragment by the distance traveled: the shortest fragment will have traveled farthest, the next shortest will be just behind, and so on.

Note: You will observe six bands (Figure 14D-3). The band closest to the origin may appear to be diffuse—it is actually composed of pieces of DNA of two different sizes, 27,491 and 23,130 bp. For graphing purposes, you will use a base-pair size of 23,130. Two additional bands, 564 and 125 bp, are usually not observed. The larger (564 bp) usually does not contain enough DNA to be visible using methylene blue stain; the smaller (125 bp) usually runs off the end of the gel.

Figure 14D-3 *Agarose gel (photo not to scale) from electrophoresis of fragments produced by restriction endonuclease digestion of lambda DNA. Lane 1, digestion with EcoRI; lane 2, digestion with HindIII; lane 3, undigested lambda DNA.*

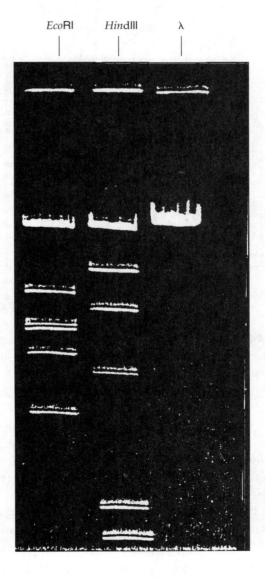

6. Use Figure 14D-4b at the end of the lab to graph your results. [The horizontal (X) axis of semilog paper is divided into a linear scale; the vertical (Y) axis is divided into a logarithmic scale.] Mark the X-axis at 1 cm, 2 cm, 3 cm, 4 cm, and 5 cm. Label this axis "migration distance."

7. Size in base pairs is plotted along the Y-axis. Assume that the first section or cycle of semilog paper represents 0 to 1,000 bp, the second represents 1,000 to 10,000 bp (see Figure 14D-4a). (On semilog paper, each section along the vertical axis is used to represent an increase by the power of 10.) On the Y-axis, mark the approximate position of each of the phage lambda *Hind*III standard fragment sizes in base pairs.

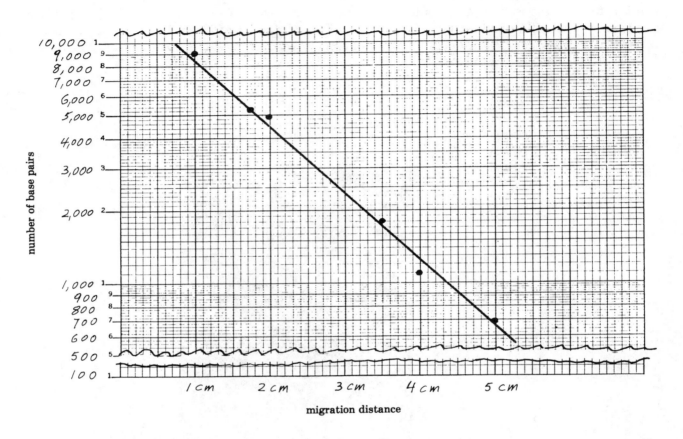

Figure 14D-4a *Example of a standard curve used to determine DNA fragment size. Note: The electrophoresis running time specified in your experiment is different from the running time used to generate the example standard curve shown here, so your standard curve will differ from this one.*

8. Each band on your gel of the *Hind*III digest should correlate with one of the fragment sizes in Table 14D-1. To plot your curve, locate the base-pair length you marked for each fragment on the Y-axis, then move horizontally along the X-axis according to the distance the fragment has traveled. When you have plotted all your points, draw a straight line that fits as close as possible to all the points (although it will not intersect all of them). This line describes the trend of the data and will serve as what is called a *standard curve*, similar, but not identical, to the one shown in Figure 14D-3.

9. Use this standard curve to determine the sizes of the fragments of phage lambda DNA digested with *Eco*RI. You should observe five bands (Figure 14D-3). Measure the migration distance for each *Eco*RI fragment. Locate that distance on the X-axis of your graph and use a ruler to extend a line upward until it crosses your standard curve. Mark the point where the lines cross, and use a ruler or the edge of a piece of paper to find where this point lies on the Y-axis, which gives you the number of base pairs.

a. *What is the relationship between DNA fragment size and rate of travel through the agarose gel?*

_____

_____

Expected *Eco*RI fragment sizes in base pairs are listed in Table 14D-2. Compare your observed results with the expected sizes by entering the base-pair sizes you observed beside the corresponding expected fragment size. *Note:* This technique is not exact—you should expect as much as a 10% to 15% error.

Table 14D-2   *Eco*RI Fragment Sizes for Phage Lambda DNA

Expected	Observed
21,226	
7,421	
5,643	
4,878	
3,530	

👁 **PART 2   Constructing Restriction Maps**

A **restriction map** shows the location of each restriction site (place where the restriction endonuclease "cuts" the DNA) in relation to other sites. A restriction map of a viral or bacterial chromosome can be constructed by comparing the sizes of DNA fragments produced when the chromosomal DNA is digested by a combination of restriction enzymes. First, individual enzymes are used to cut the DNA into fragments of a certain size. A mixture of the same enzymes is then used to cut the DNA into fragments of different sizes (Figure 14D-5a). By determining fragment size and sequencing (establishing the order of) the overlapping fragments, the restriction sites on the DNA can be mapped in relation to the linear sequence of the DNA fragments (Figure 14D-5b). Note that "kb" indicates kilobase pairs; 1,000 base pairs.

Figure 14D-5   *(a) Enzyme A cuts at one end of the DNA molecule while enzyme B cuts at the other end. The fragments could NOT be arranged in sequence as 3/4/5 kb because enzyme B produced two pieces of 4 kb and 8 kb and there is no way to get an 8-kb piece unless the 3-kb piece is next to the 5-kb piece. (b) For this reason the fragments must be arranged as 3/5/4 kb.*

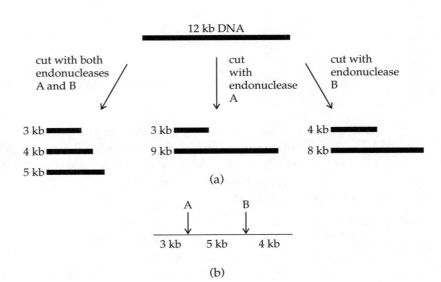

||||| **Objectives** ||||||||||||||||||||||||||||||||||||

☐ Construct restriction maps of circular plasmid DNAs.

☐ Construct restriction maps of linear DNA molecules.

||||| **Procedure** |||||||||||||||||||||||||||||||||||||||||||

1. A circular plasmid has been cut with restriction enzymes H, B, and E alone and in combination. Electrophoresis of restriction fragments produces the gel shown below. Base pair (bp) sizes are shown on the scale to the left of the gel. Estimate the size of each fragment using this scale. Start with enzyme H and assume that it produces two fragments.

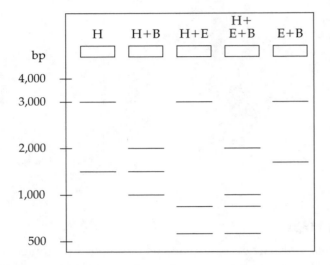

   a. *How many restriction sites were present for enzyme H?* _____

2. Draw a circle in the space below and locate the enzyme H restriction sites in relation to approximate sizes of the restriction fragments.

3. Continue to add restriction sites at the appropriate locations by determining the approximate sizes of the restriction fragments in the H+B lane on the gel. Follow with each additional set of fragments until you have established a complete restriction map; a restriction map for which the correct fragment sizes would be produced by digestion with restriction endonucleases H, E, and B.

4. Next, a linear piece of DNA is cut using enzymes H, B, and E alone and in combination. Electrophoresis of restriction fragments produces the gel shown on page 14-21. Base-pair sizes are shown on the scale to the left of the gel. Estimate the sizes of each fragment using this scale. Start with enzyme H and assume that it produces two fragments.

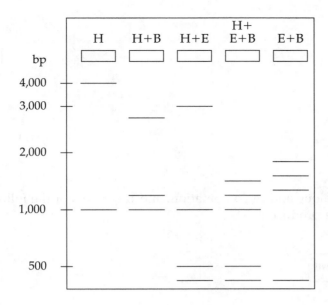

b. How many restriction sites were present for enzyme H? _____

5. Draw a line in the space below and indicate the position of the restriction site(s) for enzyme H.

6. Now continue to add restriction sites at the appropriate places along the line by estimating the fragment sizes produced by the other enzymatic digestions represented on the gel. By sequencing the overlapping fragments, you will create a complete restriction map for the linear DNA molecule.

PART 3   Mapping the Bacteriophage Lambda (λ) Chromosome

Bacteriophage lambda (λ) is a temperate phage; it can replicate autonomously or can convert *E. coli* to the **lysogenic** cycle by inserting into the *E. coli* chromosome as a prophage. Phage λ exists as a double-stranded DNA molecule of 48,502 base pairs. It can be either a linear or circular molecule because each end of the chromosome has a single-stranded tail (called the COS site). The tails are complementary (like sticky ends), allowing the linear molecule to easily convert to a circle. Because of its relatively small size, restriction enzyme digestion of phage λ DNA can be used to construct a restriction map.

Objectives

☐ Construct a restriction map of phage λ DNA from restriction digest fragment sizes.

☐ Use electrophoretic data to construct a map of phage λ DNA.

Procedure

1. Lambda DNA (48,502 bp) is cut using restriction enzymes *Afl*II (from *Anabaena flos-aquae*) and *Apa*I (from *Acetobacter pasteurianus*), as shown on the following page.

$$\overset{\downarrow}{}$$
*Afl*II –C–T–T–A–A–G

     –G–A–A–T–T–C–
$$\underset{\uparrow}{}$$

$$\overset{\downarrow}{}$$
*Apa*I –G–G–G–C–C–C–

     –C–C–C–G–G–G–
$$\underset{\uparrow}{}$$

The enzymes are used both alone and in combination, and fragments of the following approximate sizes (in bp) are produced:

*Afl*II	*Apa*I	*Afl*II + *Apa*I
5,872	10,086	2,532
6,078	38,416	3,546
6,540		5,872
30,012		6,540
		30,012

Use these data to establish the sequence of fragments and the restriction sites of *Afl*II and *Apa*I on λ DNA. Record this sequence as a linear restriction map (see Figure 14D-5) of bacteriophage λ on the line below.

|——|

42,502 bp

2. In Part 1 of the exercise you determined the restriction fragment sizes for lambda DNA digested by *Hind*III and *Eco*RI. Using the *Hind*III map (Figure 14D-6) for practice, determine the sizes of the known fragments (the length of DNA in bp between two restriction sites) by subtracting the restriction site bp designation at the left of the fragment from the bp designation on the right. For example, the fragment marked (*) is 25,157 bp − 23,130 bp = 2,027 bp in length. Record this known size below the *Hind*III map on the lines provided (_____ bp).

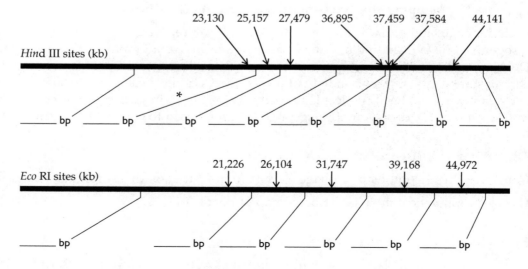

Figure 14D-6 *Lambda DNA restriction enzyme sites. The locations of "cuts" (arrows) are indicated in base pairs (bp) from the origin.*

Using your data from Part 1, compare your restriction fragment sizes determined by electrophoresis (see Tables 14D-1 and 14D-2) with those in the restriction-site maps of λ DNA in Figure 14D-6.

a. Are there any fragments shown on the map that are missing from your gel? _____

How could this be the case? _____

3. Now, using your own electrophoresis data for *Eco*RI fragments (Part 1), compare the restriction fragment sizes you determined by electrophoresis to the actual *Eco*RI fragment sizes recorded on the bacteriophage λ restriction map in Figure 14D-6.

b. How do the fragment sizes for EcoRI that you determined from your electrophoresis standard curve compare to those of known length as determined from the EcoRI map in Figure 14D-6?

_____

4. Use your *Eco*RI data from Part 1 to construct your own λ DNA map. By matching your fragment sizes (as determined after electrophoresis from your standard curve) to the known data for *Eco*RI (Figure 14D-6), you should be able to establish the correct order of the fragments. Some bands of similar size may migrate together.

├──┤

(λDNA)

Recall that the base sequences of the fragments of DNA produced by restriction enzyme digests can be determined by several means. This makes it possible to develop a complete genetic sequence of the chromosome. For larger pieces of DNA, a procedure called **"chromosome walking"** can be used. Two different restriction digests are used to cut identical pieces of DNA into fragments. Each of these DNA fragments is introduced into a bacterium via a plasmid vector. Each bacterium then clones its fragment as it replicates and forms a colony. Thus, each bacterial colony contains many copies (clones) of the same fragment, and

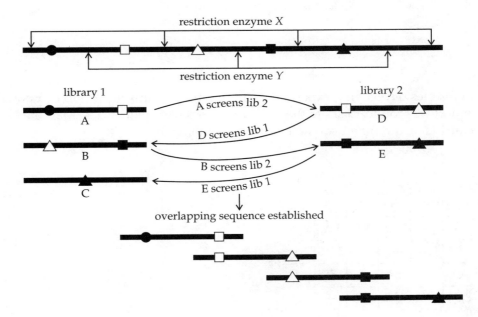

Figure 14D-7 *Chromosome walking. Restriction enzyme fragments are introduced into plasmids for cloning. All fragments formed by digesting DNA with Enzyme X form a library 1. All fragments formed by digesting DNA with Enzyme Y form library 2. By searching for complementary ends, hybridizing probes from library 1 clones with library 2 clones, it is possible to determine the linear sequence of DNA on a chromosome.*

there are many colonies, each containing a different "clone" fragment. The group of cloned colonies that develops from the fragments created by one restriction enzyme represents a **library.** The two resulting libraries can be used as probes to screen each other for overlapping complementary sequences; the right-hand end of one piece complements the left-hand end of the next piece and the right-hand end of the second piece complements the left-hand end of the next piece as if "walking" down the chromosome (Figure 14D-7).

Laboratory Review Questions and Problems

1. Ascospores of the fungus *Sordaria* are haploid. Why is this an advantage in studying the genetics of the organism?

2. Two genes for shell color, genes *A* and *B*, are on the same arm of a chromosome in a rare species of clam. Mating yellow-shelled individuals produced 800 yellow-shelled clams and 225 orange-shelled clams. Orange shells result from recombination of alleles in crossing-over events that occurred during meiosis and gamete production. What is the map distance between genes *A* and *B*?

3. There are four genes, *A*, *B*, *C*, and *D*, on a chromosome that you wish to map. These are the recombination frequencies among these genes: $B \times D = 4\%$, $B \times C = 10\%$, $D \times A = 2\%$, $C \times A = 16\%$, $C \times D = 14\%$. Map the chromosome.

4. In a series of breeding experiments among frogs, a linkage group composed of genes *A*, *B*, *C*, and *D* was found to show the following crossover frequencies. Map the chromosome. (Use the matrix like a Punnett square to show recombinations.)

	A	B	C	D
A	—	10%	4%	9%
B	10%	—	6%	19%
C	4%	6%	—	13%
D	9%	19%	13%	—

5. You are trying to map a newly isolated bacterial chromosome. After allowing different mutant strains to conjugate for different lengths of time, you test whether a certain gene (for example, leu^-) that was previously nonfunctional in the mutant is now functional (leu^+) due to transfer

of the gene from the donor strain and subsequent recombination. Only recombinants will grow on minimal medium. Use these data to map the *leu*, *pro*, *bio*, and *thi* genes on the chromosome. (*Note:* When the bacterial chromosome breaks open for replication and transfer, the F factor is split. The leading edge of the chromosome being transferred is on the *right* in the diagram below.)

		Growth on Minimal Medium			
Donor	Recipient	5 min	10 min	30 min	45 min
leu⁺ bio⁻ × *leu⁻ bio⁺*		Yes	Yes	Yes	Yes
bio⁺ pro⁻ × *bio⁻ pro⁺*		No	Yes	Yes	Yes
pro⁺ leu⁻ × *pro⁻ leu⁺*		No	No	Yes	Yes
pro⁻ leu⁺ × *pro⁺ leu⁻*		Yes	Yes	Yes	Yes
bio⁻ thi⁺ × *bio⁺ thi⁻*		No	No	No	Yes

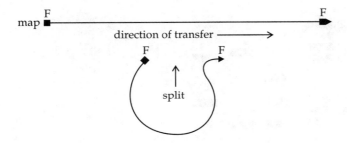

6. When samples of DNA are subjected to electrophoresis, why should the samples be loaded onto the gel at the negative pole of the electrophoresis apparatus?

7. Restriction endonucleases are used by bacteria as a form of "protection." Explain how this occurs and why the endonucleases do not destroy the bacterial cells themselves.

8. A circular bacterial DNA plasmid is cut using two restriction endonucleases. Restriction enzyme A yields a single linear molecule of 45,000 base pairs (4.5 kilobase pairs, or 4.5 kb). Enzyme B produces two restriction fragments of 1.2 kb and 3.3 kb. A combination of enzymes A and B produces three restriction fragments of 1.2 kb, 1.3 kb, and 2.0 kb. Map the plasmid, showing restriction sites for A and B and relative fragment lengths.

How many recognition sites are present for each restriction enzyme?

9. A length of human DNA from chromosome 2 has been cut by two restriction enzymes, A and B. Electrophoresis is carried out using *Hind*III fragment markers for a standard. The resulting gel is shown below. Note that enzyme A cuts the DNA into two pieces, and enzyme B cuts the DNA into two pieces, but of different sizes. When A and B are used together as a double digest, three pieces are produced.

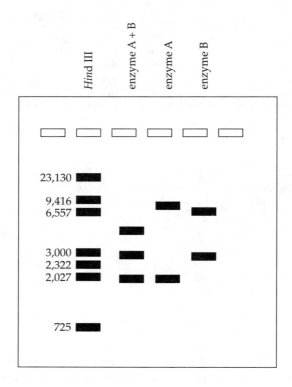

Use the *Hind*III fragments to make a standard curve, using Figure 14D-4b. Then determine the lengths of the fragments produced by enzymes A and B alone and A and B in combination.

To determine the base sequence of the fragments, you first need to know their order in the single piece of human DNA (this is like knowing the letters in three words but not knowing the order of the words in a sentence). In trying to find the order, you are mapping the gene. This is called a restriction map. How are the pieces of DNA ordered in the original piece of human DNA?

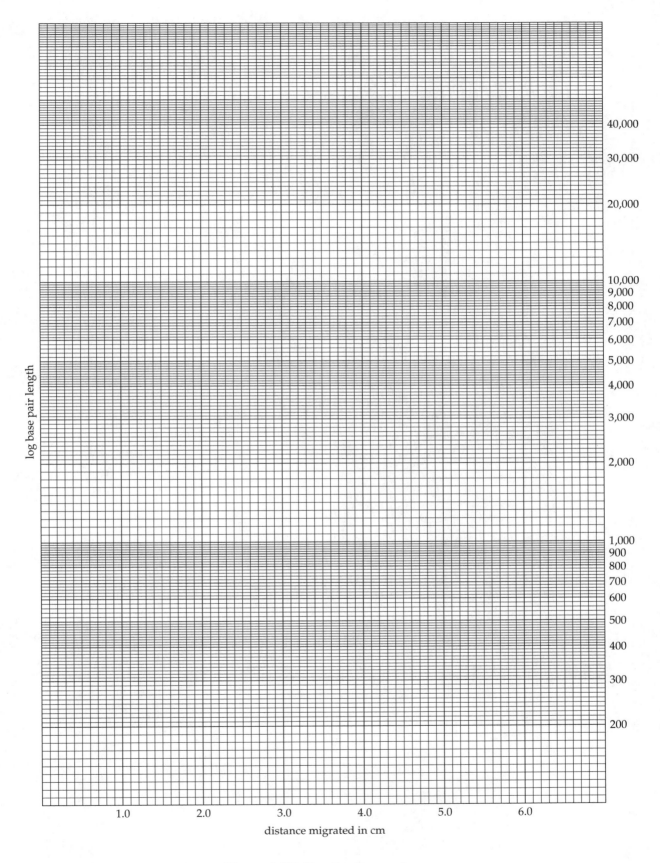

Figure 14D-4b *Semilog graph paper.*

Human Genetic Traits

OVERVIEW

All people are recognizably human, but no one is exactly like anyone else, not even an identical twin. The basis for the similarity and the reasons for the diversity that coexist in all species have puzzled and intrigued people for thousands of years. Recently, experiments with a wide variety of plants, animals, and microorganisms have yielded detailed knowledge of how traits are passed from one generation to the next, how genetic information is decoded and expressed during development, and how genetic variability can account for gradual evolution.

STUDENT PREPARATION

To prepare for this laboratory, read the text pages indicated by your instructor. Familiarizing yourself in advance with the information and procedures covered in this laboratory will give you a better understanding of the material and improve your efficiency. Work through Exercise A before coming to the laboratory.

EXERCISE A   Human Cytogenetics

Many human hereditary defects caused by chromosomal abnormalities may be identified by examining human chromosomes from cells that have been arrested in metaphase of mitosis—a stage when chromosomes are very short and compact. Leukocytes (white blood cells) or fetal cells obtained by amniocentesis or chorionic villus sampling are often used for diagnosis.

The cells are cultured (to increase their number), treated with a chemical that disrupts the mitotic spindle apparatus, and placed in a hypotonic salt solution to swell their nuclei. The mixture is then centrifuged (to increase the concentration of cells) and transferred to a glass slide. As a drop of the cell

Figure 15A-1   *G-banded chromosomes. This pattern is produced by using Giemsa stain. Bands do not represent genes, but they do serve as markers for locating genes and gene families.*

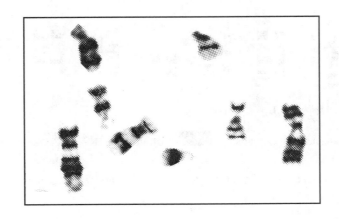

suspension hits the slide, the nuclei break open and the chromosomes spread apart; usually chromosomes from a single cell remain in an identifiable group. The cells are then stained, sometimes using special procedures that result in banded chromosomes (Figure 15A-1).

The "metaphase spread" produced by a single cell is then photographed. The photograph can be cut apart and homologous chromosomes can be arranged in pairs according to size, location of the centromeres, and length of the chromosome arms. Chromosome pairs are arranged in a specific order and labeled. The result is called a **karyotype.**

▍▍▍▍▍ Objectives ▍▍▍▍▍▍▍▍▍▍▍▍▍▍▍▍▍▍▍▍▍▍▍▍▍▍▍▍▍▍▍▍▍▍▍▍▍▍

☐ Match and order pairs of human chromosomes to make a karyotype.

▍▍▍▍▍ Procedure ▍▍▍▍▍▍▍▍▍▍▍▍▍▍▍▍▍▍▍▍▍▍▍▍▍▍▍▍▍▍▍▍▍▍▍▍▍▍

1. Figure 15A-2 is a diagrammatic representation of G-banded human chromosomes (only one chromosome from each homologous pair is represented). Figure 15A-3b (page 15-20) is a metaphase preparation of banded chromosomes. Cut out these chromosomes and match them to the chromosomes shown in Figure 15A-2. Match homologous chromosomes by size, length of arms, and location of the centromere. Place the homologous pairs together above corresponding numbers in Figure 15A-3a (page 15-3). A sample karyotype, Figure 15A-4, will also assist you in matching chromosome pairs by size and banding.

2. Note that the X chromosome has a single thick band on its upper end and four bands on its lower end. The Y chromosome is very small and has a single thick band on the tips of its arms at one end.

3. Once you have matched the chromosomes, tape them onto the blank karyotype sheet (Figure 15A-3a).

   a. Are these chromosomes from a male or a female? _____

Bring the completed karyotype with you to the laboratory.

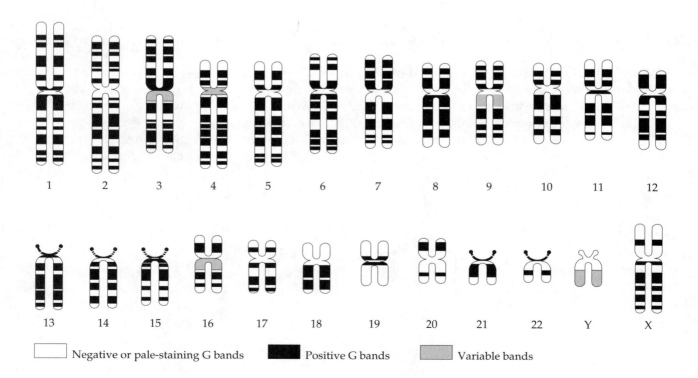

Figure 15A-2 *Diagram of G-banded human chromosomes.*

Figure 15A-3a *Match homologous chromosome pairs from Figure 15A-3b (page 15-20) and attach them in appropriate spaces above. Use the karyotype (Figure 15A-4) to help you match the pairs by size and banding patterns.*

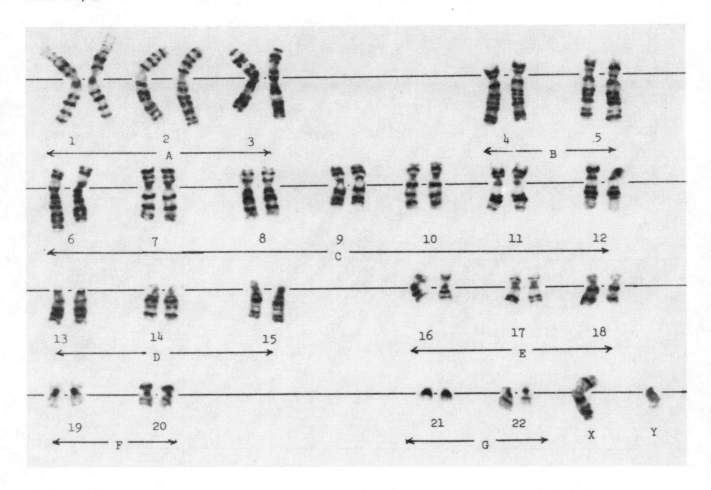

Figure 15A-4   *Karyotype of G-banded human chromosomes.*

✔ **EXERCISE B**   **X and Y Chromosomes**

An individual's sex chromosomes can be identified more simply than by karyotyping. Cells scraped from the inside of the mouth can be stained with a dye specific for DNA. If the individual is female, a darkly staining **Barr body,** or sex-chromatin body, appears near the nuclear membrane (Figure 15B-1). This structure is one of the *X* chromosomes that has condensed; only one *X* chromosome remains active in females. No such body is found in male cells.

Figure 15B-1   *Part of a cell from the squamous epithelium of a female, showing nucleus with Barr body (top).*

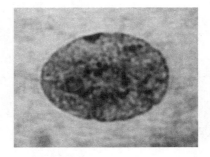

a. How many X chromosomes are found in male cells? _____

Both males and females ultimately have just one active X chromosome, which provides the correct amount of X chromosome products required for the normal functioning of cells in both males and females. In cases where more than two X chromosomes are present (see Exercise D, Part 1), the number of Barr bodies in a cell is $N - 1$, where N is the number of chromosomes in the cell.

While Barr-body stains reveal the number of X chromosomes present, they do not reveal anything about the presence or number of Y chromosomes. Instead, a special stain, acridine orange, is used to stain the Y chromosome, which then stands out as a very bright fluorescent structure in the cell.

Objectives

☐ Describe how, without karyotyping, an individual's sex chromosomes composition can be determined.

Procedure

Examine the two slides on demonstration. Both slides show cheek scrapings, one from a female and the other from a male. In the space provided below, draw what you see on these slides. Determine which slide shows tissue from a female and which shows tissue from a male.

✔ 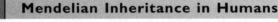 **EXERCISE C**　Mendelian Inheritance in Humans

✔ **PART I**　Monohybrid Crosses in Humans

Several human traits may be used to demonstrate Mendel's law of segregation. The traits you will study are controlled by a single gene that can occur in two forms (alleles). One allele is dominant, the other is recessive. A person's phenotype is determined by the combination of alleles present. An individual can be homozygous dominant, heterozygous, or homozygous recessive.

Objectives

☐ Describe the genetic basis for several human traits evident in the classroom population.

☐ Determine the extent to which individuals are different from one another based on the presence or absence of selected genetic traits.

Procedure

For each of the following 11 traits, determine your phenotype and enter it in Table 15C-1. Based on your phenotype, record your possible genotypes for each of the traits. Remember that if you have a recessive characteristic you must have both recessive alleles, but if you have a dominant characteristic you may be either homozygous dominant or heterozygous. If you have a dominant phenotype, you may have no way of knowing whether you carry a recessive allele. In this case, use a dash (—) to represent the unknown second gene.

With the help of your laboratory instructor, compile the data for your section. Calculate the percentages for each trait.

1. **Dimpled chin.** A cleft in the chin is a dominant trait. (*D*, allele for dimpling; *d*, allele for absence of a dimple.)

2. **Free ear lobe.** In most people the ear lobes hang free (dominant allele, *E*), but in a person with two recessive alleles (*e*), the ear lobes are attached directly to the head. Use a mirror or the opinion of your classmates to determine your phenotype.

3. **Widow's peak.** The action of the dominant allele (*W*) results in a hairline that forms a distinct point, known as a widow's peak, in the center of the forehead. The recessive allele (*w*) produces a continuous hairline. Omit this tabulation if a gene for baldness has had some effect on your hairline.

4. **Ability to taste PTC.** Some persons detect a distinct bitter taste in small concentrations of the chemical phenylthiocarbamide (PTC), while others do not taste it. A dominant allele *T* confers the ability to taste this chemical; those who are homozygous for the recessive allele *t* are nontasters. Place a PTC paper strip on your tongue and allow it to remain there for about 10 seconds. If you are a taster you will know it. If you have any doubt about your ability to taste the substance, you are a nontaster.

5. **Interlocking fingers.** When the fingers are interlocked, some people will almost invariably place the left thumb on top of the right (dominant allele *F*), whereas others will place the right over the left (recessive allele *f*).

6. **Bent little finger.** A dominant allele *B* causes the last joint of the little finger to bend inward toward the fourth finger (*b* is the recessive allele for a straight finger). Lay both hands flat on the table, relax your muscles, and note whether you have a bent or a straight little finger.

7. **Hitchhiker's thumb.** This characteristic, more precisely called distal hyperextensibility of the thumb, can be determined by bending the distal joint of the thumb back as far as possible. While there tends to be some degree of variation, certain individuals can bend it back until there is almost a 90-degree angle between the two joints. This characteristic is an effect of a recessive allele *h* (dominant allele, *H*).

8. **Long palmar muscle.** A person homozygous for a recessive allele *l* has a long palmar muscle that can be detected by examination of the tendons running over the inside of the wrists. Clench your first tightly and flex your hand. Now feel the tendons. If there are three, you have the long palmar muscle. If there are only two tendons (the large middle one will be missing) you do not have this muscle. Examine both wrists—if you find this trait in one or both wrists you have two recessive alleles. If not, you have the dominant allele *L*.

9. **Pigmented irises.** When a person is homozygous for the recessive allele *p*, there is no pigment in the front part of the eyes and a blue layer at the back of the iris shows through, resulting in blue eyes. A dominant allele of this gene, *P*, causes pigment to be deposited in the front layer of the iris, thus masking it blue to varying degrees. Other genes determine the exact nature and density of this pigment, thus there are brown, hazel, violet, green, and other eye colors. Here, you are concerned only with the presence or absence of such pigment.

10. **Mid-digital hair.** Some people have hair on the second (middle) joint of one or more of the fingers, while others do not. The complete absence of hair on this joint for all fingers is due to a recessive allele *m* and the presence of hair is due to a dominant allele *M*. There seem to be a number of alleles determining whether hair will grow on one, two, three, or four fingers. This hair may be very fine, so you should use a hand lens to look carefully on all fingers before deciding whether this hair is present on any one of your fingers, indicating the presence of dominant allele *M*.

11. **Second (index) finger shorter than the fourth.** This is a characteristic that appears to be sex-influenced. Use the symbol S^S for a shorter second finger and the symbol S^L for a longer second finger. Tabulate your results by sex, since the frequency should vary by sex.

   a. *For the traits observed, did you find that the dominant alleles were expressed most often in your laboratory section?* _____

   b. *For which traits was this not true?* _____

   _____

   c. *How do you explain your results?* _____

   _____

Table 15C-1   Genetic Traits

Characteristic	Your Phenotype	Your Possible Genotypes	Data for Your Laboratory Section	
			Number of Each Phenotype	Percentage
1. Dimpled chin (D)				
Nondimpled chin (d)				
2. Free ear lobes (E)				
Attached ear lobes (e)				
3. Widow's peak (W)				
No widow's peak (w)				
4. Taster of PTC (T)				
Nontaster (t)				
5. Left thumb on top (F)				
Right thumb on top (f)				
6. Bent little finger (B)				
Finger not bent (b)				
7. Hitchhiker's thumb (h)				
Normal thumb (H)				
8. Long palmar muscle (l)				
Two tendons only (L)				
9. Pigmented iris (P)				
Unpigmented iris (p)				
10. Mid-digital hair (M)				
No mid-digital hair (m)				
11. Shorter second finger (S^S)			♂	♂
Longer second finger (S^L)			♀	♀

👁 **PART 2**   **How Individual Is Each Individual?**

⦙⦙⦙⦙⦙ Procedure ⦙⦙⦙

1. Pick a member of your class to serve as an "individual."

2. Everyone in the class should stand up.

3. Have the "individual" call out his or her phenotype for each of the traits studied. As each phenotype is called out, all those who do not have that phenotype should sit down.

   a. *How many characteristics must be considered before the "individual" stands out as a unique individual?* _____

   b. *In the United States, a great mixing of genotypes has taken place through immigration and intermarriage. In a country where there has been little immigration, would you expect an individual to stand out sooner or later than occurred in your class?* _____

4. Instead of considering the inheritance of only one of the traits just studied, consider the inheritance of two of these traits. All of these traits are unlinked, thus Mendel's principle of independent assortment applies.

   c. *What is the phenotype of a person with the genotype EETT?* _____

   d. *What alleles would be present in gametes produced by this individual?*

   _____

   e. *What is the phenotype of a person with the genotype eett?* _____

   f. *What alleles would be present in gametes produced by this individual?*

   _____

   g. *If the two homozygous individuals above (EETT and eett) produce offspring, what would be the expected genotypes and phenotypes of their offspring?*

   Genotypes _____ *Phenotypes* _____

   h. *Assume that two individuals heterozygous for both of these traits (EeTt) marry and produce offspring. What would be the expected genotypes and phenotypes of their offspring?*

   Genotypes _____ *Phenotypes* _____

✔ **EXERCISE D**   **Chromosomal Abnormalities—Nondisjunction and Translocation**

Mitosis and meiosis are usually very exact processes that result in the correct distribution of chromosomes to the daughter cells. However, mistakes occasionally result in an abnormal number of chromosomes or pieces of chromosomes in the daughter cells. Usually, zygotes with abnormal chromosomal compositions are spontaneously aborted. However, if the combinations are not lethal at an early stage in development, the phenotype and viability of the resulting individual may be seriously affected.

Recall that during meiosis, homologous chromosomes synapse during prophase I and then separate from each other during anaphase I (Figure 15D-1a). Sometimes a pair of chromosomes may adhere so tightly that they do not pull apart during anaphase I. This will result in one of the daughter cells receiving duplicate chromosomes and the other receiving none of that type of chromosome (see Figure 15D-1b). This failure of chromosomes to separate is called **nondisjunction.** Nondisjunction may also occur during the second meiotic division if the chromatids of a dyad chromosome do not separate from each other (Figure 15D-1c).

Figure 15D-1 *Meiosis in humans (n = 23). (a) Normal meiosis. (b) Nondisjunction at meiosis I. (c) Nondisjunction at meiosis II. Numbers indicate the number of chromosomes present in the cell after each division.*

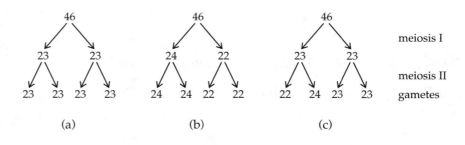

Other chromosomal abnormalities are caused by chromosome breakage and redistribution of chromosome parts. Actual breakage of the chromosome can be caused by a variety of factors, including ionizing radiation and certain drugs and chemicals. The transfer of a portion of one chromosome to another, usually nonhomologous, chromosome is called a **translocation.** The abnormal chromosome may then be passed on from parent to offspring. Some cases of Down syndrome are caused by a translocation in which a large portion of chromosome 21 becomes attached to chromosome 15.

ⅠⅠⅠⅠⅠ Objectives ⅠⅠⅠⅠⅠⅠⅠⅠⅠⅠⅠⅠⅠⅠⅠⅠⅠⅠⅠⅠⅠⅠⅠⅠⅠⅠⅠⅠⅠ

☐ Define and distinguish between nondisjunction and translocation.

☐ Explain the chromosomal basis of Klinefelter's, Turner's, triple-X, Jacob's, and Down syndromes.

✔ **PART Ⅰ** **Nondisjunction: Sex Chromosomes**

Normal meiosis in a human female results in the production of egg cells with 22 autosomes and one X chromosome. Normal meiosis in a human male results in the production of two types of sperm: one half with 22 autosomes and one X chromosome and one half with 22 autosomes and one Y chromosome.

a. *If an X-containing egg is fertilized by an X-containing sperm, an XX zygote will be formed. Will this individual be male or female?* _____

b. *How many Barr bodies (sex-chromatin bodies) will the zygote have (see Exercise B)?* _____

c. *If an X-containing egg is fertilized by a Y-containing sperm, an XY zygote will be formed. Will it be male or female?* _____

d. *How many Barr bodies will it have?* _____

If nondisjunction of the X chromosomes occurs during meiosis in a human female, some of the eggs will contain two X chromosomes and others will contain no X chromosome. In a male, nondisjunction can result in four kinds of sperm: nondisjunction during meiosis I will result in sperm with both the X and Y chromosomes and sperm with no sex chromosome, designated *0*. Nondisjunction during meiosis II may result in sperm with *XX* and sperm with no sex chromosome (*0*) or in sperm with *YY* and sperm with *0*. Gametes produced by nondisjunction usually fertilize, or are fertilized by, normal gametes.

ⅠⅠⅠⅠⅠ Procedure ⅠⅠⅠⅠⅠⅠⅠⅠⅠⅠⅠⅠⅠⅠⅠⅠⅠⅠⅠⅠⅠⅠⅠⅠⅠⅠⅠⅠⅠⅠⅠⅠⅠⅠⅠ

Several syndromes resulting from nondisjunction in humans are listed below.

Triple-X syndrome The individual will develop into a normal-appearing female, but may be sterile. She may also be mentally retarded.

Klinefelter's syndrome The individual will develop as a male. During early development he appears normal, but abnormalities become apparent at puberty. Testes do not fully develop. The person is usually taller than average, his muscular development may be somewhat feminine, breast development may occur, and his voice may be higher-pitched than normal. Although the individual may develop a "female" appearance, fluorescent cell staining always reveals the presence of a Y chromosome.

Turner's syndrome The individual will be female. She appears normal during early development, but at puberty does not menstruate, breasts do not develop, and no eggs are produced by the ovaries.

Table 15D-1   Nondisjunction in a Human Female

Abnormal Gamete (egg)	Normal Gamete (sperm)	Zygote (genotype)	Expected Sex	Number of Barr Bodies	Name of Syndrome
XX	X				Triple-X syndrome
XX	Y				Klinefelter's syndrome
0	X				Turner's syndrome
0	Y				Nonviable, not seen

Table 15D-2   Nondisjunction in a Human Male

Abnormal Gamete (egg)	Normal Gamete (sperm)	Zygote (genotype)	Expected Sex	Number of Barr Bodies	Name of Syndrome
X	XX				Triple-X syndrome
X	XY				Klinefelter's syndrome
X	0				Turner's syndrome
X	YY				Jacob's syndrome

Jacob's syndrome The individual appears to be sexually normal.

Note that all of these syndromes, except Jacob's syndrome, may result from nondisjunction in either the male or the female parent.

Based on the information given in this and previous exercises, complete Tables 15D-1 and 15D-2.

✔ **PART 2   Nondisjunction: Autosomes**

Autosomal nondisjunction involves chromosomes other than the sex chromosomes. The incidence of Down syndrome caused by autosomal nondisjunction of chromosome 21 increases with the age of the mother (although evidence suggests that the father may sometimes be responsible for providing the extra chromosome).

a. What explanations might be proposed for the relationship between increasing maternal age and the increasing

frequency of Down syndrome? _____

_____

Down syndrome—trisomy 21 Three number 21 chromosomes are present. Individuals are characterized by a fold of the upper eyelid, short stature, broad hands, stubby feet, a wide, rounded face, a large tongue, and mental retardation.

Trisomy 18 This syndrome involves an extra chromosome 18. Individuals are characterized by a misshapen skull, eye problems, overlapping fingers, heart defects, feeding problems, and severe mental and developmental retardation.

⦚⦚⦚ **Procedure** ⦚⦚⦚

Work in pairs. Obtain a blank karyotype form and a copy of a photograph of human chromosomes. The chromosomes may be banded or simply stained with Feulgen stain. Carefully cut out the individual chromosomes. Using as a guide the sample karyotype that you prepared in Exercise A, arrange each homologous pair of chromosomes on the blank karyotype form. (Remember that the comparison of bands is helpful.) Do not fasten the chromosomes until they have been checked by your instructor. Keep all scraps until you have identified each chromosome. Now tape the chromosomes to the karyotype form. Identify the karyotype you have made. The possible choices are normal female; trisomy 21 (Down syndrome) male; trisomy 21 (Down syndrome) female; trisomy 18 male or female.

✔ 👁 **EXERCISE E | Constructing a Human Pedigree**

Genetic analysis of the inheritance pattern for a specific human trait often requires collecting information about a family's genetic history and using it to construct a pedigree chart, which traces the occurrence of particular characteristics through several generations, or "lines." Such pedigrees enable genetic counselors to derive facts about a couple's genetic makeup and thereby accurately calculate the chances that their offspring will inherit detrimental genes.

The pedigree chart provides a systematic and convenient method for recording data. One set of symbols and arrangements commonly used in pedigree construction is given in Table 15E-1. Individuals are designated by specific symbols, ◯ for female and ▢ for male. Phenotypes are represented by shading the symbols. Familial relationships must also be indicated on pedigree charts: marriages, offspring, siblings, and identical or fraternal twins. For easy reference, symbols are often numbered, and birth or death dates are recorded beside the symbols.

||||| **Objectives** |||||||||||||||||||||||||||||||||||||||

☐ Construct a human pedigree from case history information.

||||| **Procedure** |||||||||||||||||||||||||||||||||||||||||

1. Read the true case history that follows. Analyze and complete the pedigree chart in Figure 15E-1 using the symbols given in Table 15E-1.

Case History 147*

Jane M. requested genetic counseling because she and her husband were planning to begin a family and were concerned because Jane's brother, Brian, was moderately retarded.

The family history indicated that Brian was the third of four children. The pregnancy was uncomplicated, but delivery required the use of forceps. Shortly after his birth, the family noticed that Brian was fussy and was slow to crawl and sit. After a neurological evaluation, Brian's parents were told that he had brain damage, probably due to the forceps delivery. Brian had attended special schools and now, at the age of 22, lived at home and had no major physical problems.

Jane's and Brian's mother, Mrs. S., indicated that her daughter, Carol, had a 6-year-old daughter who required a special education program. Another daughter, Nancy, had a son, 4 years old, considered to be developmentally delayed. Brian's mother denied that any of her three daughters, Jane, Carol, or Nancy, were mentally retarded or had any learning disability, but it was noted during counseling that Brian's mother, as well as Jane and Carol, appeared to have below-average intelligence.

After genetic screening tests, Brian was diagnosed as having a condition known as "fragile X" syndrome. This X-linked recessive condition can be identified by karyotype studies of cells cultured in media deficient in specific nutrients. Under these conditions, the X chromosome breaks very close to the end of its long arm and releases a fragment. Males who carry the fragile X chromosome are usually severely retarded. If the carrier of the fragile X chromosome is female, it is likely that, in the heterozygous condition, the presence of a normal X chromosome will prevent severe mental retardation. Protection by the normal X chromosome is not complete, however, for heterozygous female carriers usually exhibit mild forms of mental retardation.

After further screening of family members, it was explained to the family that Brian's mother carried the fragile X chromosome, as did Jane and Carol, but Nancy's karyotype did not reveal the fragile X chromosome. Women who carry the fragile X are at 50 percent risk of having affected sons and 50 percent risk of having girls who are "carriers." Nancy would not be at risk for having affected children and fragile X was not the explanation for her child's developmental delay.

Further investigation revealed additional information about the family background as noted in the partial pedigree (Figure 15E-1).

2. Examine Figure 15E-1. On a separate sheet of paper, write a paragraph describing the family history prior to that for Mrs. S.

---

*Case history information was provided by Dr. Nina Caris, Department of Biology, Texas A & M University.

Table 15E-1 Symbols Used in Pedigree Analysis

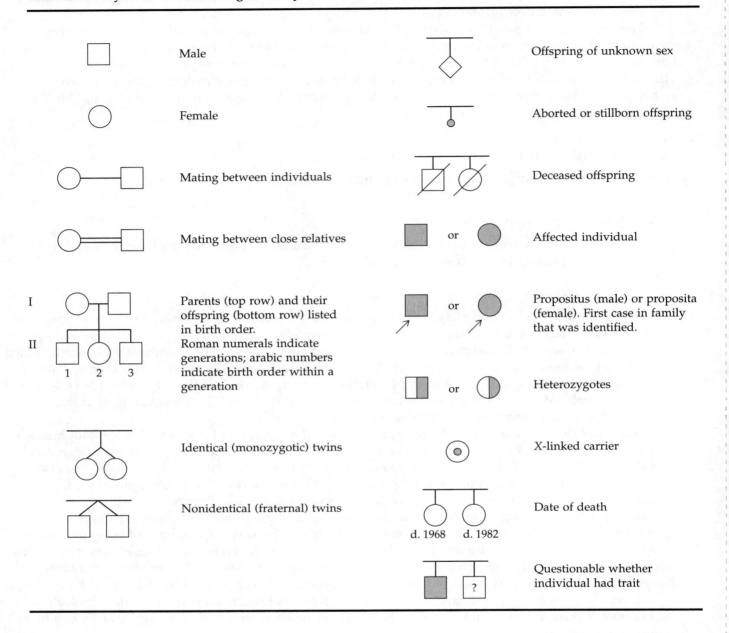

Male

Female

Mating between individuals

Mating between close relatives

I Parents (top row) and their offspring (bottom row) listed in birth order.
II Roman numerals indicate generations; arabic numbers indicate birth order within a generation

Identical (monozygotic) twins

Nonidentical (fraternal) twins

Offspring of unknown sex

Aborted or stillborn offspring

Deceased offspring

or   Affected individual

or   Propositus (male) or proposita (female). First case in family that was identified.

or   Heterozygotes

X-linked carrier

Date of death

d. 1968   d. 1982

Questionable whether individual had trait

Figure 15E-1 *Partial pedigree for the fragile X syndrome in Mrs. S.'s generation and in the generation preceding hers. Complete the chart by drawing the symbol for Mrs. S. in the position indicated, then add symbols (see Table 15E-1) for the family members described in the case study.*

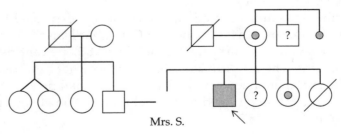

Mrs. S.

3. Complete the pedigree (Figure 15E-1) for Mrs. S., her children, and grandchildren according to the information given in the family history.

   a. *How would you counsel Jane? And, if Carol sought genetic counseling about having further children, how would you counsel her? (Write your answer on a separate sheet.)*

   b. *If Carol's daughter marries and decides to have children, what would you tell her if she sought your counsel? (Write your answer on a separate sheet of paper.)*

   c. *What would you infer about the genotype of Mrs. S.'s grandparents?* _____

   d. *What might have been the genotype of Mrs. S.'s uncle?* _____

EXERCISE F   Forensic Science: DNA Fingerprinting

In human DNA, there are many long sequences of base pairs that are similar in all individuals. Some of these sequences code for proteins, but others do not. (In fact, because most of the human genome does *not* code for protein, a large amount of variation can occur without consequence.) Mutations in coding regions may randomly eliminate restriction sites (places where restriction enzymes cut DNA) or form new ones. Thus, if the same restriction enzyme is used to cut the DNA from two individuals, the resulting fragments from corresponding allelic regions may be of different lengths. These different-length pieces are called **RFLPs** (pronounced "riflips"), **restriction fragment length polymorphisms** (Figure 15F-1a).

Restriction fragments of different lengths (RFLPs) can be separated on a gel by electrophoresis to produce a **DNA fingerprint;** the pattern of bands on the gel will be different for each individual (except for identical twins). The steps involved in analyzing DNA fingerprints, used to identify or match individuals (Figure 15F-2), can be summarized as follows:

1. Once RFLPs are separated on a gel, the gel is treated with an alkaline solution to denature the DNA. The two strands of the DNA separate, forming single strands.

2. A sheet of nitrocellulose paper or a nylon membrane is blotted onto the gel to pick up the single-stranded DNA fragments.

3. Artificially synthesized, radioactive, single-stranded, DNA fragments called **probes,** complementary in sequence to portions of specific RFLPs, are applied to the blotting sheet. They base-pair only with complementary sequences. Excess probe is rinsed off.

4. The blotting sheet is used to expose X-ray film. Since only radioactive materials will expose the film, only those sequences that have base-paired with the radioactive probes will show up on the film. The resulting pattern is called a DNA fingerprint.

Since mutations in DNA are cumulative over time, more closely related individuals and groups can be recognized by their genetic fingerprints: similar banding patterns of RFLPs. Thus, DNA fingerprinting can be used not only to identify genetic material from the same individual, but also to establish paternity and to discern evolutionary relationships.

For added precision, especially when forensic evidence is needed, RFLP analysis is accompanied by another type of DNA fingerprinting that analyzes the **variable number of tandem repeats (VNTRs)** present in genomic DNA. Within the human genome, short nucleotide sequences of 5 to 10 base pairs may be repeated over and over again (in tandem, and often head to tail) to form longer sequences of 20 to 200 base pairs. Often these tandem nucleotide sequences occur in front of or behind single-copy genes that code for proteins.

The sequence of base pairs in tandem repeats shows little variation from individual to individual, but the *number* of repeats can vary greatly, leading to different fingerprint patterns among individuals. Differences in length among arrays of tandem repeats result from unequal crossing-over during meiosis. VNTRs protect genes from damage by allowing for some "slippage" in front of and behind a gene being "cut out" of one chromosome and "pasted into" or recombined with another during crossing-over. As a consequence, the lengths of tandem repeat arrays located close to a particular allele on maternal and paternal homologous chromosomes belonging to the same individual also may differ (Figure 15F-3).

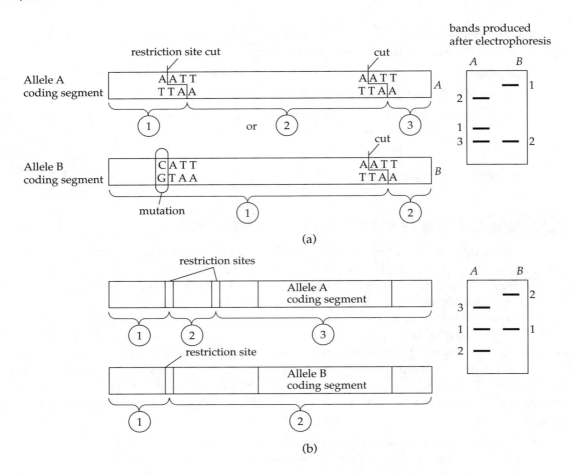

Figure 15F-1 *DNA fingerprinting using RFLP sequences. Mutations in DNA can result in eliminating or forming new restriction sites where restriction endonucleases can cut the DNA to form fragments of varying lengths (RFLPs).*

(a) Elimination of a restriction site within the coding sequence of an allele due to a base-pair change (converting allele A to allele B) results in only two fragments being formed during enzymatic digestion, rather than three. Since the fragments vary in length, a different fingerprint is obtained for alleles A and B.

(b) If restriction sites in a DNA sequence preceding an allele are affected, rather than the allele itself, differences can also be detected from DNA fingerprints. Often, a unique fingerprint pattern results from a mutation in a noncoding sequence found in front of a DNA sequence (allele) suspected of causing a specific genetic disorder. If this pattern is consistently found among those with the genetic disorder (pattern B in this case), this fingerprint pattern can be used for diagnostic purposes even though identification of the gene is not conclusive and its specific location is not known.

To obtain a DNA fingerprint using VNTR regions, the DNA is cut with restriction enzymes that do *not* cut within the tandem repeat area. Instead, they cut in front of and behind the repeat sequence. The polymerase chain reaction (PCR) is used to augment the number of copies of the isolated repeat areas. (For accuracy, several different repeat areas are analyzed by cutting with different restriction enzymes and are then amplified using PCR.) The PCR reaction uses short DNA primers to synthesize many copies of the VNTR regions to be used for electrophoresis and fingerprinting (Figure 15F-3). PCR makes it possible to obtain the required amount of DNA from even the smallest traces of blood, semen, and other tissues and cells of an individual.

Figure 15F-2 *DNA fingerprinting. (a) DNA is broken into fragments by restriction endonucleases. (b) The fragments are separated by gel electrophoresis, then the gel is treated by heat or an alkaline solution to separate the strands. (c) A sheet of nylon or nitrocellulose film is used to "blot" the gel and pick up DNA. (d) When radioactive DNA probes are added to the sheet, they base-pair only with complementary strands in various regions, producing a DNA fingerprint. (The sheet is then rinsed to wash away any probes that do not hybridize so that these pieces will not interfere with the pattern.) (e) The blotting sheet is allowed to expose X-ray film. The radioactive probes will identify the RFLPs of interest— only these will be visible on the film.*

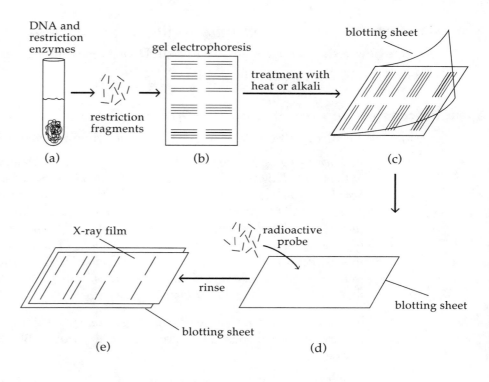

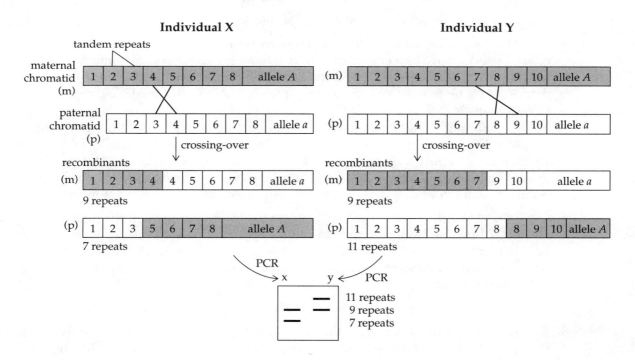

Figure 15F-3 *VNTR differences in homologous chromosomes. The number of tandemly repeated sequences of DNA associated with particular genes on a chromosome differs widely among individuals. These differences are generated by unequal crossing-over, so that even maternal and paternal homologues of the same individual, as well as different individuals, have different numbers of sequences associated with the same allele.*

||||| **Objectives** |||||||||||||||||||||||||||||||||||||||||

☐ Describe how restriction enzymes are used in DNA fingerprinting.

☐ Interpret results from a DNA fingerprint.

||||| **Procedure*** |||||||||||||||||||||||||||||||||||||||||||

A crime has been committed. A brief scenario of the crime is given below.

A woman jogging through a mid-city park at dusk was accosted by a man who dragged her into a stand of trees and bludgeoned her with a baseball bat. The man fled when interrupted by a fellow jogger who heard the woman's screams. The jogger who witnessed the crime saw the man flee from the scene but could not identify the suspect for certain because it was fairly dark at that time of evening. In the police lineup, the witness tentatively identified a suspect, but the man standing next to him also looked a lot like the alleged attacker. Fortunately, the victim had pulled some hair from the attacker's head. DNA was extracted from the papillae of the hairs and a DNA fingerprint was developed to compare the DNA from the evidence (hairs), and the two suspects, X and Y. (Electrophoresis of PCR-amplified VNTRs was used to create the needed forensic evidence.)

1. Insert a gel comb into an electrophoresis gel tray and cast an agarose gel as described in Laboratory 14, Exercise D. (Add 2 drops of Carolina Blu stain to 50 ml of agarose, if instructed to do so.)

2. When the gel has solidified (about 10 minutes), place it in the gel box so that the comb is at the negative (black) end.

3. Fill the gel box with TBE (tris-borate EDTA) buffer until the buffer covers the gel. (Add Carolina Blu stain to the buffer if instructed to do so.)

4. Gently remove the gel comb and check for bubbles.

5. Add DNA samples to the wells using a microcapillary pipette. Load the contents of each tube—Suspect X-1, Suspect X-2, Evidence 1, Evidence 2, Suspect Y-1, and Suspect Y-2—into separate wells. Use a clean pipette for each. Position the pipette above the well to load. *Do not put the pipette tip into the well* or you may punch a hole in the gel! (See Laboratory 14, Exercise D, for details.)

    Note: During PCR, two different primers were used to produce multiple copies of the suspects' DNA and the evidence DNA. Primer 1 was used to produce the DNA of Suspect X-1, Evidence 1, and Suspect Y-1. Primer 2 was used to produce the DNA of Suspect X-2, Evidence 2, and Suspect Y-2. To match a suspect to crime scene evidence, the PCR products from Primer 1 must match Evidence 1 and the products from Primer 2 must match Evidence 2.

6. Close the electrophoresis apparatus, connect the leads, and turn on the power supply to 80 volts. Allow the electrophoresis to run for approximately 2 hours or until the bromophenol blue band nears the end of the gel (see Laboratory 14, Exercise D, for details). If Carolina Blu stain has been added to the agarose and buffer, you will also see faint blue bands in the "fingerprint" pattern.

7. Turn off the power supply, disconnect the leads, and open the gel box.

8. Remove the gel and stain it using the procedure described by your instructor (staining procedures vary with the preparation of agarose), using either methylene blue or Carolina Blu Final Stain.

9. Analyze the gel by placing it on a light box or transilluminator. You may wish to use plastic wrap or transparency film to make "copies" of the gel and its bands for analysis. (See Laboratory 14, Exercise D, for details.)

---

*This exercise is adapted from the Carolina Biological Supply Kit 21-121-, PCR Forensic Simulation Kit.

a. Which "suspect" DNA matches the "evidence" DNA? Who committed the crime?

_____

b. If identical twins had been included among the suspects, what would you have observed in the DNA fingerprint pattern? _____

Laboratory Review Questions and Problems

Use a separate sheet of paper to answer the following questions. Show all of your work. Indicate all genotypes and phenotypes, where appropriate. If more than one type of offspring is produced, indicate the proportion of each type.

1. A man with attached ear lobes marries a woman who is heterozygous for free ear lobes. What types of offspring and what proportion of each type would be expected from this mating?

2. Polydactyly is an inherited human trait in which the affected individual has extra fingers or toes. This defect results from the dominance of the defective gene (*P*) over the normal allele (*p*).

a. If a man with polydactyly (assume that he is heterozygous) marries a normal woman, what are the possible genotypes of their children?

b. What are the phenotypes of these genotypes?

c. If two individuals heterozygous for this trait marry, what are the possible genotypes of their children?

d. What are the phenotypes of these genotypes?

3. Formation of Barr bodies takes place when a zygote has already divided to form a multicellular embryo. Because of this, the body of a female can be a "mosaic of sex-linked phenotypes." Explain.

4. Color blindness is an *X*-linked recessive trait. The phenotypes within a family pedigree are shown below. Determine the genotypes of each individual.

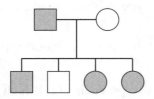

5. A color-blind man marries a woman whose father was color-blind but whose mother had normal vision. Would it be possible for the couple to have a color-blind son? A color-blind daughter? What is the probability of each of these events happening?

6. The inheritance of human blood types is dependent on multiple alleles. The ABO blood system, characterized by three alleles, is commonly used to determine the suitability of donors and recipients for blood transfusions.

Normally, the body does not store antibodies for proteins it has never encountered. The ABO blood system, however, is an exception to this rule. In this system, antibodies to other blood types occur naturally whenever the antigen (A or B) is not present. An individual with type A blood possesses type A antigen and type B antibody; an individual with type B blood possesses type B antigen and type A antibody; type O blood contains neither the A nor the B antigen, but both antibodies; type AB blood contains both antigens and neither antibody. As a result, if, for instance, a type A person receives type B blood, the B antigen will be agglutinated by the B antibodies present in the type A blood. Antibodies in the donor's blood are generally of little consequence because they are so diluted in the recipient's blood.

The general rule in transfusions is never to allow an individual to receive an antigen that does not occur in his or her blood. Since type O blood contains neither antigen A nor antigen B, this blood type is considered as the "universal donor" and may be transfused to any blood type. To study the inheritance of the ABO alleles, the symbols I and i with superscripts are used, since more than two alleles must be considered (capital and lowercase letters will not suffice). I^A, I^B, and i represent alleles of antigens A and B and the allele for blood type O (which produces no antigens), respectively.

An individual homozygous for type A blood ($I^A I^A$) will produce I^A gametes; a heterozygote ($I^A i$) will produce I^A and i gametes. (Remember that in this case the genotype cannot be determined from the phenotype.) A similar situation holds for type B individuals. Individuals with type AB blood will produce both I^A and I^B gametes. Only i gametes will be produced by an individual with type O blood. Remember that each parent contributes only one allele to its offspring, and if the parent produces more than one type of gamete, each allele has the same probability of being passed on to the offspring.

a. What types of gametes will a woman with type AB blood produce?

b. What types of gametes will a man with type O blood produce?

c. If these two people were to marry and have children, what would be the possible genotypes of their children's blood?

d. Which blood types could their children have?

e. Which blood types could not be inherited by the children of these parents?

7. The pattern of inheritance of blood groups has some practical application in medico-legal cases involving disputed parentage.

a. If a child has type A blood, could both parents be of type O? Why?

b. If the mother is type A and the father is type B, would blood tests help to determine whether a particular child belonged to them? Why?

c. If a woman has type O blood and her child has type A blood, could a man with type B blood possibly be the father of the child? Why?

d. If both husband and wife have type AB blood, what possible blood types could their offspring have?

8. A child of blood type O has a mother whose blood type is also O. Which of these men could be the father of this child: a man of blood type A, one of blood type B, or one of blood type O? Recall that blood tests reveal only the phenotype of the individual, not the genotype, so an individual with type A blood may carry the alleles for types A and O. (Refer to question 6.) Based on the results of blood typing alone, can you rule out any of the three men as possible fathers? Why or why not?

Conclusive evidence can be gained by DNA fingerprinting, using a sample of DNA from each of the three men and from the child and the mother. The child's DNA fingerprint should contain bands that are also present in the mother's fingerprint. If additional bands are present, they must be represented in the father's fingerprint. The following results are obtained after a restriction enzyme digest, followed by autoradiography.

Key to bands:

1 Mother

2 Child

3 Possible father 1, blood type A

4 Possible father 2, blood type B

5 Possible father 3, blood type O

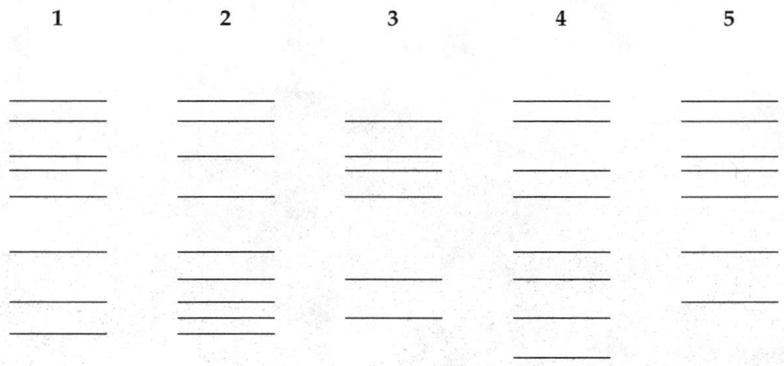

1	2	3	4	5

Which of the three men is the father of the child? How do you know?

9. List the steps involved in DNA fingerprinting. Why would there be an advantage to using more than one probe when developing a fingerprint?

10. A cemetery next to a river has been flooded after a spring of continuous rain. Many of the coffins have ruptured and bones are scattered about. Members of three families are buried in the cemetery. The families would like to reclaim the bones of their ancestors and place them in a proper burial place. As a forensic scientist, you have been asked to assist. How might your proceed to determine to which families the bones belong?

Figure 15A-3b *Cut out chromosomes and use Figures 15A-2 and 15A-4 to help you match homologous chromosome pairs. Arrange these pairs above the appropriate positions in Figure 15A-3a.*

DNA Isolation

OVERVIEW

To most students of biology, DNA is an abstraction. You can memorize the names and structures of the nitrogenous bases and know all about the history of DNA's discovery, but until you actually handle DNA, it remains a strange and mysterious substance.

The purpose of this laboratory is to give you firsthand experience with DNA by isolating it from plant tissue. (Optional exercises for isolating DNA from animal or bacterial cells are also provided.) You will start with whole onions and end with a relatively pure preparation of DNA, containing literally billions of genes. Once isolated, the DNA can be stored in alcohol or dried out. It will be possible for you to hold in your hands the key to an organism's development and structure.

STUDENT PREPARATION

Prepare for the laboratory by reading the text pages indicated by your instructor. Familiarizing yourself in advance with the information and procedures covered in this laboratory will give you a better understanding of the material and improve your efficiency. Isolating DNA will require the entire period, so be prepared to begin immediately.

EXERCISE A   DNA Isolation Procedure

DNA can be isolated from onion cells using several techniques. The procedure in Part 1 requires the use of chloroform and yields DNA that is fairly clean (chromosomal proteins associated with the DNA have been removed). Part 2 introduces a quick method for isolating DNA without the use of chloroform, but yields are not as clean. The DNA may also be sheared into shorter lengths that are more difficult to recover.

If all steps of the procedures in Parts 1 and 2 are followed carefully and the molecular structure of DNA remains intact, the genetic material of onions will precipitate as a thick, stringy white mass that may be spooled out on a glass rod. Isolation of DNA involves three basic steps:

1. **Homogenization** Before DNA can be released from the nuclei of onion tissue, the cell walls, plasma membranes, and nuclear membranes must first be broken down. This is done by homogenizing the onion tissues in a blender. Detergents in the homogenizing medium help to solubilize membranes and denature proteins.

2. **Deproteinization** Chromosomal proteins must be stripped from the DNA. The proteins can then be denatured and precipitated from the homogenate containing the DNA.

3. **Precipitation of DNA** When ice-cold ethanol is added to the homogenate, all components of the homogenate stay in solution—except DNA, which precipitates at the interface between the alcohol and homogenate layers.

If the DNA is damaged during the isolation procedure, it will still precipitate, but as a white fuzzy mass that cannot be collected on a glass rod.

ⅠⅠⅠⅠⅠ Objectives ⅠⅠⅠⅠⅠⅠⅠⅠⅠⅠⅠⅠⅠⅠⅠⅠⅠⅠⅠⅠⅠⅠⅠⅠⅠⅠⅠⅠⅠⅠⅠⅠⅠⅠⅠⅠⅠⅠ

☐ Become familiar with the physical properties of DNA by isolating it from living tissue.

☐ Learn the purpose of each step in the isolation procedure as it relates to the physical and biochemical characteristics of the genetic material.

👁 PART I   DNA Isolation (*Chloroform*)

ⅠⅠⅠⅠⅠ Procedure ⅠⅠ

Work in pairs throughout the DNA isolation procedure. Do not proceed to a subsequent step until *both* of you agree that everything to be done in the present step has been completed correctly. The DNA molecule is easily degraded, or broken down, so it is important to follow all instructions closely. If available, wear gloves to prevent nucleases (present on your skin) from contaminating the glassware.

HOMOGENIZATION

1. Dice a medium-sized onion into cubes no larger than 3 mm. (This step may already have been done for you to save time.)

2. Obtain a mass of 50 g of diced onion. Transfer *all* of this material to a 250-ml beaker.

3. Add 100 ml of homogenizing medium to the diced onion and incubate the beaker in a 60°C water bath for 15 minutes (No longer!). This heat treatment softens the onion tissue and allows penetration of the homogenization solution. It also denatures many enzymes that could interfere with the isolation procedure. The detergent-like action of the sodium lauryl sulfate present in the homogenizing medium helps to dissolve cell membranes and denature proteins.

4. Quickly cool your preparation to 15 to 20°C in an ice bath (a slush of ice and water). This step prevents the denaturation of DNA.

5. Pour your cooled preparation into a blender and fasten the lid. Homogenize for 45 seconds at low speed, followed by 30 seconds at high speed. Homogenization breaks open the cells and releases their contents (carbohydrates, proteins, fats, and nucleic acids). The NaCl/ sodium citrate buffer stabilizes the DNA by forming a Na$^+$ shell around the negatively charged phosphates of the DNA. The citrate inactivates the DNase that would otherwise break down the DNA.

6. Pour the homogenate from the blender into a 1-liter beaker. Allow it to stand in an ice bath for 15 to 20 minutes.

7. Filter the homogenate through four thicknesses of cheesecloth into a 500-ml beaker, taking care to leave the foam behind.

   a. Where is the DNA at this point? _____

DEPROTEINIZATION

8. Pour 80 ml of 95% ethanol into a clean 250-ml flask and pack ice completely around it. The alcohol must get very cold before it is used in the procedure (step 16). (Your instructor may provide you with alcohol that has been kept in a freezer; pack this alcohol in ice to keep it cold.)

9. Pour *only* 50 ml of your filtered homogenate into a clean 250-ml flask.

10. Use a 5-ml pipette to gently add 2 ml of chloroform to the homogenate.

> **CAUTION**: *Your laboratory instructor will demonstrate how to do this safely!*

11. *Gently* swirl the contents of the flask. The purpose of this step is to increase the contact between the chloroform layer and the homogenate layer, but not to mix them (Figure 16A-1). Swirling too vigorously may result in the formation of an emulsion in which no layers are discernible. (If this occurs, see the procedure given below for treatment of an emulsion.)

    b. *What is the white material in the interface layer between the homogenate and the chloroform?*

    _____

Figure 16A-1 *Denatured protein collects at the interface between the two nonmiscible layers of homogenizing medium and chloroform.*

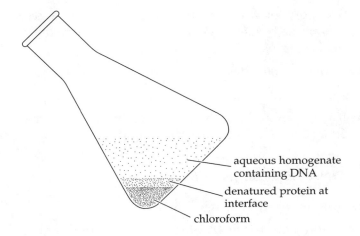

aqueous homogenate containing DNA

denatured protein at interface

chloroform

12. Carefully pour the homogenate into another clean 250-ml flask, leaving the chloroform and protein layers behind. Rinse out the flask from which you poured the homogenate so that you may use it as the clean flask when this step is repeated.

13. Repeat steps 10 through 12 four more times. By the fifth time, much less denatured protein will be separating from the homogenate. If, at any time, an emulsion is formed, follow the procedure given below for treatment of the emulsion.

TREATMENT OF EMULSIONS

If you have swirled your flask too vigorously and created an emulsion, you will have to centrifuge the preparation.

A. Pour the emulsion into a centrifuge tube. Leave about 2 cm unfilled at the top of the tube. Use water to prepare a balance tube of equal volume. The balance tube may be another group's preparation if it is exactly the same volume as yours. Your preparation and its balance tube should be placed opposite one another in the centrifuge head.

B. Advance the centrifuge slowly from one step to the next (a few seconds at each step) until top speed is attained. If the centrifuge begins to shake violently, turn it off immediately and check the balance of your tubes.

C. Centrifuge for 5 minutes at top speed. When you turn the centrifuge off, let it coast to a stop. *Do not open the lid until it has stopped. Under no circumstances should you try to stop the moving head with your hand.*

D. Carefully pour the homogenate from the tube into a clean 250-ml flask, leaving the chloroform and protein layers behind.

Repeat steps 10 through 12 for the remaining number of times, as required by step 13.

14. After the fifth deproteinization treatment, gently pour the homogenate into a 250-ml beaker. It is very important that the homogenate not be contaminated by chloroform, so it may be necessary to leave a small amount of homogenate behind when you pour it off from the chloroform and protein layers this final time.

PRECIPITATION OF DNA

15. Place your beaker with its deproteinized homogenate in an ice bath. Let it cool until it reaches 10 to 15°C.

16. Slowly add ice-cold ethanol (prepared in step 8) down the side of the beaker until the white, stringy DNA precipitate appears. It may not take all 80 ml of alcohol to precipitate the DNA.

17. Spool out, or wind up, the stringy DNA onto a glass rod by rotating the rod in *one direction only* in the beaker of DNA. Continue to rotate the rod as you move it in large circles through the beaker.

   c. *Why is it important always to rotate the rod in the same direction?*

   _____

18. If you want to keep the DNA, gently ease it off the end of the glass rod into a vial filled with 50% ethanol. Be sure the cap is tight enough to prevent leakage.

◉ PART 2   DNA Isolation (*No Chloroform*)

ⅠⅠⅠⅠⅠ Procedure ⅠⅠⅠⅠⅠⅠⅠⅠⅠⅠⅠⅠⅠⅠⅠⅠⅠⅠⅠⅠⅠⅠⅠⅠⅠⅠⅠⅠⅠⅠⅠⅠⅠⅠⅠⅠⅠⅠ

Work in pairs throughout the DNA isolation procedure. Do not proceed to the next step until *both* of you agree that everything to be done in the present step has been completed correctly. The DNA molecule is easily degraded, or broken down, so it is important to follow all instructions closely. If available, wear gloves to prevent nucleases (present on your skin) from contaminating the glassware.

Your instructor will perform steps 1–3. After these preparation steps have been completed, proceed with steps 4 through 8.

1. Cut a yellow onion into wedges and place them in a blender.

2. Add 100 ml of chilled buffer/detergent solution. Homogenize the mixture at low speed for 45 seconds, then at high speed for 30 seconds. Homogenization breaks open the onion cells and releases their contents (carbohydrates, proteins, fats, and nucleic acids).

3. Filter the homogenized mixture through cheesecloth (place the filter in a funnel) into a beaker on ice. Make sure everything is kept cold!

4. Each student should obtain approximately 4 ml of the onion preparation in a clean test tube.

5. Add 2 ml of meat tenderizer to the solution in the test tube and mix the contents slightly with a Pasteur pipette.

   a. *What color is the solution?* _____

   b. *What does the meat tenderizer do?* _____

   _____

6. Use a Pasteur pipette to slowly add an equal volume (approximately 6 ml) of ice-cold 95% ethanol down the side of the test tube. Tilt the tube slightly. You should see a distinct layer of ethanol over the colored filtrate.

7. Gently swirl the Pasteur pipette at the interface of the two layers. This process is called "spooling" the DNA. Always rotate the pipette in the same direction. The stringy, slightly

gelatinous material that attaches to the pipette is DNA. (If the DNA has been damaged, it will still precipitate, but as white flakes that cannot be collected on the glass rod.)

c. *Why does isolated DNA appear stringy?* _____

_____

8. If you want to keep the DNA, gently ease it off the end of the pipette into a vial of 50% ethanol. Cap the vial tightly.

EXERCISE B   Isolation of DNA from Animal Cells *(Optional)**

The procedure for isolating DNA from animal cells is similar to that for isolating DNA from plant cells (Exercise A). In this exercise, you will isolate DNA from mammalian testicular tissue. Tissue from the testes is a good choice for DNA isolation because of its high nucleus-to-cytoplasm ratio, which prevents lipid or protein from interfering with the procedure. Another advantage is that fresh tissue can be obtained without sacrificing the organism: testes are readily available from your local veterinarian or from spay/neuter clinics.

Objectives

☐ Isolate DNA from animal tissue.
☐ Use a colorimetric test to identify DNA.

PART I   Isolation of DNA

Procedure

1. Chill, on ice, a tube of homogenization buffer.
2. Slice or mince a small piece (2 to 3 mm^3) of testis tissue and place it in a cold tissue homogenizer. Add 1 ml of homogenization buffer and two to three drops of sodium lauryl sulfate solution.
3. Homogenize the tissue over ice, using a tissue homogenizer attached to a variable-speed electric drill (a chilled mortar and pestle may be used instead, but a drill is preferable).
4. Transfer the slurry into a centrifuge tube. Obtain a 2 M solution of NaCl. To the volume of slurry in the tube add twice that volume of 2 M NaCl (an equal volume of chloroform can be used instead of NaCl, but, in this case, be sure to use a fume hood). Stopper the tube and shake vigorously for at least 2 minutes.

   a. *Has the color or viscosity of the contents in the tube changed?* _____ *If so, how?*

   _____

5. Centrifuge the tube at least 5 to 7 minutes at top speed in a clinical centrifuge. If 2 M NaCl was used, the tube will contain a precipitate; the liquid portion will contain the DNA. If chloroform was used, the tube will contain three layers following centrifugation; the top layer contains the DNA.
6. Decant the portion of the liquid containing the DNA into a chilled beaker.
7. Slowly add two volumes of cold ethanol to the liquid containing the DNA. Use a glass rod to spool out the whitish DNA fibers (scoring the end of the glass rod with a diamond pencil or file will make it easier to spool out the DNA).

The DNA can be redissolved in 0.1 M EDTA, or, more slowly, in a salt solution (4% NaCl).

*This exercise was developed by Peggy O'Neill Skinner, Bush School, Seattle, Washington.

👁 **PART 2** **Colorimetric Detection of DNA**

You can demonstrate that the whitish material you have isolated is, in fact, DNA by using a colorimetric test; **diphenylamine** will turn blue when it reacts with DNA.

‖‖‖‖ **Procedure** ‖‖‖‖‖‖‖‖‖‖‖‖‖‖‖‖‖‖‖‖‖‖‖‖‖‖‖‖‖‖‖

1. Pour 3 ml of 4% NaCl into a clean test tube.

2. Use an applicator stick to remove some of your isolated DNA from the glass stirring rod and redissolve this material in the NaCl.

3. Mark the tube with an "I" to indicate that it contains your *isolated* material.

4. Place 3 ml of a DNA standard solution into a second test tube and 3 ml of water into a third test tube. Mark these with an appropriate indication of their contents. These tubes will provide you with a standard and a control with which to compare your results.

5. Add 3 ml of diphenylamine reagent to each tube.

6. Using a beaker of water, boil the three tubes for 15 minutes. Cover the tops of the test tubes with marbles and be sure to use boiling chips in the beaker of water.

7. Compare colors in the tubes and record your results. Tube I _____;

   DNA standard _____; (H_2O) _____

   a. *Why did you use a standard in this experiment?* _____

   _____

   b. *Why does the standard not serve as a control in this experiment?* _____

   _____

✔ 👁 **EXERCISE C** | **Isolation of DNA from Bacteria (*Optional*)**

Isolation of DNA from the bacterium *Escherichia coli* is similar to isolation of DNA from plant cells (Exercise A). Since *E. coli* are prokaryotic cells and do not contain nuclei, it is fairly simple to isolate DNA from ruptured cells.

‖‖‖‖ **Objectives** ‖‖‖‖‖‖‖‖‖‖‖‖‖‖‖‖‖‖‖‖‖‖‖‖‖‖‖‖‖‖‖‖‖‖‖

☐ Isolate DNA from bacterial cells.

☐ Determine the absorbance characteristics of DNA.

✔ **PART I** **Isolation of DNA**

‖‖‖‖ **Procedure** ‖‖‖‖‖‖‖‖‖‖‖‖‖‖‖‖‖‖‖‖‖‖‖‖‖‖‖‖‖‖‖‖‖‖‖

1. Suspend 2 g of bacterial paste in 25 ml of 0.15 M NaCl and 0.1 M EDTA. (Use a screw-top glass tube if you do not have access to a clinical centrifuge for later steps.)

2. Add 1 ml of lysozyme solution.

3. Incubate in a water bath at 37°C for 30 minutes.

4. Add 2 ml of 25% sodium lauryl sulfate.

5. Incubate at 50°C for 10 minutes or until the mixture is clear.

6. Add 7.5 ml of 5 M $NaClO_4$ and stir.

7. Add an equal volume (35 ml) of a chloroform/isoamyl alcohol mixture (50:1).

8. Shake well.

9. If you have access to a refrigerated centrifuge, centrifuge the mixture at 10,000 rpm for 10 minutes. Alternatively, you may use a clinical centrifuge. Spin at top speed. A protein pellet will form at the interface between the aqueous buffer and chloroform/isoamyl alcohol solutions. Remove the clear aqueous upper phase, which contains the DNA, and save it in a beaker on ice. If you do not have access to a centrifuge, allow the solution to settle out (this will take approximately 45 minutes), then pipette out the clear aqueous phase (upper layer) and place it in a large glass test tube or beaker.

10. Add two volumes of cold 85% ethanol slowly down the side of the beaker or tube containing the DNA solution.

11. Stir with a glass rod (acid-washed or heated and cooled to remove any nucleases). Stir gently and spool out the DNA. Scoring the end of the glass rod with a diamond pencil will make it easier to spool out the DNA.

PART 2   Measuring Absorbance of DNA

If an ultraviolet spectrophotometer is available, you can test the purity of the DNA you have isolated. The DNA can be dissolved and absorbance readings can then be taken at 280, 260, and 230 nm. For nucleic acids, absorbance is maximum at 260 nm and minimum at 230 nm. Most proteins have a strong absorption at 280 nm. Thus, the higher the ratio of the absorbance at 280 to the absorbance at 260 ($A_{280/260}$), the higher the protein content.

Procedure

1. Use a wooden applicator stick to remove some of the isolated DNA from the glass rod.

2. Redissolve the isolated DNA in 5 ml of saline citrate buffer.

3. See Laboratory 4 for directions on using the spectrophotometer. To prepare an absorption spectrum for DNA, first determine the absorption at 160 nm. If it is "off scale," dilute your sample. Then determine the absorbance at 220 nm, and thereafter at 10-nm increments up to 300 nm. Graph the absorption spectrum on a separate piece of graph paper. (For instructions on graphing absorption spectra, see Laboratory 4.)

4. If a pure DNA standard is available, repeat the procedure in step 3 using this standard. Plot this curve on the same graph.

   a. How do the two absorption spectra compare? _____

   _____

5. From your graph, determine the $A_{280/260}$ ratio for both the DNA you have isolated and the pure DNA. Remember that proteins absorb strongly at 280 nm.

   b. Which sample is purer (contains a smaller amount of protein contamination)?

   _____

PART 3   Colorimetric Detection of DNA

You can use a colorimetric test to demonstrate that you have isolated DNA. Follow the directions in Exercise B, Part 2.

PART 4   Preparing a Standard Curve

If time permits, use dilutions of the DNA standard (known concentrations in μg/ml) to prepare a standard curve (see Laboratory 4) by reacting samples of specific concentrations of DNA with diphenylamine. Use the spectrophotometer to read absorbance at 660 nm. To determine concentration without a spectrophotometer, simply compare the color of the unknown to the colors of the DNA standard dilutions. Plot the standard curve on a separate piece of graph paper. From this standard curve you should be able to determine the DNA concentration in your sample. DNA concentration = _____ μg/ml

Laboratory Review Questions and Problems

Some of the following questions can be answered from your laboratory experience. Other questions may require that you use the knowledge gained from lectures or from reading the text.

1. What did you learn about the properties of DNA during this laboratory period?

2. As your text explains, scientists once believed that proteins constituted the genetic material because of the great number of variations that are possible in their composition and structure. Why would these qualities be important in the genetic material? How does DNA meet this requirement?

3. What structural characteristic of DNA allows it to be spooled out on a glass rod? Why is it not possible to spool out precipitated proteins? (*Hint:* Compare the relative lengths of DNA and protein molecules.)

4. Construct a flow chart that illustrates the steps you took in isolating DNA from onion tissue. At each step, briefly indicate the purpose of the procedure or the solution used.

5. If you wanted to determine the DNA concentration in the tissue you used for the extraction procedure, how might you proceed? (*Hint:* If you did not perform Exercise C, read through Parts 2 and 4 of Exercise C and apply what you learn to the procedure you used in Exercise A.)

DNA—The Genetic Material: Replication, Transcription, and Translation

OVERVIEW

Exact, yet variable and mutable—these are the characteristics of our genetic material, **DNA (deoxyribonucleic acid).** DNA is composed of subunits called **nucleotides** which bond together to form long polynucleotide strands. When nucleotides in one strand pair specifically (by hydrogen bonding) with nucleotides in a second strand, a double-stranded molecule—the DNA helix—is formed.

Each DNA nucleotide consists of a sugar (deoxyribose), a phosphate, and a nitrogenous base. An enormous amount of information is encoded in DNA using only four nitrogenous bases (adenine, guanine, cytosine, and thymine) in DNA nucleotides. Variability in DNA results from the arrangement (or sequence) of nucleotide bases along the polynucleotide strands. This sequence is transmitted faithfully, as exact copies, through DNA synthesis (**replication**) at each cell division, and from generation to generation, in all organisms. Occasionally, however, changes (**mutations**) occur in the nucleotide sequence; these are the basis of evolution.

According to the "central dogma of molecular biology," DNA does not act directly, but rather codes for the synthesis of **RNA (ribonucleic acid)** molecules in a process called **transcription.** These RNA "messages" are decoded in the process of protein synthesis (**translation**). Thus DNA regulates cell activity and determines the phenotype of organisms by determining the type of proteins produced by the cell (Figure 17-1).

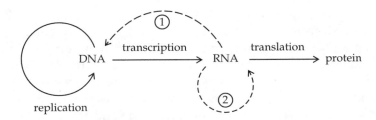

Figure 17-1 *The "central dogma of molecular biology" states that DNA is transcribed to form messenger RNA which can the be translated into protein. In this way, information stored in DNA is encoded in RNA and then is decoded to form a protein product. However, it is now known that exceptions to the central dogma exist: certain RNA viruses* ① *can produce DNA by reverse transcription (RNA → DNA) and* ② *can also replicate their RNA molecules (RNA → RNA).*

STUDENT PREPARATION

Cut out the "nucleotides" on both the green and blue sheets of paper distributed by your instructor. Keep these in two separate envelopes. Also, cut out the yellow "amino acids." Bring these materials to the laboratory with you.

Familiarize yourself with the processes of replication, transcription, and translation by reading the text pages indicated by your instructor. Familiarizing yourself in advance with the information and procedures covered in this laboratory will give you a better understanding of the material and improve your efficiency.

✔ **EXERCISE A**   **Replication**

The DNA molecule is composed of two strands of nucleotides (polynucleotides) hydrogen-bonded together and twisted to form double-stranded DNA. The double-stranded DNA helix is regular, linear, and stable because small nucleotide bases called **pyrimidines** always pair specifically with larger nucleotide bases called **purines.** Thymine (T) and cytosine (C) are the pyrimidines and adenine (A) and guanine (G) are the corresponding purines. Adenine always pairs with thymine, forming two hydrogen bonds (A=T), and cytosine always bonds with guanine, forming three hydrogen bonds (G≡C) (Figure 17A-1).

Figure 17A-1 *The purine and pyrimidine bases present in the nucleotides of DNA and RNA.*

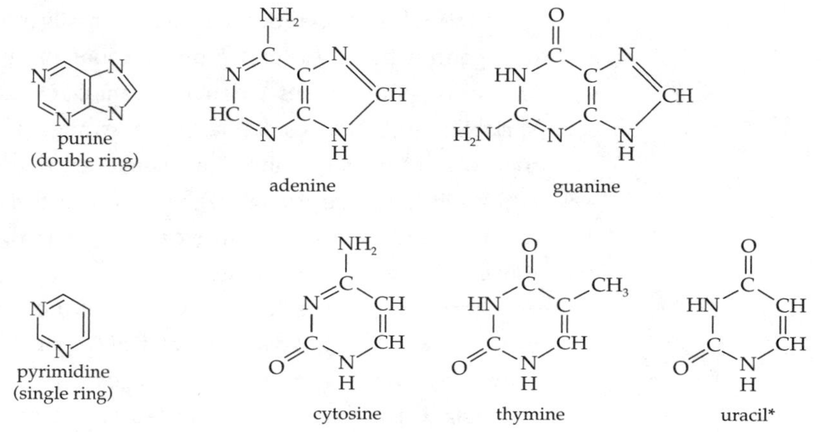

*Replaces thymine in RNA molecules; the CH_3 group present in thymine is absent in uracil.

DNA polynucleotide strands have beginnings and ends, just like sentences. At one end of each strand is a nucleotide bearing a phosphate group that is linked to carbon number 5 (5′ carbon) of the sugar deoxyribose. This is called the 5′ end. At the other end of the chain, a hydroxyl (—OH) group extends from carbon number 3 (3′ carbon) of the deoxyribose in the last nucleotide. This end is called the 3′ end. Similarly, within the DNA molecule, each bond between two adjacent nucleotides in the polynucleotide strand is formed between the 3′ hydroxyl of one nucleotide and the 5′ phosphate of the next. These bonds are called 3′ → 5′ phosphodiester bonds (Figure 17A-2).

One polynucleotide chain in the double-stranded molecule always runs in a 5′ → 3′ direction while the other runs in a 3′ → 5′ direction: the chains are said to be arranged **antiparallel** to each other (Figure 17A-3). You may not think this is important, but before you read a book you must know which direction to read. Before DNA can work, it too must know its directions.

Replication of DNA is **semiconservative.** (*Semi-*, like *hemi-*, means half.) It is possible to break the hydrogen bonds of double-stranded helical DNA molecules, separating the two polynucleotide strands. Because the base pairing is specific (A=T and G≡C), each single strand can then serve as a **template** or

Figure 17A-2 *Nucleotides are added to a DNA chain, one at a time, by attaching the 5′ phosphate of an incoming nucleotide to the 3′ hydroxyl of the last nucleotide in the lengthening DNA chain. A phosphodiester bond is formed and inorganic pyrophosphate (PP_i) is released. The overall direction of synthesis is 5′ → 3′ for the new polynucleotide strand.*

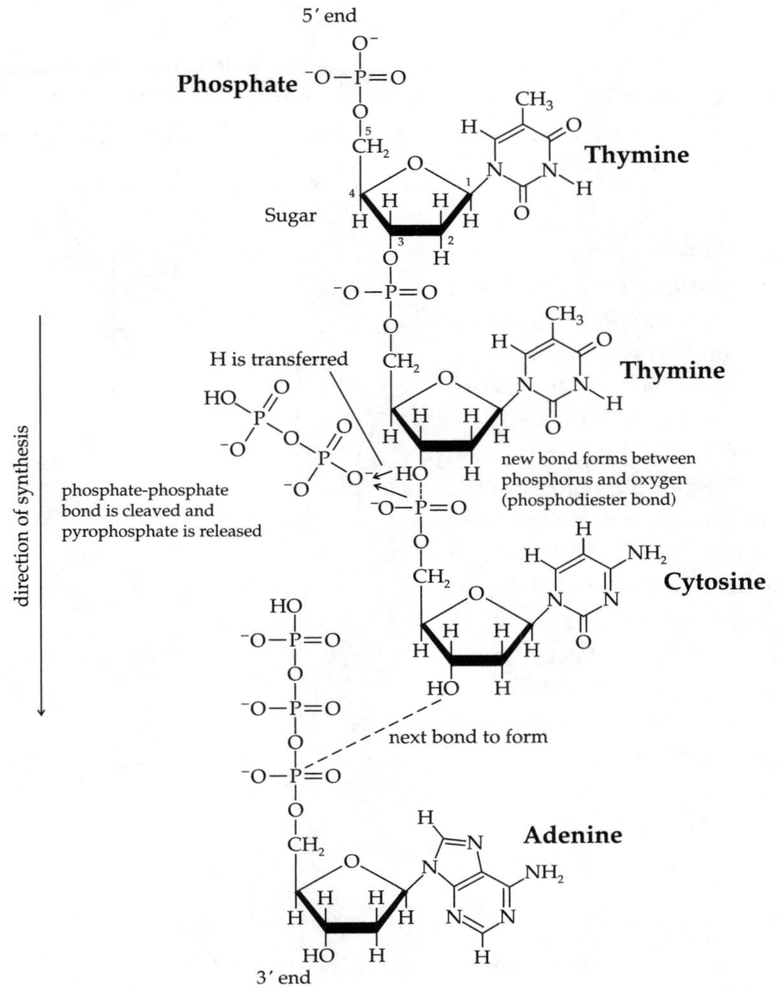

Figure 17A-3 *Polynucleotide strands in the double-stranded helical DNA molecule run antiparallel to one another. Dotted lines represent hydrogen bonds.*

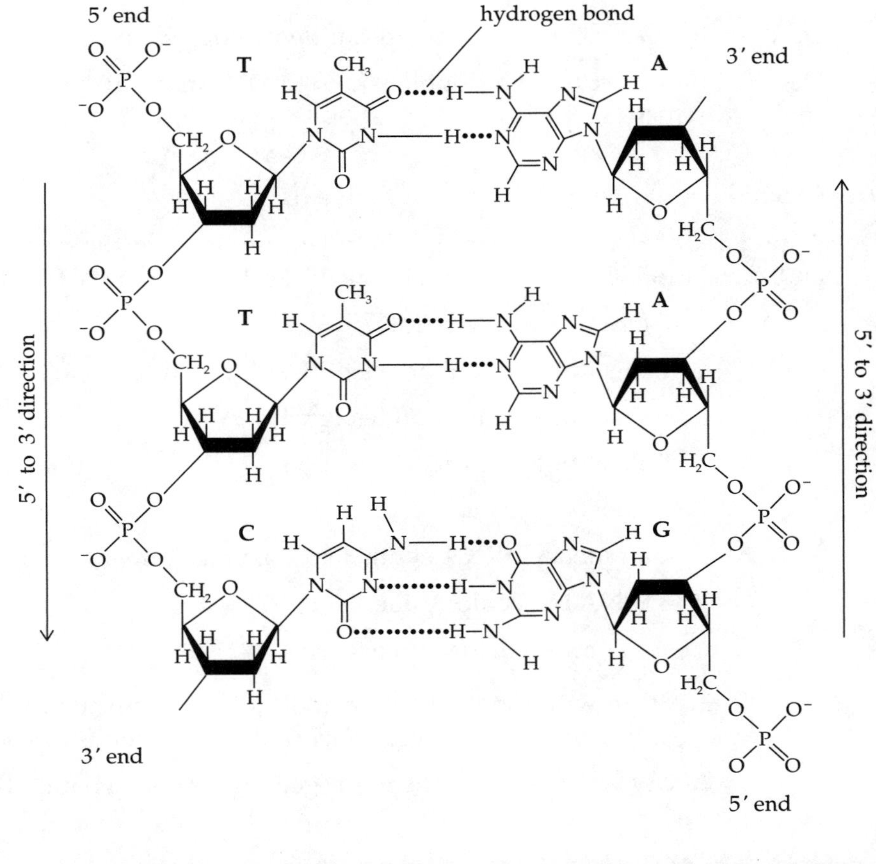

pattern for the formation of a **complementary strand.** Two new double-stranded DNA molecules are produced—each composed of an old polynucleotide strand and a newly synthesized polynucleotide strand. In other words, each of the "daughter" DNA molecules is half new and half old (Figure 17A-4).

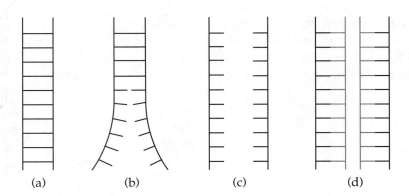

Figure 17A-4 *DNA replication is semiconservative. (a) Nucleotides in DNA are specifically paired and held together by hydrogen bonds. (b) During replication, the hydrogen bonds break and (c) the two original halves of the DNA molecule can serve as templates for the synthesis of complementary strands made from new nucleotides. (d) The two "daughter" molecules of DNA are duplicates of the original DNA. One polynucleotide strand of each molecule consists of nucleotides from the original DNA molecule; the other strand is composed of newly synthesized DNA.*

(a)   (b)   (c)   (d)

||||| **Objectives** ||||||||||||||||||||||||||||||||||||||

☐ Describe the process of semiconservative replication.

☐ Explain how the structure of the DNA molecule (including hydrogen bonding and base-pairing specificity) makes semiconservative replication possible.

☐ Describe how semiconservative replication duplicates the parent DNA molecule.

☐ Describe the structure of a double-stranded DNA molecule and explain the nature of its antiparallel-stranded structure.

||||| **Procedure** ||||||||||||||||||||||||||||||||||||||||||

A model kit has been prepared to help you understand the basic mechanisms of replication, transcription, and translation. Check to see that your kit includes the following materials:

        1 DNA molecule (white)

        1 sheet of deoxyribonucleotides (blue): dA, dG, dT, dC (*d* indicates *deoxy*ribonucleotide)

        1 sheet of ribonucleotides (green): A, G, U, C

        1 sheet of amino acids (yellow)

        1 ribosome (black)

        4 aminoacyl-tRNA synthetase enzymes (green)

        4 tRNA molecules (blue)

        1 ATP molecule (orange)

1. Begin with the white DNA molecule. Cut out the two strips and paste them together as indicated on the strips. Label the 5′ and 3′ ends of each of the two strands.

2. Use scissors to cut the hydrogen bonds and separate the two nucleotide chains labeled I and II.

3. Using the blue nucleotides, semiconservatively replicate the DNA. Line up the blue nucleotides in the proper order along each of the original DNA strands (Figure 17A-5a).

a. *Which four bases are present in the nucleotides used to synthesize DNA?*

_____

Figure 17A-5 *Replicating a DNA molecule. (a) Blue nucleotides (shown here in shading) complementary to those in the white DNA are lined up properly. (b) Nucleotides are taped to a piece of transparent tape to simulate polymerization of a polynucleotide strand in a 5′ → 3′ direction.*

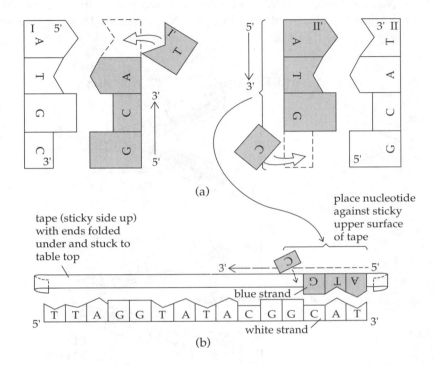

4. Obtain a piece of transparent adhesive tape. Stick one end to the laboratory bench, then turn the tape over, keeping its sticky side up; turn the other end under and stick it to the bench (Figure 17A-5b). Attach each new nucleotide, letter side up, to the sticky side of the tape, aligning the straight bottom edge of each nucleotide along the straight edge of the tape so that most of the nucleotide covers the tape. Each blue nucleotide should correspond to its complementary nucleotide in the white strand (Figure 17A-5b). Be sure to synthesize the new DNA strand in the proper direction—starting at the 5′ end of each *new* strand (opposite the 3′ end of the original white DNA strand), you should add one nucleotide at a time until you reach the 3′ end of the new chain. This will be opposite the 5′ end of the original white DNA strand. Use the same method to tape together the nucleotides of the other new strand.

This process of bonding one nucleotide to the next within a lengthening or "growing" strand is called **polymerization** and is accomplished by an enzyme, **DNA polymerase.** Do *not* tape the blue and white strands to each other.

5. Indicate which new strand was made using strand I DNA as a template and which was made using strand II DNA as a template by writing I′ and II′ on them (Figure 17A-5a). Mark the 5′ and 3′ ends of each newly synthesized strand.

b. *What types of bonds are made between nucleotides within the new strand?*

_____

For the purposes of making a model, you have kept the blue and white DNA strands separated. However, during replication, nucleotides in the original strand are bonded to complementary nucleotides in the new strand as it is being synthesized.

c. *What types of bonds join the template and complementary polynucleotide strands together?*

_____

   d. *Compare the new strands (blue) with the old strands (white). To which white strand is the blue*

   strand I' identical? _____ *Complementary?* _____

   6. Save these molecules to show your laboratory assistant, who will be coming around to check
   your work.

Note that the piece of DNA you have been working with is fairly short and represents only a small part of
a replication "bubble" within a longer DNA molecule that is being replicated *bidirectionally* (Figure 17A-6a).

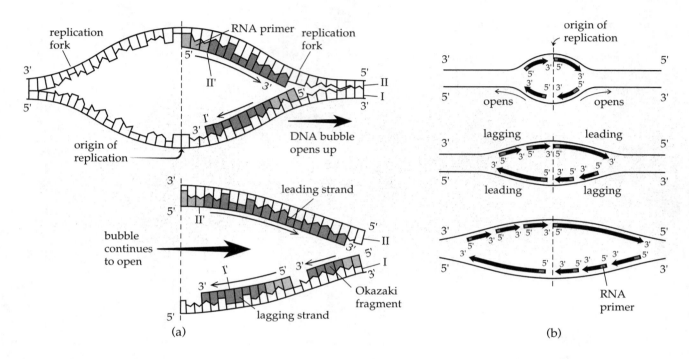

Figure 17A-6   *As a DNA replication bubble opens up bidirectionally, one
strand of DNA at each replication fork can be used as a template for continuous
synthesis of complementary DNA in a 5' → 3' direction, while the other DNA
strand serves as a template for the discontinuous synthesis of a complement in
short 5' → 3' segments that are eventually linked together.*

*(a) In this replication fork, strand II serves as the complement for continuous
synthesis of II' in a 5' → 3' direction, corresponding to the direction in which the
bubble is expanding. This strand is called the "leading strand." On the other
side, however, the complement to strand I is being made discontinuously. Short I'
pieces are synthesized in a 5' → 3' direction that is opposite to that of bubble
expansion. These pieces (called Okazaki fragments) will eventually be linked
together by the enzyme DNA ligase. This new strand, synthesized more slowly
and in small pieces, is called the "lagging strand." Small RNA primers are
necessary to start the synthesis of both leading and lagging strands.*

*(b) Note that leading and lagging strands reverse in "top-bottom"
orientation at opposite ends of the replication bubble because of the directions in
which the opposite ends of the bubble are opening.*

✔ 👁 **EXERCISE B**   **Transcription and Translation**

DNA directs the synthesis of proteins through the processes of **transcription** (DNA → RNA) and **translation** (RNA → protein). Replication, however, does *not* necessarily precede transcription and translation.

Although the genetic material, the DNA, ultimately codes for all the RNA and protein made by a cell, DNA is not *directly* involved in protein synthesis, or translation. DNA functions as a reference library from which books do not circulate. There is always an intermediate step between DNA and protein: the **messenger RNA (mRNA).**

For each particular protein to be synthesized, the DNA nucleotide sequence is first transcribed into mRNA. In eukaryotes, where a nuclear membrane separates the DNA from the cytoplasm, the mRNA is modified and then moves from the nucleus to the cytoplasm and attaches to ribosomes, where the machinery for protein synthesis is found. In prokaryotes, ribosomes can attach to mRNA even while it is being synthesized, and translation can begin immediately.

IIIII **Objectives** IIIIIIIIIIIIIIIIIIIIIIIIIIIIIIIIIIIIIIII

☐ Differentiate between the chemical structures of RNA and DNA.

☐ Distinguish between the structures and roles of tRNA, rRNA, and mRNA.

☐ Describe the process of transcription.

☐ Describe the role of aminoacyl-tRNA synthetase enzymes.

☐ Distinguish between the functions of amino acid activation and charging of tRNAs.

☐ Define codon and explain why the genetic code is called a triplet code.

☐ Read the genetic code and identify the amino acid that corresponds to a given codon.

☐ List, in sequence, the steps involved in translation.

☐ Describe the events of elongation, translocation, and termination.

☐ Indicate the role of AUG, UGA, UAA, and UAG codons.

✔ **PART I**   **Transcription—RNA Synthesis**

Using the I' strand of blue DNA made from the I strand of white DNA, you will now synthesize a molecule of messenger RNA (mRNA) using the green ribonucleotides (Figure 17B-1). *Note:* You will be transcribing only one strand of the double-stranded DNA molecule. This strand serves as the template for mRNA synthesis. The messenger RNA made complementary to this strand of the DNA encodes the same information (sequence of nucleotides), with U substituted for T, as the nontranscribed strand of the original double-stranded DNA (often called the "sense" strand because it is the information in the nontranscribed DNA strand that makes "sense" and will ultimately determine the sequence of amino acids in the protein). The enzyme **RNA polymerase** is responsible for the polymerization reactions of transcription.

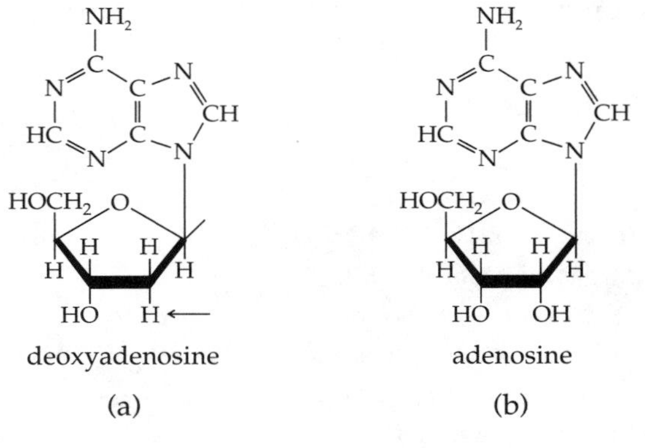

Figure 17B-1 *You will notice that the green nucleotides have the same shape as the blue nucleotides. The only difference in their structure is that* deoxyribonucleotides *are missing an oxygen on carbon number 2 (carbon 2') of the sugar molecule (a). The sugar is deoxyribose in the nucleotides of DNA (a) and ribose in the nucleotides of RNA (b).*

deoxyadenosine

(a)

adenosine

(b)

IIIII **Procedure** IIIIIIIIIIIIIIIIIIIIIIIIIIIIIIIIIIIIIIIII

1. Line up the green ribonucleotides complementary to the I' template DNA strand and tape them together, as in Exercise A, to form a polynucleotide strand. Like DNA, mRNA is always

synthesized in a 5 → 3' direction. Make sure that you tape the nucleotides together in the proper direction to simulate the process of polymerization.

2. Label the 5' and 3' ends of the messenger RNA you have made.

   a. *What do 5' and 3' refer to?* _____

3. Record the nucleotide sequence of your mRNA.

   5' end _ _ _ _ _ _ _ _ _ _ _ _ _ _ _ _ _ _ _ 3' end

4. Each group of three nucleotides in a messenger RNA is called a **codon.** Using brackets, identify the codons of your mRNA sequence as written above. (Assume the first codon begins with the first nucleotide at the 5'end.)

   b. *What is the importance of these codons?* _____

   _____

PART 2   Translation—Protein Synthesis

You are now ready to use your mRNA molecule to synthesize a protein. The first step in this process of translation is the **activation** of the amino acids, the addition of adenosine monophosphate (AMP) to amino acids so that they can be attached to tRNA molecules. (AMP is one part of the two-part orange ATP molecule in your kit.) Special enzymes called **aminoacyl-tRNA synthetases** (the large green molecules in your kit) then "charge" or bind specific "activated" amino acids onto the proper transfer RNAs (tRNAs)—one for each amino acid. Charged transfer RNAs carry amino acids to the ribosomes and serve as "adapters" between the code built into the nucleotide sequence of the mRNA and the sequence of amino acids in the protein. (Keep in mind that amino acids and nucleotides are two very different kinds of molecules that must be paired during the process of protein synthesis. Just as you use an adapter to put a three-pronged plug into a two-pronged outlet, a tRNA molecule pairs amino acids and nucleotides.)

Each tRNA (there are four blue tRNA molecules in your kit) has a nucleotide triplet, an **anticodon,** at a specific site in the molecule's three-dimensional structure. Eventually, this anticodon sequence will pair with a specific codon (also a nucleotide triplet) on an mRNA molecule that is being translated. Amino acids are attached to another specific site at one end of the tRNA molecule. As the anticodons of tRNA molecules pair, one at a time, with the sequence of codons in mRNA, the proper amino acids are aligned in the order dictated by the sequence of mRNA nucleotide triplets. Peptide bonds are formed between the amino acids to synthesize the polypeptide chain coded for by the mRNA.

Procedure

1. Find the four green aminoacyl-tRNA synthetase enzymes in your kit (see Figure 17B-2). Place one of the enzymes on your desk.

Figure 17B-2 *Each aminoacyl-tRNA synthetase enzyme contains a binding site for the unique R group of a specific amino acid and a binding site for the anticodon of a specific tRNA. In this way, a specific amino acid will be attached to a prescribed tRNA. A third binding site for the adenosine portion of AMP is also present.*

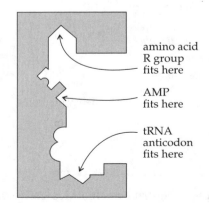

amino acid
R group
fits here

AMP
fits here

tRNA
anticodon
fits here

2. Find the proper amino acid that fits into the enzyme. The enzyme is shaped so that the anticodon bases of a tRNA fit into one end and the R group of a specific amino acid fits into the other. This is how a particular tRNA molecule ends up carrying its specific amino acid: each aminoacyl-tRNA synthetase enzyme contains binding sites for a specific amino acid and a specific tRNA (Figure 17B-2). Do *not*, however, insert the amino acid into the enzyme at this time.

ATP is used to "activate" the amino acid—that is, to convert it to a higher-energy form—so that it *can* be inserted into the enzyme and then attached to its tRNA. When this happens, ATP is split to form AMP and inorganic pyrophosphate (PP_i). The AMP, attached to the amino acid, fits into a groove on the enzyme (Figure 17B-3).

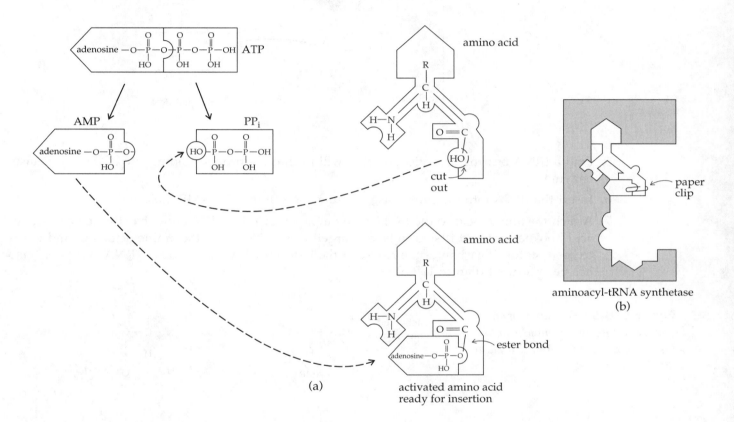

Figure 17B-3 *(a) An amino acid is activated by the splitting of ATP, resulting in the formation of AMP, inorganic pyrophosphate (PP_i), and the energy required to attach the activated amino acid to the aminoacyl-tRNA synthetase enzyme (b). During its activation, the amino acid loses an —OH group, which is incorporated into the pyrophosphate. (The nonionized forms of the amino acid, ATP, AMP, and PP_i are shown here.)*

3. Find the ATP molecule (orange). Cut the —OH group from the carboxylic acid group on the end of the amino acid you are working with. Use a paper clip to attach AMP at the same site (Figure 17B-3). Note that an ester bond is formed between the carbonyl group (C=O) of the amino acid and the phosphate group of AMP. The amino acid is now activated.

4. Insert the amino acid into the aminoacyl-tRNA synthetase enzyme.

Now tRNA can enter the synthetase enzyme (Figure 17B-4). Each tRNA has the same set of three nucleotides (CCA) on its 3′ end. This is where the amino acid will be attached. Since each aminoacyl-tRNA synthetase enzyme has a special binding site for a particular anticodon, only one tRNA can enter the enzyme to bind with the already-bound amino acid. The adenine nucleotide (A) of the 3′ CCA of tRNA fits into the same groove in the synthetase enzyme as does the adenosine of AMP. When AMP is released, the A residue of the CCA can be attached to the amino acid (Figure 17B-4).

Figure 17B-4 *The amino acid attaches to the 3′ hydroxyl of the A nucleotide at the CCA end of tRNA.*

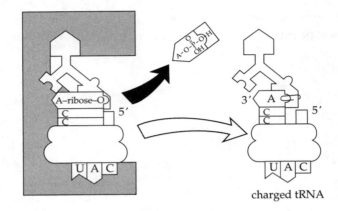

5. Find a tRNA bearing an anticodon that will fit into the proper binding site of the synthetase enzyme.

6. Insert the tRNA into the synthetase enzyme and remove the AMP molecule.

7. Attach the amino acid to the tRNA using a paper clip. The tRNA now has the correct amino acid hooked to it and is said to be a **charged** tRNA. Note that the amino acid is bound to the oxygen on the 3′ carbon of the ribose of the 3′-terminal A nucleotide of tRNA. An ester bond has been formed (Figure 17B-5).

Figure 17B-5 *Formation of an ester bond linking an amino acid to the 3′ hydroxyl of the A nucleotide at the CCA end of tRNA.*

8. The tRNA carrying its appropriate amino acid now breaks loose from the aminoacyl-tRNA synthetase molecule to participate in the process of protein synthesis. The synthetase enzyme can be used repeatedly. Remove the charged tRNA from the enzyme.

9. Now charge your three other tRNAs with amino acids, as in steps 3–8. Once all four tRNA molecules are charged, you are ready to begin the process of protein synthesis.

During protein synthesis, messenger RNA attaches to a small ribosomal subunit. In most bacteria, a short sequence of nucleotides in the 16s RNA of the small ribosomal subunit binds to a special sequence of nucleotides (the Shine-Dalgarno sequence) in the mRNA near the "start" site for protein synthesis. In eukaryotes, the 5' cap present on all mRNAs is involved in recognition of and binding to the small ribosomal subunit.

10. Attach your messenger RNA molecule to the ribosome by sliding it up through the right-hand slit. (For convenience, the large and small ribosomal subunits are already attached to one another, but this is not the case in the living cell.) Position the first two codons between the two slits, with the **AUG** or **initiation codon** on the left and the second mRNA codon on the right. (Although the AUG codon is the first one in the mRNA you are using, this is not usually the case in cells. The nucleotides to the 5' side of the AUG—to its left, or upstream—represent the **leader sequence,** much like the leader on a movie film or VCR tape. Once the mRNA is bound, the AUG codon will be in register at the correct site for translation.) You will translate the message in the familiar 5' → 3' direction (Figure 17B-6).

Figure 17B-6 *Attaching mRNA to the ribosome. When you first attach the mRNA, codon 1 is AUG, the initiation codon.*

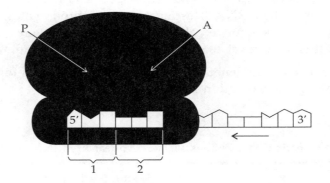

Now, a charged tRNA, carrying an amino acid, will pair with the AUG initiation codon of mRNA. The 3' end of the tRNA anticodon will pair with the 5' nucleotide of mRNA—the tRNA and mRNA are antiparallel. (Two nucleotide strands, no matter how short, can interact only when they are antiparallel.) The combination of tRNA + mRNA + small ribosomal subunit is called the **initiation complex** (Figure 17B-7).

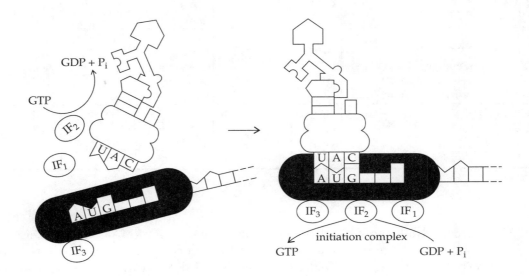

Figure 17B-7 *Formation of the initiation complex.*

A protein "factor" (small protein molecule) IF_3 is involved in binding AUG and in the attachment of mRNA to the small subunit during formation of this initiation complex. Factors IF_1 and IF_2-GTP are also involved in the attachment of the first tRNA molecule. (GTP, guanosine triphosphate, is used as a source of energy for most steps in protein synthesis.)

11. Find the blue tRNA that has an anticodon complementary to the AUG initiation codon. Pair it to the mRNA on the ribosome.

The large ribosomal subunit now attaches to the small ribosomal subunit of the initiation complex. Hydrolysis of the GTP of IF_2-GTP is required for this step. The large subunit is configured to form two major sites of activity: the **P site** (for peptidyl-tRNA), where new peptide bonds are formed, on the left, and the **A site** (for amino-acyl-tRNA), where new charged tRNAs arrive, on the right (Figure 17B-6). A third site, the **E site** (not shown), is occupied for a short time by the CCA end of the tRNA about to be ejected from the P site following removal of its amino acid during peptide bond synthesis.

12. Using the genetic code table (Table 17B-1), look up the initiation codon, AUG. Note that both methionine (Met) and formylmethionine (fMet) are specified for this codon. In prokaryotes, the initiation codon, AUG, always specifies fMet; Met is used in response to internal AUG codons in the mRNA. This means that, in prokaryotes, all proteins originally begin with fMet (which can be removed at a later time). In eukaryotes, methionine is also used for initiation, but it is not formulated. Write fMet on the yellow amino acid attached to the blue tRNA.

Important: **The genetic code specifies amino acids for codons in messenger RNA. Never look up the anticodon in the genetic code table.**

Table 17B-1   The Genetic Code: Codons as They Appear in mRNA

		Second Nucleotide				
		U	**C**	**A**	**G**	
First Nucleotide	**U**	UUU ⎤ phenylalanine UUC ⎦ UUA ⎤ leucine UUG ⎦	UCU ⎤ UCC ⎥ serine UCA ⎥ UCG ⎦	UAU ⎤ tyrosine UAC ⎦ UAA —— stop UAG —— stop	UGU ⎤ cysteine UGC ⎦ UGA —— stop UGG —— tryptophan	U C A G
	C	CUU ⎤ CUC ⎥ leucine CUA ⎥ CUG ⎦	CCU ⎤ CCC ⎥ proline CCA ⎥ CCG ⎦	CAU ⎤ histidine CAC ⎦ CAA ⎤ glutamine CAG ⎦	CGU ⎤ CGC ⎥ arginine CGA ⎥ CGG ⎦	U C A G
	A	AUU ⎤ isoleucine AUC ⎦ AUA — methionine AUG ⎡ methionine (start) ⎣ formylmethionine (start)	ACU ⎤ ACC ⎥ threonine ACA ⎥ ACG ⎦	AAU ⎤ asparagine AAC ⎦ AAA ⎤ lysine AAG ⎦	AGU ⎤ serine AGC ⎦ AGA ⎤ arginine AGG ⎦	U C A G
	G	GUU ⎤ GUC ⎥ valine GUA ⎥ GUG ⎦	GCU ⎤ GCC ⎥ alanine GCA ⎥ GCG ⎦	GAU ⎤ aspartic acid GAC ⎦ GAA ⎤ glutamic acid GAG ⎦	GGU ⎤ GGC ⎥ glycine GGA ⎥ GGG ⎦	U C A G

(right margin: **Third Nucleotide**)

13. A second tRNA now attaches to the ribosome-tRNA-mRNA complex. This tRNA fits into the A site of the large ribosomal subunit and its anticodon is complementary to the second mRNA codon (Figure 17B-8). Insert the second tRNA, carrying amino acid 2, into the A site on the mRNA-ribosome complex. Which amino acid is carried by this second tRNA? Write the name on the amino acid.

Figure 17B-8 *The large ribosomal subunit attaches to the initiation complex. Note the P and A sites in the large subunit (see Figure 17B-6). A second charged tRNA pairs with the codon located at the A site. The elongation factors shown here (Tu, Ts) are those present in bacteria (prokaryotes).*

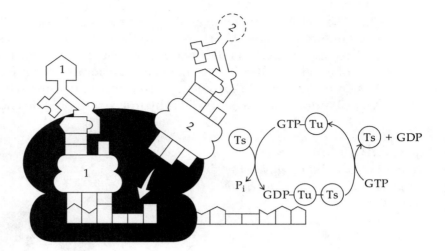

A **peptide bond** is now formed between the —C—O— of amino acid 1 and the —NH$_2$ of amino acid 2. When this happens, the bond between the first tRNA (in the P site) and its amino acid breaks. Amino acid 1 is now held by the peptide bond to amino acid 2 on the second tRNA (Figure 17B-9). This step is called **peptidyl transfer.**

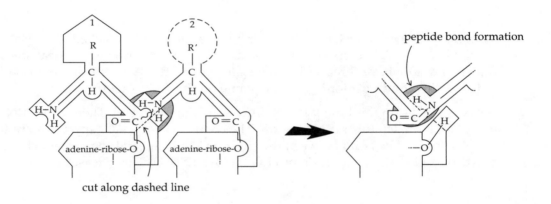

Figure 17B-9 *Formation of a peptide bond between two amino acids. The enzyme peptidyl transferase catalyzes the reaction. The H from —NH$_2$ of amino acid 2 is transferred to the oxygen in the 3' position on the A residue of the first tRNA, restoring it to a 3' —OH group.*

14. Attach the two amino acids together with a piece of tape. Using your scissors, remove the extra H on —NH$_2$. A peptide bond has been formed and the protein chain is now two amino acids long.

Each time a tRNA carrying an amino acid is added and a peptide bond is formed, the chain gets longer. Thus, this process is known as **elongation.** A protein elongation factor complexed with GTP aids in insertion of charged tRNAs into the A site on the ribosome. In prokaryotes, the elongation factor is a protein complex of Tu-GTP and Ts (which reactivates Tu after hydrolysis of GTP). In eukaryotes, elongation factors EF$_1$ and EF$_{1\beta}$ are used. Hydrolysis of GTP to GDP and P$_i$ provides the energy for elongation (Figure 17B-8). The peptidyl transfer reaction is accomplished by an enzyme complex (**peptidyl transferase**) which is part of the large ribosomal subunit. However, recent evidence suggests that rather than ribosomal proteins possessing the enzymatic activity necessary for this reaction, it is RNA that is responsible: the RNA acts as a **ribozyme** (an RNA molecule with enzymatic activity).

15. The ribosome now moves along the message in a $5' \rightarrow 3'$ direction. Place your fingers on the mRNA and move the ribosome to the right. The tRNA associated with the first codon is released since it is no longer bound to its amino acid. The process of ribosome movement from one codon to the next is known as **translocation** (Figure 17B-10). Another elongation factor (G in prokaryotes, EF_2 in eukaryotes) complexed with GTP is involved in the movement of the ribosome. Once again, GTP is hydrolyzed to provide energy for this movement.

Figure 17B-10 *Translocation and insertion of a new charged tRNA. The elongation factor (G) is that present in prokaryotes.*

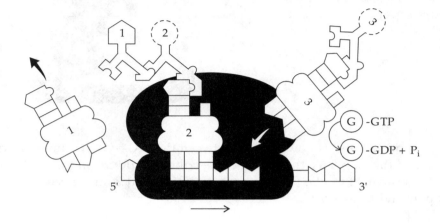

16. Now, tRNA 2 is on your left with the first two amino acids attached to it. Match the anticodon of tRNA 3, carrying amino acid 3, to the next codon and repeat steps 14 and 15. You should now have three amino acids attached to tRNA 3. Which amino acid has been added in this third position? Write the name on the amino acid.

17. Repeat steps 14–16 until the last mRNA codon is in the A site. There is no tRNA having an anticodon to match this mRNA codon. The codons UAA, UAG, and UGA are **termination** or "stop" **codons.** The bond between the tRNA in the P site and the protein chain attached to it is hydrolyzed with the addition of H_2O (Figure 17B-11). This releases the newly synthesized

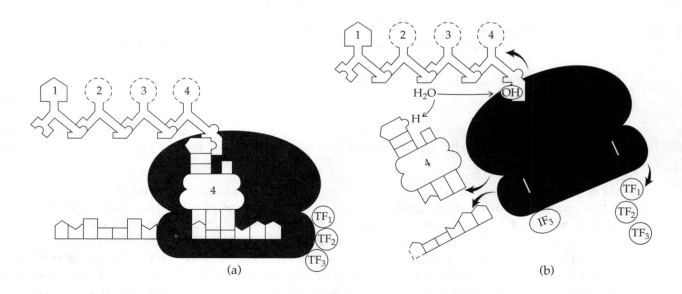

Figure 17B-11 *Chain termination. A termination or "stop" codon on mRNA is located in the A site and peptidyl transferase hydrolyzes the bond between the last amino acid of the protein and tRNA.*

protein. Several protein termination factors (TF) aid in the recognition of "stop" codons and termination of the peptide chain.

18. Release your protein chain, which should now be four amino acids long. The tRNA in the P site is also released, and the ribosomal subunits and mRNA separate.

Note that as a ribosome moves across a message, additional ribosomes can attach to the freed codons (codon 1, 2, 3, etc.). Each of these ribosomes can then serve as a site to start the synthesis of a protein. In this way, several molecules of a protein can be made simultaneously from one message. A complex of several ribosomes attached to a messenger RNA is called a **polysome.**

19. Make sure that you have identified all amino acids. Have all peptide bonds been formed correctly? Your laboratory instructor will check your work to see that you have completed the peptide chain correctly.

EXERCISE C Point Mutations in DNA

Point mutations are small changes in the DNA, such as base substitutions, base additions, and base deletions, but they may have profound effects on the protein formed by the gene, depending on where the mutations occur.

IIIII Objectives IIIIIIIIIIIIIIIIIIIIIIIIIIIIIIIIIIIIIIIIIIIIII

☐ Determine the effect on the amino acid sequence of a point mutation (base substitution, base addition, or base deletion) in DNA.

PART 1 Base Substitutions—Possible Effects

IIIII Procedure IIIIIIIIIIIIIIIIIIIIIIIIIIIIIIIIIIIIIIIIIIIIII

Refer back to the blue DNA strand that you used to make the messenger RNA for your model (Exercise B, Part 1).

a. Assume that a base substitution has occurred such that the ninth nucleotide has been changed from an A to a G. What is the sequence of the nucleotides in the third codon of the mRNA now?

_____

b. What amino acid, if any, does this codon specify? _____

c. What effect will this have on the protein formed from the DNA? _____

_____

d. If the ninth nucleotide had been changed from an A to a C, what effect would this have had on the protein formed? _____

e. If the ninth nucleotide had been changed from an A to a T, what effect would this have had on the protein formed? _____

f. Which of the base substitutions specified in questions a, d, and e would be most likely to cause the production of a defective protein? _____

PART 2 Base Substitution Resulting in Sickle-Cell Anemia

Sickle-cell anemia is a genetic disease caused by a base substitution in the DNA of the gene coding for one of the polypeptides that makes up hemoglobin, the oxygen-carrying molecule of the red blood cells.

A normal individual has two alleles for the production of normal hemoglobin. The red blood cells of this individual will have the typical "doughnut shape." An individual with **sickle-cell anemia** has two alleles (i.e., is homozygous) for the production of sickle-cell hemoglobin. The presence of sickle-cell hemoglobin causes red blood cells to take on sickle shapes, causing difficulty in passing through small blood vessels. These cells may clump and clog the blood vessels to the internal organs, thus depriving them of needed oxygen. Individuals with sickle-cell anemia may often be pale, tired, and short of breath. They may have pain in their arms, legs, back, and abdomen, and their joints may swell. Their low resistance to infections can trigger severe worsening of their condition and may eventually be fatal. (The anemia results from the fragility of the red blood cells.)

IIIII Procedure IIIIIIIIIIIIIIIIIIIIIIIIIIIIIIIIIIIIIIIIIII

1. Examine the prepared slide of normal human blood that is on demonstration. Draw a representative area of the slide in the space below.

2. Examine the prepared slide of blood from a person with sickle-cell anemia. Draw a representative area of the slide in the space below.

An individual who has one allele for sickle-cell hemoglobin and one normal allele produces both normal and sickle-cell hemoglobin; the amount of normal hemoglobin is enough to prevent the individual from becoming anemic. This individual is classified as a **carrier** since he or she can pass the recessive allele to offspring. In most ways, a carrier is normal and healthy. However, when the individual is in an environment that is low in oxygen, some red blood cells will undergo sickling. Such individuals are said to have **sickle-cell trait** (*not* sickle-cell anemia).

Sickle-cell hemoglobin differs from normal hemoglobin by only one amino acid in each of two β chains. Normal hemoglobin contains glutamic acid but sickle-cell hemoglobin contains valine.

a. *What are the codons for glutamic acid?* _____

b. *What are the codons for valine?* _____

c. *Based on this information, what base substitution had to occur in the DNA in order to produce the allele for*

   sickle-cell hemoglobin? _____

Valine is a nonpolar amino acid, whereas glutamic acid is an acidic amino acid. Valine produces hydrophobic areas on the β chains that interact with hydrophobic areas on the β chains of other hemoglobin molecules, causing the hemoglobin molecules to clump. This produces the deformed red blood cells characteristic of sickle-cell anemia.

PART 3   Frame-Shift Mutations: Base Additions and Deletions

Refer back to the blue DNA strand that you used to make your mRNA (Exercise B, Part I). Assume that an A has been added between the fourth and fifth nucleotides. Indicate the sequence of the bases in the mRNA which will be transcribed from the DNA.

5′__ __ __ __ __ __ __ __ __ __ __ __ __ __ __ __3′

In the space below, indicate the sequence of amino acids in the protein coded for by this mRNA.

a. *How does the above sequence of amino acids compare with the protein formed from the original DNA?*

_____

b. *What effect would the removal of a nucleotide from the original DNA have on the protein formed?*

_____

_____

Laboratory Review Questions and Problems

1. If DNA contained only the bases adenine and thymine, how long a code word would be necessary to enable coding for each of 20 different amino acids?

2. A particular DNA base sequence transcribed into messenger RNA is TTATCTTCGGGAGAGAAAACA. (a) If reading begins at the left, what amino acids are coded by this sequence? (*Note:* The initiation sequence is disregarded in this example.)

   (b) If proflavine treatment caused the deletion of the first adenine nucleotide on the left, what changes would occur in the first six amino acids coded by this sequence?

3. Streisinger and co-workers studied amino acid sequences in the lysozyme protein produced by the T4 phage. One sequence is Lys-Ser-Pro-Ser-Leu-Asn-Ala, but as a result of a deletion of a single nucleotide and subsequent insertion of another nucleotide, this amino acid sequence was found to change to Lys-Val-His-His-Leu-Met-Ala. Using the codons in Table 17B-1, determine the nucleotide sequences that produced (a) the original amino acid sequence and (b) the subsequent changes.

4. A single (+) strand of DNA (base composition: A, 21 percent; G, 29 percent; C, 29 percent, T, 21 percent) is replicated by DNA polymerase to yield a complementary (−) strand. The resulting duplex DNA is then used as a template by RNA polymerase, which transcribes the (−) strand. Indicate the base composition of the RNA formed.

5. a. Given the following DNA molecule, write the sequence of the messenger RNA synthesized from the upper strand.

b. Indicate the 5' and 3' ends of the message.

c. Groups of three letters on the mRNA molecule are called _____.

d. What is the special significance of the *first* group of three letters in the message?

e. What is the special significance of the *last* group of three letters in the message?

6. a. In the diagram, label the indicated parts.

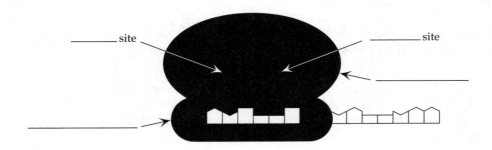

_____ site   _____ site

_____

_____

b. Label the 5' and 3' ends of the messenger RNA in the diagram.

c. The triplet of tRNA nucleotides responsible for insertion of the correct amino acid into a protein chain by complementing with a triplet of mRNA is known as the

_____.

d. What are the last three nucleotides always found at the 3' end of tRNA?

e. Use the space below to draw two tRNA molecules with attached amino acids. Two amino acids will be joined by a _____ bond to form a dipeptide. The enzyme responsible for this reaction is _____. Explain how the bond between the two amino acids is formed and show the final result by drawing an additional diagram.

7. You have the following DNA strand. Synthesize a protein from this strand. (Recall that a leader sequence may precede the AUG initiation codon.)

3' AGATTACTCGAGCCGGGTAATCGGC 5'

mRNA

Protein

8. Make a strand of DNA complementary to the DNA strand in question 7. Mark the new strand's 5' and 3' ends. Now synthesize mRNA and a protein from this strand. Is the message the same as in question 7? Is the protein the same? (Recall that mRNA is read in a 5' → 3' direction!)

   Complementary DNA

   mRNA

   Protein

9. You have synthesized the following protein: fMet-Pro-Asp-Gly-Thr. You accomplished this in a cell-free system containing tRNA molecules with the anticodons listed below:

   3' CCG 5'

   5' UGU 3'

   5' CGG 3'

   5' CAU 3'

   3' CUG 5'

   mRNA

   Construct the double-stranded DNA molecule from which this protein was synthesized. Show all of your reasoning.

   DNA (2 strands)

Molecular Genetics: Recombinant DNA

OVERVIEW

During the last decade there has been a technological revolution in the field of molecular genetics. Scientists can now explore and "engineer" changes in the genomes of a variety of organisms by obtaining pieces of DNA molecules and **recombining** them in different ways.

One of the key developments in **recombinant DNA technology** was the discovery of special enzymes called **restriction endonucleases.** These "restriction enzymes" have been isolated from a variety of prokaryotic organisms, especially bacteria. They protect bacteria by restricting foreign DNA, particularly viral DNA, from entering and functioning within cells. Restriction enzymes cut the foreign DNA at specific base sequences (restriction sites). The small pieces are then easily destroyed by other bacterial enzymes. Scientists have learned to use the same restriction enzymes as "molecular scissors" to cut all types of DNA molecules into smaller segments at specific locations.

The small pieces of DNA snipped from bacterial cells, fruit flies, frogs, or even humans can be recombined with other DNA. Often, the pieces are inserted into viruses that have been disabled or into bacterial **plasmids** (small, double-stranded DNA molecules located outside the bacterial chromosome). The plasmids or viruses act as **vectors** or carriers to transfer the DNA into the cell of a host—perhaps another bacterial cell or a eukaryotic cell. Bacterial host cells and some eukaryotic cells can multiply to form **clones** (a collection of copies of themselves) that can express the new genetic information and make new gene products. The cells have been **transformed** and may even express a new phenotype as a result of the added gene products. Commercially we can produce large quantities of rare proteins or other specific gene products, such as insulin or growth hormone, using recombinant DNA techniques. **Gene therapy,** the transfer of beneficial genes into the human body is also possible.

In this laboratory, you will investigate some of the basic principles of genetic engineering. Plasmids containing specific fragments of foreign DNA will be used to transform *Escherichia coli* cells, conferring both antibiotic (ampicillin) resistance and lac^+ phenotype (ability to metabolize lactose) to recipient cells.

STUDENT PREPARATION

Prepare for this laboratory by reading the text pages indicated by your instructor. Familiarizing yourself in advance with the information and procedures covered in this laboratory will give you a better understanding of the material and improve your efficiency.

EXERCISE A   Bacterial Transformation: Constructing Recombinant Plasmids

The bacterium *Escherichia coli* (*E. coli*) is an ideal organism for genetic manipulation and has been used extensively in recombinant DNA research. It is a common inhabitant of the human colon and can easily be grown in standard nutrient mediums.

The single circular chromosome of *E. coli* contains 5 million DNA base pairs (1/600th the total amount of DNA in a human cell). In addition, the cell contains small, circular, *extrachromosomal* (outside the chromosome) DNA molecules called **plasmids.** These fragments of DNA, 1,000 to 200,000 base pairs in length, also carry genetic information. Some plasmids replicate only when the bacterial chromosome replicates and usually exist only as single copies within the bacterial cell. Others replicate autonomously and often occur in as many as 10 to 200 copies within a single bacterial cell. Certain plasmids, called R plasmids, carry genes for resistance to antibiotics such as ampicillin, kanamycin, or tetracycline.

In nature, genes can be transferred between bacteria in three ways: conjugation, transduction, or transformation. **Conjugation** is a mating process during which genetic material is transferred from one bacterium to another "sexually" different type. (See Laboratory 14, Exercise C.) **Transduction** requires the presence of a virus to act as a **vector** (carrier) to transfer small pieces of DNA from one bacterium to another. **Bacterial transformation** involves transfer of genetic information into a cell by direct absorption of the DNA from a donor cell.

Through the process of bacterial transformation, a bacterium can acquire a new trait by incorporating and expressing foreign DNA. In the laboratory, the DNA used most commonly for transformation experiments is bacterial plasmid DNA. These plasmids often carry a gene for antibiotic resistance. The presence of the antibiotic-resistance gene makes it possible to **select** bacteria containing the plasmid of interest; the bacteria that contain the plasmid will grow on a medium that contains the antibiotic, whereas bacteria lacking the plasmid will not be resistant to the antibiotic and will die.

Transformation can occur naturally, but the incidence is extremely low and is limited to a relatively few bacterial strains. During the growth cycle of these strains, there exists a short period of time when the bacteria are most receptive to uptake of foreign DNA. At this stage the cells are said to be **competent.** (Competence to absorb DNA usually develops toward the end of the logarithmic growth phase, just before cells enter the stationary phase in culture.) The mechanism by which competence is acquired is not completely understood, but in the laboratory, the competent state can be induced by treating bacterial cells with divalent cations such as Ca^{2+} and Mg^{2+}.

In this exercise, you will simulate the construction of a recombinant plasmid. Plasmids can transfer genes such as those for antibiotic resistance which are already a part of the plasmid, or plasmids can act as carriers for introducing foreign DNA from other bacteria, plasmids, or even eukaryotes into bacterial cells. Restriction endonucleases are used to cut and insert pieces of foreign DNA into the plasmid vectors (Figure 18A-1).

Each restriction endonuclease "recognizes" a specific DNA sequence (usually a 4- to 6-base-pair sequence of nucleotides) in double-stranded DNA and digests phosphodiester bonds at specific sites in the sequence (recall that phosphodiester bonds link one nucleotide to the next in a DNA polynucleotide chain). If circular DNA is cut at only one site, an open circle results. If the restriction endonuclease recognizes two or more sites on the DNA molecule, two or more fragments will result. The length of each DNA fragment corresponds to the distance between restriction sites (restriction sites flank the fragment at its ends). Some restriction endonucleases cut cleanly through the DNA helix at the same position on both strands to produce fragments with blunt ends. Other endonucleases cut specific nucleotides on each strand to produce fragments with overhangs or "sticky ends" (Figure 18A-2). Using the same restriction endonuclease to cut DNA from two different organisms produces complementary sticky ends, which can be realigned in a "template–complement" manner, thus recombining the DNA from the two sources (Figure 18A-2).

In bacteria, restriction enzymes provide protection by breaking and destroying the DNA of invaders, such as that of bacteriophage viruses. However, since the recognition sites for restriction endonucleases also occur within the bacterial DNA itself, bacteria have a mechanism for preventing their own restriction enzymes from digesting their own DNA. For each restriction endonuclease produced by a bacterium, there is a corresponding enzyme that methylates the bacterial DNA at that enzyme's specific recognition

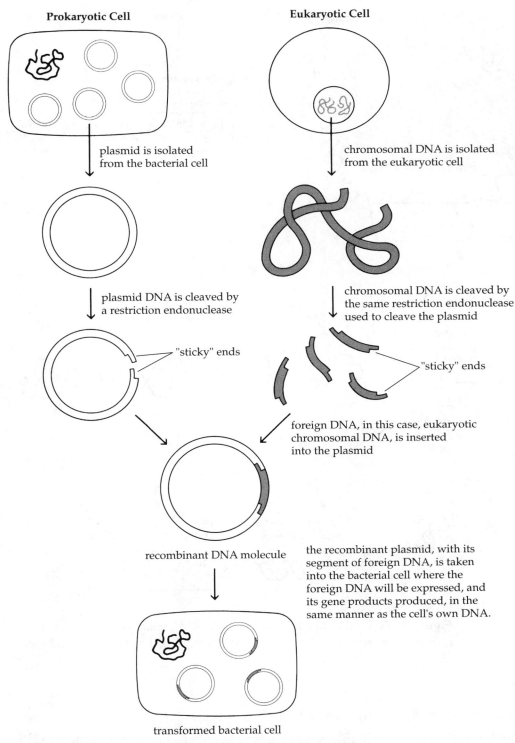

Prokaryotic Cell

Eukaryotic Cell

plasmid is isolated
from the bacterial cell

chromosomal DNA is isolated
from the eukaryotic cell

plasmid DNA is cleaved by
a restriction endonuclease

chromosomal DNA is cleaved by
the same restriction endonuclease
used to cleave the plasmid

"sticky" ends

"sticky" ends

foreign DNA, in this case, eukaryotic
chromosomal DNA, is inserted
into the plasmid

recombinant DNA molecule

the recombinant plasmid, with its
segment of foreign DNA, is taken
into the bacterial cell where the
foreign DNA will be expressed, and
its gene products produced, in the
same manner as the cell's own DNA.

transformed bacterial cell

Figure 18A-1   *A typical recombinant DNA experiment.*

site: addition of a methyl group to the nucleic acid base prevents a close association from forming
between the restriction endonuclease and the recognition site. In this way, bacteria can break down the
DNA of invaders while protecting their own genetic material from destruction.

ıııı Objectives ıııııııııııııııııııııııııııııııııııııı

☐ Describe how to construct a recombinant DNA plasmid.

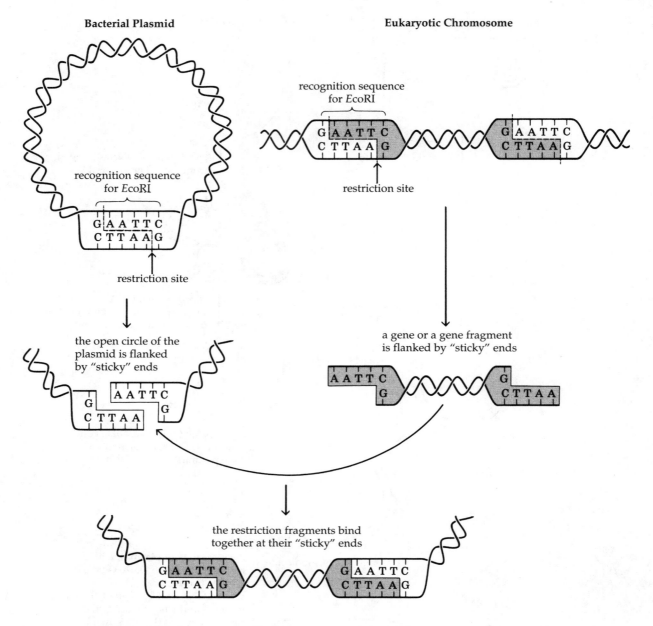

Bacterial Plasmid

Eukaryotic Chromosome

recognition sequence
for *Eco*RI

restriction site

recognition sequence
for *Eco*RI

restriction site

the open circle of the
plasmid is flanked
by "sticky" ends

a gene or a gene fragment
is flanked by "sticky" ends

the restriction fragments bind
together at their "sticky" ends

Figure 18A-2 *An example of how a bacterial plasmid and a fragment of eukaryotic chromosomal DNA are cleaved by the* EcoRI *endonuclease and then recombined.*

Procedure

The restriction endonuclease *Hind*III, isolated from *Haemophilus influenza*, recognizes the following restriction site and produces fragments with "sticky ends."

cut
↓
5' –A–A–G–C–T–T– 3'
3' –T–T–C–G–A–A– 5'
↑
cut

1. The shaded pieces of DNA in Figure 18A-3 (page 18-15) represent a segment of DNA from a human chromosome. The white piece of DNA represents a circular *E. coli* plasmid. Tape the nucleotide strands together as indicated in order to form a circular plasmid of bacterial DNA and a long linear strand of eukaryotic DNA.

2. Use a pair of scissors to cut the DNA of both the *E. coli* plasmid and the human DNA sequence as they would be cleaved by *Hind*III.

3. Insert the human DNA into the plasmid and tape the fragments together.

   a. *DNA ligase is used to join the fragments. What do you notice about the 3' and 5' ends of the restriction fragments as they are recombined?* _____

4. Suppose you wish to insert a second gene into the plasmid. You have *Eco*RI and *Bam*HI restriction endonucleases available. They cleave the DNA as follows:

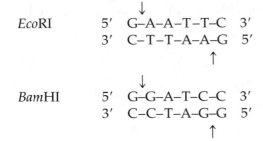

   *Eco*RI   5' G–A–A–T–T–C 3'
               3' C–T–T–A–A–G 5'

   *Bam*HI   5' G–G–A–T–C–C 3'
               3' C–C–T–A–G–G 5'

   Design a method for recombining the second gene with your plasmid.

   b. *Which restriction endonuclease would you use?* _____ *Why?* _____

   _____

   c. *What would be the required characteristics of the human DNA sequence?*

   _____

   _____

5. Use the "empty" DNA fragment in Figure 18A-3 to construct an appropriate human DNA fragment and insert it into the plasmid.

👁 **EXERCISE B** | **Rapid Colony Transformation with pAMP: Ampicillin Resistance***

Normally, *E. coli* cells are destroyed by the antibiotic ampicillin. In this exercise, you will induce competent *E. coli* cells to take up the plasmid pAMP, which contains a gene for ampicillin resistance. Only *E. coli* cells that have been transformed will be able to grow on agar plates containing ampicillin. Thus we can **select** for transformants: those cells that are not transformed will be killed by ampicillin; those that have been transformed will survive.

⊪⊪⊪ Objectives ⊪⊪⊪⊪⊪⊪⊪⊪⊪⊪⊪⊪⊪⊪⊪⊪⊪⊪⊪⊪⊪⊪⊪⊪⊪⊪⊪⊪⊪⊪⊪

☐ Discuss the principles of bacterial transformation.

☐ Describe how to prepare competent *E. coli* cells.

☐ Outline the general procedure for gene transfer using plasmid vectors.

☐ Carry out the transfer of the antibiotic gene *Amp^r* and describe how to select for transformed cells that contain the *Amp^r* gene.

Exercise B was developed by Dr. David Micklos, DNA Learning Center, Cold Spring Harbor Laboratory, and Dr. Greg Freyer, Columbia University College of Physicians and Surgeons.

||||| **Procedure** |||||||||||||||||||||||||||||||||||||

Formulate a hypothesis on which to base an investigation of how *E. coli* cells can be transformed by the pAMP plasmid.

HYPOTHESIS:

NULL HYPOTHESIS:

What do you **predict** will happen when ampicillin-sensitive E. coli cells are transformed by pAMP?

What is the **independent variable** in this investigation?

What is the **dependent variable** in this investigation?

Use the following procedure to test your hypothesis.

1. Use a sterile micropipette to add 250 μl of ice-cold 0.05 M $CaCl_2$ to two Eppendorf micro-centrifuge tubes.

2. Sterilize an inoculating loop by flaming it and then cool it by sticking it into the agar plate in an area where no bacteria are growing. Use the sterile inoculating loop to transfer a large (3-mm) colony of *E. coli* to one of the tubes. Be careful not to transfer any agar.

3. Vigorously tap the loop against the wall of the tube to dislodge the cell mass.

4. Suspend the cells immediately by vigorous pipetting using a 100-μl micropipette with a sterile tip or a sterile plastic transfer pipette.

5. Mark this first tube "(+)" and return it to the ice.

6. Repeat steps 2 to 5 for the second tube. Mark the tube "(−)."

7. Use a sterile inoculating loop to transfer 1 loopful (10 μl) of pAMP plasmid directly into the cell suspension in tube (+). At the correct angle, you will be able to see the plasmid solution form a film across the loop (much like what happens on a toy bubble-maker loop). Immerse the loop in the (+) cell suspension and mix well. Be sure to introduce the plasmid solution directly into the cell suspension—do not touch the wall of the tube as you insert the inoculating loop. Mix by tapping the tube with your finger.

8. Return the tube to ice for 15 minutes.

9. While the tubes are incubating, obtain two LB agar and two LB/Amp agar (LB agar containing ampicillin) plates. Label one LB agar plate "LB+" and the other "LB−." Label one LB/Amp plate "LB/Amp+" and the other "LB/Amp−." Mark your name on the lids.

10. A brief pulse of heat facilitates entry of foreign DNA into the *E. coli* cells. Heat-shock cells in both the (+) and (−) tubes by placing the tubes in a 42°C water bath for 90 seconds. (Tubes can be floated on the water by making an appropriate-sized hole in the center of a thin piece of Styrofoam to suspend the sample tube.) It is essential that cells be given a sharp and distinct shock, so work quickly.

11. Immediately return cells to ice for 2 minutes.

12. Use a sterile micropipette to add 250 μl of Luria broth to each tube. Mix by tapping with your finger and set at room temperature for recovery. Let sit for 10 minutes. During this period, the *Amp^r* gene, newly introduced into the transformed cells, codes for the synthesis of β-lactamase (an enzyme that destroys the antibiotic properties of ampicillin by cleaving its

β-lactam ring). The transformed cells are now resistant to ampicillin: they possess the gene whose product renders the antibiotic ineffective.

13. Place 100 μl of (+) cells onto the "LB+" plate and 100 μl of (+) cells onto the "LB/Amp+" plate. Place 100 μl of (−) cells onto the "LB−" plate and 100 μL of (−) cells onto the remaining "LB/Amp−" plate.

14. Immediately spread the cells using a sterile spreading rod. (Remove the spreading rod from ethanol and briefly pass it through a flame. Cool by touching it to the agar on a part of the dish away from the bacteria. Spread the cells and once again immerse the rod in alcohol and flame it.) Repeat the procedure for each plate.

15. Allow plates to set for 5 minutes. Tape your plates together and incubate *inverted* overnight at 37°C.

16. After 12 to 24 hours, indicate on which plates you observe growth.

    LB− _____   LB/Amp− _____

    LB+ _____   LB/Amp+ _____

    a. What is the purpose of the (−) plates? _____

    b. Why was no growth observed on the LB/Amp− plates? _____

Do your results support your hypothesis? _____ *Your null hypothesis?* _____

*What do you **conclude** about the ability of ampicillin resistance to be transferred from one bacterium to another?*

---

17. Transformation efficiency is expressed as the number of antibiotic-resistant colonies per microgram of pAMP.

    a. Determine the total amount of pAMP used: _____ μg. [You used 10 μl of pAMP (0.005 μg/μl); see step 7.]

    b. Determine the concentration of pAMP (in μg/μl) in the total suspension of cells plus Luria broth used for recovery (250 μl $CaCl_2$ + 10 μl pAMP + 250 μl Luria broth; see steps 1, 7, 12): _____ μg/μl

    c. Determine the total amount of pAMP in the 100-μl spread on the plate (see step 13): _____ μg pAMP/100 μl.

    d. Count the number of colonies on the plate: _____ colonies. (If there are too many, divide the plate into quarters, count one quarter, and multiply by 4.)

    e. Divide the number of colonies by the amount of pAMP in the 100 μl of cell suspension spread on the plate (step c) to give colonies/μg pAMP (use scientific notation): _____ colonies/μg pAMP. This is the transformation efficiency.

👁 **EXERCISE C** | **Transformation of *E. coli* with pBLU: The *lac*⁺ Phenotype**

In this exercise, you will work with the plasmid pBLU.™* In addition to the gene for ampicillin resistance (*Amp^r*), this plasmid carries a gene for production of the enzyme β-galactosidase (Figure 18C-1). This is one of the enzymes necessary for the complete breakdown of lactose, a carbohydrate that can be used in place of glucose as a source of nutrition for bacteria if glucose is unavailable. Restriction enzymes were used to insert the β-galactosidase gene into the pBLU plasmid.

---

*The pBLU™ plasmid was developed by Dr. Greg Freyer, Columbia University, College of Physicians and Surgeons, expressly for Carolina Biological Supply Company. Exercise C is adapted from the work of Dr. David Micklos, DNA Learning Center, Cold Spring Harbor Laboratory, and Dr. Greg Freyer.

Figure 18C-1 *The pBLU plasmid.™ A restriction map for the pBLU plasmid shows sites for cleavage by restriction enzymes used to insert the β-galactosidase gene into a plasmid carrying the Amp^r gene.*

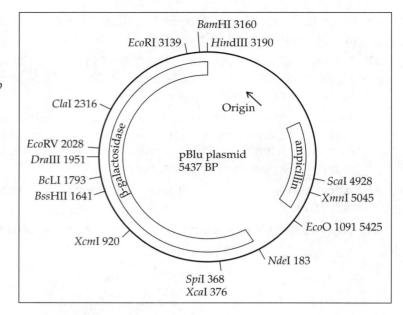

The host cells used for the study are bacterial cells, *E. coli* strain JM101. These bacteria are *not* resistant to ampicillin. They are also *unable* to metabolize lactose (they are *lac⁻*), due to a mutation in the *lac z* gene—the gene in the *lac* operon (Figure 18C-2) that produces β-galactosidase. JM101 cells, however, can acquire resistance to ampicillin (be transformed) by insertion of the pBLU plasmid carrying the *Amp^r* gene. And, at the same time, JM101 cells can acquire an undamaged copy of the β-galactosidase gene from the pBLU plasmid, transforming the cells to *lac⁺*. Transformed (*lac⁺*) cells can use lactose if it, instead of glucose, is supplied in the agar medium. *Lac⁺* cells can also use X-gal, a substitute for lactose that, when broken down by β-galactosidase, turns blue. As the transformed cells multiply to form **colonies,** the colonies will appear blue. The expression of the blue phenotype verifies that the gene for β-galactosidase (the *lac z* gene), contained in the pBLU plasmid, is being expressed.

Note that *selection* of transformed cells is actually based on their acquired ampicillin resistance and not on their ability to digest lactose. The agar used to select transformed cells contains ampicillin (in addition to X-gal), and only transformed cells (cells that now contain the *Amp^r* gene also carried on the plasmid) are able to grow. Selection for the *lac⁺* phenotype would require that the bacteria be grown on a medium containing *only* lactose as a food source, so that only those cells that are transformed to *lac⁺* could survive. This is not the case in this experiment, since some glucose is present in the agar medium.

||||| Objectives |||||||||||||||||||||||||||||||||||||

☐ Describe how to "engineer" a plasmid to include a piece of foreign DNA that confers the *lac⁺* phenotype to transformed cells.

☐ Carry out the transformation of *E. coli* using a plasmid that confers both ampicillin resistance and the *lac⁺* phenotype.

||||| Procedure ||||||||||||||||||||||||||||||||||||||||

Work in pairs. Formulate a hypothesis on which to base an investigation of how *E. coli* cells can be transformed by the pBLU plasmid.

HYPOTHESIS:

Figure 18C-2 *The* lac *operon.*
(a) The gene for production of
β-galactosidase is the lac z *gene. This*
is one of three structural genes in the
lac *operon. In combination with the*
lac y *and* lac a *genes, the* lac *operon*
produces the enzymes necessary for the
breakdown of lactose. The operon
consists of a **promoter** *region for the*
binding of RNA polymerse (as well as
a binding site for catabolite activator
protein/cyclic AMP, or CAP-cAMP,
which enhances transcriptional
activity) and an **operator** *site where*
repressors can bind. The lac *operon is*
under negative control by an active
repressor that binds to the operator
region where it blocks the movement of
RNA polymerase. (b) The operon will
remain "off" when glucose is readily
available and lactose concentrations are
low. (c) The lac *operon turns "on"*
when allolactose (a derivative of
lactose) acts as an inducer to inactivate
the repressor (causing the repressor to
fall off the operator region). This occurs
when lactose concentration in the
medium is high and glucose
concentration is low. RNA polymerase
can then transcribe a polycistronic
messenger RNA that codes for the
three gene products of the lac *operon.*

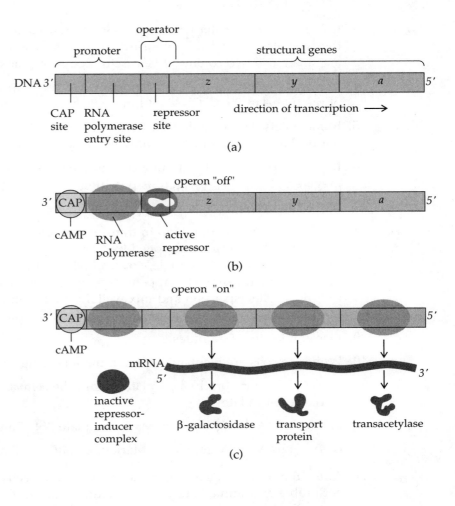

NULL HYPOTHESIS:

What do you **predict** will happen when JM101 E. coli cells that are not able to use lactose as a food source are transformed by pBLU?

What is the **independent variable** in this investigation?

What is the **dependent variable** in this investigation?

Use the following procedure to test your hypothesis.

1. Obtain two sterile plastic test tubes (with caps) to serve as "transformation tubes." Mark one "(+)pLBU" and the other "(−)pBLU."

2. Pipette 250 μl of ice-cold 0.05 M $CaCl_2$ into each of the two plastic tubes (this will be used to make the cells competent) and place both tubes on ice. (Everything must be kept cold!)

3. Using a sterile transfer (inoculating) loop, transfer a large colony of *E. coli* JM101 cells from the agar plate supplied by your instructor into one of the tubes containing $CaCl_2$. (If colonies are small, use two colonies.) Be careful not to dig the loop into the agar! If you do, start over.

4. Immerse the loop in the $CaCl_2$ solution. Tap the loop vigorously against the sides or bottom of the tube—in the $CaCl_2$—until the cells have been dislodged.

5. Immediately, use a sterile transfer pipette to break up the lump of cells. Work the cells in and out of the *tip* of the pipette until, when held up to the light, no clumps of cells are visible.

6. Repeat steps 3, 4, and 5, adding cells to the second "transformation tube." Be sure to disaggregate cell clumps by vigorous pipetting.

7. Make sure to place both tubes back into the ice.

8. Use a sterile inoculating loop to transfer one loopful (approximately 10 μl) of pBLU plasmid solution (0.005 μg/μl as supplied by Carolina Biological Supply Company) into the (+)pBLU tube only. At the correct angle, you will be able to see the plasmid solution form a film across the loop (much like what happens on a toy bubble-maker loop). Immerse the loop in the (+)pBLU cell suspension and mix well. Be sure to introduce the plasmid solution directly into the cell suspension—do not touch the wall of the tube as you insert the inoculating loop.

9. Place on ice for 15 minutes.

10. While the cells are incubating, obtain the following set of agar plates:

   a. Two LB (Luria broth) agar plates. Mark the *bottom* of one plate "(+)" and the bottom of the second plate "(−)".

   b. Two LB/Amp agar plates. Mark one plate "(+)" and the other "(−)" as above.

   c. Two LB/Amp/X-gal plates. Mark one plate "(+)" and the other "(−)" as above.

11. After 15 minutes on ice, you must heat-shock the cells to assist with plasmid uptake. Remove both tubes and immediately place the tubes into 42°C water for 90 seconds. Immediately return the tubes to ice for *at least* 1 minute before proceeding.

   a. *Why do you think you must also heat-shock the cells in the (−)pBLU tube when no plasmid was added to the tube?* _____

   _____

12. Use a sterile transfer pipette to add 250 μl (0.25 ml) of sterile Luria broth to each of the two tubes. Tap with your finger, gently, to mix and let the tubes stand at room temperature (place in a test tube rack) for 10 minutes. This is the recovery period—it will give cells, if transformed, a chance to start producing β-lactamase (see Exercise B) so that when exposed to ampicillin they will be able to degrade the antibiotic.

13. Use a sterile transfer pipette to add 100 μl of (+)pBLU cells to each of the three plates marked (+). Be careful *not* to touch the tip of the pipette to the agar—if you do, discard the pipette and obtain a clean one. Be sure to use an aseptic technique: only lift the lid above the plate—do not take it off or lay it down. You do not want air-borne bacteria and fungal spores to settle on your plates.

14. Use a second sterile transfer pipette to add 100 μl of the cell suspension from (−)pBLU to each of the three plates marked (−). Follow the same procedures and cautions given in step 13.

15. Use a sterile glass "spreader" to spread the cells across the surface of the agar plates. Dip the spreader in alcohol, and briefly pass it through the flame from a Bunsen burner or alcohol lamp. Always allow the alcohol to "burn off." Lift the lid on one of your Petri dishes and cool the spreader by placing it on the surface of the agar away from the cells—don't be surprised if it sizzles. When cool (but do *not* touch it with your fingers), use the spreader to

distribute the cells over the surface of the plate by gently rubbing back and forth at various angles. Lower the lid gently and return the spreader to the alcohol. (Do not flame the spreader before placing it back into the alcohol.) Repeat this procedure for the remaining dishes. If two spreaders are available, one partner should spread the cells on the (+) dishes, while the other partner spreads cells on the (−) dishes. Always put the spreader back into alcohol, and reflame it between using it on different Petri dishes.

16. Allow plates to stand for 5 minutes and then bundle the six plates into a stack. Tape the plates together and place *inverted* (top side downward) in a 37°C incubator. Incubate for 12 to 24 hours. If an incubator is not available, incubate at room temperature—it will simply take longer for the cells to grow and reproduce and for you to get your results!

b. *Which plates will serve as experimental control plates?* _____ *Do you expect to see cells*

   growing on these plates? _____ *Why or why not?* _____

c. *Why did you put both (+)pBLU and (−)pBLU cells on LB agar plates?*

   _____

d. *If growth occurs on both LB agar plates, what does this tell you?* _____

   _____

e. *If growth does not occur on either LB agar plate, what might you conclude?*

   _____

17. Indicate in Table 18C-1 what you expect to see, using G for growth and NG for no growth, on your plates after they have been incubated.

Table 18C-1   pBLU Transformation

Plate	Cells	Growth (G) or No Growth (NG)
LB	(+)pBLU	
LB	(−)pBLU	
LB/Amp	(+)pBLU	
LB/Amp	(−)pBLU	
LB/Amp/X-gal	(+)pBLU	
LB/Amp/X-gal	(−)pBLU	

Next Day

18. After incubation, record the number of colonies growing on the experimental LB/Amp/X-gal plate. If there are too many colonies to count, divide the plate into quarters using a marking pen. Count the number of colonies in one quarter and multiply by 4. Number of colonies:

   _____. Color development indicates that the β-galactosidase gene is functioning and the cells have been transformed to lac^+.

   f. *What color are the colonies?* _____

   If colonies are large, only their centers may be blue. X-gal is rapidly depleted from the medium as the colony grows.

   g. *Did you see any small white colonies at the edges of the blue colonies?* _____

These white colonies are feeder colonies. Often, the destruction of ampicillin by the trans-formed bacteria forms an area around the colony where nontransformed cells can grow. The nontransformed cells, however, will appear white. *h. Why?* _____

_____

19. Note the results from the control plates.

i. Did cells grow on the LB agar plates? _____ *Why or why not?*

_____

j. Did cells grow on the LB/Amp plates? _____ *Why or why not?*

_____

k. Explain the reasons for growth of the transformed cells on the LB/Amp/X-gal plates. Why are the cells resistant to ampicillin? _____

Do your results support your hypothesis? _____ *Your null hypothesis?* _____

*What do you **conclude** about the ability of transformed JM101. E. coli cells to use lactose and X-gal?*

20. Transformation efficiency is expressed as the number of antibiotic-resistant colonies per microgram of pBLU.

a. Determine the total amount of pBLU used: _____ μg. (You used 10 μl of pBLU (0.005 μg/μl); see step 8.)

b. Determine the concentration of pBLU (in μg/μl) in the total suspension of cells plus Luria broth used for recovery (250 μl $CaCl_2$ + 10 μl pBLU + 250 μl Luria broth; see steps 2, 8, 12):

_____ μg/μl

c. Determine the total amount of pBLU in the 100-μl spread on the plate (see step 13):

_____ μg pBLU/100 μl

d. Count the number of colonies on the plate: _____ colonies. (If there are too many, divide the plate into quarters, count one quarter, and multiply by 4.)

e. Divide the number of colonies by the amount of pBLU in the 100 μl of cell suspension spread on the plate (step c) to give colonies/μg pBLU (use scientific notation): _____ colonies/μg pBLU. This is the transformation efficiency.

Because transformation is limited to those cells that are competent, increasing the amount of plasmid used does not necessarily increase the probability that a cell will be transformed. A sample of competent cells can usually be saturated with small amounts of plasmid, and excess DNA may actually interfere with the transformation process.

l. How does the transformation efficiency of pBLU compare with that of pAMP (Exercise B)?

_____

Laboratory Review Questions and Problems

1. You are given the following assignment by a biotechnology firm. Transfer gene *A* from a mouse chromosome to plasmid pBR322 of *E. coli*. What would be your first consideration in choosing one or more restriction endonucleases to cut the plasmid and the chromosome? (*Hint:* Various genes on the eukaryotic chromosome are flanked by various nucleotide sequences.) Gene *A*, by the way, is responsible for production of a protein that could be economically important if it could be produced in bulk.

2. You have successfully completed a transformation experiment. There are 800 colonies on your plate. You used 50 μl of a solution containing *E. coli* cells mixed with 1×10^{-3} μg/ml of plasmid DNA to inoculate the test plate. What was the transformation efficiency in this experiment?

3. Assume that a bacterial plasmid carries the gene for resistance to the antibiotic kanamycin. Using restriction enzyme A, you open the plasmid and insert a segment of a biologically important gene isolated from a mouse. The gene was excised from the chromosome as part of a fragment cut from whole DNA by using the same restriction enzyme A. After conducting the appropriate steps in a typical bacterial transformation, you plate the transformed cells (+) and control cells (−) on LB agar containing kanamycin and on LB agar alone.

   a. What do you *expect* to see? Indicate this on the plates below:

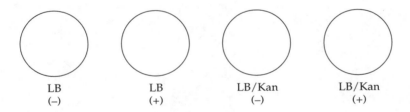

| LB | LB | LB/Kan | LB/Kan |
| (−) | (+) | (−) | (+) |

   b. What you actually observe is no growth on either LB/Kan plate, but growth on both LB plates. You try the experiment again using a different restriction enzyme B. This time you get growth of transformed (+) cells on LB/Kan but no growth of control (−) cells on LB/Kan. You get growth of both transformed (+) and control (−) cells on LB plates. How might you explain these observations? Propose a map for the bacterial plasmid and the restriction sites for restriction endonucleases A and B.

4. You are working in a recombinant DNA laboratory and are asked to clone a gene from a very rare strain of bacteria. The gene produces an important protein used in the oil industry to clean up oil spills. You need large amounts of this product. Outline the steps you would take to get an ordinary bacterium such as *E. coli* (which does *not* normally make the protein) to produce large amounts of this protein.

Figure 18A-3 *Directions: Cut out the top white strand (the "plasmid") and position tab Y under X to form a circle; tape ends together. To form the eukaryotic "chromosome," cut out the two long shaded strands and tape tab B over tab A; cut out the short shaded strand and tape tab D over tab C.*

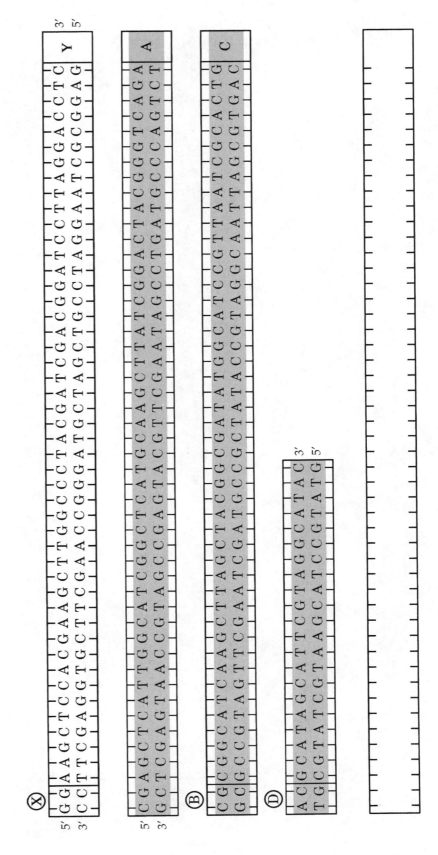

Genetic Control of Development and Immune Defenses

OVERVIEW

The hereditary material within fertilized eggs, or zygotes, of animal organisms holds the key to both structure and function. DNA guides, through time, the structural and functional development of a single cell into an embryo and its morphogenesis into an adult. Even aging is a result of genetic programming.

Biologists are just beginning to understand the genetic control of development. Advanced technologies as well as molecular and recombinant DNA techniques have made it possible to discover more and more about the genomes of organisms and to map the entire genomes of some. Common patterns of gene function are beginning to emerge, suggesting that some day, perhaps, we will be able to explain the mysteries of animal development.

Protection for the organism is also built into the genetic code. A newborn mammal possesses a full set of genetic information for the synthesis of a vast number of different immunoglobulins (antibodies). During development of the B cells of our immune system, however, each cell is modified to produce only a single type of antibody. How could the limited amount of DNA in a single organism produce millions of different types of antibodies? Research in recent years has shown that functional immunoglobulins are assembled from DNA segments that are initially separated and then rearranged during B cell development. Once again DNA exhibits its programming functions—DNA holds the secrets not only for "what" it produces but for "how" and "when" it produces its products.

During this laboratory you will have the opportunity to explore the development of sea urchin, frog, and chick embryos. You will also have an opportunity to investigate antibody activity using the precipitin ring test.

STUDENT PREPARATION

Prepare for this laboratory by reading the text pages indicated by your instructor. Familiarizing yourself in advance with the information and procedures covered in this laboratory will give you a better understanding of the material and improve your efficiency.

PART I DEVELOPMENT

All living things must be able to reproduce and develop. In animals, gametes produced by the process of meiosis unite during fertilization to form a single diploid cell, the zygote. The processes of cell division, cell movement, cellular differentiation, and morphogenesis result in the development of a multicellular embryo that will grow to form an adult.

✔ **EXERCISE A**   **Fertilization and Early Development in Sea Urchins**

Understanding how DNA regulates cell division, differentiation, and morphogenesis begins with observing the developing embryo. Sea urchins have long been the developmental biologist's favorite organism of study because they are relatively simple to obtain and culture in the laboratory. The cells of the developing sea urchin are also fairly transparent, providing us with a limited ability to "look inside" the embryo.

Early development of the sea urchin is under the genetic control not only of the zygote's DNA but of messenger RNA (mRNA) stored in the egg during its development. These messages include maternal mRNAs synthesized from maternal DNA prior to the meiotic events of oogenesis—mRNAs made from DNA that may not be included in the egg itself. Thus, the story of development begins before fertilization.

The unfertilized sea urchin egg is surrounded by a **vitelline membrane** that lies just above the surface of the cell's plasma membrane (plasmalemma). Within the cytoplasm, yolk granules (sea urchin eggs are microlecithal—they have very little yolk) can be observed. In addition to other cytoplasmic determinants and stored mRNAs, small **cortical granules,** composed of proteins and mucopolysaccharides, lie just beneath the plasmalemma (in the outer rim or cortex of the egg) (Figure 19A-1a). When a single sperm enters the egg plasmalemma, the membrane potential quickly changes as a wave of depolarization spreads from the site of sperm entry. This reaction is often referred to as a "fast block to polyspermy," since no additional sperm can gain entry following the change in membrane potential. Release of Ca^{2+} ions from the egg's endoplasmic reticulum, in response to G protein, causes the cortical granules in the egg's cortex to fuse with the plasmalemma (Figure 19A-1b).

The cortical granules discharge their contents into the space between the plasmalemma and the vitelline membranes. The excess mucopolysaccharide now present in the perivitelline space lowers the water potential of that area, and water flows in. This causes the perivitelline space to increase in diameter, making it appear as if the vitelline membrane is lifting off the surface of the zygote. Addition of proteins to the vitelline membrane hardens (or "tans") it as it is transformed into a **fertilization membrane** (Figure 19A-1c). Formation of the fertilization membrane (often called the "slow block to polyspermy") offers additional protection against multiple sperm entry. Hundreds of sperm can usually be observed still attached to the old vitelline membrane, now the fertilization membrane. These will be removed by the action of enzymes released from the cortical granules.

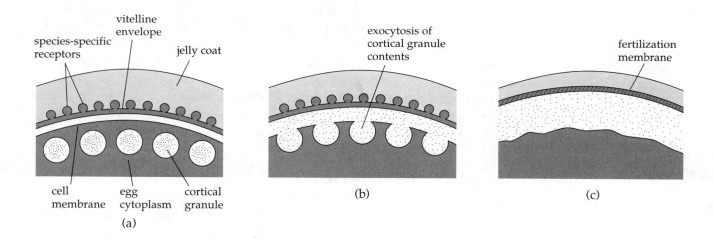

Figure 19A-1   *Formation of the fertilization membrane in the sea urchin. (a) The surface layers of an unfertilized egg include a jelly coat; the vitelline membrane or envelope, which bears species-specific receptors for sperm; and the cell membrane (plasmalemma). (b) Contact with a sperm triggers Ca^{2+} release and the fusion of cortical granules with the plasmalemma. (c) Cortical granules release their contents into the perivitelline space, causing the vitelline membrane to rise from the surface to form the fertilization membrane.*

Following fertilization, the stored mRNAs are responsible for directing protein synthesis during the early stages of development—cleavage, blastula formation, and gastrulation. In addition to the mRNA and protein already present in the fertilized egg, newly synthesized protein products are responsible for establishing regional differences in the egg cytoplasm. These proteins or **cytoplasmic determinants** are responsible for establishing the cytoarchitecture of the unfertilized egg, reorganizing the cytoplasmic elements in response to fertilization, controlling the direction of the first cleavage divisions, and establishing the axis of the embryo.

In this exercise, you will have the opportunity to observe the earliest stages of sea urchin development—beginning with fertilization. You will also have available later-stage embryos for observation of gastrulation and formation of pluteus larvae.

||||| Objectives |||||||||||||||||||||||||||||||||||||||||

☐ Describe the processes of fertilization and cleavage division in the sea urchin.

☐ Observe formation of the fertilization membrane as it expands to surround the sea urchin.

||||| Procedure |||||||||||||||||||||||||||||||||||||||||

1. Inject 0.1 to 1.0 ml of 0.5 M KCl intracoelomically through the soft tissue on the oral surface of the animal. Shake the animal gently after injection to distribute the KCl. (For some urchins, KCl injections will not work, but electrical stimulation is effective. This method requires an adjustable physiological stimulator set to a frequency of 60 Hz. Use 6 to 10 volts AC for smaller urchins and 20 to 30 volts AC for larger urchins. One electrode should be placed directly on the test (body surface) near the gonopore region and the other on a wet cotton ball or tissue on the oral surface.)

2. Gamete shedding should begin minutes after injection. (If you use electrical stimulation, gamete shedding will continue as long as the animal is stimulated, thus you can collect only the amount of material needed.) Once gamete shedding begins, the sex of the urchin can be determined. Semen is gray-white or cream colored. Eggs vary in color, depending on the species, and may be pale yellow, yellow-orange, or reddish.

3. To collect eggs, invert the female over a beaker of seawater and lower the urchin until it is covered by the seawater. Eggs will stream from the gonopores and fall to the bottom of the beaker. If electrical stimulation is used, place the female, after stimulation, right side up (gonopores up) in a beaker of seawater—the urchin should be completely covered with seawater.

4. Using a wide-mouth pipette (a turkey baster is useful), remove eggs that have settled to the bottom of the beaker and resuspend them in clean seawater.

5. Let the eggs settle, decant the seawater, and replace it with clean seawater. (This step removes any traces of coelomic fluid, which might interfere with the fertilization process.) The eggs are now ready to use.

6. It is best to collect sperm in the "dry" condition for maximum concentration. Place the male in a finger bowl of seawater, but make sure that the level of the water is below the gonopores. As the whitish semen is extruded from the gonopores, collect it with a Pasteur pipette. Store the semen in a test tube on ice (it will remain viable for several hours).

7. Return the urchins to a separate aquarium (if the animals continue to shed gametes, they may trigger other animals in the aquarium to shed gametes).

8. Obtain a depression slide and place a sample of egg suspension into the concave depression. Cover with a glass coverslip. Observe the eggs using the 10× and 40× objectives. If you wish to observe the jelly coats, add a drop of 0.1 Janus green before applying the coverslip.

9. Obtain a second depression slide and place another sample of the egg suspension into the concave depression. Cover with a coverslip, but leave a small opening on one side for the

addition of sperm in step 10. Place the slide on the microscope stage and use the 10× objective to locate the eggs.

10. Use a Pasteur pipette to place a drop of concentrated sperm suspension on a clean glass slide. Add two drops of seawater. Mix with a toothpick and, using the blunt end of the toothpick, introduce some sperm suspension into the egg suspension on the slide. Observe immediately.

11. Watch for the fertilization membrane to form, then turn off the light on your microscope or remove the slide from the microscope and gently place it in a safe place (away from heat and hot light) for later observation. If the seawater begins to evaporate from your depression slide, add more using a Pasteur pipette.

12. Fusion of pronuclei usually occurs 30 to 45 minutes after gamete fusion. The first cleavage division occurs 60 to 90 minutes after gamete fusion, depending on the species. Check your slide in approximately 45 minutes and then at 10- or 15-minute intervals until you observe the first cleavage division.

13. Your instructor may have additional samples of embryos in other stages of development for you to examine. Prepare wet-mount depression slides of these. See if you can identify four- or eight-cell embryos, early blastulae, hatching blastulae, or gastrula embryos. In the sea urchin, the larval form is called a **pluteus larva.**

14. For a composite photographic description of sea urchin development, refer to Figure 19A-2.

Figure 19A-2 *Sea urchin development. (a–h) As the egg divides, the cells become progressively smaller, so that by the blastula stage they are barely distinguishable. (i) Gastrulation begins with the formation of the blastopore, then (j, k) secondary mesenchyme cells break loose to migrate along the inner surface of the blastula. These form long pseudopodia that (l) help to "pull" the forming gut toward the opposite side of the embryo. (m) Spicules form from primary mesenchyme cells within the developing pluteus larva (n).*

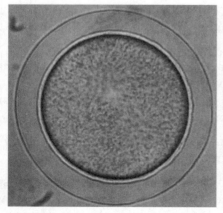
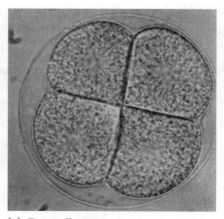
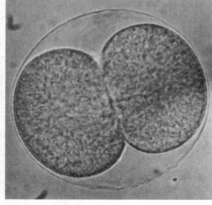
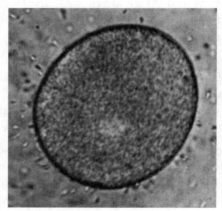
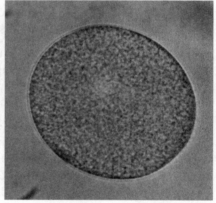

(a) Numerous spermatozoa can be seen surrounding an unfertilized egg.

(b) Fertilized egg; the fertilization membrane has just begun to form. The light area slightly above the center is the diploid nucleus.

(c) The fertilization membrane is fully formed. The egg has begun to divide; if you look closely, you can see that there are two nuclei.

(d) The first division.

(e) Four-cell stage.

(continued)

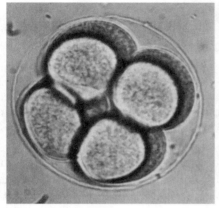

(f) Eight-cell stage.

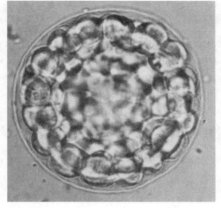

(g) The blastocoel forms.

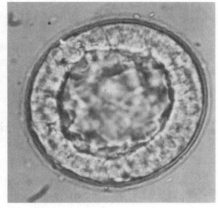

(h) The mature blastula.

0.1 mm

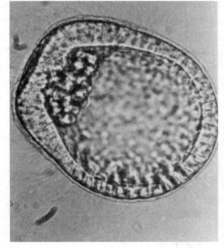

(i) The beginning of gastrulation; the blastopore has begun to form at the upper left, and cells near the blasto-pore have begun to migrate across the blastocoel.

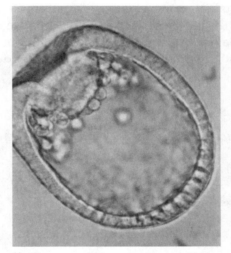

(j) The outer cell layer begins to fold inward at the blastopore, forming the archenteron.

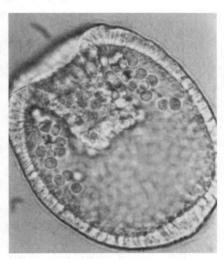

(k) The outer layer of cells continues to move across the blastocoel.

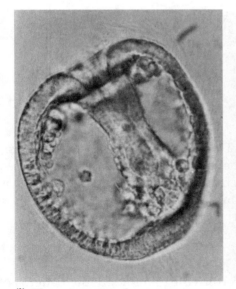

(l) The mature gastrula.

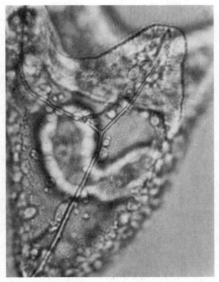

(m) Gastrula cells differentiate and or-ganize to form the pluteus larva.

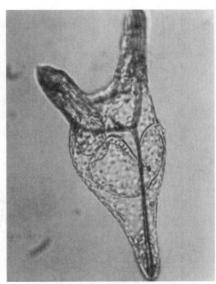

(n) Within 48 hours after fertilization, the egg has developed into a free-swimming multicellular organ-ism, the pluteus.

✓ **EXERCISE B** | **Cleavage**

In order for a fertilized egg or zygote to become a multicellular organism, the zygote must divide by mitosis. During early development, this process of division is known as **cleavage.**

The fertilized egg is not a uniform sphere. Differential concentrations of cytoplasm and yolk (if present) can affect the cleavage process. The upper portion of the egg, usually richest in cytoplasm, is known as the **animal pole,** and the lower portion of the egg, containing more yolk, as the **vegetal pole** (Figure 19B-1a). The first plane of cleavage is vertical, bisecting both the animal and vegetal poles (Figure 19B-1b).

Depending upon the amount of yolk in the egg, the planes of cleavage may pass all the way through the zygote (holoblastic cleavage, typical of cells with small to medium amounts of yolk; sea urchin and frog) or through only a part of the zygote (meroblastic cleavage, typical of cells with large amounts of yolk; chicken).

A second cleavage division typically occurs at a right angle to the first, producing four cells. The third cleavage division cuts horizontally to form eight cells, four on the top and four on the bottom (Figure 19B-1c, d). The cells produced during these cleavage divisions are known as **blastomeres.** If the blastomeres in the top "tier" lie directly above those in the bottom tier, the pattern of cleavage is said to be **radial,** a pattern characteristic of echinoderms and chordates (deuterostomes).

Figure 19B-1 *Eggs that contain a large amount of yolk in one hemisphere cleave unequally. The first two cleavages (b, c) split the egg through the poles. The third cleavage separates the yolkier (vegetal) part from the upper, less yolky (animal) part. The four cells in the animal hemisphere are much smaller than the four in the vegetal hemisphere.*

(d) The pattern of cleavage shown here is radial: the top four blastomeres are directly above the bottom four.

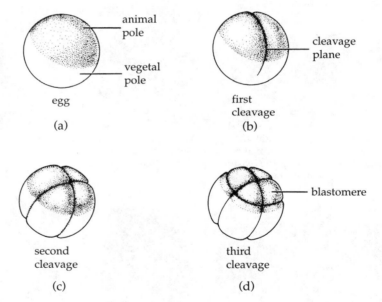

In the sea urchin, the fourth cleavage results in the formation of 16 blastomeres of three different sizes. Eight medium-size mesomeres are the product of the division of the four blastomeres in the animal hemisphere. The lower four blastomeres (vegetal hemisphere) produce four large macromeres and four small micromeres (Figure 19B-2). As a consequence of this cleavage pattern, cytoplasmic determinants are distributed in an unequal manner, laying the groundwork for future development (see Exercise C).

▮▮▮▮▮ **Objectives** ▮▮▮

☐ Describe how the amount and distribution of cytoplasm within a fertilized egg influence the patterns of cleavage.

☐ Describe how cleavage occurs in the starfish, frog, and chick.

Figure 19B-2 *Fourth cleavage division in sea urchin. An unequal fourth cleavage division results in 16 blastomeres of three different sizes. A top tier of eight medium-size blastomeres, called mesomeres, will eventually give rise to ectodermal structures. The four large macromeres will develop into endodermal structures associated with the gut. The four smallest blastomeres, micromeres, will produce mesodermal structures.*

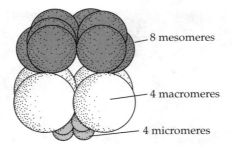

— 8 mesomeres

— 4 macromeres

— 4 micromeres

||||| Procedure ||||||||||||||||||||||||||||||||||||||||||||

1. Obtain a composite slide of starfish (*Asterias*) development. This slide will have all stages of echinoderm development represented, including unfertilized eggs, fertilized eggs, and cleavage stages (Figure 19B-3). Some of the unfertilized eggs contain a **germinal vesicle**, a large swollen nucleus containing the nucleolus (appears as a black dot). The germinal vesicle breaks down at the end of prophase I and the egg appears as an opaque sphere during the remainder of its maturation.

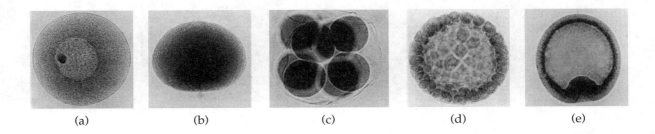

(a)   (b)   (c)   (d)   (e)

Figure 19B-3 *Embryonic development of the starfish: (a) germinal vesicle; (b) zygote with polar body; (c) 8-cell stage; (d) blastula; (e) early gastrula, blastopore apparent.*

Locate an unfertilized egg with a germinal vesicle and one without. Label a sheet of paper "Starfish Development" and draw these two examples of an unfertilized egg on the top half of the sheet. Label the nucleus and nucleolus.

a. *Why is the nucleus of the developing zygote so big?*_____

2. Find two-cell, four-cell, and eight-cell embryos on your slide of *Asterias* development. Notice that all of the cells in this embryo are the same size. On your sheet of paper, draw what you observe.

b. *Is the mass of a four-cell or an eight-cell embryo any larger than the mass of the zygote?*_____

c. *What is the effect of cleavage on cell size?*_____

*On embryo size?*_____

If live sea urchin material is available from Exercise A, observe the first cleavage division and other early cleavage stages. Compare your observations of starfish development with the stages of sea urchin development shown in Figure 19A-2.

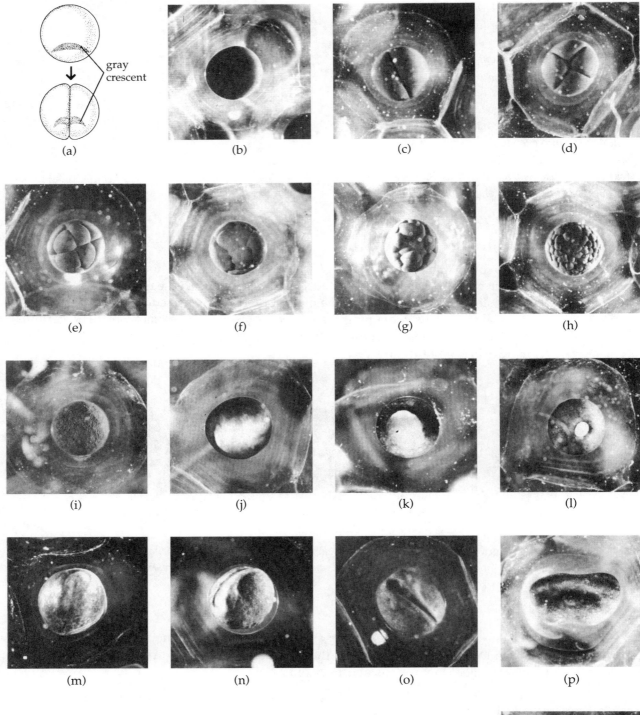

Figure 19B-4   *(a) The gray crescent of the frog zygote is bisected by the first cleavage plane so that each blastomere will contain animal-pole cytoplasm, gray-crescent cytoplasm, and vegetal-pole cytoplasm.*

Embryonic development of a frog: (b) egg; (c) 2-cell stage; (d) 4-cell stage; (e) 8-cell stage; (f) 16-cell stage; (g) 32-cell stage; (h) early blastula; (i) late blastula; (j) blastopore (the curved line where the dark region meets the light region); (k) early yolk plug; (l) late yolk plug; (m) neural plate; (n) early neural groove; (o) late neural groove; (p) neural tube; (q) tailbud stage.

  d. What do you observe? Are the developmental stages similar? _____

_____

3. Examine a fertilized frog egg (preserved specimen) using a dissecting microscope. The frog egg shows how the arrangement of cytoplasm influences the placement of the first cleavage plane. In the frog, fertilization results in a shifting of the pigmented cytoplasm of the animal pole, establishing a grayish, crescent-shaped area (the **gray crescent**) on one side of the zygote (Figure 19B-4a). This area must be cut in half (bisected) by the first cleavage plane, thereby establishing the right and left halves of the future embryo. Use dissecting needles to move the egg around until you see the gray crescent. Do not poke the egg.

4. Now examine an embryo in the two-cell stage. *e. Does the cleavage plane cut the gray crescent in half?* _____

   Interestingly, if you separated these two halves of the embryo, each would develop into a normal tadpole (twins). If, however, you artificially divided the embryo into two halves (by slowing constricting it with a fine hair loop) so that all of the gray crescent material was contained in one blastomere, only the blastomere containing the gray crescent material would develop normally.

5. Examine a sagittal section through an early cleavage stage of the frog zygote. Notice the dark pigment at the surface of the animal pole of the egg. Are the top blastomeres the same size as the lower blastomeres? _____ Label a sheet of paper "Frog Development" and draw the representative section at the top of the sheet.

✔ **EXERCISE C   Formation of the Blastula**

Repeated cleavages will result in formation of a hollow ball of cells called a **blastula.** The cavity inside the blastula is called the **blastocoel.** Even as early as this blastula stage, groups or layers of cells are already destined to become particular organs or organ systems; these layers of cells are known as **presumptive germ layers.** The major germ layers and their derivatives are listed in Table 19C-1.

Table 19C-1   Germ Layers and Their Derivatives

Germ Layer	Derivative
Epidermal ectoderm	Skin
Neural ectoderm	Brain, spinal cord, and neural crest cells
Chordamesoderm	Notochord and spinal disks in some organisms; ganglia
Mesoderm	Skeleton, circulatory system, excretory system, and parts of organs belonging to other systems
Endoderm	Gut and associated outpocketings

   In the sea urchin, blastomeres that will give rise to cells of germ layers are already laid out at the 16-cell stage. A **fate map** can be assigned. Mesomeres will give rise to ectodermal structures including the cilia that develop on the blastula's surface. Macromeres will give rise to endodermal structures. Micromeres will be responsible for formation of the body cavity, many internal organs, and the skeletal elements (**spicules**) of the embryo (see Figures 19A-2 and 19B-2).

||||| **Objectives** |||||||||||||||||||||||||||||||||||||

☐ Describe the structure of a typical blastula.

☐ Relate the structure of the frog blastula to the establishment of the dorso-ventral and antero-posterior axes of the embryo.

||||| **Procedure** ||||||||||||||||||||||||||||||||||||||

1. Reexamine the composite slide of starfish development. Locate an early blastula stage. The cells will be large enough to see and will appear as a dark ring of cells surrounding a lighter center. Remember that these are whole mounts and you are looking through a hollow sphere (see Figures 19A-2 and 19B-3). Draw your observations on the sheet of paper labeled "Starfish Development." Label blastomeres and blastocoel.

    a. Why does the center appear to be lighter? _____

    _____

2. Locate a later blastula stage. The cells of the blastula are now so small that it will be difficult to distinguish individual cells (see Figures 19A-2 and 19B-3). Again, the outer rim appears to

    be dark. *b. Why?* _____

    At this time, the cells of the blastula are covered by cilia which allow the blastula to spin and move. Draw your observations. Label blastomeres and blastocoel.

3. Examine a sagittal section through a frog blastula. Note the placement of the blastocoel.

    c. Why is the blastocoel found closer to the animal pole of the embryo?

    _____

    d. How does the size of the animal-pole blastomeres compare with that of the vegetal-pole blastomeres?

    _____

    Draw the sagittal section on your sheet of paper labeled "Frog Development," and label blastomeres and blastocoel.

✔ **EXERCISE D** | **Gastrulation**

As a blastula, most of the cells of the embryo (including those destined to become cells of internal organs and tissues) are on the outside of the hollow ball, and it is obvious that some of the sheets of cells must move, or migrate, to the inside of the blastula. This process of cell movement is called **gastrulation.**

In some organisms, such as the starfish and sea urchin, gastrulation is accomplished simply by buckling or pushing inward, forming a depression or **blastopore.** Endoderm and mesoderm reach the inside of the embryo in this manner. In other organisms, such as the frog or the chick, cells migrate to the interior by way of the blastopore, the portion of the embryo that will eventually contribute to the development of the anus (Figures 19A-2 and 19D-1). Mesoderm, endoderm, and chordamesoderm (notochord material) migrate to the inside. With the change in position of the germ layers, the blastocoel of echinoderms (starfish and sea urchin) and chordates (frog) is eventually obliterated and a new cavity, the **archenteron,** is formed within the gastrula. The archenteron is the primitive gut of the embryo (see Figures 19A-2, 19B-4, and 19D-1).

Note that during sea urchin development, secondary mesenchyme cells help to "pull" the endoderm of the newly forming archenteron (gut) toward the opposite side of the embryo, where it will fuse with the outer layer, or ectoderm, to form the mouth. (Hence the term "deuterostome" or second mouth, the anus forming at the blastopore or first opening.) See Figures 19A-2 and 19D-1.

Gastrulation in the starfish and the sea urchin

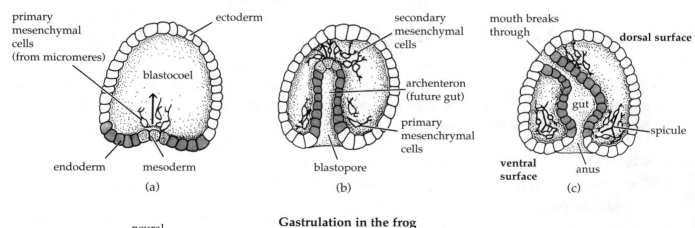

primary mesenchymal cells (from micromeres)

ectoderm

blastocoel

endoderm   mesoderm

(a)

secondary mesenchymal cells

archenteron (future gut)

primary mesenchrymal cells

blastopore

(b)

mouth breaks through

dorsal surface

gut

spicule

ventral surface   anus

(c)

Gastrulation in the frog

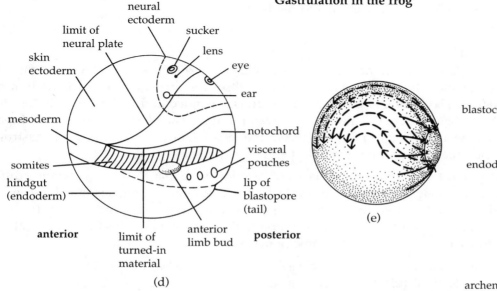

neural ectoderm

limit of neural plate

skin ectoderm

mesoderm

somites

hindgut (endoderm)

sucker

lens

eye

ear

notochord

visceral pouches

lip of blastopore (tail)

anterior   limit of turned-in material   anterior limb bud   **posterior**

(d)

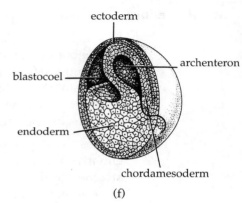

(e)

ectoderm

archenteron

blastocoel

endoderm

chordamesoderm

(f)

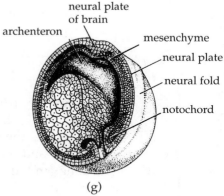

neural plate of brain

archenteron

mesenchyme

neural plate

neural fold

notochord

(g)

Figure 19D-1   *In some organisms, such as the sea urchin and starfish, gastrulation occurs as a simple buckling and pushing in of cells on one side of the embryo (a). This area of buckling becomes the blastopore, the entrance to the archenteron, or primitive gut (b). The archenteron will be lined by endoderm. The primary mesenchymal cells that arise from the mesoderm at the leading edge of the archenteron will accumulate in a ring near the vegetal pole, where they will form skeletal rods (spicules). Ectoderm will cover the entire embryo. The archenteron breaks through on its anterior end to form the mouth opening of the gut, and coelomic pouches bud from the gut (c).*

In other organisms, such as the frog, gastrulation occurs by an inward migration of cells by way of the blastopore (d, e). Sheets of ectoderm, mesoderm, and endoderm move around until they are in the correct position for organ formation. Immigration of cells and involution into the interior continue (f). The neural plate stretches in an anterior-posterior direction across the dorsal surface of the embryo (g). The edges of the neural plate will turn upward to form the neural folds along the neural groove, and eventually the entire structure will tubulate to form the brain and spinal column (h).

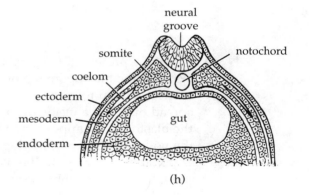

neural groove

somite

coelom

ectoderm

mesoderm

endoderm

notochord

gut

(h)

a. Why do the mesoderm, endoderm, and future notochord need to be on the inside?

_____

b. To what organs or organ systems do these germ layers give rise?

_____

Ectoderm and neural ectoderm remain on the outside of the frog or chick embryo. Eventually, the neural ectoderm will form a tube that is to become the brain and nerve cord, while the epidermal ectoderm spreads to cover this structure and the rest of the embryo.

‖‖‖ Objectives ‖‖‖‖‖‖‖‖‖‖‖‖‖‖‖‖‖‖‖‖‖‖‖‖‖‖‖‖‖‖‖‖‖‖‖‖‖‖‖

☐ Compare and contrast the structural components of the blastula and gastrula frog embryo.

☐ Describe the purpose of gastrulation.

☐ Compare the process of gastrulation in the starfish and the frog.

‖‖‖ Procedure ‖‖

1. Use high power (40×) to examine the composite slide of starfish (*Asterias*) development. Locate both early and late gastrula embryos. You will see the **archenteron** beginning to form as an inpocketing. The front end of the archenteron is expanded and will eventually fuse to the side of the embryo where the mouth will break through. The blastopore (region of inpocketing) becomes the anus. Refer to Figures 19A-2 and 19B-3. Draw the early and late gastrula embryos on the sheet of paper labeled "Starfish Development."

2. Locate a folded and twisted **bipinnaria** larva of *Asterias*. Note that the gut has become differentiated into an **esophagus, stomach,** and **intestine.** Two **coelomic pouches** may be seen attached to the gut wall. These are mesodermal and will eventually expand to form the mesodermally lined coelomic cavity. Draw a bipinnaria larva on your sheet of paper. Label the esophagus, stomach, intestine, and coelomic pouches.

   c. Recall from your study of diversity that echinoderms are enterocoelous coelomates. Explain how this description reflects the developmental events observed on the starfish composite slide.

   _____

   _____

3. Compare the structure of the starfish bipinnaria larva to that of the sea urchin pluteus larva (Figure 19A-2). If live specimens of sea urchin pluteus larvae are available from Exercise A, use a depression slide to observe that structure and movement. Observe the skeletal elements (spicules) and internal organs of the digestive system. Do you observe any coelomic pouches forming? In the space below, describe what you observe.

4. Study the preserved frog embryos on demonstration. Using the dissecting microscope, examine the blastopore region at the center of the lower edge of the gray crescent. Cells on the exterior stream toward the blastopore to turn inward (Figures 19D-1e and 19B-4j, k) and then spread out on the inside of the gastrula to take up their proper places. As cells converge on the blastopore, it appears first to form sides and then a ventral lip that completely encircles the yolky cells (the "yolk plug") of the interior (Figure 19B-4l).

   Note the changes in the blastopore region in eggs of increasing ages. Identify the yolk plug. Draw these embryo stages on the sheet of paper labeled "Frog Development."

5. Use scanning and low powers (4× and 10×) to observe a sagittal section through a late gastrula or **yolk plug** stage of the frog embryo. Identify the yolk plug and archenteron. Using Figure 19D-1, try to identify the cells of the notochord, ectoderm, and endoderm of the gastrula.

EXERCISE E   Neurulation

During this stage of development, a strip of neural ectoderm on the outside of the dorsal surface of the embryo (Figure 19B-4m) turns upward to form a **neural tube** (Figure 19D-1h). The underlying mesoderm and notochord tissue induce formation of the neural tube. The folds of tissue forming the tube are the **neural folds** and the groove between them is the **neural groove.** Eventually the anterior end of the neural tube will expand to form the **brain;** the spinal cord will develop posterior to the brain.

Aggregations of mesoderm (mesodermal **somites**) behind the brain and alongside the spinal cord will form vertebrae (back bones) that protect and enclose the spinal cord. Mesodermal somites also give rise to dorsal skeletal muscles and to the dermis of the skin (Figure 19D-1d). The presence of somites is an indication of the segmented nature of vertebrate embryos. Biologists have recently shown that a series of genes control the development of segmentation in all segmented embryos, whether in fruit flies, mice, or in humans. These genes, called **homeotic genes,** often work in a cascade with other genes that control the basic head-to-tail and anterior-to-posterior architecture of the embryo. Within each homeotic gene, a special sequence or **homeobox** of 180 base pairs codes for a protein that is 60 amino acids long. This is a DNA regulatory protein that can "turn on" other genes involved in segmentation and segment identity (whether wings, antennae, legs, and so forth, are attached). The same homeobox is found in all homeotic genes of segmented organisms—its base sequence has been **conserved** throughout evolution.

The gut will tubulate during this later stage of development, and the ventral unsegmented mesoderm will split to form a mesodermally lined coelom (Figure 19D-1h).

ⅠⅠⅠⅠⅠ Objectives ⅠⅠⅠⅠⅠⅠⅠⅠⅠⅠⅠⅠⅠⅠⅠⅠⅠⅠⅠⅠⅠⅠⅠⅠⅠⅠⅠⅠⅠⅠⅠⅠⅠⅠⅠⅠ

☐ Describe the process of neurulation.

☐ Describe the function of the chordamesoderm.

☐ Relate the development of the brain and the spinal column to the development of the neural tube and neural folds.

ⅠⅠⅠⅠⅠ Procedure ⅠⅠⅠⅠⅠⅠⅠⅠⅠⅠⅠⅠⅠⅠⅠⅠⅠⅠⅠⅠⅠⅠⅠⅠⅠⅠⅠⅠⅠⅠⅠⅠⅠⅠⅠⅠ

1. Examine a cross section of a frog late neurula. Draw this on your "Frog Development" sheet. Identify and label the neural tube, notochord, somite mesoderm, gut cavity, epidermis, and mesoderm lining the coelom.

2. Examine a preserved late-neurula frog embryo. Iddentify the neural groove and neural folds.

3. Examine later stages of frog embryos (Figure 19B-4).

   a. *What happens to the shape of the embryo after the sides of the neural tube close?*

   _____

EXERCISE F   Development of the Chick

Note that the same body-forming movements occur during gastrulation in the development of chickens as in frogs, except that the end product of the process is three flat layers of cells rather than three concentric rings in a sphere (Figure 19F-1). At the beginning of gastrulation, the disk of cells (blastodisc) on the top of the egg is composed of two layers—the upper layer containing ectoderm, neural ectoderm, notochord, and mesoderm tissue; the lower layer containing endoderm. During gastrulation, notochord and mesoderm cells are moved from the top layer to form a middle layer.

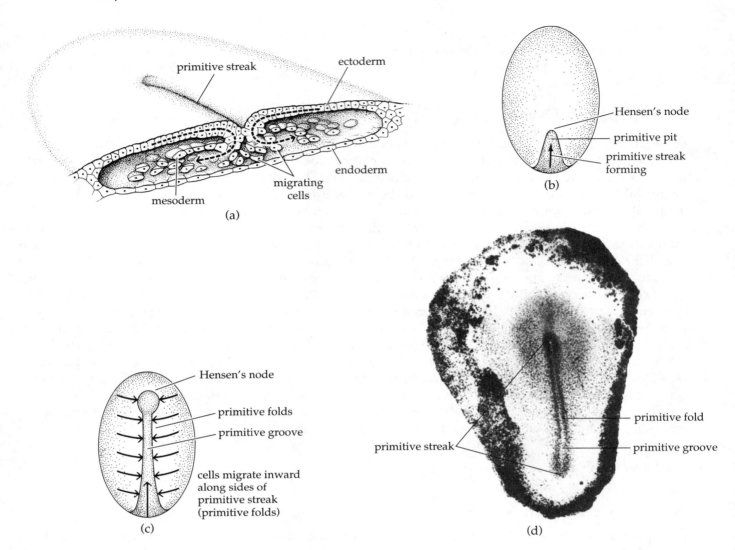

primitive streak

ectoderm

endoderm

migrating cells

mesoderm

(a)

Hensen's node

primitive pit

primitive streak forming

(b)

Hensen's node

primitive folds

primitive groove

cells migrate inward along sides of primitive streak (primitive folds)

(c)

primitive streak

primitive fold

primitive groove

(d)

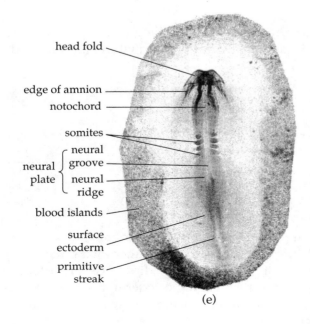

head fold

edge of amnion

notochord

somites

neural groove

neural plate

neural ridge

blood islands

surface ectoderm

primitive streak

(e)

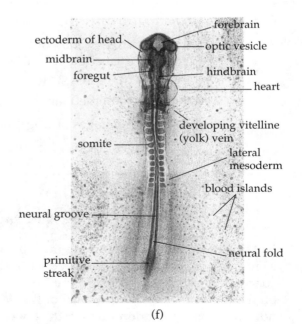

ectoderm of head

midbrain

foregut

forebrain

optic vesicle

hindbrain

heart

developing vitelline (yolk) vein

somite

lateral mesoderm

blood islands

neural groove

primitive streak

neural fold

(f)

Figure 19F-1 *(Opposite.) Development of the chick blastodisc. (a) Cells migrate into and away from the primitive streak to form the mesoderm and some endoderm. (b, c) Formation of the primitive streak. Cells converge toward the midline of the blastodisc. (d) The chick embryo at 16 hours. The rounded mound of cells (Hensen's node) at the anterior end of the primitive streak represents an accumulation of chordamesoderm cells (notochord and cephalic mesoderm) that will invaginate. As cells move to the interior, the streak will recede toward the posterior end of the embryo, and the notochord cells will be pulled out into a strip in the middle layer (between the mesoderm that has already moved inward from the surface layer). Neural ectoderm cells will be pulled back as a sheet of tissue on the outside overlying the notochord. (e) The chick embryo at 24 hours. (f) The chick embryo at 33 hours. During the process of neurulation, the underlying mesoderm and notochord induce the formation of a neural tube in the strip of neural ectoderm lying above them.*

As cells migrate to form the underlying tissue layer, they converge toward the posterior portion of the midline of the blastodisc (Figure 19F-1c). There they sink downward (similar to the way cells sink inward at the blastopore of the frog gastrula). This area, at first triangular in appearance, is called the **primitive streak.** As more cells move into the area, the primitive streak appears to lengthen, forming the **primitive groove** bordered by **primitive folds,** formed by accumulations of cells converging on the area in preparation for invagination. If you made a tube out of the flat layers of the chick after completion of gastrulation, the structure would look surprisingly like a frog gastrula.

ııııı **Objectives** ıııııııııııııııııııııııııııııııııııııı

☐ Compare gastrulation and neurulation in the chick and the frog.

ııııı **Procedure** ıııııııııııııııııııııııııııııııııııııı

1. Observe a prepared slide of a 16-hour-old chick and identify the primitive groove and primitive folds.

2. Observe later stages of chick development (Figure 19F-1d, e, f). Note that the neural folds and mesodermal somite blocks can easily be seen very early in the development of a chick (17 to 29 hours). Remember that the chick must tubulate in order for its "flat" layers of tissues to form a circular gut tube. The head is the first area to tubulate, and it appears to be lifted up off the surface of the blastodisc (Figure 19F-1f).

✔ **EXERCISE G** **Formation of Extraembryonic Membranes in the Chick**

As the chick continues to develop inside the egg, special vascularized membranes develop to serve the respiratory, protective, nutritional, and excretory needs of the embryo.

In the chicken embryo, four membranous sacs are formed by the growth and folding of the sheets of ectoderm, mesoderm, and endoderm at the edges of the blastodisc. These **extraembryonic** membranes are continuous with embryonic tissue, but lie outside (hence "extraembryonic") the embryo (Figure 19G-1a). The **amnion** is formed around the embryo and fills with fluid to serve a protective function. The vascularized **chorion** lies beneath the shell and serves a respiratory function by picking up oxygen that has diffused through the shell and transporting it to the embryo. The **allantois,** which invaginates from the hindgut, serves as a repository for wastes and eventually fuses with the chorion to aid in respiration. The **yolk sac** invaginates from the gut cavity and becomes highly vascularized by a network of vitelline vessels that pick up nutrients from the yolk and transport them to the embryo (Figure 19G-1b).

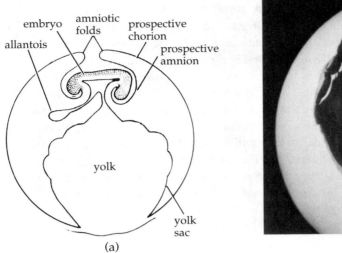

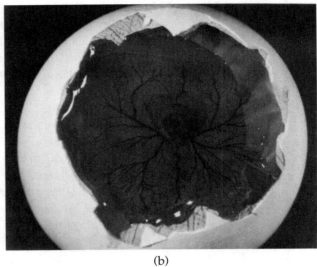

(a)

(b)

Figure 19G-1 *(a) Extraembryonic membranes of the chick. (b) The yolk sac vascularized by a network of vitelline vessels.*

⫶⫶⫶⫶⫶ Objectives ⫶⫶⫶⫶⫶⫶⫶⫶⫶⫶⫶⫶⫶⫶⫶⫶⫶⫶⫶⫶⫶⫶⫶⫶⫶⫶⫶⫶⫶⫶⫶⫶⫶

☐ List and give the functions of the four extraembryonic membranes.

⫶⫶⫶⫶⫶ Procedure ⫶⫶⫶⫶⫶⫶⫶⫶⫶⫶⫶⫶⫶⫶⫶⫶⫶⫶⫶⫶⫶⫶⫶⫶⫶⫶⫶⫶⫶⫶⫶⫶⫶⫶⫶

1. Work in pairs. Obtain an incubated 96-hour-old or 5- to 6-day-old chicken egg. Place the egg on a towel so that one side faces upward. The embryo will rotate to the top side after 3 to 5 minutes. Meanwhile, fill a small finger bowl with warm (37°C) chick Ringer's solution.

2. Gently crack the egg on the lower surface and allow the egg yolk and embryo to float out of the shell into the Ringer's solution. Do not jerk the two halves of the shell apart or separate them too quickly. It will take some time for the chorion to separate from the shell membranes of the older eggs.

3. Identify the yolk sac with its vessels. How far has the network of vitelline vessels extended. These vessels will eventually cover the entire surface of the yolk.

4. With a blunt instrument, push on the embryo. You will see that it is floating within a sac—the amnion.

5. Look for the allantois. It will appear as a small bubble emerging from the posterior end of the embryo.

6. With a pair of forceps, lift up the outermost tissue layer lying above the embryo. This is the chorion. How far does it appear to extend?

PART II   IMMUNE RESPONSES

Consider that a mammalian organism can make more than 10 million different types of antibodies. However, there is not enough DNA in the organism's entire genome to code for all of these molecules. How can we account for the variety of antibodies, or **immunoglobulins,** that protect our bodies?

The answer lies in the observation that the DNA of eukaryotic chromosomes can rearrange itself, making new coding combinations from old sequences. Each antibody molecule is made up of two **heavy** (long) **chains** and two **light** (short) **chains.** A portion of each chain, the **variable region** (containing approximately 100 amino acids), is responsible for antibody diversity and function (Figure 19II-1).

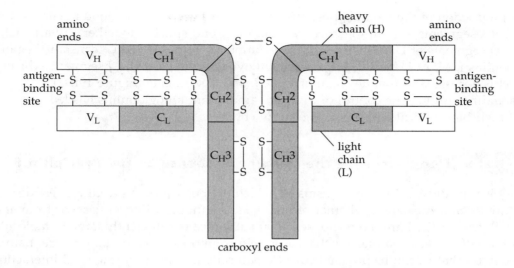

Figure 19II-1   *The immunoglobulin molecule. An antibody (immunoglobulin) molecule has two light (L) and two heavy (H) polypeptide chains, each of which has variable (V) and constant (C) regions. The heavy and light chains are connected to each other by disulfide bonds.*

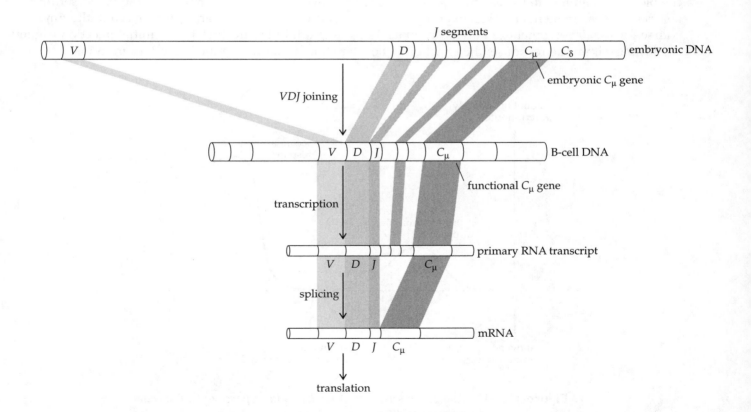

Figure 19II-2   *Rearrangement of the DNA segments of the genes for the heavy chains of immunoglobulins. The V, D, J, and C segments are joined to form the gene. After transcription, the primary RNA transcript is spliced to remove the transcripts of introns and extra J segments. The mature mRNA is then ready for translation.*

During production of the variable portions of heavy or light immunoglobulin chains, various combinations of DNA segments are put together by "selecting" and splicing together *V* (variable), *D* (diversity), and *J* (joining) segments of DNA. These are then combined with selected *C* (constant) regions for immunoglobulins I, M, A, D, and E. Introns are removed, as well as extra *J* segments, during splicing. The final product is a single and unique chain of *V*, *D*, *J*, and *C* segments (Figure 19II-2).

Understanding this genetic control of diversity among immunoglobulins has led us to a better understanding of their biochemical and physiological functions.

EXERCISE H   Demonstrating the Immune Response by the Precipitin Ring Test*

Immunology is the study of the mechanisms by which the body protects itself against disease-causing microbes. Any foreign substance (whether of microbial origin, including viruses, or from another species of animal) will trigger the immune response if that substance is sufficiently large (a molecular weight of at least 10,000), is sufficiently complex (such as a protein composed of at least 20 different amino acids), and is biodegradable—that is, can be broken down by naturally occurring biochemical interactions. Some large synthetic molecules, such as many plastics, are nonbiodegradable—tend not to interact with or be broken down by substances within the body—and therefore do not normally trigger an immune response. Foreign substances that cause an immune response are called **antigens.**

When antigens enter the body (for example, when you are injected with a flu vaccine), they set off a complex series of reactions which are demonstrable in a variety of ways. One of the easiest ways to demonstrate the immune response is to do a blood **serum titer analysis.** A titer is a measure of the amount of antibodies in the blood serum that can react with a specific antigen. (Serum is the liquid part of the blood remaining after cells and fibrin have been removed. Antibodies circulate in the blood serum.) If a series of measurements is taken over a period of several weeks after the antigen is injected, the time course for the development of antibodies looks like the graph in Figure 19H-1. The initial injection triggers the primary immune response; the effect of a "booster shot" is shown by the secondary immune

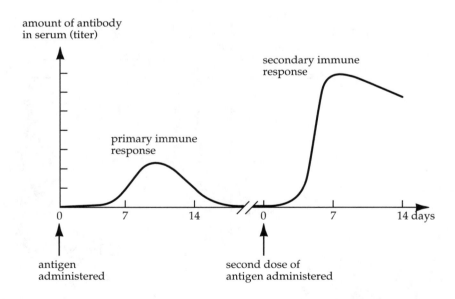

Figure 19H-1   *Changes in serum antibody levels in response to two injections of an antigen administered at different times.*

---

*This exercise was developed by Dr. Fred Stutzenberger and Jzuen-Rong Tzeng, Clemson University, Clemson, South Carolina.

response—it is greater and longer-lasting than the primary response. This secondary response is specific for the challenging antigen—serum will not react this way with other antigens that the body has not yet experienced (a phenomenon known as **immunomemory**).

a. *Why do some types of vaccines require a "booster shot"?* _____

_____

b. *Which cells are responsible for the secondary immune response?*

_____

c. *Explain how this response occurs.* _____

_____

Antibodies are **immunoglobulins** (globular-shaped proteins with an immune function) which can often be found in other body fluids, such as saliva, breast milk, and respiratory secretions, as well as in serum. Although immunoglobulins are found in several different forms, depending on their location in the body, all have similar structures (see Figure 19II-1). When serum antibodies combine with their specific antigen (Figure 19H-2), they form a complex **immunoprecipitate.** If this reaction is carried out in a test tube, the immunoprecipitate will form a ring at the interface where the antigen and antibody meet—hence the name "ring test."

The ring test is a useful qualitative method for the rapid detection of either antigen or antibody. It involves carefully overlaying a solution of antibody (antiserum) with a solution of antigen so that a sharp liquid interface is formed. Evidence of a positive reaction is the formation of a cloudy precipitate at the interface. Either antigen or antibody can be detected quickly in amounts as small as 1.0 μg of protein if care is taken in carrying out the test. However, the ring test is not a highly quantitative method.

Figure 19H-2 *Binding of antibodies to antigens. Because each antibody molecule has two antigen-binding sites, a single antibody can bind to antigens on two different cells or particles, causing them to stick together (agglutinate). Phagocytic white blood cells then consume these larger masses of foreign particles and antibodies.*

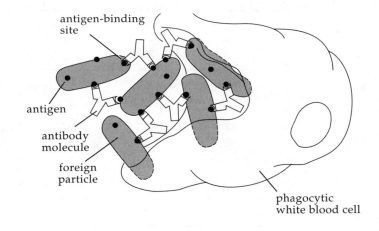

antigen-binding site

antigen

antibody molecule

foreign particle

phagocytic white blood cell

ııııı **Objectives** ıııııııııııııııııııııııııııııııııııı

☐ Describe the relationship between antigen and antibody.

☐ Describe the primary and secondary immune responses.

☐ Use the precipitin ring test to demonstrate the immune response.

ııııı **Procedure** ııııııııııııııııııııııııııııııııııııııı

1. Place five microculture tubes (numbered 1 through 5) in a rack. With a Pasteur pipette, introduce anti-BSA antiserum into tubes 1 and 5 to a height of about 10 mm. Be sure to note the approximate volume of the liquid in the Pasteur pipette. Using a separate pipette for each

solution, introduce a similar amount of buffer into tube 2 and of normal serum into tubes 3 and 4, as indicated in Table 19H-1.

2. Gently overlay each volume of solution with an equal volume of the buffer or antigen indicated in Table 19H-1. *Be careful not to mix these two layers of solution.* This can be avoided by slightly slanting the tube and carefully introducing the buffer or antigen solution along the wall of the tube near the surface of the bottom solution, then gradually pulling up the Pasteur pipette. Place the rack of tubes in front of the light source and observe after 10 to 15 minutes at room temperature.

Table 19H-1   Ring Test for Detection of an Antigen (BSA) or an Antibody (anti-BSA)

	Tube Number*				
	1	2	3	4	5
Top layer	B	Ag	B	Ag	Ag
Bottom layer	Ab	B	N	N	Ab
Reaction					

*B = buffer, Ag = antigen, Ab = antiserum, N = normal serum.

3. A positive reaction is indicated by formation of a milky white ring at the interface. Record reactions in Table 19H-1.

   d. *Explain why you had a positive reaction or no reaction in:*

   Tube 1 _____

   Tube 2 _____

   Tube 3 _____

   Tube 4 _____

   Tube 5 _____

   e. *How might the ring test be used to check whether an individual demonstrates immunity to a particular antigen, perhaps to a disease such as measles or chicken pox?*

   _____

   f. *How is this reaction similar to the type of reaction that would occur if a person with type B blood received a transfusion of type A blood by mistake?* _____

Laboratory Review Questions and Problems

1. Fill in the following table to summarize the major events of early development.

	Formation of the Blastula	Early Gastrula	Late Gastrula	Neurula
Processes occurring				
Type of structure formed				
Characteristics of structure				
Significance of stage				

2. Compare and contrast major developmental events in the sea urchin, frog, and chick by completing the following table.

	Sea Urchin	Frog	Chick
Type of egg			
Type/pattern of cleavage			
Distinguishing characteristics of blastula			
How gastrulation occurs			
Events of neurulation			
Distinguishing characteristics of later development			

3. What are cytoplasmic determinants? How do they affect early development, including cleavage of the zygote?

4. How can maternal DNA affect the development of an embryo when the maternal genes are not included in the egg produced as a result of meiosis?

5. What are homeotic genes? How do they control the development of segmented organisms?

6. How is the heavy chain of an immunoglobulin molecule of the M type constructed?

7. How can you explain the tremendous variability among immunoglobulins if DNA does not have enough genes to code for each molecule individually?

8. What are antigens and antibodies? What types of cells make antibodies?

9. Why would it be important to be able to determine whether a person had antibodies to a specific antigen? Can you think of a medical crisis in which this knowledge would be helpful?

10. Describe the function of each of the following extraembryonic membranes in the chicken embryo: amnion, chorion, allantois, yolk sac.

11. What is the function of the vitelline blood vessels in the chicken embryo?

12. In development of the human embryo, the trophoblast is like the chorion, surrounding the embryo in its allantois. The trophoblast becomes the placenta during development of the fetus. How does the function of the human placenta compare to the chorion of the chicken embryo?

13. During human development, a rudimentary allantois is formed but is never used. Why would it not be used? How are the functions that the allantois performs in the chicken embryo carried out during human fetal development?

Dueling Alleles

Dueling Alleles is a computer simulation that can be used to explore the effects of genetic drift and selection for many more generations than is possible in the course of Laboratory 20.

- **Dueling Alleles** allows you to investigate the effects of varying population sizes, ranging from 1 to 500 mating pairs, and can simulate hundreds of generations in a few minutes.

- By choosing a small population size, you will be able to observe how alleles can be driven to extinction by violent fluctuations in allelic frequency caused by genetic drift. Larger simulated populations result in a more gentle drift.

- By adjusting the degree of selection against different genotypes, you can observe both the rapid elimination of a deleterious dominant allele and how recessiveness or heterozygote advantage can maintain an allele for many generations, even when selection against it is severe.

Dueling Alleles allows you to create populations that increase, decrease— or become extinct!

The Genetic Basis of Evolution

OVERVIEW

To the population geneticist, evolution means changes in the frequencies of the **alleles,** alternative forms of the same gene, or genotypes, in the gene pool. The potential of a population to experience evolutionary change as it adapts to its environment depends upon the amount of variation within the population's gene pool. **Natural selection,** the differential reproduction of phenotypes, is one of the chief agents of evolutionary change. Evolution depends on natural selection operating in a population having a variety of alleles that produce a variety of genotypes.

STUDENT PREPARATION

Prepare for this laboratory by reading the text pages indicated by your instructor. Familiarizing yourself in advance with the information and procedures covered in this laboratory will give you a better understanding of the material and improve your efficiency.

✔ **EXERCISE A**   **Understanding Variation**

It is important to understand that variation in phenotype is due not only to genetic variation but also to environmental effects. Variation may be either **continuous,** characterized by small gradations between individuals, or **discrete,** showing clear-cut differences. The continuous variations in human anatomical traits you will measure in this exercise are subject to both genetic and environmental influences. It is not known to what extent natural selection influences these traits.

IIIII **Objectives** IIIIIIIIIIIIIIIIIIIIIIIIIIIIIIIIIIIIII

☐ Understand continuous variation through examination of some human traits.

☐ Make a quantitative analysis of continuous variation within a population.

✔ **PART I**   **Measuring Cephalic Index**

The **cephalic index** is a ratio that relates the breadth of a skull to its length. This measure is used by anthropologists to compare head shape in human populations and by paleontologists to study fossilized skulls.

IIIII **Procedure** IIIIIIIIIIIIIIIIIIIIIIIIIIIIIIIIIIIIII

1. Work in pairs. Use a set of calipers and a meter stick to measure the breadth and length of your partner's head in centimeters. Breadth is the maximum width of the head measured

above and behind the ears (Figure 20A-1a). Length is the distance from the bulge in the forehead (glabella), just above the nose, to the bony protrusion at the base of the skull (opisthocranium) (Figure 20A-1b). Round your measurements to the nearest whole number.

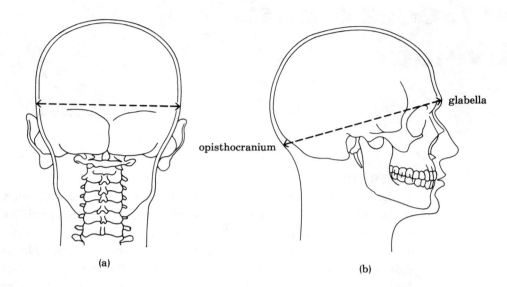

Figure 20A-1 *Measuring (a) head breadth and (b) head length.*

2. To obtain the cephalic index, divide the breadth of the skull by its length and multiply the quotient by 100. (Keep in mind that the cephalic index in humans is not well correlated with either brain size or intelligence.) Use the space below for your calculations.

3. Your partner should now calculate your cephalic index.

4. Write both cephalic indexes on the blackboard. Designate your sex by M or F.

5. On the bar graph in Figure 20A-2, mark cephalic index measurements by placing an **X** for each member of the class in the column directly above the appropriate cephalic index range on the horizontal axis. Use different colored pencils or pens to mark measurements for males and females.

   a. *What is the average cephalic index for the class?* _____ *For males?* _____
   For females? _____

   b. *For which sex is the range of measurements greater? (The range is the distance between the highest and the lowest values in a group of measurements.)* _____

(If there is disagreement, your instructor may use a statistical calculation to decide the question.)

Figure 20A-2 *Bar graph for class cephalic index data.*

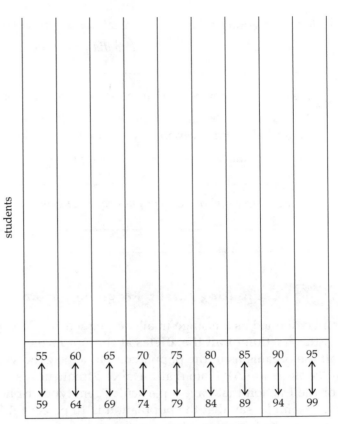

students

cephalic index ranges (cm)

| 55 ↑ 59 | 60 ↑ 64 | 65 ↑ 69 | 70 ↑ 74 | 75 ↑ 79 | 80 ↑ 84 | 85 ↑ 89 | 90 ↑ 94 | 95 ↑ 99 |

✔ **PART 2** **Measuring Relative Sitting Height**

Another highly variable trait in humans is the length of the torso relative to overall stature.

‖‖‖‖ **Procedure** ‖‖‖

1. Work in pairs. Have your partner sit as erect as possible on the lab table while you measure the distance from the top of the head to the top of the lab table in centimeters.

2. Now have your partner stand against the wall of the classroom. Mark a point just above the top of the head, then measure from the floor to your mark.

3. To obtain relative sitting height, divide your partner's sitting height by his or her overall height and multiply the quotient by 100.

4. Now your partner should repeat these measurements and calculations for your relative sitting height.

5. Write both relative sitting heights on the blackboard. Designate your sex by M or F.

6. On a separate sheet of paper, construct a bar graph like the one in Figure 20A-2. The intervals of five along the horizontal axis should begin with the multiple of five that includes the lowest measurement in the class and proceed, in fives, up to the multiple of five that includes the highest measurement. Use different colored pencils to mark measurements for males and females in the appropriate sitting-height category. Add this graph to your laboratory manual.

    a. *What is the average relative sitting height for the class?* _____ *For males?* _____

       For females? _____

    b. *For which sex is the range of measurements greater?* _____

c. Is there a greater range in relative sitting heights among the males or among the females in your class? _____ How did you arrive at your conclusion? _____

_____

(Again, your instructor may verify this conclusion with a statistical calculation.)

d. Recall that in Laboratory 13 you examined the genetic trait of corn seed color, a discrete (discontinuous) phenotypic variation. Compare the variation in corn seed color with the variations you observed in this exercise. Describe the differences. _____

_____

e. What do these differences tell you about the genes involved in the inheritance of different traits?

_____

_____

👁 EXERCISE B | Estimating Allelic Frequency from a Population Sample

Evolution can be defined as a change in allelic frequencies. **The allelic frequency of allele A in a population is the fraction of all the alleles at the A/a locus in the population's gene pool that are A.** For example, if all the organisms in the population have the genotype Aa, then the frequency of allele A is 0.5 (because *half the alleles* are A). Note that the allelic frequency is *not* the fraction of the organisms that have an A allele, or the fraction that have a particular genotype or phenotype.

As another example, say that the population is 10 percent AA, 60 percent Aa, and 30 percent aa. Then we would compute the frequency of A as follows:

> **All** the alleles of the AA organisms are A, so the contribution of the AA organisms to the frequency of A is (1.0)(0.1).

> **Half** of the alleles of the Aa organisms are A, so the contribution of the heterozygotes is (0.5)(0.6).

> **None** of the alleles of the aa organisms is A, so the contribution of the aa organisms is (0)(0.3).

Thus, the frequency of $A = (1.0)(0.1) + (0.5)(0.6) + (0)(0.3) = 0.4$.

Note that if the population has only A and a alleles at the gene (A/a) locus, then any allele that is not A is a. The frequency of the a allele is $1-$ the frequency of the A allele. Therefore, in the population above, in which the frequency of A was 0.4, the frequency of a must be $1 - 0.4 = 0.6$.

In situations in which there are only two alleles at a locus and one is dominant and one is recessive, we define two variables, p and q:

> $p =$ **the frequency of the dominant allele** (0.4 in the example above)

> $q =$ **The frequency of the recessive allele** (0.6 in the example above)

Then,

> $p + q = 1$

a. A population consists of 20 percent AA, 30 percent Aa, and 50 percent aa organisms. What is the frequency of the A allele in this population? _____

The **Hardy-Weinberg rule** predicts what will happen to allelic frequencies and the proportion of genotypes in the population under a very restrictive set of conditions:

> **1.** There is no mutation of the alleles.

> **2.** Mating is random: no genotype has a mating advantage, and every genotype is equally likely to mate with any other genotype.

3. The population is very large.

4. Immigration and emigration do not occur.

5. There is no selection: all genotypes have an equal chance of surviving and reproducing.

These conditions are almost never true in nature. However, if they were true, the population is said to be at **Hardy-Weinberg equilibrium.** In a population at Hardy-Weinberg equilibrium:

1. **Evolution does not occur, because allelic frequencies never change.**

2. **Genotypic frequencies can be predicted from allelic frequencies.**

We will be focusing on this second consequence of the Hardy-Weinberg equilibrium in this exercise.

A **genotypic frequency** is the fraction of the population that is a particular genotype. For example, if a population is 20 percent *AA*, 20 percent *Aa*, and 60 percent *aa*, the genotypic frequencies are 0.2 *AA*, 0.2 *Aa*, and 0.6 *aa*.

If we *cannot* assume Hardy-Weinberg equilibrium, there is no necessary relationship between allelic frequencies and genotypic frequencies. For example, the *allelic frequencies* of *A* and *a* are both 0.5 in all the populations listed in the table below.

	AA	*Aa*	*aa*
Population 1	0.00	1.00	0.00
Population 2	0.50	0.00	0.50
Population 3	0.20	0.60	0.20
Population 4	0.25	0.50	0.25
Population 5	0.40	0.20	0.40

However, if a population is at Hardy-Weinberg equilibrium, where the frequency of $A = p$ and the frequency of $a = q$, we will always see the following genotypic frequencies:

	AA	*Aa*	*aa*
Hardy-Weinberg equilibrium population	p^2	$2pq$	q^2

For example, the population described above, where $p = 0.5$ and $q = 0.5$, if the population is at Hardy-Weinberg equilibrium, we would expect to find the frequency $AA = (0.5)^2$ or 0.25; the frequency $Aa = 2 (0.5)(0.5) = 0.50$; and the frequency $aa = (0.5)^2 = 0.25$.

b. Among the five populations listed in the table above, are there any that might be at Hardy-Weinberg equilibrium?

_____ *Which one(s)?* _____

If there are only two possible alleles at a locus, the only possible genotypes are the two homozygotes and the heterozygote. Therefore, the frequencies of these genotypes must add up to 1.0. This gives us the **Hardy-Weinberg equation:**

$$p^2 + 2pq + q^2 = 1$$

If we know all of the genotypic frequencies in a population, we can determine if these *observed* genotypic frequencies agree with what would be *expected* if the population were at Hardy-Weinberg equilibrium. For example, if we are told that a population's genotyopic frequencies are 0.1 *AA*, 0.2 *Aa*, and 0.7 *aa*, we can determine whether or not the population is in Hardy-Weinberg equilibrium by a three-step procedure.

Step 1 Determine allelic frequencies from genotypic frequencies.

p (frequency of A) = (1.0)(0.1) + (0.5)(0.2) + (0.0)(0.7) = 0.2

q (frequency of a) = 1 − p = 1 − 0.2 = 0.8

Step 2 Determine *expected* genotypic frequencies if the population is at Hardy-Weinberg equilibrium.

Expected frequency of $AA = p^2 = 0.04$

Expected frequency of $Aa = 2pq = 0.32$

Expected frequency of $aa = q^2 = 0.64$

Step 3 Compare observed and expected (if in equilibrium) genotypic frequencies.

	AA	*Aa*	*aa*
Observed	0.10	0.20	0.70
Expected	0.04	0.32	0.64

There is a fairly serious disagreement between observed and expected here, so we would probably conclude that this population is *not* at Hardy-Weinberg equilibrium. If we wanted a more exact determination of agreement, we could do a statistical test with chi-square (Appendix I).

c. *A population has 10 percent* AA, *45 percent* Aa, *and 45 percent* aa *organisms. Does this population seem to be at Hardy-Weinberg equilibrium?* _____

One final application of the Hardy-Weinberg rule is using it to estimate allelic frequencies where the genotypic frequencies are partially unknown *but the Hardy-Weinberg equilibrium can be assumed.* For example, say *EE* and *Ee* individuals have normal earlobes, but *ee* individuals have a distinctive shape to their earlobes. We know that 16 percent of the North American population shows this latter phenotype (*ee*), and we are willing to assume Hardy-Weinberg equilibrium. We can use two steps to estimate the frequencies of the alleles and the genotypes in the population:

Step 1 Determine allelic frequencies.

q^2 (frequency of *ee*) = 0.16

$q = \sqrt{0.16} = 0.40$

$p = 1 - q = 0.60$

Step 2 Determine expected genotypic frequencies if the population is at Hardy-Weinberg equilibrium.

Expected frequency of $EE = p^2 = 0.36$

Expected frequency of $Ee = 2pq = 0.48$

Expected frequency of $ee = q^2 = 0.16$

Note that if we are unwilling to assume Hardy-Weinberg equilibrium, we cannot do any of this. Without Hardy-Weinberg equilibrium, the Hardy-Weinberg equation does not apply, and although we still know that 16 percent of the population is *ee*, the normal-earlobe phenotypes might be all *EE*, all *Ee*, or any mixture of *EE* and *Ee*. Therefore, it is important to realize when you can and cannot do certain computations.

- You can always calculate allelic frequencies from a complete list of genotypic frequencies, *whether or not* you assume Hardy-Weinberg equilibrium.

- You can compute genotypic frequencies from allelic frequencies *only* if you can assume Hardy-Weinberg equilibrium.

- You can estimate allelic and genotypic frequencies from the fraction of homozygous recessives in a population *only* if you can assume Hardy-Weinberg equilibrium.

In this exercise you will use the class as a sample population and will assume Hardy-Weinberg equilibrium. You will estimate the frequencies of alleles of a gene controlling the ability to taste the chemical phenylthiocarbamide (PTC). A bitter-taste reaction to PTC is evidence of the presence of a dominant allele, in either the homozygous condition (frequency = p^2) or the heterozygous condition (frequency = $2pq$). The inability to taste the chemical at all depends on the presence of a homozygous recessive genotype (frequency = q^2). To find the frequency of the PTC-tasting allele in the population, you must find p; to find the frequency of the non-PTC-tasting allele, you must find q.

ⅠⅠⅠⅠⅠ Objectives ⅠⅠⅠⅠⅠⅠⅠⅠⅠⅠⅠⅠⅠⅠⅠⅠⅠⅠⅠⅠⅠⅠⅠⅠⅠⅠⅠⅠⅠⅠⅠⅠⅠⅠⅠ

☐ Estimate allelic frequencies for a specific trait within a sample population and compare these with estimated frequencies for the entire population.

ⅠⅠⅠⅠⅠ Procedure ⅠⅠⅠⅠⅠⅠⅠⅠⅠⅠⅠⅠⅠⅠⅠⅠⅠⅠⅠⅠⅠⅠⅠⅠⅠⅠⅠⅠⅠⅠⅠⅠⅠⅠⅠ

1. Your instructor will provide you with PTC taste test papers. Tear off a short strip and press it to the tip of your tongue. A few PTC tasters report a salty, sweet, or sour taste rather than a bitter taste. These individuals are considered tasters for the purpose of this experiment.

2. Calculate a decimal number representing the proportion, or frequency, of tasters ($p^2 + 2pq$) by dividing the number of tasters in the class by the total number of students in the class. Obtain a decimal number representing the frequency of nontasters (q^2) by dividing the number of nontasters by the total number of students. In Table 20B-1, record these numbers under the appropriate headings.

   a. *Why is the term 2pq included in the taster phenotype?* _____

   _____

Table 20B-1 Phenotypic Proportions of Tasters and Nontasters of PTC and Frequencies of the Determining Alleles

	Phenotype		Allelic Frequency	
Sample	Tasters ($p^2 + 2pq$)	Nontasters (q^2)	p	q
Class population				
North American population	0.55			

3. Assuming that your class is in Hardy-Weinberg equilibrium, use the Hardy-Weinberg equation to determine the frequencies (p and q) of the two alleles. The frequency q can be calculated by

taking the square root of q^2. Since $p + q = 1$, you can calculate the frequency p of the taster allele. Record your data in Table 20B-1.

4. Use the data on the North American population given in the table to calculate the expected frequencies (p and q) of the two alleles. Record these in Table 20B-1.

b. *How do the sample frequencies for the class compare with the frequencies for the population at large?*

_____

_____

Note: You may want to determine whether the difference between the allelic frequencies (p and q) for your class and the known frequencies for the North American population is significant. Use the chi-square test (Appendix I).

c. *How do you explain any discrepancy in these frequencies?* _____

_____

✔ **EXERCISE C** **The Founder Effect as an Example of Genetic Drift**

As indicated in Exercise B, the Hardy-Weinberg principle states that five conditions must be met for allelic frequencies to remain constant. Briefly stated, these are the conditions required: (1) no mutation, (2) random mating, (3) large population, (4) no migration, (5) no selection. Under these conditions, evolution will not occur. *Why not?* _____

_____

These conditions are so restrictive that they are almost never met in nature, even for short periods. The consequence of this is that allelic frequencies are constantly changing. In Exercises C and D you will explore two cases in which allelic frequencies change. In this exercise, the change in allelic frequencies, genetic drift, is caused by small population size (the **founder effect**). In Exercise D you will discover how **migration** can also cause fluctuations in allelic frequencies.

Genetic drift is the change in allelic frequencies that results from the random outcome of matings. An analogy would be a coin toss. If 100 people each tossed coins at the same time, we would be very surprised if 50 percent of the coins turned up heads and 50 percent tails on toss after toss. The same chance deviations from predicted frequencies occur for combinations of alleles. For example, even if the frequency of both *B* and *b* is 50 percent in a population, and *BB*, *Bb*, and *bb* zygotes are all equally viable, in every "round" of mating, the percentage of zygotes with *BB*, *Bb*, and *bb* genotypes will undergo chance fluctuations from their calculated frequencies of 25 percent, 50 percent, and 25 percent.

Drift is exaggerated in small populations such as founder populations. A **founder population** is a small segment of a population that splinters off from a main population. Often a founder population, because of its small size, has a gene pool that differs from that of the parent stock; the incidence of a rare inherited condition may be greater in the founder group than in the parent group.

If we continue the coin-toss analogy, but with only two people (a founder population) tossing coins, the frequency of heads will fluctuate from 0 percent to 50 percent to 100 percent. Likewise, small populations experience larger chance fluctuations in allelic frequencies.

One last coin-toss analogy should help to confirm the importance of genetic drift. Say that we have a rule stating that *if all coins in the toss turn up heads, then all future coins will be made into heads on both sides, and tails will never be seen in the "population" again.* If there are 100 coins in the toss, there is little chance that all 100 coins will land heads up at the same time. But if there are only two, then within a very few tosses both coins will come up heads, and tails will vanish forever.

This is similar to the dynamics of genes in a population. If the population is *small*, there is a good chance that random fluctuations will, over many generations, result in the *loss* of an allele when its frequency randomly fluctuates to zero. We say that the allele has become "extinct" and that the alternative allele has been "fixed" in the population. Unless migration or mutation brings the extinct allele back into the population, it has disappeared forever. Thus, even if all other conditions for the Hardy-Weinberg equilibrium hold, when the population is small, genetic drift can cause rapid and permanent changes in gene frequencies.

One practical consequence of drift is the damaging effects of such "genetic bottlenecks" in, for example, the small population of an endangered species. Drift may cause a loss of many alleles and a drastic reduction in genetic diversity. Even if the population recovers in numbers of individuals, its reduced ability to adapt to changing conditions may assure its eventual extinction.

ⅡⅠⅠⅠⅠ Objectives ⅠⅠⅠ

☐ Observe the founder effect using a simulation in which a founder population becomes isolated from its parent population.

☐ Define genetic drift and explain why it occurs when small populations become isolated from parent populations of organisms.

☐ Explain why drift produces more severe changes in allelic frequencies in smaller populations.

ⅡⅠⅠⅠⅠ Procedure ⅠⅠ

1. Divide the class into groups of four students. Each group should obtain a jar containing beads of two colors. The two colors represent the two alleles (*A* and *a*) of a particular gene. Your instructor will give you the exact count for each color and tell you which color bead produces the recessive trait that you will follow in the founder population. [Beads representing the recessive (and in this case, rare) allele will be fewer in number than those representing the dominant allele.] Compute the frequency of each color (*p* and *q*) by dividing the number of beads of each color by the total number of beads. Record the allelic frequencies in the spaces below "Parent Generation" in Table 20C-1.

2. Each pair of individuals in your group should pick beads randomly from the jar (without looking), placing a total of 10 beads in each of 10 cups to represent founder populations of 5 individuals (10 beads; 2 alleles per individual). Count the number of rare alleles in each cup and record your data in Table 20C-1. Record class data in the same table, indicating for each group the number of cups that contain 0, 1, 2, 3, or more rare alleles.

3. Compute the percentage of populations (cups) containing 0, 1, 2, 3, or more rare alleles by dividing the total number of cups in each column by the total number of cups in the class. Record these percentages in Table 20C-1.

4. Examine the class data.

   a. *Are there any founder populations (cups) in which the rare alleles are absent?* _____

   b. *In what percentage of all populations is the rare allele absent?* _____

   c. *Are there any founder populations (cups) in which the rare allele appears in higher proportions than in the parent population?* _____

   d. *In what percentage of all populations is the allelic frequency the same as in the parent population?*

   _____

   e. *From your data, what do you observe about the allelic frequencies of small founder populations compared with those of parent populations?* _____

   _____

5. From the class data, determine the range in allelic frequencies for the rare allele. (What are the highest and lowest allelic frequencies when all the cups in the class are considered?) Record your data in Table 20C-2 for founder populations of 5 individuals.

Table 20C-1   Data for Comparison of Allelic Frequencies in Parent Population and Founder Populations

Parent Generation $p =$ _____; $q =$ _____	Number of Founder Populations (Cups) Containing the Following Numbers (0–6) of Rare Alleles							
	0	1	2	3	4	5	6	
Group 1								
Group 2								
Group 3								
Group 4								
Group 5								
Group 6								
Group 7								
Group 8								
Group 9								
Group 10								
Group 11								
Group 12								
Group 13								
Group 14								
Group 15								
Group 16								
Total no. of cups with indicated no. of rare alleles								Total cups _____
Percent of all populations with indicated no. of rare alleles								

Table 20C-2   Range of Allelic Frequencies in Founder Populations

Population Size (number of individuals)	Lowest q	Highest q
5		
10		
25		
50		

6. Transfer your data for founder populations of 5 individuals that contain zero rare alleles (the column under "0" in Table 20C-1) to the first column in Table 20C-3.

Table 20C-3   Effect of the Size of Founder Populations (5, 10, or 25) on the Frequency of Rare Alleles

	Number of Founder Populations with 0 Rare Alleles for Populations of 5, 10, or 25 Individuals		
	5	10	25
Group 1			
Group 2			
Group 3			
Group 4			
Group 5			
Group 6			
Group 7			
Group 8			
Group 9			
Group 10			
Group 11			
Group 12			
Group 13			
Group 14			
Group 15			
Group 16			
Total no. of cups or groups of cups with zero rare alleles			
Percent of all populations with zero rare alleles			

7. Now you will increase the size of the founder population to 10. Randomly (so you may want students from another group to help) choose pairs of cups from your set of 10 cups. Each pair of cups represents a founder population of 10 individuals (20 beads; 2 alleles per individual).

   f. Do you have any populations with rare alleles? _____

8. Collect class data for the number of populations with zero rare alleles. Record the data in Table 20C-3 in the column under "10."

   g. What percentage of populations (of 10 individuals) in the class have zero rare alleles? _____ *Record this in Table 20C-3. Is this more or less than observed when founder populations were smaller (only 5 individuals)?* _____

9. From the class data, determine the range in allelic frequency for the rare allele in populations of 10 individuals. Record your data in Table 20C-2.

10. Randomly group your 10 cups into two founder populations of 5 cups each to represent founder populations of 25 individuals (50 beads; 2 alleles per individual). Collect class

data for the number of populations with zero rare alleles. Record your data in Table 20C-3 under "25."

    h. Now, what percentage of populations (of 25 individuals) in the class have zero rare alleles?

    _____ *Record this in Table 20C-3. Is this more or less than observed when populations were*

    smaller? _____

11. From the class data, determine the range in allelic frequency for the rare allele in populations of 25 individuals. Record your data in Table 20C-2.

12. If you consider all 10 cups as a single population of 50 individuals, what are the allelic frequencies for *A* and *a* in the population?

    $A(p) =$ _____

    $a(q) =$ _____

13. From the class data, determine the range in allelic frequency for the rare allele in populations of 50 individuals. Record your data in Table 20C-2.

    i. From your data and observations, what do you conclude about the relationship between the size of a founder population and the likelihood that alleles may be "lost" (become extinct) due to drift when

    founder populations become isolated from parent populations? _____

    _____

    j. From your data and observations, what do you conclude about the size of a founder population and the likelihood that it will resemble the parent population (will have similar allelic frequencies)?

    _____

    _____

    k. What could be the consequences of drift for a small population? _____

    _____

◉ EXERCISE D   The Role of Gene Flow in Similarity Between Two Populations

Gene flow is the term used by population geneticists to describe the movement of alleles from the gene pool of one population to the gene pool of another. Gene flow occurs as **emigration** (individuals leaving the population) or **immigration** (individuals entering the population). When gene flow takes place randomly between two populations, that is, without respect to the phenotype of the immigrants and emigrants, then the frequencies of alleles in the two populations approach equality, and because of this the two populations become more genetically similar.

An analogy may help explain the role of gene flow in making two populations more similar. Imagine several small ponds, some of which are sheltered from the sun and are cold, and some of which are in the open and are warm. If all the ponds are interconnected by underground pipes and there is a constant exchange of water between them, the temperatures in each pond will become very similar. However, if the water exchange stops, each pond's temperature may vary greatly from that of the other ponds. Likewise, when populations have even a low level of gene flow due to interbreeding, their allelic frequencies will tend to be similar and fluctuations in allelic frequencies will be moderated. (Even in founder populations, separated but not totally isolated from parent populations, fluctuations in allelic and genotypic frequencies may be moderated if some interbreeding with the parent population is still going on.)

In all cases, however, if interbreeding stops, each isolated population will feel the full effects of drift; allelic frequencies may fluctuate drastically, and alleles may be lost. Thus, it is small *isolated* populations that usually give rise to *new species*.

IIIII **Objectives** IIIIIIIIIIIIIIIIIIIIIIIIIIIIIIIIIIIIIIIIIII

☐ Demonstrate, using a simulation, the effect of gene flow on the genetic similarity of two populations that exchange individuals.

IIIII **Procedure** IIIIIIIIIIIIIIIIIIIIIIIIIIIIIIIIIIIIIIIIIIIIIII

On the lab table are two trays, labeled 1 and 2, containing beads that represent pairs of alleles—*A*, the dominant allele, and *a*, the recessive allele—in the gene pools of two neighboring populations. Note that the proportions of the two colors of beads in the two trays are different, indicating that the two populations are genetically different.

1. Work in pairs. Count the number of beads of each color in tray 1 and calculate the frequencies (p_1 and q_1) of the two alleles represented by the two different colored beads. Enter these frequencies as p_1 and q_1 in Table 20D-1. Repeat this procedure for tray 2 and record the initial frequencies as p_2 and q_2.

2. Shake the two trays to distribute the beads randomly. Without looking at the beads, draw (one at a time) 10 beads from tray 1. At the same time your partner will draw 10 beads (also one at a time) from tray 2. Transfer your beads to tray 2 *at the same time* that your partner transfers beads to tray 1.

  a. *Explain how these transfers of beads are related to immigration and emigration of individuals*

     between two populations. _____

  _____

3. Count the beads in each tray, calculate the frequencies of the two alleles for the two trays, and record these frequencies for the first round of gene flow in Table 20D-1.

Table 20D-1 Changes in Allele Frequencies of Two Populations Experiencing Gene Flow

Round of Gene Flow	Tray 1		Tray 2		Differences Between Trays	
	p_1	q_1	p_2	q_2	$p_1 - p_2$	$q_1 - q_2$
One						
Two						
Three						
Four						
Five						

4. Repeat steps 2 and 3 four more times, each time recording the allele frequencies for both trays in Table 20D-1 for the appropriate round.

5. Complete the rest of the data table by subtracting corresponding allele frequencies ($p_1 - p_2$ and $q_1 - q_2$) for the two trays.

  b. *How does the difference in the final frequency of allele* A *(that is,* $p_1 - p_2$*) compare with the*

     difference in the original frequency (first line of table)? _____

  _____

    c. On which round of gene flow did the allele frequencies in the trays show the greatest change?

    _____ *Is this what you expected?* _____ *Why?* _____

    _____

    d. On which round was the genetic makeup of the two gene pools most nearly the same? _____

    How would you expect the two gene pools to compare at the end of 10 rounds of gene flow?

    _____

    At the end of 20 rounds? _____

    _____

    e. What does this simulation indicate about the effect of gene flow on the genetic makeup of

    neighboring populations? _____

    _____

    f. In each of the following cases, explain how the results of the simulation would have been changed.
    Only 4 beads, instead of 10, are transferred between trays on each round.

    _____

    Beads are transferred from tray 1 to tray 2, but are not transferred from tray 2 to tray 1.

    _____

👁 EXERCISE E | The Effect of Selection on the Loss of an Allele from a Population

Evolution is a change in frequencies of alleles and genotypes in populations over time. This change can be brought about by mutation, drift (see Exercise C), immigration or emigration (see Exercise D), and **selection.**

    Selection refers to the differential rate of reproduction of different genotypes in a population. Each individual in a population, however, is the sum of *all* of its genotypes for *all* characteristics. Their cumulative expression gives rise to an individual's **phenotype:** the expression of many different genes, often working in a coordinated fashion. Those individuals whose phenotype is best adapted to their environment are more "fit"—they have a greater chance to survive and reproduce, contributing more offspring, and thus more alleles of certain types, to the next generation.

    A single allele rarely determines **fitness** or a "winning" phenotype. If an African antelope has superior genes for every characteristic except watchfulness for lions, its phenotype will have reduced fitness if lions are very active in its habitat. All of its "superior" genes will not be passed on to the next generation because they are bound up in one package with its nearly lethal "predator watchfulness" characteristic. Thus, it is differential rates of reproduction of different phenotypes, resulting from interactions of organisms with their environments, which gives rise to changes in the relative frequency of alleles and genotypes in a population—an overall process that biologists call **natural selection.**

    In a population, genotypes that reduce the fitness of the phenotype are said to be "selected against." Given the reduced reproductive potential of the individuals carrying these genotypes, deleterious alleles responsible for the genotypes should slowly decrease in frequency and eventually disappear from the gene pool of the population. This might raise the question of why any deleterious alleles are left after millions of years of natural selection. Several factors tend to slow or stop the allele extinction process:

    1. Elimination of deleterious alleles is slow, perhaps requiring hundreds or even thousands of generations, *except* when the selective disadvantage of the genotype is severe.

    2. Mutation or immigration might introduce new copies of deleterious alleles.

    3. The "disadvantaged" genotype may be selected against only when it is abundant, and may even enjoy a selective advantage when it is rare (frequency-dependent selection).

4. Recessive deleterious alleles may not experience any selective disadvantage when they are masked by the dominant allele.

5. Heterozygotes may experience a selective advantage over either homozygote, and thus both alleles will persist.

In this exercise, you will explore recessiveness and heterozygote advantage as factors that tend to maintain deleterious alleles in a population.

PART I   The Effects of Recessiveness on Deleterious Alleles

Deleterious alleles are those that have a detrimental effect on any organism that contains them as part of its genome. Most deleterious alleles exist as the recessive form of a gene at a particular locus. Let us say that the *B* gene exists as two alleles, *B* and *b*, and *b* is deleterious. The homozygous recessive genotype is most selected against because two copies of the maladaptive gene are present. In many cases, the homozygous recessive condition (*bb*) is lethal. Although the homozygous dominant genotype is unaffected, the heterozygote may also be affected, depending on how much the dominant allele is able to mask the effects of the deleterious recessive allele.

If the *Bb* heterozygote suffers no selective disadvantage (*B* masks *b*), then the *b* allele may persist for long periods in the population by "hiding out" in the *Bb* heterozygote. This heterozygote "refuge" will also become more and more effective as the *b* allele becomes increasingly rare. This occurs because the *b* allele will experience selective disadvantage only when a *bb* offspring occurs, and this will happen only when two heterozygotes mate. If the heterozygotes are affected (*B* does not mask *b*) and the heterozygotes become rare, then almost all matings of heterozygotes will be with *BB* organisms and the *b* allele will persist.

This part of the exercise will illustrate this principle. A population is isolated in an area where a particular allele (*b*) has become deleterious. First, you will determine the persistence of the deleterious *b* allele when *B* does not mask *b*; the *B* allele is only incompletely dominant over *b*, and *bb* and *Bb* experience selective disadvantage. Then you will do the same exercise for when *B* masks *b*; the *B* allele is completely dominant and only *bb* is at a disadvantage.

IIIII Objectives IIIIIIIIIIIIIIIIIIIIIIIIIIIIIIIIIIIIIIIIIIII

☐ Determine the relationship between persistence of an allele in a population and its "fitness."

☐ Determine the effects of selection on the persistence of deleterious alleles in a population.

A   Allele *B* Does Not Mask Allele *b*

In this simulation, all *bb* individuals die before reproducing, and there is a 50 percent chance that *Bb* individuals will die.

IIIII Procedure IIIIIIIIIIIIIIIIIIIIIIIIIIIIIIIIIIIIIIIIIIII

1. Divide the class into groups of four students. For each group, students will work in pairs. Each pair of students should obtain a cup. Place into the cup 16 beads of one color representing the common (dominant) allele and 4 beads of a different color representing the rare (recessive) allele, for a total of 20 beads. Your instructor will tell you which colors represent the two kinds of alleles (*B* and *b*) of a particular gene. (If selected by pairs, each pair of alleles would represent an individual of a particular genotype.)

2. The sum of *all* beads (40) in the two cups of both pairs of students represents the total gene pool of zygotes making up the parent population. This parent population is composed of half males and half females; one pair of students represents the 10 females in the population while the other pair of students represents the 10 males.

   a. *If there are 20 beads representing each sex, why are there only 10 individuals of each sex and not 20?* _____

You will consider this to be a *pre-reproductive* population since some zygotes representing particular genotypes (all *bb* and some *Bb*) may never mature to reproductive age (they die early). In the space below, compute the frequency of each allele (*p* and *q*) in the total pre-reproductive parent population by dividing the total number of beads of each color in both cups of the pre-reproductive parent population by the total number of beads in the population:

$$\text{Allele frequency} = \frac{\text{total number of one color in both cups}}{40}$$

Record the allelic frequencies of the pre-reproductive parent generation in Table 20E-1.

Table 20E-1   Effects of Recessiveness on the Persistence of a Deleterious Allele When *B* Does Not Mask *b*

| Generation | Frequency of Alleles | | Frequency of Genotypes | | |
	p (B)	q (b)	p^2 (BB)	$2pq$ (Bb)	q^2 (bb)
Parent					
F_1					
F_2					

3. Each pair of students should now pick pairs of beads from the cup randomly (close your eyes) and line up the 10 bead pairs on the laboratory table. One pair of students has 10 females and the other pair of students has 10 males. These represent the genotypes of the parent individuals. Compute the genotypic frequencies:

$$\text{Genotypic frequencies} = \frac{\text{number of beads of a particular genotype}}{20}$$

Record the observed genotypic frequencies for all 20 bead pairs representing individuals of the pre-reproductive parent population in Table 20E-1.

4. The parent individuals will now pair up and reproduce to form an F_1 population of offspring. First, however, you must eliminate all potential parents that do not reach reproductive maturity. Since all *bb* individuals die before reaching reproductive age, eliminate these individuals by placing the beads in a discard container. Since there is only a 50% chance that heterozygotes will live to reproductive maturity, for each heterozygote, flip a coin to determine if it will live to reproduce (heads = lives; tails = dies).

Rules for Fate of Individuals When *B* Does Not Mask *b*

Individual	Fate
BB	All live
Bb	Flip a coin to determine if individuals will live (heads = lives; tails = dies)
bb	All die

5. Rearrange the remaining "reproductive" parents in a row, left to right. A group of four students should now have two rows of bead pairs—a row of females and a row of males. If there are more individuals of one sex than the other, eliminate the excess bead

pairs at the right end of your row to even up the number of males and females (in this population, males and females pair-bond for life).

6. Now, pair each male (2 beads) with a female (2 beads). Determine the genotypes of the individuals in each mating pair. If the individual is *BB*, you know that he or she will donate a *B* allele to form an offspring. If an individual is *Bb*, flip a coin to determine which allele is contributed to the new offspring, *B* or *b* (heads = *B*; tails = *b*). Each mating pair should make *four* offspring in this manner, males and females donating one allele at a time in a gamete (it takes two gametes to form a zygote, thus each new individual has two alleles represented by a pair of beads). Obtain beads from the reserve containers on the table to make offspring. Line up the F_1 generation individuals on the table as they are created.

7. After all parent individuals have reproduced to form F_1 individuals, be sure to remove all parent individuals from the table, placing the beads in the discard container.

8. Count the number of beads of each color and the total number of all beads in the F_1 generation to calculate the allelic frequencies (*p* and *q*) for the F_1 generation. Record these values in Table 20E-1.

   b. How do these frequencies compare with the allelic frequencies of the parent generation?

   _____

9. Record the observed genotypic frequencies of the pre-reproductive F_1 generation.

   c. What trend do you see as you compare the genotypic frequencies of the parent and F_1 generations?

   _____

   _____

10. Repeat steps 5–8 using the F_1 generation to produce an F_2 generation. Record all data in Table 20E-1.

   d. What trend do you observe in allelic frequencies for the B *does not mask* b *experiment?*

   _____

   e. Why do you think this trend occurs? _____

   f. Are you data representative of the class data? _____

11. Put all beads that remain on the table into the discard container.

B Allele *B* Masks Allele *b*

In this simulation, *B* masks *b*, so all *BB* and *Bb* individuals live and are reproductively successful. Only *bb* individuals die.

וווו Procedure וו

1. Return to step 1 in Part 1A, and start with a new population.

2. Repeat steps 1–10, but use the following rules for *B* masks *b* when determining which individuals will or will not live to reproductive maturity.

Rules for Fate of Individuals when *B* Masks *b*

Individual	Fate
BB	All live
Bb	All live
bb	All die

3. Record all data for parental, F_1, and F_2 allelic frequencies and genotypic frequencies in Table 20E-2.

Table 20E-2   Effects of Recessiveness on the Persistence of a Deleterious Allele when B Masks b

Generation	Frequency of Alleles		Frequency of Genotypes		
	p (B)	q (b)	p^2 (BB)	$2pq$ (Bb)	q^2 (bb)
Parent					
F_1					
F_2					

g. What trend do you observe in allelic frequencies for the B masks b experiment?

_____

h. Why do you think this trend occurs? _____

_____

i. Why do recessive deleterious alleles that are masked by a dominant allele persist longer in a population than alleles that cannot be masked by other alleles? _____

_____

4. Compare your observations for B masks b with those for B does not mask b.

j. Did you see the effect described in question h in your data (b persisting longer when masked by B)?

_____

k. If recessiveness will help a deleterious allele persist longer in a population, would recessiveness speed or hinder the spread of a beneficial allele? _____

Why? _____

_____

5. Put all beads that remain on the table into the discard container.

PART 2   Heterozygote Advantage

Another way in which deleterious alleles are preserved is if heterozygotes have a selective advantage over either of the homozygotes (assuming that only two alleles are possible at a locus). In the extreme case in which both homozygotes (say bb and BB) die before reproducing and only the heterozygote survives, both the B and b alleles will persist in the heterozygotes and both bb and BB individuals will continue to be produced from Bb × Bb matings, driving the allelic frequencies toward $p = 0.5$ and $q = 0.5$. *Why?*

_____

A classic example of heterozygote advantage is the persistence of sickle-cell anemia in Africa. Sickle-cell anemia is a severe blood disease present in ss individuals. Ss individuals have some mild sickle-cell symptoms, but they are also resistant to malaria. Normal (SS), nonanemic individuals may suffer disability or death if they become infected with malaria. In areas of eastern Africa where malaria is prevalent, heterozygotes have a selective advantage of 26 percent over normal individuals, which seems to preserve the sickle-cell allele (s) in the gene pool. In some places, 45 percent of the population are heterozygotes. However, in areas where malaria is not common, Ss individuals have no advantage and the sickle-cell allele is eliminated. It is estimated that 300 to 350 years ago, 22 percent of the slaves in the

American South carried the sickle-cell allele; but by the early 1950s, only 9 percent of the black American population carried the allele.

In this simulation, you will demonstrate the preservation of a deleterious allele by heterozygote advantage. You will also observe the disappearance of a deleterious allele when the heterozygous condition is no longer advantageous.

||||| Objectives |||||||||||||||||||||||||||||||||||||

☐ Explain heterozygote advantage and how it may maintain alleles in a population.

||||| Procedure |||||||||||||||||||||||||||||||||||||

1. Follow the same procedures used in Exercise E, Part 1. Remember that all couples must produce *four* viable offspring. However, new rules for heterozygote advantage apply for reproduction.

Rules for Fate of Individuals in Heterozygote Advantage

Individual	Fate
BB	Flip a coin to determine if individuals will live (heads = lives; tails = dies)
Bb	All live
bb	All die

2. Repeat steps 1–10 of Part 1A. Record all data for parental, F_1, and F_2 allelic frequencies and genotypic frequencies in Table 20E-3.

Table 20E-3   Effects of Heterozygote Advantage on Allelic Frequencies

Generation	Frequency of Alleles		Frequency of Genotypes		
	p (B)	q (b)	p^2 (BB)	$2pq$ (Bb)	q^2 (bb)
Parent					
F_1					
F_2					

a. Compare the allelic frequencies of the parent population with those of the F_1 and F_2 generations.

_____

_____

b. Does the recessive allele ever disappear? _____ Why or why not? _____

_____

c. Toward what values of p *and* q *is the population in this simulation moving?*

_____

Why? _____

d. Could either p *or* q *be greater than 0.5 in this simulation?* ————

Why or why not? _____

e. Compare the genotypic frequencies of the parent population with those of the F_1 *and* F_2 *generations.*

_____

_____

f. Is your population typical of most of the populations in the class? ————

g. Explain how heterozygote advantage can maintain two or more alleles in a population.

_____

_____

h. Suggest why a deleterious allele that persists because of heterozygote advantage disappears once the

heterozygote advantage is removed. _____

_____

_____

EXTENDING YOUR INVESTIGATION: SELECTION PRESSURE

Selection pressure is often placed on certain **prey** phenotypes by **predators.** Predators may be more successful at capturing prey with coloration or behavior that attracts attention and less successful at capturing those that blend into their habitat (cryptic coloration). For instance, variations in the color or markings of certain insects or their larvae affect predation by birds.

You will use yarn of different colors to simulate "wooly worms" of different phenotypes distributed on the grass at the test site. You will be given 10 minutes to collect as many pieces of yarn as you can. If collection of yarn is random, like flipping a coin, then the class should collect nearly equal numbers of all colors of yarn. If there are significant differences in the colors collected, it is possible that the differences can be explained by a process of selection. Formulate a hypothesis that can be used to explore the relationship between "wooly worm" phenotype and selection.

HYPOTHESIS:

NULL HYPOTHESIS:

What do you **predict** will happen to the "wooly worm" population being preyed upon in this experiment?

Identify the **independent variable.**

Identify the **dependent variable.**

Use the following procedure to test your hypothesis.

PROCEDURE:

1. Work in pairs. At the test site, collect as many pieces of yarn as possible in the 10 minutes allotted.

2. Return to your laboratory and count the number of pieces of each color of yarn collected. Record the numbers in column A of Table 20-I. Total all the numbers of column A and enter the sum at the bottom of the column.

3. The expected number for each color—that is, the number we would expect if the number collected were affected only by chance—can be determined as follows:

$$\frac{\text{Total number collected}}{\text{Number of colors used}} = \text{number expected for each color}$$

Enter this number in column B for every color. (It will be the same for each color.)

Table 20-I   Chi-Square Calculations

Color	A No. Observed (collected)	B No. Expected (by chance)	C Observed − Expected	D (Observed − Expected)^2	E (Observed − Expected)^2/ Expected
1.					
2.					
3.					
4.					
5.					
6.					
7.					
8.					
9.					
10.					
11.					
12.					
13.					
14.					
15.					
Total number observed =		Sum of chi-square values =			

4. Use Table 20-I to determine the chi-square value for each color (see Appendix I for an explanation of the chi-square test).

a. Subtract the number in column B from the number in column A and enter the result in column C. Numbers may be positive or negative.

b. Square each number in column C and enter it in column D (negative numbers become positive).

c. Divide each number in column D by the number in column B.

d. Add all the numbers in column E to get the sum of chi-square values. Enter this number in the table.

RESULTS: You have completed the chi-square calculations. You can now check your results against the chi-square distribution (Table 20-II) to determine the probability that chance alone was responsible for the numbers of colored yarn pieces collected.

Degrees of Freedom. This will always be 1 less than the total events (colors) observed. For instance, if 10 colors of yarn were collected (observed), 9 degrees of freedom will be used.

Probability. The columns marked $p = 0.99, 0.95, 0.50, 0.05, 0.01$, and 0.001 refer to probabilities that your observations differ from those expected based on the actual numbers of yarn pieces of each color distributed. For instance, if 10 colors were used (9 degrees of freedom) and you obtained a chi-square value of 8.5, you could conclude that the numbers observed vary from expected numbers about 50 percent of the time. Thus, chance alone could cause such a variation and the null hypothesis is supported. However, if you obtain a chi-square value of 17 (16.919) or higher, which gives a probability of 0.05 (or less), this indicates that the numbers collected or observed would show such a variation *by chance alone* only about 5 percent of the time. In other words, there is only 1 chance in 20 that the variation you observed is by chance alone. The smaller the probability, the less likely it is that the results are due to chance. In general, researchers consider a 0.05 probability as a minimum for considering results to be due to factors other than chance. However, remember that even if you obtain a probability level of 0.05, there is still 1 chance in 20 that your results are due only to chance.

From your results, describe the effects of selection on your population of "wooly worms."

Do your results support your hypothesis?

Your null hypothesis?

Was your prediction correct?

What do you **conclude** about the effects of varying phenotypes on their selection?

Which colors of yarn were subjected to positive selection pressure and which to negative selection pressure?

Why?

If the environment remains constant over time, how will gene frequencies be affected in future generations?

How could you increase the level of certainty that the results you obtained are due to selection and not to chance? (*Hint:* Think about Mendel's experiments with peas, and especially about the experimental setup he used to obtain the 9:3:3:1 ratio.)

Table 20-II   Chi-Square Distribution*

Degrees of Freedom	p = 0.99 (99 in 100)	p = 0.95 (95 in 100)	p = 0.50 (50 in 100)	p = 0.05 (5 in 100)	p = 0.01 (1 in 100)	p = 0.001 (1 in 1,000)
1	<0.001	0.004	0.455	3.841	6.635	10.827
2	0.020	0.103	1.386	5.991	9.210	13.815
3	0.115	0.352	2.366	7.815	11.345	16.286
4	0.297	0.711	3.357	9.488	13.277	18.465
5	0.554	1.145	4.351	11.070	15.086	20.517
6	0.872	1.635	5.348	12.592	16.812	22.457
7	1.239	2.167	6.346	14.067	18.475	24.322
8	1.646	2.733	7.344	15.507	20.090	26.125
9	2.088	3.325	8.343	16.919	21.666	27.877
10	2.558	3.940	9.342	18.307	23.206	29.588
11	3.053	4.575	10.341	19.675	24.725	31.264
12	3.571	5.226	11.340	21.026	26.217	32.909
13	4.107	5.892	12.340	22.362	27.688	34.528
14	4.660	6.571	13.339	23.685	29.141	36.123
15	5.229	7.261	14.339	24.996	30.578	37.697

High probability that results are due to chance   Low probability that results are due to chance

*p is the probability that results could be due to chance alone. The numbers in parentheses below each value of p restate p in terms of chance. For example, at a p value of 0.01, chances are 1 in 100 that results are due to chance.

Laboratory Review Questions and Problems

1. What is an allelic frequency?

2. If you have a complete list of genotypic frequencies for a population, can you calculate allelic frequencies? Always? (Consider populations that *are* and *are not* in Hardy-Weinberg equilibrium.)

3. In a population that is 10 percent *AA*, 20 percent *Aa*, and 70 percent *aa*, what is the frequency of allele *A*? What is the frequency of *a*?

4. If *A* and *a* are the only alleles at the gene *A* locus, does the population in question 3 seem to be at Hardy-Weinberg equilibrium? If not, list some possible reasons why.

5. How would you tell if a population is at Hardy-Weinberg equilibrium?

6. A simple Mendelian trait has two alleles, *D* and *d*. If a population is in Hardy-Weinberg equilibrium and is 49 percent homozygous dominant, what percentage is heterozygous? What are the allelic frequencies in the population?

7. If you know the allelic frequencies for a particular gene (e.g., one that determines the degree of attachment of earlobes) in your class, can you calculate the genotypic frequencies for your class? Why or why not?

8. If you know allelic frequencies for a population, can you always calculate genotypic frequencies? Why or why not?

9. If you know the fraction of homozygous recessive individuals in a population, can you always estimate the allelic and genotypic frequencies for that population? What conditions must be met by the population in order to do this? (Compare your answer with that for question 6.)

10. In a population that is in Hardy-Weinberg equilibrium, the frequency of the *d* allele is 0.4. What fraction of the population is *DD*? Why is it important to mention that this population is in Hardy-Weinberg equilibrium?

11. What five conditions must be met for the Hardy-Weinberg equation to predict the genetic makeup of a population accurately? Why do we consider the Hardy-Weinberg prediction to be a "null" hypothesis?

12. When you relax (or violate) any of the conditions necessary for Hardy-Weinberg equilibrium, what happens to allelic frequencies in a population?

13. What is genetic drift? Why can drift usually be ignored in a large population, but never in a small population?

14. Explain why nonrandom mating and migration can disrupt Hardy-Weinberg equilibrium.

15. Can a population evolve solely due to genetic drift? Explain.

16. In Exercises B through E, which Hardy-Weinberg criteria did you violate? In each case, what happened to allelic frequencies?

Exercise B

Exercise C

Exercise D

Exercise E

17. In Exercise E, Parts 1A, 1B, and 2, which genotypes were favored? Why?

Part 1A

Part 1B

Part 2

18. To illustrate Hardy-Weinberg principles, a biology textbook describes the case of a homozygous recessive fatal disease that occurs in 4 percent of a population. Students are asked to use the Hardy-Weinberg equation to compute p and q for this population. Is this an appropriate use of the Hardy-Weinberg equation? Why or why not?

Genetic Basis of Evolution II— Diversity

OVERVIEW

Evolution is one of the major unifying themes in biology. Organismal diversity (Laboratories 22 to 27) is a consequence of ongoing evolution.

The study of biological diversity is called **systematics.** Systematists reconstruct the evolutionary history or **phylogeny** of a group of related organisms, making it easier to classify the diversity of fossils as well as organisms alive today. Phylogenetic classification schemes group organisms into assemblages based on common ancestry. Traditionally, morphological, embryological, and fossil evidence have been compared to determine common lineage. Recently, molecular evidence from analysis of proteins, DNA, and RNA have also been gathered and used for building phylogenies.

In this laboratory, you will investigate the phylogenetic relationships among members of a group of ungulate mammals ("hoofed" herbivores). Electrophoretic analysis of structural differences in the protein lactate dehydrogenase (LDH) will provide evidence for determining evolutionary relationships among cows, goats, sheep, and horses.

STUDENT PREPARATION

Prepare for this laboratory by reading the text pages indicated by your instructor. Familiarizing yourself in advance with the information and procedures covered in this laboratory will give you a better understanding of the material and improve your efficiency. Complete Exercise A before coming to the laboratory.

✔ **EXERCISE A** | **Understanding Evolutionary Classification**

The goals of systematics, a branch of evolutionary biology, are to determine **relationships** among organisms, preferably based on genealogy or relationship by descent (phylogeny), and to develop useful **classifications** that reflect these relationships. The organization of species into **taxonomic groups** reflects a hierarchy of these phylogenetic relationships—from broad inclusive taxonomic groups to smaller and smaller branches of the phylogenetic tree. An understanding of relationships helps us ask relevant biological questions and understand the origins and adaptations of organisms making up our world. An understanding of classification helps us organize the diversity ("biodiversity") in our global environment and is necessary for the successful management of our total ecosystem.

Every organism is classified using a hierarchy of categories—kingdom, phylum or division, class, order, family, genus, and species (with some intermediate levels). Classification defines the position of a species within a broader phylogenetic spectrum of taxonomic groups.

A group of organisms at any level in the hierarchy (species, genus, family, etc.) is called a **taxon** (plural, taxa). A taxon is **monophyletic** if a single ancestor has given rise to all species within that group and to no other species in any other group. (Note that monophyly is relative. In one sense, all birds may be regarded as monophyletic; but so may all vertebrates or all animals. The concept of monophyly is most useful within genera or families.) Ideally, all taxa *should* be monophyletic. A taxon is **polyphyletic** if its members are derived from more than one ancestor. (It is not always possible to separate species phylogenetically, particularly in less known taxa or in groups that do not reproduce sexually, because reproductive isolation is the criterion for separating biological species. Certain taxa may be "dumping grounds" or may contain species placed there provisionally pending further study.) A taxon is **paraphyletic** if it includes the group and its common ancestor but excludes all other groups that may also have come from the common ancestor. (We know that birds and mammals trace their ancestry to somewhat different groups of reptiles; monophyly would suggest that all three "classes" be placed together. Our common practice of separating reptiles, birds, and mammals, by criteria you will use in Exercise A, provides a useful classification in spite of being "paraphyletic.")

A major goal of systematics is to determine phylogenies and to arrange taxa into monophyletic groups related by descent.

ⅠⅠⅠⅠⅠ Objectives ⅠⅠ

☐ Identify levels of phylogenetic classification.

☐ Distinguish among monophyletic, polyphyletic, and paraphyletic classifications.

☐ Understand approaches by systematists to determine relationships and construct classifications.

ⅠⅠⅠⅠⅠ Procedures ⅠⅠ

1. Use information in your text to classify humans.

Kingdom _____

Phylum _____

Subphylum _____

Class _____

Subclass _____

Order _____

Family _____

Genus _____

Species _____

Be sure that the species name consists of *two* words—the genus name and the species (or "specific") epithet (an adjective)—providing a **binomial nomenclature** as first practiced consistently by Linnaeus. Note that the species epithet is an adjective and can *never* be used alone. Thus *Homo* is humans' generic name. The species name is *Homo sapiens*. (Where the complete binomial species name has been used and there is no possibility of confusing *Homo* with any other generic name, you may see the species name abbreviated *H. sapiens*. But never, under any circumstances, can humans be named using only the specific epithet "*sapiens*.")

2. In Figure 21A-1, circle all monophyletic groups and label them "M."

3. Circle all polyphyletic groups and label them "P."

4. Circle all paraphyletic groups and label them "A."

5. In the hypothetical phylogenetic tree of primates (Figure 21A-2), circle all monophyletic, polyphyletic, and paraphyletic groups and label them as in steps 2–4.

6. Within the Vertebrata (a subphylum of the Chordata), there are several classes of living organisms that are familiar to you. List the major traits or characteristics by which you would

Figure 21A-1 *Hypothetical phylogeny of species groups.*

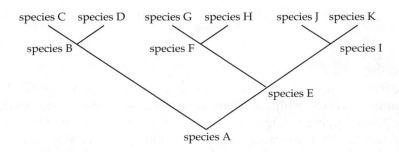

Figure 21A-2 *Hypothetical phylogeny of primates.*

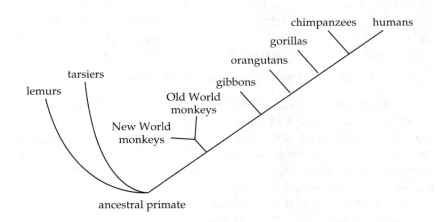

recognize the following taxa. Use information from your text to group the listed organisms into their appropriate classes.

Cartilaginous fishes (sharks, skates, rays, chimaeras) _____

_____

Bony fishes (e.g., herring, trout, goldfish) _____

_____

Amphibians (frogs, salamanders) _____

_____

Reptiles _____

_____

Birds _____

_____

Mammals _____

_____

a. *What features are common to all of these groups of vertebrates?* _____

   _____

b. *What features distinguish among the groups and allow you to place* all *species in their appropriate*

   classes? _____

   _____

c. *Which of these features do you suppose have a common genetic/developmental basis that verifies a*

   true relationship? _____

d. Which of these features are similar but may not be related by ancestry?

_____

_____

The observations and inferences you have just made are very similar to those made by systematists in determining relationships and constructing classifications. However, it is important that **homologous traits**, those having a common genetic and developmental substrate or background, be distinguished from characteristics that resemble each other (**analogous traits**) or that provide a similar function (**homoplastic traits**). The forelimbs of fish (involved in swimming), birds (flying), and mammals (walking or running) are all homologous, even though they have different shapes and functions (and in the case of fishes, may have quite different structures). The wings of birds and beetles are, however, only analogous, even though in both cases they are involved in flying. Lungs and gills are homoplastic—both are involved in gas exchange, but they are unrelated structures.

7. Return to your list of vertebrates above and circle traits that you believe are homologous.

Early biologists classified organisms by similarities. After Darwin's publications and a general understanding of the concept of natural selection, attempts were made to refine classifications along phylogenetic lines. This trend toward a "new systematics" developed in the 1940s with publications by Dobzhansky, Mayr, Stebbins, Simpson, and many others. Systematists attempted to use a combination of traits to construct phylogenies, but the resolution of conflicts often depended on relatively arbitrary decisions (not necessarily wrong).

In the 1950s, with the advent of computer technology, a group of systematists employed numerical techniques to determine "similarities" among species—a number of characters were measured and "crunched" to construct a matrix that allowed species to be arranged by similarity (but not necessarily by relationship—no attempt is made to distinguish among homologies and analogies). Thus, the school of **phenetics** developed. With modern computers, numerical techniques have become important in all phases of systematics (particularly in the analysis of DNA and molecular data), but phenetic concepts have not satisfied our desire to understand phylogeny.

Another approach to determining relationships among organisms is based on "clades"—evolutionary branches—and the school of phylogenetic systematics developed by Hennig is commonly known as **cladistics.** Cladists determine branching of evolutionary lines by novel homologies unique to all descendants contained within that branch—so-called "synapomorphies" or shared derived characteristics. Cladists consider only branching, not the degree of evolutionary divergence among the organisms they study. This and some of the assumptions made by cladists may diminish the value of this approach and limit its ability to produce stable classifications.

Modern **evolutionary systematists** use a combination of numerical techniques and all of their understanding of the biology of a group of organisms (including both branching and divergence) in an attempt to construct realistic **phylograms** (phylogenetic trees) that can be used to produce a useful and stable classification. They recognize, for example, that modern birds as a group are uniquely defined by feathers and endothermy and have more in common with each other than with their reptilian ancestors. It is "useful" to recognize birds as a higher category even though birds and reptiles are related by descent.

8. Using the groups of vertebrates listed in step 6, draw a "cladogram" (dichotomous branches only) and a "phylogram" to express your understanding of the different approaches used by systematists.

Cladogram   Phylogram

Systematists have studied and named about 1.6 to 1.7 million species to date. Estimates of the total number of species present on the planet range from 5–6 million to 30 million or more.

e. Why is it vital for us to study and catalog this diversity immediately?

_____

_____

Each species contains a unique genetic library. *f. Why is it necessary for us to manage our environment and planet to preserve as much of this genetic diversity as possible?*

_____

EXERCISE B | Electrophoretic Analysis of LDH in Ungulate Mammals*

As species evolve, subtle changes occur in macromolecules within organisms. These changes cause alterations in the structures of proteins, RNA, and DNA that can be studied using electrophoretic and other molecular techniques. In particular, enzyme proteins vary in composition in various organisms, giving rise to dissimilar banding patterns using electrophoresis. Systematists using molecular techniques expect differences among organisms to reflect the length of time that the lines have been distinct and the degree of divergence between those lines. They assume that small "neutral" mutations occur at a *constant* rate and that the greater the interspecies differences between chromosomal DNA, mitochondrial DNA, ribosomal RNA, and common, highly conserved proteins (such as the cytochromes), the longer the populations have been separated. Unfortunately, evolution does not always proceed at a constant rate, and so-called molecular clocks defined by this approach must be realistically calibrated. Molecular studies have been very useful and can be coupled with other systematic approaches, often with similar results.

The analysis of cytochrome *c* is probably the most noted application of these molecular techniques (molecular systematics). Human cytochrome *c* differs from the cytochrome *c* of yeast in 44 of the 104 amino acid residues. In contrast, the amino acid sequences of cytochrome *c* of humans and chimpanzees are identical. This evidence has helped confirm the ancestral link between humans and other primates.

In this exercise, you will compare the properties of a specific protein, lactate dehydrogenase, in different ungulate mammals. Ungulates are hoofed, herbivorous animals that can be divided into two major orders: the artiodactyls and the perissodactyls. Artiodactyls, having an even number of toes on each foot, include pigs, camels, cattle, sheep, and goats. Perissodactyls, with an odd number of toes, include rhinoceroses, present-day horses, and zebras. The evolution of ungulates is shown in Figure 21B-1.

Figure 21B-1 *Mammalian phylogenetic tree.*

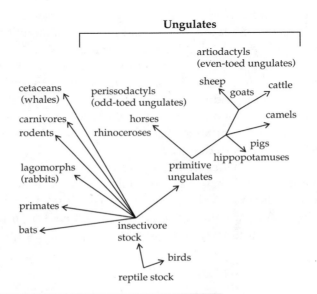

Adaption of Experiment 106, Protein Fingerprinting: Teaching Modern Biology in the Laboratory, Modern Biology, Inc. © John N. Anderson, 1994. Chemicals for this experiment can be purchased from Modern Biology, Inc., 111 North 500 West, West Lafayette, IN 47906; phone 1-800-733-6544.

Lactate dehydrogenase (LDH) is an enzyme that catalyzes the conversion of pyruvic acid (pyruvate) to lactic acid (lactate) (Figure 21B-2). This protein is produced by most tissues but is especially prevalent in muscle. LDH is also found in blood serum, the noncellular portion of blood, as a result of normal cell death with release of cellular enzymes into the bloodstream.

Figure 21B-2 *Conversion of pyruvate to lactate. Pyruvate, the end product of glycolysis, can accept hydrogen electrons from NADH + H^+, reoxidizing the NADH + H^+ to NAD^+.*

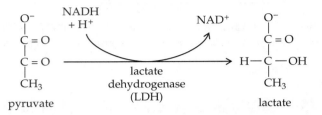

Lactate dehydrogenase production is controlled by two genes. Each gene is monoallelic (it has only one form) and produces a single type of peptide (monomer). Monomeric peptides combine to produce tetramers. LDH is a tetramer, and in mammals it exists as five isoenzymes (different molecular forms of the same enzyme). If all of the LDH peptides are of one type, the enzyme exists as the LDH-1 isoenzyme or the LDH-5 isoenzyme, depending on which gene produced the peptide monomers. The remaining three isoenzymes result from different combinations of the peptides produced by the two genes. If we call the two genes and their peptide products **M** (for muscle) and **H** (for heart), the LDH isoenzymes are designated as follows:

MMMM	LDH-5
MMMH	LDH-4
MMHH	LDH-3
MHHH	LDH-2
HHHH	LDH-1

The order of listing the peptides is inconsequential; MMMH is the same as MHMM.

All five LDH isoenzymes catalyze the same reaction, but they differ in their amino acid composition and net charge. These differences make them easily separable by electrophoresis. LDH-5 is the lightest (lowest molecular weight) and will therefore travel the farthest on a gel. LDH-1 is the heaviest and will remain closer to the origin.

While all mammals are capable of producing all five isoenzymes, the amount of each isoenzyme produced is controlled by regulatory genes that control the activities of the two LDH-producing genes. Therefore, not all isoenzymes will be detectable on a gel even if they are present. (Some may occur in concentrations too low to react with the stain.) Furthermore, LDH isoenzymes composed of the same combination of monomers are not identical among different mammalian species. There are antigenic differences among mammals, indicating differences in amino acid composition. Therefore LDH from two different mammalian species will not necessarily band in the same place on the gel. It is these differences, combined with the number of different bands observed, that make it possible to determine evolutionary trends among the mammals tested.

In this exercise, you will separate the serum proteins from different ungulates (cow, sheep, goat, and horse) by electrophoresis, and then use a special staining technique to identify the LDH isoenzymes. The results from each serum sample will appear as a distinct gel fingerprint of LDH activity, with different isoenzymes showing different banding patterns. A comparison of the LDH fingerprints should enable you to determine the basic evolutionary relationships among the ungulates tested.

ııııı **Objectives** ııı

☐ Use electrophoretic evidence to distinguish among related species of organisms.

☐ Determine the phylogenetic relationships among ungulate mammals.

ııııı **Procedure** ıııııııııııııııııııııııııııııııııı

Formulate a hypothesis for this investigation.

Hʏᴘᴏᴛʜᴇsɪs:

Nᴜʟʟ ʜʏᴘᴏᴛʜᴇsɪs:

What do you **predict** will be the outcome of this investigation?

What is the **independent variable**?

What is the **dependent variable**?

Use the following procedure to test your hypothesis.

1. Work in pairs. Place the gel comb into the end slots of the gel carrier and seal off the ends with tape to make a gel box.

2. Pour 50 ml of 1.2% agarose into the gel box until an even layer of approximately one-eighth of an inch of agarose covers the bottom of the box.

3. Allow the gel to cool for approximately 15 minutes.

4. Remove the comb and tape from the gel box, and place the gel box into the electrophoresis unit so that the gel wells are at the negative electrode (black) end.

5. Pour enough tris-citrate buffer into the electrophoresis unit to cover the gel completely.

6. Load 15 μl of each type of serum into the sample wells as indicated in Figure 21B-3. The serum has been combined with a loading dye (bromphenol blue and glycerol).

Figure 21B-3   *Loading order for ungulate sera samples.*

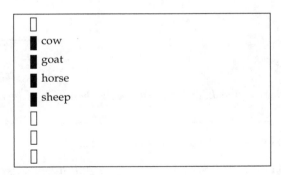

7. Conduct electrophoresis at 100 volts until the loading dye has migrated to within 2 cm of the positive electrode end of the gel.

8. *Just* before completion of the electrophoresis run, prepare a fresh stain solution by combining the solutions listed in Table 21B-1. This solution couples the LDH-catalyzed reaction (pyruvate $\rightarrow$ lactate) to a second reaction that produces a formazan compound, which is brown in color and collects at the sites of LDH activity. Therefore, this stain selectively excludes all proteins in the serum except LDH, allowing the simple differentiation of LDH patterns among the samples tested. (You will have enough stain for only your own gel.)

Table 21B-1   Chemical Composition of the LDH-Staining Solution

Solution	Amount (ml)
Tris-HCl buffer	30
Lilactate	6
NAD	1.3
MTT	0.3
PMS	0.5

9. Remove the agarose gel from the electrophoresis unit and gel box, and place it into a staining dish. (Warning: Be certain that the power has been turned off and the electrodes removed from the power supply before attempting to remove the gel.)

10. Cover the gel with the staining solution you have prepared, and place the gel into a 37°C incubation chamber for 30 minutes. (The incubation chamber may be an incubator or a water bath equipped to support staining dishes.) *The staining reaction should occur in the dark,* so be sure to replace the lid on the chamber.

11. After incubation, pour the staining solution into the waste beaker. Cover the gel with the 10% acetic acid destain solution, and observe the banding patterns.

12. Sketch the LDH banding patterns on your gel.

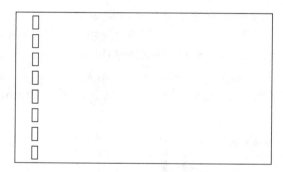

a. *Which organisms show the most similar banding patterns?* _____

_____

b. *Does this agree with your predictions?* _____

Do your results support your hypothesis? _____ *Your null hypothesis?* _____

From the results obtained in this exercise, what do you **conclude** *about the evolution of ungulates?*

_____

_____

c. *Are the ungulates a monophyletic or polyphyletic taxon?* _____ *How do you know?*

_____

Laboratory Review Questions and Problems

1. Define the following terms and give an example for each.

   Phylogeny

   Taxon

   Binomial nomenclature

   Monophyletic taxon

   Polyphyletic taxon

   Paraphyletic taxon

   Cladistics

   Evolutionary systematics

   Homology

   Analogy

   Homoplasy

   Phylogeny

   Classification

   Systematics

2. The gel shown below represents the electrophoretic pattern of a protein from four different but related organisms.

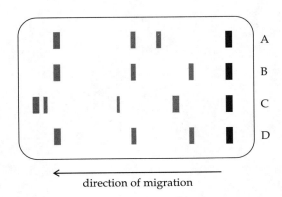

direction of migration

a. Which two organisms are most closely related?

b. Which organism is least related to the others, although it shares a common ancestor?

c. In the space below, construct a possible phylogenetic tree showing the evolutionary relationships among organisms A, B, C, and D.

Diversity— Kingdoms Eubacteria, Archaebacteria, and Protista

OVERVIEW

Biologists currently recognize three *domains* of living things: **Bacteria, Archaea,** and **Eukarya.** The domain Bacteria is comprised of one kingdom, **Eubacteria** (true bacteria). The domain Archaea is also composed of one kingdom, **Archaebacteria** (ancient bacteria). The domain Eukarya includes all other living organisms: kingdoms **Protista** (unicellular animals and plants, slime molds, and algae), **Fungi, Plantae** (multicellular plants), and **Animalia** (multicellular animals).*

Among the diverse organisms in these kingdoms you will find two major types. **Prokaryotes** are unicellular (single-celled) organisms characterized by the absence of a membrane-bound nucleus and membrane-bound organelles. In contrast, **eukaryotes** have both a well-formed nucleus and many other types of membrane-bound organelles.

Prokaryotes and eukaryotes differ also in the chemical composition of their cell walls (if present), the organization of their genetic material, and the structure of their flagella. The kingdoms **Eubacteria** and **Archaebacteria** are composed of prokaryotic organisms. Eukaryotic organisms are found in the other four kingdoms.

STUDENT PREPARATION

Prepare for this laboratory by reading the text pages indicated by your instructor. Familiarizing yourself in advance with the information and procedures covered in this laboratory will give you a better understanding of the material and improve your efficiency.

PART I   DOMAINS BACTERIA AND ARCHAEA

Prokaryotes represent the oldest and simplest living things. Prokaryotes are clearly different from eukaryotes, but ordering them taxonomically into subgroups of organisms according to anatomical and physiological affinities is difficult. Based on comparison of DNA and RNA sequences, it appears that prokaryotes diverged early in evolutionary history into two distinct lineages now recognized as domains, Bacteria and Archaea (Figure 22I-1).

Belonging to the domain Archaea, the kingdom Archaebacteria is composed of four phyla of organisms that live at environmental extremes: methanogens (bacteria that use H_2 to reduce CO_2 to

---

*Because scientists are continually learning more about the structure and origins of organisms, the assignment of organisms within a particular classification may change. Indeed, the number of phyla and even the number of kingdoms have changed during the past decade. It is possible that your instructor will adopt a different classification scheme than that used in the following exercises, but the principle of classification will remain the same—the grouping of organisms according to similarities that indicate evolutionary relationships.

Figure 22I-1 *According to present thinking, the domains Eukarya and Archaea share a common ancestry and thus are more closely related to each other than to the domain Bacteria.*

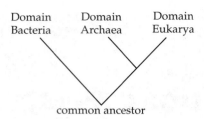

methane), thermoacidophiles or hyperthermophiles (bacteria that live where it is both hot and acidic), extreme halophiles (salt-loving bacteria), and thermoplasma. Belonging to the domain Bacteria, the kingdom Eubacteria is composed of 12 phyla, among which are gram-negative, gram-positive, and photosynthetic bacteria. These include myxobacteria, rickettsias, desulfovibrio, purple nonsulfur bacteria, rhodopseudomonas, purple sulfur bacteria, spirochetes, actinomycetes, clostridia, and mycoplasmas.

Bacteria are virtually ubiquitous. Not only do bacteria surround you, they thrive within you. Most of these bacteria are nonpathogenic and actually keep your body free from more harmful bacteria through competition. However, conditions sometimes make the difference between a pathogen and a nonpathogen. For example, *Staphylococcus aureus*, a common bacterium, can cause severe infections if introduced into an open cut. The same bacterium is also thought to be an agent in toxic shock syndrome.

Other bacteria play a large role in the environment and in the economy. Some bacteria are important as decomposers, recycling dead material into components required by living organisms. In sewage treatment plants, they promote the breakdown of solid wastes. Many foods, such as vinegar, sour cream, yogurt, and cheeses, are made with the help of bacteria. Certain antibiotics are produced by bacteria. Some bacteria can even be used to clean up oil spills. Bacteria are also a main focus of genetic and biochemical research.

✔ **EXERCISE A** | **Morphology of Bacteria**

Most bacteria may be classified into one of three major morphological groups: rods (bacilli), spheres (cocci), or spirals (spirilla). You can observe the morphology of these groups both microscopically and macroscopically, by observing growth forms.

▮▮▮▮▮ **Objectives** ▮▮

☐ Recognize the different morphological forms of bacteria.

▮▮▮▮▮ **Procedure** ▮▮

1. Check the demonstration microscopes and observe the basic morphological forms of bacteria (Figure 22A-1). Draw the bacterial forms and identify each in the space that follows.

Figure 22A-1 *Morphology of bacteria.*

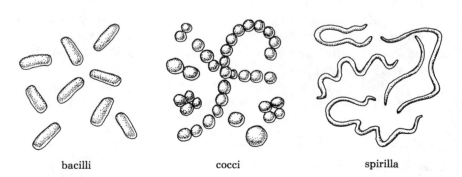

bacilli   cocci   spirilla

Demonstration A   Demonstration B   Demonstration C

Type _____   Type _____   Type _____

a. *Strep throat is caused by streptococcal bacteria. What shape would you expect these bacteria to have?*

   _____

b. *The prefix "strep" means strip or chain. How would you expect streptococcal bacteria to be*

   arranged? _____

c. *How are the bacterial cells arranged in the demonstration slides you have observed? As single cells?*

   Clusters? Strands? A _____ *B* _____ *C* _____

2. Gram stain is a specific stain (purple) used to test for gram-positive bacteria. In these bacteria, the cell wall contains a small amount of lipid combined with a thick layer of a peptide/sugar material (peptidoglycan). The peptidoglycan molecules trap Gram stain and the cell walls retain a purple color. The cell walls of gram-negative bacteria contain a great deal of lipid but little peptidoglycan and the purple Gram stain washes out. A second stain, safranin, stains the gram-negative bacteria pinkish-red. What color were the bacteria on the demonstration slides? Were they gram-positive or gram-negative?

Color	Gram + or Gram −
A	
B	
C	

d. *Are all coccal bacteria gram-positive?* _____ (Hint: *Review the slides on demonstration.*)

3. Streptococcal bacteria are gram-positive. Physicians use Gram stain to identify specific bacteria.

   e. *What color would Gram-stained streptococcus be?* _____

4. Penicillin is usually an effective antibiotic for treating infections caused by gram-positive bacteria. Penicillin affects the ability of bacterial cells to synthesize peptidoglycan.

   f. *Would penicillin be an effective way to treat strep throat?* _____

5. You were given an agar-filled Petri dish with instructions to expose it to a potential source of bacteria in your environment, for example, keys, money, shoes; air in various locations (inside the house, in a bathroom; outside, in moist or dry environments); your hands before washing, after washing with regular soap, after washing with an "antibacterial" hand soap; the foot of a

cat, dog, or bird. Examine your dish and those of your classmates. *Do not remove the lids.* Note the number of bacterial colonies in each plate. These will have the appearance of small, shiny masses. The "fuzzy" masses that may appear on some plates are fungi. Theoretically, each colony of bacteria or fungi arose from a single bacterium or fungal spore (though closely adjacent individual colonies may coalesce and appear as one). The appearance of a bacterial colony on agar is used in classification. Record your observations of the number of different colonies on each of four plates, and the shape, color, texture, and size of each type of colony.

Plate 1 _____

Plate 2 _____

Plate 3 _____

Plate 4 _____

👁 EXERCISE B   Characteristics of Bacteria: Sensitivity to Antibiotics

Since the number of basic bacterial morphologies is limited, identification of species is largely dependent on biochemical and physiological characteristics. Composition of the cell wall, DNA or RNA sequencing, sensitivity to antibiotics, mode of metabolism, and mode of reproduction can all be used to distinguish among bacterial types. These characteristics are the topics of this exercise and Exercises C through D.

The bacterial cell wall and ribosomes represent two major sites for the action of antibacterial agents. As you have already learned, many of the common antibiotics treat bacterial infections by inhibiting the synthesis of the essential structural polymer, peptidoglycan, in the bacterial cell walls. Recall that penicillin is an example of one of these antibiotics. But how effective are other antibiotics against gram-positive bacteria? And what about gram-negative bacteria?

ⅠⅠⅠⅠⅠ Objectives ⅠⅠⅠⅠⅠⅠⅠⅠⅠⅠⅠⅠⅠⅠⅠⅠⅠⅠⅠⅠⅠⅠⅠⅠⅠⅠⅠⅠⅠ

☐ Determine the sensitivity of selected bacteria to several antibiotics.

☐ Describe the action of antibiotics.

There are two general types of antibiotics. **Narrow-spectrum antibiotics** target a limited number of bacterial species, while **broad-spectrum antibiotics** are effective against a wider spectrum of bacterial types. This exercise will show the relationship between bacterial cell-wall structure, as indicated by Gram stain, and susceptibility to different types of antibiotics.

ⅠⅠⅠⅠⅠ Procedure ⅠⅠⅠⅠⅠⅠⅠⅠⅠⅠⅠⅠⅠⅠⅠⅠⅠⅠⅠⅠⅠⅠⅠⅠⅠⅠⅠⅠⅠⅠⅠ

1. Examine the Petri dishes on display. In preparing these, the agar surfaces were first covered with a suspension of a known type of bacteria—one plate with gram-positive and the other with gram-negative. Paper disks impregnated with an antibiotic or antiseptic were then placed on top of the bacteria (Figure 22B-1). After an incubation period, bacterial growth will be

Figure 22B-1 *Bacteria growing on an agar plate demonstrate sensitivity to antibiotics. Clear areas indicate inability of bacteria to grow in the presence of an antibiotic. The diameter of the zone of inhibited growth depends on the relative diffusibility of the antibiotic and the sensitivity of the particular bacterium.*

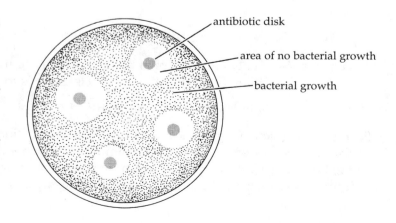

antibiotic disk

area of no bacterial growth

bacterial growth

visible unless inhibited by the substance in the disk. (This procedure is often used to determine which drug would be most effective against a particular bacterium.)

2. Record the results in Table 22B-1. List the antibiotics in the left-hand column, then record the effects of each antibiotic on each of the two types of bacteria. Use the following symbols to record your observations of the dishes: −, growth not inhibited; +, growth slightly inhibited; ++, growth greatly inhibited.

Table 22B-1   Effects of Antibiotics

Antibiotic	Effect on Growth	
	Plate 1: Gram −	Plate 2: Gram +
A.		
B.		
C.		
D.		
E.		

3. In Table 22B-2, list the antibiotics and the types of bacteria against which they are effective, and then decide whether each antibiotic has a broad or narrow range, or spectrum, of action.

Table 22B-2   Spectrum of Antibiotic Activity

Antibiotic	Bacteria Affected	Spectrum
A.		
B.		
C.		
D.		
E.		

a. Under what circumstances might a doctor prescribe a narrow-spectrum antibiotic?

_____

A broad-spectrum antibiotic? _____

EXTENDING YOUR INVESTIGATION: HOW EFFECTIVE IS YOUR SOAP?

Why do you use the brand of soap that you do? Is your soap effective in clearing bacteria from your hands or face? Would water be just as effective? Is medicated or deodorant soap better than "just plain soap"? Is alcohol an effective disinfectant? What about commercial disinfectants such as Lysol or Mr. Clean? When antibacterial substances are used on living organisms, they are called **antiseptics**. Those used to clean surfaces are called **disinfectants**.

Formulate a hypothesis concerning the effectiveness of an antiseptic or disinfectant of your choice.

HYPOTHESIS:

NULL HYPOTHESIS:

What do you **predict** will be the effect of the soap or the antiseptic or disinfectant solutions you have chosen?

What is the **independent variable?**

What is the **dependent variable?**

The following materials will be available to you: an *S. aureus* (gram-positive) culture or plate, an *E. coli* (gram-negative) culture or plate, detergents, soaps, and disinfectants (or bring your own). Use these materials to design an experiment based on the general procedure that follows.

PROCEDURE: Your instructor may provide you with agar plates that have been inoculated with bacteria. If not, inoculate the plates as in steps 1–3. Then proceed to testing your selected detergents, soaps, or antiseptics (steps 4–10).

1. Wipe your laboratory area clean with a 10% Clorox solution.

2. Obtain an agar plate and a broth culture of *S. aureus* or *E. coli*.

3. Prepare a lawn of bacteria as follows:

   a. Lift the lid of the agar plate and hold it above the plate.

   b. Use a sterile cotton swab to remove some bacterial culture from one of the flasks. Make sure the cotton is saturated, but not dripping.

   c. Apply the bacteria to the agar plate in a zig-zag motion over the entire surface, as shown below (a).

   d. Rotate the plate 90° and repeat this motion (b).

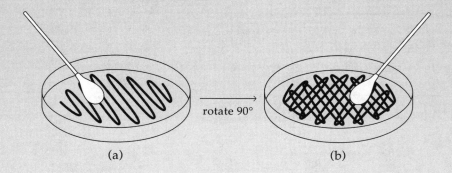

rotate 90°

(a)   (b)

   e. Dispose of the cotton swab in the container provided.

4. Pour small amounts of your selected liquid antiseptics or disinfectants into shallow dishes or other containers. If you are testing a solid material such as a bar of soap or a soap powder, make a liquid solution.

5. Using a pair of forceps, submerge a paper disk in a solution. Drain off excess liquid by touching one end of the disk to a paper towel.

6. Place the paper disk on your agar plate for 2 minutes.

7. While the disk is on the agar, mark the undersurface of the agar plate with a circle to show the placement of the disk and the name of the substance tested.

8. Repeat this procedure for three more disks impregnated with different solutions. Remember to mark and label the agar plate for each disk position. Remove the disks and dispose of them in the container provided.

9. Incubate the dishes, upside down, until the next laboratory period. Be sure that your name is on the culture plate.

10. After you complete your work, wash your work area with 10% Clorox. Also, wash your hands with soap and water!

RESULTS: A clear area around the position of a paper disk indicates that bacterial growth has been inhibited. Quantify your results by measuring the diameter of this region and noting it in the following table.

Test Substance	Zone of Inhibition (Diameter)

Do your results support your hypothesis?

Your null hypothesis?

What do you **conclude** about the effectiveness of the disinfectant, antiseptic, or soap solutions you tested?

EXERCISE C   Nitrogen-Fixing Bacteria

Although all organisms require nitrogen for the production of proteins and nucleic acids, eukaryotic organisms cannot use atmospheric nitrogen (N_2) directly. It must first be converted to a compound such as ammonia (NH_3). Only bacteria and cyanobacteria are capable of **nitrogen fixation,** the conversion of N_2 into NH_3. This capacity is the basis of a significant relationship between bacteria and eukaryotes.

Legumes, the family of plants to which peas and beans belong, form a symbiotic relationship with bacteria of the genus *Rhizobium*. *Rhizobium* live in the root nodules of the legumes and supply the plants with usable nitrogen by converting atmospheric N_2 into NH_3. The plants supply *Rhizobium* with sugar, the product of photosynthesis.

||||| **Objectives** |||||||||||||||||||||||||||||||||||||||

☐ Recognize *Rhizobium* bacteria on the root nodules of legumes.

||||| **Procedure** |||||||||||||||||||||||||||||||||||||||

1. Examine the two groups of soybean plants provided. One group was inoculated with *Rhizobium*.

    a. *Which group of plants is more robust?* _____

    b. *Examine the roots. Which group of plants has root nodules?* _____

    c. *Which group of plants has a symbiotic relationship with* Rhizobium?

    _____

2. Observe *Rhizobium* bacteria on a demonstration slide if available. Make a sketch in the space below.

    d. *Morphologically, how would you classify* Rhizobium *bacteria?* _____

✔ **EXERCISE D**   **Bioluminescent Bacteria**

Some marine bacteria are capable of producing light in the presence of oxygen. Several marine fishes are equipped with bizarre light organs inhabited by these bacteria. The fish provide the bacteria with nutrients and oxygen and use the light produced by the bacteria primarily to attract prey.

Bioluminescence in bacteria is the result of a reaction between a *luciferin* (an aldehyde compound), reduced flavin mononucleotide, or $FMNH_2$ (FMN is a component of the electron transport chain), and O_2 in the presence of the enzyme *luciferase*.

||||| **Objectives** |||||||||||||||||||||||||||||||||||||||

☐ Observe the bioluminescent properties of bacteria.

||||| **Procedure** |||||||||||||||||||||||||||||||||||||||

1. Your instructor has prepared a broth culture of the bioluminescent marine bacterium *Vibrio fisheri*. Obtain a tube of the culture and place a rubber stopper in the end of the tube.

2. In a darkened room or closet, shake the tube vigorously.

    a. *What do you see?* _____ *What color does the light appear to be?*

    _____

    b. *Why did you have to shake the tube?* _____

    _____

3. Stop shaking the tube. c. *What happens?* _____

    _____

    d. *Which part of the tube darkens first?* _____ *Why?*

    _____

    e. *How is the supply of oxygen normally maintained in the seas?* _____

    _____

✓ **EXERCISE E**   **Diversity and Structure of Cyanobacteria**

Traditionally, cyanobacteria have been called "blue-green algae." However, they are not algae, but prokaryotes: gram-negative photosynthetic bacteria. Cyanobacteria contain chlorophyll *a* (as do photosynthetic eukaryotic green plants and algae), but it is characteristically masked by blue, red, and purple pigments. These pigments (**phycobilins**) enhance light absorption by the cells and serve as nitrogen reservoirs. All cyanobacteria are unicellular, but individual cells are commonly attached to each other by a gelatinous sheath, thus producing filaments or colonies. Like some other bacteria, many cyanobacteria are able to fix atmospheric nitrogen.

Cyanobacteria share with other bacteria the ability to inhabit the most inhospitable locations on earth, such as hot springs and bare rocks. Cyanobacteria can be desiccated for many years yet resume growth when water is again present.

IIIII **Objectives** IIIIIIIIIIIIIIIIIIIIIIIIIIIIIIIIIIIIIIIIIII

☐ Recognize several types of cyanobacteria and compare these to other types of bacteria.

☐ Recognize several cell types common to cyanobacteria.

IIIII **Procedure** IIIIIIIIIIIIIIIIIIIIIIIIIIIIIIIIIIIIIIIIIIII

Work in pairs. Prepare and examine material on a wet-mount slide and then exchange slides with your partner. Depending on the material available, examine one or more examples of cyanobacteria. Refer to Table 22E-1, and add further information to the table where necessary. Make sketches if instructed to do so, adding labels where possible.

Most cells of cyanobacteria are structurally undistinguished, but a few specialized cell types can be recognized. **Heterocysts** are round or oval, *clear* cells that allow cyanobacteria to fix atmospheric nitrogen. Many cyanobacteria can fix nitrogen when they are in an anaerobic environment, but heterocysts are

Table 22E-1   Characteristics of Cyanobacteria

Type	Morphology	Distinctive Features
Nostoc (a)	Filaments of round cells; gelatinous sheath surrounds filament.	Can combine in large gelatinous balls containing hundreds of filaments. Reproduce by fission or fragmentation.
Cylindrospermum (b)	Filaments of rectangular cells; length greater than width.	**Heterocysts** at the ends of filaments function in nitrogen fixation. Reproduce by fission or fragmentation. **Akinetes** are special sporelike reproductive cells resistant to adverse environmental conditions.
Oscillatoria (c)	Filaments of rectangular cells covered by a sheath; width greater than length.	Oscillate, seek specific conditions in water. Reproduce by fragmentation only. **Hormogonia** are short fragments between dead cells where fragmentation takes place.

(continued)

Table 22E-1   *(continued)*

Type	Morphology	Distinctive Features
Anabaena akinete heterocyst ated (d)	Barrel-shaped vegetative cells held in a gelatinous matrix.	Heterocysts are integral or terminal and function in nitrogen fixation. Reproduce by fission or fragmentation. Akinetes are dispersed among vegetative cells.
Gleocapsa (e)	Spherical cells; single or groups of 2 to 8; each cell surrounded by its own sheath; colony surrounded by sheath.	Can fix nitrogen despite absence of heterocysts. Reproduce by fission.

Sketches

necessary for aerobic nitrogen fixation. **Akinetes** are generally larger, usually oval, densely packed, sporelike reproductive cells that are resistant to adverse conditions.

a. *What are the basic cyanobacterial cell shapes?* _____ *How are these individual cells combined to form colonies and filaments?* _____

b. *Which of the specialized cell types can you recognize in each of the types of cyanobacteria?* _____ *What does the presence of these cells (heterocysts, akinetes) indicate about the environment of these organisms?* _____

c. *Cyanobacteria grow prolifically in streams and lakes with low oxygen levels and high nutrient concentrations. How might the presence or absence of cyanobacteria be used as an index of pollution in lakes?* _____

d. *How can you determine, from your microscope observations, whether cyanobacteria are prokaryotes or eukaryotes?* _____
Name three differences between prokaryotes and eukaryotes.
_____
_____

e. *Cyanobacteria have a peptidoglycan cell wall. Would you expect them to be sensitive to antibiotics? Explain.*
_____

f. *Prokaryotes have membrane-bound organelles. Did the cyanobacteria that you examined contain chloroplasts?* _____ *Chlorophyll?* _____

PART II   KINGDOM PROTISTA

With the kingdom Protista,* we begin our study of eukaryotes. The cells of eukaryotic organisms contain both a nucleus and membrane-bound organelles. All other eukaryotic organisms (including fungi, plants, and animals) probably originated from the primitive protists.

For our purposes, protists can be divided into three broad groups, usually based on modes of nutrition.

Protozoa Unicellular heterotrophs, typically animal-like.

Fungus-like protists (slime molds) Sometimes referred to as the "lower fungi" because they may be multinucleate, as are fungi, during some part of their life cycle. They are classified with protists because of their similarities to protozoans.

Algae Unicellular and multicellular plant-like organisms.

PROTOZOA

Protozoans are unicellular organisms. Most are motile. Protozoans can be found in free-living and parasitic forms and in freshwater or marine environments.

✔ **EXERCISE F**   **Identifying Protozoans**

There are many phyla of protozoans. Some of the most common forms are represented below. They can be distinguished by body form and mode of locomotion (Table 22F-1).

||||| **Objectives** |||||||||||||||||||||||||||||||||||||||||||||||

☐ Identify and classify representative protozoans.

||||| **Procedure** ||||||||||||||||||||||||||||||||||||||||||||||||

Observe material using prepared slides or, if fresh material is available, make temporary wet-mount slides according to directions in Table 22F-1. Label structures and make notes or sketches of any identifying characteristics or behaviors.

Table 22F-1   Characteristics of Protozoans

Phylum/Representative	Method of Observation	Mode of Locomotion
Zoomastigophora *Trypanosoma* (a)	Study a prepared slide.	Flagellar movement. A single **flagellum** is united basally with the body of cell by an undulating membrane. Amoeboid extensions (**pseudopodia**) are also found in many flagellates. *Trypanosoma gambiense* is the causative agent of African sleeping sickness.

(continued)

*Originally, only unicellular organisms were assigned to the kingdom Protista, but in recent years it has been suggested that the kingdom be expanded to include some multicellular organisms—the multicellular algae and fungus-like organisms that lack some of the important characteristics of true fungi. The name Protoctista has been proposed for this "expanded kingdom."

Table 22F-1   *(continued)*

Phylum/Representative	Method of Observation	Mode of Locomotion
Rhizopoda 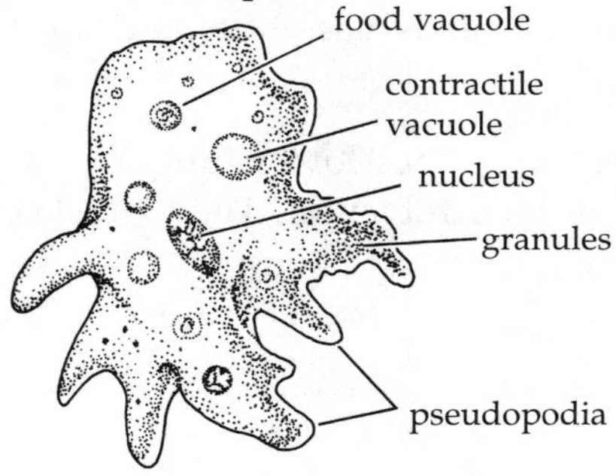 *Amoeba* (b)	Examine living amoebas. You can see the organism on the bottom or side of the culture dish. Remove an amoeba with a pipette and place it on a glass slide. Observe without a coverslip. Adjust light. If motion does not occur, add coverslip.	Amoeboid movement—pseudopodia. Cytoplasmic extensions change in size.
Ciliophora 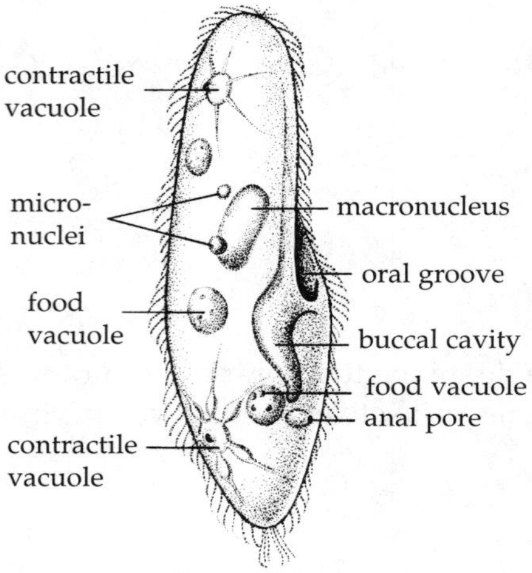 *Paramecium* (c)	Examine living paramecia. Place a small drop of culture medium on a clean glass slide. Mix in a drop of Protoslo (methylcellulose). Cover with a coverslip and observe. Is ciliary movement coordinated? _____ Can you see food vacuoles? _____ Do you see any contractile vacuoles? _____ How often do they fill and empty? _____	Ciliary movement. Cilia have the same internal structure as flagella, but are shorter.
Apicomplexa 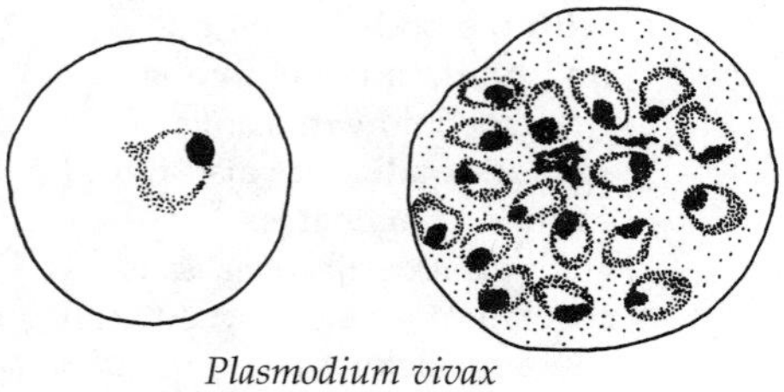 *Plasmodium vivax* (d)	Study a prepared slide. Locate sporozoites. See life cycle, Figure 22F-1.	Nonmotile phases predominate. Blood parasites.

(continued)

Table 22F-1 *(continued)*

Phylum/Representative	Method of Observation	Mode of Locomotion
Actinopoda 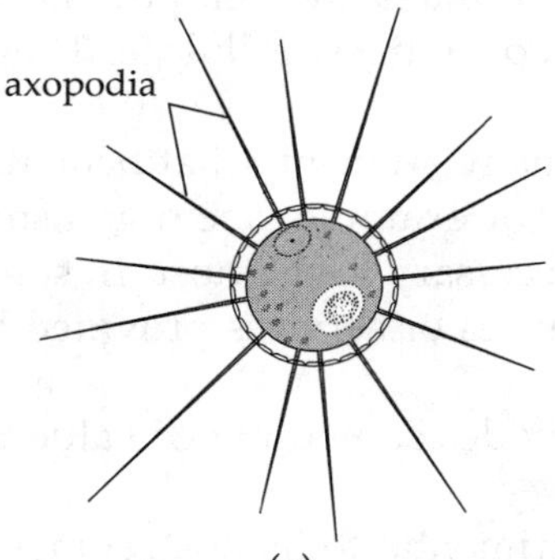 axopodia (e)	Not studied in this laboratory.	Slender pseudopodia called **axopodia** (reinforced by microtubules) help the organisms to float and feed. Phylum includes heliozoans and radiolarians.
Foraminifera 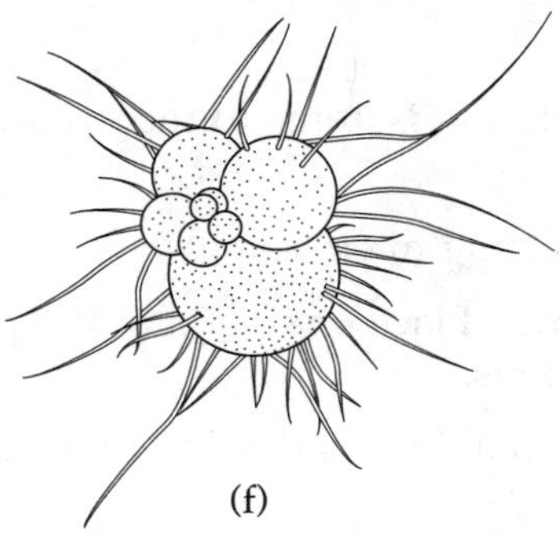 (f)	Not studied in this laboratory.	Foraminiferans (forams) are named for their porous shells. Strands of cytoplasm extend from holes in the shell for swimming and feeding.

Figure 22F-1 *Life cycle of a* Plasmodium.

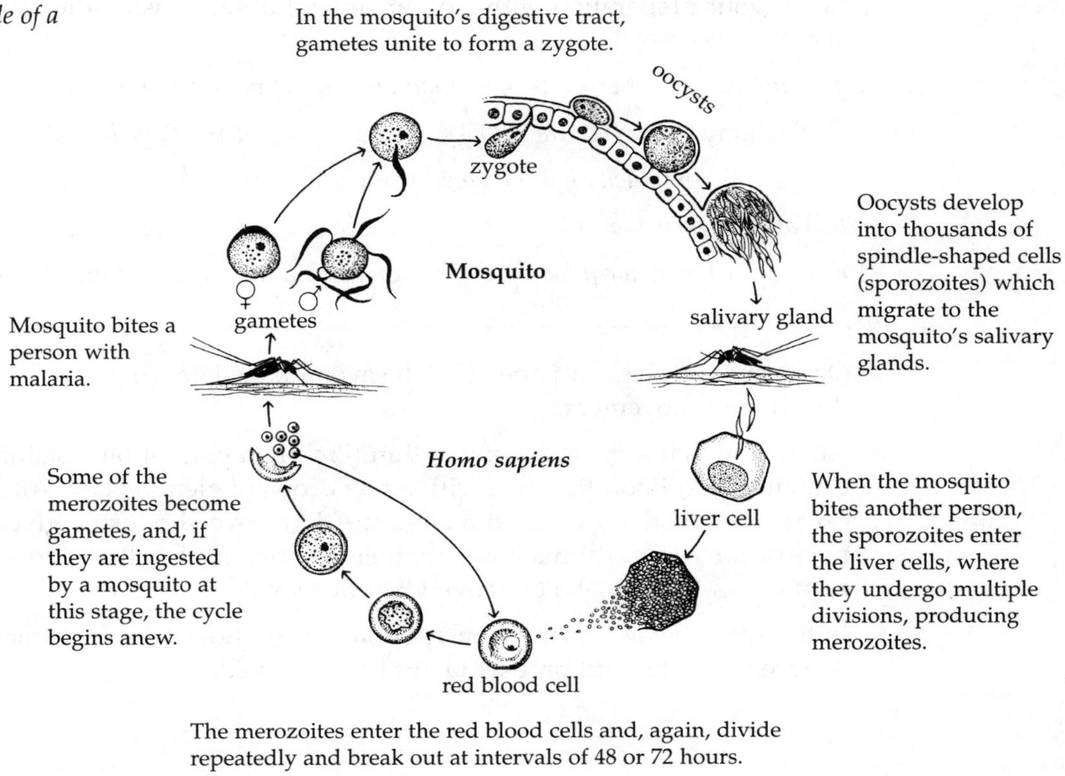

In the mosquito's digestive tract, gametes unite to form a zygote.
oocysts
zygote
Oocysts develop into thousands of spindle-shaped cells (sporozoites) which migrate to the mosquito's salivary glands.
Mosquito
♀
♂
gametes
salivary gland
Mosquito bites a person with malaria.
Homo sapiens
Some of the merozoites become gametes, and, if they are ingested by a mosquito at this stage, the cycle begins anew.
liver cell
When the mosquito bites another person, the sporozoites enter the liver cells, where they undergo multiple divisions, producing merozoites.
red blood cell
The merozoites enter the red blood cells and, again, divide repeatedly and break out at intervals of 48 or 72 hours.

How does Plasmodium's *lack of motility affect the mechanism required for infection in humans?* ______________________

____

👁 EXERCISE G | Symbiosis in the Termite: A Study of Flagellates

A very good source of protozoans is the gut of a termite. Termites are well known for their ability to eat their way through wooden buildings. However, these organisms cannot, on their own, digest the cellulose of wood, even though it is the major constituent of their diet. The guts of termites are well populated by flagellated protozoans, and at least some of these have the enzyme required to digest cellulose. The termite–flagellate relationship is an example of **symbiosis.**

Symbiosis is the close living relationship of two organisms. A symbiotic relationship that benefits both organisms, as is seen in the termites and flagellates, is called **mutualism.** For example, one organism may obtain nourishment from another and in turn provide a nutrient that is necessary to the host. In some mutualistic relationships, each organism becomes so dependent on substances or services provided by the other that neither can survive alone; this is *obligatory mutualism.*

In another type of symbiosis called **commensalism,** one organism provides something of value to the second, but is neither harmed nor helped by the relationship.

The most extreme form of symbiosis is **parasitism.** The parasite is detrimental to its host, sometimes to the point of causing the host's death.

ⅠⅠⅠⅠⅠ Procedure ⅠⅠⅠ

1. Use a dropper to place a drop of insect Ringers (a saline solution isotonic to the tissues of the termite) onto a glass slide.

2. Place a termite into the Ringers and observe using the dissecting microscope.

3. Use two dissecting needles or fine forceps to open the termite. Place one point at the posterior end and one at the anterior end and pull in opposite directions.

4. The intestine is long and tubular. Locate it and move the remaining material to one side.

5. Use the dissecting needles to tear open the intestine.

6. Cover your preparation with a coverslip and observe using the compound microscope at 4×
   and 10× .

   a. *What do you observe? Is more than one type of organism present?* _____ *Do you see*

   Trichonympha *(Figure 22G-1)?* _____ *How many flagella are present on each organism?*

   _____ *In which phylum would you place these organisms?* _____

   b. *Describe their movements.* _____

   c. *Suggest the major function of* Trichonympha *in termites. (Hint: Consider the diet of the termite.)*

   _____

7. On a separate sheet of paper, sketch several flagellates. Include flagella and indicate the direction of movement.

8. Allow your slide to air dry into a thin film, then place it on a staining trough or other suitable container and flood the slide with a few drops of giemsa stain. After 15 minutes, rinse off the excess with distilled water and air dry. Add a coverslip, using glycerol, and examine under the microscope again. Estimate the number of different flagellates present. Compare these to Figure 22G-1 and make tentative identifications.

9. In the space below, record your speculations on how symbiotic relationships such as that between termites and protists might have evolved.

Figure 22G-1 *Some flagellated protists found in the termite gut.*

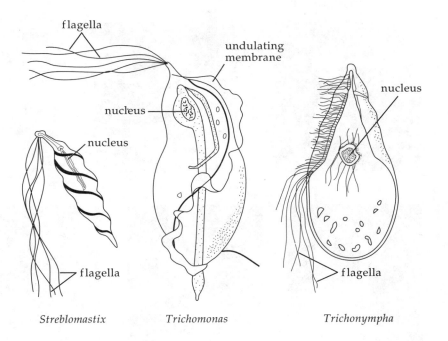

Streblomastix   *Trichomonas*   *Trichonympha*

d. What significant ecological role do termites and their symbionts perform (aside from destroying human-built wooden structures)? _____

_____

e. Is their feeding niche shared by other organisms? _____ Name some examples.

_____

FUNGUS-LIKE PROTISTS

✔ **EXERCISE H** **Plasmodial Slime Molds**

The slime molds are sometimes referred to as the "lower fungi" because they are multinucleate during part of their life cycle. Also, there are some ways in which slime molds resemble amoebas.

Some slime molds, known as **plasmodial slime molds** (phylum Myxomycota), are multinucleate masses of streaming protoplasm (Figure 22H-1). Others, the **cellular slime molds** (phylum Dictyostelida

Figure 22H-1 *The plasmodial slime mold,* Physarum.

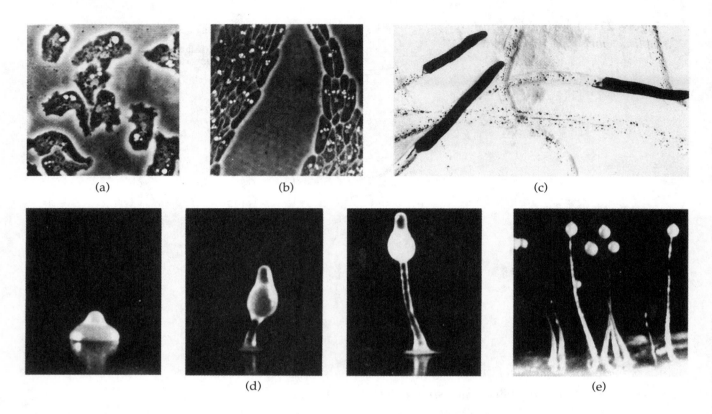

(a)   (b)   (c)

(d)   (e)

Figure 22H-2 *Life cycle of a cellular slime mold,* Dictyostelium
discoideum. *(a) Haploid (n) amoebas in the feeding stage. (b) Amoebas
aggregating. (c) Migrating slugs (pseudoplasmodia). (d) At the end of
migration, each pseudoplasmodium (2n) gathers together and begins to rise
vertically, differentiating into a stalk and fruiting body (e). Meiosis restores the
haploid condition.*

and Acrasida), have bodies, or plasmodia, composed of aggregates of small cells called amoebas. These
amoebas retain their identity as individual cells (Figure 22H-2).

‖‖‖ **Objectives** ‖‖‖‖‖‖‖‖‖‖‖‖‖‖‖‖‖‖‖‖‖‖‖‖‖‖‖‖‖‖‖‖‖‖

☐ Describe the structure of a typical plasmodial slime mold, *Physarum.*

☐ Determine behavioral responses of *Physarum* to heat, light, and overcrowding.

‖‖‖ **Procedure** ‖‖‖‖‖‖‖‖‖‖‖‖‖‖‖‖‖‖‖‖‖‖‖‖‖‖‖‖‖‖‖‖‖‖‖‖‖

    1. Examine the well-developed body (**plasmodium**) of *Physarum* on demonstration. This multi-
nucleate mass of protoplasm ingests food particles as it moves across the agar surface of the
Petri dish. In the natural environment, plasmodia can be found on rotting stumps or fallen
logs, usually underneath the bark.

    2. Study the reproductive structures (**sporangia**) of those species on demonstration. The onset of
unfavorable conditions—for example, the lack of water or food—triggers the formation of
sporangia.

       a. Are these reproductive structures plant-like or animal-like? _____

    3. Obtain a piece of filter paper containing dried plasmodial material. Wet the paper and place it
on the agar in the middle of a Petri dish.

4. Place two or three oat flakes around the material.

5. Take this preparation home with you and observe what happens during the next week.

   b. When the slime mold runs out of space, what happens? _____

   _____

   c. Keep the slime mold in one place and observe the direction in which it moves. To what stimulus does it seem to be responding? _____

   d. Punch a hole in the plasmodium. What happens to it? _____

   e. Slightly warm one-half of the dish. What happens to the slime mold?

   _____

   f. Illuminate one-half of the dish with your desk lamp and cover the other half with aluminum foil.

   Does Physarum *react to light? Why or why not?* _____

6. Summarize your experiments and their results on a separate sheet of paper. Be sure to include the hypotheses that determined the experimental procedures you used. Hand in your report, if assigned.

✓ EXERCISE I   Water Molds

Oomycetes, the "egg fungi" (phylum Oomycota), get their name from their sexual reproduction cycle in which large nonmotile eggs are produced inside a special structure called an **oogonium** (Figure 22I-1). Egg fungi are also called by several common names, including water molds, algae-like fungi, and downy mildews.

Figure 22I-1 *(a) Asexual reproduction in the water mold* Achlya ambisexualis *is characterized by production of motile flagellated zoospores from sporangia located on the tips of the hyphae. (b) Sexual reproduction is characterized by formation of large nonmotile eggs within an oogonium (a type of female gametangium) to which tubal outgrowths on the antheridia (male gametangia) fuse, allowing sperm to enter the oogonium and fertilize the eggs. Zygotes develop thick coats and are called oospores.*

(a)

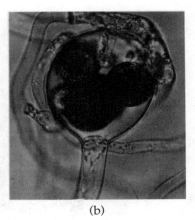
(b)

Unlike other fungi, with cell walls composed of chitin, the cell walls of oomycetes are made up of cellulose. Another distinction is that the spores formed by oomycetes during asexual reproduction are flagellated, a distinctly protistan characteristic.

You are probably more familiar with the oomycetes than you realize. Since these molds can attack diseased or dying fish, you may have experienced a problem with *ick*—a disease caused by water molds—

in your aquarium. One oomycete, *Phytophthora,* was responsible for the potato famine in Ireland in the 1850s. Another, *Plasmopara,* almost destroyed the French wine industry.

||||| **Objectives** ||||||||||||||||||||||||||||||||||||||

☐ Identify the structural characteristics typical of Oomycota, and distinguish between the structures of asexual and sexual reproductive phases.

☐ Explain the basis for the name "Oomycota."

||||| **Procedure** |||||||||||||||||||||||||||||||||||||||

1. In the demonstration area, you will find water molds growing on dead seed. Note the cottony mass of filaments, or hyphae, which constitute the mycelium.

   a. *Are these oomycetes parasitic or saprophytic?* _____

2. Examine (at 10× and 40×) a wet-mount slide of a portion of the water mold mycelium on demonstration. Look for denser areas at the tips of the hyphae. These individual cells, called **zoosporangia,** produce motile zoospores asexually.

3. Examine a prepared slide of the water mold *Saprolegnia.* Identify oogonia, antheridia, and zoosporangia. Use the space below to draw and label your observations.

ALGAE

A number of distinct lines of simple, photosynthetic eukaryotes, the **algae,** evolved more than 450 million years ago. All modern algae have chlorophyll *a* as their main photosynthetic pigment, and they also have accessory pigments. Distinctions are made among divisions of algae based on the type of accessory photosynthetic pigments; the nature of the stored food reserve; the composition of the cell wall, if present; and whether the body (**thallus**) is unicellular or multicellular.

✔ **EXERCISE J**   **Studying and Classifying Algae**

You will study six phyla of algae. Representatives range in size from microscopic to extremely large. Both freshwater and marine species exist.

||||| **Objectives** ||||||||||||||||||||||||||||||||||||||

☐ Identify types of algae based on morphological characteristics.

||||| **Procedure** |||||||||||||||||||||||||||||||||||||||

Obtain fresh material or prepared slides to study representatives of the six divisions of algae. Refer to Table 22J-1 for characteristics and methods of observation.

Table 22J-1   Characteristics of Algae

Phylum/Representative	Characteristics/Method of Observation
Euglenophyta* *Euglena* (a)	Unicellular. True eye-socket algae. **Flagellum** attached within reservoir, distinct orange-red eyespot adjacent to the flagellum. A flexible protein layer (pellicle) rather than a rigid cell wall allows the organism to change its shape. Many bright green chloroplasts. Locomotion: swimming, creeping, or floating. <u>Method of Observation</u> Make a wet mount of *Euglena.* Mix a drop of Protoslo with culture.
Chrysophyta *Diatoms* (b)	Diatoms only: Unicellular or chains of rod (**pennate**) or circular (**centric**) shapes. Cell walls of silica with numerous holes. Walls make two overlapping halves (**thecas**) that fit together like the halves of a Petri dish. Cells are brownish yellow. Locomotion: attached, gliding, or floating. <u>Method of Observation</u> Make a wet mount of diatomaceous earth, if available, or use material collected from stream rocks.
Pyrrophyta (Dinophyta) *Peridinium Gymnodinium* (c)	Unicellular. Spinning flagellates. All members are biflagellate and motile. One flagellum wraps around the middle of the cell and allows it to spin; another flagellum trails and pushes the cell along. Cell wall composed of many interlocking plates, giving an armored appearance. Brownish color. Locomotion: floating or swimming. <u>Method of Observation</u> Study prepared slide of *Peridinium.*
Phaeophyta (brown algae) *Fucus Laminaria Ectocarpus* (d)	Multicellular, small to massive. Organisms called rockweeds, kelps, and brown seaweeds. All marine. Body differentiated into **blade, stipe, holdfast,** and, in some, **air bladder floats.** Cell walls contain mucilage, algin, used commercially as additive to food and cosmetics and as a thickener for some ice cream. Almost all are attached to the bottom (**benthic**) in coastal marine environments. <u>Method of Observation</u> Study preserved specimens.

(continued on next page)

Table 22J-1   *(continued)*

Phylum/Representative	Characteristics/Method of Observation
Rhodophyta (red algae) *Corallopsis Dasya Gigartina* (e)	Multicellular filaments to medium-sized seaweeds. Red-colored. Predominantly marine. More branched than brown algae. Cell walls in certain species contain mucilage—carrageenan or agar used to give a smoother, thicker texture to many milk products and to make bacterial growth media Almost all are benthic and attached. Method of Observation Study preserved specimens.
Chlorophyta (green algae) *Chlamydomonas* *Spirogyra* *Gonium* *Volvox* *Zygnema* *Stigeoclonium* *Ulva* (see below)	Unicellular, colonial, filamentous. Green color, starch inside plastids. Many different forms adapted for attachment on benthic substrate or for swimming or floating in planktonic environments. Mostly freshwater types, some marine. Method of Observation Study fresh specimens or prepared slides as available.

*May be included among the flagellated protozoans.

✔ **EXERCISE K**   **Diversity Among the Green Algae: Phylum Chlorophyta**

It is thought that the ancestor of land plants was a green alga. Several evolutionary trends are obvious among the **Chlorophyta,** including:

* Increase in size accompanied by **cell differentiation.** Within a group of cells, certain cells have specific functions; individual cells do not act independently.

* Sexual reproduction. Among the algae are three types of sexual reproduction (listed from the most primitive to the most advanced):

Isogamy Male and female gametes look exactly like (isogametes); both are motile.

Anisogamy Also called heterogamy. Male and female gametes look alike except that the female gamete (egg) is larger; both are motile.

Oogamy The male gamete (sperm) is small and motile. The female gamete (egg) is large and nonmotile.

▮▮▮▮▮ **Objectives** ▮▮▮▮▮▮▮▮▮▮▮▮▮▮▮▮▮▮▮▮▮▮▮▮▮▮▮▮▮▮▮▮▮▮▮▮▮▮▮

☐ Compare and contrast the representatives of the phylum Chlorophyta.

☐ Describe the progression in complexity of form observed among the green algae.

IIIII **Procedure** IIIIIIIIIIIIIIIIIIIIIIIIIIIIIIIIIIIIIIIIIII

Use prepared slides or, if fresh material is available, make temporary mounts of the following organisms. Observe the progression in size and complexity illustrated in the green algae. Sketch the organisms in the spaces provided.

Chlamydomonas (class Chlorophyceae)
Unicellular thallus.

Gonium (class Chlorophyceae)
Spherical colony made up of 4 to 32 cells, depending on the species.

Volvox (class Chlorophyceae)
Spherical colony made up of 500 to 50,000 cells, depending on the species.

Zygnema (class Chlorophyceae)
Simple, unbranched filament (cell division occurs in a single plane).

Stigeoclonium (class Chlorophyceae)
Branched filament (cell division occurs in two planes).

Oedogonium (class Chlorophyceae)
Simple unbranched filament with netlike chloroplast. Oogamous.

Ulva (class Ulvophyceae)
"Sheets" that are two cells thick (cell division occurs in three planes).

Ulothrix (class Ulvophyceae)
Simple unbranched filament with only the basal cell differentiated into a holdfast. Isogamous.

Spirogyra (class Charophyceae)
Unbranched green alga with one or more ribbonlike chloroplasts helically arranged. Isogamous.

Desmids (class Charophyceae)
Unicellular (some multicellular). Cell wall is in two sections with a narrow construction, the isthmus, between them.

✔ **EXERCISE L** **Recognizing Protists Among the Plankton**

Plankton is a general term for small (mostly microscopic) aquatic organisms found in the upper levels of water where light is abundant. Plankton includes both plant-like photosynthetic forms (*phytoplankton*) and animal-like heterotrophic forms (*zooplankton*). A sample from enriched natural water, such as a fish pond, is an excellent source of algae and protozoans, as well as microscopic animals.

ıııı **Objectives** ıııııııııııııııııııııııııııııııı

☐ Identify diverse types of flagellates, ciliates, sarcodines, and algae.

ıııı **Procedure** ıııııııııııııııııııııııııııııııııııı

1. Place a small drop of the plankton sample on a slide and add a coverslip. Your instructor will provide some illustrations of types of organisms you are likely to see.

2. Identify as many organisms as possible. Various types of algae (diatoms, desmids, and, possibly, filamentous green algae) may be visible. Some of the flagellates you find may belong to the algal division Euglenophyta rather than Zoomastigophora, but they, too, illustrate the way in which flagellates, in general, move. Study this movement carefully.

   a. *Can you find any algae?* _____ *To which divisions might these algae belong?*

   _____

   How do you distinguish between algae of different divisions?

   _____

   b. *Are there any sarcodines?* _____ c. *Are there any ciliates?* _____

   d. *Can you find any multicellular rotifers? (These are members of the phylum Rotifera, one of the animal phyla.)* _____

3. Sketch representatives in the space provided below.

   e. *What might be the role of the plankton in the food chain of an ocean or lake?*

   _____

   f. *Both phytoplankton and zooplankton can migrate to various depths in a lake or ocean. What might cause these organisms to surface or to move to greater depths within a lake or ocean environment?*

   _____

Laboratory Review Questions and Problems

1. List three characteristics that distinguish prokaryotes from protists.

2. Why are the cyanobacteria included with the bacteria?

State the differences that distinguish cyanobacteria from green plants.

3. Fill in the table below, giving the distinguishing characteristics of each protozoan phylum and an example of a representative organism studied in the laboratory.

Protozoan Phylum	Distinguishing Characteristics	Example
Zoomastigophora		
Rhizopoda		
Ciliophora		
Sporozoa		

4. Slime molds and water molds are included among the Protista rather than in the kingdom Fungi. Why?

5. Fill in the table below, giving the distinguishing characteristics of each algal phylum and an example of a representative organism studied in the laboratory.

Algal Phylum	Distinguishing Characteristics	Example
Euglenophyta		
Chrysophyta		
Pyrrophyta		
Phaeophyta		
Rhodophyta		
Chlorophyta		

6. *Euglena* can be classified among the algae or among the flagellated protozoans. Explain why this organism is so difficult to classify.

7. A friend comments that you have more in common with an archaean than with a bacterium. Is this true? Why?

Diversity—Fungi and the Nontracheophytes

OVERVIEW

Multicellularity allows for great increases in the size of organisms and for the specialization of their parts. The advantages of multicellularity have allowed organisms belonging to the kingdoms Fungi and Plantae to develop the specific structural modifications, life styles, and unique reproductive mechanisms required to adapt to and succeed in a variety of habitats.

During this laboratory period you will examine representatives of the **Fungi** and begin your study of the Plantae by examining the **nontracheophytes**—plants that lack special vascular tissues for the distribution of water, minerals, and photosynthetic products.

STUDENT PREPARATION

Prepare for this laboratory by reading the text pages indicated by your instructor. Familiarizing yourself in advance with the information and procedures covered in this laboratory will give you a better understanding of the material and improve your efficiency.

Last week you were given three plastic bags and a slice of bread. You exposed the pieces of bread to different environmental conditions (heat, cold, light, or dark) for approximately 6 days. You should also have formulated a hypothesis stating how you expected the experimental conditions you chose to affect fungal growth.

In Exercise A, Extending Your Investigation (p. 23-4), state your hypothesis and summarize your method of treatment (experimentation), results (observations), and conclusions supporting or refuting your supposition. Bring your experiment on fungal growth to the laboratory and place your materials on the demonstration table for other students to observe.

On campus or in a nearby woods, you may be able to find examples of fungi, lichens, and mosses. Bring them to class for identification. If you find something you do not recognize, bring it too. You and your classmates can try to identify unknown plant organisms during this laboratory period.

PART I   KINGDOM FUNGI

Members of the kingdom Fungi were long classified among the plants, but they are so unlike any other plant group that taxonomists now assign them to a separate kingdom. Like other eukaryotes, fungi probably originated from an ancestral heterotrophic protist. Most fungi are multinucleate or multicellular, although some, including the yeasts, are uninucleate and unicellular. Unlike plants, the fungi lack chlorophyll. They have cell walls composed of cellulose or **chitin.**

Because fungi lack chlorophyll and therefore cannot manufacture their own food, they feed either on decomposing organic matter (**saprophytes**) or on living organisms (**parasites**). Fungi maximize their contact with sources of nutrients by making branched or unbranched threadlike filaments called **hyphae,** which form a spreading mass called a **mycelium.**

Both asexual and sexual reproduction are common among the fungi and result in the production of **spores.** Elaborate mechanisms of spore production, including formation of many different, and sometimes bizarre, types of **fruiting structures,** protect fungal spores and promote their dissemination to new habitats.

Fungi can be grouped into four phyla based on their basic structure and patterns of reproduction, particularly sexual reproduction. The four phyla include: the Zygomycota, zygospore fungi; the Ascomycota, sac fungi; the Basidiomycota, club fungi; and the Chytridiomycota, chytrids. Because the imperfect fungi (deuteromycetes) are probably polyphyletic (all other phyla of fungi are monophyletic), they are not assigned phylum status.

EXERCISE A   Phylum Zygomycota

If you leave a piece of bread or other bakery product covered and at room temperature for a while, a fuzzy gray or black mold will appear. This fungus, *Rhizopus stolonifer,* a common bread mold, is representative of the **zygomycetes,** members of the phylum Zygomycota. This group of fungi is characterized by the formation of **zygospores**—special resting structures composed of a zygote surrounded by a thick protective wall. The zygospores germinate to form a fruiting structure, the **sporangium,** which produces spores by meiosis. Accumulations of these dark-colored sporangia give *Rhizopus* its gray-black color.

ııııı **Objectives** ıııııııııııııııııııııııııııııııııııı

☐ Identify the structural characteristics typical of zygomycetes, and distinguish between asexual and sexual reproductive structures.

☐ Explain the basis for the name "Zygomycota."

ııııı **Procedure** ıı

1. Locate the following structures in the life-cycle diagram of *Rhizopus stolonifer* (Figure 23A-1) and familiarize yourself with the function or importance of each structure.

    Hypha Strands (hyphae) composing the fungal body are multinucleate (**coenocytic,** containing many haploid nuclei) in most zygomycetes. Some hyphae arch upward and are called **stolons.** These form hyphal **rhizoids** wherever they touch the substrate (surface on which the fungi are growing).

    Mycelium The fungal body composed of a mass of hyphae.

    Sporangium The structure responsible for producing spores, either asexually by mitosis from hyphal cells (**sporangiospores**) or sexually by meiosis from the **zygospore** (meiospores). A long stalk raises the sporangium above the surface, where air can disperse the spores.

    Gametangium Haploid cell formed at the tip of a hypha. Gametangia fuse to produce the zygote (Figure 23A-2a).

    Zygospore Zygote formed by the fusion (**syngamy**) of two gametangia and covered by a thick protective wall. Since the zygote is not derived from the fusion of true gametes (egg and sperm) but from the fusion of gametangia, it is called a zygospore. The zygospore germinates to form a sporangium which produces spores by meiosis (Figure 23A-2b).

2. *Without opening the dishes,* use a dissecting microscope to identify as many structures as possible in the available living material. Note the black "dots" on older parts of the mycelium.

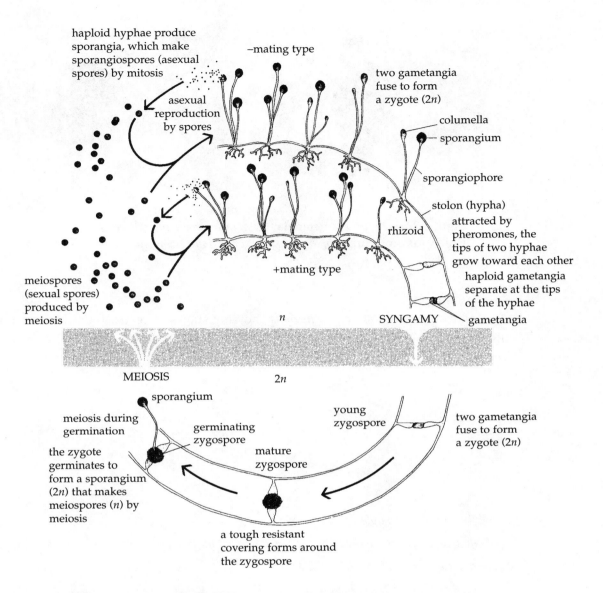

Figure 23A-1 *Life cycle of the zygomycete* Rhizopus stolonifer. *Sexual reproduction occurs between different mating strains, traditionally referred to as the + and − mating strains, even though they are morphologically alike. Asexual reproduction occurs by the mitotic production of spores within haploid sporangia.*

These are sporangia. Observe the upright stalks (**sporangiophores**) bearing the sporangia. Locate the stolons and rhizoids. Draw and label these structures in the space below.

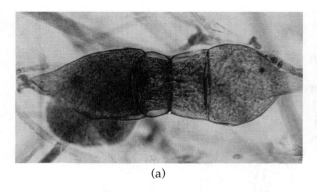
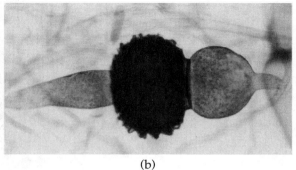

(a)   (b)

Figure 23A-2   (a) Gametangia, *the gamete-producing structures of*
Rhizopus stolonifer. *(b) A zygospore, or sexual resting spore (the dark mass in
the center).*

3. Examine a prepared slide of *Rhizopus* showing stages of sexual reproduction. Locate the
gametangia and zygospores. Draw and label these structures in the space below.

a. *Is the mycelium haploid (n) or diploid (2n)?* ———————

b. *What types of products (n or 2n) are produced by mitosis?* ———————— *By meiosis?* ————————

c. *Are true gametes formed?* ————————

d. *By what process are gametangia formed?* ————————————————————————

4. On Figure 23A-1, circle and label the portion of the life cycle involving sexual reproduction;
now do the same for the portion of the life cycle involving asexual reproduction.

EXTENDING YOUR INVESTIGATION: CONDITIONS FOR FUNGAL GROWTH

Last week you were asked to choose experimental conditions for growing fungus on bread and to
formulate a hypothesis about how these conditions would affect fungal growth. What was your
hypothesis?

HYPOTHESIS:

NULL HYPOTHESIS:

What did you **predict** would happen under the conditions that you chose?

What was the **independent variable?**

What was the **dependent variable?**

What experimental procedure did you use to conduct your investigation? Describe it in the space below.

PROCEDURE:

OBSERVATIONS AND RESULTS:

What type of mold did you find growing on the bread? Was *Rhizopus* present? Examine the slices of bread brought to the laboratory by your classmates. Each person treated the bread in a different experimental manner. In the chart that follows, indicate the conditions tested by your classmates and state the results, growth or no growth.

Condition Tested	Growth or No Growth

Based on these data, what conditions favor the growth of mold or fungi?

From your observations, why do you suppose that molds are so common?

Do your results and those of your classmates support your hypothesis?

Your null hypothesis?

EXERCISE B   Phylum Ascomycota

Ascomycetes, or sac fungi, are often referred to (along with the basidiomycetes) as "higher fungi" because their hyphae are made up of uninucleate cells partitioned by cell walls (septate), whereas the hyphae of Zygomycota are coenocytic (multinucleate).

Ascomycetes include some familiar fungi such as morels, truffles, *Sordaria, Neurospora,* and yeasts. One member of this division, *Claviceps,* produces a mycelium known as ergot on grasses such as rye grass. Perhaps one of the most well known ergot-derived products is LSD. Yeasts produce ethyl alcohol by the process of fermentation and thus are important in the production of beer and wine. Chestnut blight, Dutch elm disease, apple scab, brown rot in fruit, and powdery mildew infections are serious plant diseases caused by the ascomycetes. There are approximately 30,000 known species of ascomycetes.

Asexual reproduction in the ascomycetes occurs by the mitotic production of haploid **conidiospores,** which are produced in long chains at the end of specialized hyphae called **conidophores.**

Sexual reproduction in most ascomycetes results in the development of an **ascus** (derived from the Latin word for "sac") containing four to eight **ascospores** (four are produced by meiosis; additional spores are the result of mitosis). A group of asci are usually found together in a fruiting body known as an **ascocarp** (Figure 23B-1). The ascocarp occurs in a variety of shapes, by which different ascomycetes can be identified. Species that have an ascocarp are collectively called *euascomycetes* (true ascomycetes); those without ascocarps are called *hemiascomycetes* (half ascomycetes).

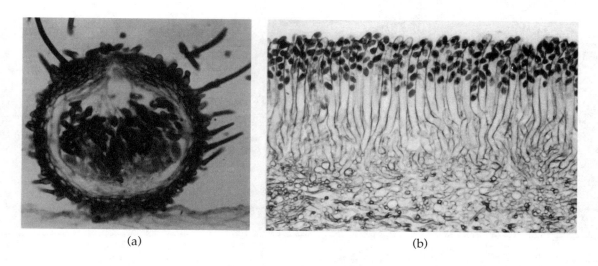

(a)   (b)

Figure 23B-1   *(a) Ascocarp of* Chaetomium erraticum *showing enclosed asci and ascospores. Note the small pore at the top. (b) A section through a layer of asci on the inner surface of an ascocarp of* Morchella.

ııııı Objectives ıııııııııııııııııııııııııııııııııııııı

☐ Identify the structures typical of ascomycetes.

☐ Distinguish between mechanisms of asexual and sexual reproduction.

☐ Explain the basis for the name "Ascomycota."

ııııı Procedure ıı

1. Use a dissecting microscope to examine living material from *Sordaria,* or observe prepared slides using the 40× objective on your microscope. In the space below, sketch and label the fruiting bodies (ascocarps), asci, and ascospores of the ascomycete *Sordaria.*

2. Remove several of the small, black, round fruiting bodies from a culture of *Sordaria* and prepare a wet-mount slide. Press on the coverslip with a small cork to squash the ascocarps and release the asci (Figure 23B-2). Study the slide at high power (40× objective). In the space next to Figure 23B-2, draw and label your observations.

Figure 23B-2   *Sordaria ascospores within asci.*

a. *How many ascospores are present in each ascus?* _____

b. *By what division process (or processes) were the ascospores produced?*

_____

3. Obtain a drop of yeast culture and make a wet-mount slide. Notice the buds on some of the cells. Instead of having a hyphal filament, yeasts are unicellular and can reproduce asexually by budding or sexually by the production of asci (Figure 23B-3a, b).

Figure 23B-3   *Yeasts. (a) Budding cells of bread yeast,* Saccharomyces cerevisiae. *(b) Asci with ascospores of* Schizosaccharomyces octosporus.

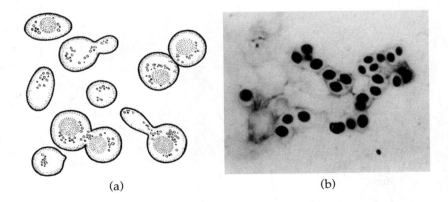

(a)   (b)

4. Examine a prepared slide of *Schizosaccharomyces* on demonstration. Identify an ascus containing ascospores (Figure 23B-3b).

c. *How many ascospores are present in each ascus?* _____

✔ **EXERCISE C** | **Phylum Basidiomycota**

The **basidiomycetes,** or club fungi, have a septate mycelium, as do the ascomycetes, but they differ from the ascomycetes in having sexual spores (**basidiospores**) borne *externally* on a club-shaped structure, the **basidium,** instead of within a sac.

The fruiting body of the basidiomycetes is the **basidiocarp.** The mature basidiocarp develops a large number of pores or gills on its underside. These gills contain numerous club-shaped basidia, single cells that produce basidiospores (Figure 23C-1). Basidiocarps come in many different sizes, shapes, and colors. Some, such as the mushrooms with which you are familiar, are edible.

Rusts, smuts, puffballs, toadstools, and shelf fungi are also members of the phylum Basidiomycota. Many species form **mycorrhizal** associations with plants, in which case a symbiotic relationship develops between fungal hyphae and plant roots, providing both the plants and fungi with important nutritional elements.

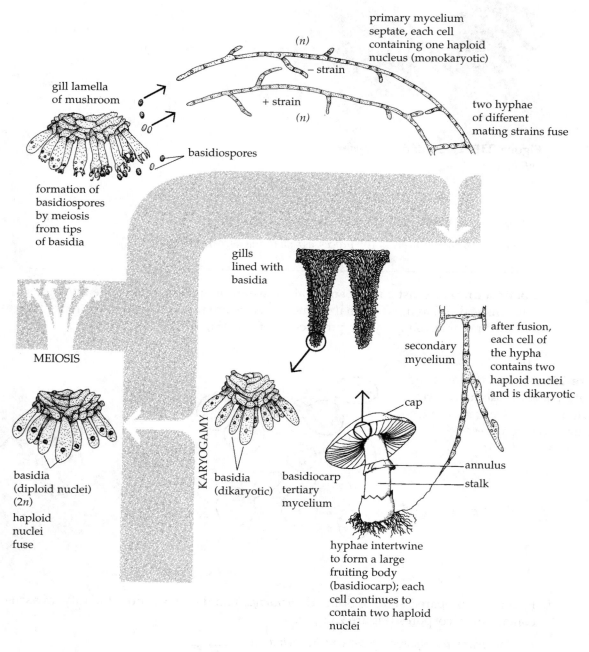

Figure 23C-1 *Generalized life cycle of a basidiomycete.*

IIIII **Objectives** IIIIIIIIIIIIIIIIIIIIIIIIIIIIIIIIIIIII

☐ Identify the structures typical of basidiomycetes.

☐ Describe how a "mushroom" is formed.

IIIII **Procedure** IIIIIIIIIIIIIIIIIIIIIIIIIIIIIIIIIIIIII

1. Obtain a fresh edible mushroom, *Agaricus,* and examine it carefully. Identify the parts described below and locate these on the life-cycle diagram (Figure 23C-1).

   Cap The umbrella-shaped portion of the fruiting body (basidiocarp).

   Gills Radiating strips of tissue (**lamellae**) on the undersurface of the cap; basidia form on the surface of the gills.

   Basidia Club-shaped, spore-producing structures on the surface of the gills (Figure 23C-2a).

   Basidiospore A spore produced by meiosis on the outside of a basidium (Figure 23C-2b).

   Stalk The upright portion of the fruiting body that supports the cap—a mycelium composed of many intertwined hyphae.

   Ring (or **annulus**) A membrane surrounding the stalk of the fruiting body at the point where the unexpanded cap was attached to the stalk.

Figure 23C-2 *(a) Section through the gills of* Coprinus. *The dark margins constitute the layer of developing basidia. (b) Mature basidiospores attached to a basidium.*

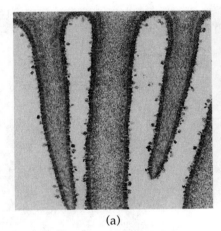

(a)

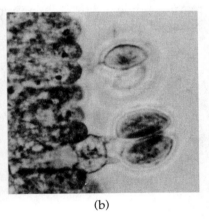

(b)

2. Remove a small portion of several gills. While holding them together, cut a very thin cross section with a razor blade. Put sections in a drop of water on a slide and cover this with a coverslip. You should be able to see basidia on the surface of the gills. Study your slide at high power (40× objective).

   a. *Can you see the hyphae making up the thickness of the gills?* _____ *Are they septate?* _____

      Are the basidia club-shaped? _____

3. For a clearer view of the reproductive structures of basidiomycetes, examine a prepared slide of a cross section through the cap of the basidiomycete *Coprinus.* Find basidia and basidiospores (Figure 23C-2a, b). In the space below, draw and label your observations.

4. Note the diversity of structures in other examples of basidiomycete fruiting bodies on demonstration. Draw several of these in the space below.

✔ EXERCISE D   Phylum Chytridiomycota

Chytridiomycota is the most ancient of the fungal phyla (Figure 23D-1). Chytrids are often classified among the protists, along with the Oomycetes and other water molds. This is because they have flagellated spores and gametes—the only flagella present in the kingdom Fungi. Molecular evidence indicates, however, that they are monophyletic with the fungi. They were probably the first to diverge from a common fungal ancestor. All fungi in the other branch lost their flagella.

Chytrids are either parasitic or saprobic. Most live in fresh water or in damp soil, but some are marine. Some are unicellular, while others take the form of branched chains. They can reproduce sexually or asexually. *Allomyces*, a common chytrid, displays alternation of generations. See Figure 23D-2.

Figure 23D-1   *Chytridium confervae, a common chytrid. You can see the slender rhizoids extending downward.*

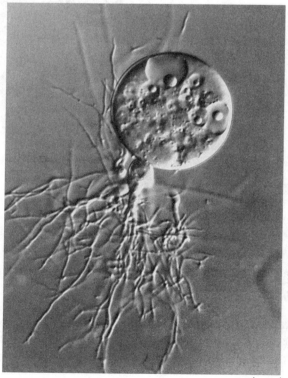

10 μm

▐▐▐▐▐ Objectives ▐▐▐▐▐▐▐▐▐▐▐▐▐▐▐▐▐▐▐▐▐▐▐▐▐▐▐▐▐▐▐▐▐▐

☐ Observe life-cycle stages of chytrids.

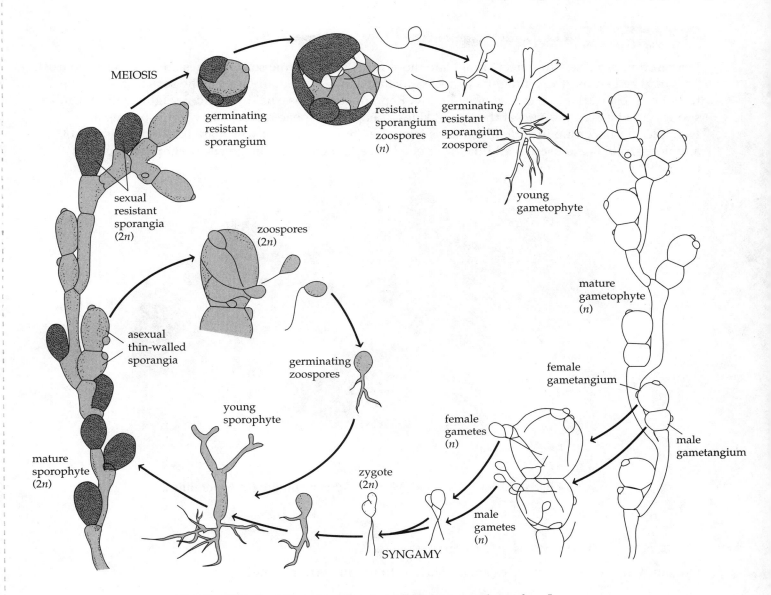

Figure 23D-2 *Life cycle of the chytrid* Allomyces arbusculus. *In
Allomyces, haploid zoospores germinate on the seeds or dead plant material
available. They form male and female gametangia, which produce haploid gametes
by mitosis. Both types of gametes have flagella. The female gametes produce a
chemical that attracts the male gametes, and they fuse to form a diploid zygote.
Cell divisions give rise to a small diploid organism that can then produce diploid
flagellated zoospores, which germinate to form more diploid organisms.
Eventually the diploid organisms form a thick wall around themselves and
become resting zygospores. Meiosis within the resting zygospores produces
haploid zoospores, and the cycle begins again.*

ııııı **Procedure** ıı

1. Examine living cultures of *Allomyces* on demonstration. Identify the resting spores.

2. Remove a sample of the lake water and examine it under the microscope (40×).

   a. Do you see any flagellated zoospores? _____

✔ **EXERCISE E** | **The Deuteromycetes**

Deuteromycetes, the "imperfect" fungi, are those fungi in which the sexual stages are not known to exist. This may be because these fungi have not been completely studied, or because the sexual stages have truly been lost during the course of evolution. Reproduction is asexual by **conidia** (Figure 23E-1). Unlike spores, conidia are produced at the tips or sides of haploid hyphae rather than within sporangia. Examples in this group include blue molds and green molds, some of which are important sources of antibiotics; others are used to add flavor or odor to certain cheeses. The genus **Trichophyton** causes athlete's foot.

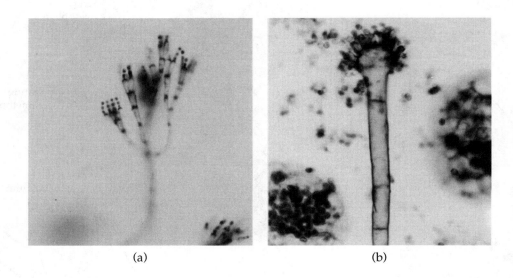

(a)   (b)

Figure 23E-1   *The conidiophores of (a)* Penicillium; *(b)* Aspergillus.

ııııı **Objectives** ıııııııııııııııııııııııııııııııııııı

☐ Explain why the deuteromycetes are considered to be "imperfect" fungi.

ııııı **Procedure** ıııııııııııııııııııııııııııııııııııııı

Examine a prepared slide of *Aspergillus* or *Penicillium*. Study the slide using the compound microscope (40× objective) and identify conida. Notice that most of the conidia dispersed as the slide was made. Air currents perform this task in nature. In the space below, make a labeled sketch of your observations.

a. *Since the imperfect fungi are not known to have a sexual cycle, are conidiospores produced by mitosis or meiosis?*

_____

b. *Of what economic importance is the mold* Penicillium *to the medical industry?*

_____

To the food industry? _____

👁 **EXERCISE F**   **Identification of Collected Fungi**

||||| **Objectives** ||||||||||||||||||||||||||||||||||||||||||

☐ Classify a variety of local fungal specimens by phylum or type.

||||| **Procedure** ||||||||||||||||||||||||||||||||||||||||||

1. Now that you have completed your study of the fungi, try to determine the type of fungus you collected. Give the reasons for your decision.

2. On a 3" × 5" card, write your name, the date, the type of fungus collected, and the place where it was found, and describe the substrate upon which the fungus was growing. Place the card and your material in the demonstration area. Examine the other specimens and decide whether or not you agree with your classmates' identifications. If you disagree with a particular identification, locate the "collector" and see if you can come to an agreement.

   a. *List several ways in which fungi are beneficial to humans.* _____

   _____

   b. *List several ways in which fungi are harmful to humans.* _____

   _____

✔ **EXERCISE G**   **Diversity Among the Lichens**

Lichens are distinct organisms that are actually two organisms in one. The body is made up of certain genera of green algae or cyanobacteria that embed themselves in the mycelium of a fungus (usually an ascomycete or a basidiomycete) and live symbiotically with it. The fungus is the dominant (most prominent) of the two organisms. Thus lichens are usually studied with the kingdom Fungi.

Lichens are found on tree trunks, rocks, and arctic mountain tops, to name just a few locations. Often lichens are the first colonists on bare, rocky areas.

||||| **Objectives** ||||||||||||||||||||||||||||||||||||||||||

☐ Recognize lichens in nature and identify their growth forms.

||||| **Procedure** ||||||||||||||||||||||||||||||||||||||||||

1. Note the growth forms of lichens on demonstration. See if you can identify the following three types (Figure 23G-1a–c): **crustose,** closely encrusting bodies; **foliose,** leafy bodies; **fruticose,** shrubby, branching bodies.

2. Examine a prepared slide of lichen thallus showing algal cells surrounded by fungal hyphae (Figure 23G-1d). In the space below, draw and label your observations.

3. Did you or others in your class collect any lichens?

   a. *Do your lichen collections (or lack or them) reflect the air quality in your community or are they simply indicative of the habitats surveyed for the collections?* _____

   _____

   b. *In which Kingdom would you classify lichens: with the algae?* _____ *with the fungi?*

   _____

   c. *Which lichen characteristics would you use for classification?* _____

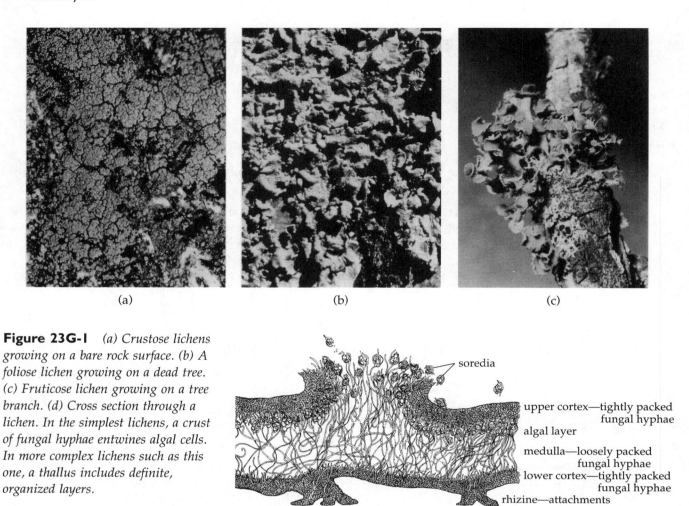

(a)   (b)   (c)

Figure 23G-1 *(a) Crustose lichens growing on a bare rock surface. (b) A foliose lichen growing on a dead tree. (c) Fruticose lichen growing on a tree branch. (d) Cross section through a lichen. In the simplest lichens, a crust of fungal hyphae entwines algal cells. In more complex lichens such as this one, a thallus includes definite, organized layers.*

soredia

upper cortex—tightly packed fungal hyphae
algal layer
medulla—loosely packed fungal hyphae
lower cortex—tightly packed fungal hyphae
rhizine—attachments

(d)

PART II   KINGDOM PLANTAE

NONTRACHEOPHYTES

Members of the kingdom Plantae are photosynthetic autotrophs. Those in existence today can be grouped into 12 phyla. Most are land plants that developed the following adaptations to make the transition from an aquatic to a terrestrial environment:

* A waxy coating or cork layer that retards water loss on plant parts located above ground.

* Pores (called **stomata**) in the aboveground parts for gas exchange.

* Multicellular reproductive organs (**gametangia** and **sporangia**).

* Retention of the fertilized egg within the female gametangium so that the young sporophyte plant is protected.

Not long after the transition to land, plants diverged into two lines—one gave rise to the **nontracheophyte plants** (including mosses, hornworts, and liverworts) and the other to the **tracheophyte plants** (including ferns, gymnosperms, and angiosperms). (*Note:* The term "nonvascular plants" has long been used to describe the mosses, liverworts, and hornworts, but it is misleading because some mosses, unlike liverworts and hornworts, have limited amounts of vascular tissue. The term "bryophytes" has also been used to describe these three phyla of nontracheophytes, but this term is now used to refer to the mosses only.)

In general, tracheophytes (vascular plants) have specialized vascular tissues: **xylem** for the transport of water and minerals upward to the plant body and **phloem** for the distribution of photosynthetic products.

Nontracheophytes absorb moisture mainly through aboveground structures and depend on diffusion for transport.

All land plants have the same type of life cycle involving an alternation of generations. A haploid **gametophyte** plant produces haploid (*n*) gametes by mitosis. These gametes fuse to form a diploid (2*n*) **zygote,** which then grows into a diploid plant, the **sporophyte,** which produces spores (*n*) by meiosis. The haploid spores then develop into haploid gametophytes and the cycle begins again (Figure 23II-1).

Among the nonvascular plants, the gametophyte generation is the most conspicuous and occupies the dominant part of the life cycle.

Before you continue, make sure you understand the importance of the last two paragraphs. Check your understanding by answering the following questions.

a. *In plants, what process produces gametes?* _____

b. *What is the process that produces gametes in animals such as humans?* _____ *Is this the same process that produces gametes in plants?* _____

c. *Why is it beneficial to plants to produce spores in addition to gametes?*

_____

d. *What process produces spores?* _____

e. *Why don't humans produce spores?* _____

f. *What is the dominant part of the life cycle in the nontracheophytes?*

_____

Figure 23II-1   *Generalized life cycle, known as alternation of generations, characteristic of land plants.*

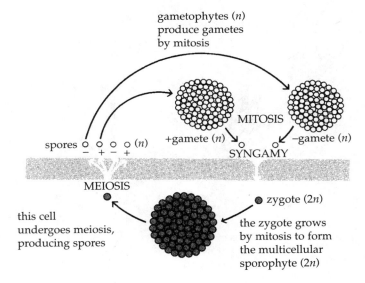

gametophytes (*n*) produce gametes by mitosis

MITOSIS

spores ○ ○ ○ ○ (*n*)   +gamete (*n*)   −gamete (*n*)
         − + − +   SYNGAMY

MEIOSIS

this cell undergoes meiosis, producing spores

zygote (2*n*)

the zygote grows by mitosis to form the multicellular sporophyte (2*n*)

EXERCISE H   **Nontracheophytes—Mosses, Liverworts, and Hornworts**

Although basically terrestrial, nontracheophyte plants are restricted to moist habitats such as creek banks and moist woods. Some are even aquatic, though none are marine. Members of this group are small plants that have structures resembling roots, stems, and leaves, but because most nontracheophytes lack the vascular tissues typical of most land plants (phloem and xylem), they do not, strictly speaking, have "true" roots, stems, or leaves.

A distinct alternation of generations occurs. Both gametophyte and sporophyte are multicellular and visible to the eye, with the gametophyte being more prominent.

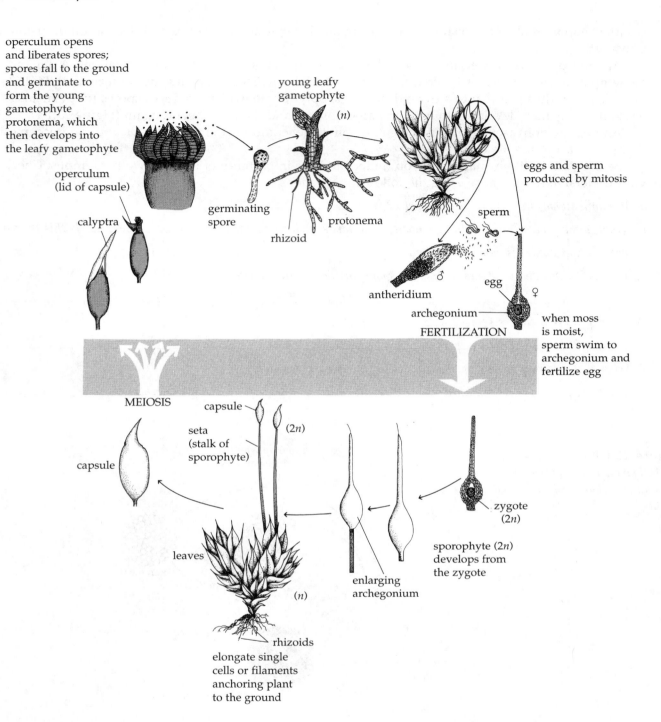

operculum opens and liberates spores; spores fall to the ground and germinate to form the young gametophyte protonema, which then develops into the leafy gametophyte

operculum (lid of capsule)

calyptra

germinating spore

young leafy gametophyte (n)

rhizoid

protonema

eggs and sperm produced by mitosis

sperm

antheridium ♂

archegonium

egg ♀

FERTILIZATION

when moss is moist, sperm swim to archegonium and fertilize egg

MEIOSIS

capsule

seta (stalk of sporophyte)

capsule

capsule (2n)

leaves

(n)

rhizoids elongate single cells or filaments anchoring plant to the ground

enlarging archegonium

zygote (2n)

sporophyte (2n) develops from the zygote

Figure 23H-1 *Life cycle of the moss. The gametophyte generation is the soft, green carpet you walk on in the woods. Gametes are produced by mitosis within specialized gametophyte structures called archegonia (produce eggs) and antheridia (produce free-swimming biflagellated sperm).*

If you examine moss closely at certain times of the year, you will see the sporophyte generation—small hairlike projections topped by capsules (sporangia)—rising above the green leaves of the gametophytes. Spores produced within the sporangia by meiosis fall to the ground and germinate, giving rise to a filamentous or platelike protonema—the new gametophyte generation.

Mosses (phylum **Bryophyta**), liverworts (phylum **Hepatophyta**), and hornworts (phylum **Anthocerophyta**) are the three phyla of nontracheophyte plants.

ııııı **Objectives** ıııııııııııııııııııııııııııııııııııııı

☐ Name the major structures of a moss.

☐ Describe the life cycle of a typical bryophyte.

ııııı **Procedure** ııı

1. Examine fresh moss (phylum Bryophyta) sporophyte and gametophyte material and identify the following structures. Refer to the moss life-cycle diagram (Figure 23H-1) and be sure that you understand the relative importance and function of each structure.

   Gametophyte The leafy green plant of the haploid generation.

   "Leaves" Bladelike structures spirally or alternately arranged around the axis of the moss gametophyte.

   Rhizoids Rootlike structures anchoring the gametophyte.

   Protonema Haploid structure produced by the germinating spore, which gives rise to the gametophyte.

   Sporophyte The body of the diploid generation, consisting of a foot, stalk (**seta**), and capsule (Figure 23H-2).

   Capsule (sporangium) The top portion of the moss sporophyte within which spores are produced. Spores are released through the lid (**operculum**) of the capsule.

   Spores Haploid reproductive structures responsible for the asexual portion of the moss life cycle.

   a. *Why do mosses need a moist environment to reproduce sexually? (Refer to the moss life cycle, Figure 23H-1.)* _____

   _____

Figure 23H-2 *Spore-bearing setae of the hairy moss,* Pogonatum brachyphyllum. *(b) Capsule of the sporophyte of a moss, with the calyptra (the enlarged archegonium) totally removed, revealing the lid, or operculum, of the capsule.*

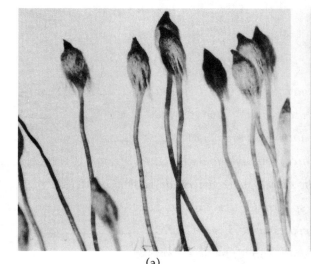

(a)   (b)

Figure 23H-3 *Gametangia of a moss, Mnium. (a) Longitudinal section through female gametangia (archegonia) with eggs. (b) Longitudinal section through male gametangia (antheridia) containing male gametes, sperm.*

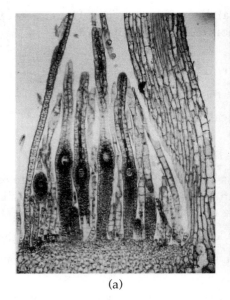
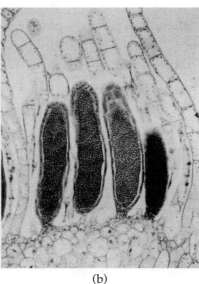

(a)   (b)

b. *How (or when) do mosses growing in dry environments reproduce?*

_____

_____

2. Examine prepared slides of moss antheridia and archegonia (Figure 23H-3) and identify the following structures. (Refer to the moss life-cycle diagram to be sure that you understand their relationship to other gametophyte and sporophyte structures.) Draw and label your observations in the space below.

Antheridium The male reproductive organ in which sperm develop.

Sperm The motile (flagellated) male gamete produced in an antheridium.

Archegonium Female gametangium in which the egg develops.

Egg Nonmotile female gamete produced in an archegonium.

c. *Is water necessary for fertilization?* _____

d. *The archegonium produces* _____.

e. *The antheridium produces* _____.

f. *The archegonium and antheridium are part of the* _____ *generation.*

g. *Is the gametophyte generation haploid or diploid?* _____

h. *The sporophyte structure produces* _____.

i. *Are spores haploid or diploid?* _____

j. *The process that produces spores is* _____.

k. *What do you notice about the shapes of the antheridia and archegonia?*

_____

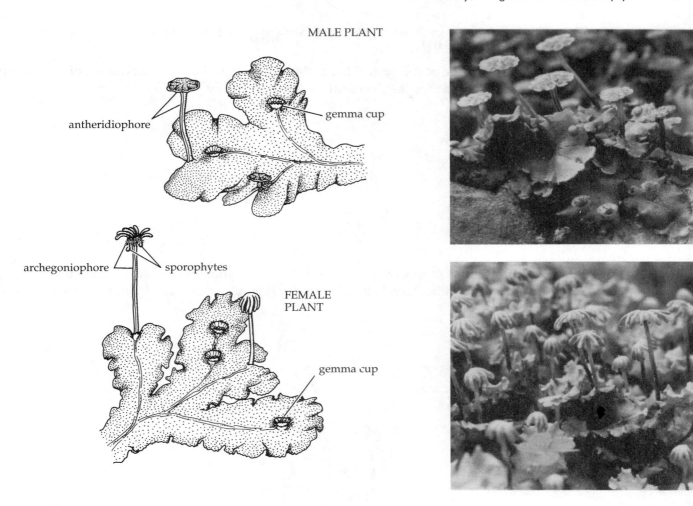

Figure 23H-4   *Gametophyte and sporophyte of the leafy liverwort*
Marchantia. *The male and female gametophyte plants bear raised gametangia.*
Antheridia, which produce sperm, appear on the male plant; archegonia, which
produce eggs, appear on the female plant.

l. How would you describe the relationship between their shapes and their functions?

3. If liverworts (phylum Hepatophyta) are available, examine them carefully (Figure 23H-4). At
   first glance, a liverwort will probably not resemble any other kind of plant you have seen
   before. Notice that the plant body is not differentiated into recognizable roots, stems, or
   leaves. The type of structure you see is called a **thallus.** Liverworts assume one of two
   different forms: **thallose,** in which the plant is flat, ribbonlike, and dichotomously branched;
   or **leafy,** in which the thallus is lobed and leaflike in appearance (Figure 23H-4). More liverworts are leafy than thallose, but the species that you will see is a type of thallose liverwort.

4. Obtain a living liverwort, *Marchantia,* and examine it with a dissecting microscope. The plant
   body is the gametophyte—as in the moss, the gametophyte generation in the liverwort is the
   dominant one. The gamete-producing structures, archegonia and antheridia, appear as small
   palm-tree-like structures on the thallus of male and female plants. On the lower surface of the
   thallus, you will find rootlike structures (**rhizoids**) with which the liverwort adheres to its
   growth site and obtains nutrients. Also, look for **cupules** on the upper surface of the thallus.

These cups contain specialized reproductive structures called **gemmae** used for asexual reproduction (Figure 23H-4).

5. Using your knowledge of the typical life cycle of a moss and your observations of *Marchantia*, draw a typical life cycle for the liverwort in the space below.

6. If hornworts (phylum Anthocerophyta) are available in your laboratory, examine their structure and identify the parts belonging to the sporophyte and gametophyte generations (Figure 23H-5).

Figure 23H-5 *Gametophyte of* Anthoceros, *a hornwort, showing attached sporophytes.*

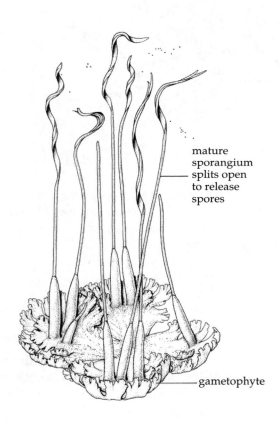

mature sporangium splits open to release spores

gametophyte

Laboratory Review Questions and Problems

1. Fill in the following table to summarize the differences among phyla in the kingdom Fungi.

Phylum	Distinguishing Characteristics	Type of Hyphae	Type of Reproductive Structures
Zygomycota			
Ascomycota			
Basidiomycota			
Chytridiomycota			

2. Why are the deuteromycetes ("fungi imperfecti") not included among the phyla of fungi?

3. Why are the chytrids (phylum Chytridiomycota) unusual among the fungi?

4. Lichens are usually studied with the kingdom Fungi. Is there another kingdom with which they might be studied? Explain.

5. Why can lichens be used as an example of symbiosis?

6. Why is it not correct to refer to mosses as "nonvascular plants"? How are the following terms related: bryophytes, nontracheophytes, nonvascular plants?

7. Draw a typical life cycle for a moss. Which generation is the conspicuous generation?

8. Define the following terms and indicate whether each term can be used to describe mosses, liverworts, or both.

Rhizoids

Sporangium

Thallus

Gemmae

Archegonium

Antheridium

Spores

Diversity— The Tracheophytes (Vascular Land Plants)

OVERVIEW

Despite the variety of their forms and the diversity of the environments to which they have become adapted, all vascular land plants, or tracheophytes, share some characteristics that reveal a common evolutionary organ.

Alternation of generations is a characteristic common to all land plants. As you examine the evolutionary progression of land plants, from nonvascular to vascular forms, you will notice a continuous reduction in the size and complexity of the gametophyte until it ultimately becomes greatly reduced and nutritionally dependent upon a much larger, more complex sporophyte.

Also associated with the evolutionary trend toward sporophyte dominance is a change from the **homosporous** condition (the simplest vascular plants produce only one type of spore) to the **heterosporous** condition (more advanced vascular plants produce two different types of spores—megaspores, which develop into female gametophytes, and microspores, which develop into male gametophytes). Another factor essential to the success of tracheophytes on dry land is the change from unprotected zygotes to the protection of the embryo sporophyte within a seed. In addition, modern tracheophytes contain networks of vascular strands (xylem and phloem). Leaves in the most primitive vascular plants have only single vascular strands.

During this laboratory period, you will study representatives of the many types of vascular plants, both with and without seeds.

STUDENT PREPARATION

To prepare for this laboratory, read the text pages indicated by your instructor. Familiarizing yourself in advance with the information and procedures covered in this laboratory will give you a better understanding of the material and improve your efficiency.

PART I   TRACHEOPHYTES (VASCULAR PLANTS) WITHOUT SEEDS

✔ **EXERCISE A** | **Examining Seedless Tracheophytes**

The tracheophytes without seeds include the following phyla: the Psilophyta, whisk ferns; the Lycophyta, club mosses; the Sphenophyta, horsetails; and the Pterophyta, ferns. Most of the members of the first three phyla are extinct; their characteristics are summed up in Table 24A-1.

Extinct Lycophyta once loomed as giants in the forests of the Carboniferous period. All living lycopods, however, are small. *Lycopodium* species are commonly called ground pine or club moss. An entire patch of *Lycopodium* may be connected by a single **rhizome,** a characteristic that once made this plant a favorite for wreath-making (collection is now prohibited in most states).

Table 24A-1   Vascular Plants, Divisions Psilotophyta, Lycophyta, and Sphenophyta

Phylum	Characteristics	Method of Observation

Psilotophyta
Rep. *Psilotum* (whisk fern)

	Sporophyte lacks both leaves and roots. Dichotomously branched triangular stem with small scalelike appendages in place of leaves.
	    Rhizomes and **rhizoids** in place of roots. Vascular tissue present (see below).
	    Spores produced in **trilobed sporangia** (spore-producing organs). Homosporous.
	
	phloem xylem
	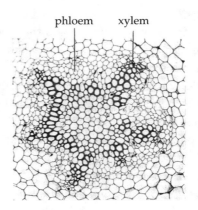
		Study preserved and living material on demonstration. Label sporangia, rhizomes, and rhizoids on the diagram below. Observe vascular tissue (phloem and xylem) in a cross section of a *Psilotum* stem, if available.
		
		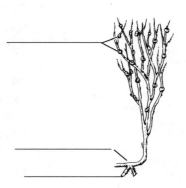

Lycophyta
Rep. *Lycopodium* (club moss)

	Leaves considered to be true leaves since they contain vascular tissue. Sporangia produced in the **axils** of leaves called **sporophylls;** these form conelike structures called **strobili.** Roots and rhizomes are present.
	    Lycopodium is homosporous. *Selaginella* is heterosporous.
	    In the *Selaginella* strobilus below, note the microsporangia and megasporangia.
	
	microsporangia with microspores megasporangia with megaspores
	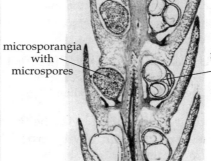
		Identify specimens of *Lycopodium* and *Selaginella* on demonstration. Identify microphyllous leaves (sporophylls), strobili, roots, and rhizomes. Label these structures on the diagram below.

(continued)

Table 24A-1   *(continued)*

Phylum	Characteristics	Method of Observation
Sphenophyta Rep. *Equisetum* (horsetail) 	Branched, ribbed stem with well-defined **nodes** and **internodes**. **Microphyllous leaves** occur regularly at each node. **Sporangia** produced within the **strobilus.** Homosporous.	Observe specimens on demonstration. Label the node, internodes, microphyllous leaves, and strobilus on the diagram below.

There is only one living genus of Sphenophyta—*Equisetum*, commonly called horsetail or scouring rush. Its ribbed, photosynthetic stems contain silica, providing an abrasive quality useful for scrubbing cookware while camping.

Alternation of generations is characteristic of all seedless vascular plants. In all four phyla the sporophyte is the conspicuous phase and the gametophyte is small and subterranean. Both sporophyte and gametophyte remain nutritionally independent of one another.

Seedless vascular plants require water for the sperm to reach the nonmotile egg. Therefore, one of the key events in the early invasion of land by these plants was the development of durable spores that could be dispersed over areas of dry land by wind. When all of the spores produced by a plant look alike, the plant is said to be **homosporous.** If two kinds of spores are produced, the plant is **heterosporous.** It was among certain seedless vascular plants that the heterosporous condition first appeared.

ⅠⅠⅠⅠⅠ Objectives ⅠⅠ

- ☐ Distinguish between roots, rhizoids, and rhizomes.
- ☐ Describe the life cycle of a fern; distinguish between the gametophyte and sporophyte generations.

PHYLA PSILOPHYTA, LYCOPHYTA, AND SPHENOPHYTA

The presence of primitive conducting tissues (**vascular tissues**) consisting of xylem and phloem allows us to classify the whisk ferns, club mosses, horsetails, and ferns as tracheophytes. Refer to Table 24A-1 and examine the representatives of the Phyla Psilophyta, Lycophyta, and Sphenophyta that are available in the laboratory. *a) Which plants have photosynthetic leaves?* _____ *b) Where does photosynthesis occur in the Psilophyta?* _____ *c) Which plants have vascularized leaves?* _____

PHYLUM PTEROPHYTA

The obvious parts of ferns are leaves that may be large and are usually delicate and deeply divided. Stems usually grow either horizontally along the soil surface or underground (**rhizomes**), although some "tree ferns" with vertical stems are found in the tropics. Although ferns are commonly found in moist areas, a few are thick-leaved and adapted to living in hot, dry environments.

The fern life cycle is characterized by an alternation of generations in which the sporophyte generation is more conspicuous than the gametophyte generation. The large green leaves we recognize as ferns are the sporophyte generation. (*What was the conspicuous generation in the moss?*) The sporophyte generation is diploid (2*n*). *a. What does this generation produce?* _____ *By what process?*

_____

Terrestrial ferns are homosporous, producing only one kind of spore. Fern leaves bear **sporangia** that produce haploid (*n*) spores by meiosis. Sporangia usually appear in clusters as rusty-looking patches called **sori** on the underside of a leaf or in large aggregations on a modified leaf (Figure 24A-1).

Figure 24A-1 *Types of sori. (a) Bare sori with spores exposed in* Polypodium virginianum. *(b) Sori are located along the margins of the leaf blades, which are rolled back over them in* Pellaea glabella.

 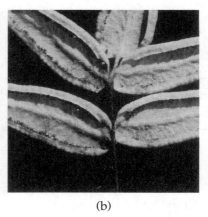

(a)   (b)

The haploid gametophyte develops from a germinating spore and is inconspicuous. The young gametophyte is a heart-shaped structure (**prothallus**) bearing **archegonia** (containing eggs) and **antheridia** (producing sperm). The nonmotile egg remains in the archegonium, and sperm swim to it. Water (rain or dew) is necessary for fertilization. The diploid sporophyte then develops from the fertilized egg or zygote (2*n*). The sporophyte and gametophyte are nutritionally independent (Figure 24A-2).

IIIII Procedure IIIIIIIIIIIIIIIIIIIIIIIIIIIIIIIIIIIIIIIIIIII

1. Study the herbarium sheets and living fern sporophytes on demonstration. Identify the following parts of the mature sporophyte and study their function as part of the fern life cycle (Figure 24A-2).

   Leaf Fronds or sporophylls bearing sori—the most conspicuous part of the sporophyte.

   Sorus Cluster of **sporangia** that produce haploid spores by meiosis.

   Rhizome Underground horizontal stem of the sporophyte.

   Adventitious roots Roots extending from the rhizome.

2. Obtain a portion of a fern frond or leaf with sori on the underside. Heat the sporangia composing the sorus by placing a lamp near the frond. A single sporangium has a row of specialized cells (known as the **annulus**) which runs two-thirds of the way around its edge. When dry conditions cause water to evaporate from the annulus, changes in the size of its cells cause it to spring back sharply, releasing some spores, and then rapidly snap back, catapulting the remaining spores. Watch for the annulus to move violently as it disperses spores.

   b. What does this mechanism of spore release indicate about spore dissemination?

   _____

   _____

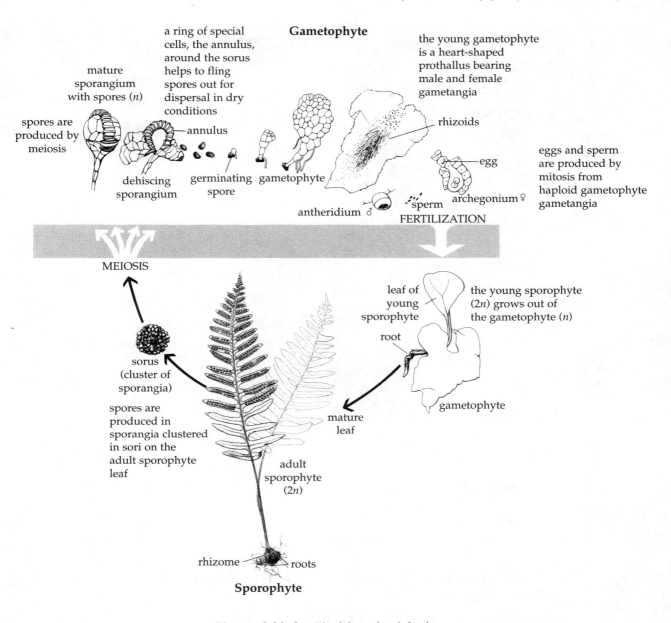

Figure 24A-2 *The life cycle of the fern.*

3. Prepare a wet mount of a living heart-shaped fern gametophyte (prothallus) or obtain a prepared slide of a young fern gametophyte (whole mount). Observe the specimen using the 10× and 40× objectives of your microscope. Identify the following parts and study their function as part of the fern life cycle (Figure 24A-2).

Rhizoids Specialized cellular filaments extending from the lower surface of the gametophyte into the substrate.

Antheridium Sperm are produced within the antheridia located on the lower surface of the gametophyte among rhizoids. When an adequate supply of water is present, mature antheridia burst and release sperm (Figure 24A-3a).

Sperm Flagellated male gametes produced in antheridia.

Archegonium This flask-shaped organ contains the female gamete (egg) and is located on the lower surface of the gametophyte (Figure 24A-3b). Sperm swim down the neck of the

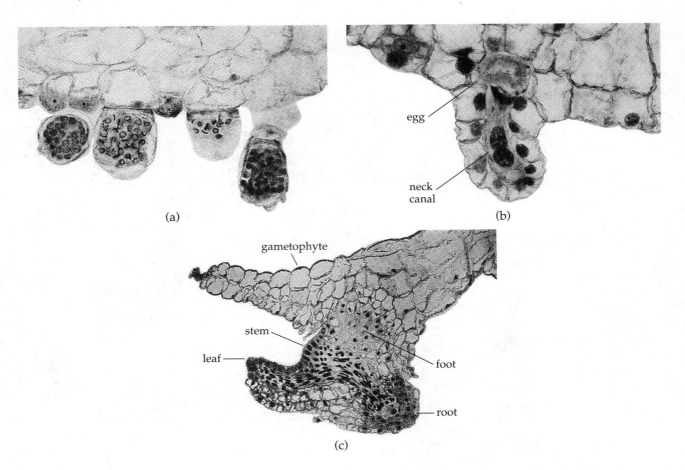

Figure 24A-3 *(a) Antheridia with nearly mature archegonia on the prothallus of* Osmunda. *(b) Archegonia. (c) Young sporophyte growing out of the gametophyte.*

archegonium to fertilize the egg, forming the zygote, which develops directly into the young embryo sporophyte.

Eggs Female gametes produced in archegonia.

c. How many cell layers are there in the young prothallus? _____

4. Obtain a prepared slide of a gametophyte with an attached sporophyte, or use living material (Figure 24A-3c). Fertilization has occurred several days earlier. The resulting sporophyte consists of one or more leaves and a *root.*

d. Is the new plant haploid or diploid? _____

e. What will be the fate of the gametophyte? _____

PART II TRACHEOPHYTES (VASCULAR PLANTS) WITH SEEDS—GYMNOSPERMS AND ANGIOSPERMS

What is the ecological advantage of a seed? Seeds provide a mechanism for a plant to survive during adverse environmental conditions, such as lack of water and extreme cold. Seed plants achieved prominence during a time when the earth's climate became colder and drier following the warm and humid Carboniferous period.

Gymnosperms are plants with "naked" or unprotected seeds, and **angiosperms** are plants with enclosed, protected seeds. In gymnosperms, seeds develop directly on the surface of the scales of the female cone; in angiosperms, seeds are enclosed in a fruit.

The evolutionary trend toward **sporophyte dominance** is most strikingly apparent among gymnosperms and angiosperms. In angiosperms, the mature female gametophyte is often as small as seven cells (one of which is the egg) and the mature male gametophyte may consist of only three cells (two of which are sperm). The inconspicuous gametophyte generation is completely dependent upon the sporophyte generation for nutrition.

Understanding the life cycles of heterosporous seed plants is made easier by first studying a generalized life cycle (Figure 24II-1). Keep in mind that the particulars may differ for individual gymnosperms and angiosperms.

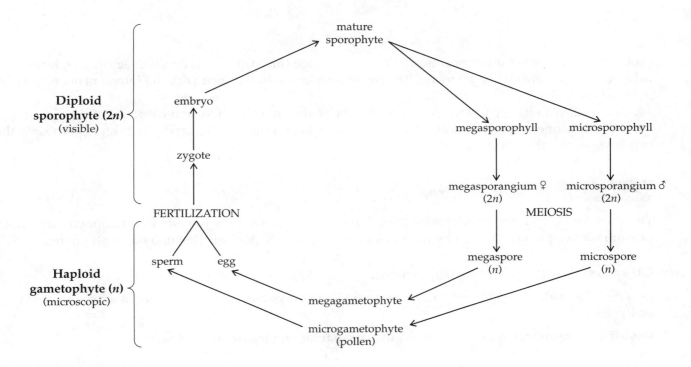

Figure 24II-1   *Generalized life cycle of the vascular plants. This life cycle is characterized by an alternation of sporophyte and gametophyte generations.*

The types of spores produced by heterosporous seed plants, called **megaspores** (n) and **microspores** (n), are produced by meiosis from **megasporangia** ($2n$) and **microsporangia** ($2n$). Within the megasporangium (*nucellus*), one megaspore divides many times to develop into the multicellular female gametophyte (*megagametophyte*) containing, as a rule, two archegonia. Inside each archegonium, an egg develops.

Microspores develop into the male gametophyte (**microgametophyte**). The mature microgametophyte is the **pollen grain,** which can be dispersed by the wind. After landing on an ovule, the pollen grain produces a special structure called a **pollen tube** that transports the sperm produced within the pollen grain into the vicinity of the egg. (In contrast to the nontracheophytes and seedless vascular plants, the heterosporous land plants do not require water for fertilization; thus, these plants thrive in a greater diversity of environmental conditions.)

The egg, developed from the megaspore, remains inside the female megasporangium (Figure 24II-2), often borne on a leaflike **megasporophyll** of the sporophyte plant. The fertilized egg develops into the young embryo within the megasporangium, which, in seed plants, is covered by one or two layers of

Figure 24II-2 *Sectional view of the ovule of the ancient seed plant* Eurystoma angulare, *showing the spatial relationship of the integument, the megasporangium, and the megaspore.*

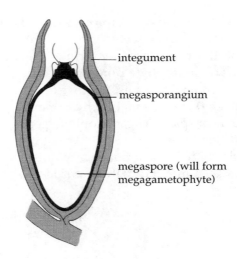

integument

megasporangium

megaspore (will form megagametophyte)

protective tissue called **integuments.** The entire structure (megaspore within the megasporangium covered by integuments) is the **ovule.** The ovule develops into the **seed** with the integuments serving as the **seed coat.**

The seed, with its specialized structures for the nutrition, protection, and dispersal of the young sporophyte, is one of the most important innovations to arise among the land plants and is probably the major reason for their dominance.

✔ **EXERCISE B** | **Gymnosperms**

There are four phyla of **gymnosperms** with living representatives: the phylum Coniferophyta, conifers; phylum Cycadophyta, cycads; phylum Ginkgophyta, ginkgos; phylum Gnetophyta, gnetophytes.

ııııı **Objectives** ıı

☐ Describe the relationship between microsporophyll, microsporangium, microspore, microgametophyte, and pollen.

☐ Describe the relationship between ovule, megasporangium, megaspore, and seed.

ııııı **Procedure** ıı

1. Examine the branches and cones from different types of conifers on demonstration in your laboratory. Most conifers are large evergreen trees or shrubs with needlelike or scalelike leaves. Confiers can be identified by the length, appearance, and number of modified leaves held together in a single group attached to the stem. For each of the conifers you study, write a brief description that distinguishes it from the others (Table 24B-1). (*Note:* Your instructor may provide you with a dichotomous key to help you identify the conifers.

Table 24B-1   Types of Conifers on Demonstration

Conifer	Characteristics

2. Locate a male cone (staminate or pollen cone) from one of the conifers on demonstration.

a. How does it differ from the typical (female) pine cone? _____

Carefully remove one of the scales (**microsporophylls**) that aggregate to form the cone. Observe with a dissecting microscope. Note the two attached microsporangia (Figure 24B-1).

3. Place the microsporophyll in a drop of water on a slide and carefully crush it to release the **microspores** from the microsporangium. A mature microgametophyte is known as a pollen grain (Figure 24B-2).

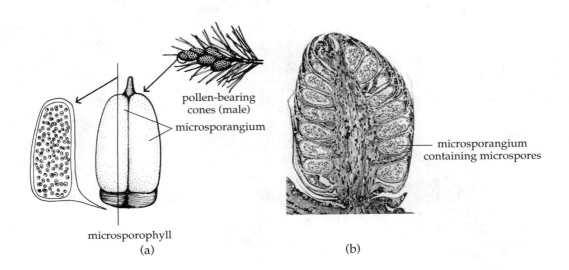

Figure 24B-1   *(a) Microsporophyll with attached microsporangia containing developing microspores. (b) Longitudinal view of a pollen-producing cone, showing microsporophylls and microsporangia containing mature pollen grains.*

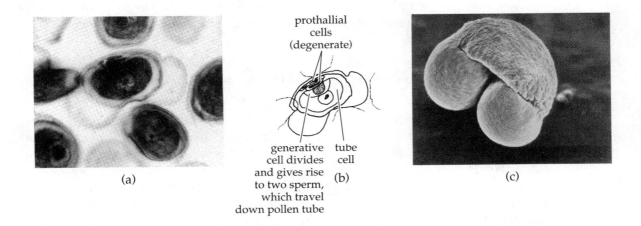

Figure 24B-2   *(a) Immature pollen grain. (b) The pollen grain consists of several cells; the generative cell will eventually divide to give rise to sperm cells. (c) Scanning electron micrograph of a pine pollen grain. When the pollen grain germinates, the pollen tube emerges from the lower end of the grain, between the wings.*

4. View a prepared slide of a longitudinal section of a staminate cone and look for microsporo-phylls, microsporangia, and microspores.

5. Examine a female cone (ovulate or seed cone) and locate the ovules (Figure 24B-3). The seeds of gymnosperms are said to be naked because they are produced on the surface of scales constituting the ovulate cone. At maturity, each scale bears two seeds or ovules containing embryos. Locate and examine a seed.

   b. How are conifer seeds dispersed? _____

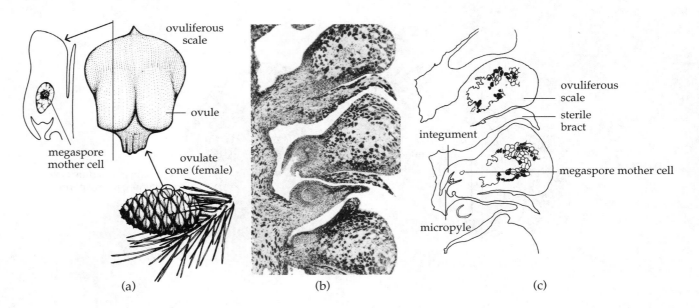

Figure 24B-3   *(a) Ovules on the scale of an ovulate cone. (b) Cross section of ovuliferous scales. (c) The megaspore mother cell divides to form the megaspores. The micropyle opening allows the pollen tube to enter so that fertilization can occur later.*

6. View a prepared slide of a longitudinal section of an ovulate cone. The **ovule** consists of **integuments** containing the **megasporangium** (Figure 24B-3). The **megaspore mother cell** develops within the megasporangium and gives rise to **megaspores** by meiosis. Locate these structures on your slide.

7. *Optional.* If assigned by your instructor, review the life cycle of pine in Figure 24B-4. Write the letter that follows each structure described in the caption in the appropriate box in the life-cycle diagram.

8. Examine living and preserved cycads (phylum Cycadophyta) and ginkgos (phylum Ginkgo-phyta) on demonstration in the laboratory. Phylum Ginkgophyta is represented by only a single living species, *Ginkgo biloba*. This tree was once a favored landscape planting for city sidewalks. However, the seeds have a fleshy outer layer that contains a foul-smelling compound, butyric acid. The putrid, slimy messes (not to mention the smell) created by the prolific production and dropping of seeds onto city sidewalks soon discouraged the planting of female *Ginkgo* trees. You will find the most recently planted *Ginkgo* trees are male.

   Fewer than a dozen living genera of Cycadophyta exist today; they occur naturally only in the tropics or subtropics. Note the fernlike leaves and short, thick stem. Also study the cones if present. Cycads and ginkgos are *dioecious*—pollen and seed cones are found on different plants. Look for ovules in female cones and microsporangia in male cones.

   c. How do representatives of these other gymnosperm phyla resemble the conifers?

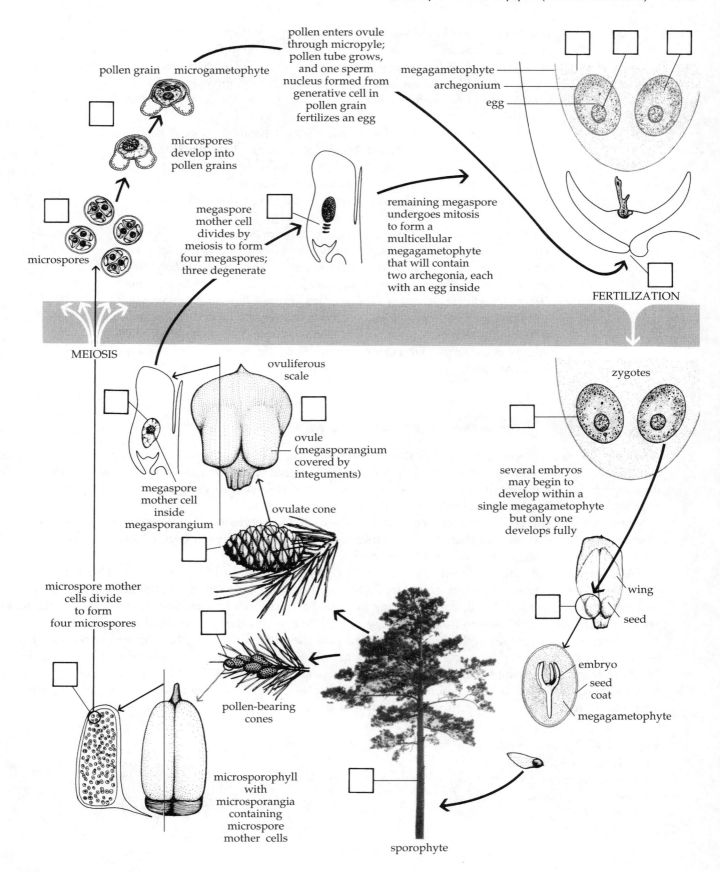

pollen grain   microgametophyte

pollen enters ovule
through micropyle;
pollen tube grows,
and one sperm
nucleus formed from
generative cell in
pollen grain
fertilizes an egg

megagametophyte
archegonium
egg

microspores
develop into
pollen grains

megaspore
mother cell
divides by
meiosis to form
four megaspores;
three degenerate

remaining megaspore
undergoes mitosis
to form a
multicellular
megagametophyte
that will contain
two archegonia, each
with an egg inside

microspores

FERTILIZATION

MEIOSIS

ovuliferous
scale

zygotes

ovule
(megasporangium
covered by
integuments)

megaspore
mother cell
inside
megasporangium

ovulate cone

several embryos
may begin to
develop within a
single megagametophyte
but only one
develops fully

microspore mother
cells divide
to form
four microspores

wing
seed

embryo
seed
coat

megagametophyte

pollen-bearing
cones

microsporophyll
with
microsporangia
containing
microspore
mother cells

sporophyte

Figure 24B-4   *Beginning at the bottom of the diagram: The* **immature**
sporophyte (gymnosperm seedling) develops into the **mature sporophyte**

(**a**). *Male cones* (**b**) *are usually produced in clusters of up to 50 or more at the tips of low branches. The male cones produce pollen within the* **microsporangia** (**c**) *that occur in pairs on the microsporophylls (leaves) of the cone. Microspore mother cells within the microsporangia produce* **haploid microspores** (**d**) *that develop into four-celled immature* **pollen grains** (**e**). *At this stage, the pollen grains are shed.*

After landing in the vicinity of a micropyle of a female cone, one of the cells of the pollen grain, the tube cell, will develop into a pollen tube during the germination of the pollen grain. Another of the cells will divide to form two sperm cells as the pollen grain matures. A mature pollen grain is the mature male microgametophyte. **Female cones** (**f**) *are generally larger than male cones and consist of woody scales containing paired* **ovules** (**g**). *The ovule is a megasporangium covered by integuments. Inside each megasporangium is a* **megaspore mother cell** (**h**), *which produces four haploid megaspores by meiosis. Three of these cells degenerate* (**i**) *and the remaining one divides mitotically to produce a multicellular* **megagametophyte** (**j**) *that will contain two* **archegonia** (**k**), *each containing an egg* (**l**). *Pollen enters the ovule through the micropyle* (**m**) *opening in the integument. One sperm nucleus unites with the egg to fertilize it. Generally, all eggs are fertilized and several embryos* (**n**) *begin to develop within a single megagametophyte, but only one develops fully. Each scale bears only two seeds* (**o**), *which separate and are dispersed by the wind.*

✔ | **EXERCISE C**   **Angiosperms (Phylum Angiospermae)**

Angiosperm means "covered seed." Following fertilization, seeds produced by angiosperms develop from ovules and are covered by a structure called the **ovary** located within the **flower.** The ovary eventually develops into a **fruit** containing one or more seeds. Enclosed seeds, flowers, and fruits are all unique to angiosperms, which are, at present, the dominant plants on earth.

Angiosperms can be divided into two classes: the **dicots** (dicotyledons) and **monocots** (monocotyledons). These names refer to the fact that the dicot embryo has two seed leaves or **cotyledons** (leaflike parts of the embryo), while monocot embryos have only one.

Among monocots are the familiar grasses, lilies, irises, orchids, cattails, and palms. Dicots include many herbs and almost all shrubs and trees (other than conifers). The many distinctions between monocots and dicots will be discussed during later laboratories, which examine the structure of the leaves, stems, and roots of angiosperm plants. Here, you will turn your attention to the flower.

A **flower** can be considered a highly modified and specialized shoot: a stem tip where modified leaves, the petals and sepals, occur "bunched together." Floral parts and leaves have many similarities. Working from the outside toward the inside, similar parts are grouped as whorls (Figure 24C-1). Monocots usually have flower parts arranged in threes (or multiples of three). Dicots usually have flower parts arranged in fours or fives (or multiples of four and five).

Figure 24C-1 *(a) Generalized flower parts. (b) Similar parts are grouped in whorls. In monocots, whorled structures occur in groups of three. In dicots, they occur in groups of four or five.*

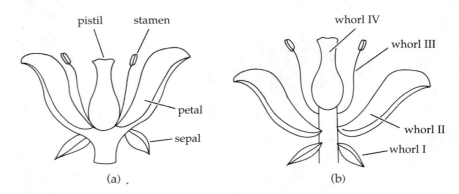

(a)   (b)

‖‖‖‖ Objectives ‖‖‖‖‖‖‖‖‖‖‖‖‖‖‖‖‖‖‖‖‖‖‖‖‖‖‖‖‖

☐ Distinguish between monocot and dicot flower types.

☐ Describe the structures of the male and female parts of a flower.

☐ Differentiate between pollination and fertilization.

‖‖‖‖ Procedure ‖‖‖‖‖‖‖‖‖‖‖‖‖‖‖‖‖‖‖‖‖‖‖‖‖‖‖‖‖‖‖‖

1. Obtain a flower from those provided. *a. Is your flower a monocot or dicot?* _____

2. Identify the parts of the flower in the following list. Locate as many parts as you can by simply looking at the flower. Then, using forceps and a dissecting microscope, start at the first whorl of the flower and work inward, locating each of the parts. Label these structures on the flower diagram (Figure 24C-2). Be sure that you understand the function of each part.

   Peduncle Narrow stalk that bears a flower or group of flowers (**inflorescence**). The stalk of an individual flower in an inflorescence is called a **pedicel** (Figure 24C-3).

   Receptacle Part of the flower stalk to which floral parts are attached.

Figure 24C-2 *Parts of a flower. Note: The receptacle is located at the base of the sepals and is not visible in this figure.*

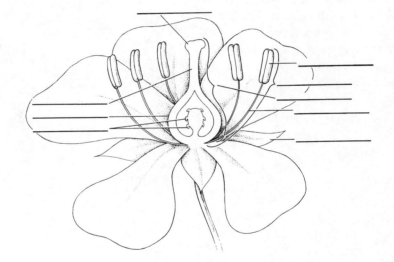

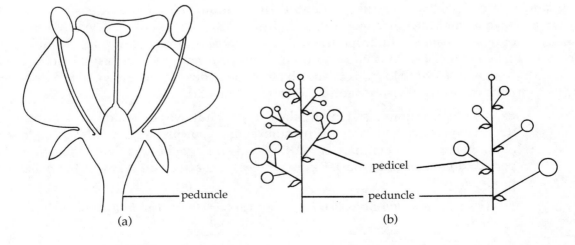

Figure 24C-3 *(a) A single flower attached at the peduncle. (b) Inflorescences supported by peduncles; each flower is attached to the peduncle by a pedicel.*

Sepals Outer leafy parts of the flower collectively called the **calyx.** Sepals usually enclose the outer flower parts, protect the bud, and later surround the ovary.

Petals Often brightly colored, forming the conspicuous inner whorl of flower parts. Collectively, petals are called the **corolla.**

Stamens Produce pollen grains, the male gametophytes. The gametes, sperm, are not produced until after pollination, as in gymnosperms. Stamens are actually microsporophylls consisting of a **filament** and a two-lobed **anther** containing the microsporangia.

Carpels Carpels may be individual (free) or fused. The pistil (composed of one or more carpels) is differentiated into a lower part, the **ovary,** and an upper part, or **stigma,** which receives pollen. The **style** connects the ovary to the stigma. The portion of the ovary to which the **ovules** are attached is called the **placenta** (Figure 24C-4).

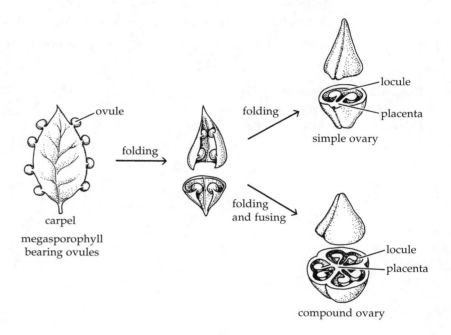

Figure 24C-4 *Presumed evolutionary development of simple and compound ovaries. A leaf-shaped carpel with ovules along its edge (left) folded in on itself, and the edges fused to form a simple ovary. Compound ovaries were formed by the fusing of separate infolded carpels.*

In angiosperms, pollination results in **double fertilization:** one sperm nucleus fuses with the egg cell to form a 2*n* **zygote** and the other sperm nucleus fuses with the two polar nuclei to form a polyploid (usually 3*n* or 5*n*) **endosperm.** Endosperm provides a source of stored energy for the developing embryo. The fertilized ovule then develops into a seed. In monocots, endosperm persists, forming a food reserve to be used during germination. Most dicots absorb all of the endosperm during embryo development, storing nutrients in fleshy cotyledons for use during germination.

3. Examine pollen grains by tapping the anther of a flower over a drop of sucrose solution on a microscope slide. Add a coverslip and examine your preparation, first using low power (10× objective) and then high power (40× objective). Keep this slide until the end of the laboratory period (do not let it dry out). Before you leave, check to see if any of the pollen grains have produced pollen tubes.

   b. Why did you mount the pollen in sucrose solution rather than in water alone?

   _____

4. *Optional.* If assigned by your instructor, study the life-cycle diagram for angiosperms in Figure 24C-5. Write the letter that follows each structure or process described in the caption in the appropriate box in the diagram.

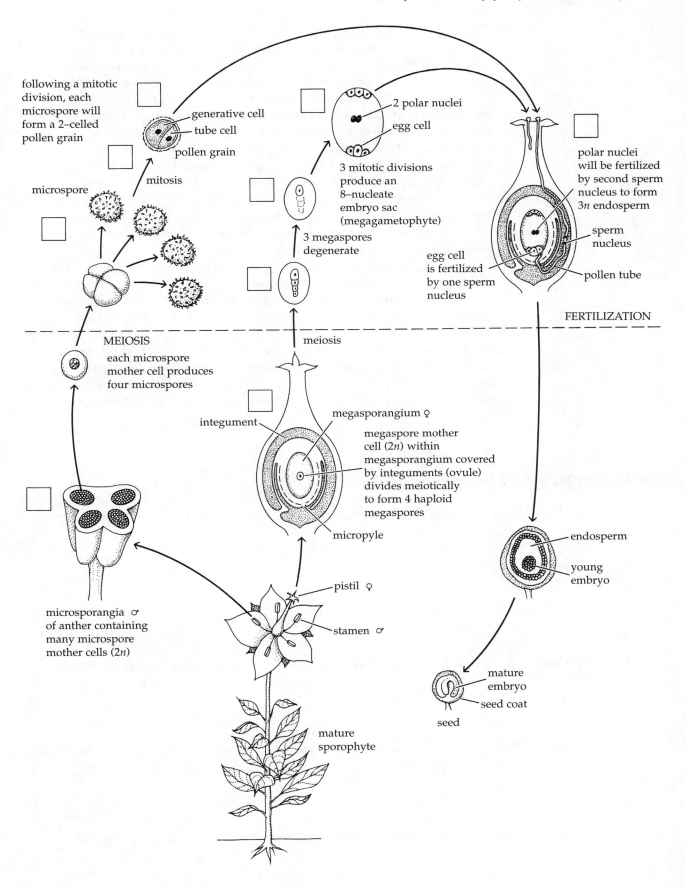

following a mitotic division, each microspore will form a 2–celled pollen grain

generative cell
tube cell
pollen grain

microspore

mitosis

2 polar nuclei
egg cell

3 mitotic divisions produce an 8–nucleate embryo sac (megagametophyte)

polar nuclei will be fertilized by second sperm nucleus to form 3n endosperm

sperm nucleus

pollen tube

egg cell is fertilized by one sperm nucleus

3 megaspores degenerate

FERTILIZATION

MEIOSIS

each microspore mother cell produces four microspores

meiosis

integument

megasporangium ♀

megaspore mother cell (2n) within megasporangium covered by integuments (ovule) divides meiotically to form 4 haploid megaspores

micropyle

endosperm

young embryo

microsporangia ♂ of anther containing many microspore mother cells (2n)

pistil ♀

stamen ♂

mature embryo
seed coat
seed

mature sporophyte

Figure 24C-5   *(page 24-15) Beginning in the center of the diagram: Within the **ovary** of the flower pistil, one or several **ovules** are attached to the ovary wall. Each of these ovules contains a megasporangium covered by **integuments** (**a**). Inside the megasporangium is the **megaspore mother cell** which produces four **megaspores** by meiosis (**b**), three of which degenerate (**c**). The remaining megaspore enlarges and undergoes three mitotic divisions to form eight nuclei. Of these, one nucleus develops a cell wall and becomes the egg; upon fertilization, two other nuclei (polar nuclei) will develop into a food source for the embryo (**d**). This structure represents the mature female gametophyte—the microscopic gametophyte generation in the sporophyte-dominated life cycle of seed plants.*

*Microsporogenesis takes place within the **anther** of the flower. Microspores are produced by meiosis within the pollen sacs (microsporangia) of the anther (**e**). Each microspore develops a resistant outer wall and the specialized structures common to **pollen grains** of that species (**f**). The microspore also divides mitotically (**g**) to form two cells, the **tube cell** and the **generative cell** (**h**). Most pollen grains are at this stage when released. A mature pollen grain containing these two cells represents the male **microgametophyte**—one of the only visible parts of the gametophyte generation in the sporophyte-dominated life cycle of angiosperms.*

*Pollination occurs when wind or insects carry the pollen to the stigma of the flower. After pollination, the tube nucleus forms a **pollen tube**. The pollen tube grows down through the style of the pistil and enters the ovary and then the ovule through an opening in the integuments (**micropyle**) (**i**). Fertilization can then occur.*

Laboratory Review Questions and Problems

1. Compare the life cycle of a moss (Laboratory 23) with that of a fern. Which is the visible generation of each?

2. How do the tracheophytes (vascular land plants) differ from the nontracheophytes?

3. Do whisk ferns, club mosses, horsetails, and ferns have vascular tissue? Seeds? Based on your answer, how do we group these phyla in terms of a general category?

4. In what group of land plants do we first observe true roots, stems, and leaves?

5. What three major advances are obvious among the vascular plants with seeds?

6. Are ferns heterosporous or homosporous? _____ Gymnosperms? _____ Angiosperms? _____

7. In the table below, indicate whether each structure is haploid (*n*) or diplois (2*n*), gametophyte or sporophyte.

Structure	*n* or 2*n*	Gametophyte or Sporophyte
Fern archegonium		
Moss antheridium		
Fern leaf		
Moss "leaflet"		
Pollen grain of pine		
Megagametophyte of angiosperm		
Microsporangium of angiosperm		
Pine tree		
Flower		

8. Flowers play an essential role in sexual reproduction of angiosperms. What types of insects or animals pollinate flowers? _____ What do these flowers look like? _____ What types of plants are pollinated by wind? _____

Diversity—Porifera, Cnidaria, and Wormlike Invertebrates

OVERVIEW

Multicellular organisms, or **metazoans,** comprise all of what are generally included in the kingdom Animalia (single-celled animals or protozoans are included in the kingdom Protista). Metazoans with poorly defined tissues and no internal organs are called **parazoans.** Two phyla, Placozoa and Porifera, are composed of parazoans. All other metazoans are **eumetazoans**—animals with internal organs and a digestive cavity with at least one opening, the mouth.

Figure 25-1 *Evolutionary "family tree" of the animal kingdom. (Adapted from Robert D. Barnes,* Invertebrate Zoology, *5/e, p. 61, Saunders, Philadelphia, 1986.)*

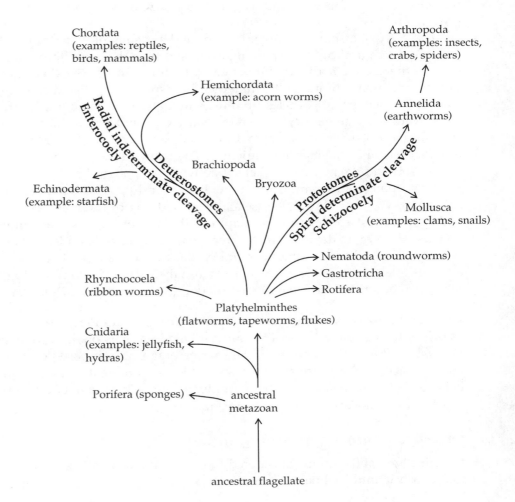

The phylogenetic relationships among various groups of metazoan organisms within the kingdom Animalia can best be studied by visualizing an evolutionary "family tree" that positions animals according to their relatedness to other organisms. Comparisons of anatomical, biochemical, physiological, and developmental similarities, together with fossil evidence, allow systematists to clarify phylogenetic relationships and to construct this "family tree." The tree presented in Figure 25-1 is one, but by no means the only, possible view of evolutionary relationships among the major animal phyla.

During this laboratory, you will survey six phyla of **invertebrates**—animals generally characterized as lacking a "backbone," or vertebral column. Our survey will focus on their diversity of form and function, proceeding from the relatively simple sponges (phylum Porifera) to the phylum Cnidaria, and then to wormlike animals in the phyla Platyhelminthes, Rhynchocoela, Nematoda, and Annelida.

STUDENT PREPARATION

Prepare for this laboratory by reviewing the text pages indicated by your instructor. Familiarizing yourself in advance with the information and procedures covered in this laboratory will give you a better understanding of the material and improve your efficiency.

EXERCISE A   Phyla Porifera and Placozoa

Phylum Porifera The sponges, phylum Porifera, are the least complex of all multicellular animals. They are **asymmetrical,** with poorly defined tissues and no organs. In fact, if the cells of a sponge are separated, the cells become amoeboid and reaggregate and redifferentiate into a new sponge without regard to their previous roles.

The body of a sponge is organized around a system of water canals. Water is drawn through small pores into a central cavity, the **spongocoel,** and then flows out through a larger opening, the **osculum.** Cells of the sponge body are differentiated by function. Flattened **epithelial cells** cover the outer surface to form the **pinacoderm.** On the inner surface, special flagellated cells called **choanocytes,** or "collar cells," strain extremely small particles from the water and thus serve in **filter-feeding.** In the middle jellylike layer, wandering **amoebocytes** secrete a skeleton composed of calcium carbonate ($CaCO_3$), silicon dioxide (SiO_2), or a protein called spongin. Calcareous and siliceous sponges are hard due to the presence of tiny rodlike skeletal elements called **spicules.** The natural sponges you might buy for bathing or to wash your car are soft and are made of a skeletal network of spongin fibers.

Most sponges are marine, but a few live in fresh water. As adults, all are **sessile** (attached to a substrate). They can reproduce asexually by budding or fragmentation and sexually by production of eggs and sperm. Most sponges are **hermaphroditic** (or monoecious): each individual has both male and female gonads. The zygote develops into a free-swimming, flagellated larva—a free-swimming hollow ball of flagellated cells that resembles the embryonic blastula of other organisms. When the larva settles and attaches to a substrate, the external cells lose their flagella and move to the interior in a process of cellular reorganization much like that of gastrulation (see Laboratory 19) in other animals.

Phylum Placozoa A second group of parazoans, phylum Placozoa, is usually included with the sponges as one of the early branches of the animal kingdom. This phylum contains only two species. One of these, *Trichoplax adhaerens*, is a multicellular marine organism composed of two irregularly shaped, flattened layers of cells that enclose an inner layer of loose, contractile mesenchyme cells. Originally, *Trichoplax* was believed to be the larval form of an unidentified adult animal, but scientists now believe that it may represent the most primitive living metazoan.

ⅠⅠⅠⅠⅠ Objectives ⅠⅠⅠ

☐ Describe why sponges are considered to represent a level of organization between that of a colony and that of a true multicellular organism.

☐ List the cell types present in the body of a sponge and describe their functions.

☐ Explain how sponges feed.

☐ Differentiate between mineralized and proteinaceous sponge skeletons.

 Procedure

1. Examine the examples of sponges on demonstration.

a. Which of the forms displayed have mineralized skeletons (calcium or silicon salts)?

_____

b. Which have fibrous (protein) skeletons? _____

2. Obtain a piece of sponge from one of the samples of *Grantia* and use your dissecting microscope to search for the tiny pores (**ostia**) through which water is taken into the body cavity (**spongocoel**) (Figure 25A-1).

c. What is the function of the larger pore (osculum)? _____

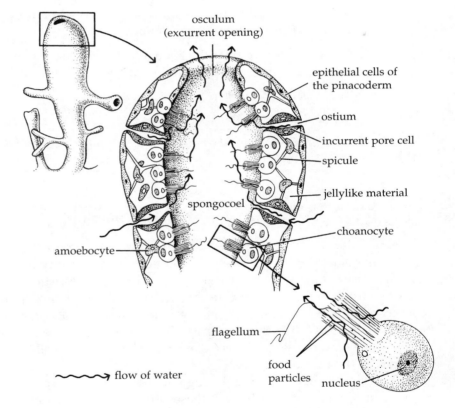

Figure 25A-1 *A simple sponge. Choanocytes (collar cells) with their flagella help maintain a flow of water into the central cavity of the sponge through small pores, the ostia. Particles of food and organic debris are filtered from the water by the choanocytes, becoming trapped in the delicate cytoplasmic collar surrounding the base of the cell's flagellum. Amoebocytes then carry food particles to nonfeeding cells; digestion takes place within the cells and not within the body cavity. Water is released through a larger opening, the osculum.*

The asconoid body form, shown here, is the most primitive among the sponges. In asconoid sponges, choanocytes line a single, unpartitioned spongocoel. In other sponges (synconoid type), choanocytes line canals that extend from the spongocoel. In leuconoid sponges, the choanocytes are distributed along the surfaces of numerous smaller chambers that branch off canals leading from the spongocoel. This arrangement allows leuconoid sponges to become very large and to exhibit great variety of shape.

3. In a prepared slide of *Grantia,* identify the choanocytes (collar cells) lining the spongocoel.

    d. What functions do these cells perform?

    _____

    e. What type of structure do you see extending from the choanocytes? _____

    f. How would this structure aid the choancytes in carrying out their main functions?

    _____

    g. Describe the region that separates the two cellular layers of the sponge.

    _____

    h. Are there any cells in this region? _____

4. On a separate sheet of paper, make a drawing of *Grantia.* Indicate the direction of water flow through the sponge. Label the following structures: choanocyte, amoebocyte, flagellum, ostium, osculum. Insert the drawing into your laboratory manual.

5. Obtain a small piece of *Spongilla* and place it on a glass slide, but do not add a coverslip. View the slide using the light microscope at scanning power (4×).

6. Remove the slide from the microscope and add a drop of bleach solution or vinegar to the sponge. Let the slide sit for 5 minutes.

7. Place a coverslip over the sponge material. Use a paper towel to soak up any excess fluid before returning the slide to the microscope. View at 10× and 40×.

    i. Describe what you see. _____

    j. What do you think these structures are? _____

    k. Why did you add the vinegar or bleach solution? What did it do to the sponge material?

    _____

👁 EXERCISE B   Phyla Cnidaria and Ctenophora

Phylum Cnidaria Organisms in the phylum Cnidaria, and all those that are higher on the phylogenetic tree, have distinct cell layers and are **symmetrical.** Symmetry implies a higher degree of complexity and organization than the asymmetrical organization characteristic of the sponges (Figure 25B-1).

    Organisms in the phylum Cnidaria are generally radially symmetrical. They are **diploblastic,** composed of two true tissue layers, the outer **epidermis** and the inner **gastrodermis,** separated by a gelatinous matrix called the **mesoglea.** The mesoglea may be thin or relatively thick and may be either cellular or without cells. Cnidarians are named for special cells called **cnidocytes,** which contain stinging organelles, the **nematocysts.**

    Two body forms are found among cnidarians—the **polyp** and the **medusa** (Figure 25B-2). A single species may exhibit one or both of these body forms. Polyps may be free-living or attached to a substrate, whereas medusas are swimming forms. In colonial cnidarians, polyps and medusas may live together and share the functions of food gathering (by polyps) and reproduction (by medusas).

    Both the polyp and medusa have tentacles armed with cnidocytes for capturing food and gathering it into their mouths. Most cnidarians feed on zooplankton, small animals and larvae that move passively with water currents. Food is ingested and wastes are voided through the mouth, the only opening into the digestive cavity. A digestive cavity with a single opening is called a **gastrovascular cavity.** In cnidarians, it also serves a circulatory function since its branches are close to all tissues. Neurons usually form a network of fibers at the interface of the epidermal and gastrodermal layers. This "nerve net" controls the limited behaviors permitted by the longitudinal epidermal fibers and the water-filled gastrovascular cavity that acts as a supporting **hydrostatic skeleton.**

Figure 25B-1 *(a) If a radially symmetrical animal is bisected in a particular plane, the shapes of the sections will be the same. (b) If a bilaterally symmetrical organism, which has dorsal, ventral, anterior, posterior, and right and left sides, is sectioned in different planes, the sections will not always be the same shape.*

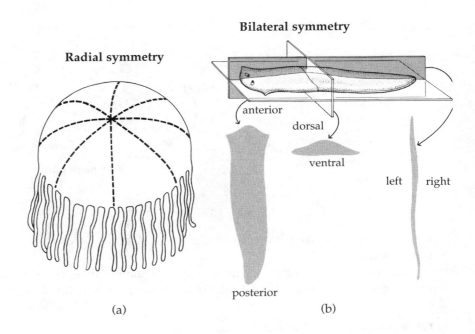

Radial symmetry

Bilateral symmetry

anterior

dorsal

ventral

left   right

posterior

(a)   (b)

Figure 25B-2 *Two body forms are found among the cnidarians, the polyp (a) and the medusa (b). A single species may exhibit one or both of these body forms during its life cycle.*

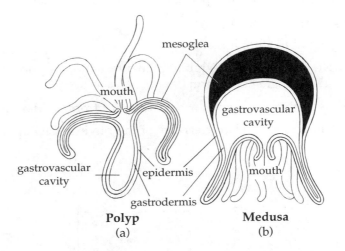

mesoglea

mouth

gastrovascular cavity

mouth

gastrovascular cavity

epidermis

gastrodermis

Polyp
(a)

Medusa
(b)

Cnidarians are found in both marine and freshwater environments. Reproduction occurs both asexually by budding or fragmentation and sexually by the production of eggs and sperm. Free-swimming, ciliated **planula larvae** are characteristic of most cnidarians.

Representatives of the three classes of cnidarians include the **hydrozoans,** the jellyfishes (**scypho-zoans**), and the corals and sea anemones (**anthozoans**). About 9,000 living species are known today, and there is also a rich fossil record of this phylum extending back to the Cambrian period.

Phylum Ctenophora Ctenophores probably evolved from the cnidarians or, at least, both groups share a common ancestry. Ctenophores, commonly called jellies or sea walnuts, are diploblastic and have a globular, medusoid-like body with a thick, transparent mesoglea that contains fibers, amoebocytes, and muscle cells. No nematocysts are present, except in a single species.

The spherical ctenophore body is biradially symmetrical. The mouth is on the lower side and the body is divided into equal sections by eight rows of ciliated "combs" (transverse plates of long, fused cilia), the combs arranged one behind the other to form comb rows (Figure 25B-3). In many comb jellies, two long tentacles protrude from epidermal pouches on the side opposite the mouth. Ctenophores are carnivorous and can evert their tentacles to trap small planktonic organisms by

Figure 25B-3 *A ctenophore.*

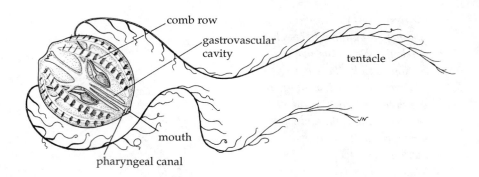

means of special adhesive cells (colloblasts). Ctenophores that lack tentacles simply catch food with their mouths.

The phylum Ctenophora contains approximately 50 species, all of which are marine. They usually range in size from several millimeters to about the diameter of a golf ball.

ııııı Objectives ııııııııııııııııııııııııııııııııııııı

☐ Relate the name Cnidaria to the special structures of organisms in this phylum.

☐ Describe the general form of a polyp.

☐ Describe the general form of a medusa.

☐ Give an example of a cnidarian exhibiting an alternation of polyp and medusa forms and of a cnidarian that exhibits only a single body form.

☐ Indicate how ctenophores are similar to, and differ from, cnidarians.

☐ Describe the distinguishing characteristics of the phylum Ctenophora.

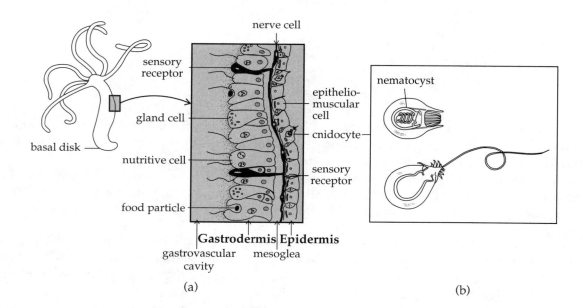

Figure 25B-4 *(a) Longitudinal section through a* Hydra *showing two tissue layers and specialized cells, including cnidocytes. (b) The cnidocyte is filled with a nematocyst that discharges a filament in response to a chemical or mechanical stimulus.*

ⅠⅠⅠⅠⅠ Procedure ⅠⅠⅠⅠⅠⅠⅠⅠⅠⅠⅠⅠⅠⅠⅠⅠⅠⅠⅠⅠⅠⅠⅠⅠⅠⅠⅠⅠⅠⅠⅠⅠⅠⅠⅠⅠⅠ

1. Study the live hydrozoan, *Hydra*, using a dissecting microscope. Note the general **polyp** body plan (Figure 25B-4). In this species, no medusa is present during the life cycle. Reproduction is either asexual by budding or sexual by production of eggs and sperm. Identify the mouth, tentacles, and basal disk. Touch one of the tentacles gently with a dissecting needle. What happens?

2. Watch the *Hydra* move (be patient and avoid jarring your microscope). Locomotion is mainly by contraction of muscle fibers in the outer tissue layer. The basal disk detaches from the substrate, and nematocysts help "glue" tentacles to the substrate so the polyp can somersault. In some cases, an air bubble in cells of the basal disk allows the *Hydra* to float. *Hydra* uses regular bursts of contraction and extension to sample the environment and search for food.

3. What color is your *Hydra*? Some species are green, which might suggest that they should be included among the plants. Symbiotic green algae in the cells lining the gastrovascular cavity, however, are the source of the green color. This type of symbiosis is not uncommon in cnidarians.

4. Recall that cnidarians such as *Hydra* have stinging organelles called nematocysts. What do they use these for? Have you seen any evidence that these structures were needed for

   locomotion? _____   Do you think they would be of use in capturing prey? _____

   Hydra eat small protozoans and larvae of other organisms or even other small organisms such as *Daphnia* or brine shrimp. Formulate a hypothesis that predicts how *Hydra* will behave when it comes into contact with prey (food). Is there a preference for certain types of food; large or small, dead or alive, fast moving or slow moving?

   HYPOTHESIS:

   NULL HYPOTHESIS:

   What do you **predict** will happen when *Hydra* is fed?

   Design an experimental procedure to test your hypothesis. Identify the **independent variable** in this investigation.

   Identify the **dependent variable** in this investigation.

   PROCEDURE:

RESULTS: Determine how *Hydra* feeds. Is there any evidence that *Hydra* uses stinging cells to obtain food? Does body shape change during feeding? How do the tentacles function? Are all tentacles used or just a few?

Do your results support your hypothesis?

Your null hypothesis?

Was your prediction correct?

What do you **conclude** about the food preferences of *Hydra?*

5. Examine a prepared slide of *Obelia,* also a hydrozoan, under low power (10×). *Obelia* is a **colonial** cnidarian with a life cycle that alternates between an asexual polyp and a sexual medusa (Figure 25B-5). The colony is composed of two types of polyps connected by a branching "stem." Feeding polyps have tentacles, whereas reproductive individuals do not. Nutrients are supplied to the reproductive polyps through the stem. Identify the feeding polyps and the reproductive polyps.

Figure 25B-5 Obelia, *a representative colonial cnidarian containing both feeding polyps and reproductive polyps. When mature, medusas are released from the reproductive polyps. As they swim, medusas release gametes that unite to form a diploid zygote. The zygote develops into a swimming larval form, the planula, which eventually settles onto a substrate and attaches itself. There it differentiates into a new colony of polyps.*

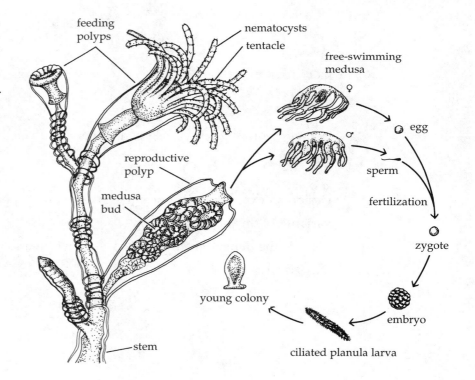

6. Adjust your microscope to high power (40×) and examine the nematocysts located within cnidocytes along the tentacles of a feeding polyp. These tiny, hairlike barbs are used to snare and paralyze small prey. Immobilized organisms are then taken into the gastrovascular cavity through the mouth opening between the tentacles.

7. Notice the two distinct tissue layers in *Obelia*. Gastrodermal cells lining the digestive cavity are separated from epidermal cells covering the outer surface of the polyp by a jellylike layer, the mesoglea. This layer contains wandering amoebocytes, which secrete materials that compose the mesoglea.

8. Examine a reproductive polyp at high power. Does it contain ringlets of cells along its longitudinal axis? These cells are the beginnings of medusas and are called, collectively, medusa buds.

   a. *What part do medusa buds play in the life cycle (Figure 25B-5)?* _____

   b. *Of what advantage is the colonial life form adopted by* Obelia? _____

   _____

9. Examine the Portuguese man-of-war (*Physalia*) on demonstration or study Figure 25B-6. This is a free-living, complex hydrozoan colony composed of both medusas and polyps. The most obvious individual in the colony is a modified medusa that forms the large, crested, gas-filled float. Feeding polyps hang from this float trailing tentacles as far as 60 feet (18 meters) from the colony. Particularly venomous nematocysts may kill invertebrates and small fish (and can be extremely painful to humans).

Figure 25B-6 *The Portuguese man-of-war is a colonial organism composed of (a) an inflated medusa and (b) many types of polyps, which have feeding and reproductive functions.*

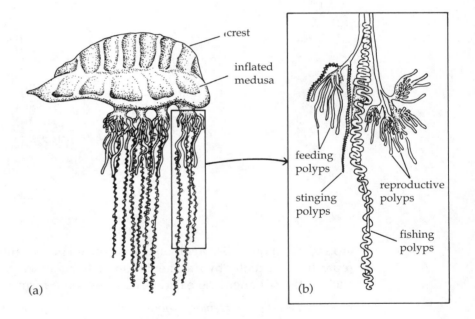

(a)   (b)

10. Study the plastic mount of the scyphozoan jellyfish *Aurelia* on demonstration (Figure 25B-7). The common mouth/anus is in the middle, surrounded by tentacles or arms. Gonads surround pouches of the gastric system whose radial canals extend to the edge of the umbrella (as a sort of gastrovascular circulatory cavity). The jellylike mesoglea forms the mass of the tissue making up the bell.

    Located at the outer margin of the bell are smaller tentacles and sense organs that determine position (**statocysts**). All exposed areas bear stinging cells.

    The jellyfish, like the hydrozoans, have an alternation of generations, but the medusa, which is a free-swimming jellyfish, is the predominant and most obvious form.

11. Examine the preserved and dissected sea anemones on demonstration or study Figure 25B-8. Note the internal structures displayed. A distinctive feature of anemones is the presence of a **pharynx** and sheets of tissue called **mesenteries** formed from infoldings of the body wall

Figure 25B-7 *The scyphozoan jellyfish,* Aurelia, *ventral view. Note that the margin of the bell is free of the ring of tissue (velum) present in hydrozoan medusas, which may otherwise resemble scyphozoan jelly-fish in form.*

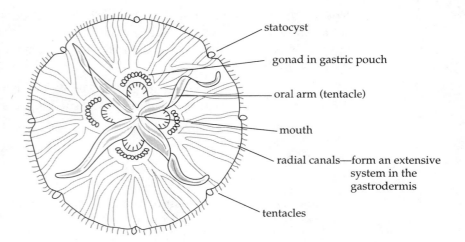

- statocyst
- gonad in gastric pouch
- oral arm (tentacle)
- mouth
- radial canals—form an extensive system in the gastrodermis
- tentacles

Figure 25B-8 *Diagrammatic representation of the internal structure of a sea anemone.*

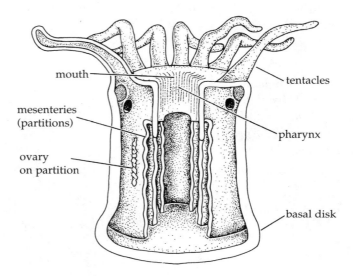

- mouth
- mesenteries (partitions)
- ovary on partition
- tentacles
- pharynx
- basal disk

around the mouth. Sea anemones feed on small fish and other invertebrates. They may reproduce asexually by fragmentation or sexually by production of gametes. **Anthozoans** such as the sea anemone exist only in the polyp form.

c. How does the sea anemone polyp resemble the hydrozoan polyp?

_____

12. Examine the corals on demonstration. Corals are also anthozoans existing only in the polyp form. A colony of polyps secretes a calcareous skeleton around the component individuals. Look closely at the coral. Do you see tiny holes? These are the holes through which the living polyps extend while feeding and reproducing. Sea fans and sea whips are "soft corals"; they are composed of a proteinaceous skeleton rather than a calcareous one.

d. How does the structure of the soft corals resemble that of the other corals?

_____

13. Study the comb jelly, *Pleurobranchia,* on demonstration or examine Figure 25B-3. Note the comb rows, mouth, pharyngeal canal, and tentacles.

e. How would you distinguish organisms in the phylum Cnidaria from organisms in the phylum Ctenophora?

_____

EXERCISE C   Wormlike Animals: Platyhelminthes, Rhynchocoela, Nematoda, Annelida

"Worms" and all remaining phyla of animals consist of organisms that are **triploblastic** and possess bilateral symmetry at some time in their life history.

The representatives of the four diverse phyla of wormlike organisms considered in this exercise exhibit many important evolutionary advances. One of these is the **mesoderm,** a third distinct embryonic tissue layer between the ectoderm and endoderm (hence the term "triploblastic"). Your study of representatives of the four "worm" phyla will show additional advances in organization and function.

PART I   Phylum Platyhelminthes: Flatworms

"Flatworms"—flattened, unsegmented worms—include **planarians,** class Turbellaria; **flukes,** class Trematoda; and **tapeworms,** class Cestoda. All members of the classes Trematoda and Cestoda are parasitic. They are **acoelomate** (have no coelom), with body organs embedded in their mesodermal tissues. Platyhelminthes exhibit the first extensive **organ-system** level of development.

Free-living flatworms, the turbellarians, are small, and most are marine, living on or in bottom sediments. Locomotion is by cilia and, in some larger flatworms, undulating muscular movements may help. The nervous system includes a small anterior ganglionic "brain" and longitudinal nerve cords. "Eyespots" consist of concentrations of pigment (melanin) that shade photoreceptive neurons (**ocelli**). The turbellarian digestive tract is a blind sac with no anus—the mouth is used for both ingestion and egestion. Primitive osmoregulatory structures, **protonephridia** (flame cells), are also present. Turbellarians are hermaphroditic and larvae are free-swimming.

Adult flukes, the trematodes, are all parasites, either external or internal. Flukes are flattened and have a ventral sucker or other adhesive organ for attaching to their host. In some trematodes, a second sucker is associated with the anterior mouth. Most flukes are hermaphroditic. The life cycle may involve one to four hosts—intermediate hosts (hosts that harbor the immature stages) may be invertebrates, but the definitive host (the host that harbors the sexually mature stage) is always a vertebrate.

Tapeworms, the cestodes, are intestinal parasites of vertebrates and are highly adapted for a hostile environment, where they nonetheless enjoy a rich food supply provided by their host. Like flukes, tapeworms are hermaphroditic, and their life cycle may involve an intermediate host in which a "bladder worm" stage encysts, awaiting ingestion by the definitive host.

IIIII Objectives IIIIIIIIIIIIIIIIIIIIIIIIIIIIIIIIIIIIIIIIII

☐ State the evolutionary advances demonstrated by members of the phylum Platyhelminthes.

☐ Describe how the tapeworm is adapted for parasitism.

IIIII Procedure IIIIIIIIIIIIIIIIIIIIIIIIIIIIIIIIIIIIIIIIII

1. The planarian *Dugesia* lives in ponds and streams under submerged rocks and logs. Place a living specimen of *Dugesia* into a Petri dish with some pond water. Examine it using your dissecting microscope. On a separate piece of paper, make diagrams of its dorsal and ventral surfaces.

2. Draw a dotted line on your diagrams to indicate the plane of symmetry. Indicate which end is anterior and which is posterior.

   a. *Which type of symmetry do flatworms exhibit?* _____

3. Place a few pieces of fresh (or dried) liver or egg yolk into your Petri dish and observe how *Dugesia* feeds.

   b. *Which organ is used in feeding?* _____ *What do you observe?* _____

   c. *How do planarians rid themselves of solid wastes?* _____

4. Cephalization (development of a head), a central nervous system, and an excretory system are first seen in the phylum Platyhelminthes. Figure 25C-1 will assist you in labeling the sensory lobes (auricles), eyespots ("eyes"), and pharynx and internal anatomy of *Dugesia*.

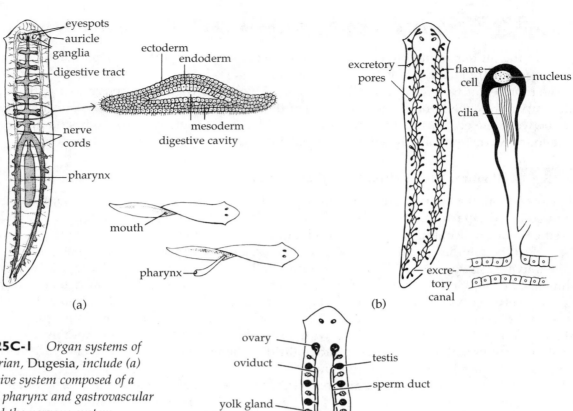

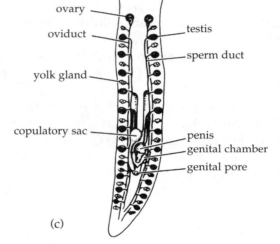

Figure 25C-1 *Organ systems of the planarian,* Dugesia, *include (a) the digestive system composed of a muscular pharynx and gastrovascular cavity and the nervous system composed of a "brain" of nerve tissue and two latero-ventral nerve cords; (b) the excretory system composed of flame cells embedded in the mesoderm and opening to the outside through excretory pores; and (c) the hermaphroditic reproductive system with male and female gonads, ducts, and specialized copulatory structures.*

5. Cephalization allows for forward movement. This is an important advance because it allows animals to hunt prey and to react, moving away or toward environmental stimuli. Sensory cells aid in these behaviors.

Sensory cells are distributed over the entire surface of *Dugesia*. They are slender pointed cells that lie with their ends projecting from the body surface between the epithelial cells. Some of the sensory cells are concentrated in the auricles and are used to gather tactile information. The two "eyes" are sensory organs specialized for light reception. Each consists of a pigmented black cup lined by sensory cells that continue as nerves to the "brain," a ganglion-like collection of nerve cells in the head. The pigmented cup shades the sensory cells from light in all directions but one and allows *Dugesia* to respond to light.

Planarians demonstrate **taxis,** locomotion either directly toward or away from a stimulus. Eyespots allow them to move either toward light (a positive phototaxis) or away from light (a negative phototaxis). Formulate a hypothesis that predicts how planaria will respond to light.

HYPOTHESIS:

NULL HYPOTHESIS:

What do you **predict** about the taxis of *Dugesia?* Is it toward light (positive) or away from light (negative)?

Identify the **independent variable** in this investigation.

Identify the **dependent variable** in this investigation.

Use the following experimental procedure to test your hypothesis. Modify the procedure if necessary.

PROCEDURE:

1. Draw a line across the middle of the bottom of a Petri dish. Mark one side "light," and the other "dark."

2. Put enough pond water in the dish to keep the animal happy.

3. Cover one half of the lid with aluminum foil.

4. Introduce a *Dugesia* anywhere in the Petri dish. Put on the lid so that the aluminum foil covers the side of the bottom marked "dark."

5. Place the dish directly under a moderate light source and leave it for about 10 minutes.

6. Remove the *Dugesia* with an eye dropper.

7. Sprinkle a small amount of carmine powder over the water and allow it to settle.

8. Gently swirl the dish so that the carmine comes in contact with all parts of the mucous trail left by the planarian on the bottom of the dish.

9. In one smooth motion, decant (pour out) the water and unattached carmine particles.

10. You now have a visual record of the movements of *Dugesia* during the experiment.

RESULTS: What do you observe about the movement of the planarian under the experimental conditions?

Do your results support your hypothesis?

What do you **conclude** about the function of the "eyes" in planaria?

Your null hypothesis?

Was your prediction correct?

6. Examine a microscope slide of the human liver fluke *Chlonorchis sinensis* (Figure 25C-2). Note the anterior sucker and the two blind intestinal sacs extending from the pharynx and esophagus. Much of the central part of the body is occupied by the uterus. The single ovary and paired testes are located in the posterior third of the body.

   d. Notice that there are no sense organs in trematodes. Why not?

   _____

   e. What is the intermediate host of Chlonorchis? _____

   The definitive host? _____

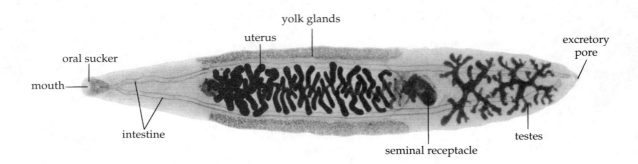

Figure 25C-2 *In the life cycle of* Chlonorchis, *snails ingest the eggs of this parasite, which then hatch and develop into intermediate larval forms within the host. The larvae develop into tadpole-like swimming forms that bore through the flesh of the snail and escape into the water. There they swim until they find a fish of the appropriate type, bore into its flesh, lose their tails, and become encysted in muscle tissue. In a human who eats raw, infected fish, the larvae encyst in liver tissue and become adults. Eggs of the parasite are excreted in the feces. If raw sewage finds its way into bodies of water, snails may continue the* Chlonorchis *life cycle by ingesting the eggs.*

7. Examine preserved specimens of **tapeworms** on demonstration. The adult tapeworm (Taenia) lives in the cavity of the human intestine and the intestine of domesticated animals such as dogs and cats. An anterior region of the "worm," the **scolex,** is modified to anchor the organism in the intestinal tissue of the host. Behind the scolex is a chain of segmentlike reproductive structures, called **proglottids.** This chain may reach a length of 20 feet or 6 meters and may consist of over 1,000 proglottids. New proglottids are constantly formed by budding in the "neck" region behind the scolex. Mature proglottids detach from the posterior end and are voided with the feces of the host. They may then be passed to other individuals through food contamination.

8. Observe a prepared slide of portions of *Taenia* at 10× to observe the general size and shape of different regions; then use high power (40×) to see the detailed structure of the scolex and proglottids (Figure 25C-3).

   f. Describe the structure of the scolex and state how it is adapted for attachment to the host.

   _____

   The disk-shaped structure surrounded by hooks at the tip of the scolex does not contain a mouth—tapeworms have no digestive system. Instead, food molecules are taken directly from the intestine of the host through the integument of the parasite.

Figure 25C-3 Taenia, *the tape-worm. (a) The scolex. (b) Internal view of a mature proglottid showing the male and female reproductive organs present in each proglottid.*

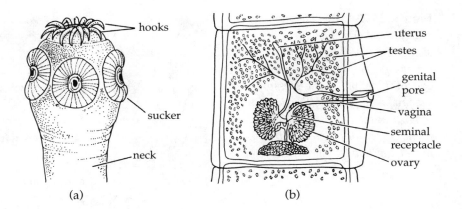

hooks

sucker

neck

(a)

uterus

testes

genital pore

vagina

seminal receptacle

ovary

(b)

g. *Tapeworms are covered by a thick cuticle secreted by the epidermis. Of what advantage would this be to an organism living inside an intestine?*

_____

h. *Do all of the proglottids on the slide appear to be about the same size and have the same internal structure?* _____ *Which proglottids were formed first?* _____

_____

i. *What are the reproductive structures apparent in the proglottids on your slide (Figure 25C-3b)?*

_____

j. *What advantage does a hermaphroditic species have over a species with separate sexes?*

_____

✔ **PART 2**   **Phylum Rhynchocoela: Ribbon Worms**

Organisms in the phylum **Rhynchocoela,** or Nemertea (nemertines), are called "**ribbon worms**" or "proboscis worms." Ribbon worms are long and often flattened, with a remarkable eversible **proboscis** usually equipped with a barb for food gathering. The proboscis is actually a long tube coiled in a body cavity called the **rhynchocoel** (it is much like the finger of a rubber glove that is pushed inside the glove, springing out when the glove is squeezed). Muscles attached to the proboscis are used to evert it (Figure 25C-4).

The phylum Rhynchocoela may be an offshoot of the Platyhelminthes; however, the ribbon worms exhibit evolutionary advances over the flatworms. For the first time in our consideration of the phyla, we observe **two digestive openings,** a **mouth** and an **anus;** and a **closed circulatory system** composed of two lateral vessels connected anteriorly and posteriorly. In addition, a **dorsal nerve cord** and two **lateral nerve cords,** somewhat similar to the nervous system of the flatworms, are found in members of this group. The ribbon worms are the first phylum to possess excretory cells, or **protonephridia,** that are truly excretory rather than simply osmoregulatory (as are the characteristic flame cells of the flatworms). Sexes are separate in most nemertines, and fertilization is external.

ııııı **Objectives** ıı

☐ State two evolutionary advances characteristic of the phylum Rhynchocoela.

☐ Describe the general structure of a ribbon worm.

ııııı **Procedure** ıı

Examine preserved specimens of ribbon worms. *Describe their general appearance.*

_____

Figure 25C-4 *(a) Diagrams of the anterior end of a ribbon worm showing the proboscis withdrawn and extended. The proboscis is rapidly extended by hydrostatic pressure and the prey is impaled by the stylet or tangled in the proboscis. (b) Diagram of the circulatory system with one dorsal and two lateral vessels.*

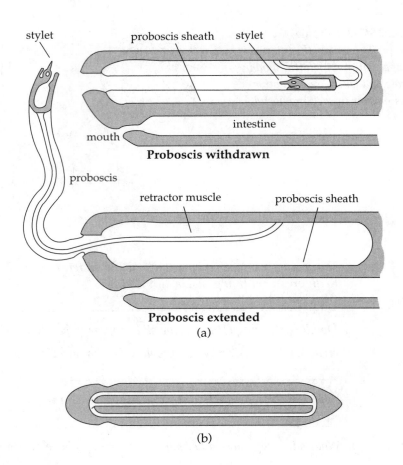

(a)

(b)

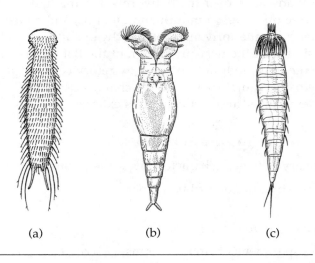

PART 3   Phylum Nematoda (Roundworms) and Other Wormlike Phyla

Gastrotrichs, rotifers, kinorhynchs, nematodes (roundworms), horsehair worms, and a few other minor wormlike groups form an assemblage of small, generally free-living animals with an anterior mouth and sense organs, but no well-defined head (Figure 25C-5).* Cilia are reduced and the body is covered with a secreted cuticle. The digestive tract is usually complete and has a specialized "pharynx." Most of these organisms have protonephridia. Many of the nematodes and rotifers have a constant number of nuclei in various organs, and mitosis ceases following embryonic development. The sexes are separate in most of

Figure 25C-5 *Representatives of wormlike phyla: (a) gastrotrich (phylum Gastrotricha), (b) rotifer (phylum Rotifera), (c) kinorhynch (phylum Kinorhyncha).*

(a)   (b)   (c)

Because of their similarities, these animals are sometimes grouped together as the Aschelminthes.

these wormlike organisms. Here, we will consider only the most diverse group of these animals, the phylum Nematoda (roundworms) with about 12,000 known species.

The **roundworms** or **nematodes** are unsegmented worms that may be either free-living or parasitic. Many free-living forms are inhabitants of the soil. Parasitic forms invade plant bodies and destroy tissues. Common animal parasites include hookworms, intestinal roundworms (*Ascaris*), *Trichinella*, and pinworms.

In contrast to the acoelomate flatworms and ribbon worms, organisms in this phylum have a type of body cavity (**coelom**). A true coelom is completely lined by mesoderm and will not be found until we progress to organisms in more advanced phyla. In nematodes, the coelom is called a **pseudocoelom** because it appears to be only partially lined by mesoderm (Figure 25C-6).

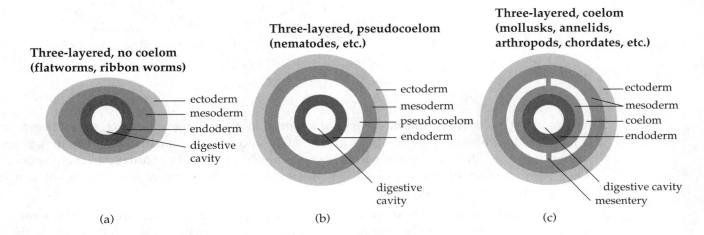

Figure 25C-6   *(a) Acoelomate animals, such as platyhelminths, have only one body cavity—the digestive cavity; body organs are embedded in mesoderm. (b) Other animals, for example, nematodes, have a second body cavity, a pseudo-coelom. (c) A true coelom (defined as a second body cavity completely lined with mesoderm) is found in annelids and almost all other higher phyla. Why is a pseudocoelom not a true coelom?*

Organisms in this phylum also have a true gut with two openings—a mouth and an anus, allowing regions of the digestive tract to assume specialized functions.

⋮⋮⋮⋮⋮ **Objectives** ⋮⋮⋮⋮⋮⋮⋮⋮⋮⋮⋮⋮⋮⋮⋮⋮⋮⋮⋮⋮⋮⋮⋮⋮⋮⋮⋮⋮⋮⋮⋮⋮⋮⋮⋮⋮⋮

☐ State two evolutionary advances in the phylum Nematoda.

☐ Describe the generalized structure of a free-living nematode.

☐ Describe the modifications for parasitism observed among the nematodes.

⋮⋮⋮⋮⋮ **Procedure** ⋮⋮⋮⋮⋮⋮⋮⋮⋮⋮⋮⋮⋮⋮⋮⋮⋮⋮⋮⋮⋮⋮⋮⋮⋮⋮⋮⋮⋮⋮⋮⋮⋮⋮⋮⋮⋮⋮

A free-living nematode, the vinegar eel *Turbatrix*, lives at the bottom of casks of vinegar, where it feeds on bacteria and yeast.

1. Place a few drops of the vinegar eel culture onto a clean microscope slide. Add one drop of 0.2% neutral red to increase the contrast of the mount and add a coverslip. Examine a vinegar eel at 10× to observe its overall shape, and then use high power (40×) to see details of its internal anatomy. If your vinegar eels are moving about too rapidly for you to see them well, touch the edge of a paper towel to the coverslip to draw off some of the culture solution.

Figure 25C-7 *(a) Apparatus used to collect nematodes in soil samples. (b) Staining a wet mount slide with methylene blue. Place a drop of stain to one side of the coverslip. Apply a small piece of paper towel to the other side (this helps "draw" the stain beneath the coverslip).*

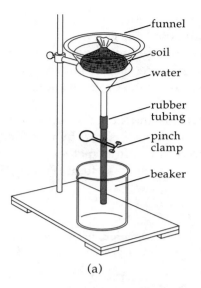

funnel
soil
water
rubber tubing
pinch clamp
beaker

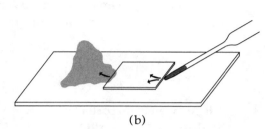

(a)   (b)

2. Beginning at the anterior (rounded) end of a worm, locate the digestive tract, which begins at a narrow slit, the mouth, and empties immediately into a featherlike pharynx. The remainder of the tract, the **intestine,** begins with a bulbous region and tapers gradually to the posterior. On a separate piece of paper, make a sketch of as much of the internal anatomy as you can see and label the structures you draw.

3. Nematodes are commonly **dioecious**—sexes are separate. Reproductive organs of the female vinegar eel may be visible. Female nematodes have a very prominent elongate **uterus** in their midsection; it may obscure much of your view of the digestive tract. The uterus may contain tiny offspring hatched within it.

   During copulation, male nematodes expel sperm from a posterior opening, the anus, from which they also excrete wastes.

4. Use high power (40×) to examine slides of the parasitic nematode, *Trichinella;* you will see adults and encysted larvae in the muscle tissue of the pig. Like other nematodes, this organism is covered by a thick protective cuticle. *Trichinella* larvae encysted in meat are ingested by a host, either pig or human. Inside the intestine of the host, larvae are released from their cysts (Figure 25C-8a) and develop into adults; they mate and produce more larvae. These larvae invade the body tissues and migrate to the muscles where they encyst. Large infections can produce considerable muscle damage leading to weakness and even death. The disease produced by this infection is called **trichinosis.**

   a. *How do the larval forms differ in appearance from adults?* _____

   b. *Why is it important that pork be cooked well, or not eaten rare?* _____

5. On a separate sheet of paper, describe the events that begin with eating infected pork and lead to the harboring of *Trichinella* parasites in muscle tissues. Make a drawing of the life cycle of *Trichinella* to assist you in writing your explanation.

   More than 50 different types of roundworms have been found in humans. *Ascaris lumbricoides* live in the intestine and can grow to more than a foot in length (Figure 25C-8b). A female can lay up to 200,000 eggs per day. *Ascaris* eggs are found in human feces.

   The worms hatch in the intestine but leave immediately, burrowing through the body to the lungs. From here they work they way up the bronchial tubes to the mouth and are then swallowed back into the stomach and return to the intestine.

   c. *List several ways in which a human could become infected by* Ascaris. _____

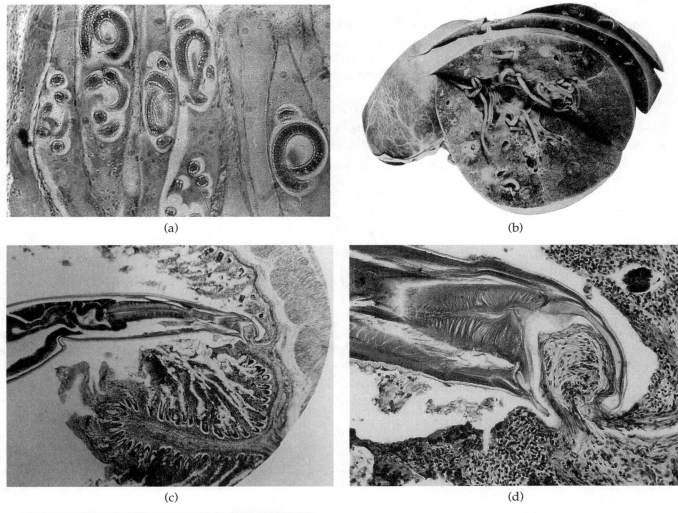

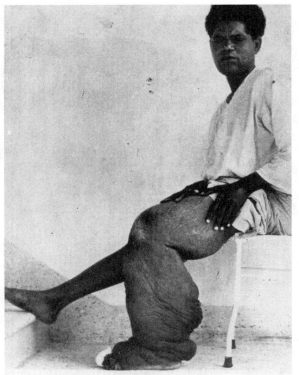

Figure 25C-8 *(a)* Trichina *cysts in the muscle of a pig. The cyst itself is harmless; the worm eventually dies. The damage is caused by millions of larvae boring through the organism before they encyst. (b)* Ascaris *in a human liver. As with* Trichina, *the greatest damage occurs when the worms migrate. As adults, they are relatively harmless unless their numbers increase to the point of blocking vital ducts or the intestine itself. Sometimes adults also wander to the liver (as here) or up through the esophagus and out through the nose (to the surprise and horror of their host!). (c) Section through hookworm biting wall of intestine. (d) Closeup of the hookworm in (c). (e) Elephantiasis caused by filaria worms.*

Hookworms attach to the lining of the intestine (Figure 25C-8c, d) and feed on blood and tissue fluids. Eggs pass from the intestine in the feces. Larvae hatching from the eggs burrow into the ground, where they eat and grow until they are capable of infecting other humans. The larvae (0.05 cm) bore through body tissue (usually through bare feet that touch contaminated soil). After entering the skin, the worms take the same path as *Ascaris*. Hookworm infections can be very debilitating, often leading to extreme weakness and stunted growth.

Filaria worms differ from *Ascaris* and hookworms. They require an intermediate host—a mosquito. Filaria live in the lymph glands of their human hosts. Adult females are 3 to 4 inches long. A female gives birth to larvae, microfilaria, that enter the blood vessels.

d. *What is the relation of lymph channels to the bloodstream?*

The larvae, if picked up by a mosquito, can be transferred to another person bitten by the same mosquito. The larvae penetrate the skin near the bite and travel to the lymph nodes where, in large numbers, they may block lymph channels, causing severe swelling of affected body parts (Figure 25C-8e). This condition is known as **elephantiasis.**

The **guinea worm** is perhaps one of the most bizarre human parasites. The female can grow to more than a meter in length but is only 0.5 mm wide. Guinea worms are common in Egypt, Arabia, India, and Africa. The worms coil into a blisterlike formation near the skin's surface. When an infected individual enters cold water, the blister opens and larvae escape. Larvae swim in the water until they find their intermediate host—*Cyclops*, a tiny crustacean about 2 mm long. Larvae develop into adults within the tissues of *Cyclops*. Ingesting *Cyclops* in unfiltered water can then spread the guinea worm to other humans.

Chemicals can be used to treat infections and doctors can extract the worms from blisters by winding them around a needle—a very painful solution.

e. *How could infections of this parasite be controlled in areas where water is contaminated? (Note that in some countries, religious beliefs dictate how water is obtained or used.)*

EXTENDING YOUR INVESTIGATION: NEMATODE DIVERSITY

One of the richest sources of nematodes, by far, is the soil. It is easy to collect soil nematodes. Do you think different types would be found in different kinds of soil? Do roundworms prefer more acidic or more basic soil? Do you find more roundworms in areas of open soil or in areas covered by vegetation? Formulate a hypothesis to investigate how different environments support different sizes of nematode populations in the soil.

HYPOTHESIS:

NULL HYPOTHESIS:

What do you **predict** you will find if you sample some soil from your local environment?

Identify the **independent variable** in this investigation.

Identify the **dependent variable** in this investigation.

Use the following procedure to test your hypothesis.

PROCEDURE:

1. Set up an apparatus similar to the one shown in Figure 25C-7a.

2. Collect soil and wrap it in a triple layer of cheesecloth to form a small sack. Secure the ends of the cheesecloth together with a rubber band.

3. Place enough water in the funnel to cover the bag of soil. Let stand for 24 hours. The nematodes will crawl into the water and will eventually sink into the neck of the funnel.

4. Obtain a plastic cup or beaker and open the pinch clamp for a few seconds in order to collect a small amount of water.

5. Make a wet-mount slide by using a single drop of water from a Pasteur pipette. *Hint:* You may wish to stain your preparation with methylene blue (see Figure 25C-7b). *Do you see any nematodes?* _____ *How many?* _____ How do they move? _____

6. Sample at least three drops of water. One milliliter contains approximately 15 drops. How may nematodes/milliliter (mL) are found in your soil sample? Record your results below.

RESULTS: Describe the diversity of nematodes present in your soil sample.

Do your results support your hypothesis?

Your null hypothesis?

Was your prediction correct?

What do you **conclude** about the presence of nematodes in the soil sample you collected?

✓ PART 4   Phylum Annelida: Segmented Worms

Phylum Annelida contains the segmented worms, almost all of which are free-living. Note that we have now reached a branching point in the family tree of animal phyla. The annelids, and all organisms above them in the phylogenetic tree, have true coeloms. The annelid coelom is formed by a splitting of the embryonic mesoderm and is said to be **schizocoelous** (*schizo*- means split). This type of coelom is also found among arthropods and mollusks. Note that these three groups are on the same branch of the phylogenetic tree (Figure 25-I). Organisms on the other side of the branch point are also coelomate, but their coelom is formed by an outpocketing, or evagination, of the primitive gut, or *enteron*. Hence the resulting coelomic cavity is said to be **enterocoelous.** This type of coelom is found among the echinoderms and chordates.

The coelom of annelids is compartmentalized into segments by **septa.** Coelomic fluid within the body cavity acts like a **hydrostatic skeleton** against which muscles work to change body shape. Like nematodes, annelids have a one-way digestive tract with a mouth, anus, and several specialized regions.

A dorsal mass of nerve cells forming a ganglion or "brain" and a ventral nerve cord provide a primitive nervous system. The circulatory system is closed, blood being confined to vessels.

Marine **polychaetes** (sand worms), **oligochaetes** (freshwater annelids and earthworms), and **leeches** are among the most common annelids. There are over 8,700 known species of annelid worms.

The polychaetes (class Polychaeta) make up the largest group of annelids. Most are marine and are an important food source for fish and crustaceans. Polychaetes have **parapodia,** fleshy appendages on the body segments. The polychaetes have well-developed sense organs on their heads, including eyes, antennae, and chemoreceptors. Some polychaetes build tubes to live in. Many of these sedentary forms use tentacles covered with cilia to trap food such as tiny animals and decaying organic matter and transport it to the mouth. Others pump water through their burrows and filter food from the water.

Most leeches (class Hirudinea) live in fresh water. They are parasitic or predaceous, feeding on tissue fluids, blood, or small invertebrates. Leeches lack the setae characteristic of other members of the phylum.

ⅠⅠⅠⅠⅠ Objectives ⅠⅠ

☐ Explain why annelids are found on a separate branch of a typical phylogenetic tree.

☐ Define "coelom" and relate this term to the other five phyla studied during this laboratory.

☐ Describe the hydrostatic skeleton of the earthworm.

☐ Explain how the circular and longitudinal muscle layers surrounding the earthworm gut allow the organism to move.

ⅠⅠⅠⅠⅠ Procedure ⅠⅠⅠⅠⅠⅠⅠⅠⅠⅠⅠⅠⅠⅠⅠⅠⅠⅠⅠⅠⅠⅠⅠⅠⅠⅠⅠⅠⅠⅠⅠⅠⅠⅠⅠⅠⅠⅠ

1. Study the anterior end of a living or preserved earthworm, class Oligochaeta. The first part of the worm is the **prostomium.** It is above and anterior to the mouth. The terminal end of the worm, bearing the anus, is the **pygidium.** Find the mouth, a crescent-shaped opening partly covered by the prostomium, and the anus, an oval opening in the posterior extremity of the last segment. A glandular area, the **clitellum,** is located near the anterior end of the worm. It is formed from several swollen segments and secretes mucus for copulation and cocoon formation.

2. Note that the surface of the body is distinctly marked by transverse grooves that encircle it. These grooves divide the body into **segments.**

   a. *Do the segmental grooves completely encircle the body in every region of the earthworm?*

   _____

   b. *How many segments are present in the specimen you observed?* _____

   Do all earthworms have the same number of segments? _____

   The body is covered by the **cuticle,** a delicate, iridescent, nonliving membrane secreted by the epidermis. The cuticle contains numerous pores that provide an exit for the mucus secreted by underlying cells. The mucus spreads over the entire surface of the cuticle and provides the proper conditions for respiration through the cuticle.

   Every segment except the first and last is provided with four pairs of chitinous bristles called **setae** that are used for locomotion. Pass your finger back and forth on the ventral and lateral surfaces of the specimen. You should be able to feel the setae projecting from the surface.

   With a sharp scalpel and a fine pair of scissors, carefully cut open the earthworm lengthwise along the dorsal surface. Pin the body wall to the dissection pan or board. Observe that the body wall is separated from the intestine, the central dark tube in the posterior part of the body, by a space, the **coelom.** Refer to Figure 25C-9 and identify and study the following parts.

   Prostomium The most anterior segment is incomplete and lies in front of the mouth, which opens beneath it into the first segment of the body.

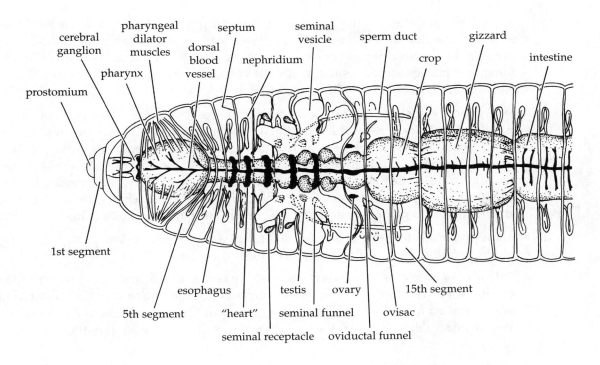

Figure 25C-9 *Internal structures of the earthworm (dorsal view). (After Robert
D. Barnes,* Invertebrate Zoology, *6/e, p. 559, Saunders, Philadelphia, 1986.)*

Pharynx Find this portion of the digestive tube in segments 1 through 6. The anterior part,
the buccal pouch or mouth cavity, occupies the first two segments. The pharynx acts as a
suction pump to ingest loose soil when dilator muscle fibers attached to the body wall
contract and expand the pharynx. Subsequently, when the mouth is closed, the muscular walls
of the pharynx contract and force the contents posteriorly.

Cerebral ganglion Look for a small whitish, bilobed structure, the so-called brain (cerebral
ganglion), embedded in the muscle tissue on the dorsal anterior surface of the pharynx.
Locate the ventral nerve cord running beneath the gut. It is attached to the cerebral ganglion
by a pair of connections encircling the esophagus.

Esophagus Observe this thin-walled narrow channel that extends from the pharynx to segment
13 or 14. Most of the esophagus is hidden by the overlying seminal vesicles and aortic loops.

Crop and **gizzard** These expanded portions of the food tube lie in segments 14 through 20,
just posterior to the seminal vesicles. The gizzard, located posterior to the crop, is more
muscular. Food accumulated in the crop is passed to the gizzard to be ground with the help of
sand and soil particles.

Intestine The remainder of the digestive tract, fairly uniform in appearance, is the intestine.

Seminal vesicles You will find these prominent whitish bodies in segments 9 through 13.
Earthworms are hermaphroditic; however, copulation, with mutual transfer of sperm, is the
characteristic method of reproduction. Two worms align with their ventral surfaces opposed
and with their anterior ends facing opposite directions. The worms are held together by
mucus secreted by the **clitellum,** a reproductive region consisting of five or six segments
beginning at about segment 32 in *Lumbricus,* and sometimes by special genital setae. Sperm,
released from the genital pores, pass down a pair of grooves on the ventral surface and enter
the seminal receptacles of the other worm (the grooves are covered over with mucus, thus
keeping the sperm from the two worms separate). Copulation takes 2 to 3 hours.

A few days after copulation, the clitellum secretes a dense cocoon material. Albumin is also secreted inside the cocoon. The secretions form a band which encircles the body and eventually begins to slip forward. Eggs from the female genital pores and sperm from the seminal receptacles are deposited inside the cocoon as it slips toward the anterior end, where the open ends seal as it is shed. Fertilization and development take place within the cocoon.

Dorsal blood vessel Identify the very fine, dark, reddish-brown tube that runs longitudinally along the mid-dorsal line of the intestine. Follow it anteriorly. In the region of the seminal vesicles, partially concealed, it gives rise to five large lateral branches, the aortic loops ("hearts") that encircle the digestive tract and attach to the ventral blood vessel. Blood moves anteriorly in the pulsating dorsal vessel and posteriorly in the ventral vessel. In the earthworm, hemoglobin is dissolved in the circulating blood. Three of the four blood pigments found in metazoan animals are present in the phylum Annelida.

3. Study a prepared slide of a cross section of the earthworm. Use low power (10×) to locate each of the following structures and label them on Figure 25C-10.

   a **Intestine** The horseshoe-shaped space at the center of the section is the cavity (lumen) of the intestine. The dark-staining tissue surrounding the lumen contains gut-epithelial cells, which absorb food molecules from the gut. Locate the dorsal fold in the intestine or **b typhlosole,** which increases the surface area for digestion and the uptake of the products of digestion.

   c **Muscle tissue** Two thin bands of muscle tissue surround the gut and are associated with the movement of food through the gut. These are of mesodermal origin.

   d **Chlorogen tissue** Special digestive cells (lighter staining cells around the intestine), performing some of the functions carried out by the liver of higher organisms, including the storage of glycogen and fat, the synthesis of urea, and the removal of silicates.

   e **Coelom** The large space bordering the digestive tract, lined by a sheet of mesoderm called the peritoneum. The fluid-filled space is called a **hydrocoel.**

   f **Nerve cord** Flattened, oval structure in the interior of the coelom. Recall from your dissection that it is a ventral nerve cord.

   g **Blood vessels** One or two large blood vessels should be apparent next to the nerve cord. A fairly large dorsal vessel can be seen above the intestine. Anteriorly, this vessel gives rise to several stout lateral loops ("hearts") that encircle the intestine.

   h **Nephridium** Inside each coelomic segment is a pair of excretory organs, the nephridia. Open ciliated funnels on the end of the nephridia filter the coelomic fluid and then carry it to the outside through excretory pores in the next most posterior segment. During the passage of fluid through these ducts, water can be reabsorbed and the contents modified as in the kidney tubule of vertebrates.

   i **Longitudinal** and **j** **circular muscle bands** Two bands of muscular tissue are located on the inside of the body wall. Increase the magnification to observe the muscle arrangement in greater detail. Identify the longitudinal bundles and circular bands of muscle tissue. Because the earthworm is segmented—its body is made up of a number of separate sections or compartments—the contraction of these muscles, acting in concert with the supporting hydrocoel present in each segment, can produce local movements that are quite different from the "lashing" activities of the vinegar eel. In general, contraction of the longitudinal muscles shortens (or bends) the worm while contraction of the circular muscles lengthens it. However, one part of the earthworm may be shortening or remain immobile while another part is lengthening. If live worms are available, observe their movements.

   k **Epidermis** Thin tissue surrounding the muscles. It secretes the **l cuticle** (body covering) and the slime trail upon which the earthworm glides. Respiration occurs by diffusion of gases through the epidermis.

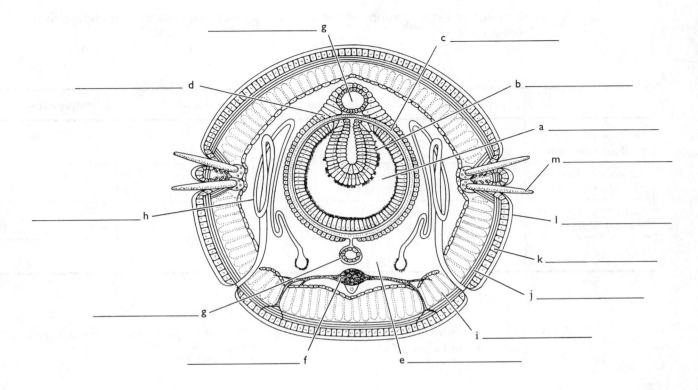

Figure 25C-10   *Label the structures visible in the earthworm cross section.*

m **Seta** Setae may be present in pairs as short stalks penetrating the integument. They are used to anchor one part of the worm's body while it moves another part forward.

c. What does the coelom of a living earthworm contain? _____

d. What is the function of its coelom? _____

4. Look at the preserved and live specimens of other representative annelids on demonstration.

Laboratory Review Questions and Problems

1. Define each of the following terms and describe how each is related to the others: Metazoa, Parazoa, Eumetazoa, and Protozoa.

2. Summarize your observations of the Cnidaria and Ctenophora by completing the following table.

	Hydrozoans	Scyphozoans	Anthozoans	Ctenophorans
Polyp or medusa as predominant form				
Presence or absence of cnidocytes				
Distinguishing characteristics				
Examples				

3. Why is the phylum of organisms containing hydrozoans, jellyfishes, corals, and sea anemones named Cnidaria?

4. What type of symmetry is characteristic of the ctenophores? Do they possess nematocysts like the Cnidaria?

5. Describe the type of symmetry and the type of body cavity in the phylum Platyhelminthes.

6. How does the function of protonephridia in ribbon worms and flatworms differ? How does the structure of the digestive tract in these two groups of organisms differ?

7. Fill in the "family tree" below to clarify your understanding of how coelom type is used to classify animals. Draw a line to divide the true coelomates from the other organisms.

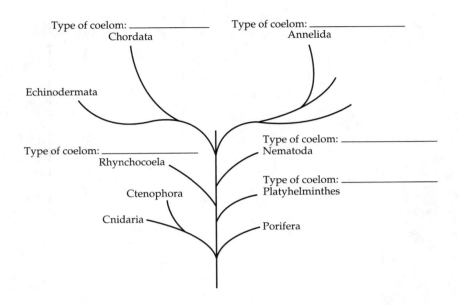

8. How are the coeloms of flatworms, ribbon worms, roundworms, and segmented worms constructed? How does this relate to the way in which the family tree of animals is constructed (see Figure 25-I)?

9. The organization of body tissues into layers allows us to distinguish between the parazoans and the eumetazoans that are diploblastic and triploblastic. On the family tree in question 7, draw circles around the groups of organisms that have the same number of tissue layers. Indicate how many tissue layers each group has, using the correct terms.

10. Symmetry differs among organisms. Draw a line on the tree in question 7 to separate asymmetrical animals from bilateral animals.

11. From your answers to previous questions and from your textbook reading, fill in the following summary table. Compare the characteristics of the major phyla you have studied and identify any evolutionary trends.

	Porifera	Cnidaria	Platyhelminthes	Nematoda	Annelida	Evolutionary Trend
Number of tissue layers						
Type of digestive cavity						
Type of coelom (if present)						
Type of reproduction (sexual/asexual)						
Larva (if discussed)						
Nervous system (form and location)						
Type of circulation (open or closed)						
Type of excretory organs (structure; osmoregulatory or excretory functions)						
Type of symmetry						

Diversity—Mollusks, Arthropods, and Echinoderms

OVERVIEW

In Laboratory 25, several phyla of multicellular animals illustrated important evolutionary advances in organization—from the loosely organized tissuelike structures found in sponges to the two-layered (diploblastic) structure characteristic of cnidarians and the three-layered (triploblastic) structure of flatworms and all other organisms with organs and organ systems. The appearance of a functional body cavity in the nematodes (roundworms) and other wormlike groups was another particularly significant evolutionary milestone. From this point on the "trunk" of the family tree of animals, invertebrates can be divided into two discrete phylogenetic lines, the **protostomes** and the **deuterostomes** (review Figure 25-1 in Laboratory 25).

Protostomes were introduced in the last laboratory with the study of annelid worms. In this evolutionary line, the coelom develops as a schizocoel. In addition, mitotic divisions during the cleavage of the egg produce daughter cells whose fate is more or less committed to the production of a specified part of the body at division (the cell division is **determinate**). Cleavage is also said to be **spiral** due to the characteristic alignment of the mitotic spindles during division. In the eight-celled embryo, the top layer of cells spirals as though it were twisted to the right or left relative to the lower layer of cells. In addition, the blastopore (the opening into the developing gut cavity of the gastrula) becomes the mouth, hence the name protostome (first mouth) for this phylogenetic line.

In deuterostomes, the coelom develops as an evagination (outpocketing) of the primitive gut, or *enteron,* and is called an enterocoel. The mitotic divisions of the fertilized egg produce cells whose fate is not determined at the time of division, hence early cell division is said to be **indeterminate.** Cleavage is **radial:** daughter cells remain aligned above one another. The blastopore becomes the posterior opening of the gut and the mouth end forms later in development—hence the name deuterostome (second mouth) for this evolutionary line that leads eventually to humans (Figure 26-1).

In the present laboratory, you will complete your study of protostomes (phylum Mollusca and phylum Arthropoda) and begin a more detailed examination of the deuterostomes with the phylum Echinodermata.

STUDENT PREPARATION

To prepare for this laboratory, read the text pages indicated by your instructor. Familiarizing yourself in advance with the information and procedures covered in this laboratory will give you a better understanding of the material and improve your efficiency. Review Figure 25-1 carefully.

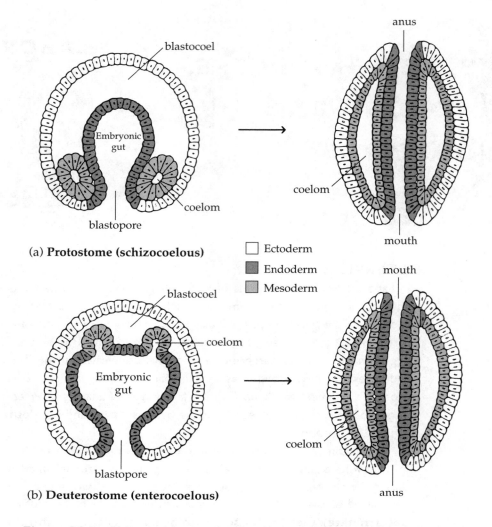

blastocoel

anus

Embryonic
gut

coelom

blastopore

coelom

mouth

(a) **Protostome (schizocoelous)**

☐ Ectoderm
■ Endoderm
▨ Mesoderm

blastocoel

mouth

coelom

Embryonic
gut

coelom

blastopore

coelom

(b) **Deuterostome (enterocoelous)**

anus

Figure 26-1   *(a) In protostomes, the mesoderm takes shape between the endo-derm and the ectoderm, in the region around the blastopore. The coelom arises from a splitting of the solid mesoderm. As the cells continue to multiply, the blastocoel—the original embryonic cavity of the blastula—is obliterated. The blastopore of protostomes becomes the mouth.*

   (b) In deuterostomes, the mesoderm originates from outpocketings of the embryonic gut. These outpocketings create cavities within the mesoderm that become the coelom. The blastopore becomes the anus, and the mouth develops elsewhere.

✔ **EXERCISE A   Phylum Mollusca**

Mollusks represent the second largest phylum, consisting of more than 50,000 living species of marine, freshwater, and terrestrial animals. They are bilaterally symmetrical, coelomate, and apparently unseg-mented.

   The general body plan of a mollusk includes three regions: the **head-foot** (used in locomotion and food capture), the **visceral hump** or visceral mass (containing the major organ systems), and the **mantle** (soft tissue that secretes the calcium-containing shell present in many mollusks).

   Mollusks have an open circulatory system with a chambered heart (one ventricle and two atria) and their blood contains an oxygen-carrying respiratory pigment, hemocyanin. Excretory organs, the **metanephridia,** drain the relatively small coelom surrounding the heart and a portion of the intestine.

Gills are present in the mantle cavity of most mollusks. Mollusks may be filter-feeders, sediment-feeders, herbivores, or carnivores.

Mollusks with shells include species that have shells formed from several plates (**chitons**), hinged shells (**bivalves,** including clams, oysters, and scallops), conical, twisted shells (**gastropods,** including snails and limpets), and reduced or internalized shells (**cephalopods,** including squids, octopuses, and cuttlefishes). (See Figures 26A-1, 26A-2, and 26A-3.)

IIIII **Objectives** IIIIIIIIIIIIIIIIIIIIIIIIIIIIIIIIIIII

☐ Distinguish between the shell types characteristic of chitons, bivalves, gastropods, and cephalopods.

☐ Explain what the markings on the inner surface of a clam shell represent.

☐ Explain how water is moved through a clam for respiratory and feeding purposes.

☐ Distinguish between an open and a closed circulatory system.

☐ Distinguish between torsion and coiling of the gastropod shell.

☐ Describe how a cephalopod makes use of the visceral mass of its foot and the mantle cavity.

✔ **PART 1** Chitons

IIIII **Procedure** IIIIIIIIIIIIIIIIIIIIIIIIIIIIIIIIIIIIIIIIIIIIII

Study the preserved chitons (class Polyplacophora) on demonstration. In most chitons, several plates or valves cover the upper body, suggesting a type of "segmentation," although this pattern is not reflected by internal structures (Figure 26A-1). These plates may be protective adaptations, allowing the organism to better fit the ocean bottom or to wrap itself into a ball. Identify the mouth and general head region. The specimen on demonstration has a well-developed foot. Locate the foot on the ventral surface. The outer covering of the body beneath the shell is the **mantle.**

   a. How many plates are present in the shell? _____

Figure 26A-1   *(a) External body plan of the chiton. (b) The chiton in cross section.*

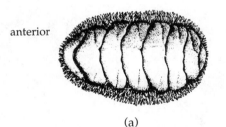
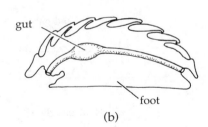

anterior

gut

foot

(a)   (b)

✔ 👁 **PART 2** Bivalves

IIIII **Procedure** IIIIIIIIIIIIIIIIIIIIIIIIIIIIIIIIIIIIIIIIIIII

1. Examine a clam (class Bivalvia) shell.

   a. Is a clam a bivalve mollusk or a gastropod? _____

   b. The concentric rings on the shells of mollusks are referred to as growth rings. Which part of the shell do you suppose forms first? _____

Why does the outside of the shell have lines? The clam shell is composed of three layers: an outer horny layer, middle prismatic layer (mostly calcium carbonate), and inner pearly layer (this is where the pearl buttons on your shirts come from). The first two layers are laid down by the cells in the edge of the mantle, so as the clam grows the mantle makes layers in the outward direction. The entire mantle lays down the pearly substance. If a parasite or foreign object gets stuck in the mantle of a clam or an oyster, concentric rings of pearly material are secreted around it and a "pearl" is produced.

2. Orient the shell so that the anterior part, called the **umbo,** or beak, corresponds to the orientation of the shell in Figure 26A-2a. As you examine features in the shell, match the letters in the text to the letters in Figure 26A-2b; write the name of each anatomical structure on the diagram.

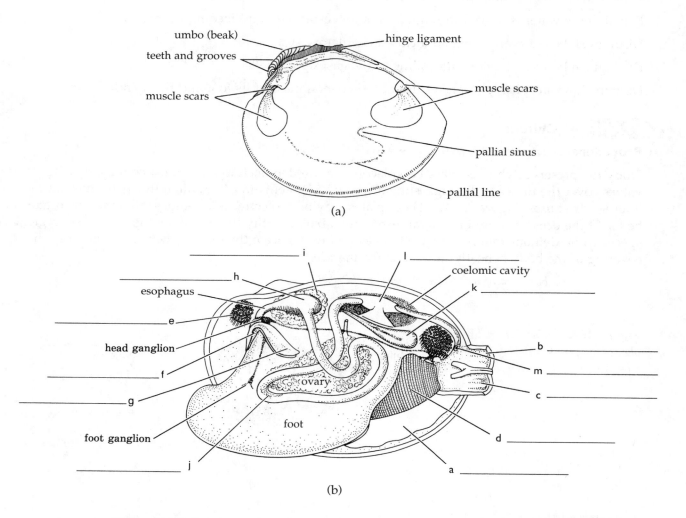

Figure 26A-2   *(a) Features of the clam shell. (b) Internal structure of the clam.*

    The faint line (**pallial line**) that follows along the margin of the shell marks the edge of the attached **mantle tissue a** inside the shell. At the posterior end of the shell, the pallial line furrows inward. The incurrent opening **b** and the excurrent opening **c** (siphons), which carry water to and from the mantle cavity, originate in this region. Water currents bathe the mantle and **gills d** to facilitate gas exchange, and water passed through the gills is filtered through **mucous nets** to remove food particles that are subsequently eaten. Trace the path of food movement in Figure 26A-2b.

3. Locate the prominent **muscle "scars" e** on the inside surface of the shell. The contraction and relaxation of the muscles (whose points of attachment you see inside the valve) are responsible for closing and opening the shell. The two halves of a bivalve shell are held together by the **hinge ligament** whose scar may be visible along the inner dorsal edge of the shell. Also along this inner edge or lip are "teeth" and grooves that fit together with the corresponding "teeth" and grooves of the opposing valve.

4. Observe a preserved clam. Identify the following structures: **mouth f, palps g, stomach h, digestive gland i, intestine j, excretory organ k, heart l,** and **anus m.**

   The saclike stomach is surrounded by the digestive gland. The intestine runs from the stomach through the foot and passes through the cavity of the heart or *pericardial cavity* (the heart is actually wrapped around the intestine). Intestinal wastes are eliminated at the excurrent pore. Excretory organs (paired) lie beneath the heart, each resembling a tube coiled back on itself. The anterior portion of each organ collects waste from the blood and pericardial cavity, and the posterior portion eliminates waste into the dorsal gill passage.

   The circulatory system is open. The heart has three chambers (one ventricle and two atria). Blood is pumped anteriorly and posteriorly in open sinuses. When the large blood sinus in the foot is engorged with blood, it swells and makes locomotion possible.

   The nervous system consists of only three ganglia (foot, head, and visceral ganglia) connected to each other by long neurons. Sensory cells containing small amounts of limestone are found in the food and are used for balancing. A small patch of cells near the visceral ganglion is sensitive to chemicals.

5. Observe the compressed, bladelike form of the large muscular **foot** used in burrowing. Movement of this organ depends on blood pressure changes and on the pedal protractors and retractors, muscles that extend from each side of the foot to the shell on the opposite side and attach near the anterior adductor muscles.

   c. *How is the foot used in burrowing?* _____

   _____

6. Examine other bivalve mollusk shells on demonstration. You are probably familiar with many of these.

   d. *List these representatives.* _____

   _____

EXTENDING YOUR INVESTIGATION: FILTER-FEEDING

Clams are filter-feeders and feed mostly on microscopic particles. If live clams are available, you can follow the ciliary action using fine carbon particles. Form a hypothesis that predicts the pathway that carbon particles will take if the clam treats them as food.

HYPOTHESIS:

NULL HYPOTHESIS:

What do you **predict** will happen when carbon particles are placed on the surface of the gills?

Identify the **independent variable** in this investigation.

Identify the **dependent variable** in this investigation.

Use the following procedure to test your hypothesis:

PROCEDURE:

1. Use a candle or several wooden matches to apply a small amount of carbon film on the outside of a beaker or test tube.

2. Insert a one-piece scalpel or cartilage knife between the shells of a live clam. Angle the blade against the inside of one side of the shell and carefully cut the adductor muscles that hold the shells closed. **Do not cut toward your hand!**

3. Open the clam and place it in a dish with just enough water (seawater for a marine clam) to fill the shell and moisten the exposed gills.

4. Use a dissecting needle or small brush to remove a few carbon particles from the smoked beaker prepared in the first step. Gently place these particles on the posterior surface of the gills.

RESULTS: Determine how the carbon particles move across the gills.

What did you observe happening to the carbon particles?

Do your results support your hypothesis?

Your null hypothesis?

Was your prediction correct?

What do you **conclude** about the way in which filter feeders like the clam obtain food?

PART 3   Gastropods

|||||| **Procedure** ||||||||||||||||||||||||||||||||||||||

1. Observe the gastropod mollusk shells (class Gastropoda) on display.

   a. *Does the shell coil to the right or to the left?* _____

   The direction of shell coiling is determined as early as the eight-cell stage in development by maternal messenger RNA contained in the egg. The coiled nature of the shell disappears in some gastropods, with the adult shell representing a single large expanded whorl as in the abalone or slipper shells.

   In some gastropods, such as terrestrial slugs and marine nudibranchs, the shell may be completely lost. Note the many ornate ridges and bumps on the shells of gastropods on display. Look for the channels and grooves in the shells that house siphons through which water is flushed to aerate the gills.

2. Examine living pulmonate land snails on display. These snails do not have gills. Instead, the mantle cavity functions like a lung.

   b. When a snail is crawling about, what parts of its body are outside the confines of the shell?

   _____

   c. Would you say that cephalization (development of the head) is more apparent in gastropods than in

   bivalves? _____ Why? _____

3. Move a blunt probe toward the snail you are observing. Which part of the body retracts into the shell first? Notice the disk-shaped plate (**operculum**) covering the opening of the shell.

   d. Does the operculum completely seal off the snail's body from its surroundings when the animal is

   disturbed? _____ Is the operculum continuous with the shell itself? _____

   e. Where does the operculum go when the snail is crawling along? _____

4. Snails and slugs secrete mucus that is laid down by the foot. The slime is protective—a slug can pass unharmed over the edge of a sharp razor—and also contains hormones (pheromones) that guide other snails along the same trail. Can you find the slime trail that is formed as the animal crawls across the substrate?

PART 4 Cephalopods

IIIII Procedure IIIIIIIIIIIIIIIIIIIIIIIIIIIIIIIIIIIIIIIIII

1. Examine the squid on display. Cephalopod mollusks (class Cephalopoda) are among the most advanced invertebrates. The name means "head-footed," for in these animals the foot, which is divided into a number of "arms," is wrapped around the head (Figure 26A-3). The cephalopod head has complex sensors and the nervous system is well developed. Cephalopods are even capable of learning complex tasks.

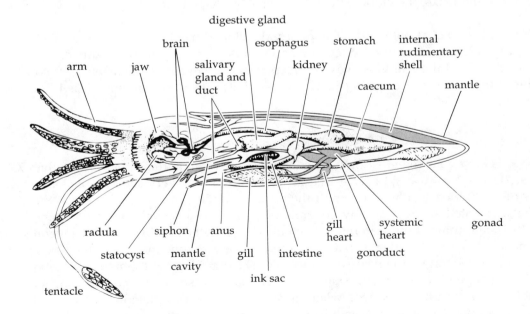

Figure 26A-3 *Internal structure of a squid. In the cephalopods, the head is modified into a circle of arms, and part of the head-foot forms a tubelike siphon through which water can be forcibly expelled. The arrows indicate the direction of water movement.*

2. The shell, characteristic of most mollusks, is only a thin vestige or "pen," buried in the body tissues of the squid. However, in the cuttlefish, a close relative of the squid, this internal shell is calcified. If you have cared for a pet bird, you may have provided this skeleton ("cuttlebone") for your pet to use in sharpening its beak and as a source of calcium. Examine a squid's "pen" and a cuttlefish's "cuttlebone" on demonstration.

3. The mantle, which secretes the "shell," is thick and muscular and serves a protective function, compensating for reduced shell size. When the mantle is relaxed, water enters the mantle cavity and, as it contracts, the edge of the mantle becomes tightly sealed to the body, forcing water out the funnel. Gently probe this area with your fingers to see these structures. When the animal is excited, the mantle contracts strongly and water is forced out of the funnel, pushing the animal in the direction opposite to that of the water jet. When attacked, the squid may emit dark ink from an ink sac opening into the funnel, thereby escaping as the predator is confused by this "smoke screen."

4. Examine the head. The squid has eyes constructed much like human eyes, although they developed in quite a different way.
    The squid has internal cartilaginous supports and cartilage protects the brain.

5. If a dissected squid specimen is available, locate the gills, ink sac, intestine, gill hearts, stomach, and systemic heart (see Figure 26A-3).

6. The octopus and chambered nautilus are also cephalopod mollusks. An external shell has been retained in the nautilus, but is lost in the octopus.

    a. What structure represents the foot in the octopus? _____

    The mantle of the octopus is fused to the body wall, and the entrance to the mantle cavity is more restricted than in other cephalopods. The octopus can swim by creating a jet of water, but typically it crawls over rocks and retreats into its den when attacked.

✔ **PART 5**   **The Scaphopoda**

This class of mollusks, the "tooth shells," have shells shaped like an elephant's task. Study the tooth shells on display.
    The shell is open at both ends. The poorly developed head bears a number of extensible filaments used to capture prey. A radula for feeding and a muscular foot for burrowing are also present. The gills have been lost and the mantle serves as a respiratory organ as water currents are constantly maintained in and out of the upper end of the shell. The lower end of the shell remains buried in the sand.

👁 **EXERCISE B**   **Phylum Arthropoda**

Arthropods are by far the most numerous and diverse of all animals, with more than 750,000 known species. Marine, freshwater, or terrestrial forms are found in every conceivable habitat due to their high degree of evolutionary adaptability and their great mobility, including, for some, the ability to fly.
    The segmented arthropod body, covered by a chitinous **exoskeleton,** is typically divided into three parts: the **head, thorax,** and **abdomen.** Each of these may be subdivided into several segments to which are attached jointed appendages that carry out a variety of functions. As arthropods grow, they shed their chitinous exoskeleton by the process of **molting.** During this growth, many arthropods may also undergo a marked change in form (**metamorphosis**). If this is the case, the larva, a feeding stage, often bears no resemblance to the adult produced when metamorphosis is completed.
    The arthropod circulatory system is open: a distinct muscular heart (Figure 26B-1) pumps blood through open spaces in the tissues—the **hemocoel.** The coelom, correspondingly, has been reduced and is represented in most arthropods only by the cavity of the gonads. The digestive tract of arthropods is well developed and modified into several distinct parts. The nervous system and associated sense organs are particularly well developed and control a variety of complex behaviors, including flight in winged insects.

Figure 26B-1 *Generalized body plan of an arthropod.*

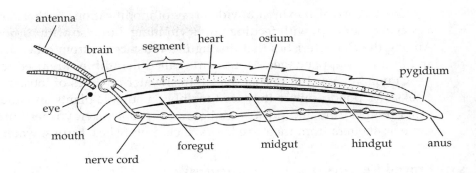

The group of organisms we collectively call arthropods appears to have diverged into at least four separate subphyla, often regarded as phyla. The subphylum Cheliceriformes (or phylum Chelicerata) includes the classes Merostoma (horseshoe crabs), Arachnida (spiders, mites, ticks, scorpions, and their relatives), and Pycnogonada (sea spiders). Chelicerates lack antennae and mandibles (jaws). Instead, the first pair of appendages, the **chelicerae,** are in the form of pincers or fangs. The subphylum (or phylum) Crustacea includes crabs, lobsters, shrimps (class Malacostraca) and barnacles (class Cirripedia), and a number of other small aquatic invertebrates. The crustacea are **mandibulates** and are characterized by **biramous** (two-branched) appendages, two pairs of antennae and mandibles (jaws), and a pair of compound eyes. Members of the subphylum Uniramia* include the insects (class Insecta), centipedes (class Chilopoda), and millipedes (class Diplopoda). Insects are **uniramous,** having only one pair of antennae and unbranched (uniramous) appendages.

A fourth group of arthropods (subphylum Trilobitomorpha or phylum Trilobita) is an extinct taxon of primitive arthropods that were abundant in the Cambrian and Ordovician Periods and are known from their rich fossil remains.

The origin of arthropods remains a mystery. Systematists originally believed that members of the phylum Onycophora, a group of wormlike animals with claw-bearing legs and a thin chitinous cuticle, represented a type of common ancestor, a "missing link" between annelids and arthropods. Ribosomal RNA evidence, however, indicates that onycophorans are true arthropods. Members of a second phylum, Tardigrada, the "water bears," are similar to the onycophorans. Most likely these two groups are similar in structure to early arthropods but do not represent common ancestral lines.

In examining representatives of three major groups of arthropods (Crustacea, Arachnida, Insecta), we will concentrate on characteristics that are particularly useful in distinguishing arthropod species: the numbers and types of appendages, the degree of external segmentation, and the types of sensory organs.

▎▎▎▎▎ Objectives ▎▎▎▎▎▎▎▎▎▎▎▎▎▎▎▎▎▎▎▎▎▎▎▎▎▎▎▎▎▎▎▎▎▎▎▎▎▎▎

☐ Distinguish among the external body plans of the crustaceans, arachnids, and insects.

☐ Discuss the importance of molting in arthropod life cycles.

☐ Describe the respiratory system of insects.

☐ Give reasons for the success of insects.

✔ **PART I**   **Crustacea: External Morphology of the Crayfish**

Crustaceans exhibit a great range of sizes, from lobsters and crabs to tiny water fleas and brine shrimp, hundreds of which can be cultured in a small laboratory beaker.

If Uniramia is considered a phylum, two subphyla are identified: Myriapoda (centipedes and millipedes) and Insecta.

Crustaceans also exhibit a wide array of modifications to their jointed appendages. These are associated not only with feeding and swimming, but also with reproduction, respiration, and burrowing. Among the characteristics that distinguish crustaceans from other arthropods are two pairs of antennae and biramous appendages. Mouthparts called **mandibles** are used to crush and grind food.

Crayfishes, used in this study, are freshwater relatives of lobsters. They patrol the bottoms of shallow creeks and ponds, where they feed mainly on aquatic plants and and on animal carcasses. When high water temperature and turbulence become a problem, crayfishes burrow into the mud. Their burrows also serve as shelters from their predators, including fishes, turtles, water snakes, and humans.

Procedure

1. Carefully remove a crayfish from the jar of preserved specimens available on the demonstration table. Place it dorsal side up in a dissecting tray (or Petri dish) and use the dissecting microscope to make your observations.

2. Note the general body plan of the crayfish. It is divided into a **cephalothorax** (anterior half) and an **abdomen** (posterior half). The dorsal portion of the cephalothorax of crustaceans is heavily mineralized (with calcium and phosphate salts) and is called the **carapace**. Running laterally across the carapace is a suture (the **cervical groove**) which marks the nearly complete fusion of the head and thorax.

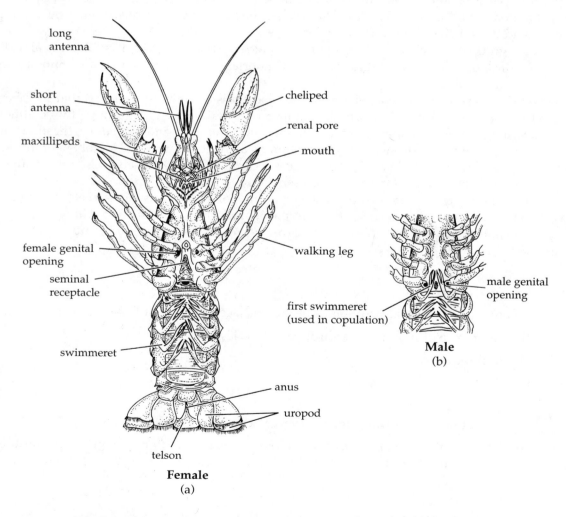

Figure 26B-2 *External features of (a) a female and (b) a male crayfish.*

3. Examine the head of your crayfish under a dissecting microscope. The carapace ends anteriorly in a protective hood, the **rostrum.** On either side of the rostrum is a compound eye, situated on a stalk that allows the eye to move. Gently use your forefinger and thumb to peel off the outer layer of the eye (the cornea). Note the many small rectangular facets that make up the structure of the compound eye.

4. Crustaceans have two pairs of antennae, used primarily as sensory appendages. Briefly compare the structure of the two types of antennae. At the base of one of the long antennae, locate the small pore that functions in the excretion of nitrogenous waste and excess water.

5. In addition to antennae, crayfishes have 17 other pairs of appendages; three pairs make up the inner jaw (mandibles and maxillae), three pairs make up the outer jaw (maxillipeds), five pairs are associated mainly with crawling and digging, and the most posterior (abdominal) six pairs are used mainly in swimming. It is best to examine these with the crayfish positioned so that the ventral side is facing you (Figure 26B-2). As you study each type of appendage, relate its structure to its function.

6. Summarize the characteristics of crustaceans in Table 26B-1.

Table 26B-1   Comparison of the External Morphology of Arthropods

Characteristics	Crustaceans	Arachnids	Insects	Myriapods
Main divisions of the body				
Main body divisions that show external segmentation				
Locomotor structures				
Other appendages				
Number of pairs of antennae				
Mouthparts				
Number and type of eyes				

✔ **PART 2**   **Arachnida: External Morphology of a Spider**

Arachnids are a group of eight-legged terrestrial arthropods that are, for the most part, fierce predators of other arthropods. The most distinctive characteristic of these carnivores is a pair of appendages, the **chelicerae,** used in procuring and eating prey. Arachnids include not only spiders, but also scorpions, harvestmen (daddy longlegs and their relatives), ticks, and mites.

Many arachnids use venom to paralyze their prey. **Silk,** another common arachnid product, may be used for prey capture, reproductive activities, and dispersal.

ııııı **Procedure** ıııııııııııııııııııııııııııııııııııı

1. Use a pair of forceps to carefully remove a preserved spider from its vial. Place it in a Petri dish containing alcohol and a cotton pad. With the ventral side resting against the cotton, orient it so that the front edge of the cephalothorax is toward you. Use the dissecting microscope to examine the spider.

2. In arachnids, as in crustaceans, the body is divided into two parts, the cephalothorax and abdomen.

   a. How can you distinguish the cephalothorax of your specimen from its abdomen?

   _____

   b. To which part of the body are the eight legs attached? _____

3. Position the spider against the cotton pad so that the anterior edge of the cephalothorax is turned upward and the spider is "looking up" into the microscope. Note that the eyes of spiders are simple rather than compound. They occur in pairs of different sizes.

   c. How many eyes can you find? _____ *How are these eyes arranged?* _____

   d. Which pair of eyes is the largest? _____ (In some spiders, the posterior pair of eyes is well back on the sides of the cephalothorax.)*

4. The first pair of appendages, the **chelicerae,** are below the first pair of eyes. Turn your specimen over onto its dorsal surface (so that you view its ventral surface as in Figure 26B-3). Gently push aside the other appendages with your dissecting needle to find where the chelicerae are attached to the body. In spiders, these stout but powerful appendages are sharply pointed to form fangs used to inject venom into prey. In other arachnids, the chelicerae are not fangs but claws.

   The mouth of a spider lies directly posterior to the chelicerae. It is a small hole that might be difficult to see because of the many hairs that surround it. These hairs are used in filtering out solid materials—spiders consume only the liquefied portion of their prey.

Figure 26B-3 *External features of a female spider.*

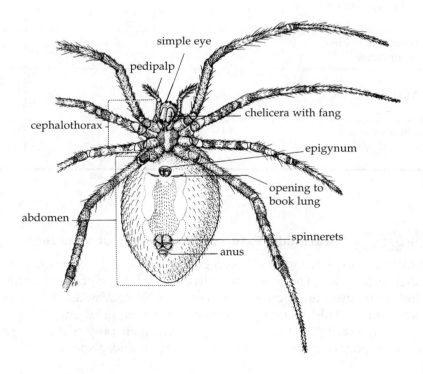

5. On either side of the chelicerae are the **pedipalps,** short leglike appendages used for a number of purposes. In male spiders, the tips of the pedipalps contain an organ (the **tarsal organ**) used to store sperm. This organ and the sensory hairs around it give the tip of the pedipalp the appearance of a tiny boxing glove.

   If your specimen does not appear to be a male, examine its abdomen for a furrowed area bordered by patches of color (Figure 26B-3). This is the **epigynum,** whose two tiny openings lead into the female reproductive tract. The male inserts its tarsal organs into the opening of the epigynum to deliver sperm.

6. On either side of the epigynum, look for the openings to the **book lungs.** (They are in the same location in males as in females.) A book lung consists of a highly folded membrane with a rich supply of blood capillaries. It is used for respiratory gas exchange.

7. At the tip of the abdomen is a pair of **spinnerets,** appendages that secrete silk from glands within the abdomen. Each pair is specialized to secrete a different type of silk suited to a specific purpose.

   e. How many spinnerets are present in your specimen? _____

   Posterior to the last pair of spinnerets is the **anal tubercle** with an anal pore at its tip. This pore is usually closed and not visible.

8. Summarize arachnid characteristics in Table 26B-1. Compare these with the structural characteristics of crustaceans.

PART 3 Chelicerata: Biochemical Taxonomy of the Horseshoe Crab (*An Optional Exercise*)*

Although *Limulus* (the horseshoe crab) appears to have many anatomical and physiological similarities with crustaceans, biochemical evidence suggests that it is more closely related to the arachnids. With which group does *Limulus* belong?

Organisms demonstrating biochemical similarities among their proteins are thought to be more closely related to each other than to other organisms. The number and sequence of amino acids and the manner in which a protein folds are all a reflection of the "recipe" in an organism's DNA. Similar proteins have similar amino acid sequences dictated by similar DNA sequences, and scientists believe that similar DNA sequences indicate a common evolutionary origin.

The more biochemically similar two organisms are, the more similar their enzymes will be. In this exercise you will test for the activity of lactic acid dehydrogenase (LDH), a multi-subunit enzyme present in the muscle tissue of all animals. During strenuous exercise, LDH catalyzes the anaerobic conversion of pyyruvic acid to the lactic acid. In this reaction, lactic acid is formed when electrons carried by NADH are accepted by the pyruvic acid:

$$CH_3CCOOH + NADH + H^+ \rightleftharpoons CH_3CHOOH + NAD^+$$

$$\underset{\text{pyruvic acid}}{\overset{\parallel}{O}} \qquad\qquad \underset{\text{lactic acid}}{\overset{|}{OH}}$$

This reaction is reversible, and in this exercise you will study the conversion of lactic acid to pyruvic acid. During this reaction, electrons are transferred from lactic acid to NAD^+ and then to phenazine metasulfate (PMS) and, finally, to dichloro-phenol-indophenol (DPIP). As DPIP accepts electrons and is reduced, its color changes from blue to clear.

The tertiary structures of LDH subunits vary slightly among different organisms. These tertiary structures are determined by the interactions of amino acids, which, in turn, depend upon primary

---

*This experiment was developed by Mary Ellen Hart and Scott E. Pattison. Funding was provided by the National Science Foundation and California State University system under the auspices of the Institute for Cellular and Molecular Biology, California State Polytechnic University, Pomona.

structures and thus upon DNA sequences. The structural variations in LDH among types of organisms result in differences in the interaction with NAD^+ and, consequently, in the degree of color change in DPIP. To investigate the degree of evolutionary similarity among organisms, you will compare LDH activity in spiders (Arachnida), crabs (Crustacea), and *Limulus* (Merostomata).

||||| Objectives |||||||||||||||||||||||||||||||||||||||||

☐ Use biochemical evidence to elucidate evolutionary relationships.

☐ Demonstrate that anatomical similarities do not always indicate close taxonomic or evolutionary relationships.

||||| Procedure ||||||||||||||||||||||||||||||||||||||||||||

1. Label four test tubes: control, *Limulus*, spider, and crab.

2. Add 3 ml of lactic acid solution to each of the four test tubes.

3. If you are using the spectrophotometer to quantitate the DPIP reaction, use a tube containing the lactic acid solution to "zero" the spectrophotometer (see Laboratory 4) at 600 nm. Record the data for all four tubes in the first row of Table 26B-2.

4. Add six drops of DPIP indicator solution to each tube. Cover each tube with a small piece of Parafilm and invert the tube to mix the solutions.

5. Record the color of each of the sample tubes in Table 26B-2. If you are using a spectrophotometer, also record the absorbance.

6. Add 10 drops of spider supernatant to the tube marked "spider" (the supernatant contains homogenized muscle tissue). Cover the tube with Parafilm and invert to mix. Immediately note the initial color and record the absorbance in Table 26B-2.

7. Add 10 drops of crab supernatant to the tube marked "crab," 10 drops of *Limulus* supernatant to the tube marked "*Limulus*," and 10 drops of distilled water to the tube marked "control." Cover each tube with Parafilm and invert to mix. Note the initial color and record absorbance in Table 26B-2.

8. At 5, 10, and 15 minutes, note the color and record the absorbance for each tube.

Table 26B-2

	Spider		*Limulus*		Crustacean		Control	
	Absorbance	/ Color	Absorbance	/ Color	Absorbance	/ Color	Absorbance	/ Color
Lactic acid solution								
Lactic acid solution + DPIP								
Lactic acid solution + DPIP + supernatant: initial reading								
5 minutes								
10 minutes								
15 minutes								

  a. Based on observed color differences or on absorbance data, is Limulus more closely related to the spiders (Arachnida) or to the crabs (Crustacea)? _____

  b. Anatomically, which group of organisms does Limulus more closely resemble? _____

  c. How would you explain the fact that anatomically, Limulus resembles one group of organisms, but biochemically, it is more closely related to another group of organisms?

  _____

  _____

✔ **PART 4** **Insecta: External Morphology of the Grasshopper**

Of all invertebrates, insects are the most diverse and the most complex. Almost 90 percent of all known animal species belong to this group. The enormous success of insects as a group can be attributed, in part, to their diverse and specialized types of locomotor and feeding appendages and their well-developed sensory organs. The most important reason for their success, however, is their capacity for flight.

ııııı **Procedure** ııı

1. Use your dissecting microscope to examine a preserved grasshopper. Orient the specimen in a Petri dish so that its head is to the left as you view it. From Figure 26B-4 and the discussion that follows, locate each of the structures indicated in the text in boldface type. On Figure 26B-4, write in the names of the structures accompanied by letters in the text.

2. The insect body is covered by a strong, protective **cuticle** secreted by cells of the epidermis. The major chemical component of this cuticle is chitin. A layer of pigment covers the chitin and gives the insect its characteristic color. On the outer surface, a waxy layer protects the insect from desiccation.

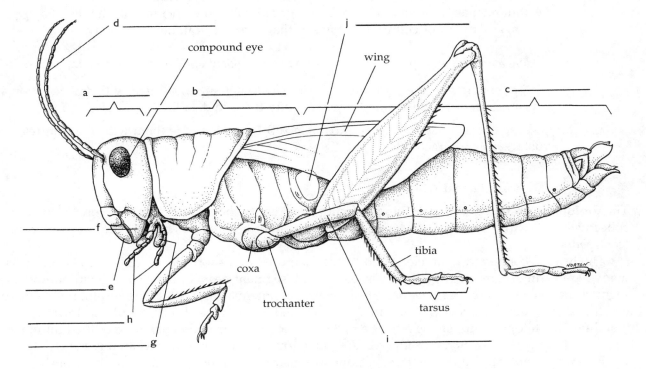

Figure 26B-4 *External anatomy of the grasshopper.*

3. The body of an insect is divided into three parts: the **head a**, **thorax b**, and **abdomen c**.

   a. How does this body plan differ from that of crustaceans and arachnids?

   _____

4. The insect head bears both sensory organs and mouthparts. Locate the **antennae d** (first segment) and the single pair of compound eyes (second segment.) Insects also have simple eyes (**ocelli**) that are much smaller than compound eyes and lack the multiple sense elements found in compound eyes. Ocelli may be very sensitive to low light levels and may also enhance the sensitivity of the compound eyes.

   b. How many ocelli do you see above the antennae? _____

5. Now, using a dissecting needle, carefully manipulate and identify the mouthparts. The third segment bears the upper lip, **labrum e**, and the fourth a pair of toothed, horny jaws, **mandibles f**, which are accompanied by accessory jaws, **maxillae**, on the fifth segment. Both the maxillae and the lower lip, **labium g**, bear sensory **palps h**, delicate, segmented appendages used in manipulating and "smelling" food.

6. Look at the dorsal part of the thorax of your grasshopper. This portion of the thorax is a hardened shield called the **prothorax**. In flying insects, the thorax bears one or two pairs of **wings**. A grasshopper has hardened forewings overlying translucent membranous hindwings. Veins in the hindwing are formed by a network of chitinous tubules that provide strength without adding much weight.

   The thorax bears three pairs of **legs,** one pair on each thoracic segment. All insects have six legs. In the grasshopper, the first two pairs are typical walking legs and the third is specialized for jumping. Identify the leg parts in Figure 26B-4. Do you see the many spinelike and toothlike structures used for grasping? The upper segment of the grasshopper hindleg, the **femur i,** is enlarged; it contains the muscles used in jumping.

7. Now examine the abdomen of your grasshopper. It is divided into several segments externally, but not internally. The first segment of the abdomen of a grasshopper bears a relatively large sense organ, the **tympanum j,** used in detecting airborne vibrations (sound). Along the side of the abdomen are a number of tiny holes, or **spiracles,** the openings of the insect's respiratory system, a network of chitin-lined tubules that spread throughout the body.

   c. How many pairs of spiracles are present on each segment of the grasshopper? _____

   Appendages on the posterior tip of the abdomen are associated with reproduction and sensory detection. You will not need to study these in detail.

8. Add the characteristics of insects to Table 26B-1 and compare them with the features found in crustaceans and arachnids.

✔ **PART 5**   **Diplopoda and Chilopoda**

The **diplopods** (millipedes) and **chilopods** (centipedes) are flightless terrestrial arthropods sometimes collectively called the **myriapods.** Myriapods may be either herbivorous (millipedes) or carnivorous (centipedes).

   The head is connected to an elongated trunk composed of a short thorax and long abdomen with many leg-bearing segments. In centipedes, each trunk segment bears one pair of legs; in millipedes each segment has two pairs of legs. The head bears a pair of antennae and sometimes ocelli, but compound eyes are almost never present. The mouth is surrounded by an upper lip (labrum) and lower maxillae and mandible. In chilopods, appendages of the first trunk segment (maxillipeds) are modified into poison claws for food gathering. Some centipedes are dangerous to humans—their venom is quite potent.

   Examine a millipede and a centipede. Compare their external body plans with those of the crustaceans, arachnids, and insects. Summarize their characteristics in Table 26B-1.

✔ **PART 6** **Insect Mobility**

The capacity for flight has provided insects with the ability to disperse themselves widely and to exploit food resources more efficiently. Flight demands a great deal of energy and oxygen. To fill this demand, insects have developed a highly efficient system for exchanging gases between cells and the environment. This is the **tracheal system,** a finely branching network of tiny tubules through which oxygen passes directly to the cells of muscle and other tissues throughout the body. The tracheal system is independent of the pumping of the heart and operates solely on the principle of diffusion (with muscular activity aiding in the ventilation of larger tracheoles). Although this system is efficient, diffusion into and out of small tracheoles limits the overall body size that an insect can attain.

a. What is the largest insect you have seen or read about? _____

In addition, the tracheal system makes insects particularly vulnerable to various diseases, particularly those caused by fungi that invade body tissues through these tubes.

IIIII Procedure IIIIIIIIIIIIIIIIIIIIIIIIIIIIIIIIIIIIIIIIIII

1. Look at the prepared slide of an insect spiracle.

   b. What is the external location and orientation of the spiracle on the insect's body?

   _____

   c. How is the chitin arranged within the spiracle? _____

   d. What might be a purpose for the tiny hairs lining the air space in the spiracle?

   _____

2. Look at the prepared slide of an insect tracheal system.

   e. How is the chitin arranged within individual tracheae and tracheoles? _____

   f. What might be the purpose of this type of arrangement? _____

   g. Describe the type of branching within the tracheal system.

   _____

3. Examine the preserved insect specimens on display in the laboratory. List those that have wings and indicate how many pairs of wings are present. Record your results in Table 26B-3.

Table 26B-3

No Wings	One Pair of Wings	Two Pairs of Wings

   h. From what division of the body do the wings extend? _____

   i. What other parts of the body do the wings cover? _____

   j. What secondary function does this suggest for the wings? _____

✔ | **EXERCISE C** | **Protostomes: Minor Groups**

There are several groups of protostome invertebrates that do not fit into any of the large phyla we have discussed. These include several phyla of wormlike organisms: the Pogonophora (the largest of which are found along hydrothermal rifts in the ocean floor), the Sipuncula or "peanut worms," and the Echiura and Pentastomida (parasites that inhabit the respiratory systems of reptiles). Tardigrades (Tardigrada), tiny soil organisms called "water bears," are often found in the water films of mosses. (These are a favorite of most biology students). Recall that the tardigrades and onycophorans (Onycophora), which resemble slugs with legs, appear to be similar to primitive arthropods, although their role as ancestral "missing links" is questionable.

Four additional phyla, the Bryozoa, Entoprocta, Phoronida, and Brachiopoda, all possess a tentacular filter-feeding apparatus called a **lophophore**. Often these phyla are grouped together as lophophorate coelomates. However, they are not closely related to one another; bryozoans and entoprocts are more closely related to deuterostomes. Bryozoans or "moss animals" include colonial and *sessile* (attached to a substrate) organisms with a mouth surrounded by tentacles. Entoprocts are similar to the bryozoans but, unlike the bryozoans, both the mouth and anus are located within the lophophore. The phoronids live within chitinous tubes in marine waters, while the brachiopods or "lamp shells" are encased by the dorsal and ventral valves of their shell (rather than left and right valves as in bivalve mollusks). Brachiopods are widely distributed as fossils of the Paleozoic and Mesozoic eras.

ıııı **Objectives** ıııııııııııııııııııııııııııııııııııı

☐ Define "protostome."

☐ Identify minor phyla by describing an organism for each.

ıııı **Procedure** ıııııııııııııııııııııııııııııııııııı

1. Observe the pictures or preserved specimens of this diverse group of organisms on display in the laboratory.

2. Make a wet mount of moss leaves (or use a prepared culture) to observe tardigrades, the "water bears."

✔ | **EXERCISE D** | **Phylum Echinodermata**

If you examine the phylogenetic tree presented at the beginning of Laboratory 25, you will see that our study of echinoderms takes us to a new branch of the tree. Recall that three major differences in early embryonic development and form typify these organisms. First, unlike the protostome organisms we have studied thus far, echinoderms are **deuterostomes**. During development, the blastopore contributes to the formation of the anus rather than of the mouth as in protostomes. Second, the cleavage pattern typical of deuterostome zygotes differs from that of protostomes: cleavage is **radial** rather than spiral. Third, the true coelom of deuterostomes is derived from a mesodermal outpocketing of the primitive gut, or **enteron;** the deuterostomes are called enterocoelous coelomates. Recall that the coelom of annelids, arthropods, and mollusks (all protostomes) in schizocoelous, formed from a splitting of a solid mass of mesoderm within the body.

The phylum Echinodermata includes four major groups of marine bottom dwellers or burrowers: the starfishes and brittle stars (class Stelleroidea); sea urchins and their relatives, the sand dollars (class Echinoidea); sea lilies (class Crinoidea); and sea cucumbers (class Holothuroidea). Echinoderms are noted for their spiny protective skins, their five-part structure, and the presence of numerous small appendages, the tube feet, which function as part of a water vascular system derived from the coelom (Figure 26D-1). The tube feet are used for locomotion, feeding, and respiration. The coelom carries out circulatory, respiratory, and excretory functions.

Figure 26D-1 *The aboral and oral surfaces of the sand dollar.*

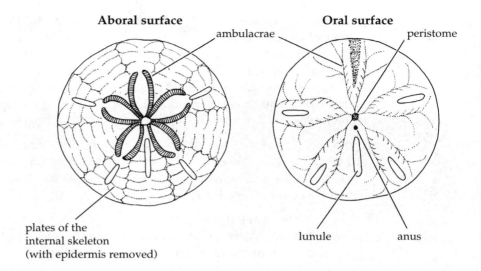

Aboral surface

ambulacrae

plates of the
internal skeleton
(with epidermis removed)

Oral surface

peristome

lunule   anus

ııııı **Objectives** ıııııııııııııııııııııııııııııııı

☐ Define "deuterostome."

☐ Explain why echinoderms form a new branch of the phylogenetic tree.

☐ Give three major characteristics of organisms grouped as echinoderms.

☐ Describe radial symmetry among the echninoderms.

ııııı **Procedure** ıııııııııııııııııııııııııııııııııı

1. Examine the echinoderms on display in the laboratory.

    a. Which forms are radially symmetrical? _____

    b. Which are bilaterally symmetrical? _____

    c. Which have arms? _____

    d. Which are spherical? _____

    e. Which are disklike? _____

2. Unlike arthropods, which have exoskeletons, echinoderms have internal skeletons. The skeleton is composed of flattened calcareous plates called **ossicles.** Spines are outward extensions of these plates and are characteristic of the echinoderms, often called the "spiny-skinned" animals. In addition to spines, some echinoderms (the Stelleroidea and Echinoidea) also have **pedicellaria** extending from their surfaces. These are small "pincers" that aid in capturing food and keeping the body surface clean. Look at the rounded surface of a sand dollar (Figure 26D-1). Do you see the faint lines indicating an array of interconnected plates? Or a separate sheet of paper, make a sketch of this part of the skeleton. In the living sand dollar, as in other echinoderms, these plates are covered by an epidermis and would not be visible. Insert your drawing into the manual.

3. The terms dorsal and ventral are not usually used to describe radially symmetrical organisms. Instead, the terms **oral** (on the same side as the mouth) and **aboral** (on the side opposite the mouth) are preferred. The mouth of radially symmetrical echinoderms is on the lower surface. Identify the oral and aboral surfaces of the sand dollar.

Radiating outward from the center of the sand dollar on both sides is a pattern of "arms" (**ambulacrae**) resembling the arms of a starfish. Along each edge of the ambulacrae is a row of tiny perforations through which the tube feet project.

Gonopores, from which eggs and sperm are shed, may be seen where the ambulacrae join at the center of the organism. The narrow elongated holes in the skeleton, called **lunules,** are channels through which food can be moved from the upper (aboral) surface to the mouth on the lower (oral) surface.

4. Examine the skeleton (**test**) of a sea urchin. Compare its form to that of the sand dollar. Like the sand dollar, the ossicles bear spines, but they are more pronounced. Depending on the species, these can be narrow and pointed, long or short, and may sometimes be very thick and heavy as in the pencil urchins.

5. Examine the oral surface of the sea urchin. At its center is an open ring, the **peristome.** In the live organism, the peristome contains the gills and is covered by a membrane. Inside the ring is a circle of large, mineralized teeth shielding the opening of the mouth. The teeth of both the sand dollar and sea urchin are part of a complex feeding apparatus called Aristotle's lantern. Look again at the oral surface of the sand dollar. Compare its central structure with that of the sea urchin.

f. How many teeth are present in the sand dollar? _____

Surrounding the peristome of living echinoids are rings of tube feet. Tube feet are also scattered over much of the oral surface in sea urchins. Can you see the holes in the test through which the tube feet project? Tube feet near the mouth are used like tiny hands to move food toward the opening. The others are used like tiny suction cups in locomotion. Tube feet are exposed directly to the water and have permeable membranes through which oxygen can diffuse. The anus of the sea urchin is near the center of the aboral surface; it may not be visible.

6. Compare the structure of the sand dollar and sea urchin with that of the sea star (starfish, class Stelleroidea) on display. On the oral surface, identify the ambulacral grooves and spines. In the living starfish, tube feet project from these grooves. Examine the aboral surface and find the **madreporite,** a small round plate placed somewhat off-center on the central disk. The madreporite is a perforated plate that guards the opening through which water enters the water vascular system (Figure 26D-2).

7. Related to the sea stars are the brittle stars. They lack the pedicellaria and skin gills (**papulae;** thin areas of the body wall projecting outward between ossicles) present in sea stars. Also, their tube feet do not have suckers and are sensory in function rather than locomotory. Examine the brittle star on display. Notice that the arms are sharply distinct from the central disk. Both sea stars and brittle stars are able to regenerate lost parts.

8. Examine a preserved specimen of the sea cucumber (class Holothuroidea). Sea cucumbers lack arms, spines, and pedicellaria. They lie on the ocean floor (looking much like cucumbers) and filter-feed by means of branched tentacles, which are really modified tube feet. Some are burrowers. The ossicles are small and are buried in the leathery body wall.

Sea cucumbers have an interesting way of discouraging predators—they eviscerate (cast out part of their internal organs). The predator is left holding the viscera while the sea cucumber escapes. Lost parts are then quickly regenerated.

9. The sea lilies (class Crinoidea) are the most primitive group of echinoderms. Some are stalked and sessile, while others (feather stars) are motile. Crinoids trap detritus and planktonic organisms in the center of their feathery arms and then move it toward the mouth through a ciliated ambulacral groove with the assistance of many suckerless tube feet. In all other groups of echinoderms, the tube feet are simply locomotory or sensory, not ingestive, in function. Examine the preserved feather star specimen on demonstration.

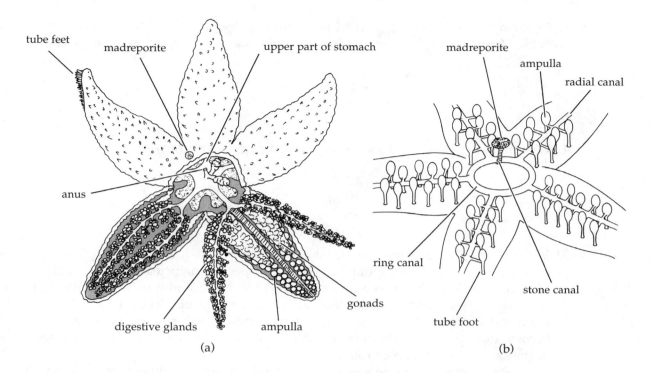

tube feet   madreporite   upper part of stomach   madreporite
ampulla
radial canal
anus
ring canal
stone canal
gonads
tube foot
digestive glands   ampulla
(a)   (b)

Figure 26D-2   *The sea star, (a) external and (b) internal anatomy. The water vascular system is filled with fluid similar to seawater, and some seawater can enter the madreporite to maintain water pressure. The internal canals—stone canal, ring canal, and radial canals—are interconnected. Radial canals terminate in a series of tube feet and bulbs (*ampullae*) separated from the canal by a valve. During locomotion, the ampullae contract and the valves close, forcing water into the tube feet, which extend and adhere to the substrate by means of mucous adhesion and suction.*

Laboratory Review Questions and Problems

1. Protostomes, including the annelids, arthropods, and mollusks, can be distinguished from deutoerostomes, including echinoderms and chordates (to be studied during the next laboratory period). Fill in the following table to summarize the differences.

	Protostomes	Deuterostomes
Type of coelom		
Type of cleavage in egg		
Type of development		
Fate of the blastopore		

2. The shells of chitons are segmented into eight plates, which may be buried in the mantle. What is the significance of this observation with respect to their placement in the annelids, arthropods, and mollusks?

3. Cephalopod mollusks are adapted for swimming by propulsion and for carnivorous eating habits. Explain how the foot and mantle cavity are modified for these purposes.

4. What unifying characteristics do you find among the arthropods?

5. If you were to assign someone the task of efficiently dividing a large, diverse group of objects or organisms into smaller groups according to specific characteristics, you might devise a **dichotomous key.** This method uses pairs (*dichot-*, two) of mutually exclusive descriptions to subdivide larger groups into successively smaller subgroups. These descriptions are based on the possession of, or lack of, particular characteristics. With each subdivision, the characteristics become more limited, until each "path" of characteristics can lead to only one specific member or the smallest identifiable group.

   For example, suppose you are given a rubber bulb, a rubber stopper, a glass rod, a glass slide, and piece of glass tubing. Your dichotomous key might look like this:

1a object rubber   go to 2
1b object glass   go to 3

    2a object squeezable   rubber bulb
    2b object solid   rubber stopper

        3a object flat   glass slide
        3b object cylindrical   go to 4

            4a object hollow   glass tubing
            4b object solid   glass rod

Notice that each alternative choice excludes the other. Each message on the right indicates the "path" to be followed through the key until the object is definitively identified.

   Devise a dichotomous key to separate the five classes of arthropods that you have studied:

1a _____

1b _____

    2a _____

    2b _____

        3a _____

        3b _____

            4a _____

            4b _____

                5a _____

                5b _____

6. What two characteristics distinguish crustaceans from other arthropods?

7. Distinguish between the general body plan of crustaceans and arachnids and that of insects.

8. Echinoderms are deuterostomes and represent a different branch of the phylogenetic tree from the other phyla you have studied. How do they differ from these other phyla?

9. What common characteristics are shared by the four classes of echinoderms you studied?

Diversity— Phylum Chordata

OVERVIEW

The phylum to which we, *Homo sapiens,* belong is the Chordata. All representatives of this group have, or are clearly related to forms that have, each of the following characteristics: (1) a flexible but incompressible supporting skeletal rod called the **notochord,** from which comes the name Chordata; (2) a **dorsal tubular nerve cord** lying above the notochord; (3) **pharyngeal pouches,** also called visceral or "gill" pouches, located in the pharynx (an anterior region of the gut); and (4) a **post-anal tail.**

The phylum Chordata includes two groups (subphyla), the tunicates and sea lancelets, that do not have backbones, and it is closely related to another phylum, the Hemichordata, containing pterobranchs and acorn worms (acorn worms were previously classified with the chordates). However, the hemichordates do not possess all of the features of chordates and probably represent an early offshoot of the line leading to the chordates. All of these "invertebrate" chordates (tunicates and sea lancelets) and the Hemichordata are referred to as **protochordates.**

A third group (subphylum) of the phylum Chordata contains organisms that have a *backbone,* a bony spinal or vertebral column that replaces most of the notochord and encases the nerve cord, and a *skull* surrounding the brain, an anterior expansion of the nerve cord. These organisms are called **vertebrates.** There are seven vertebrate classes with living representatives. The classification of hemichordates and chordates is summarized below:

Phylum Hemichordata (Stomochordata)
    Class Pterobranchia: pterobranchs
    Class Enteropneusta: acorn worms
Phylum Chordata
    Subphylum Urochordata (Tunicata): sea squirts and relatives
    Subphylum Cephalochordata (Acraniata): sea lancelets
    Subphylum Vertebrata: vertebrates
        Class Agnatha: jawless fishes
        Class Placodermi (extinct)
        Class Chondrichthyes: cartilaginous fishes
        Class Osteichthyes: bony fishes
        Class Amphibia: frogs, salamanders, and others
        Class Reptilia: turtles, snakes, lizards, crocodiles, and others
        Class Aves: birds
        Class Mammalia: mammals

Figure 27-1 *Chordates may have evolved as follows: A primitive sessile filter-feeder with ciliated tentacles may have given rise to primitive echinoderms such as sea lilies and to modern hemichordates. In the hemichordate line, multiplication of gill slits and replacement of tentacles by the pharyngeal basket as the primary feeding mechanism distinguished tunicates from their hemichordate ancestors. Chordate characteristics originated in the larvae of one line of tunicates. In a process called* paedo-morphosis *(retention of juvenile characteristics by the adult), this group lost the typical adult sessile stage and gave rise to the lancelets and more advanced free-swimming chordates. (After Alfred S. Romer and Thomas S. Parsons,* The Vertebrate Body, *6/e, p. 34, CBS, New York, 1983.)*

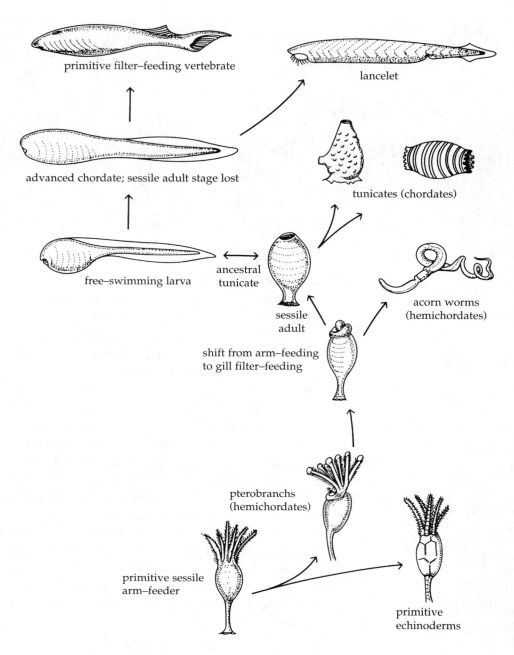

Chordates, like the echinoderms (Laboratory 26), are deuterostomes. One scenario for evolution among the deuterostomes is suggested in Figure 27-1 in which a primitive sessile feeder is shown giving rise to two lines of descent—primitive echinoderms on the one hand, and modern hemichordates and tunicates on the other. Chordate characteristics originated in the larvae of one group of tunicates. Through the loss of the typical sessile adult stage, these tunicate larvae may have given rise to lancelets and to more advanced, free-swimming chordates, including the vertebrates.

Evolutionary relationships among the vertebrate groups are shown in Figure 27-II.

STUDENT PREPARATION

Prepare for this laboratory by reading the text pages indicated by your instructor. Familiarizing yourself in advance with the information and procedures in this laboratory will give you a better understanding of the material and will improve your efficiency.

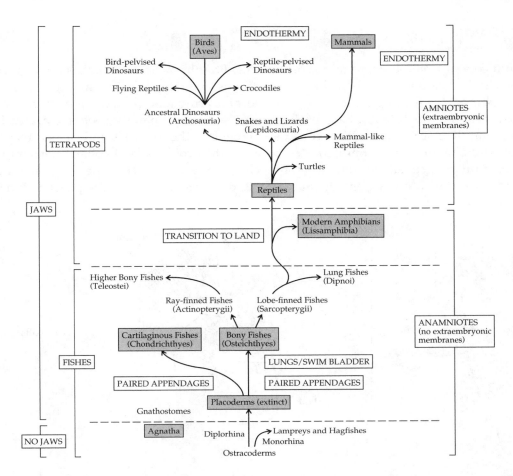

Figure 27-II *There are seven classes of vertebrates with living representatives. The first class, the **Agnatha**, includes extinct jawless fishes, the **ostracoderms**, and their specialized and aberrant living descendants, the lampreys and hagfishes. Note that modern agnathans have only a single nostril—they and their ancestors are grouped with the **Monorhina**. Other agnathans, all extinct, had two nostrils and are classified as **Diplorhina**—this is the line that must have given rise to higher vertebrates, but the transition is obscured by time. The remaining vertebrates all have jaws and can be called **gnathostomes** (gnatho-, jaw).*

*We can group together a class of diverse extinct fishes called **placoderms** and their descendants, the **Chondrichthyes**—sharks, skates, and rays, the second class of living vertebrates. In members of this group, bony tissues have generally been replaced with less dense cartilage, reducing the amount of energy required for remaining afloat.*

*The third class includes the bony fishes, or **Osteichthyes**. This line developed a lung/swim bladder, a gas-containing sac used in respiration in some members and also in maintaining neutral buoyancy so that continuous swimming is not required to remain afloat. Two lines of evolution are evident in the group. The **ray-finned fishes** include the higher bony fishes or **teleosts**, a group containing the majority of vertebrate species and most of the fishes with which we are familiar. The **lobe-finned fishes** developed*

*fleshy appendages with bone structures apparently homologous to those of the higher vertebrates. This line of bony fishes led to four-legged animals (**tetrapods**) and life on land.*

* **Amphibians** are the fourth class of living vertebrates. Modern representatives (frogs, salamanders, and caecilians) are the rather specialized descendants of diverse fossil forms that first invaded land. Although many adults in this group leave the water, eggs and larvae require an aquatic environment (or a reasonable substitute) for development. Thus amphibians are not completely terrestrial.*

* The fifth class of vertebrates, **Reptilia**, made the transition to land with the development of an egg containing the nutritional reserves required for development and extraembryonic structures necessary for food utilization, waste storage, and gas exchange. Reptiles once counted among their members dinosaurs, flying pterosaurs, swimming ichthyosaurs, and other familiar but extinct species. Several lines survive today; of these, the best known are turtles, snakes and lizards, and alligators.*

* **Birds**, the sixth class of vertebrates, are thought to have originated from small carnivorous dinosaurs or their ancestors. Recent evidence, however, suggests that birds may actually predate dinosaurs. The seventh class, **mammals**, originated from a different line of running carnivorous reptiles and diversified into forms familiar to us today.*

✔ **EXERCISE A**   **Hemichordates**

This phylum contains pterobranchs and acorn worms. **Pterobranchs** are small, sedentary, colonial marine organisms with individuals (zooids) held in a gelatinous mass (Figure 27A-1a). Each individual has a *proboscis,* a *collar* with a feeding structure (lophophore) bearing tentacles, and a body *trunk* held within a secreted covering (Figure 27A-1b). Each of these body sections contains a coelomic cavity. Food is filtered from the surrounding seawater by a mucous net secreted by the tentacles and carried by ciliary movements to the mouth. The mouth, located between the proboscis and collar, opens into a pharynx (with a pair of pharyngeal slits in some pterobranchs) that leads to a **U**-shaped digestive tract with the anus opening near the mouth. Extending anteriorly into the proboscis from the pharynx is a strand of tissue

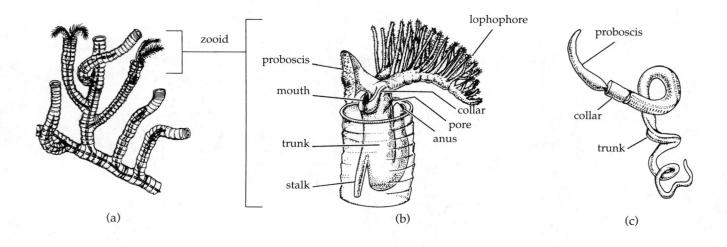

(a)   (b)   (c)

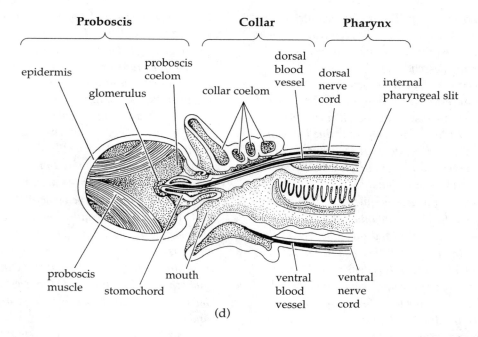

(d)

Figure 27A-1   *(a) Part of a pterobranch colony. (b) An individual pterobranch zooid* (Rhabdopleura). *(c) An acorn worm* (Balanoglossus). *(d) A mid-sagittal section of an acorn worm. [(a) (b) (c) after Alfred S. Romer and Thomas S. Parsons,* The Vertebrate Body, *6/e, p. 25, CBS, New York, 1983; (d) after Saul Wischnitzer,* Atlas and Dissection Guide for Comparative Anatomy: Lab Manual, *p. 11, W. H. Freeman, New York, 1979.]*

that resembles the notochord of chordates. However, this skeletal element originates from endodermal (not mesodermal) tissues and does not extend the length of the body as does a true notochord. Thus, it is not currently regarded as homologous to the notochord. Because of its origin near the mouth, this structure is called a **stomochord.**

Acorn worms (Figure 27A-1c, d), marine burrowing animals from 2 cm to 2 m in length, also have a stomochord, but have lost the lophophore. The proboscis has become muscular and assists in burrowing. Cilia on the surface of the proboscis collect food in a mucous net and move it to the mouth. The mouth can also be used to gather detritus from the ocean bottom. As in pterobranchs, the body is divided into three sections: the proboscis, collar (lacking the lophophore), and trunk, each with its own coelomic cavity. The anus is at the end of the body.

Although these organisms are not now placed in the phylum Chordata, the presence of pharyngeal gill slits and other developmental similarities clearly suggest a relationship with the echinoderm-chordate line.

ⅠⅠⅠⅠⅠ Objectives ⅠⅠⅠⅠⅠⅠⅠⅠⅠⅠⅠⅠⅠⅠⅠⅠⅠⅠⅠⅠⅠⅠⅠⅠⅠⅠⅠⅠⅠⅠⅠⅠⅠⅠⅠⅠⅠ

☐ Describe the organization of pterobranchs and acorn worms.

☐ Relate the evolution of hemichordates to that of chordates.

ⅠⅠⅠⅠⅠ Procedure ⅠⅠⅠⅠⅠⅠⅠⅠⅠⅠⅠⅠⅠⅠⅠⅠⅠⅠⅠⅠⅠⅠⅠⅠⅠⅠⅠⅠⅠⅠⅠⅠⅠⅠⅠⅠⅠ

1. Study the diagrams of pterobranchs in Figure 27A-1a, b. Compare the structural features of an individual zooid with those of an acorn worm, as seen in Figure 27A-1c, d.

2. Examine a preserved acorn worm (preserved or plastic mount). Note the three regions of the body: the proboscis, used in burrowing and feeding; the collar; and the trunk. The anterior section of the body also bears a number of pharyngeal slits.

3. Examine a median longitudinal section of the anterior end of an acorn worm. Identify as many of the structures shown in Figure 27A-1d as possible. The collar serves as an anchor for extension and contraction of the proboscis during burrowing. The surface of the proboscis is ciliated and is used in feeding. Try to identify the stomochord which projects into the proboscis coelom. At the anterior end of this structure is a tissue mass known as the glomerulus (it may have some excretory function). The nervous system consists of a nerve net with dorsal and ventral nerve cords—there is no real "cerebral" expansion and there are no special sensory organs. Acorn worms have a circulatory system in which blood flows anteriorly in dorsal vessels and posteriorly in ventral vessels; the blood is colorless.

   a. *What features suggest that pterobranchs and acorn worms are related?*

   _____

   _____

   b. *What structures link the hemichordates to the chordates?* _____

   _____

✔ **EXERCISE B   Tunicates**

Tunicates or **urochordates** are a relatively diverse group of marine filter-feeders. They are solitary or colonial and may be either attached to the substrate or free-floating. The most familiar forms are **ascidians** or **sea squirts.** Adult sea squirts can be found attached to pilings or rocks in the intertidal zone of our coasts. Larvae of solitary forms, however, are free-swimming, bilaterally symmetrical "tadpoles" with characteristic chordate features—a notochord extending into a post-anal tail, a dorsal tubular nerve cord, and pharyngeal pouches (Figure 27B-1a). These larvae locate an appropriate substrate, attach with a sucker at their head end, and undergo metamorphosis during which most of the characteristics of the chordate are lost. Great interest has focused on this larva because it is a "typical" chordate, whereas adult sea squirts bear little resemblance to any other representative of the phylum.

IIIII Objectives IIIIIIIIIIIIIIIIIIIIIIIIIIIIIIIIIIIIIII

☐ Describe the structures of the larva and the adult form of solitary sea squirts and relate these structures to chordate characteristics.

IIIII Procedure IIIIIIIIIIIIIIIIIIIIIIIIIIIIIIIIIIIIIIIIII

1. Obtain a slide of the tadpole larva of a tunicate and examine it under low power (10×), or study Figure 27B-1a if slides are not available.

   The larva has a tail with a **notochord** that may be visible. A **nerve cord** lies above this supporting element, directing the activities of muscle cells alongside the notochord. Anteriorly, the nerve cord expands to form a hollow cephalic vesicle with a simple light-sensitive structure and a gravity receptor.

   A mouth (which may not be open at this stage) leads into a pharynx, perforated by several gill slits. These open into a surrounding pocket called the **atrium**, which exits dorsally. An incomplete intestine is present. On a separate sheet of paper, sketch those features you can identify. Metamorphosis includes reabsorption of the tail and a rotation of the body so that the mouth opens opposite the site of attachment to the substrate.

2. Observe an adult sea squirt on demonstration (Figure 27B-1b). The outer body is covered by a **test** or "tunic." This structure contains **tunicin,** a cellulose-like material that is unique in the animal world. Your specimen is probably relatively soft and pigmented.

   Directly opposite the point of attachment is the mouth, which leads into a pharynx perforated by many pharyngeal slits. Water, propelled by cilia lining the pharyngeal basket, passes through a mucous net where food particles are trapped. This net is coiled by ciliary action then passed into a simple **U**-shaped intestine where digestion takes place. Wastes are voided through the anus into the atrium, the chamber surrounding the pharynx. Filtered water and feces are expelled from the atrium through the atrial siphon near the mouth. If a dissected specimen is available, locate the mouth and pharyngeal basket.

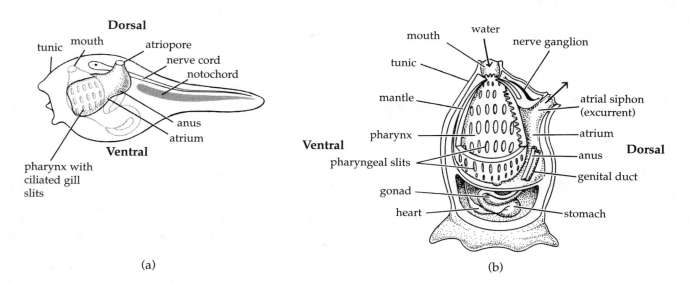

(a)   (b)

Figure 27B-1   *(a) The tunicate larva. (b) The adult form.*

Gonads may also be apparent near the base of the organism. Tunicates are hermaphroditic—individuals possess both testes and ovaries. Ducts lead to the atrium where gametes are shed at the appropriate season.

A fascinating feature of the tunicate "heart," an open tube located below the pharynx, is that it can reverse the flow of blood from dorsal to ventral and from ventral to dorsal every two or three minutes. Blood cells in some tunicates contain pigments with vanadium or niobium, rare elements concentrated from the surrounding ocean. However, these pigments probably have no respiratory function. Tunicates were once believed to have no coelom, but it is now thought that the hemocoel may represent this structure in adults.

If a sea squirt is disturbed, longitudinal muscles contract, pulling the organism toward the site of attachment and expelling a jet of water from the siphon as the pharynx collapses.

a. What feature does the adult tunicate retain that is typical of all chordates?

_____

✔ **EXERCISE C**   **Sea Lancelets**

A second small subphylum of chordates, the **cephalochordates (Acraniata),** contains small, bilaterally symmetrical marine organisms with all of the characteristic features of chordates. However, unlike all other representatives of the phylum, in cephalochordates the notochord extends all the way to the front of the head, beyond the anterior end of the dorsal nerve tube. In other chordates, the notochord ends behind the expanded part of the dorsal nerve tube or cerebral vesicle (behind the forebrain in vertebrates). Thus the name cephalochordate (*cephalo-*, head) is appropriate for this group. The most commonly studied cephalochordate is *Branchiostum lanceolatum* (formerly known as *Amphioxus*).

This group is of particular interest because it is representative of the types of organisms that could have evolved from the tunicate larva.

⦙⦙⦙⦙⦙ **Objectives** ⦙⦙

☐ Describe the structure of a lancelet and relate features of this organism to those of other chordates.

⦙⦙⦙⦙⦙ **Procedure** ⦙⦙⦙

1. Place a preserved lancelet in a small watch glass under your dissecting microscope or obtain a whole-mount prepared slide for study. On a separate sheet of paper, sketch and label the features you are able to identify. Use Figure 27C-1a as a reference.

2. Examine the external body plan of a lancelet.

    a. What kind of symmetry is exhibited by the lancelet? _____

3. The integument in adult lancelets is covered with a cuticle secreted by the outer layer of cells. In larvae, the integument is ciliated. These features may not be apparent in your specimen.

4. Try to find the **notochord.** It may be visible as a homogeneous column of tissue extending the length of the animal and located just above the pharyngeal basket. This incompressible rod is flexible.

5. The **nerve cord** is located above the notochord. If you have a whole mount, you may be able to see small pigmented areas along the side of the cord. These "eye spots" function in the reception of light but cannot form images. Anteriorly, the nerve cord is not expanded; it terminates behind the anterior end of the notochord, where a small pigmented spot may be found.

6. Above the nerve cord is a series of **dorsal fin rays** that support the dorsal fin. Along the length of the body you can find the chevron-shaped **myomeres,** clusters of longitudinal

muscle separated by sheets of tissue (myosepta). It is the contraction of these muscles on alternate sides of the larva that flexes the notochord and propels the larva through the water.

b. *Why is it important that the notochord is incompressible? (Hint: What happens when the myomeres contract?)* _____

---

7. Find the **buccal cirri** projecting around the mouth; these guard the opening of the mouth against sand and large detritus that could damage the delicate pharynx. The **wheel organ** is a ciliated structure on the inner lateral and dorsal walls of the vestibule (mouth or buccal cavity). The cilia of the wheel organ direct a vortex of water toward the mouth. The mouth leads into the pharynx which is perforated by about 200 **pharyngeal slits** leading to the atrium, a cavity surrounding the pharynx. These slits are covered by a mucous net secreted by cells lining the pharynx and, as water passes through the net directed by cilia on the walls of the pharynx, food is filtered from the water, wrapped in the mucus, and moved to the intestine. Note that a branch of the gut, the **hepatic diverticulum** (named because of some similarities with the vertebrate liver), protrudes anteriorly into the atrium, lying alongside the pharynx. Digestive wastes are voided through the anus which may be seen ventrally, near the base of the caudal fin. Water filtered through the pharyngeal slits and into the atrium leaves the body through the **atriopore,** anterior to the anus.

8. Obtain a slide of a cross section through the pharyngeal region of a lancelet and study it under scanning power (4×) using your compound microscope. Identify the structures labeled on Figure 27C-1b. Note the fin ray supporting the dorsal fin. Beneath this is the nerve cord lying over the larger notochord. Laterally, you can see sections through the chevron-shaped body musculature (myomeres), separated by connective tissue septa. Ventrolaterally, there is a **metapleural fold** on each side of the body. These folds and the dorsal fin help the lancelet control roll, pitch, and yaw when swimming (the caudal fin provides the major propulsive thrust).

9. Projecting from the lateral wall of the body into the atrium are gonads. Male and female gonads are found in separate organisms. Gametes are shed into the atrium and carried to the outside with the water passing through the pharynx to the atriopore. You may also note a section of the hepatic diverticulum in your specimen.

10. In the middle of the atrium, you will observe the pharynx tissue. The **pharyngeal bars** or "gill" bars of the pharynx bound the pharyngeal or "gill" slits used in filter-feeding. The pharynx is attached to the body in the mid-dorsal line but hangs free into the atrial cavity. Ventrally, the pharyngeal bars attach to the **endostyle.** Because of the delicate nature of the pharynx, lancelets attempt to keep larger particles out of the pharynx—when they are unsuccessful, "cough" reflexes mediated by giant nerve fibers are used to eject detritus from the mouth.

11. Relate the structures you find in the cross section to those seen in your whole mount. Be sure that you understand the spatial relations of the pharynx and atrium and how water is passed through these structures.

    In the cross section, you may be able to identify several vessels of the circulatory system. This is a closed system with large sinuses but without true capillaries. The blood is acellular and oxygen is carried in solution rather than bound to a blood pigment. There is no "heart" as such, but several major vessels, including the "ventral aorta," located under the endostyle, pulsate and assist a series of contractile bulbules (found at the base of alternate pharyngeal bars) to pass blood dorsally where it is collected by two vessels (**radices**) that join to form the "dorsal aorta." These vessels, coupled with a lateral return system of sinuses and vessels surrounding the gut and the hepatic diverticulum, suggest a circulatory pattern very similar to that of vertebrates. Near the dorsal vessels, clusters of excretory

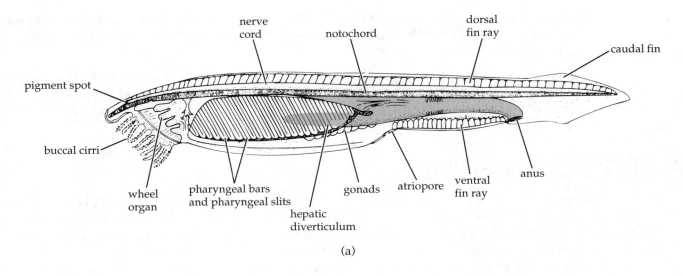

(a)

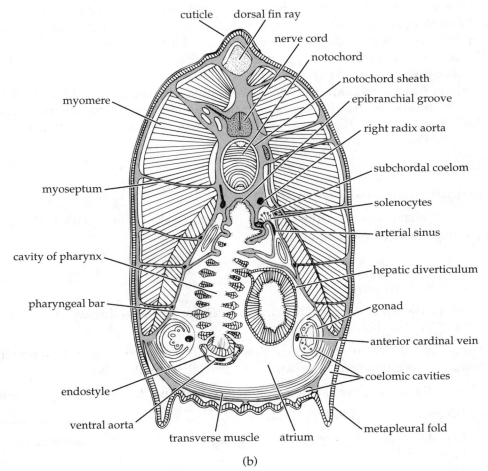

(b)

Figure 27C-1 *(a) Internal anatomy of a lancelet. These animals are filter-feeders that spend most of their lives buried in the bottom sediment of shallow marine environments. Only the "head" and mouth protrude above the bottom. Seawater taken into the pharynx passes through the pharyngeal slits. Oxygen from the seawater diffuses into the pharyngeal bars, which are richly supplied with blood vessels. A mucous net between the pharyngeal bars collects particles of debris and small organisms that are then passed from the pharynx into the gut. (b) A cross section through the pharyngeal region of a lancelet. (After Malcolm Jollie,* Chordate Morphology, *p. 7, Krieger, Melbourne, FL, 1973.)*

structures contain nephridia with flagella. These structures, called **solenocytes,** are intermediate between the protonephridia found in some larvae of smaller invertebrates and the metanephridia of larger animals.

c. *Cephalochordates show no signs of an enlargement of the nerve tube and associated sense organs usually found in the anterior end of the chordate nervous system. Why might this be so?*

_____

✔ **EXERCISE D   Vertebrates—Fishes**

The three groups of vertebrates you will study in this exercise are all fishes: the Agnatha, Chondrichthyes, and Osteichthyes. Some of the features that characterize these groups can be demonstrated by a study of the external morphology of a lamprey, a shark, and a teleost (bony fish). Also see Laboratory 37, Table 37-1, for a summary of structural and functional adaptations of vertebrates, including fishes.

The class Agnatha includes all jawless fishes. According to the fossil record, these are the oldest known fish. At one time, they had bony skeletons, but the skeletons of modern representatives, the lampreys and lungfishes, are composed of cartilage. Lampreys feed upon their prey by attaching to the skin and sucking blood from soft tissues—that is, the adult forms are parasites.

Modern Chondrichthyes include the rays and sharks (Selachii), of which the dogfish shark, *Squalus,* is one of the smallest members, and the chimeras (Holocephali).

The Osteichthyes include modern fish. Most have **bony endoskeletons.** The class includes many popular marine and freshwater fishes such as trout, bass, salmon, and tuna—all members of the higher bony fishes, the Teleostei. Unlike the cartilaginous fishes, which have external gill slits, osteichthyans have gills covered by a flap, or **operculum.**

||||| **Objectives** ||||||||||||||||||||||||||||||||||||||||

☐ State several distinguishing characteristics of each of the three classes of aquatic vertebrates by examining some features of their external morphology.

||||| **Procedure** ||||||||||||||||||||||||||||||||||||||||

1. Examine the lamprey, dogfish shark, and teleost fishes on display in the laboratory. Complete Table 27D-1 as you proceed.

2. Use a hand lens to study the skin (integument) of each specimen. Describe the surface features that characterize the skin and record these in the table.

3. Identify the **dorsal, caudal, pectoral, pelvic,** and **anal fins** on your specimens.

   a. *Which fishes have paired fins?* _____

   b. *How does the structure of the fins in the dogfish shark differ from the fin structures of the other two classes?* _____

   _____

4. Examine the head of each specimen. Look for the **external nares** (nostrils) of the dogfish shark. How are they arranged? Probe one of the nostrils with a dissecting needle. You should be able to feel a short passageway that makes no connection with the pharynx. The nares of fish are not used in respiratory gas exchange, but instead detect the presence of dissolved chemicals in the water—their role is in olfaction, the sense of smell.

5. Look for a single nostril on the surface of the head of the lamprey. Find the external nares of the teleost fish. In Table 27D-1, describe the arrangement of nares on the head of each fish.

Table 27D-1   Comparison of the External Morphology of Three Fishes

	Lamprey	Dogfish	Teleost
Features of the integument			
Number of fins (paired or unpaired)			
Location of fins			
Number of external nares			
Position of external nares			
Texture, size, and shape of the teeth			
Position of the teeth inside the mouth			
Number of gill slits			
Location of gill slits (covered or uncovered)			
Other distinguishing features observed			

6. In the adult lamprey, one set of "teeth" is on the lateral walls of the large funnel at the anteroventral end of the organism. The lamprey uses these teeth to penetrate the skin of its prey. Additional teeth surround the mouth at the rear of the funnel, and another set of teeth is on the tongue, which is rolled up inside the mouth. Feel these "horny" teeth composed of cornified epidermal cells and describe their size and shape in Table 27D-1.

7. Compare the teeth of the lamprey with the teeth of the dogfish shark.

8. Examine the teeth of the teleost fish. The teeth of both the dogfish shark and the teleost fish are made of **dentine,** a type of bone not supplied with blood vessels.

   c. *How can the relative positions of the nares and the mouth in a lamprey and a dogfish be explained in terms of their method of feeding?* _____

   _____

9. External gill slits in cartilaginous fishes occur in a lateral series near the back of the head. Water is taken in through the mouth, passed over the gills, and expelled through the gill slits. Record the number of these slits in the lamprey and the dogfish shark. In the dogfish shark, the first gill slit is reduced to a small opening, the **spiracle,** located above the angle of the jaw. In higher vertebrates, this gill slit becomes the outer and middle ear and the eustachian tube.

10. Using a blunt probe, gently move aside the operculum of the teleost fish. Sketch the arrangement of the gills in the space below.

11. When your data table is complete, and you are finished with the three specimens, cover the specimens with wet paper towels so they will not dry out.

✔ **EXERCISE E   Amphibians and Reptiles**

Amphibians are incompletely adapted to terrestrial environments. Even in those species that possess lungs, some gas exchange must occur through the skin. This requires that the skin be kept moist, a condition that prevents amphibians from living in a strictly terrestrial environment.

Amphibians also depend on water for reproduction. In many amphibians, such as the frog, eggs are laid in freshwater ponds or streams and are fertilized externally. The tadpole larva is a swimming stage that undergoes metamorphosis into an adult frog. Frogs may live on land, but they must return to the water to lay their eggs.

Unlike amphibians, most reptiles lead strictly terrestrial lives. Their tough, horny skin need not be kept moist and actually retards water loss. Reptilian eggs have hard or leathery shells and contain all the food and water needed for complete embryonic development.

See Laboratory 37, Table 37-1, for a summary of structural and functional adaptations of vertebrates, including amphibians and reptiles.

▮▮▮▮▮ **Objectives** ▮▮▮▮▮▮▮▮▮▮▮▮▮▮▮▮▮▮▮▮▮▮▮▮▮▮▮▮▮▮▮▮▮▮▮▮▮▮

☐ Compare amphibian and reptilian adaptations to living on land.

▮▮▮▮▮ **Procedure** ▮▮▮▮▮▮▮▮▮▮▮▮▮▮▮▮▮▮▮▮▮▮▮▮▮▮▮▮▮▮▮▮▮▮▮▮▮▮▮

1. Feel the skins of amphibians and reptiles on display (use live specimens if they are available).

   a. *How would you describe them?* _____

   _____

   b. *Which type of skin offers more protection from desiccation in a terrestrial environment?*

   _____

   c. *The integument of reptiles is formed from a number of epidermal plates or scales. What is the general shape of these scales?* _____

   d. *In what regions of the reptile's body do the scales become very small and numerous?*

   _____

   e. *What are the advantages of this arrangement for mobility?* _____

   _____

2. Compare the limbs and feet of the amphibians and reptiles on display.

   f. *What are the general differences?* _____

   _____

   g. *What structures of amphibian limbs betray their aquatic heritage?*

   _____

   h. *How are the amphibian and reptilian limbs oriented with respect to the sides of the body?*

   _____

   i. *For what locomotor activities would such an arrangement be best suited?*

   _____

3. Compare the arrangement of bones in the forelimbs of the turtle and the frog shown in Figure 27E-1a, b.

j. *How do the two forelimbs differ in number and kinds of bones present?*

_____

k. *How does the angle of the joint between the upper and lower part of the leg differ between the frog and the turtle?* _____

_____

4. Compare the arrangement of bones of the hindlimbs of the frog and turtle as shown in Figure 27E-1c, d.

l. *How do the hindlimbs differ in number and kinds of bones present?*

_____

m. *What is the major difference between the hindlimbs of these two animals?*

_____

n. *How is the difference in the angle of their hindlimbs related to the difference in the locomotor behavior of frogs and turtles?* _____

_____

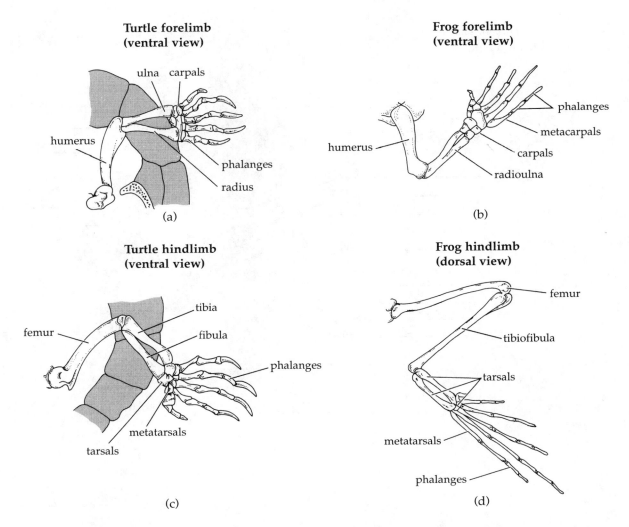

Figure 27E-1 *Comparison of the forelimbs and the hindlimbs of the turtle and the frog.*

5. If amphibian and reptile eggs are available, examine them under the dissecting microscope.

    o. How do they differ in size and texture? _____

    p. What surrounds and protects the frog egg? _____

    q. How do the types of egg laid by amphibians and reptiles relate to the mode of fertilization and the
    type of environment in which the egg must hatch? _____

    _____

✔ **EXERCISE F** | **The Avian and Mammalian Skeletal Systems**

Birds and mammals trace their ancestry to groups within the class Reptilia. Inferences about these
relationships are supported by observations of similar skeletal structures—many of the elements in both
groups are **homologous,** or related by descent. Of course, the avian skeleton shows many specializations
and constraints related to flight, and the mammalian skeleton is more generalized and adaptable to
different modes of locomotion.

    See Laboratory 37, Table 37-1, for a summary of structural and functional adaptations of vertebrates,
including birds and mammals.

Figure 27F-1 *(a) Human skeleton.*
(b, opposite) Skeleton of a bird.

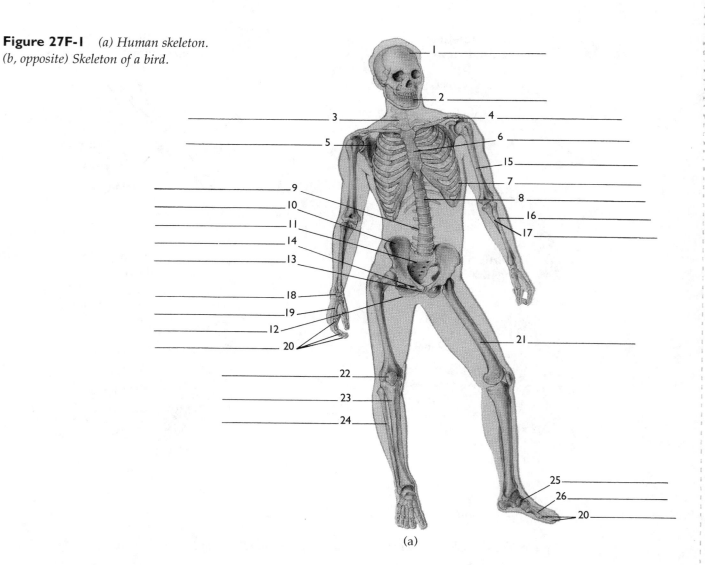

(a)

Table 27F-1   Bones of the Mammalian and the Avian Skeletons

The Axial Skeleton			
1	Skull	**8**	Thoracic vertebra
2	Mandible	**9**	Lumbar vertebra
3	Cervical vertebra	**10**	Ilium
4	Clavicle	**11**	Sacral vertebra (synsacrum in bird)
5	Scapula	**12**	Ischium
6	Sternum	**13**	Pubis
7	Ribs	**14**	Coccyx in human (caudal vertebrae in bird and in cat)

The Appendicular Skeleton			
15	Humerus	**21**	Femur
16	Radius	**22**	Patella
17	Ulna	**23**	Tibia (tibiotarsus in bird)
18	Carpal	**24**	Fibula
19	Metacarpal (carpometacarpal in bird)	**25**	Tarsal (for bird, see tibiotarsus)
20	Phalanges	**26**	Metatarsal (tarsometatarsus in bird)

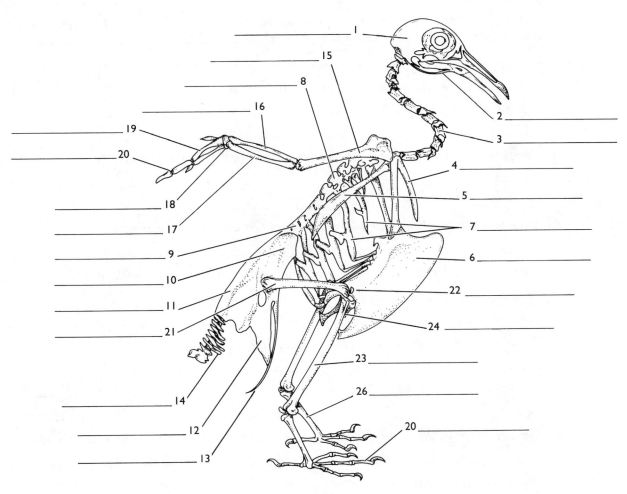

(b)

||||| **Objectives** |||||||||||||||||||||||||||||||||||||

☐ Compare a representative avian skeleton with that of a representative mammal.

☐ Relate the skeletal morphology in birds and mammals to a single ancestral pattern.

||||| **Procedure** |||||||||||||||||||||||||||||||||||

1. Study a mammalian skeleton (cat or human) and an avian skeleton (pigeon). If skeletons are not available, refer to Figure 27F-1 throughout this exercise.

2. Locate all of the bones listed in Table 27F-1 in both skeletons. Note that vertebrate skeletal systems are divided into (1) the **axial skeleton,** including the skull and spinal column, and the rib cage and sternum, and (2) the **appendicular skeleton,** including the bones of the forelimbs and hindlimbs and their connections to the axial skeleton.

3. Compare the size of the skulls of the pigeon and the mammal.

   a. *Do they appear to differ in size relative to the size of each animal's entire skeleton?*

   _____

4. Compare a cut bone from a pigeon (or chicken) with that of a mammal. Use bones of approximately the same diameter and length.

   b. *Which appears to be lighter in weight?* _____ *Explain.*

   _____

   c. *Certain bones in the pigeon contain extensions of air sacs, which are part of the respiratory system. What feature of avian bones makes this possible?* _____

   _____

5. Examine the axial skeleton of the mammal and the pigeon.

   d. *List, in order, the names of the major categories of vertebrae.* _____

   _____

   e. *Compare the number of vertebrae in each region of the vertebral column in the pigeon and the cat or human.* _____

   _____

   f. *Why do you think the pigeon has more cervical vertebrae than a mammal?*

   _____

   g. *Name the fused bones in the pigeon that are commonly known as the wishbone.*

   _____

   h. *Where are these bones located in the mammalian skeleton?* _____

   i. *What are the names of the three bones of the pelvis in mammals?*

   _____

   Name the homologous bones in the pigeon _____

   _____

   j. *Describe the orientation of the femur in relation to the pelvis in the mammal and the bird.*

   _____

   k. *How does this difference relate to the different modes of locomotion?* _____

   _____

6. Locate the bones of the pigeon's wing that correspond to the carpals, metacarpals, and phalanges in mammals. Compare the numbers of such bones in the two groups. The reduced number of bones in the pigeon's wing results from the fusion of adjacent bones.

   l.   Why would this arrangement be of advantage to the pigeon? _____

   _____

Subphylum or Class	Chordate Characteristics Present	Chordate Characteristics Missing	Invertebrate or Vertebrate	Distinguishing Characteristics	Adaptations to Environment
Urochordata					
Cephalochordata					
Agnatha					
Chondrichthyes					
Osteichthyes					
Amphibia					
Reptilia					
Aves					
Mammalia					

7. Locate the bones of the pigeon's hindlimb that correspond to the mammalian tarsals, metatarsals, and phalanges. Note that the number of these bones is reduced in the bird.

m. Suggest why the pattern of hindlimb structure in birds differs from that in mammals.

_____

_____

Laboratory Review Questions and Problems

1. Define "homologous structure." Give three examples of homologous structures studied in the laboratory.

2. Complete the table on page 27-17. List representative organisms under the appropriate group in the left-most column.

Plant Anatomy— Roots, Stems, and Leaves

OVERVIEW

As a result of evolutionary specialization among the land plants, numerous morphological and physiological differences led to the development of roots, stems, and leaves and their many modifications.

Roots anchor the plant in the ground, absorb and transport water and nutrients, and have some storage capabilities. Stems support both leaves and reproductive parts, transport water and nutrients, and sometimes have storage and photosynthetic capabilities. Leaves evolved to carry out photosynthesis. Some stems and leaves are specialized to perform a protective function: thorns, for instance, are modified stems, and spines are modified leaves. The stem and leaves collectively make up the shoot system of vascular land plants.

During this laboratory period, you will study the structure of the root and shoot systems of the plant and the many specialized types of cells and tissues that carry out the functions of roots, stems, and leaves.

STUDENT PREPARATION

Prepare for this laboratory by reading the text pages indicated by your instructor. Familiarizing yourself in advance with the information and procedures covered in this laboratory will give you a better understanding of the material and improve your efficiency.

✔ **EXERCISE A** | **Plant Tissues**

Roots, stems, and leaves are **plant organs** composed of tissues organized into **tissue systems** during primary and secondary growth. The specialized cells of plant systems develop from **meristems,** undifferentiated tissue located at sites of active growth. **Primary growth** (arising from primary meristems) lengthens the stems and roots, and **secondary growth** (arising from secondary meristems) increases their thickness.

Three tissue systems occur in all organs of the plant and are continuous from organ to organ. These are the **dermal tissue system** making up the outer covering of the plant; the **vascular tissue system** comprising the conductive tissues—xylem and phloem; and the **ground tissue system** consisting of all the "packing" cells that are neither dermal nor vascular.

The types of cells and tissues found in each of these systems are listed in Table 28A-1. Those tissues composed of only one cell type are called **simple tissues,** and those composed of more than one cell type are called **complex tissues.** Although the plant cells listed in the table have special features that distinguish one cell type from another, they have certain common features.

Table 28A-1 Plant Tissue Cell Types

Tissue or Cell Type and Function	Characteristics of Cells	Location
Parenchyma — Storage of food, usually starch; participation in metabolic processes, water balance, and wound healing.	Many-sided. May have only thin primary cell walls or both primary and secondary cell walls. Living at maturity: retain protoplasts.	Throughout plant. Mesophyll of leaves, flesh of fruits, and pith and cortex of roots and stems.
Ground Tissue System — **Sclerenchyma** — **Sclereid** Mechanical and protective functions. — **Fiber** Support function.	Short. Primary wall and thick secondary wall, generally lignified. May be living or dead at maturity. Elongate. Usually (but not always) dead at maturity (protoplast gone).	Throughout plant. Component of fruits, nutshells, and other hard coverings. Gives "gritty" texture to fruits. In vascular bundles and in bundles in leaves, stems, and fruits. Examples: hemp, jute, and flax.
Collenchyma — Support function.	Elongate. Primary walls only; thickened at corners—not lignified. Living at maturity.	On periphery of stems and along veins in some leaves, giving ribbed appearance to stem or leaf.
Vascular Tissue System* — **Xylem (Tracheary Elements)** — **Tracheid** Water conduction. — **Vessel Element** Water conduction.	Lignified xylem elements are major components of wood. Elongate, tapering, and lignified. Primary wall and secondary wall with thickenings. Membrane-covered pits concentrated on overlapping ends of cells through which water passes from cell to cell. Dead at maturity (protopalst gone). Elongate but shorter than tracheid. Primary and lignified secondary walls. In addition to pits, vessel elements have perforations, "holes," usually in the end walls of the cells, which offer less resistance to water flow than the membrane-covered pits of tracheids. Vessel elements join at the ends to form long continuous vessels. Dead at maturity (protoplast gone).	In vascular bundles or vascular tissue areas. In vascular bundles or vascular tissue areas.

Table 28A-1   *(continued)*

Tissue or Cell Type and Function		Characteristics of Cells	Location
Vascular Tissue System*	**Phloem** **Sieve-Tube Member** Food distribution.	Elongate and tapering. Primary cell wall only. Sieve areas on end walls with larger pores than side walls. Several sieve-tube members in a vertical series constitute a **sieve tube.** Living at maturity, but unique in lacking a nucleus.	In vascular bundles or vascular tissue areas.
	Companion Cell Role in movement of food into and out of sieve-tube member.	Variable shape, generally elongate. Associated with sieve-tube members and aid in transfer of food. Living at maturity.	In vascular bundles or vascular tissue areas.
Dermal Tissue System*	**Epidermis** Covering.	Variable shape; some contain a waxy substance called cutin. Specialized guard cells open and close stomata to retard water loss. **Trichomes, hairs,** root hairs are specialized epidermal cells. Living at maturity. If secondary growth occurs, epidermis is replaced by **periderm**—part of the bark on trees.	On outer surface of roots, stems, and leaves.

*This tissue system also contains sclerenchyma and parenchyma.

Typically, a living plant cell consists of a rather rigid **cell wall,** a structure not present in animal cells, and the **protoplast,** the living contents of the cell inside the cell wall. The protoplast, bounded by the plasma membrane (just inside the cell wall), is made up of the nucleus and the cytoplasm, which includes organelles such as mitochondria and plastids.

Living, active cells, those that are carrying on metabolic processes or actively dividing, have a **primary cell wall.** These are composed mostly of cellulose and are somewhat plastic (can be stretched during growth). Certain cells also have a **secondary cell wall** laid down inside the primary cell wall after the cell matures. Lamination (layering) of cellulose fibers oriented in different directions strengthens secondary cell walls, and **lignin,** a complex polymer, frequently contributes to their rigidity. After depositing the secondary wall, the protoplast dies and no further metabolic activities take place in the cell. These cells have an empty appearance. Rigid secondary walls are found mainly in cells performing a support function.

▐▐▐▐▐ Objectives ▐▐▐▐▐▐▐▐▐▐▐▐▐▐▐▐▐▐▐▐▐▐▐▐▐▐▐▐▐▐▐▐▐

☐ Describe the cell types present in the dermal tissue system and in the ground tissue system.

☐ Differentiate between the structures and functions of xylem and phloem.

▐▐▐▐▐ Procedure ▐▐▐▐▐▐▐▐▐▐▐▐▐▐▐▐▐▐▐▐▐▐▐▐▐▐▐▐▐▐▐▐▐

You can best study specialized cell types by first examining the components of some familiar plant materials. Refer to Table 28A-1 to assist you in recognizing the cell types present in these materials.

1. Strip off a piece of the outermost cell layer from a leaf of lettuce. Make a wet-mount slide and observe it at high power (40×). Identify the epidermal cells (you are looking at a surface view). In the space below, sketch your observations.

2. Make a very thin cross section of a celery petiole. Prepare a wet-mount slide. Locate **parenchyma, collenchyma,** and **sclerenchyma fibers.** The vascular tissue occurs in bundles— the "strings" that can be pulled out of the celery—and is composed of *xylem* and *phloem.* In cross section, sclerenchyma cells are found to one side of the vascular bundle and appear as a "cap." Collenchyma cells with their thickened corners are found mainly along the vascular tissue of the ribs (and are an important constituent of celery strings). In the space below, sketch and label your cross section.

3. Make a longitudinal section through one of the vascular bundles (strings) of the celery petiole and prepare a wet-mount slide. Note the elongate nature of the vascular tissue cells. Thickenings in the secondary walls of the tracheary elements should be obvious. In the space below, sketch and label your observations.

   a. What type of tissue are the tracheary elements? _____

4. Mount a small piece of pulp from a pear in a drop of water and prepare a wet-mount slide. Gently apply pressure to the coverslip with the eraser end of your pencil. Locate a cluster of **sclereids.** (These give pears their characteristic gritty texture.) Sketch your observations.

   b. What type of tissue are sclereids? _____

✔ **EXERCISE B**   **The Monocot and Dicot Angiosperm Body Plan**

Angiosperms can be divided into two classes: **monocotyledons** (monocots) and **dicotyledons** (dicots). Although these two groups have much in common, there are certain distinctive differences in the organization of their root and shoot systems. These are listed in Table 28B-1 and shown diagrammatically in Figure 28B-1.

Table 28B-1   **Characteristics of Monocots and Dicots**

Organ	Monocotyledons (one cotyledon)	Dicotyledons (two cotyledons)
Leaf	Parallel veins	Netlike pattern of veins
Stem	Bundles of vascular tissue scattered throughout ground tissue. No vascular cambium.	Bundles of vascular tissue arranged around a central core of ground tissue. Vascular cambium present.
Root	Bundles of xylem located in a ring with small bundles of phloem between pith in center stele. No vascular cambium.	Xylem centrally located in stele in cross-shaped arrangement with phloem between arms of xylem. Vascular cambium present.
Flower	Parts occur in threes or multiples of three.	Parts occur in fours or fives or multiples of four or five.

In addition, you should note that all monocots are **herbaceous.** Herbaceous plants are usually short-lived (one or two seasons), fleshy, and of limited diameter. Their tissues are derived exclusively, or at least predominantly, from **primary tissues,** tissues arising during primary, lengthening growth as a result of cell division within apical meristems at the tips of shoots and roots.

Dicots, on the other hand, can be either herbaceous or **woody.** Herbaceous dicots, like herbaceous monocots, are composed of tissues developed as a result of primary growth. Woody dicots, however, produce "woody tissues" during secondary growth that thickens their stems and roots as a result of cell division within lateral (secondary) meristems.

In this laboratory, you will confine your study to the plant body composed of primary tissues arising from primary growth. Thus you will study only herbaceous monocots and dicots. You will examine secondary tissues and secondary growth during your study of plant development in Laboratory 29.

ıııı Objectives ııııııııııııııııııııııııııııııııııııı

☐ Describe the major anatomical differences that distinguish the roots, stems, leaves, and flowers of herbaceous monocots from those of herbaceous dicts.

ıııı Procedure ıııııııııııııııııııııııııııııııııııııı

Study Figure 28B-1 closely. Continue to refer to these diagrams while organizing your observations in this exercise.

Figure 28B-1 *Diagram of the principal structures of the primary plant body of (a) a dicot and (b) a monocot. All structures below the soil surface constitute the root; those above the soil make up the shoot.*

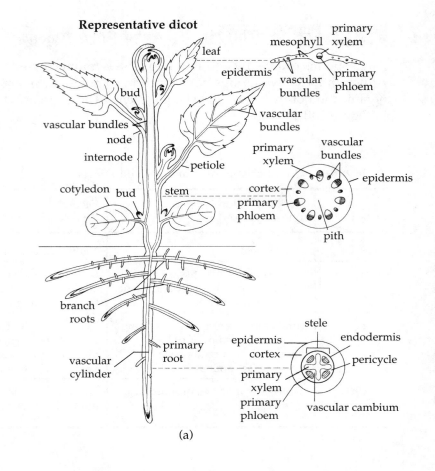

(a)

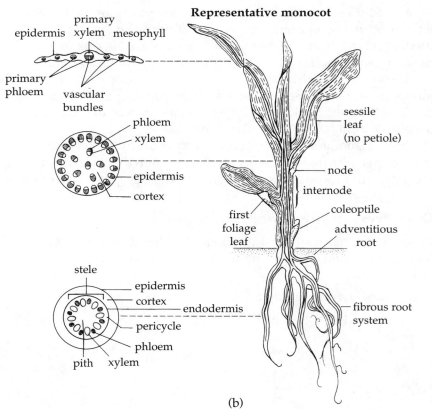

(b)

PART I   Structure of the Dicot and Monocot Root

✔ ### A. The Herbaceous Dicot Root—Buttercup (*Ranunculus*)

> **1.** Use a prepared slide to study a cross section of the matured region of a buttercup root. Keep in mind that each structure you see has a third dimension: what appears as a circle in cross section is a vertical cylinder in the plant. Beginning at the outside, locate the structures and tissues discussed in the text below. Label Figure 28B-2 by matching the letters in the text to the letters on the illustrations.

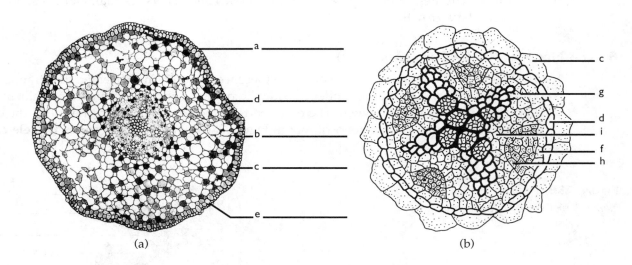

(a)   (b)

Figure 28B-2   *(a) Cross section of a dicot root (buttercup,* Ranunculus*). (b) Enlargement of the stele and associated tissues.*

> **a Epidermis** Appears as a row of thin-walled cells in the form of a circle around the outer edge of the section.
>
> **b Cortex** Consists of cortical parenchyma and endodermis.
>
> > **c Cortical parenchyma** These large, thin-walled cells make up the greater part of the root. The numerous dark bodies within the cells are starch grains. The main function of this part of the root is food storage.
> >
> > **d Endodermis** This inner layer of the cortex forms an easily distinguished "circle" of smaller cells. Surrounding each endodermal cell, in the way that a wide rubber band would wrap around the edges of a box, is a layer of waxy, impermeable material called the *Casparian strip* (you will not be able to see these structures). Since the cells are arranged edge to edge—the Casparian strip around one cell touching the Casparian strip of its neighbor—water and solutes cannot pass between the cells, but instead must move through the protoplast of the endodermal cells, either by crossing their plasma membranes or by way of plasmodesmata, before reaching the vascular tissues of the root. Because water, oxygen, and carbon dioxide pass easily through cell membranes, but many ions and other substances do not, the membranes of endodermal cells regulate the substances that pass from the root to other parts of the plant body.
>
> **e Stele** Consists of the following parts (**f** to **i**) within the endodermis:
>
> **f Pericycle** Thin-walled cells of the stele just within the endodermis. Branch roots arise from this primary meristem tissue.

g Primary xylem A cross-shaped group of cells in the center of the stele. Xylem is easily distinguished because most of the cells (vessel elements and tracheids) are large, thick-walled, and appear to lack protoplasts. The chief function of xylem is to conduct water.

h Primary phloem Strands of thin-walled cells that appear in cross section as roughly circular groups. One strand of phloem lies in each space between the "points" of the primary xylem. The main function of the phloem is to translocate sugar, a product of photosynthesis.

i Vascular cambium Meristem tissue composed of a layer of thin-walled cells separating the xylem and the phloem. As these cells divide, they produce secondary xylem and phloem tissues, which increase root diameter. This increase in diameter is called secondary growth.

✔ **B. The Monocot Root—Corn (*Zea mays*)**

1. Obtain a slide of a monocot root. Notice the presence of a **pith** in the center, the strands of primary xylem arranged in a cylinder outside the pith, and the small patches of phloem alternating with strands of xylem. The other parts of the root are similar to those of the dicot root. Label structures of the monocot root in Figure 28B-3: **epidermis a, cortex b, stele c, pith d.**

Figure 28B-3 *Cross section of a monocot root (corn, Zea mays).*

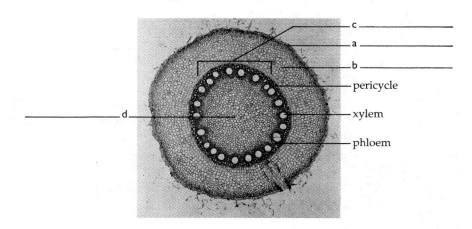

c _____
a _____
b _____
pericycle
xylem
phloem

d _____

✔ **C. Origin of a Branch (Secondary) Root—Willow (*Salix*), a Dicot**

Examine a slide of a root cross section showing the origin of branch roots.

    a. From which tissue do branch roots arise? _____

    b. What is the functional advantage of branch roots arising opposite xylem points rather than opposite strands of phloem? _____

_____

PART 2   Structure of the Dicot and Monocot Stem

✔ **A. The Herbaceous Dicot Stem—Alfalfa (*Medicago sativa*)**

1. Using a prepared slide, study the cross section of an alfalfa stem. Use low-power magnification (10×) for an overall view. Most of the tissues you see will be primary tissues. Refer to Figure 28B-4 to identify the following structures.

a Epidermis The outermost layer of cells. Stems often have epidermal hairs that appear as multicellular projections from the epidermis. These hairs may serve a protective function.

b Cortex A zone of varying width inside the epidermis. Any or all of the following tissues may be present:

> **c Parenchyma** These thin-walled cells lack chloroplasts, but have starch-containing plastids (amyloplasts).
>
> **d Chlorenchyma** Parenchyma cells with chloroplasts. Chlorenchyma cells are usually located toward the outer edge of the cortex, where light is available for photosynthesis.
>
> **e Collenchyma** The walls of these cells are thickened in the corners and provide flexible support for the stem.

f Pith The central portion of the stem; composed of loosely arranged parenchyma cells.

g Vascular bundles Strands of conductive tissue spaced in a circle around the pith and composed of the following:

> **h Phloem** Located next to the cortex. Thin-walled, irregular cells consisting mainly of sieve-tube members (joined together as sieve tubes), companion cells, and parenchyma.
>
> **i Xylem** Located next to the pith.
>
> **j Vascular cambium** This lateral meristem tissue consists of a circular layer of brick-shaped cells. The cells of the vascular cambium (j^1) that lie *between the xylem and the phloem* within the vascular bundles give rise to secondary xylem and secondary phloem. The vascular cambium (j^2) that lies *between the vascular bundles* gives rise to the parenchyma cells that form the vascular rays (see below).

Sclerenchyma fibers May be present as a crescent-shaped cap of thick-walled cells outside the phloem; the cap supports and protects the vascular bundle. You will observe this cap in the *Coleus* stem.

Vascular rays Also called *pith rays*. Parenchyma tissue between the vascular bundles; connects pith with cortex (See Figure 29F-1.)

2. Label the structures of the dicot stem in Figure 28B-4a, b.

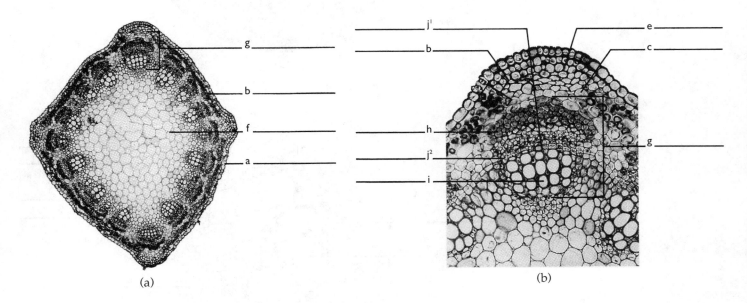

(a)   (b)

Figure 28B-4   *(a) Cross section of a dicot stem (alfalfa,* Medicago sativa*). (b) Enlargement of a portion of the same stem.*

✔ **B. The Monocot Stem—Corn (*Zea mays*)**

1. Study a prepared slide of a cross section of a young corn stem. Identify the following using Figure 28B-5 as a guide:

   a Epidermis Outside layer of cells around the stem.

   b Sclerenchyma Tissue with extremely thick-walled cells directly adjacent to the epidermis; forms the "rind" of the mature stem.

   c Parenchyma Thin-walled cells constituting the tissue that makes up the bulk of the stem; it is not divided into cortex and pith as in dicot stems.

   d Vascular bundles Scattered throughout the stem cross section. Study a single vascular bundle under high power (40×) and identify the following:

   e Phloem Located toward the outside of the vascular bundle and composed of sieve tubes and companion cells.

   f Xylem The arrangement of the vessels gives the appearance of a small face.

   g Sclerenchyma fibers Surround the vascular bundle to form the **bundle sheath.**

2. Label the structures of the monocot stem in Figure 28B-5a, b.

Figure 28B-5 *(a) Cross section of a monocot stem (corn,* Zea mays). *(b) Enlargement of a vascular bundle and surrounding tissues.*

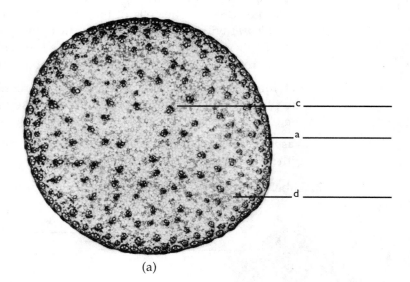

(a)

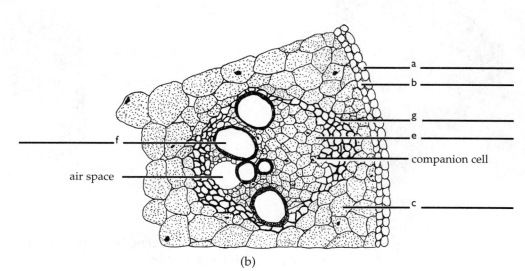

(b)

C. The *Coleus* Root and Stem

1. Work in pairs. Obtain a *Coleus* plant from your instructor. One laboratory partner should cut off a small section of one of the oldest (thickest) roots. Do all cutting on a glass slide. Take a section about 1.5 to 2.0 cm long. The other partner should obtain a piece of *Coleus* stem of about the same length (1.5 to 2.0 cm).

2. Obtain a nut-and-bolt microtome from your instructor.

3. Turn the nut until it is almost at the end of the bolt, forming a small "cup."

4. Stand the root or stem on its end in the center of the opening of the nut and, while holding the tissue upright, carefully pour melted paraffin into the cut until it fills the opening (be careful that the paraffin is not too hot when you pour it or you will cook your plant tissue). This assembly will allow you to hold the stem or root tissue upright and to cut very thin slices.

5. Hold the head of the bolt flat on the table with one hand. With the razor blade in your other hand, remove the excess stem or root and wax by "shaving" down to the nut. This technique keeps your fingers out of the way of the razor blade (Figure 28B-6).

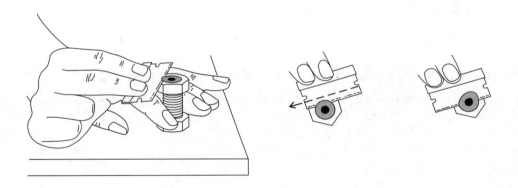

Figure 28B-6 *Using the nut-and-bolt microtome.*

6. Twist the bolt just a little, so a very thin core of paraffin and stem or root protrudes above the surface of the nut.

7. Use a slicing motion to cut this section flush with the nut.

8. Using forceps, place the slice in a Petri dish containing 50% ethanol and follow the procedure for vital staining (see instructions for vital staining in the box on page 28-12).

9. Twist the bolt a bit more to get another slice. Remember: you are trying to get the thinnest possible slice. It is better to get only part of a thin slice than a thick slice that is entirely round. As you obtain each slice, put it in a dish of 50% ethanol and stain as described in the box. Slice the entire block (or at least 10 sections) in this manner.

10. On a separate sheet of paper, make a drawing of the stem and root sections and identify and label as many cell and tissue types as possible. Insert the drawings into your laboratory manual.

    a. Is Coleus *a monocot or dicot?* _____

    b. *What characteristics of the root sections allowed you to distinguish whether* Coleus *is a monocot or dicot?*

    _____

VITAL STAINING OF PLANT TISSUE WITH TOLUIDINE BLUE O

A. Place the sections in 50% ethanol for 5 minutes.

B. Move the sections to a dish of toluidine blue O and leave them there for 5 minutes.

C. Rinse sections in a dish of distilled water. Remove any paraffin that may still be attached to the plant sections.

D. Use a drop of glycerine solution to mount the plant tissue on a slide. Cover with a coverslip and use the 10× and 40× objectives to make your observations.

c. *What characteristics of the stem sections allowed you to distinguish whether* Coleus *is a monocot or dicot?* _____

_____

d. *Did toluidine blue stain all tissues the same color?* _____ *Which tissues were stained green?*

_____ *Purple or blue?* _____

Red? _____

EXTENDING YOUR INVESTIGATION: IS IT A MONOCOT OR A DICOT?

You may have a houseplant or two in your room; or perhaps when you walk to class, you pass some weeds or plants growing wild in a field. Collect a few specimens with soft stems and bring them to the laboratory. What kinds of plants do you think most houseplants or common weeds might be—monocots or dicots? Formulate a hypothesis that addresses this question.

HYPOTHESIS:

NULL HYPOTHESIS:

Based on your knowledge of the specimens you have collected, what do you **predict** you will find when you study the structure of their stems?

Identify the **independent variable** in this investigation.

Identify the **dependent variable** in this investigation.

Design an experimental procedure to test your hypothesis. (Incorporate the "Vital Staining of Plant Tissue with Toluidine Blue O" procedures.)

PROCEDURE:

RESULTS: What types of plants (monocots or dicots) have you sampled?

Do your observations and results support your hypothesis?

Your null hypothesis?

Was your prediction correct?

What do you **conclude** about the classification of most common houseplants?

Is this a safe conclusion based on your sample size?

PART 3   Leaf Structure

✔ A. External Features of Leaves

The external morphology of leaves varies considerably within the angiosperms, and leaf characteristics afford the botanist a relatively dependable means of identifying species. Some general characteristics also provide a way of differentiating between monocots and dicots.

A typical dicot leaf is petiolate: the **petiole** attaches the **blade** of the leaf to the stem of the plant. The point of attachment is called a **node.** The leaves of monocots are more likely to exhibit the **sessile** pattern of attachment: they lack a petiole and are attached directly to the stem by a sheath.

Leaves, whether compound or simple, are arranged on the stem in one of three ways: in an **alternate** (spiral), **opposite** (in pairs), or **whorled** (three or more at a node) pattern.

Blades may differ in shape and may be single, as in **simple** leaves, or divided into leaflets, as in **compound** leaves. Leaflets on a compound leaf can arise from a single point on the petiole, as in a **palmately compound** leaf (think of the palm of your hand), or the leaflets can arise from many different points on the petiole, as in a **pinnately compound** leaf, with odd or even patterns (Figure 28B-7). Since leaflets are similar in appearance to leaves, it is sometimes difficult to tell them apart. One distinction is that leaves, but not leaflets, have buds in their *axils* (the upper angle between the leaf and the stem).

There are two principal types of venation in angiosperms. **Netted** venation is a pattern of highly branched veins characteristic of dicots. Netted veined leaves are usually further described as having a palmate or pinnate pattern of venation (Figure 28B-7). Monocots have **parallel** venation with little or no branching.

Procedure

Observe the leaves and branches on demonstration and make sketches on a separate piece of paper. Identify the leaf type (simple, pinnately compound, or palmately compound), type of venation, and leaf arrangement (alternate, opposite, or whorled). Determine whether the specimen is from a monocot or a dicot. Insert your sketches into the laboratory manual.

✔ B. The Structure of a Dicot Leaf—Privet (*Ligustrum*)

1. Examine a prepared slide of a cross section of a *Ligustrum* leaf at low power (10×) and then at high power (40×). Identify the following:

   Upper epidermis A single layer of cells often covered by a noncellular, waxy secretion (cutin) forming a separate layer, the **cuticle.**

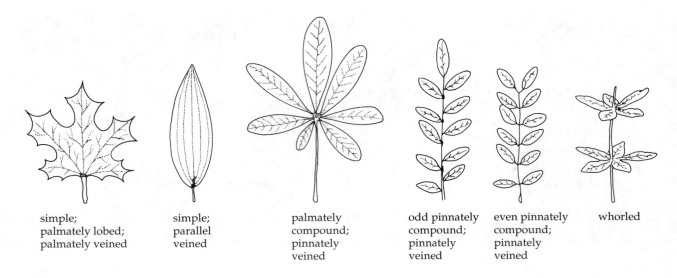

simple;
palmately lobed;
palmately veined

simple;
parallel
veined

palmately
compound;
pinnately
veined

odd pinnately
compound;
pinnately
veined

even pinnately
compound;
pinnately
veined

whorled

Figure 28B-7   *Simple and compound leaves: patterns of leaf venation and leaflet arrangement.*

a. What is the function of the cuticle?

_____

Mesophyll The middle portion of the leaf, consisting of chlorenchyma cells (parenchyma cells containing chloroplasts) distinguishable as two layers: **palisade mesophyll,** compactly arranged elongate cells with small intercellular spaces; and **spongy mesophyll,** loosely arranged cells with many large intercellular air spaces.

Veins Made up of xylem and phloem cells and mechanical support tissue.

b. What is the position of xylem with reference to the upper and lower surfaces of the leaf?

_____ *With reference to the phloem?* _____

Hint: Compare xylem and phloem position in the stem and trace these tissues into the leaf to explain xylem and phloem position in the leaf.)

Notice that in dicots the veins branch and rebranch from a central midrib. The midrib appears in cross section, while the secondary and tertiary branch veins are seen in longitudinal or oblique sections. Compare the midvein to the smaller secondary branch veins. Secondary veins are usually located between the spongy and palisade mesophyll.

Bundle sheath A layer of thin-walled, compactly arranged parenchyma cells enclosing the smaller veins. These cells are responsible for loading the products of photosynthesis into and out of the phloem and are also the site of the C_4 pathway.

Sclerenchyma fibers Surround the midrib bundle sheath, providing mechanical support.

Lower epidermis Similar to the upper epidermis, but contains more numerous stomata. (Stomata can occur on both upper and lower epidermis.) Look for guard cells as seen in the cross section. Notice that each stoma opens into an air space within the spongy mesophyll. Outgrowths or epidermal hairs (trichomes) may also be present on the lower epidermis.

c. What is the significance of the intercellular spaces of the spongy mesophyll?

_____

_____

Figure 28B-8 *Cross section of a dicot leaf (lilac, Syringa vulgaris).*

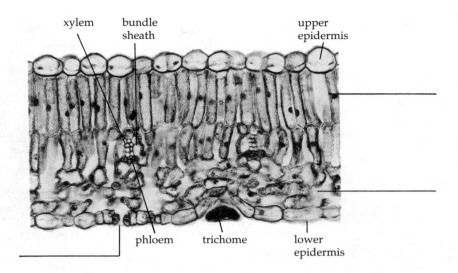

xylem   bundle sheath   upper epidermis

phloem   trichome   lower epidermis

2. Complete the labeling of the dicot leaf structures and tissues in Figure 28B-8.

✔ **C. Monocot Leaf Structure—Corn (*Zea mays*)**

1. Obtain a slide of a monocot leaf. Use Figure 28B-9 and the following discussion to identify structures and tissues. Study the slide at low power (10×) and then at high power (40×). Note the presence of thin-walled bubblelike (**bulliform**) cells in the **a upper epidermis.** When these cells are turgid the corn blade is flat and open. When they lose turgor, the leaf rolls.

   d. How does this mechanism conserve water? _____

2. Examine the stomata **b.**

   e. Are stomata on the upper epidermis, lower epidermis, or both? _____

   Are they evenly spaced? _____

   f. How does this arrangement differ from what you observed in the dicot leaf?

   _____

   Corn leaves have a distinct type of stomatal construction. The guard cells are flattened in the middle with thickened walls and have bulbous ends with thin walls. As turgor increases, the bulbous ends swell, pulling the guard cells apart. Each guard cell is attached to an accessory cell. Study a stoma using the high-power (40×) objective.

3. Examine the **mesophyll c.**

   g. How does the mesophyll in corn differ from that in privet?

   _____

4. Examine the vascular bundles or veins. Veins are parallel in monocot leaves and produce a highly regular pattern through a section. Identify **xylem d** and **phloem e.**

   h. How are xylem and phloem arranged in relation to one another?

   _____

5. Examine the **bundle sheath cells f.** These large cells have a few large chloroplasts, which contain abundant amounts of starch. Extensions of the bundle sheath are connected to sclerenchyma cells beneath the epidermis. Recall that corn is a C_4 plant with the C_4 pathway occurring in the chloroplasts of mesophyll cells and the Calvin cycle, or C_3 pathway, in the chloroplasts of bundle sheath cells.

Figure 28B-9 *Cross section of a monocot leaf (corn, Zea mays).*

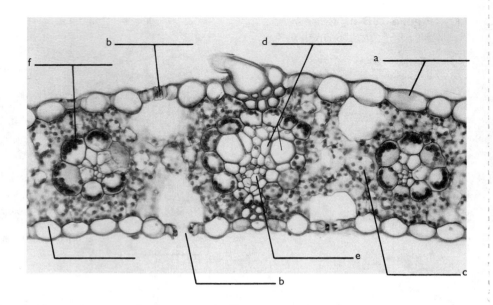

6. Label the structures and tissues of the monocot leaf in Figure 28B-9.

✔ **EXERCISE C** | **The Pine Leaf (Optional)**

Obtain a prepared slide of a cross section of a pine leaf. Identify the structures and tissues in the leaf and label Figure 28C-1.

Conifers (gymnosperms) have needlelike leaves covered by a rather thick **epidermal cuticle a.** Beneath the epidermis are one or more layers of thick, compact cells, the **hypodermis b.** The mesophyll cells are penetrated by **resin ducts c** and are separated from the vascular tissue by the **endodermis d.** One or two **vascular bundles e** are typically found in the center of the leaf surrounded by **transfusion tissue f,** which conducts materials between vascular bundles and **mesophyll g.** Note that the **stomata h** are sunken below the leaf surface—an adaptation for growth in arid climates.

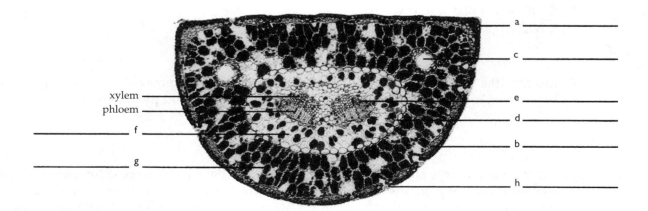

Figure 28C-1 *Cross section of a pine leaf.*

Laboratory Review Questions and Problems

Summarize your answers to questions 1–4 in the following table.

Summary of Tissue Systems, Tissues, and Cell Types

Tissue Systems	Tissues	Cell Types	Cell Functions
Dermal	Epidermis		
	Periderm		
Vascular	Xylem		
	Phloem		
Ground (or fundamental)	Parenchyma		
	Collenchyma		
	Sclerenchyma		

1. Describe and distinguish among ground tissue, vascular tissue, and dermal tissue.

2. Distinguish between epidermis and periderm. What types of cells are present in these tissues? Summarize your observations on cell types and functions in the accompanying table.

3. Distinguish between xylem and phloem. What do they have in common? How are they different? Summarize your observations on cell types and functions in the accompanying table.

4. Distinguish among parenchyma, collenchyma, and sclerenchyma. What do they have in common? How are they different? Summarize your observations on cell types and functions in the accompanying table.

5. Distinguish among cortex, pith, stele, pericycle, and endodermis. Of what tissues are they composed? What is the function of each?

6. The terms herbaceous and woody are often used to describe plants. How do these descriptions relate to the classification of plants as monocots or dicots?

7. List four major structural differences that can be used to distinguish monocots and dicots.

8. List three functions of roots.

9. List three functions of stems.

10. Fill in the following table.

	Arrangement and Location of Vascular Tissues	Presence or Absence of Stele	Presence or Absence of Pith	Special Characteristics
Dicot stem				
Monocot stem				
Dicot root				
Monocot root				
Dicot leaf				
Monocot leaf				

11. Corn is a C_4 plant. Describe the location of the light-dependent and light-independent reactions of photosynthesis relative to the structure of the corn leaf. What are bundle sheath cells and where are they located?

12. Distinguish between each of the following:

Simple leaves and compound leaves

Palmately compound and pinnately compound leaves

Netted venation and parallel venation

Angiosperm Development—Fruits, Seeds, Meristems, and Secondary Growth

OVERVIEW

Angiosperms protect their seeds inside fruits that develop from the ovary tissues of flowers. Seeds, liberated from their fruits, germinate and form the **epicotyl,** which develops into the shoot system, and the **hypocotyl,** which develops into the root system. **Apical meristems,** located at the tips of growing roots and stems, give rise to the **primary tissues** and increase the length of both stems and roots. In dicots, **lateral (secondary) meristems** give rise to **secondary tissues,** which increase the width of the stem and root, making them woody.

During this laboratory you will examine fruits, their seeds, how the seeds germinate, and how a plant develops through the activities of both apical and lateral meristems.

STUDENT PREPARATION

Prepare for this laboratory by reading the text pages indicated by your instructor. Familiarizing yourself in advance with the information and procedures covered in this laboratory will give you a better understanding of the material and improve your efficiency.

✔ **EXERCISE A** | Fruits

A **fruit** is a mature seed-containing ovary, a cluster of mature ovaries, or an ovary and closely associated tissues. Recall that the ovary is the enlarged portion of a flower carpel and contains one or more ovules. (Review flower structure, Laboratory 24, Exercise C.) Seeds develop from fertilized ovules, thus seeds are inside fruits. As the ovary develops into a fruit, the ovary wall thickens and becomes the **pericarp.** The evolution of such covered seeds marked a great evolutionary advance for angiosperms.

Fertilization most often initiates development of the fruit as well as the seed but, in some plants, pollination alone serves as the stimulus for fruit development.

Structural adaptations of various fruits have facilitated the worldwide dispersal of many plants. The structural organization of fruits reflects that of the flowers from which they develop. Fruits can be classified into three major types:

1. **Aggregate fruits** Consist of a number of enlarged multiple ovaries of a single flower, massed on or scattered over the surface of a single receptacle (the part of the flower stalk that bears the floral organs). The separate ovaries are called fruitlets. Examples include the raspberry, blackberry, and strawberry.

2. **Multiple fruits** Consist of the enlarged ovaries of several flowers more or less coalesced into one mass. Examples include the mulberry, fig, and pineapple.

3. Simple fruits Arise from the ovary (composed of a carpel or several united carpels) of a single flower. Simple fruits are divided into several categories based on the consistency of the pericarp and on structure and dehiscence (manner of opening). The two major groups include fleshy fruits and dry fruits.

a. **Fleshy fruits** The thickened pericarp sometimes becomes differentiated into three distinct layers: proceeding from the outside to the inside of the fruit, the **exocarp**, the **mesocarp**, and the **endocarp.** The development and consistency of these layers differ among types of fruit. There are several types of fleshy fruits (Figure 29A-1), including berries (examples: grapes and tomatoes), drupes (examples: cherries and peaches), pomes (examples: pears and apples), hesperidia (citrus), and pepos (squash, melons, and cucumbers).

b. **Dry fruits** Dry fruits (Figure 29A-2) are simple fruits usually classified according to whether they are **dehiscent** (split open when ripe) or **indehiscent** (do not split open when ripe). Further distinctions are made according to their mechanisms of dehiscence and other features of structure. Peas and beans are examples of dehiscent fruits. Sunflower seeds and wheat are indehiscent fruits.

ⅠⅠⅠⅠⅠ **Objectives** ⅠⅠⅠⅠⅠⅠⅠⅠⅠⅠⅠⅠⅠⅠⅠⅠⅠⅠⅠⅠⅠⅠⅠⅠⅠⅠⅠⅠⅠⅠⅠⅠⅠⅠ

☐ Define the term "fruit."

☐ Relate fruit structure to the ovary and ovules of a flower.

☐ Distinguish among aggregate, multiple, and simple fruits.

Fleshy fruits

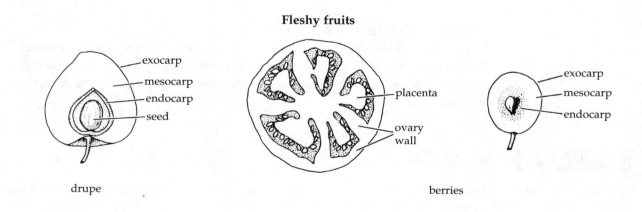

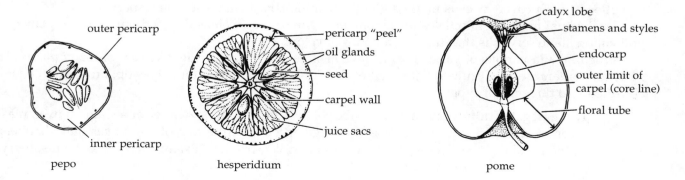

Figure 29A-1 *Examples of fleshy fruits.*

☐ Relate the structure of angiosperm flowers to their fruits.

☐ Recognize common types of fleshy fruits.

☐ Recognize specific examples of dry (dehiscent and indehiscent) fruits.

IIIII **Procedure** IIIIIIIIIIIIIIIIIIIIIIIIIIIIIIIIIIIIIIIIIIIII

1. Observe the cut peach and coconut on demonstration. The peach and the coconut are classified as **drupes:** in both fruits, the single seed is enclosed by a hard, stony endocarp (Figure 29A-1). In the peach the mesocarp is fleshy and edible, whereas in the coconut the mesocarp is fibrous. The exocarp of the peach forms the skin. The exocarp of the coconut is a hard covering that is usually removed, along with most of the mesocarp, before it reaches the produce department. The coconut you buy is the seed enclosed in the endocarp with remnants of the fibrous mesocarp still attached.

2. Cut a grape in half lengthwise. The grape is a **berry** (Figure 29A-1). The endocarp, mesocarp, and exocarp (pericarp) are fleshy. Mesocarp and endocarp are not well differentiated. The exocarp forms the skin.

3. Examine the cut apple on the demonstration table. The apple is a **pome** (Figure 29A-1). The endocarp surrounding the seeds is papery or leathery. The core constitutes the extent of the ovary, and the exocarp and mesocarp are indistinguishable. The flesh of the fruit derives from the enlarged bases of the petals and stamens that surround the ovary, not from the ovary itself.

4. Using the key to simple fruits (Table 29A-1) and the diagrams in Figures 29A-1 and 29A-2, try to identify each of the fruits on demonstration. For those fruits commonly eaten as food, identify the part of the fruit usually eaten (exocarp, endocarp, or the entire fruit). Since seed dispersal is closely associated with fruit structure, suggest the probable means of seed dispersal for each fruit. Record your observations in Table 29A-2.

Dry fruits

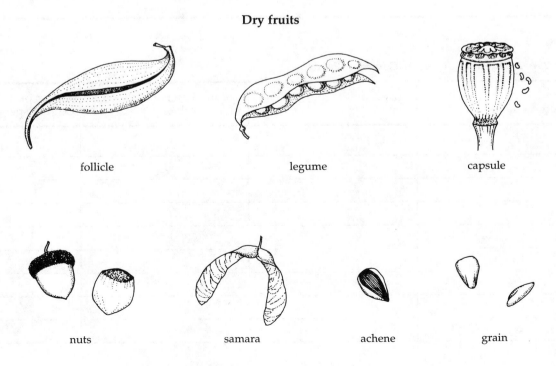

follicle   legume   capsule

nuts   samara   achene   grain

Figure 29A-2 *Examples of dry fruits.*

Table 29A-1   A Dichotomous Key to Identify Simple Fruits

1. Fruit dry at maturity 2
   2. Fruit splits open (is dehiscent) 3
      3. Fruit contains one chamber or compartment 4
         4. Fruit splits open along one side . . . . Follicle—example: milkweed
         4. Fruit splits open along two sides . . . . Legume—examples: peas, beans
      3. Fruit contains two or more chambers and splits along more
         than two sides . . . . Capsule—example: lily
   2. Fruit does not split open (is indehiscent) 5
      5. Ovary wall very hard . . . . Nut—examples: oak, hickory
      5. Ovary wall easily cracked or opened 6
         6. Ovary wall with winglike outgrowth . . . . Samara—examples: ash, maple, elm
         6. Ovary wall without winglike outgrowth 7
            7. Seed attached at only one place to the inside of
               the ovary wall . . . . Achene—example: sunflower
            7. Seed completely attached to ovary wall so that
               the two cannot be separated . . . . Grain—example: wheat, rice, corn
1. Fruit fleshy at maturity 8
         8. Fruit containing only one seed (located
            inside the pit) . . . . Drupe: the exocarp is skinlike; the
                                     mesocarp, fleshy; and the endocarp,
                                       stony—examples: cherry, peach, almond,
                                       olive
         8. Fruit containing several seeds 9
           9. Fruit has a firm leathery or hard rind 10
              10. Fruit has a hard rind and no
                  sections; rind inseparable . . . . Pepo—examples: cucumber and squash
              10. Fruit has leathery rind and many
                  sections; rind separable . . . . Hesperidium: citrus fruits
           9. Fruit has a peel that is not leathery and
            may be eaten 11
              11. Fruit has a core . . . . Pome—examples: pear, apple
              11. Fruit more or less fleshy
                throughout; seeds may be
                eaten . . . . Berry—examples: grape, pepper, tomato

Table 29A-2   Identifications of Common Fruits

Plant (common name)	Fruit Type	Portion Eaten (if applicable)	Means of Seed Dispersal
1.			
2.			
3.			
4.			
5.			
6.			
7.			
8.			
9.			
10.			

✔ EXERCISE B   Seed Structure

Angiosperm and gymnosperm seeds are integumented ovules. The integuments form the seed coat. The ovule itself contains the embryo sac within which double fertilization occurs. One sperm nucleus from the pollen grain unites with the egg to form a zygote. The other sperm nucleus unites with two polar nuclei to form a triploid primary endosperm nucleus. The zygote divides mitotically to form the embryo. The primary endosperm nucleus divides mitotically to form the endosperm, food for the embryo.

ⅡⅡⅡ Objectives ⅡⅡⅡⅡⅡⅡⅡⅡⅡⅡⅡⅡⅡⅡⅡⅡⅡⅡⅡⅡⅡⅡⅡⅡⅡⅡⅡⅡ

- ☐ Identify the parts of a monocot and a dicot seed and describe the function of each.
- ☐ Describe the function of endosperm and the cotyledons.
- ☐ Compare the mechanisms of endosperm storage in monocots and dicots.

✔ PART I   Examining the Dicot Bean Seed

ⅡⅡⅡ Procedure ⅡⅡⅡⅡⅡⅡⅡⅡⅡⅡⅡⅡⅡⅡⅡⅡⅡⅡⅡⅡⅡⅡⅡⅡⅡⅡⅡ

1. Obtain a water-soaked bean (*Phaseolus vulgaris*) seed. Examine the **seed coat** and note the **hilum,** the former point of attachment of the ovule to the ovary.

2. Gently remove the seed coat with your fingernails. Integuments around the ovule thicken and harden into this protective covering.

3. Pull the two cotyledons apart. The embryo consists of three parts:

   Cotyledons Leaflike food storage organs (seed leaves). In most dicots, the endosperm is absorbed during embryonic development and the seeds develop two fleshy, food-storing cotyledons.

   Hypocotyl Located below the attachment of the cotyledons, this portion of the embryo will develop into the embryonic root, or **radicle.**

   Epicotyl Located above the attachment of the cotyledons, this portion of the embryo will develop into the embryonic shoot. The first true leaves or **plumule** develop from the epicotyl.

4. In Figure 29B-1a, label the seed coat **a**, cotyledon **b**, hilum **c**, hypocotyl **d**, epicotyl **e,** first leaves or plumule **f**.

PART 2   Examining the Monocot Corn Seed

1. Examine the water-soaked kernels of corn (*Zea mays*). The kernel is a one-seeded fruit with the **pericarp** fused with the seed coat.

2. Cut a kernel lengthwise, bisecting the embryo. The **single cotyledon** or seed leaf is the broad surface of the embryo pressed against the starch-rich **endosperm.** In monocots, the cotyledon performs an absorbing rather than a food-storing function. The root tip (at the end of the embryonic root or radicle) is enclosed by a protective sheath called the **coleorhiza.** The shoot tip is enclosed in a conical sheath called the **coleoptile.**

3. In Figure 29B-1b, label the pericarp **a**, endosperm **b**, cotyledon **c**, coleoptile **d**, epicotyl **e**, hypocotyl **f**, coleorhiza **g**.

   a. What is the difference between the functions of the cotyledon in corn and in the bean?

   _____

   b. Which part of the corn seed contains the food source for the developing embryo?

   _____

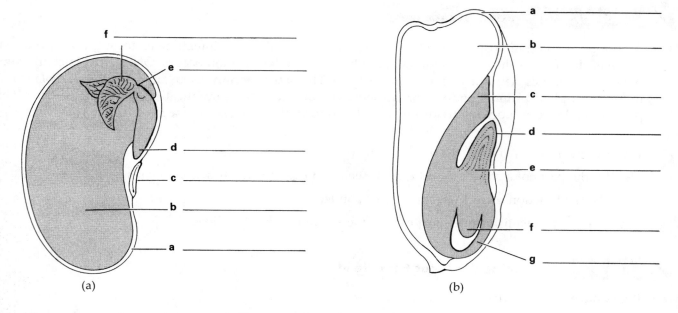

Figure 29B-1   *(a) The bean seed, a dicot. (b) A seed of corn, a monocot.*

👁 **EXERCISE C** | **Found a Peanut**

The peanut, a dicot, provides excellent material with which to check your understanding of fruits, seeds, and embryos.

||||| **Objectives** |||||||||||||||||||||||||||||||||||

☐   Diagram and discuss the functions of the parts of a peanut.

||||| **Procedure** ||||||||||||||||||||||||||||||||||||||

1. Obtain a peanut. *a. What type of fruit is it? (Should we be calling the peanut a fruit?)* _____
   b. Is the peanut really a nut? _____ *Why or why not?* _____
   _____

   c. What part of a fruit does the "shell" of a peanut represent? _____

2. Open the peanut. *d. On how many sides does it split open? (This should give you a clue for answering question a.)* _____

3. Remove the edible part. *e. What part of the fruit is the brown skin on the surface of the parts you eat?* _____ *f. What is the part of the fruit that you are used to calling the "nut?"* _____

4. Open the seed carefully. *g. Is the peanut a monocot or a dicot?* _____
   h. What does each half of the seed represent? _____

5. Look carefully for the embryo at the end of the seed. You should be able to see the hypocotyl, epicotyl, and some little leafy-looking structures, the first true leaves of the stem.
   i. Why are these little leafy structures not the cotyledons? _____
   _____

👁 EXERCISE D | Seedling Development

As seed germination occurs, the first structure to emerge from most seeds is the **radicle** (embryonic root), formed from the hypocotyl. The continuation of this root is called the **primary root. Branch roots** (lateral roots) develop on the primary root. In monocots, the primary root is short-lived and the root system of the adult plant develops from **adventitious roots** that arise from the first **node** (place on the stem where leaves are attached). Lateral roots are then produced from the adventitious roots.

The way in which the shoot emerges from the seed differs among various plants. In some plants, such as beans, the cotyledons are carried above the soil by the hypocotyl. This first forms a hook and then straightens out to lift up the cotyledons, which become photosynthetic. The food stored in the cotyledons is digested and transported to areas of the growing seedling. The cotyledons shrink and eventually fall off the stem.

In other dicot plants, such as the garden pea, the epicotyl forms a hook that carries the plumule (epicotyl plus first foliage leaves) above the ground while the cotyledons remain below ground. In corn the root and shoot emerge from the protective coleoptile (the first seedling leaf), and the single cotyledon remains below ground.

ⅠⅠⅠⅠⅠ Objectives ⅠⅠⅠⅠⅠⅠⅠⅠⅠⅠⅠⅠⅠⅠⅠⅠⅠⅠⅠⅠⅠⅠⅠⅠⅠⅠⅠⅠⅠⅠⅠⅠⅠⅠ

☐ Distinguish between the germination processes in beans, peas, and corn.

☐ Describe the process of seedling development by stating the fates of the hypocotyl, epicotyl, cotyledons, shoot apex, and root apex.

👁 PART I | Comparing Germination in Beans, Peas, and Corn

ⅠⅠⅠⅠⅠ Procedure ⅠⅠⅠ

1. Examine the different stages of germinating bean seedlings on demonstration. In the bean, a dicot, seedling emergence is due to expansion of the hypocotyl. Note in particular the **hypocotyl arch** (hook).

   a. *What is the advantage of having the hypocotyl arch pull the cotyledons up through the soil, rather than push them up?* _____

   _____

   Once the arch breaks through the soil surface and is exposed to light, it straightens out to hold the cotyledons and epicotyl in an upright position. Identify the following parts on the live germinated bean seedling: seed coat, cotyledon, first leaves, hypocotyl, epicotyl, and primary root. Refer to Figure 29D-1a.

   b. *What does the hypocotyl become?* _____

2. Examine the germinated pea seedlings on demonstration. In the pea (a dicot) seedling, the cotyledons remain underground. The epicotyl forms a hook which then straightens out and pulls the plumule above ground. Identify the following parts on the live germinated pea seedling: seed coat, cotyledon, first leaves, hypocotyl, epicotyl, and primary root. Refer to Figure 29D-1b.

3. Study the different stages of germinated corn on demonstration. Identify the coleoptile. Note that it is initially a closed tubular structure that grows to the soil surface. Once the coleoptile emerges from the soil, it ceases to grow in length, splits, and exposes the rolled leaves within. The shoot apex is at the base of the underground coleoptile. The "stalk" of the seedling is formed of rolled leaf bases. Only after the corn plant is several inches in height does the stem start to grow. The single cotyledon remains below ground. The radicle, originally enclosed in the coleorhiza, forms the primary root. Adventitious roots will later develop from the stem (Figure 29D-1c).

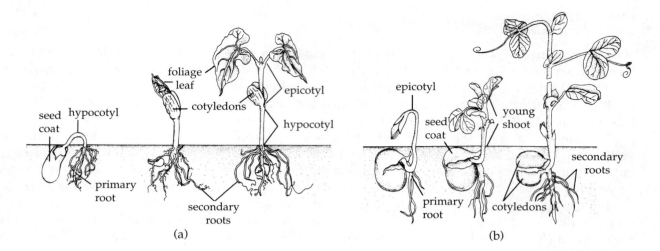

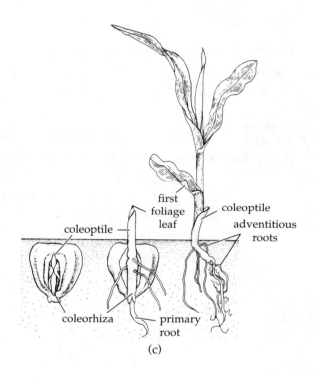

Figure 29D-1   *(a) During seedling emergence in the dicot bean, the hypocotyl expands and forms an arch that lifts the cotyledons up and out of the ground. (b) During germination of the dicot pea seedling, cotyledons remain below ground and the epicotyl forms a hook to pull the plumule above ground. (c) During germination of the monocot corn seed, the single cotyledon remains below ground and the shoot emerges from the coleoptile.*

4. Identify the leaf, coleoptile, primary root, and adventitious roots of the germinated corn seedling.

5. Cut a germinating corn seedling in half and add Lugol's solution (I₂KI) to it. Add I₂KI to the ungerminated seed examined earlier (Exercise B) and compare the color reaction.

    c. Which test is positive? _____

    d. What has happened to the food reserves of the endosperm in the germinated seedling? _____

    _____

6. In the following table (Table 29D-1), list the above-ground and below-ground structures you observed in the bean, pea, and corn seedlings, including the function of each. What are the major differences in the germination processes for each seed type?

Table 29D-1   Above and Below Ground Structures

	Bean	Pea	Corn
Above ground			
Below ground			

PART 2   Observing the Germination and Development of Seeds

Observe the development of bean, pea, corn, and other seeds in a germination chamber.

Procedure

1. Fill a clear plastic cup with wet soil or wrap a moist piece of blotter paper around the inside surface of the sides of the cup and pour an inch of water into the cup.

2. Place seeds against the plastic about an inch below the soil or between the blotter and the cup. You should be able to see the seeds. Cover with a Petri dish lid.

3. Observe the seeds for two weeks (or longer). Remember to keep the soil moist. Keep the blotter paper moist by keeping a one-fourth inch of water in the bottom of the cup. The blotter paper will act as a wick and soak up the water.

4. In Table 29D-2 record your observations over 10 days.

Table 29D-2   Seed Germination

Type of Seed	Time until Emergence of Epicotyl	Time until Emergence of Hypocotyl	Fate of Cotyledons	Fate of Hypocotyl and Epicotyl	Day 5 Length of Hypocotyl and Epicotyl General Appearance	Day 10 Length of Root and Shoot General Appearance

EXERCISE E   Studying the Stem Tip and Root Tip

The developing shoot (epicotyl) and root (hypocotyl) are dependent upon the presence of **apical meristems**—the growing points or areas of mitotic activity at the tips of the shoot and root. These are responsible for primary growth and the laying down of primary tissues.

Three primary meristematic tissues are present in the root and shoot tip. (1) The **protoderm** is the outermost layer of cells along the surface of the apical tissue and is responsible for production of the epidermis. (2) The **procambium** forms lengthwise columns of tissue that will differentiate into vascular bundles composed of xylem and phloem. (3) The **ground meristem** adds cells to the remaining space, giving rise to the cortex and the pith. Development of the stem and root from these meristematic tissues occurs through cell division, cell enlargement, and cell differentiation.

ⅠⅠⅠⅠⅠ Objectives ⅠⅠ

☐ Distinguish between apical and lateral meristems in the root and shoot.

☐ Identify the types of tissues formed by apical meristems and describe their location in the young root and shoot.

✔ **PART Ⅰ**   **Examining the Stem Tip**

ⅠⅠⅠⅠⅠ Procedure ⅠⅠ

1. Locate the apical buds at the tips of all branches on a living *Coleus* plant.

2. Look at the **axil** of each leaf between the leaf base and the stem axis and identify the **axillary bud.** When branches develop, axillary buds become apical buds.

3. Use low power (10×) to examine a prepared slide of the apical bud of *Coleus* (longitudinal section), then switch to high power (40×). Locate each of the following structures and label them in Figure 29E-1.

Figure 29E-1   *Stem tip of the* Coleus *plant.*

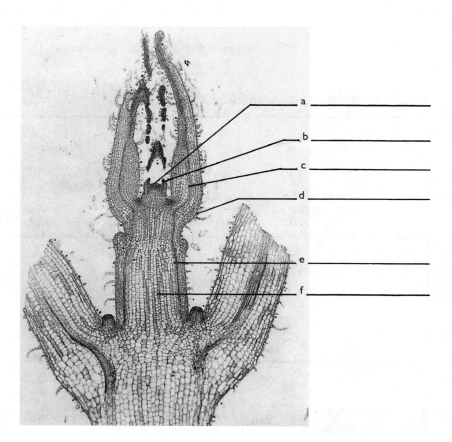

a _____

b _____

c _____

d _____

e _____

f _____

    a Meristematic cells Meristematic cells are small with relatively large nuclei and dense cytoplasm, and are stained more intensely than surrounding nonmeristematic cells. Note the mound of meristematic cells at the stem tip and in the regions along the sides of the stem in the axils of the young leaves.

   b Leaf primordia The size of these rudimentary leaves increases from the tip downward. Notice that leaf primordia are visible only on alternate nodes, and intervening nodes appear as mounds of tissue.

   a. Is the arrangement of Coleus *leaves alternate or opposite?* _____

   c Vascular tissue Strands of vascular tissue are seen in lengthwise view along the sides of the stem and in older leaves.

   d Protoderm Outermost layer developing into the epidermis.

   e Procambium Located inside the protoderm; gives rise to the primary vascular tissue.

   f Ground meristem Inner cells giving rise to ground tissue, including the cortex.

4. Examine the head of a cabbage or a brussels sprout that has been cut lengthwise through the center. Compare it with the *Coleus* stem tip.

   b. What may the entire head be called? _____

 **PART 2**   **Examining the Root Tip**

||||| **Procedure** |||||||||||||||||||||||||||||||||||||||||||||

1. Place a young radish or rye seedling in a large drop of water on a slide.

2. Use a razor blade to cut off the grain and shoot of the seedling and carefully lower the cover glass over the root. Study the following structures under low power (10×) and label them in Figure 29E-2.

   a Root cap Located at the tip of the root, covering the apical meristem. Cells of the root cap are usually loosely arranged and often have broken away from the root.

   b Apical meristem Located behind the root cap; a **region of cell division, c** Cells are tightly packed and appear dark.

   d Region of elongation Region behind the apical meristem where cells elongate. Region ends at level of first root hairs.

Figure 29E-2 *Root tip.*

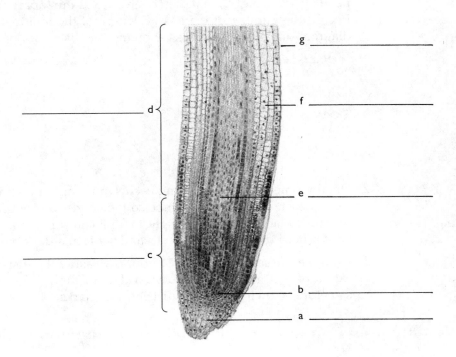

e Procambium Innermost area of the root; matures to form the central vascular cylinder.

f Ground meristem Middle tissue layer; forms the cortex.

g Protoderm Outer layer of cells; differentiates into epidermis.

EXTENDING YOUR INVESTIGATION: GROWING LONGER

The elongation of cells in the region of elongation produces most of the increase in length of the root. Do you think that there is any increase in length beyond this region? To test your answer, formulate a hypothesis that predicts what will happen to the length of the region of elongation relative to the region of maturation in the growing root.

HYPOTHESIS:

NULL HYPOTHESIS:

Identify the **independent variable** in this experiment.

Identify the **dependent variable** in this experiment.

If you observe the root tip of a germinating pea seed over a period of 24 hours, what do you **predict** you will observe?

Use the following procedure to test your hypothesis.

PROCEDURE:

1. Obtain three pea (*Pisum*) seedlings of equal size with straight roots about 2 to 3 cm long. One at a time, lay each seedling flat on a moist paper towel and then, with a fine indelible ink pen (e.g., Sharpie™), make marks exactly 2 mm apart along the entire length of the root. Record the number of marks and length of the root in Table 29E-1. In the space below, diagram one of the seedlings showing the distribution of the marks.

   Pea Seedling   *Pea Seedling after 24 Hours*

2. Obtain 3 pieces of glass tubing each 4 cm long with an inner diameter of 5 mm. Insert the root of each seedling into a tube so the cotyledons of each seedling are resting on one end of the tube. Place the tubes upright in a plastic cup containing 25 ml of water. Cover the cup with a Petri dish or plastic wrap and set it in a dark, warm place.

3. Examine the seedlings after 24 hours. Measure the distances between the marks and the total extent of root growth. Record your data in Table 29E-1. Draw the same seedling (see above) as before, showing the new distribution marks.

RESULTS: What was the result of your experiment? Which portion of the root elongated?

Do your data and results support your hypothesis?

Your null hypothesis?

Was your prediction correct?

What do you **conclude** about the way in which root growth occurs?

Table 29E-1   Root Length of Germinating Pea Seedlings

Pea Seedling	Number of Marks 2 mm Apart	Total Length of Root at Start (mm)	Root Length after 24 Hours	
			Distances Between Marks	Total Growth (mm)
1				
2				
3				

✓ **EXERCISE F**   **Secondary Growth of Angiosperms—The Woody Stem (*Tilia*)**

We have observed that all vascular plants have primary meristems in two locations: stem tips (buds) and root tips. These are responsible for primary growth and production of primary tissues. Leaves are limited to primary growth, but the stems and roots of woody plants (not herbaceous plants) can thicken as a result of **secondary growth.**

There are two secondary meristems in woody plants: the **vascular cambium** and the **cork cambium.** The vascular cambium consists of a row of cells that lies between the xylem and phloem of the vascular bundles and continues between individual vascular bundles, thus forming a continuous cylindrical sheath of cells within the stem. The meristematic cells divide and produce new cells along both interior and exterior faces of the vascular cambium. Cells produced along the interior face differentiate into secondary xylem, whereas those produced along the exterior face differentiate into secondary phloem (Figure 29F-1).

During each growing season a new layer of secondary xylem is added to the inside, and a new layer of secondary phloem to the outside of the vascular cambium. Xylem layers build up and form the woody core of the plant. The xylem cells are dead and contain **lignin** to strengthen their cell walls. In contrast, secondary phloem does not build up: older layers of phloem, as well as the cortex and epidermis formed during primary growth, are continuously sloughed off the stem and new protective tissues are formed by the **cork cambium.** Because the **bark** layers external to the vascular cambium are continuously sloughed, these outer regions do not add significantly to the cumulative thickness of the woody stem and root: wood is primarily secondary xylem.

Figure 29F-1 *Cross section of a 3-year-old stem, showing annual growth layers. On the perimeter of the outermost growth layer of xylem is the vascular cambium, encircled by a band of secondary phloem. The primary phloem and also the cortex will eventually be sloughed off. The tissues outside the vascular cambium, including the phloem, constitute the bark.*

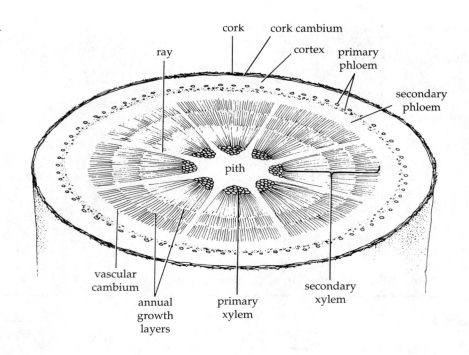

ııııı Objectives ııııııııııııııııııııııııııııııııı

☐ Describe how wood is formed.

☐ Distinguish between summer wood, spring wood, heartwood, and sapwood.

☐ Describe how to determine the age of a tree.

☐ Describe the importance of secondary meristems to the woody plant.

ııııı Procedure ııııııııııııııııııııııııııııııııııı

1. Examine a prepared slide of a woody stem (*Tilia*, basswood). Locate the following structures and label them in Figure 29F-2.

   a Pith Located in the center of the stem, just as in the herbaceous dicot stem of alfalfa (Laboratory 28).

   Vascular tissue Xylem. The closely spaced bundles of **secondary xylem b** form a wide, continuous ring around the pith. Clusters of **primary xylem** cells **c** are located toward the interior around the pith.

   d Vascular cambium Situated at the outer edge of the xylem, these thin-walled cells give rise to the secondary xylem and phloem and the **vascular rays e**, ribbonlike aggregates of two to three cells extending radially through the xylem and widening to form triangular areas toward the outside.

   f Annual rings In temperate zone woody plants, the secondary xylem cells formed by vascular cambium early in the growing season when more moisture is present, called **spring wood g**, have a larger diameter than those formed later, called **summer wood h**. The contrasting cell types in spring wood and summer wood produce the annual ring, an indication of the xylem produced during one growing season. You can determine the age of a tree by counting its annual rings.

   i Vascular tissue Phloem. This appears as wedge-shaped areas between widened vascular rays. The banded appearance is the result of layers of thick-walled **phloem fibers j** interspersed with layers of **sieve tubes** and **companion cells k**.

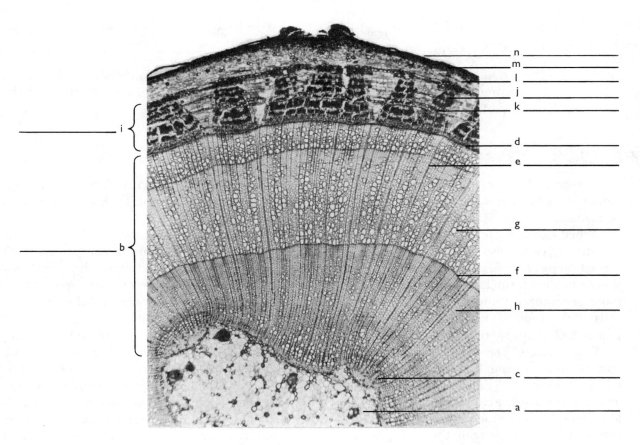

Figure 29F-2 *Cross section of* Tilia *stem.*

a. *How old is the basswood section you are examining?* _____

b. *In the tropics, rainfall and temperature are the same year round. Do you think that trees grown in the tropics would have annual rings?* _____

c. *Which kind of wood within an annual ring is* always *located to the inside?* _____
   Why? _____

l **Cortex** Located outside the phloem, the cortex varies in thickness and is somewhat distorted by the pressure exerted by increasing amounts of xylem and phloem tissue. Parenchyma, collenchyma, and fibers are present.

m **Cork** Located just beneath the **epidermis n**. Cork is formed by the cork cambium, which develops in the outer cortex about the time that second-year growth occurs.

Wood All tissues outside the vascular cambium are known as *bark*. Tissues lying inside the cambium are referred to as *wood*. As trees grow older, the older xylem cells lose their ability to conduct water due to the accumulation of resin substances and other metabolic waste products within the cells. This wood is called **heartwood**. The younger conductive wood surrounding the heartwood is called **sapwood**. Heartwood is often darker than sapwood.

2. Examine the tree cross sections on demonstration. Identify bark, cambial region, annual rings, spring wood, heartwood, sapwood, summer wood, rays, and pith.

d. Which structures of woody stems are included in the bark? _____

In wood? _____

3. The wood samples on demonstration have been cut transversely, radially, and tangentially. Determine how each has been cut in relation to growth rings and rays. Note the characteristic wood pattern of various cuts.

EXERCISE G   Plant Tissue Culture*

Small pieces of tissue, **explants,** removed from a plant and grown in a specialized medium can give rise to **clones,** new individuals having the same genetic makeup as the parent plant. In this experiment, you will culture explants in vitro to produce clones.

When explants are removed from the parent plant, the excised tissues respond to being "wounded" by stimulating cell division to form a protective layer of wound tissue, sometimes accompanied by the formation of **callus** tissue. Callus tissue is often described as undifferentiated; the cells of explants and those of callus formations undergo "dedifferentiation" during which they become less specialized and more generalized (however, not *all* cells reach the same state of generalization). The "dedifferentiated" cells, with proper stimulation by hormones, particularly auxins and cytokinins, can regenerate entire plants. Cells capable of giving rise to all the structures of a mature plant are said to be **totipotent.**

Cauliflowers, *Brassica oleracea,* a member of the mustard family, can be used to demonstrate plant tissue culture or cloning techniques. The "head" of the cauliflower is composed of a mass of apical meristems within floral buds. The buds can be induced to produce leafy shoots by treatment with a mixture of the auxin **indoleacetic acid** (IAA) and the cytokinin **kinetin.** Once shoots have formed, they are transferred to a second growth medium without hormones, where they develop roots. Rooted shoots can then be placed in soil to propagate the multiple copies or clones of the original parent plant. Development from explant to plant requires 6 weeks.

Procedure

Ideally, this procedure should be carried out under a laminar-flow hood. However, plant tissue culture can be done almost anywhere if simple precautions are taken to maintain aseptic conditions. Choose an area where there is little or no air movement. Clean your work area with a 1% bleach solution. Plan to have all necessary tools nearby and accessible with a minimal amount of movement. *Do not* lean over the work surface (hair is a good source of bacteria). Carry out all manipulations *in front of you.*

1. Break apart or cut cauliflower into 0.5-inch to 1-inch sections.

2. Place three or four pieces in a small jar or Erlenmeyer flask containing 100 ml of tap water and 3 drops of dishwashing detergent (Joy™). Shake periodically for 5 minutes. Drain off the liquid and rinse the flask and the cauliflower well in tap water.

3. Add 50 ml of 70% ethanol and shake for 1 minute.

4. Drain off the ethanol (into a waste beaker) and add 100 ml of 10% bleach. Shake gently for 15 minutes.

5. Drain off the bleach solution. Using sterile forceps that have been dipped in 95% ethanol, flamed, and cooled, transfer the cauliflower pieces to a jar containing 100 ml of sterile distilled water. Swirl for 2 minutes.

6. Carefully drain off the water and, using sterile forceps, transfer the pieces of tissue to a sterile Petri dish.

---

*Adapted from Janice H. Haldeman and Jane P. Ellis, "Using cauliflower to demonstrate plant tissue culture,"
The American Biology Teacher, *vol. 50, no. 3, March 1988.*

7. Use a sterile scalpel to make slices of the explant tissue (approximately 1 cm by 1 cm and 3 mm in thickness). Make sure that each slice contains bud tissue.

8. Use sterile forceps to transfer four slices of tissue to a Petri dish containing Medium A (Murashige and Skoog Minimal Organic Medium) supplemented with sucrose but without hormones. Seal the edges of the dish with a strip of paraffin film.

9. Place the dish of explants 12 to 15 inches below a fluorescent light source connected to a timer set to provide 16 hours of light and 8 hours of dark. Culturing should take place at room temperature (26 ± 2°C).

Next Laboratory Period

1. Check for dead or contaminated explants: those that appear fuzzy (covered by fungus) or slick (covered by bacteria) should be discarded. Healthy explants will show bud development and some evidence of "greening" (tips may redden due to the production of anthocyanins).

2. Using aseptic technique, transfer healthy explants to a sterile jar containing Medium B (same as Medium A, but supplemented with 2.5 mg/l kinetin and 8.0 mg/l IAA).

3. Return explants to the light source and allow approximately 2 weeks for shoots to develop.

Rooting New Shoots

As shoots form, you will notice some root development. These roots have not formed from bud apical meristem tissue but from the vascular cambium of the non-bud tissue of the explant. These first roots do not serve the newly formed shoots and must be removed so that the shoots themselves can form roots.

1. Use sterile forceps to transfer a shoot cluster to an empty sterile Petri dish.

2. Use a sterile scalpel to cut shoots at the base, separating them so that each shoot includes at least two leaves and a bud. Prepare at least six shoots.

3. Place three shoots in a sterile jar of Medium A, gently pushing the end of the shoot into the medium so that the stem stands up vertically.

4. Return the explants to the same light source. Root development will take place in approximately 1 week.

Transplanting

When roots (three to four) ranging from 0.5 to 1.5 cm in length have developed, the "clones" can be transferred to soil. Aseptic technique is no longer critical.

1. Carefully remove rooted shoots from the Medium A jars and rinse with tap water to remove all traces of Medium A.

2. Obtain a pot of soil. Make a hole in the soil with your pencil and place the plant into the hole (the roots and approximately one-third of the stem should be buried). Press the soil firmly around the base of the stem.

3. Water thoroughly and place a plastic bag over the pot to retain moisture. Return the plants to the light source or place them on a windowsill in indirect light.

ⅢⅢⅠ Alternative Procedure ⅢⅢⅢⅢⅢⅢⅢⅢⅢⅢⅢⅢⅢⅢⅢⅢⅢⅢⅢⅢⅢⅢⅢⅢⅢⅢⅢ

Alternatively, your instructor may ask you to culture tobacco leaf disks. In this case, swab a tobacco leaf with 95% ethanol and place it in a sterile Petri dish containing a 20% solution of bleach. After 10 minutes rinse the leaf three times with sterile distilled water and use a sterile cork borer (#4) to cut leaf disks from the tissue between veins. Transfer the leaf disks aseptically to flasks of sterile nutrient medium and gently tap them down onto the surface of the medium using a sterile, round-end glass rod. Place the covered flasks under the same light source used for the cauliflower explants. A callus will form within 3 to 4 weeks, followed by development of roots and shoots (4 to 8 weeks) plus buds and flowers (8 to 12 weeks).

Laboratory Review Questions and Problems

1. Strictly speaking, what is a fruit?

2. Why is a peanut actually a fruit rather than a nut?

3. Distinguish between simple, multiple, and aggregate fruits.

4. Distinguish between fleshy and dry fruits.

5. Identify the following fruits as simple, multiple, or aggregate. For simple fruits, use the dichotomous key in Table 29A-1 to further classify the fruit type.

Fruit	Simple, Multiple, or Aggregate	Type of Simple Fruit
Plum		
Strawberry		
Peanut		
Pineapple		
Banana		
Fig		
Tomato		
Coconut		
Orange		

6. When parents tell children to "eat their vegetables," what do they actually mean? What is a vegetable?

7. For the following "vegetables," identify the parts we consume: roots, stems, leaves, flowers, or fruits. Review Laboratory 28 and refer to your textbook for assistance.

"Vegetable"	Plant Part Eaten	"Vegetable"	Plant Part Eaten
Carrot		Spinach	
Irish potato		Squash	
Sweet potato		Lettuce	
Celery		Onion	
Broccoli		Corn	

8. Fill in the following table to summarize some characteristics of monocots and dicots.

		Monocot	Dicot
SEED	Number of cotyledons		
	Presence of plumule		
	Presence of coleoptile		
	Presence of coleorhiza		
STEM	Presence of vascular cambium (lateral meristem)		
	Presence of primary meristem		

9. Distinguish between each of the following:
Ovary, carpel

Fruit, ovary

Seed, ovule

Indehiscent, dehiscent

Hypocotyl, epicotyl

Monocot, dicot

Branch roots, adventitious roots

Primary meristem, secondary meristem

Wood, secondary xylem

Spring wood, summer wood

Heartwood, sapwood

Bark, cork

10. Which tissues compose the bark of a tree? Why does "girdling," cutting completely through the bark around the circumference of the trunk, kill a tree?

Water Movement and Mineral Nutrition in Plants

OVERVIEW

The amount of water needed daily by plants for photosynthesis, cell growth, and maintenance is very small, yet plants require large volumes of water to live. Amazingly, over 90 percent of the water moved throughout plant tissues by the process of **translocation** is lost to the air by **transpiration** (loss of water vapor from the plant surface) or **guttation** (loss of liquid from the ends of vascular tissue at the leaf margins). During this laboratory period you will study some of the factors involved in the movement of water and minerals throughout the plant.

STUDENT PREPARATION

Prepare for this laboratory by reading the text pages indicated by your instructor. Familiarizing yourself in advance with the information and procedures covered in this laboratory will give you a better understanding of the material and improve your efficiency. Review Laboratory 8, Exercise D, A Look at Osmosis, and be sure to familiarize yourself with the following terms: *water potential, pressure potential,* and *osmosis.*

✔ **EXERCISE A** | **Observing Stomata**

Stomata are minute openings bordered by **guard cells** in the epidermis of leaves and stems. It is through these openings that gases pass and water evaporates as photosynthesis, respiration, and transpiration occur.

Stomata open and close as a result of changes in turgor pressure within guard cells. Light-activated proton pumps in guard cell membranes actively transport H^+ ions out of the guard cells and, in turn, are responsible for the uptake of potassium ions (K^+) into these cells. An increase in the K^+ concentration inside guard cells causes their water potential (ψ) to become more negative, and water from the surrounding mesophyll cells (which have a more positive water potential) moves into the guard cells. As water accumulates and turgor increases within these cells, they inflate and bulge outward. The stomatal opening increases in size as the guard cells swell (Figure 30A-1).

Stomata usually remain open during the day and close at night, balancing the need for photosynthesis with that for conserving water. When mesophyll cells of the leaf are actively photosynthesizing, the amount of CO_2 present in air spaces between mesophyll cells decreases rapidly, signaling guard cells to open. On hot days, water depletion may cause loss of turgor in the guard cells and the stomata will close. Abscisic acid, produced by mesophyll cells of drought-stressed plants, can also signal guard cells to close during the day. In both cases photosynthetic rates will decrease, but this is a necessary trade-off. C_4 plants and succulent plants that live in hot, dry environments accumulate CO_2 in their leaves in the form of organic acids. Their stomata can remain closed during the day to conserve water while photosynthesis is made possible by release of CO_2 from the organic acids.

Figure 30A-1 *The mechanism of stomatal movements. (a) A closed stoma. The kidney-shaped guard cells are close together. Note the microfibrils that loop around the guard cells radially. (b) When water enters the guard cells, the microfibrils prevent them from expanding in circumference, so they expand in length and push apart at their attached ends.*

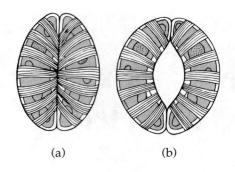

(a)   (b)

IIIII Objectives IIIIIIIIIIIIIIIIIIIIIIIIIIIIIIIIIIIIIIIII

☐ Describe how water transpires from the surface of a leaf.

☐ Diagram a stoma and label its parts.

IIIII Procedure IIIIIIIIIIIIIIIIIIIIIIIIIIIIIIIIIIIIIIIIIIII

1. Obtain a leaf from the plant provided. Use forceps to remove a small piece of the lower epidermis. Make a wet-mount slide. Alternatively, paint the leaf surface with clear nail polish and, when it is dry, use a piece of clear adhesive tape to remove the "polish peel" from the leaf and mount it directly onto the slide. Examine the peel using the compound microscope at low power (10×). Count the number of stomata visible in one field of view. Move the slide and repeat the count. Record your data in Table 30A-1.

2. Repeat step 1 using a piece of upper epidermis from the same leaf. Record your data in Table 30A-1.

Table 30A-1   Distribution of Stomata

Name of Plant Used: _____		
	Number of Stomata in One Field of View:	
	Lower Epidermis	Upper Epidermis
	_____	_____
	_____	_____
Average	_____	_____

3. Switch to high power (40×) and, in the space provided next to the table above, prepare a drawing of several epidermal cells, guard cells, and stomata. Label these parts on your drawing.

    a. *Are stomata more numerous on one surface than the other? _____ If so, which one?*

    _____

    b. *Compare your counts with those of a classmate who used a different plant. Are they the same?*
    _____ *How do you explain this?* _____

    _____

Elodea, a common aquarium plant whose leaves are only two or three cells thick, has no stomata.

c. *How does gas exchange occur?* _____

4. Mount another piece of epidermis (upper or lower) on a microscope slide, but this time mount it in a 20% sucrose solution. (If you use the nail polish technique, soak your leaf in sucrose for 5 minutes and then blot it dry prior to making the "polish peel.") Add a coverslip and use a compound microscope to make your observations. You may need to use high power (40×).

   d. *Do the stomata appear to be different?* _____ e. *Are they opened or closed?* _____

   f. *Why?* _____

   g. *What is the shape of the guard cells?* _____

5. On the left side of the space provided below, diagram a stoma treated with sucrose.

6. Remove the coverslip from your slide and place the *same* piece of epidermis into a drop of water on a clean slide. Observe your slide immediately.

   h. *Have the guard cells changed in shape?* _____ *Are they open or closed?* _____

   i. *Why?* _____

7. In the remaining space above, diagram a stoma as it appears after the change from the sucrose solution to water. Label the guard cells.

✔ EXERCISE B | Guttation

Guttation is the loss of water from the ends of veins at the tips and margins of leaves. It will occur only when soil moisture levels are high and the relative humidity is 100 percent. Under these conditions, transpiration, the loss of water vapor, is slow or absent, and a buildup of pressure in the roots (root pressure) forces water up the xylem. The water exudes through special openings called **hydathodes** at the tips and margins of leaves (*not* through stomata). Much of the dew you see on grass in the early morning is not water condensed from the air, but rather water from inside leaves leaving by the process of guttation.

|||||| Objectives ||||||||||||||||||||||||||||||||||||||||||||||

☐ Explain how guttation occurs in a leaf.

|||||| Procedure ||||||||||||||||||||||||||||||||||||||||||||||

Examine the young grain seedlings on demonstration. Droplets on the leaves are the result of guttation.

   a. *Where on the leaves do the droplets of the water of guttation appear?* _____

✔ 👁 EXERCISE C | Transpiration

Transpiration, the loss of water vapor by plant parts, occurs mainly through the stomata of the leaves. Water can move up a plant to the leaves by being pushed from the bottom or pulled from the top. **Root pressure** is not sufficient to push water all the way from the bottom to the top of a large plant such as a tree. It is more likely that water is pulled up through the plant body by the **cohesion–tension mechanism.**

As water leaves the intercellular spaces between mesophyll cells in the leaf and evaporates through the stomata, it is replaced by water from within the cells themselves. Since water moves out of these cells freely but solutes do not, the solute concentration within the mesophyll cells increases and the water potential (ψ) of the cells decreases. Water will then move into the mesophyll cells from surrounding cells with higher water potentials—for example, the cells of the xylem. Thus, as a result of transpiration, a gradient of differences in water potential from the xylem to the air outside the leaf is formed and water tends to be "pulled" upward. The **cohesion** of water molecules (one hydrogen-bonding to another) and their **adhesion** to the walls of the xylem cells cause the water to be pulled up as a continuous column.

The upward transpiration pull on the fluid within the xylem causes a **tension** (negative pressure) to form, pulling the walls of the xylem inward (you can actually measure the decrease in stem diameter of a plant on a hot sunny day, when the transpiration rate is very high). Tension, since it is "negative" pressure, causes water potential in the xylem to decrease. The decrease in water potential, transmitted through the column of fluid in the xylem, all the way to the roots, causes water to move from the soil across the cortex of the root and into the xylem of the stele, once again moving from an area of higher water potential to an area of lower water potential.

The opening of stomata, which allows transpiration to occur, is also required for the entry of CO_2 used in photosynthesis. A balance must be maintained between the two processes, transpiration and photosynthesis, by regulating the opening and closing of stomata.

ııııı Objectives ııı

☐ Determine how environmental conditions affect the rate of transpiration.

✔ **PART I**

ııııı Procedure ıı

Examine the flasks (covered with beakers) in the demonstration area: flask 1, no *Coleus* shoot; flask 2, *Coleus* shoot; and flask 3, *Coleus* shoot with leaves coated with petroleum jelly.

a. *In which beaker or beakers have water droplets formed?* _____

b. *Which of the beakers serves as a control?* _____ *How?* _____

_____

c. *Where did the water that has condensed on the inside of the beaker come from? Be specific.*

_____

d. *What was the purpose of the petroleum jelly treatment?*

_____

👁 **PART 2**

Any environmental condition that increases evaporation will increase transpiration. Your laboratory instructor will demonstrate the method for assembly and use of a simple **potometer** (Figure 30C-1), an apparatus used to measure water transpiration in a plant shoot.

ııııı Procedure ııı

1. Work in pairs. Each pair of students will determine the rate of transpiration using one of four treatments (to be assigned by your laboratory instructor):

   A. Establish a control by running the experiment under room conditions.

   B. Simulate wind by placing the plant about 2 meters from a fan. Use a low or medium setting (too much wind will cause stomata to close).

Figure 30C-1   *A sample potometer.*

C. Increase the humidity (vapor pressure) by spraying the plant with water and covering it with a plastic bag.

D. Increase the temperature of both the leaf and the air and increase the light intensity by placing the plant a prescribed distance from a flood lamp. (Ask your instructor for assistance—this distance will vary with different types of plants; the usual distance is 1.25 to 1.5 meters.)

Formulate a hypothesis on how the rate of transpiration is affected by the altered environmental conditions assigned to you.

HYPOTHESIS:

NULL HYPOTHESIS:

What do you **predict** will happen to the transpiration rate in your experiment compared with the rates for other treatments?

What is the **independent variable**?

What is the **dependent variable**?

2. Now determine transpiration rate. Set up your potometer as demonstrated. Choose the plant you will use and push the stem into the hole in the rubber stopper of the potometer top. Put petroleum jelly (Vaseline) around the base of the plant on the top side of the rubber stopper. Fill the potometer bottle *all the way to the top* with water. Make a fresh cut on the bottom of the stem (do this under water if possible) and immediately push the stopper (including plant and pipette) into the potometer bottle. Make sure to push hard to get a complete seal around the rubber stopper. The water should have risen up the pipette, past the last mark, and the level should not be dropping too quickly.

3. Allow the plant to equilibrate to the experimental conditions for 10 minutes. In Table 30C-1, record the water level in the pipette of the potometer at the end of 10 minutes as "(a) ml at start." After an additional 10 minutes, take a second reading and record your data in Table 30C-1; 10 minutes later, take a third reading and again record your data. (Your instructor will indicate whether you should take more readings.)

Table 30C-1   Transpiration Experiment Data

Experimental Conditions _____	Water Loss per 10-minute Interval
(a) ml at start _____	(b) − (a) _____
(b) ml at 10 minutes _____	(c) − (b) _____
(c) ml at 20 minutes _____	Average _____

4. After completing your last reading, remove all of the leaves from the plant that you used for your experiment. Using a balance, determine the combined mass (in grams, g) of all the leaves.

5. Determine the transpiration rate as average milliliters (ml) of water loss per 10 minutes.

   Convert this to water loss per hour (ml/hr). _____ ml/10 minutes; _____ ml/hr

6. Express the transpiration rate as the average amount of water loss in milliliters per hour per gram of leaf material.*

   $$\text{Transpiration rate per gram of leaf tissue} = \frac{\text{transpiration rate (ml/hr)}}{\text{mass of leaf tissue (g)}}$$

   Transpiration rate: _____ ml/hr per gram

7. Pool the class data and determine the average transpiration rate for each of the four treatments.

8. Using Figure 30C-2, make a bar graph to represent the pooled class data. Choose an appropriate scale for the vertical axis.

Do your results support your hypothesis? _____ *Your null hypothesis?* _____

Was your prediction correct? _____

*What can you **conclude** about the rate of transpiration under your assigned conditions?*

a. Under which condition was the rate of transpiration greatest? _____

b. Consider all four conditions and explain how each condition caused an increase or decrease in transpiration.

_____

_____

_____

You may want to determine the transpiration as a function of surface area (rate per square meter of leaf tissue). Estimate the total leaf surface area in square meters (m^2). To do this, divide the total mass (g) of all leaves stripped from the plant (step 4) by the mass (g) of 1 m^2 of leaf tissue. (Your instructor will give you the latter value for the type of plant being used.) For total surface area, multiply by 2 to account for both surfaces of a leaf. Calculate transpiration rate per square meter by dividing the rate (ml/hr) by the surface area (m^2) of leaf tissue of your plant.

Figure 30C-2 *Class data for effects of environmental conditions on transpiration rate.*

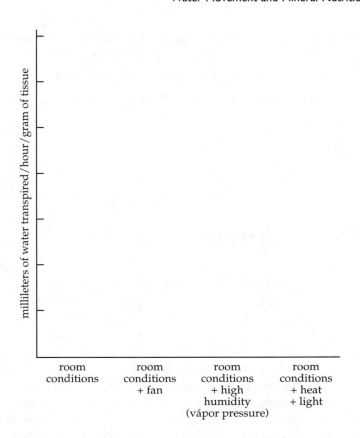

c. *Is it possible that one of the test plants might transpire more than another of equal size under the same conditions?* _____ *Explain.* _____

_____

d. *Do you think that the number or size of the leaves might affect the transpiration rate?* _____
   Why? _____

_____

e. *How would the number and distribution of stomata affect the transpiration rate?* _____

_____

✔ **EXERCISE D**   **The Pathway of Water Movement Through a Plant**

Water moves from the roots to the stem and then into leaves and flower parts through the xylem of the vascular bundles. Thus there is a continuous flow of water from the lower parts of plants to the upper parts.

⁞⁞⁞⁞ Objectives ⁞⁞⁞⁞⁞⁞⁞⁞⁞⁞⁞⁞⁞⁞⁞⁞⁞⁞⁞⁞⁞⁞⁞⁞⁞⁞⁞⁞⁞⁞⁞⁞⁞⁞

☐ Determine the pathway of water movement in a plant.

⁞⁞⁞⁞ Procedure ⁞⁞⁞⁞⁞⁞⁞⁞⁞⁞⁞⁞⁞⁞⁞⁞⁞⁞⁞⁞⁞⁞⁞⁞⁞⁞⁞⁞⁞⁞⁞⁞⁞

   1. Examine the carnation and celery plants that have been placed in one or several containers of food coloring. Note the cut surface of the celery petiole.

      a. *Does the dye appear only in certain tissues?* _____

2. Use your dissecting microscope to examine a cross section of the celery. In the space below, make a sketch showing the location of the water-conducting tissue.

b. Recall your study of root, stem, and leaf anatomy. Certain tissues are continuous through these three

organs. Through which tissue does water move? _____

EXERCISE E | **Plant Mineral Nutrition**

In addition to carbon dioxide and water, plants need a variety of minerals to form the organic molecules needed for life. These minerals are usually obtained from the soil and enter the plant with the water absorbed through the roots. The major mineral nutrients (**macronutrients**) required are nitrogen (N), phosphorus (P), potassium (K), calcium (Ca), magnesium (Mg), and sulfur (S). In addition to these, plants also require smaller amounts of **micronutrients** such as iron (Fe), manganese (Mn), copper (Cu), and zinc (Zn). The general effects of the lack of any of these nutrients are reviewed in Table 30E-1.

The macronutrients N, P, and K are used in large amounts by growing plants and can therefore be rapidly depleted from the soil. Consequently, these are the three most common components of fertilizer. In fact, types of fertilizer are usually designated according to their percentage content of N, P, and K. For example, a common garden fertilizer, 5-10-10, contains 5% nitrogen, 10% phosphorus, 10% potassium, and 75% filler.

The purpose of this exercise is to determine how deficiencies in macronutrients and micronutrients affect plant growth. For this experiment, you will grow sunflower seedlings hydroponically. **Hydroponics** is a method of growing plants in which soil is not used. Instead, plants are placed in a liquid environment (medium), and growth occurs through the uptake of nutrients from the liquid via the roots.

||||| **Objectives** |||||||||||||||||||||||||||||||||||||

☐ Describe how deficiencies in nitrogen, phosphorus, potassium, calcium, magnesium, sulfur, and iron affect plant growth.

||||| **Procedure** |||||||||||||||||||||||||||||||||||||

Work in pairs. As many as eight treatments may be used in this experiment. Your instructor will assign you to the treatment you will be using. Each represents a different mixture of macronutrients and micronutrients in a hydroponic growth medium:

       Treatment 1 Control (contains all macronutrients and micronutrients)

       Treatment 2 Lacks potassium (K)

       Treatment 3 Lacks nitrogen (N)

       Treatment 4 Lacks iron (Fe)

       Treatment 5 Lacks phosphorus (P)

       Treatment 6 Lacks calcium (Ca)

       Treatment 7 Lacks magnesium (Mg)

       Treatment 8 Lacks sulfur (S)

An aquarium is assigned to each treatment. It contains the hydroponic medium for that treatment.

After reviewing the information in Table 30E-1, formulate a hypothesis about the role of macronutrients and micronutrients in promoting plant growth. (Consider all treatments.)

Table 30E-1   Dichotomous Key to Mineral Deficiencies in Plants

A. Effects localized on older leaves or generalized to the whole plant.

    B. Local, occurring as mottling or chlorosis (loss of chlorophyll) with or without necrotic spotting (brown spotting) of the lower (older) leaves; little or no drying of lower leaves.

        C. Lower leaves curved or cupped under with yellowish mottling or necrotic spots at tops and margins; leaf drop. _____ **Potassium**—Maintains osmotic balance of cells; responsible for osmotic changes in guard cells and regulation of the opening and closing of stomata; required in protein synthesis and in some enzymatic reactions.

        C. Lower leaves chlorotic between the principal veins at leaf tips; leaf margins of a light green to white color; typically, no necrotic spots. _____ **Magnesium**—Constituent of chlorophyll; involved in numerous enzymatic reactions; stabilizes ribosomes.

    B. General, also yellowing and drying or "firing" of lower leaves.

        C. Plant light green, lower leaves yellow, drying to light brown color; plants dwarfed. _____ **Nitrogen**—Constituent of amino acids, proteins, nucleic acids, chlorophyll, and hormones.

        C. Plant dark green or may have purple tint; leaves narrow in proportion to length; plants dwarfed. _____ **Phosphorus**—Important as a constituent of nucleic acids, phospholipids, ATP, coenzymes, and some proteins; also involved in energy metabolism.

A. Effects localized on terminal growth, young leaves, and buds.

    B. Dieback involving the terminal bud is preceded by peculiar distortions and necrosis at the tips or bases of young leaves making up the terminal growth; root meristems also develop abnormal growth; growth stunted. _____ **Calcium**—Constituent of cell walls; constituent (as calcium pectate) of the middle lamella.

    B. Terminal bud remains alive; chlorosis of upper or bud leaves, with or without necrotic spots; veins either light or dark green.

        C. Young leaves with necrotic spots scattered over chlorotic leaf; smallest veins tend to remain green, producing a checkered effect. **Manganese**—Constituent and activator of enzymes involved in photosynthesis, respiration, and nitrogen metabolism.

        C. Young leaves without necrotic spots; veins either light or dark green.

            D. Young leaves are light green, never white or yellow; leaves do not dry up; veins are light green or of the same shade as interveinal tissue. _____ **Sulfur**—Constituent of proteins and coenzyme A.

            D. Young leaves or all leaves become yellowish; principal veins characteristically darker green than tissue between the veins. _____ **Iron**—Needed in energy transfer molecules in respiration and photosynthesis; also involved in chlorophyll synthesis.

HYPOTHESIS:

NULL HYPOTHESIS:

What do you **predict** will happen to the growth of plants deprived of certain macronutrients or micronutrients?

What is the **independent** variable?

What is the **dependent** variable?

1. Record your treatment type: _____

2. Remove five sunflower seedlings from the stock container. Plant the five sunflower seedlings in the aquarium assigned to your treatment.

    Each aquarium contains a styrofoam float that has been divided into a grid system. Your instructor will assign you five locations within this grid to plant your seedlings. Record these locations:

    _____   _____   _____   _____   _____

    To plant a seedling, remove the cotton plug from the grid location. Carefully insert the seedling through the hole so that its roots extend beneath the foam and its stem is standing above the float. (Do not lift the float out of the tank or push it beneath the solution!) Cotton should be repacked or twisted loosely around the stem to keep the plant from slipping through the hole.

3. Collect day 1 ("time zero") biomass data for the sunflower seedlings. To do this, remove two seedlings from the stock container. Using a balance, determine the mass of each plant and record these values in Table 30E-2.

Table 30E-2   Data on Seedling Mass

Week	Seedling	Mass (g)
Time zero (day 1)	1	
	2	
1 Week growth	1	
	2	
2 Weeks growth	1	
	2	

4. Collect day 1 ("time zero") chlorophyll content data for the sunflower seedlings. To determine the chlorophyll content of the two plants (used in step 3), they must be crushed. Cut one plant into tiny pieces (use the entire plant, including roots!) and put these into 30 ml of 90% acetone in a mortar. Grind the plant pieces with a pestle 50 times, then filter the acetone through a double layer of cheesecloth into a 50-ml beaker. (Be sure to use only 50 strokes—this will standardize your procedure.) Pour this solution into a clinical centrifuge tube, leaving approximately 2 cm of the tube unfilled. Repeat this procedure with the other plant. Label the centrifuge tubes "1" and "2," to correspond to the seedling they contain, and place both centrifuge tubes into a clinical centrifuge opposite one another. **Be sure to record the locations of each of your tubes.** Once other students have loaded their tubes into the centrifuge, spin the samples using setting 6 for 20 minutes. After centrifugation is complete, decant the supernatant into a Spectronic 20 tube or cuvette until it is filled to within approximately 2 cm of the top.

For measuring absorbance, use a blank containing 7.5 ml of 90% acetone to adjust your Spectronic 20 to 100 percent transmittance and 0 percent absorbance. Measure the absorbance of both chlorophyll samples at 663 nm. Record all data in Table 30E-3.

Table 30E-3   Data on Chlorophyll Content

Week	Seedling	Absorbance (663 nm)	Chlorophyll Content (mg/l)	Standardized Chlorophyll (mg/g)
Time zero (day 1)	1			
	2			
1 Week growth	1			
	2			
2 Weeks growth	1			
	2			

5. Determine chlorophyll content in milligrams per liter (mg/l) using the following formula:*

    Chlorophyll content (mg/l) = absorbance at 663 nm $\times$ 13.4

6. Record the chlorophyll content for each of your seedings in Table 30E-3. Standardize your chlorophyll measurements for each seedling by factoring in the biomass of the plants. Use the following formula:

$$\text{Standardized chlorophyll content (mg/g)} = \frac{\text{chlorophyll content (mg/l)} \times 0.030 \text{ (l)}}{\text{seedling biomass (g)}}$$

7. Record the standardized chlorophyll content values in Table 30E-3 and give your data, along with the biomass measurements in Table 30E-2, to your laboratory instructor to be added to a cumulative class database. Be sure your names and treatment group are included with your data.

8. During the next two weeks, at the beginning of each laboratory period, make observations (height, color, leaf side, wilting) on the plants in each treatment tank. Record these observations in Table 30E-4 and give copies of this information to your laboratory instructor. (Again, be sure your names and treatment group are included with your data.)

9. Repeat steps 3–7 for a pair of your seedlings in week 1 and for another pair in week 2. Record the biomass data in Table 30E-2 and the chlorophyll content data in Table 30E-3 for each week. Give all of your data to your instructor (include your names and treatment group).

10. Data for each treatment will be collected for all laboratory periods in week 1 and again in week 2. At the ends of weeks 1 and 2, your instructor will perform a chi-square median statistical analysis of differences in biomass and chlorophyll measurements for all treatments. Results for each week will be distributed to you. Be sure to familiarize yourself with the chi-square median analysis (Appendix I) so that you will understand how to interpret your data and come to a conclusion.

---

*Use the constant 13.4 as determined by O. T. Lind, Handbook of Common Methods in Linnology, p. 132, C. V. Mosby, St. Louis, MO, 1979.

Table 30E-4 Appearance of Plants after 1 and 2 Weeks

Week	Treatment	Observations
1 Week growth	Control	
	– K	
	– N	
	– Fe	
	– P	
	– Ca	
	– Mg	
	– S	
2 Weeks growth	Control	
	– K	
	– N	
	– Fe	
	– P	
	– Ca	
	– Mg	
	– S	

a. Which plants demonstrated the greatest amount of growth (increase in biomass)? _____ Why? _____

b. Did you see a color change in any of the plants? _____ Describe any changes. _____

c. Why did these changes occur? _____

d. In those plants with color changes, was their growth also affected? _____ If so, how? _____

e. Which plants demonstrated the greatest chlorophyll content? _____ _____

   Why? _____

f. Were there differences in biomass and chlorophyll content between control and experimental plants for all experimental treatment groups? _____ For your treatment? _____

g. If there was a difference, was it statistically significant according to the results from the chi-square median test? For all experimental plants? _____ For your treatment? _____

For your treatment? _____ Was your prediction correct? _____

Do your results support your hypothesis? _____ Your null hypothesis? _____

What can you **conclude** about the role of macronutrients and micronutrients in controlling plant growth? For all treatments? _____

Laboratory Review Questions and Problems

1. You cut some flowers and leave them beside your sink for several hours before putting them into a vase. When you finally arrange the flowers, some of the stems are too long and you cut them and immediately place them in the vase. The next day you notice that many of the flowers are wilted, but those that you cut a second time are fine. How could you explain this given what you know about the mechanism of transpiration?

2. Leaves of plants growing in shady forests tend to be large and luxuriant, while those of plants growing in sunny grasslands tend to be narrow with little surface area. Explain in terms of transpiration.

3. One factor that causes stomata to open is the depletion of CO_2 in the air spaces of the leaf. How does this fact relate to the process of photosynthesis?

4. On a hot, dry day with temperatures above 30 to 35°C, cellular respiration in leaf cells increases. In these conditions, what would happen to the stomata? Why?

5. Some plants, including cacti and succulents, open their stomata at night and close them during the day. Why? How would they fix CO_2 for photosynthetic processes? [*Hint:* These plants use crassulacean acid metabolism (CAM).]

6. Suggest how each of the following leaf modifications would be of advantage to a plant in the designated type of environment.

Modification	Environment	Advantage
Sunken stomata	Arid, windy	
Leaves modified as spines	Sunny, arid	
Stomata only on upper epidermis	Leaves float on water surface	

7. Both magnesium and iron deficiencies result in a failure to form chlorophyll. How would you expect plants deficient in these nutrients to look? If you completed the mineral nutrition experiment (Exercise E), did it support this expectation?

8. Leguminous plants obtain nitrogen in the form of NH_4^+. What is the name of this process of reducing N_2 to NH_4^+? What is responsible for this process? Why is it beneficial to rotate plantings of leguminous and nonleguminous crops?

9. Mycorrhizal fungi associated with the roots of certain plants help to transfer phosphorus and other relatively immobile nutrients such as zinc, manganese, and copper into the plant roots. What is responsible for this increased transfer?

Seedling

Seedling is a simulation of plant competition and plant physiology. **Seedling** simulates the growth of a crop plant in diverse outdoor environments or in a growth chamber with constant, user-controlled conditions.

- **Seedling** allows you to manage a growing plant by taking the sugar it synthesizes each day and allocating it either to add leaves, increase stem height or stem diameter, or grow more roots. Your choices can create a vigorous and competitive plant or one that cannot survive dehydration, shading from other plants, or windstorms. However, beware: Even well-grown plants can be cropped and trampled by cows that sometimes visit this simulation!

- **Seedling** can also simulate three growth-chamber experiments: the effects of interplant distance on competition, plant responses to differing temperature and light conditions, and the influence of temperature and humidity on transpiration.

This is your opportunity to experience the challenges that plants must face to grow, compete, or die!

Plant Responses to Stimuli

OVERVIEW

In plants, hormones are organic molecules that coordinate growth and development. Hormones are synthesized in one region of the plant and **translocated** (moved) to other regions where they cause a physiological response. Various types of responses may occur depending on hormone interactions and the plant organ affected.

There are five groups of hormones: (1) **auxins,** (2) **gibberellins** (3) **cytokinins,** (4) **abscisic acid,** and (5) **ethylene.** The effects of the hormones, either stimulatory or inhibitory, depend upon the hormone concentration, the tissue affected, and the developmental status of the tissue. Light, temperature, gravity, day length, and other external factors also play important roles in the growth responses of plants. Some growth responses are visible within a few days; others may take weeks. During this laboratory period you will examine the effects of auxins and gibberellins on plant growth and development.

STUDENT PREPARATION

Prepare for the laboratory by reading the text pages indicated by your instructor. Familiarizing yourself in advance with the information and procedures covered in this laboratory will give you a better understanding of the material and improve your efficiency.

👁 **EXERCISE A** | **Auxins**

👁 **PART I** | **Bud Inhibition and Apical Dominance**

Auxins occur naturally in several slightly different chemical forms. The most abundant is indole-3-acetic acid (IAA), often referred to simply as "auxin," which is synthesized in buds of young stems and leaves, embryos, seeds, and fruits. Once synthesized in a bud, IAA is translocated toward the base of the plant, the concentration of IAA remaining highest in the growing tips of the stem and decreasing toward the roots. IAA causes the lengthening of cells in the elongation region of a growing shoot (the area just behind the apical meristem). However, if concentrations of IAA increase beyond a certain optimum level, lengthening of the stem is inhibited rather than stimulated.

The product of IAA in the apical bud at the end of a growing shoot also inhibits the development of lateral buds. As a result of this **apical dominance,** a plant will appear to grow upward rather than outward. A simple experiment can be used to investigate the effects of IAA on stem growth.

⦚⦚⦚⦚⦚ Objectives ⦚⦚⦚⦚⦚⦚⦚⦚⦚⦚⦚⦚⦚⦚⦚⦚⦚⦚⦚⦚⦚⦚⦚⦚⦚⦚⦚⦚⦚⦚⦚⦚⦚

☐ State the effects of auxin on the nature and form of stem growth in branching plants.

☐ Describe how the presence or absence of an apical bud influences the growth form of a plant.

ııııı **Procedure** ıııııııııııııııııııııııııııııııııııııı

First Week

1. Work in groups of four. Obtain a pot containing 4-week-old *Coleus* plants.

2. Leave two plants with apical buds intact. Mark them with small tags labeled "Apical Bud." All tags should be loosely tied.

3. Remove the apical buds from each of two plants. Mark them with small tags labeled "Removed."

4. Remove the apical buds from each of two additional plants and apply a small amount of lanolin paste to the decapitated surface. Mark these plants with tags labeled "Removed— Lanolin."

5. Remove the apical buds from each of two other plants and apply a small amount of lanolin paste containing 5,000 parts per million (ppm) IAA to the decapitated surface. Mark these plants with tags labeled "Removed—IAA."

6. In each pot, place a label with your names and laboratory section (do not write on pots). Give the pot to your instructor, who will keep it in a sunny spot for the next week.

   Form a hypothesis about the effects of IAA on stem growth:

   HYPOTHESIS:

   NULL HYPOTHESIS:

   What do you **predict** will happen to the eight plants used for this experiment?

   What is the **independent** variable?

   What is the **dependent** variable?

Second Week

Seven days after the decapitation procedure, measure to the nearest millimeter the length of the axillary buds (buds located in the angle formed by the leaf and stem) or branches that have developed. Record your data in Table 31A-1.

Table 31A-1   Apical Dominance and Bud Inhibition

Treatment	Average Bud on Branch Length (millimeters)
Intact plants	
Debudded plants	
Debudded plants + lanolin	
Debudded plants + lanolin and IAA	

a. Did branching occur in plants with an apical bud? _____ b. Did branching occur in plants without an apical bud and without lanolin or IAA treatment? _____ c. Did lanolin substitute for the apical bud (that is, did it produce the same effect)? _____ d. Did lanolin and IAA substitute for the apical bud? _____

e. What is the purpose of the lanolin treatment? _____

What do you **conclude** from this experiment? _____

_____

f. Does IAA cause bud inhibition? _____ Do your results support your hypothesis? _____
Your null hypothesis? _____

g. In your experiment, what was the reason for leaving two plants intact? _____

_____

h. Gardeners commonly "pinch out" the apical bud of a plant. What effect does this practice have on plant form, and why do gardeners do it? _____

_____

👁 PART 2   Leaf Abscission

Leaves of most perennial flowering plants have a relatively short life span. Produced in the spring, they serve a photosynthetic function throughout the summer. With the onset of fall, leaves begin to **senesce** (age).

During senescence, nutrients and reusable molecules are returned to the stem, and enzymes begin to break down the walls of cells near the base of the leaf petiole, forming an **abscission layer.** The eventual abscission (drop) of leaves following senescence is the result of the formation of this abscission layer. Continued breakdown of cell walls loosens contacts between cells in the abscission layer, while the formation of a protective layer of cells below the abscission area further isolates the petiole from the stem. Eventually, the only connection between the cells of the petiole and the stem is a strand of vascular tissue. Complete abscission of the leaf from the stem then occurs from the stress of wind and gravity.

Auxin, along with abscisic acid, plays a role in leaf abscission. When leaves are young, the leaf end of the petiole maintains a higher auxin concentration than the stem end. As a leaf ages, less auxin is produced at the leaf end of the petiole and eventually the stem end has a higher concentration than the leaf end. This change in auxin concentration causes leaf abscission. Abscisic acid, isolated from leaves (and fruits), accelerates the abscission process.

▮▮▮▮▮ Objectives ▮▮▮▮▮▮▮▮▮▮▮▮▮▮▮▮▮▮▮▮▮▮▮▮▮▮▮▮▮▮▮▮▮▮▮▮▮▮▮

☐ State the effects of auxin on leaf abscission.

▮▮▮▮▮ Procedure ▮▮▮▮▮▮▮▮▮▮▮▮▮▮▮▮▮▮▮▮▮▮▮▮▮▮▮▮▮▮▮▮▮▮▮▮▮▮▮

1. Work in groups of four. Obtain a *Coleus* plant. From near the apexes of three different stems, select one young leaf and cut off only the blade portion of the leaf. Do *not* cut the petiole from the plant (Figure 31A-1).

2. To the tip of one of the petioles, apply lanolin containing 5,000 ppm IAA. To the second petiole tip, apply only lanolin. To the third, apply nothing.

3. Label each petiole with a paper reinforcement ring or small piece of tape on the stem or adjacent leaf indicating the treatment applied. In each pot, place a label with your names and laboratory section (do not write on pots). Return the *Coleus* to your instructor. After one week, make observations and record the data (petiole on or off) in Table 31A-2.

Figure 31A-1 *Location of cut on*
Coleus *leaf.*

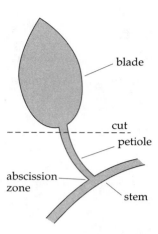

blade

cut

petiole

abscission
zone

stem

What hypothesis can you test using this experimental method?

HYPOTHESIS:

NULL HYPOTHESIS:

What do you **predict** will happen to the three petioles?

What is the **independent variable**?

What is the **dependent variable**?

a. *Does IAA promote or inhibit the formation of the abscission layer?* _____

b. *Review your results and state where the auxin controlling the development of the abscission layer is produced.*

_____

Explain your reasoning. _____

_____

c. *What was the experimental reason for placing lanolin on one petiole?*

_____

_____

Table 31A-2   Leaf Abscission

Treatment	Observations
Debladed	
Debladed + plain lanolin	
Debladed + lanolin with IAA	

What can you **conclude** about the role of auxins in leaf abscission? _____

_____

Do your results support your hypothesis? _____

Your null hypothesis? _____

👁 EXERCISE B | Gibberellins

Gibberellins are a group of naturally occurring plant steroid hormones. There are more than 90 different forms of gibberellins, although GA_3 (gibberellic acid) is the most prevalent in flowering plants.

Gibberellic acid is synthesized in meristematic regions (young leaves, shoot tips, and root tips) and is translocated in the xylem of vascular bundles. Very low levels are found in mature roots, stems, and leaves. Gibberellic acid is also synthesized in seeds.

In young plants, gibberellic acid, like auxin, promotes cell elongation. Shoot elongation and bolting (rapid growth) are under direct control of gibberellic acid. In seeds, gibberellic acid hastens germination by enhancing cell elongation and thus elongation of the embryonic root, so that it can penetrate the seed coat. Additionally, gibberellic acid produced by the embryo stimulates product of hydrolytic enzymes that digest the seed's endosperm, making food reserves available for growth.

Some plants, such as dwarf pea plants, lack the ability to synthesize GA_3, but remain sensitive to exogenous application of this hormone. In this exercise, the effects of gibberellic acid can be demonstrated using dwarf pea plants (Figure 31B-1).

ⅼⅼⅼⅼⅼ Objectives ⅼⅼⅼⅼⅼⅼⅼⅼⅼⅼⅼⅼⅼⅼⅼⅼⅼⅼⅼⅼⅼⅼⅼⅼⅼⅼⅼⅼⅼⅼⅼⅼⅼⅼⅼⅼ

☐ State the effects of gibberellic acid on stem elongation in plants.

ⅼⅼⅼⅼⅼ Procedure ⅼⅼⅼⅼⅼⅼⅼⅼⅼⅼⅼⅼⅼⅼⅼⅼⅼⅼⅼⅼⅼⅼⅼⅼⅼⅼⅼⅼⅼⅼⅼⅼⅼⅼⅼⅼⅼ

First Week

1. Work in groups of four. Each group of students should obtain two pots containing dwarf pea plants.

2. Measure the height of the stems in each pot (in millimeters) and count the number of nodes along the stem. When you measure stem height, measure from the soil surface to the terminal bud. (Be careful not to confuse the petiole of a leaf with the stem—the terminal bud of the stem may not be the highest point of the plant.) Average your data and record the averages in the columns marked "Initial Treatment" in Table 31B-1.

Figure 31B-1 *Effects of gibberellic acid on growth of dwarf pea stems.*

Table 31B-1   Shoot Length in Response to Gibberellins

Treatment	Initial Treatment: Stem Height (millimeters)	7 Days: Stem Height (millimeters)	Initial Treatment: Number of Nodes	7 Days: Number of Nodes
GA_3				
Control				

3. Add several drops of 100 mg/l GA_3 solution directly to the apex of the plants in one pot. In a similar manner, add the control solution (water plus Tween-20) to the plants in the second pot. (Both solutions contain Tween-20, a wetting agent that assists in the absorption of GA_3 by the pea plants.)

4. In each pot, place a label indicating the treatment, your names, and laboratory section (do not write on pots). Return the pots to your instructor.

Form a hypothesis that provides a tentative explanation for the effects of GA_3 on dwarf pea plants.

HYPOTHESIS:

NULL HYPOTHESIS:

What do you **predict** will happen to the plants treated by this experimental procedure?

What is the **independent variable**?

What is the **dependent variable**?

Second Week

After one week, measure the height of the stem of each plant in each pot and count the number of nodes present. Compare and record the average stem height and number of nodes for treated and control plants. When you measure the stem height, remember to measure from the soil level to the tip of the terminal bud. (Be careful not to confuse a petiole with the stem.)

a. *Did treatment with GA_3 have an effect on stem elongation?* _____ b. *Did GA_3 influence the number of nodes?* _____ c. *Did GA_3 influence the length of internodes?* _____

Do your results support your hypothesis? _____

Your null hypothesis? _____

What can you **conclude** *about the effects of gibberelic acid?* _____

_____

d. *Based on your results, what do you suppose are the differences between the amounts of GA_3 produced in dwarf plants and in normal plants?* _____

_____

e. Dwarfism in peas is a genetic characteristic. How would you relate this fact to the presence or absence of GA₃?

_____

_____

✔ **EXERCISE C** | **Tropisms**

Tropisms are directional growth movements that occur in response to stimuli. Growth toward the stimulus is said to be a **positive response;** growth away from the stimulus is called a **negative response.** Movements occur because of unequal growth, resulting in a bending toward or away from the stimulus. Unequal growth appears to be the result of redistribution of auxin or some other hormone.

👁 **PART I**   **Gravitropism (Geotropism)**

Gravitropism is directional plant growth in response to gravity. Stems tend to grow up (negative gravitropism), whereas roots tend to grow down (positive gravitropism). Branches and leaves usually show intermediate responses (**plagiotropisms**).

Gravitropic responses occur in two steps. First, something within the plant must detect gravity (this "something" is thought to be starch grains called **statoliths**). Then differential growth must occur—one side of the plant part growing faster than the other, causing the plant part to bend. If a plant stem is placed in a horizontal position, starch grains fall toward the lower surface of the stem and seem to cause an increase in the auxin concentration in that region. The result is greater cell elongation in the lower surface of the stem than in the upper surface, causing an upward bending of the stem.

In roots, statoliths are found in the root cap. If a plant is placed in a horizontal position, statoliths cause an increase in concentration of a growth inhibitor called **abscisic acid** (ABA) on the lower surface of the root cap. Inhibition of growth on the lower surface causes a downward bending of the root.

a. Describe the differences between the gravitropisms of stems and roots.

_____

_____

||||| **Objectives** ||||||||||||||||||||||||||||||||||||||||||||

☐ Differentiate between the actions of auxin and abscisic acid (ABA).

☐ Describe how auxin causes negative gravitropic effects in stems.

☐ Describe how ABA causes positive gravitropic effects in roots.

✔ **A. Negative Gravitropism in Stems**

Two chicken gizzard plants (*Iresine* sp.) of approximately the same age and size were placed in the dark for 24 hours. Plant A remained in a vertical position. Plant B was placed in a horizontal position. In the space below, sketch these two plants.

Plant A   Plant B

b. *What must have happened to the distribution of auxin in Plant B?*

_____

c. *How does this distribution result in the response observed in Plant B?*

_____

d. *Why were the plants placed in the dark?*

_____

B. Positive Gravitropism in Roots: The Role of Seed Position

Procedure

First Week

1. Obtain three corn seeds, a Petri dish, and several paper towels.

2. Moisten a single paper towel and fold it so that it fits within the bottom of the Petri dish.

3. Place three seeds onto the paper towel oriented as indicated in Figure 31C-1a. Note the position of the corn seeds.

Figure 31C-1 *(a) Proper position of corn seeds to show positive gravitropism in roots. (b) Draw your results.*

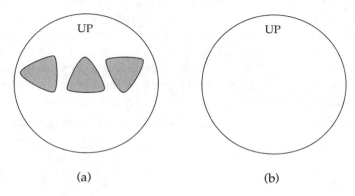

(a)   (b)

4. Moisten a second towel and place it over the seeds. Cover with the top of the Petri dish, secure with tape, and label with your name and section. *Note:* The Petri dishes will be stored vertically. With this in mind, write "UP" on the lid of the dish as shown in Figure 31C-1a so the seeds will be stored in the correct horizontal direction.

Do you think gravity combined with the action of endogenous auxin will affect the development of roots as the seeds germinate? Formulate a hypothesis that addresses the question.

HYPOTHESIS:

NULL HYPOTHESIS:

What do you **predict** will happen to the three seeds?

What is the **independent variable**?

What is the **dependent variable**?

Second Week

After 1 week, observe the general direction in which the primary roots have grown and sketch your results in Figure 31C-1b.

e. *How does the position of the seed affect the gravitropic response of the corn root?*

_____

   Explain. _____

f. *Do roots exhibit positive or negative gravitropism?* _____

g. *How might plants benefit from this gravitropic response?* _____

   From the results of this experiment, what do you **conclude** *about the role of gravity in controlling root growth during development?* _____

_____

   Do your results support your hypothesis? _____

   Your null hypothesis? _____

Third Week

After 2 weeks, observe the general direction of coleoptile growth and sketch your results in Figure 31C-1b.

h. *Does the coleoptile demonstrate a positive or negative gravitropic response?* _____
   Compare this with your observations on stem gravitropism above.

C. Positive Gravitropism in Roots: The Role of the Root Tip

Procedure ▮▮▮▮▮▮▮▮▮▮▮▮▮▮▮▮▮▮▮▮▮▮▮▮▮▮▮▮▮▮▮▮▮▮▮▮▮▮

First Week

    1. Obtain four germinated bean seedlings with primary roots about 5 mm long. Place the seedlings on a moist paper towel so they do not dry out.

    2. Obtain a razor blade and carefully cut the roots as follows (Figure 31C-2a):

       Seedling 1. Remove the entire root cap (the first millimeter of tissue).

       Seedling 2. Remove one-half of the root cap on one side.

       Seedling 3. Remove one-half of the root cap on the opposite side.

       Seedling 4. No treatment.

Figure 31C-2 *(a) Setup of germinated bean seedlings to show the role of the root tip in gravitropic responses of roots during seed germination. (b) Draw your results.*

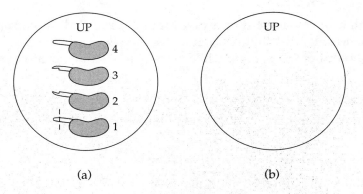

(a)   (b)

    3. Place the seedlings on a paper towel in a Petri dish as described in in Part 1B above, but make sure that all the roots are oriented in the same direction (Figure 31C-2a).

4. Cover the seedlings as described above and label the plate with your name and section. Mark "UP" as shown in Figure 31C-2a to ensure correct storage.

Formulate a hypothesis about the roles of the root cap and ABA in controlling root growth.

HYPOTHESIS:

NULL HYPOTHESIS:

What do you **predict** will happen to the roots of the four seeds? Why? _____

What is the **independent variable**? _____

What is the **dependent variable**? _____

Second Week

Observe the direction of root growth in each seedling. Diagram your results in Figure 31C-2b.

i. Is the root cap required for gravitropism in roots? (Compare seedlings 1 and 4.) _____

j. What happened to seedlings 2 and 3? _____

k. Did you find a relationship between the direction of bending and the side of the root with remaining root cap? ___

Explain your **results.** *What do you* **conclude** *about how ABA might be involved in controlling root growth?*

_____

_____

Do your results support your hypothesis? _____
Your null hypothesis? _____

👁 PART 2 Phototropism

The growth of plants toward or away from light is called **phototropism.** Phototropism occurs because light influences the distribution of auxin in the region of elongation beneath the stem tip. Stems exhibit positive phototropic responses, while roots demonstrate negative phototropic responses.

ⅠⅠⅠⅠⅠ Objectives ⅠⅠⅠⅠⅠⅠⅠⅠⅠⅠⅠⅠⅠⅠⅠⅠⅠⅠⅠⅠⅠⅠⅠⅠⅠⅠⅠⅠⅠⅠ

☐ Describe how differential accumulation of auxin in a stem affects how the stem responds to light.

ⅠⅠⅠⅠⅠ Procedure ⅠⅠⅠⅠⅠⅠⅠⅠⅠⅠⅠⅠⅠⅠⅠⅠⅠⅠⅠⅠⅠⅠⅠⅠⅠⅠⅠⅠⅠⅠ

1. Examine the plants on demonstration. Note how the stems have bent toward the light (Figure 31C-3).

a. In what region of the stem does the bending occur? _____

b. Do the plants exhibit positive or negative phototropism? _____

(a)   (b)

Figure 31C-3 *Phototropism in (a) germinating wheat seedlings (light at left) and (b) a young tomato plant (light from above).*

c. *Recall that auxin influences cell elongation. Based on what you have learned about other tropic responses, suggest how auxin is involved in stem bending.*

_____

_____

2. Sketch several cells in Figure 31C-4b to represent relative changes in size of cells on each side of the stem after exposure to light from one direction.

Figure 31C-4 *(a) Representation of cells in stem before exposure to a unilateral (one-sided) light source. (b) Diagram some cells to show relative cell length after exposure to a unilateral light source.*

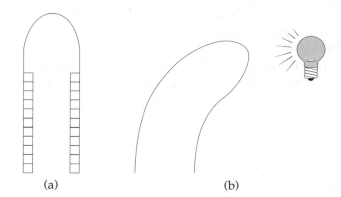

(a)   (b)

✔ **EXERCISE D**   **Light-Induced Germination**

Some plants regularly inhabit newly disturbed areas, such as cleared fields, while others grow only in areas that are already covered by vegetation. Some seeds that grow in disturbed areas require light to induce germination and will not germinate if covered by soil or other vegetation. In these *light-sensitive seeds,* the pigment **phytochrome** is the photoreceptor (light receiver).

||||| **Objectives** ||||||||||||||||||||||||||||||||||||||||||||

☐ State the role of the plant pigment phytochrome in light-induced growth responses.

☐ Explain how gibberellic acid affects the process of germination in light-sensitive seeds.

✔ **PART I**   **The Role of Phytochrome**

Procedure ████████████████████████████████████

On demonstration are two varieties of lettuce seeds: Great Lakes and Grand Rapids. Fifty seeds of each variety were allowed to germinate in the light and 50 seeds of each variety were allowed to germinate in the dark. Count the number of germinated seeds in each category and enter the results in Table 31D-1.

a. *Based on these data, which variety has a light requirement for germination?* _____

b. *Was the other variety light-inhibited?* _____

c. *Which variety has light-sensitive seeds?* _____ *Which has light-insensitive*

   seeds? _____

Table 31D-1   Light Induced Germination in Lettuce Seeds

	Light		Dark	
Variety	Grand Rapids	Great Lakes	Grand Rapids	Great Lakes
Total number of seeds	50	50	50	50
Number germinated				
Percent germinated				

👁 **PART 2**   **The Forms of Phytochrome**

From your results in the previous experiment, you have determined which variety of seeds is light-sensitive. In this experiment you will use the light-sensitive seeds to determine which *wavelength* of light is responsible for inducing germination.

Do you think that different wavelengths (different colors) of light affect germination in light-sensitive seeds? Formulate a hypothesis offering a tentative explanation about what might happen to seeds illuminated by different wavelengths of light.

HYPOTHESIS:

NULL HYPOTHESIS:

What do you **predict** will happen to seeds exposed to different wavelengths of light?

What is the **independent variable**?

What is the **dependent variable**?

Procedure ████████████████████████████████████

1. Fifty seeds were germinated in different colors of light. Count the number of germinated seedlings in each of the pots and determine the percent germination for each type of light. Record the data in Table 31D-2.

Table 31D-2   Effect of Wavelengths of Light on Germination

	A	B	C	D	E	F
Light conditions	Light	Dark	Blue	Green	Red	Far red
Total number of seeds	50	50	50	50	50	50
Number germinated						
Percent germinated						

2. Next, plot these data on the graph in Figure 31D-1. What you have generated is called an **action spectrum** for the photoreceptor pigment, phytochrome. This pigment is a protein and is responsible for absorbing the light required to trigger germination. Phytochrome has two forms: one absorbs red light and the other absorbs far-red light.

   a. *Which form of phytochrome induced germination in the light-sensitive lettuce seeds?*

   _____

   b. *Does the action spectrum indicate that red light is the only color absorbed by phytochrome?*

   _____

   From your results, what do you **conclude** *about how light of different wavelengths affects germination?*

   _____

   _____

   Do your results support your hypothesis? _____ Your null hypothesis? _____

   c. *Based on the action spectrum, what color is phytochrome? _____*

Figure 31D-1   *Effects of different wavelengths of light in inducing germination of lettuce seeds.*

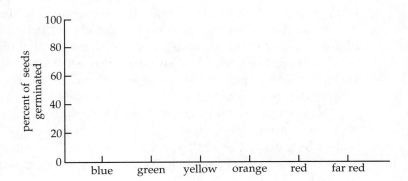

PART 3   Germination and Gibberellins (*Optional*)

Some seeds require low temperatures, long days, or red light to germinate. In some species, the application of gibberellic acid can substitute for these dormancy-breaking conditions.

||||| Procedure |||||||||||||||||||||||||||||||||||||||||||||

1. Prepare four Petri dishes by placing a square of blotter paper in each.

2. In two dishes, soak the blotter paper with water. Mark the dishes "H_2O." In the other two dishes, soak the blotter paper with a solution of gibberellic acid. Mark these dishes "GA_3."

3. Place 20 light-sensitive lettuce seeds on the filter paper in each of the four dishes. Cover the dishes.

4. Wrap one of the "H₂O" dishes and one of the "GA₃" dishes in aluminum foil. The other two dishes will remain unwrapped.

5. Place the dishes in very low light. Keep moist. After 1 to 2 weeks of incubation, determine the percent germination and record your data in Table 31D-3.

Table 31D-3   Germination and Gibberellins

Treatment	Percent Germination
H_2O/foil	
H_2O/light	
GA_3/foil	
GA_3/light	

a. Did seeds germinate in the dark? _____ Under what conditions? _____

b. What effect does gibberellic acid have on the germination of light-sensitive seeds?

_____

EXERCISE E   Photoperiodism

Everyone is aware that certain plants bloom only during particular seasons. But how do plants "know" when to bloom? They use several environmental cues, including temperature. However, most plants determine their seasonal responses by detecting the length of the day. The phenomenon is called **photoperiodism** and depends on the pigment phytochrome.

Plants are categorized as **long-day,** flowering only when light periods are longer than a critical length; **short-day,** flowering only when light periods are shorter than a critical length; or **day-neutral,** flowering independently of day length.

As noted in the previous exercise, phytochrome has two forms: a red-absorbing form (P_r) and a far-red-absorbing form (P_{fr}). Phytochrome is synthesized in the P_r form and can be converted to P_{fr} when it absorbs a photon of red light (in daylight, red wavelengths predominate over far red). And P_{fr} can be converted back into P_r when it absorbs a photon of far-red light. P_{fr} can also revert spontaneously to P_r in the dark.

During the day, the interconversion of $P_r \leftrightarrow P_{fr}$ reaches an equilibrium: approximately 60 percent of the phytochrome is in the P_{fr} form at noon on a sunny day. At night, the level of P_{fr} steadily declines as it is destroyed or converted to P_r.

P_{fr} inhibits flowering in short-day plants; when the night is long enough, a critical amount of P_{fr} is removed so the plant can flower. (If a short flash of red light interrupts the night, however, flowering is inhibited. Why?) In long-day plants, P_{fr} promotes flowering; if the night is short enough, sufficient P_{fr} is left at the end of the night to promote flowering (Figure 31E-1). Thus, it is the dark-period length (not the light-period length) that is important in determining flowering response, and phytochrome is responsible for mediating this activity. (It is generally agreed, however, that the control mechanism is more complex than phytochrome conversion alone.)

▌▌▌▌▌ **Objectives** ▌▌▌▌▌▌▌▌▌▌▌▌▌▌▌▌▌▌▌▌▌▌▌▌▌▌▌▌▌▌▌▌▌▌▌▌▌

☐ Describe how the length of the night affects flowering in long-day, short-day, and day-neutral plants.

Figure 31E-1 *At dusk most phy-tochrome is in the P_{fr} form.*

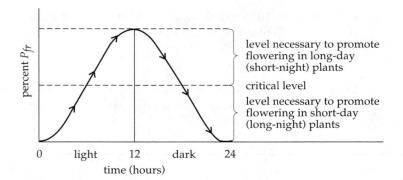

level necessary to promote flowering in long-day (short-night) plants

critical level

level necessary to promote flowering in short-day (long-night) plants

ⅢⅢ **Procedure** ⅢⅢⅢⅢⅢⅢⅢⅢⅢⅢⅢⅢⅢⅢⅢⅢⅢⅢⅢⅢⅢⅢ

Observe each of three species of plants on demonstration. One plant of each pair was exposed to 8 hours of light and 16 hours of dark; the other to 16 hours of light and 8 hours of dark. In Table 31E-1, record the presence or absence of flowers or flower buds and determine whether each species is long-day, short-day, or day-neutral.

Table 31E-1 Photoperiodism

Species	Photoperiod (hours)		Flowering (+ or −)	Category Long-Day, Short-Day, or Day-Neutral
	Light	Dark		
A	8	16		
	16	8		
B	8	16		
	16	8		
C	8	16		
	16	8		

a. For each species of plant A, B, or C, relate the conversion of $P_{fr} \rightleftarrows P_r$ to the light-dark cycle and the effect on flowering.

A. _____

_____

B. _____

_____

C. _____

_____

Laboratory Review Questions and Problems

1. Normal plant development depends upon the interaction of external (environmental) factors and internal (chemical) factors. List at least four examples of each.

2. The word "hormone" comes from the Greek *hormaein,* which means "to excite." Why is this not a completely accurate description of the role that hormones play in controlling plant responses?

3. Explain how differential accumulation of auxin in a root and a stem apparently affects how the root and stem respond to gravity.

4. From your data and observations, complete the following table showing the mechanism of action and plant response for auxins and gibberellins (the table is partly filled in to help you organize your thoughts).

Hormones	Mechanism	Plant Response
Auxins	1.	1. Gravitropism
	2. Produced in apical bud	2.
	3.	3. Leaf or fruit drop
	4. Light influences auxin distribution	4.
Gibberellins	1.	1. Stem lengthening, bolting
	2. GA$_3$ substitutes for effects of red light to break seed dormancy	2.

5. Three other groups of hormones are important in the control of plant responses. From your reading, complete the following table.

Hormone(s)	Mechanism	Plant Response
Cytokinins	1.	1. Shoot growth; increase in callus tissue
Ethylene	1. Produced by fruits	1.
	2.	2. Leaf and fruit abscission
Abscisic acid	1.	1. Prevention of water loss from leaves
	2.	2. Acceleration of abscission

6. Explain how auxins might interact with cytokinins, ethylene, or abscisic acid (often antagonistically) to control plant responses.

7. Explain the relationship between the P_r and P_{fr} forms of phytochrome. Which of the two is the active form that induces a biological response?

8. Two groups of plants are exposed to a light cycle of 8 hours of light and 16 hours of darkness. The plants in group A normally flower under these conditions, but those in group B do not. In another experiment, groups of plants of the same two species are exposed to the same light cycle, but the dark hours are interrupted by flashes of red light every hour, then the results are reversed: plants in group A do not flower, but those in group B do flower. Explain these results.

Animal Tissues

OVERVIEW

In animals, groups of similar cells specialized to perform the same function are called **tissues.** There are four general classes of tissues: **epithelial, connective, muscle,** and **nervous** tissues. In many cases, several of these tissues associate to carry out a particular function. This structural and functional unit of associated tissues is called an *organ.* An organ such as the stomach is encased, both inside and out, with epithelia, and contains connective, muscle, and nervous tissues within its walls. A group of organs that interact to carry out a particular process is called an *organ system.* The digestive system, for example, is made up of the esophagus, stomach, intestines, liver, and pancreas.

During this laboratory period you will examine the morphology of various types of tissues making up the organs and organ systems of animals.

STUDENT PREPARATION

Prepare for this laboratory by reading the text pages indicated by your instructor. Familiarizing yourself in advance with the information and procedures covered in this laboratory will give you a better understanding of the material and improve your efficiency.

✔ **EXERCISE A**   Epithelial Tissue

Epithelial tissues cover the various internal and external surfaces of the body. These layers of close-fitting cells form an effective barrier between the body and its surroundings. In some cases, such as in the sense organs, epithelial cells serve the function of responding to environmental stimuli. Other epithelial cells are responsible for absorption, secretion, and excretion and are often found in the walls of tubular glands, such as those that produce sweat, tears, and saliva.

Three major types of epithelial tissue can be distinguished according to cell shape: **squamous, cuboidal,** and **columnar.** These may exist as a single layer of cells (**simple**) or as more than one layer of cells (**stratified**). Occasionally, the cells seem to be stratified, but closer examination shows that there is only a single layer of cells; these are called **pseudostratified** epithelial cells (Figure 32A-1). The sheets of epithelial cells generally rest on a **basement membrane,** made up of collagen fibers that the cells secrete as supporting structures. This membrane separates the epithelial cells from underlying tissues.

||||| **Objectives** ||||||||||||||||||||||||||||||||||||||||

☐ Distinguish between squamous, cuboidal, and columnar epithelia.

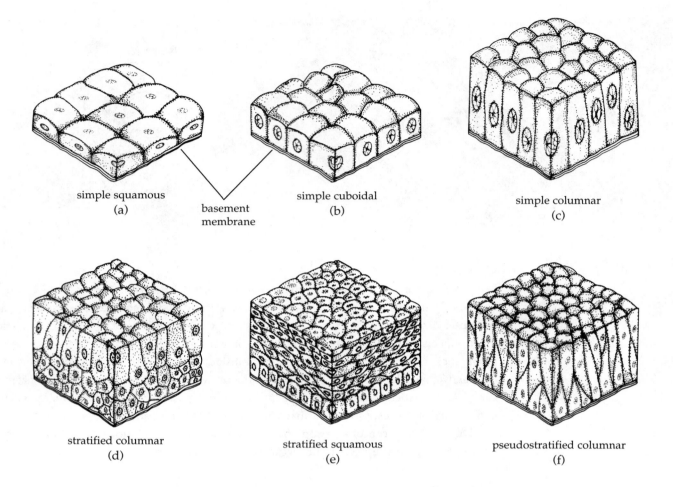

Figure 32A-1 *Epithelial cells may be (a) squamous, (b) cuboidal, or (c) columnar. These can be arranged in single sheets (a), (b), (c), or in stratified multiple layers (d), (e). In pseudostratified epithelia, each cell is connected to the basement membrane (f).*

✔ **PART I   Squamous Epithelium**

Simple squamous epithelium consists of a single layer of flattened, polygonal cells that is easily permeable to gases and liquids. This type of tissue is found inside the lung alveoli (air sacs), in the blood vessels, and in the lining of the heart. It is also found in the membranes lining the abdominal and thoracic cavities. **Stratified squamous epithelium** is found in areas of the body where additional protection is required, for example, on the skin and in the mouth, the esophagus, the anus, and the vagina (Figure 32A-2).

||||| Procedure |||||||||||||||||||||||||||||||||||||||||||

1. Obtain a small piece of shed skin (epidermis) from a frog. You may find this tissue floating in the bowel or loosely attached to the frog.

2. Make a temporary wet-mount slide using frog Ringer's solution. Add a drop of methylene blue stain before covering the tissue with a coverslip. Observe the slide at high power (40×).

   a. What shape are the cells? _____ *b. How are they arranged?* _____

3. On the upper third of a separate sheet of paper, draw and label your observations.

Figure 32A-2 *Stratified squamous epithelium.*

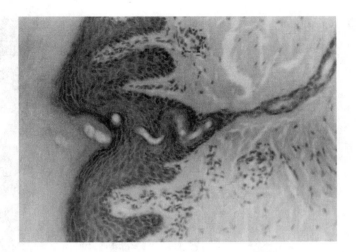

4. Repeat step 2 using a very thin piece of skin from the webbing between the toes of a pithed frog. Carefully scrape this area with a scalpel and remove the tissue with a pair of forceps. Alternatively, you may use a prepared slide of frog skin.

    c. How do these cells differ from those of the shed skin? _____

    Do you see cells containing darkly stained objects? These are **chromatophores,** pigment-containing cells in the underlayer of the skin called the dermis. The skin is composed of several layers, including the epidermis and underlying dermis.

    d. Why did you not see chromatophores in the shed skin? _____

    _____

    e. How does the size of a chromatophore cell compare with that of an epithelial cell?

    _____

5. Use the middle third of your sketch sheet to draw and label your observations of the cells in the webbing.

6. Use high power (40×) to examine a prepared slide of the stratified squamous epithelium from the esophagus. On the lower third of the sheet of paper, draw your observations. Insert the sheet into the laboratory manual.

    f. Are the cells of the esophageal epithelium ciliated? _____ *g. Do they differ in shape from the*

    simple squamous cells of the frog's skin? _____

    _____

✔ **PART 2** **Columnar Epithelium**

Columnar epithelium consists of a layer of closely packed, columnar cells in which the length of each cell greatly exceeds its width. This type of epithelium is found lining the intestinal lumen of vertebrates, where it is involved in the secretion of some digestive enzymes and in the uptake of the products of digestion. Some cells (goblet cells) are specialized for the secretion of mucus—a substance containing the protein mucin—which provides mechanical protection of the gut lining and buffers the contents of the lumen. Columnar epithelium also makes up the lining of the respiratory tract and reproductive organs, where the cells are ciliated (Figure 32A-3).

a. What might be the function of cilia in the reproductive system? _____

_____

b. What might be the function of cilia in the respiratory system?

_____

Figure 32A-3 *Columnar epithelium.*

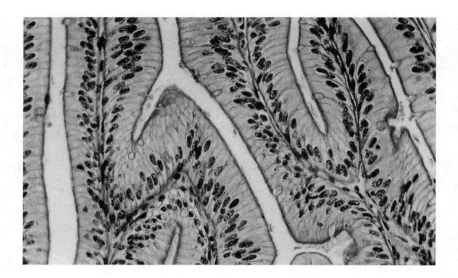

||||| Procedure |||||||||||||||||||||||||||||||||||||||

1. Use high power (40×) to observe a prepared slide of columnar epithelium from the lining of a human kidney tubule. On the upper half of a separate piece of paper, draw a section of the tissue. Label the cell nuclei and the basement membrane.

   c. *What shape are the cells?* _____

   d. *How are the cells arranged?* _____

✔ **PART 3**   **Cuboidal Epithelium**

When seen in cross section, **cuboidal epithelial** cells are shaped like cubes (Figure 32A-4). These cells often serve a secretory function and are associated with the walls of ducts and glands.

Figure 32A-4 *Cuboidal epithelium.*

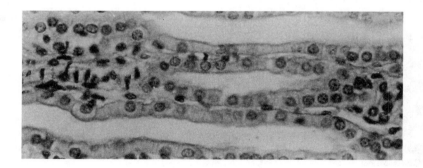

||||| Procedure |||||||||||||||||||||||||||||||||||||||

1. Use high power (40×) to observe a prepared slide of cuboidal epithelium from the human thyroid gland. On the lower portion of your sheet of paper, draw a section of this tissue; include several cells and the basement membrane. Label as many structures as you are able to identify.

   a. *How are the cells arranged?* _____

2. Work in groups of four. Obtain a live frog that has been cooled down in an ice bath, and place the frog on its back on a board covered with moist cheesecloth. Gently pull the lower jaw downward so that the mouth is opened wide; secure the upper jaw by tying a piece of string loosely around the jaw and the board. Tie the lower jaw open in the same manner.

3. Obtain some fine cork particles and position them on the roof of the mouth between the two eyes. Place the frog under a dissecting microscope and observe the movement of particles over a period of time.

    b. Do the particles move in a particular direction? _____

    c. How far do the particles move in 1 minute? _____

4. Using a sharp scalpel or fine tweezers, remove a small sample of tissue from the *surface* of the roof of the mouth. Use frog Ringer's solution to make a temporary wet-mount slide of the tissue. Study the ciliated epithelial cells at high power (40✕). You may need to adjust the illumination.

    d. Do the cilia beat rhythmically? _____   *e. Do all cilia beat in the same direction?* _____

    f. Record your observations. _____

5. Release the frog and return it to the tank.

✔   **EXERCISE B**   **Connective Tissue**

Connective tissues bind other tissues together, giving substantial form and support to the organs.

Cells of connective tissue are usually widely separated by a **matrix,** an intercellular space filled with fluid, fibers, and solid materials. The relative abundance and arrangement of the cells and the character of the matrix determine the properties of a particular kind of connective tissue.

The **mesenchyme cell** is a type of connective tissue cell found in abundance during the early development of most animals. Mesenchyme cells are undifferentiated and relatively unspecialized, giving rise to all of the connective tissues of an embryo and all of the various cell types of adult connective tissue. (Connective tissues are defined as those that have a mesenchymal origin in the embryo.) In the adult, connective tissue cells remain relatively unspecialized, and most can transform from one type to another. Included among the connective tissues are the following: cartilage, bone, loose connective tissue, dense connective tissue (ligaments and tendons), adipose (fat) tissue, and blood.

ııııı **Objectives** ıı

☐ Describe the structure of cartilage.

☐ Compare hyaline cartilage with elastic cartilage.

☐ Compare and contrast the structure of cartilage with that of bone.

☐ Distinguish between loose and dense connective tissue.

☐ Describe the structure of adipose tissue.

☐ Describe the structure of red blood cells.

✔   **PART I**   **Cartilage**

Cartilage—or gristle, as it is often called—is a connective tissue commonly having a dense fibrous matrix giving it a "rubbery" consistency.

The skeletal system of vertebrates is composed of cartilage and bone. In the embryonic development of vertebrates, cartilage serves as part of the initial supporting skeleton. As the organism matures, most of the cartilage is gradually replaced by bone in a process called **ossification** [see the discussion of endochondral (replacement) bone in Part 2 of this exercise]. Once the last of the cartilage in a bone has disappeared, the bone can no longer increase in length. This does not mean, however, that bones are nonliving and cannot change shape.

Some cartilage in the human body is never replaced by bone. One example is **hyaline cartilage** found at the ends of long bones such as the humerus of the arm and the femur and tibia of the leg

(Figure 32B-1). Hyaline cartilage is also found at the end of the nose, in the major cartilage rings of the trachea and bronchi of the respiratory system, and in the rib cartilages. Another type of cartilage, **elastic cartilage,** occurs in the ear and in the trachea; elastic fibers are embedded in the matrix.

Distributed throughout the firm but elastic matrix of cartilage are numerous **lacunae** ("lakes") containing cartilage cells (**chondrocytes**), which secrete the matrix. Note that cartilage cells shrink somewhat during the processes required for the preparation of slides. Thus, you will not see the cells filling as much of the lacunar space as they do in living tissue.

Figure 32B-1 *Hyaline cartilage showing lacunae.*

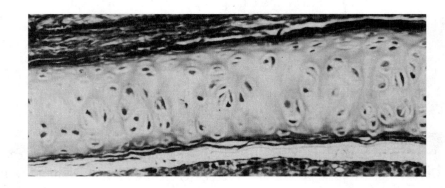

Procedure ||||||||||||||||||||||||||||||||||||||||||||||||

1. Use high power (40×) to examine slides of human elastic and hyaline cartilage. On a separate piece of paper, draw a section of each tissue and label the **matrix, lacunae,** and **chondrocytes.**

   a. In which tissue are the chondrocytes closest together? _____

   b. Do you see any fibers or other types of cells in the matrix of hyaline cartilage? _____ *Of elastic cartilage?* _____

   c. Why might the matrix of hyaline cartilage contain elastic fibers? _____

   As we age, or following injury, cartilage may begin to degenerate. Often this occurs in joints that are used to bend and lift, such as elbows, knees, and neck. As a consequence, **osteoarthritis** develops.

PART 2   **Bone**

Bone consists of a dense matrix of calcium salts and proteins secreted by bone-forming cells called **osteoblasts.** Protein is laid down as minute fibers that contribute strength and resiliency to bone; mineral salts contribute hardness. [However, the earliest bone found in fossil agnathan fishes (and in many modern teleost fishes) was acellular.] Cellular bone, containing mature osteoblasts called **osteocytes,** is not a fixed, unchanging tissue. It is constantly being remodeled by **osteoclasts,** multinucleated cells that break down the bone structure, and the continued synthetic activity of osteoblasts. The dynamic nature of cellular bone also enables it to serve as a calcium reserve: birds use calcium from their skeleton in the production of egg shells; mammals may also draw on skeletal reserves during the prenatal development of their young and in the production of milk.

Bone tissue is of two types: **dermal bone,** which forms by means of calcium deposition within the dermis of the skin, and **endochondral bone,** which replaces the cartilage of the embryo as the organism grows. Dermal bone is prevalent in lower vertebrates, where it forms like those scales and plates that covered many early and primitive fishes. In bony fishes (the group that includes trout, perch, and salmon) and in terrestrial vertebrates, dermal bone is found only in elements of the skull and pectoral (forelimb) girdle. The shapes of the long bones of adult vertebrates are prepatterned in the cartilage of the embryo. Although cartilage continues to provide support during the period of growth, most cartilage is gradually replaced by endochondral bone, through the process of ossification, as the animal matures. Endochondral bone predominates in the skeleton of higher vertebrates.

The long bones of higher vertebrates are not solid structures. Rather they are constructed to provide maximum strength with minimum weight. This is accomplished by surrounding the outer portion of the bone with heavy, **compact bone** to provide strength. A **marrow cavity** in the center of the bone shaft and **spongy bone** at the ends of the long bones reduce weight (Figure 32B-2). In mammals and their immediate reptilian ancestors, most dinosaurs, and a scattering of other vertebrates, the plates (**lamellae**) of secondary bone and osteocytes are arranged concentrically to form **Haversian systems.** In the process of bone formation, mature osteoblasts (osteocytes) become trapped in lacunae, microscopic cavities formed by their secretions of bone matrix. The lacunae are linked together by a series of small channels (**canaliculi**) that combine to form a large system of canals (**Haversian canals**). Many of the Haversian canals are large enough to contain blood vessels, which deliver oxygen and minerals to the living osteocytes. If you have ever observed a broken bone, you probably noted that it was quite bloody. Bone is a living tissue and must be supplied with nutrients and oxygen by the circulatory system, as is true for all the tissues of the body (Figure 32B-3).

The strength of bones depends on the rate at which osteoclasts reabsorb calcium and phosphorus and the rate at which osteoblasts replace these vital elements. In individuals with strong bones, osteoblasts synthesize bone materials more rapidly than osteoclasts break them down. Bones subject to disuse, especially in older people, tend to become weakened. **Osteoporosis,** a condition characterized by weakened bones, is common among postmenopausal women. Steroid hormone therapy combined with intake of extra calcium and vitamin D, a treatment that has been shown to stimulate osteoblast activity, is a common regimen among aging women. As your parents age, you might notice that they seem to "shrink" over the years. Bone loss occurs more rapidly in spongy bones. Vertebrae are composed of spongy bone, and the spinal column compresses with time and age.

Details of the cellular architecture of mammalian long bone can be observed by grinding a section of compact bone until it is extremely thin and mounting it on a slide for inspection under the microscope. Note that cellular elements are lost in this process.

Figure 32B-2   *The outer area of a long bone is composed of compact bone. Spongy bone occurs at the ends, and the marrow cavity extends through the center of the bone shaft. The surface of the bone is covered by a thin fibrous layer, the* periosteum *peri-, around;* osteon, *bone), to which muscles are attached by tendons.*

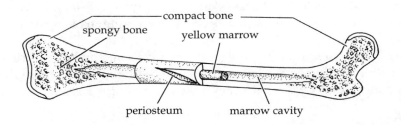

spongy bone

compact bone

yellow marrow

periosteum

marrow cavity

iiiii **Procedure** iiiiiiiiiiiiiiiiiiiiiiiiiiiiiiiiiiiiiiiiiiiiii

1. Use high power (40×) to study a slide of human compact bone. On a separate sheet of paper, draw a portion of the bone. Label the structures shown in Figure 32B-3b: the Haversian canals, canaliculi, lacunae, and lamellae.

    a. Suggest how a broken bone heals. _____

    _____

Bone first appeared, together with cartilage, in the earliest fishes found in the fossil record. In the evolution of fish, the replacement of cartilage by bone in adults was favored in those groups that developed a swim bladder/lung, a structure that can change the body's density relative to that of water. A fish can ascend or descend in the water by simply adding gas or removing gas from its swim bladder,

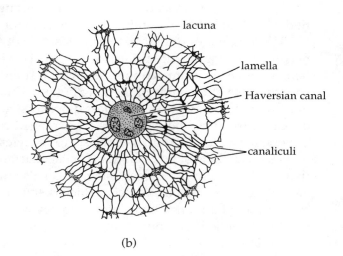

(a)   (b)

Figure 32B-3   *(a) A bony matrix is deposited in concentric cylinders, or*
lamellae, *around the* Haversian canal, *which contains blood vessels and nerves.*
Each bone cell within its lacuna is connected to adjacent bone cells and to the
*Haversian canal by cellular extensions occupying minute canals (*canaliculi*)*
in the matrix. Oxygen carried in the blood vessels of the Haversian system is
transported to the bone cells by the cytoplasmic extensions, and wastes are
returned to the Haversian system along the same route. (b) Diagrammatic
representation of the Haversian system.

thus reducing the energy required to change or maintain its position. In the evolution of those fish that
lack a swim bladder—for example, the extinct armored fishes (the placoderms) and the elasmobranchs
(the group that includes sharks, skates, and rays)—dense bone was replaced by lighter cartilage, thus
reducing the body's overall density. Some cartilaginous fishes retain bone in their vertebrae, where
support is needed.

 Among vertebrates, the replacement of cartilage by bone during maturation to adulthood may have
been extremely important in the transition to life on land, where the body of a large, active animal, no
longer supported by the buoyancy of water, requires a strong internal support structure.

✔ **PART 3**   **Loose and Dense Connective Tissue**

Connective tissue is responsible for the attachment of skin to muscle, muscle to bone (**tendons**), and bone
to bone (**ligaments**).

 Loose connective, or **areolar, tissue** connects organs and tissues (Figure 32B-4). **Dense connective**
tissue is more compact and shows a more or less orderly arrangement of its composite fibers, depending
on its function. Such tissue is found in the periosteum covering bones, in the dermis of skin, and as part
of the ligaments and tendons used for support.

⁞⁞⁞⁞⁞ **Procedure** ⁞⁞⁞

 1. Use high power (40×) to examine a prepared slide of areolar tissue. On the top half of a sheet
  of paper, draw what you observe. Identify fibroblasts, elastic fibers (yellow and branched),
  and collagen fibers (white and unbranched).

  a. Do you see any white blood cells? _____ *What purpose do you think is served by these cells*

  wandering around in the connective tissues of the body? _____

  _____

Figure 32B-4 *Loose connective tissue is composed of two types of fibers, white collagenous fibers and yellow elastic fibers, embedded in a semisolid ground substance. Fibroblasts secrete both the matrix and fibers. White blood cells (macrophages) may also be seen wandering through the tissue.*

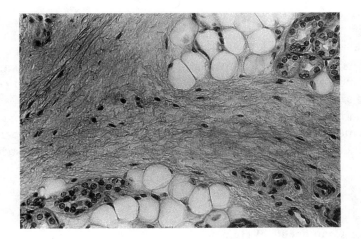

2. Examine a prepared slide of a human tendon. On the lower half of your sheet of paper, draw your observations. Label fibroblasts and bundles of elastic fibers.

 PART 4 **Adipose (Fat) Tissue**

Fat, or **adipose tissue,** is included among the connective tissues because of its mesenchymal origin and its association with other connective tissue elements (Figure 32B-5).

Figure 32B-5 *Adipose tissue with fat removed so that the large central vacuole of each cell is visible. Small, darkly stained nuclei are evident in the cytoplasm surrounding the vacuoles.*

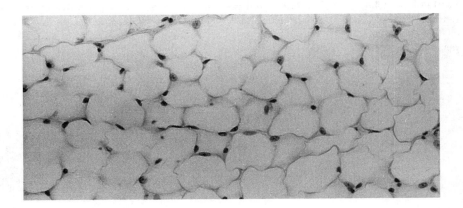

IIIII Procedure IIIIIIIIIIIIIIIIIIIIIIIIIIIIIIIIIIIIIIIIIIIII

1. Use high power (40×) to examine a prepared slide of cat adipose tissue. You will notice the enlarged fat cells containing fat globules, found in the center of each cell in a single large vacuole. The living contents of the cell form a shell around the fat globules. With weight loss, the fat globules become smaller, but the cell itself is not destroyed; it just waits to become filled with fat again. In addition, extra fat cells may form if excess food is available. On a separate sheet of paper, draw a section of the adipose tissue and label the nucleus, cytoplasm, and fat vacuole of the fat cell.

2. Examine a prepared slide of a section of adipose tissue from which the fat has been removed.

   a. *Compare these cells with the adipose tissue observed in step 1.*

   _____

   _____

✔ **PART 5** **Blood**

Blood is usually classified as a connective tissue because its cellular components (**erythrocytes, leukocytes,** and **platelets**) are suspended in an intercellular fluid matrix (**plasma**). However, since connective tissue is strictly defined as being of mesenchymal origin and not all blood cells are of mesenchymal origin, blood could instead be classified as a specialized tissue.

Mammalian blood consists of approximately 55 percent plasma (the intercellular fluid matrix) and 45 percent formed (cellular) elements. Plasma consists of water (90 percent) plus dissolved substances—plasma proteins, enzymes, digested food material, metabolic wastes, hormones, antibodies, and certain gases.

Table 32B-1   Components of Human Blood

Cells	Description	Structure
Red blood cells (erythrocytes) 3.5–5.5 million cells per milliliter	The most numerous cell type. Faintly biconcave disks. No nuclei. Contain the pigment hemoglobin, essential in the transportation of oxygen to cells and tissues. Less than normal quantities of red blood cells or hemoglobin causes anemia, a condition in which the cells and tissues are deprived of oxygen.	erythrocytes
White blood cells (leukocytes) 6,000–9,000 cells per milliliter	Nucleated disks of an amoeboid shape. Move by means of pseudopods. These cells prevent infection and the clogging of blood vessels by ingesting bacteria and other foreign materials.	
Granulocytes	Neutrophils, eosinophils, and basophils transform into phagocytic macrophages. Contain cytoplasmic granules.	eosinophil neutrophil basophil
Agranulocytes	B- and T-lymphocytes are responsible for the immune response. Cell nucleus rounded. No cytoplasmic granules. Monocytes engulf foreign material, including bacteria at sites of infection.	lymphocyte monocyte
Platelets (thrombocytes) 150,000–400,000 cells per milliliter	Platelets are responsible for blood clotting. These are cell fragments that rupture on the rough edges of torn blood vessels and release materials that initiate blood clotting.	platelets

||||| **Procedure** |||||||||||||||||||||||||||||||||||||||

1. Obtain a prepared slide of a human blood smear. Use high power (40×) to identify the cell types listed in Table 32B-1 and draw your observations on a separate sheet of paper. Be sure your drawings reflect the relative proportions and sizes of the different blood cell types. Label erythrocytes, leukocytes, and platelets.

    a. *How many different types of white blood cells do you find?* _____ *What are these types?*

    _____

    b. *Compare the structure of white blood cells and red blood cells.*

    _____

    _____

    c. *What is the general shape of the red blood cells?* _____

2. Obtain a sample of blood from a pithed frog. Cut through the skin and muscle along the centerline of the abdomen to expose the superficial ventral abdominal vein. Nick the vein and place a drop of blood on a clean glass slide. Place one drop of frog Ringer's solution (or 0.85% saline) onto the same slide and mix with the blood. Add one drop of methylene blue stain and observe the blood cells at 40×. Alternatively, you may use a prepared slide of frog blood to make your observations.

    d. *Are frog blood cells different from yours?* _____ e. *Do they have nuclei?* _____ f. *Do your red blood cells have nuclei?* _____

    g. *What distinguishes the red blood cells of humans from those of frogs?*

    _____

✔ **EXERCISE C   Muscle Tissue**

Most muscle movements result from the contraction of elongated cylindrical or spindle-shaped muscle cells called **muscle fibers,** each of which contains many microscopic elongated, parallel **myofibrils.** These fibrils are composed of the proteins actin and myosin. If the myofibrils have the appearance of alternating dark and light cross-bands (or striations) under the light microscope, the muscle fibers are classified as striated muscle tissue. Skeletal muscle and cardiac (heart) muscle are examples of striated muscle. Smooth muscle fibers also contain myofibrils but do not exhibit these cross-striations. Striations represent the alignment of actin and myosin fibers and play a fundamental role in muscle contraction.

||||| **Objectives** ||||||||||||||||||||||||||||||||||||||||

☐ Distinguish between smooth, skeletal, and cardiac muscle.

☐ Compare and contrast the functions of the three types of muscle.

☐ Describe the locations in which smooth, skeletal, and cardiac muscle are found.

✔ **PART I   Skeletal Muscle**

Skeletal muscle (also called **striated** or **voluntary** muscle) makes up the muscle masses that are attached to and move the bones of the body. In skeletal muscle, several elongated cylindrical cells are fused to form a multinucleate unbranched muscle fiber. The nuclei lie in the periphery of the fiber (Figure 32C-1a, b).

||||| **Procedure** ||||||||||||||||||||||||||||||||||||||||

1. Use high power (40×) to examine a prepared slide of skeletal muscle and make a drawing of this tissue on the upper third of a separate sheet of paper. Indicate the position of nuclei and label the cross-striations.

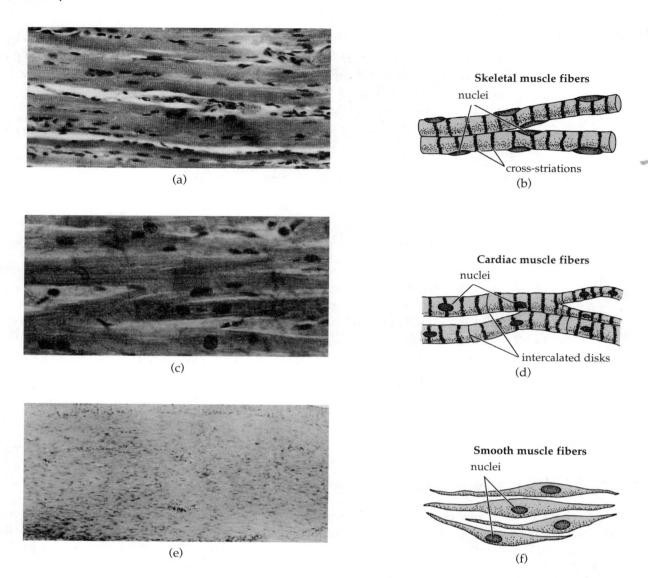

Figure 32C-1   *(a) (b) Skeletal muscle composed of fused cells producing multi-nucleate, unbranched fibers with cross-striations. (c) (d) Cardiac muscle composed of fused cells forming branched fibers with individual nucleated cells separated by intercalated disks. Cross-striations are also present. (e) (f) Smooth muscle with elongated single muscle cells (fibers) having central nuclei and no cross-striations.*

2. Use your observations to fill in Table 32C-1.

a. *What do the cross-striations represent?* _____

_____

b. *Why are they important?* _____

_____

✔ **PART 2**   **Cardiac Muscle**

Cardiac muscle is found only in the wall of the heart. The cells are cylindrical and striated. Unlike other muscle tissues, the fibers are branched. Individual cells are separated by transverse membranes called **intercalated disks,** which appear as dark bands crossing the fibers. Nuclei are centrally located (Figure 32C-1c, d).

ⅠⅠⅠⅠⅠ Procedure ⅠⅠⅠⅠⅠⅠⅠⅠⅠⅠⅠⅠⅠⅠⅠⅠⅠⅠⅠⅠⅠⅠⅠⅠⅠⅠⅠⅠⅠⅠⅠⅠⅠⅠⅠ

1. Use high power (40×) to examine a prepared slide of cardiac muscle and make a drawing of this tissue in the middle third of your sheet of paper. Indicate the position of nuclei and label the cross-striations and intercalated disks.

    a. *What is the significance of the intercalated disks in the functioning of heart muscle?*

    _____

2. Use your observations to fill in Table 32C-1.

Table 32C-1   Muscle Tissue Characteristics

Characteristic	Skeletal	Cardiac	Smooth
Shape of muscle fiber			
Number of nuclei per fiber			
Location of nuclei			
Function			
Location			

✔ **PART 3**   **Smooth Muscle**

Smooth muscles are found chiefly in the wall of the alimentary, urinary, and genital tracts and in the walls of arteries and veins. Smooth muscle fibers are spindle-shaped cells with no cross-striations, but with faint longitudinal striations. A single, elongated, central nucleus is present in each fiber (Figure 32C-1e, f).

ⅠⅠⅠⅠⅠ Procedure ⅠⅠⅠⅠⅠⅠⅠⅠⅠⅠⅠⅠⅠⅠⅠⅠⅠⅠⅠⅠⅠⅠⅠⅠⅠⅠⅠⅠⅠⅠⅠⅠⅠⅠⅠ

1. Use high power (40×) to examine a prepared slide of smooth muscle and draw a section of this tissue on the lower third of your sheet of paper. Label a nucleus and a muscle fiber.

    a. *Are actin and myosin present in smooth muscle?* _____

2. Remove a flap of skin from the thigh of a pithed frog.

3. Cut a *small* piece of muscle from the thigh. Make a wet-mount slide using Ringer's solution. Use dissecting needles to tease the tissue apart; add a drop of methylene blue.

4. Cover the tissue with a coverslip. If the coverslip does not lie flat, remove the coverslip, tease the tissue further, and try again. Use high power (40×) to study the shape and size of the cells.

    b. *Describe the shape and appearance of the muscle cells.* _____

    _____

    c. *What type of muscle tissue is represented?* _____

5. Use your observations to fill in Table 32C-1.

✔ **EXERCISE D** | **Nervous Tissue**

The basic functional unit of the nervous system is the individual nerve cell or **neuron.** The function of the neuron is the conduction of a nerve impulse (the action potential, an electrochemical signal) to another neuron or to a muscle fiber or gland. However, only about 10 percent of the cells in the nervous system are neurons. The remainder are glial elements and sheath cells, which isolate, assist, and support neurons functionally and sustain them metabolically.

Neurons consist of three major sections: the cell body (soma), dendrites, and the axon (Figure 32D-1). The **cell body** contains the nucleus and most of the cellular organelles. Cellular extensions or fibers that conduct the nerve impulse toward the cell body are called **dendrites. Axons** conduct the nerve impulse away from the cell body. The number of nerve fibers extending from the cell body depends on the particular neuron and its location within the nervous system. The junction of two neurons is called a **synapse.**

Neurons of the central nervous system maintain a background of spontaneous activity, monitor and control a variety of physiological functions, receive information and integrate it with genetically determined functions or learned responses, and initiate motor activities. Various structural and functional adaptations are responsible for this coordinated control.

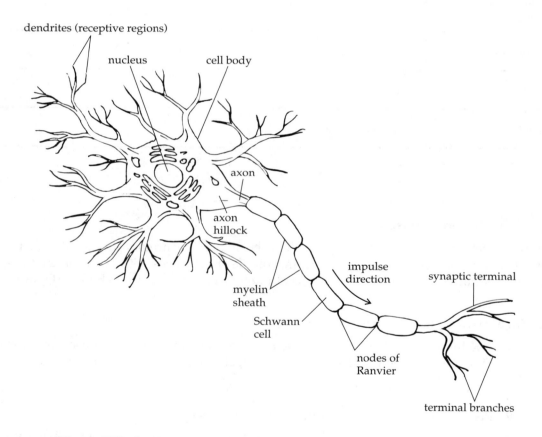

Figure 32D-1   *A typical motor neuron with its cell body, dendrites, and axon. The axon is encased within a myelin sheath produced by Schwann cells.*

||||| **Objectives** |||||||||||||||||||||||||||||||||||||||||

☐ Draw and label the major parts of a nerve cell, including the axon, dendrites, and cell body.

☐ Describe the function of each part of a nerve cell.

||||| **Procedure** ||||||||||||||||||||||||||||||||||||||||||

Obtain a prepared slide of motor neurons from the spinal cord and examine it at high power (40×). On a separate sheet of paper, draw a single neuron and label all parts: the cell body, nucleus, axon, and dendrites.

a. Can you see an entire motor neuron on this slide? _____ *b. Why or why not?* _____

_____

c. A single motor neuron extends from its cell body in the spinal cord to the most distant extremity of your body. Where would you find the nerve endings of the longest cell in your body? _____

Laboratory Review Questions and Problems

1. What are the four groups of tissues studied in today's laboratory?

2. Which group is composed of the most diverse cell types?

3. What is common to all types of connective tissue? Describe any variations in components among the different connective tissues.

4. Where in the body do you find loose connective tissue? Dense connective tissue?

5. Use a separate sheet of paper to prepare a summary table listing the six types of connective tissues, the specific types of cells in each tissue type, one place in the human body where the tissue is found, the appearance of the tissue, and its major function.

6. Stratified squamous epithelium regenerates rapidly by division and is found in areas that are subject to abrasion. Simple squamous epithelium is leaky and is usually found in areas where exchange occurs across a surface. Cuboidal and columnar epithelia contain a lot of cytoplasm and usually occur where secretory products are made. Predict which type of epithelium you would find in each of the following:

Esophagus

Lungs

Blood vessels

Kidney tubules

Intestine

7. How does dermal bone differ from endochondral bone?

8. When you break a bone it hurts and usually there is some bleeding. Why?

9. How is bone continuously "re-worked"?

10. How does the matrix of cartilage differ from that of bone?

11. Compare the three types of muscle tissue using the following criteria: striated or unstriated; voluntary or involuntary; uninucleate or multinucleate fibers; branched or unbranched fibers; spindle-shaped or rectangular fibers.

12. Why is blood usually considered a type of connective tissue? What are the major functions of red blood cells? White blood cells?

13. Draw a typical nerve cell and label the cell body, axon, and dendrites. Do all nerve cells have this same structure?

Introduction to the Study of Anatomy, and the External Anatomy and Integument of Representative Vertebrates

OVERVIEW

This laboratory introduces the topic of **anatomy**, the field of science that studies the structure and arrangement of the parts of an organism. Laboratories 33 through 36 examine the anatomical features of four vertebrates, focusing on structural adaptations that illustrate evolutionary relationships and phylogenetic trends among the vertebrates. The four animals you will study are a cartilaginous fish (a shark), an aquatic vertebrate that has retained many ancestral features; an amphibian (a frog), a tetrapod illustrating the transition of vertebrates from water to land; a reptile (a turtle), a vertebrate featuring fully terrestrial adaptations; and a mammal (a rat), possessing many of the adaptations found in the class to which we belong. (Since the phylogenetic "tree" diagram, Figure 27-II in Laboratory 27, will be important to your understanding of the relationships among these vertebrate groups, you may find it convenient to move page 27-3 to this part of the manual for easy reference.) These four laboratories will give you an overview of vertebrate anatomy and experience with dissection techniques.

STUDENT PREPARATION

Prepare for this laboratory by reading the pages indicated by your instructor. Familiarizing yourself in advance with the information and procedures covered in this laboratory will give you a better understanding of the material and improve your efficiency.

If dissection equipment is not provided, obtain the following items and bring them with you to each laboratory: one one-piece scalpel or cartilage knife, one pair of dissection scissors, one probe, and two dissection needles. Do *not* use a two-piece scalpel during these exercises.

PART 1   INTRODUCING DISSECTION: ANATOMICAL LOCATIONS

Multicellular organisms, including vertebrates, are composed of groups of interacting and cooperating organs, often with common developmental origins, which form organ systems such as the integument, the digestive system, and the respiratory system. The digestive system, for example, includes the pharynx, esophagus, stomach, intestine, liver, and other organs. Each organ is made up of one or more types of tissue (see Laboratory 32).

A variety of terms is used to describe the position of body parts or organs (or parts of organ systems) and their location in relation to specified reference points. It is important to understand these terms so that you can follow dissection directions.

Study Figure 33-I, referring to the terms in Table 33-I. Notice that there is some ambiguity in using terms such as "anterior" or "ventral" in humans because of their upright posture.

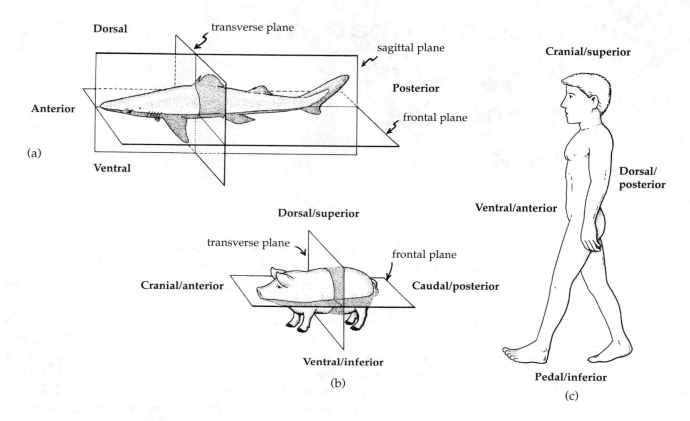

Figure 33-I   *Lateral view of (a) shark, (b) pig, and (c) human.*

TECHNIQUES OF DISSECTION

Most vertebrate specimens used in dissection are preserved in a 10% solution of formaldehyde (formalin). This preservative hardens tissues and maintains their definition. However, continued storage in formalin may make the specimens less pliable and more difficult to dissect. Formalin is also irritating to the eyes, nose, and skin, and prolonged exposure is suspected to be carcinogenic. For these reasons, after specimens have been preserved in formalin, they are washed and stored in other solutions. Your instructor will tell you how to care for the specimens used in your laboratory and will inform you of any special precautions. We recommend the following:

- Always wear disposable plastic gloves to protect your hands from the solution used to preserve or store your specimen. Contact with the preservative may cause some loss of feeling in your fingers, but sensation will return within a few hours.

- Rinse your specimen with fresh water whenever you retrieve it for use and when you open the body cavity.

- Do *not* wear contact lenses. (If you rub your eyes or get any preserved tissue in them, flush your eyes with running water for 15 minutes and see a doctor as soon as possible.)

Table 33-I   Definition of Terms Used to Describe Position or Direction in Vertebrates

Position or Direction	Definition
Anterior	In the direction toward which the organism faces; pertaining to the head end (in humans, equivalent to *ventral*)
Posterior	In the direction away from which the organism faces; pertaining to the end opposite the head (in humans, equivalent to *dorsal*)
Dorsal	Toward the vertebral column; the upper surface of the animal or organ; the back
Ventral	Toward the chest and abdomen; the underside of an animal or organ
Cranial	Toward the head
Caudal*	Toward the tail
Pedal	Toward the feet
Superior	Toward the upper parts of the body (in humans)
Inferior	Toward the lower parts of the body (in humans)
Lateral	Away from the midline of the body
Medial	Toward the midline of the body
Superficial	At or near the surface
Distal	Away from the point of reference
Proximal	Near the point of reference

*The word *caudal* is sometimes used to mean "toward the feet" in the human. But caudal literally means "toward the tail," and the human vestigial tail (the coccyx) is in the region of the pelvis, not the feet. For this reason we have defined caudal to mean "toward the tail" and pedal to mean "toward the feet."

- Wear old clothes for these laboratories, since some of the substances you will encounter may cause stains that are difficult to remove.

The goal of dissection is to separate body parts so that their size, shape, connections, and gross structure can be examined. To avoid damaging connections and disrupting anatomical continuity, use the scissors and scalpel as little as possible. The major use of scissors in dissection is to cut through surfaces to reach the structures beneath. When possible, try to plan cuts so that the edges can be brought back together and body parts can be reconstructed for review. The scalpel should not be used to make incisions; rather, use it as a cartilage knife to remove supportive tissues when necessary. Your fingers will be the most useful tools for separating organs and tissues. A blunt probe or dissecting needles may also be used for finer dissections.

UNDER NO CIRCUMSTANCES SHOULD ORGANS BE REMOVED FROM THEIR POSITION IN THE BODY WITHOUT EXPLICIT INSTRUCTIONS. Structures damaged prematurely may interfere with or be lost to future study. In many cases, it is advisable to work on one side of the body only, preserving the other side for later study. (This is particularly important in the case of muscle dissections, which destroy many of the blood vessels that will be traced during a later laboratory.)

You will use your assigned representative vertebrate for several laboratories, so it is important to treat it properly. If any body parts show signs of drying out during the laboratory period, wrap them with moist paper towels. After each laboratory, be sure that your specimen is stored properly. If the specimen is returned to a common bucket, be sure the animal is covered with fluid. If the specimen is to be stored in a plastic bag, moisten several towels with water and wrap the extremities and any skinned parts of the animal with the wet towels. Any parts of the animal that are to be discarded should be placed in a designated trash receptacle. Be sure to wash your hands with soap before leaving the laboratory.

ORGAN SYSTEMS OF VERTEBRATES

When we study vertebrates, it is convenient to group together organs with similar functions. These organs may show close anatomical and developmental interrelations with each other—as do many of the digestive organs—but structures serving the same function may also show major anatomical differences. For example, gills and lungs are organs of gas exchange but function in different environments—water versus air—and they are very different in structure. We will use the scheme presented in Table 33-II to study the functional associations of organs.

PART 2   INTRODUCTION TO THE REPRESENTATIVE VERTEBRATES

✔ EXERCISE A   **Life History and External Anatomy of Four Representative Vertebrates**

||||| **Objectives** |||||||||||||||||||||||||||||||||||||||

- ☐ Examine and compare the body forms of representative vertebrates.
- ☐ Locate the important external features of each specimen and be able to use these as landmarks for subsequent studies.
- ☐ Describe the major features of the life history and reproduction of the four representative vertebrates.

||||| **Procedure** |||||||||||||||||||||||||||||||||||||||

Work in groups of four.

1. Put on plastic gloves.

2. One person in each group should obtain a shark, one a frog, one a turtle, and one a rat. Place your specimen on a dissecting pan.

3. As you dissect your specimen, share your observations with the other members of your group. Since the objective of this series of laboratories is to learn about evolutionary relationships and phylogenetic trends, your success in these laboratories will require that you learn as much as possible from each other's dissections. *Note*: When these labs are completed, your instructor may give you a "practical" exam; such a test requires that you be able to locate and identify structures in the specimens you have dissected. To help you to remember and review your observations, keep notes and make sketches on separate sheets. Add these to your laboratory manual.

Shark

The dogfish, *Squalus acanthias*, is a small shark in the class Chondrichthyes (cartilaginous fishes, a group that includes sharks and rays). Adult sharks and rays, unlike most other vertebrates, have a skeleton composed of cartilage, not bone. In the early stages of development, the skeletons of all vertebrates are made up of cartilage, a relatively lightweight, flexible material. In most vertebrates (including the ancestors of cartilaginous fishes), during the course of growth, most cartilage is gradually replaced by harder, denser bone tissue, which provides stronger structural support in the adult animal. However, for marine and aquatic animals, an increase in body density causes a decrease in buoyancy. The density of the ancestral fish body was higher than that of water, requiring these fish to expend relatively large amounts of energy just to keep themselves from sinking. During the course of evolution, chondrichthyian fish adapted by ceasing to replace cartilage with bone during maturation, thus reducing their overall density and also the energy required to maintain their position in the water.

In the evolutionary line leading to bony fishes (a group that includes perch, trout, salmon, and many other familiar fish), animals developed a swim bladder, an organ that contains air, thereby reducing the animal's overall density, increasing its buoyancy, and again decreasing the amount of energy required to keep from sinking to the bottom.

Table 33-II   Vertebrate Organ Systems

System	Function
I. Integument (skin)	The outer covering of an organism. Provides a protective and regulatory interface with the environment.
II. Behavioral systems	Structures participating primarily in externally directed activity (behavior).
A. Sensors (affectors)	Nerve endings and sense organs that monitor both the external and internal environments.
B. Control systems*	Receive input from the sensors or feedback from other organs; produce output to structures that give rise to movement or other responses (effectors). (Adaptive responses may depend on both genetic elements and learning from past experience.)
1. Nervous system	Composed of the brain, spinal cord, and peripheral nervous system; these structures respond rapidly to input.
2. Endocrine system	An assemblage of tissues and organs with several different embryonic origins; respond to input by releasing their products into the circulatory system. These relatively slow responses may reinforce activities initiated by the nervous system.
C. Effectors	Tissues and organs that respond to stimuli to produce: movement (cilia, muscles, and muscles and bones working together), glandular secretions (gland cells and glands), light (photophores), or electrical potentials (electric organs).
III. Physiological systems	Organs participating primarily in physiological (internally directed) activities.
A. Digestive system	Processes food to provide energy and building materials for the organism.
B. Respiratory system	Functions in gas exchange; includes gills, lungs, and, in some vertebrates, the body surface.
C. Circulatory system	Carries nutrients, wastes, gases, and information-containing molecules; also has a role in disease prevention and immunological responses; serves all living tissues of the body.
D. Urogenital system	Consists of kidneys, whose functions are osmoregulation and the excretion of nitrogenous wastes, and gonads, which produce the gametes. Although these organs have very different functions, they are associated with one another during embryonic and later development.

*Note that control systems are also involved in regulating adaptive internal adjustments (physiological processes) and that both behavioral and physiological systems function to maintain an internal balance called homeostasis. During our study we will try to understand how an organism's behavior and physiology respond to external changes.

LIFE HISTORY

Dogfishes live in marine waters off the Atlantic and Pacific coasts of North America. They prefer water temperatures close to 10°C and migrate north in the spring and south in the fall, often in schools of thousands. They feed on herring, other small fishes, and squid and are themselves part of the human diet in many parts of the world. Like other sharks, dogfishes retain large amounts of the nitrogenous waste product, urea, as part of their osmoregulatory strategy. Urea retained in the body fluids helps to maintain the fluids as either isosmotic or hyperosmotic relative to seawater, thus preventing water loss to the salty ocean.

a. Explain what would happen if a shark's body fluids were hypoösmotic to seawater.

_____

Unlike more primitive sharks, dogfishes retain large, yolky eggs in their oviducts for a lengthy gestation period of 22 months, and give birth to live young; they are said to be **viviparous.** (Animals that do not retain fertilized eggs until hatching—they lay their eggs—are **oviparous.**)

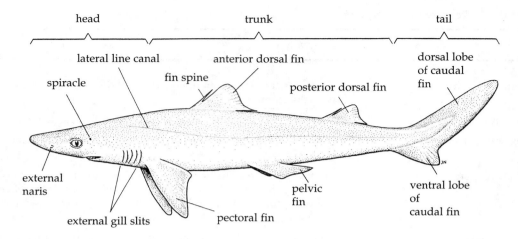

Figure 33A-1   *External anatomy of the dogfish shark.*

IIIII **Procedure** IIIIIIIIIIIIIIIIIIIIIIIIIIIIIIIIIIIIIIIIIIIII

1. Examine the shark. The major divisions of the shark's body are the **head,** including the gill region, the **trunk,** and the **tail.** The body shape is streamlined or **fusiform** (Figure 33A-1).

2. Pry open the mouth (if it is necessary to cut through the angle of jaw to see the teeth, cut on the animal's left side) and *carefully* feel the edges of the jaws. (The teeth are actually modified scales, derived from the dermis of the skin, and are similar to those covering the body surface.) Several rows of teeth may be apparent; the largest teeth are found on the margins of the jaws.

3. Find the **cloacal opening.** The **cloaca** is a chamber that receives products from the digestive system and from the excretory and reproductive systems.

    b. *Which of the other vertebrates dissected in this laboratory have a cloaca?*

    _____

    c. *Humans do not have a cloaca. What are some of the anatomical differences that distinguish the*

    human urogenital system from that of animals with a cloaca? _____

    _____

4. Find the **external gill slits.** These are openings from the pharynx (also part of the digestive system) to the exterior. Water is normally taken in through the mouth, passes to the pharynx where capillaries of the internal vascular gills provide a surface for gas exchange, then leaves through the external gill slits. The **spiracles,** one behind each eye, are modified external openings of the anterior gill slits which, during the course of evolution, have been displaced dorsally by the development of the jaw. (In higher vertebrates, structures derived from the spiracles developed into the eustachian tube and middle ear.) In bottom-dwelling cartilaginous fishes such as skates and rays, water for respiration enters through the spiracles, since the mouth often rests on the bottom.

5. Examine the paired **external nares** (on the lower portion of the snout). Use the blunt end of a probe to explore these structures. Note that they are strictly external and do not open into the mouth cavity; the nares contain the olfactory epithelium, which mediates the sense of smell.

6. Find the paired eyes. Note the fleshy, nonmotile lids above and below the eyes (in some sharks, the lower lid is movable and can cover the eyeball to protect it during feeding).

7. Press gently on the skin covering the dorsal area of the snout. Note the jellylike material extruded from numerous **ampullary organs of Lorenzini** located in this region. Ampullae contain sensors of electroreception.

   d. *How would an aquatic organism use the ability to sense electrical fields?*

   _____

8. The **lateral line canal** can be seen as a thin, light line running along the side of the body. Sensory receptors of the lateral line provide the dogfish with the ability to sense movements in the surrounding water.

9. Identify the two dorsal fins, the caudal fin, and the paired pectoral and pelvic fins, all of which are used in locomotion. Examine the pelvic fins carefully. In males, the medial border of each fin is modified to form a **clasper** with a deep groove along its border. Folds of skin overlap this groove and close it off from the surrounding water. Anteriorly, the groove connects with a siphon sac which can fill with seawater. When seawater is pumped out through the groove, it mixes with sperm from the cloaca and travels through the groove into the cloaca of the female during **copulation.** In the dogfish, fertilization is internal.

   e. *Why?* _____

   Note that the pelvic fins of the female are broad and unmodified.

   f. *What is the sex of your specimen?* _____

10. Feel the surface of the skin. Its unique texture is due to **placoid scales,** which originate in the dermis and have a structure, and evolutionary origin, similar to that of teeth (see Exercise B).

    g. *Placoid scales are part of what organ system?* _____

11. Turn to the Summary, page 33-13.

Frog

The bullfrog, *Rana catesbiana,* is a large frog in the class Lissamphibia (modern amphibians). It is an anuran—"without tail." Other living amphibians include the salamanders (urodeles) and the wormlike caecilians of the tropics. Frogs have hind legs and a pelvic girdle that are highly specialized for jumping.

LIFE HISTORY

The large genus *Rana* has over 400 species worldwide and about 27 species in North America. Adult bullfrogs live in shallow, still waters from Nova Scotia to central Florida along the East Coast and west through the Great Plains. In colder climates, adults may burrow into the mud and pass the winter in an inactive, torpid state (hibernation).

When females are ready to mate, they locate a vocalizing male. A male mounts the female's back and clasps her tightly (an embrace known as **amplexus**). This action stimulates the female to lay hundreds (or thousands) of eggs which, as they emerge, are mixed with sperm from the male. Fertilization is external. Embryos develop from jelly-coated eggs and hatch as aquatic tadpoles (a larval form with external gills, no paired appendages, and a long tail). After two or more years, the bullfrog tadpoles undergo **metamorphosis,** a process that takes several months. During this time the gills become internal and are later functionally replaced by lungs. Fore and hind limbs appear and become functional locomotor organs, and the tail is reabsorbed. Tadpoles feed on plants and decaying animal material; adults eat insects, other invertebrates, and small fishes.

▮▮▮▮▮ Procedure ▮▮▮

1. Examine a specimen of the bullfrog (if you are studying a different species, your instructor will identify it and tell you about its range and life history). The major divisions of the body are the **head,** which extends to the shoulder region, and the **trunk** (Figure 33A-2).

Figure 33A-2 *External anatomy of the adult bullfrog.*

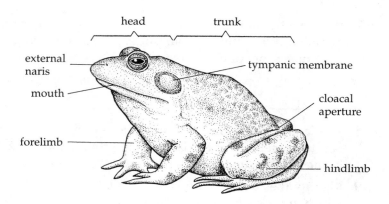

2. Locate the **mouth** and the **cloacal opening.** The **cloaca** is a chamber that receives products from the digestive system and from the excretory and reproductive systems.

   a. Which of the other vertebrates dissected in this laboratory have a cloaca?

   _____

   b. Humans do not have a cloaca. What are some of the anatomical differences that distinguish the

   human urogenital system from that of animals with a cloaca? _____

   _____

3. Find the paired **external nares.** Pry open the mouth and find the **internal nares** opening anteriorly into the roof of the mouth cavity (cut through the angle of the left side of the jaw if necessary). The nares provide a pathway for the movement of air between the outside and the lungs. The olfactory epithelium responsible for the sense of smell is located along this pathway.

   c. Are teeth present on the jaws? Describe them. _____

   _____

   d. How does the structure of the frog's nares differ from that of the shark's nares?

   _____   _____

   What is the significance of this difference? _____

   _____

4. Find the eyes. Both the upper and lower eyelids are nonmotile, but a third eyelid, the **nictitating membrane,** protects the surface of the eye. Locate this membrane if possible (it is better seen in the living frog). One of the extrinsic muscles of the eye is used to pull the eye deep into its orbit, further protecting it.

5. Behind the eyes, find the paired **tympanic membranes** of the ear. In frogs (and turtles) this membrane forms an "eardrum" flush with the surface of the body, separating the middle ear from the environment. In some reptiles and in birds and mammals, the tympanic membrane is recessed, separating the middle ear from the outer ear (the passage to the outside). The middle ear and outer ear develop, both phylogenetically and embryonically, from the first gill slit or spiracle (see step 4, page 33-10). (In the shark, there is no middle or outer ear; the inner ear is not highly modified for the perception of "sound" and the spiracle remains as an open gill slit.)

   The tympanic membrane **transduces** sound, changing airborne vibrations to mechanical vibrations that can be carried by the **columella,** a bone derived from the hyoid arch (part of the pharyngeal skeleton in sharks and ancestral vertebrates), to the inner ear where the sensors associated with hearing are located. In bullfrogs, the tympanic membranes show **sexual dimorphism**—they are larger in males than females.

    e. Is your specimen a male or a female? _____

    f. What major sensory system found in the shark is absent in the adult frog?

    _____

6. Compare the **forelimbs** and **hindlimbs.**

    g. How do the forelimbs and hindlimbs differ? _____

    h. What is the significance of this difference?

    _____

    i. How do the limbs of frogs differ from those of sharks?

    _____

7. Touch the skin. Can you feel scales? In most adult amphibians, the skin is moist and is richly vascularized.

    j. What are the functions of this moist, vascular organ? _____

    _____

8. Turn to the Summary, page 33-13.

Turtle

Turtles, descendants of the earliest reptiles, form an order within the class Reptilia. Turtles of the genus *Pseudemys,* whose members are commonly called cooters or sliders, are a favorite choice for dissection. The genus ranges from the United States to Argentina and is found in aquatic habitats including rivers, ditches, lakes, and ponds.

    Male pond sliders (*P. scripta*) have long nails on their forelimbs and their shells are flatter than those of females; the females of the species are generally larger. Males have a long grooved penis that carries sperm to the cloaca of females; fertilization is internal. Usually, several shelled eggs are laid at one time. Unlike endothermic vertebrates (birds and mammals), in which the sex of the offspring is determined genetically, **sex determination** in turtles (and some other reptiles) depends upon the incubation temperature of the eggs. In turtles, higher temperatures favor females and lower temperatures, males. (The reverse is true in lizards and alligators.)

a. Why would temperature-dependent sex determination not be a good idea in birds and mammals?

_____

    Sliders are largely vegetarian.

    Other turtles available for dissection include painted turtles, *Chrysemys,* or snapping turtles, *Chelydra.* Your instructor will give you information about these turtles if they are used.

⁞⁞⁞⁞⁞ Procedure ⁞⁞

1. Examine the preserved turtle. The major divisions of the body are the **head** borne on a long **neck,** the **trunk,** and the **tail** (Figure 33A-3). The major portion of the body, the trunk, is covered dorsally by a **carapace**—a combination of the axial skeleton and bony plates overlaid by scales or **scutes** (see Exercise B). The arrangement and characteristics of these scutes vary in different species and can be used for identification. Ventrally, the turtle is covered by a **plastron**—also a combination of bony plates and scales. Note the bony bridges that connect the plastron and carapace. A tail protrudes from the back portion of the body.

    b. What is the sex of your turtle? _____

Figure 33A-3 *External anatomy of a turtle.*

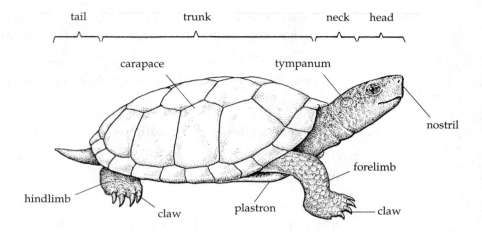

2. Locate the **mouth** and **cloacal opening.** The **cloaca** is a chamber that receives products from the digestive system and from the excretory and reproductive systems.

    c. *Which of the other vertebrates dissected in this laboratory have a cloaca?*

    _____

    d. *Humans do not have a cloaca. What are some of the anatomical differences that distinguish the*

       human urogenital system from that of animals with a cloaca? _____

       _____

3. Find the paired **external nares.** Cut through the angle of the jaw on the left side and pry open the mouth. Locate the **internal nares** opening into the roof of the mouth cavity. Airborne molecules passing through this channel stimulate cells of the olfactory epithelium, which mediates the sense of smell. Are teeth present on the jaws? Are teeth present in other reptiles?

    e. *What are the differences between the structures of the mouth cavity of the turtle and of the frog?*

       _____

4. Examine the eyes. Note the upper and lower eyelids and find the **nictitating membrane** in the anterior corner of the eye. Find the **tympanic membrane** (tympanum), the exterior covering of the **middle ear** on each side of the head just above and behind the angle of the jaw. This structure transduces sound waves in the air into mechanical vibrations that are transmitted to the **inner ear,** where they are transduced into nerve impulses.

5. Examine the **forelimbs** and **hindlimbs.** Note the platelike scales of the skin. Feel them. Is any bone associated with them? Are they slimy? (Reptiles lack integumentary mucous glands.) Examine the **claws** on each foot. Feel them. Like the scales covering the shell, the claws are composed of keratin.

    f. *How many digits are on the forelimb?* _____ *On the hindlimb?* _____

6. Turn to the Summary, page 33-13.

Rat

The laboratory rat is a rodent classified in the class Mammalia, a group distinguished by the presence of mammary glands and hair. Specimens available for dissection in the laboratory will probably be either white (albino) or hooded rats.

   Like all mammals, rats are **viviparous,** giving birth to live young. Unlike sharks, in which the embryonic yolk sac serves as a membrane for nutrient and gas exchange, mammals possess a true **placenta,** which provides for intimate contact of maternal and fetal blood and facilitates the transfer of nutrients, wastes, and gases between mother and fetus. Also, unlike most other vertebrates, laboratory rats (and many other domesticated animals, removed from rigorous natural selection and the timing cues of their natural environment) remain capable of reproduction at all times of the year. Female rats have an estrus cycle of about four days (this means that they are receptive and can mate with a male at four-day intervals, if not already pregnant). A male will mount a receptive female and insert his penis into her vagina, depositing sperm which travel through the uterus to the oviducts where eggs (ova) are fertilized. These fertilized eggs begin to divide and become embedded in the horns of the uterus. Gestation takes about 21 days. Young are born naked with their eyes closed. Parental care includes **nursing**—the provision of milk—by the mother until the young are sufficiently mature to eat solid food. Rats are omnivorous—they eat plant and animal materials, live, dead, or decaying.

Most, if not all, vertebrates prefer certain environmental temperatures. Only birds and mammals, however, are capable of regulating their internal body temperatures within a narrow range. They are said to be **endothermic** (in contrast to **ectothermic** animals, which must to some degree conform to the temperature of the environment in which they live). In mammals, hair aids in reducing heat loss by trapping a shell of air around the body, thus providing an effective insulating blanket. Insulation is also one of the functions of the feathers of birds.

⁞⁞⁞⁞⁞ Procedure ⁞⁞

1. Examine a preserved rat. Note that the major divisions of the body are the **head, neck, trunk, and tail** (Figure 33A-4). The trunk is divided into the **thorax** in the region of the rib cage and the lower or more posterior **abdomen** behind the ribs.

2. Locate the **anus** beneath the tail. Note that most mammals do not have a cloaca—the digestive system opens separately from the urogenital ducts of the excretory and reproductive systems. If your animal is a male, find the large **scrotal sacs** at the base of the tail and the **penis** in front of the anus. If your specimen is a female, locate the **vagina** in front of the anus.

    a. *What is the sex of your rat?* _____

3. Find the paired **external nares** at the tip of the snout. These lead to a series of chambers (including one in which olfactory sensory cells are located) above a **palate,** the roof of the mouth cavity, separating the nasal passageways from the mouth cavity anteriorly. The **internal nares** open into the pharynx behind the mouth cavity. Feel the bony palate.

    b. *What might be the advantage of this partitioning of the mouth cavity?*

    _____

4. Examine the mouth. Part the lips and examine the teeth. Cut through the angle of the jaw on the left side and open the mouth to see the teeth. Note that the large teeth at the front of the

Figure 33A-4 *External anatomy of the rat.*

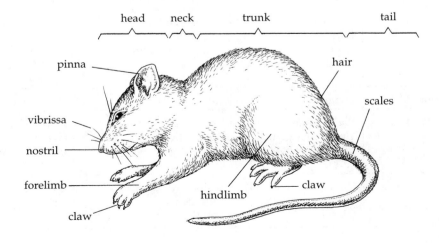

mouth, the incisors, are chisel-like and modified for cutting. More posterior teeth are used in grinding. In most submammalian vertebrates, the teeth are all similar or **homodont.** The type of dentition that includes several distinct types of teeth is called **heterodont.** Modifications of mammalian teeth are related to the types of food eaten in each group. The differences in types of teeth and the numbers of each type are distinctive enough to be used to classify mammals.

c. *Where are the teeth of the rat located with respect to the margins of the mouth?*

_____

d. *How does the positioning of teeth differ in the rat and the shark?* _____

_____

e. *Are the teeth located in bony sockets in rats?* _____ *In sharks?* _____

5. Examine the eyes. Note that movable upper and lower eyelids are present. Can you find a remnant of the reduced **nictitating membrane** in the inner corner of the eye? The **external ears** are associated with a flap of skin, the **pinna,** which can be moved in different directions to pick up sounds selectively. These sounds travel through the canal of the **outer ear** to the tympanic membrane bounding the **middle ear,** which is recessed. (Where is this membrane in the frog?) A series of three small bones transmits vibrations to the **inner ear.** Note the sensory hairs, called **vibrissae,** on the flexible snout.

6. Study the **forelimbs** and **hindlimbs.** Note the **claws** found on each.

f. *How many digits are on each forelimb?* _____ *Hindlimb?* _____

7. Examine the skin on the body and tail. Note the scalelike epidermis of the tail and the relation of these scales to hair. Is there a pattern to their association? (One theory holds that in the

Table 33A-1   Features of Representative Vertebrates

Feature	Shark	Frog	Turtle	Rat
Habitat	Aquatic	Transitional (aquatic larva)	Secondarily aquatic	Terrestrial
Body divisions	Head, trunk, tail	Head, trunk	Head with neck; trunk (in "shell")	Head with neck; trunk, tail
Digestive system	Mouth, cloaca	Mouth, cloaca	Mouth, cloaca	Mouth, anus
Teeth	Homodont	Homodont	Absent	Heterodont, in sockets
Respiratory system	Internal gills	External gills (larva) Lungs, skin (adult)	Lungs	Lungs
Sense organs	External nares Eyes No middle or inner ear	Nares to mouth Eyes Tympanum (middle ear), inner ear	Nares to mouth Eyes Tympanum (middle ear), inner ear	Nares to pharynx Eyes External, middle, and inner ear
Locomotion	Swim (fusiform body)	Jump, swim	Swim, walk	Walk, run
Integument	Placoid scales Unicellular glands	Moist, vascularized Multicellular glands	Horny skin Glands reduced (scent glands)	Hair Sweat and sebaceous glands; mammary glands
Reproduction	Viviparous, internal fertilization (clasper)	Oviparous, external fertilization (amplexus)	Oviparous, internal fertilization (penis)	Viviparous, internal fertilization (penis)

evolution of mammals, hair first appeared as sensory structures associated with epidermal scales. The hair on the tail of rats is too sparse to serve as insulation; instead, it continues to serve a sensory function.)

SUMMARY

Working with your partners, study Table 33A-1. Ask the other members of your group to show you the major features of their organisms and to teach you the important aspects of the reproduction and the life history of those organisms.

We are now ready to begin our study of the organ systems of the four representative vertebrates. We will begin by examining the integumentary system (skin) in this laboratory.

✔ **EXERCISE B   Integument—A Dynamic Interface**

The outer tissue layers of vertebrates form the skin or integument which provides an interface between the internal and external environments. The integument functions in the following ways:

Protection

Mechanical   Provides a barrier against abrasion, injury, attack, parasites, pathogens.

Chemical   Forms a buffering barrier to passage of water and ions; can also include the production of mucous and other secretions.

Visual   Provides pattern or color for camouflage or warning, for attraction and communication.

Light   Intercepts ultraviolet radiation and prevents its penetration into deeper layers of the body.

Locomotion   Ciliary movement of aquatic larvae; transmission of the physical force required for movement; scales, pads, or claws provide attachment to surfaces.

Respiration   Gas exchange between external environment and blood vessels close to the surface in some species.

Secretion   Glandular secretions modify the interface with the environment (e.g., a mucous coat "streamlines" certain fishes); may provide mechanical protection or may be toxic; may provide species identification or function as behavioral cues (pheromones).

Excretion   Ammonia and simple wastes diffuse across the integument in some vertebrates.

Water uptake   Some amphibians can take up water through their skin.

Synthesis   Formation of vitamin D (with sunlight).

Heat exchange   Heat gain or loss affected by the presence of pigments and by other epidermal structures such as sweat glands, hair, feathers.

Energy storage   Fat deposits.

Monitoring the environment   Sensors (receptors).

Behavior   Effectors include muscles, glands, and pigment cells.

The skin of vertebrates is composed of two layers, the outer **epidermis** and the deeper **dermis.**

EPIDERMIS

The deepest layer of the epidermis, the basal layer, usually consists of a single layer of cells. It is this layer of the epidermis that is most mitotically active; the new cells produced in the basal layer are pushed up through the other layers of the skin to replace cells worn away or shed from the surface.

In many aquatic organisms such as the shark, the epidermis is relatively thin and all of its cells are living (Table 33B-1). In vertebrates that made the transition from water to land, the epidermis is thicker

Table 33B-1   Comparison of the Skins of the Vertebrates Studied in This Laboratory

Trait	Shark	Frog	Turtle	Rat
Epidermis	Living, thin	Dead, thin	Dead, thick	Dead, thick
Specializations	None	Some keratin	Highly keratinized epidermal scales (scutes)	Hair, composed of keratin
Glands	Unicellular mucous glands	Multicellular mucous and granular glands	Few glands	Sebaceous, sweat, and mammary glands
Dermis				
Chromatophores	Present	Present	Present	Present; melanophores invade epidermal structures
Dermal bone	Absent	Reduced, becomes part of skeleton	Redeveloped, as dermal plates of carapace and plastron in association with skeleton	Reduced, becomes part of skeleton
Dermal scales; teeth	Placoid scales; teeth	No dermal scales; small teeth	No dermal scales or teeth	No dermal scales; teeth in sockets

and cells die and become infiltrated with a horny, proteinaceous material, **keratin,** as they move toward the surface (away from sources of nourishment). This process of keratin deposition is called **cornification;** it reduces the permeability of the skin to water and protects the organisms from desiccation. The outer **cornified layer** of the skin of terrestrial animals is composed of dead, cornified cells.

Life on land led to further modification of the epidermis, including specialized, highly keratinized structures such as the **epidermal scales** of reptiles and birds, the **feathers** of birds, and the **hair** of mammals. Feathers and hair trap an insulating layer of air around the body and are important in the evolution of **endothermy** in birds and mammals.

Within the epidermis, glands perform a variety of functions. In aquatic vertebrates, numerous **mucous glands** secrete the viscous mucus that protects the surface of the epidermis and reduces the friction of the body in water. Mucus may also retard the loss of water to the environment. In addition to mucous glands, amphibians also have **granular glands,** which produce a variety of watery secretions—some of them toxic (Indians of the Amazon basin use toxins from amphibians in making their poison arrows). Animals with a fully terrestrial life style generally have fewer epidermal glands—in part, to prevent the loss of water in secretions. Reptiles, for example, have only a few glands, which function in species recognition, reproduction, or defense. In birds, a single oil gland at the base of the tail is used to condition feathers. The major epidermal glands of mammals include sebaceous glands, two types of sweat glands, and mammary glands. **Sebaceous glands** are associated with hair follicles and produce oily secretions (**sebum**) that soften and condition the hair and skin surface. **Eccrine sweat glands,** found in humans and some other mammals, produce watery secretions (sweat) important in evaporative cooling. These glands open onto the surface of the skin between hairs. **Apocrine sweat glands** are associated with hair follicles; in humans, these glands are found in the armpits and pubic region but are not involved in thermoregulation. Apocrine secretions contain cellular debris which, when decomposed by bacteria, produces characteristic odors. **Mammary glands** are specialized integumentary glands found only in mammals; these glands produce milk for the nourishment of young.

DERMIS

The dermis is separated from the epidermis only by a thin basement membrane secreted by the basal layer of the epidermis. The dermis is composed of connective tissue, mainly collagen and elastin fibers in a gel-like matrix. In addition, the dermis contains blood vessels, nerves, sensors, and fat (adipose tissue).

Pigment cells, **chromatophores,** are a conspicuous element of the dermis. In most vertebrates, the most common type of pigment cell is a **melanophore** containing the brown pigment melanin. Some vertebrates also have lipophores containing red, orange, or yellow pigments. Chromatophores are concentrated in the upper layer of the dermis and, in endotherms, portions of these cells sometimes bud off and invade epidermal structures to provide a fixed pigment pattern to feathers and hair. However, in a variety of ectothermic vertebrates, chromatophores are capable of changing their size and shape and thus the pattern of coloration. If all of the melanophores, for example, are contracted so that their pigment (melanin) is highly concentrated in small, widely separated dots, colors and patterns provided by other pigment cells and structures are evident. When the melanin is dispersed to cover a wide area of the skin, however, the color is darkened and other colors and patterns are hidden. These changes may be mediated directly by light, indirectly by endocrine changes, or rapidly through nerve impulses—thus qualifying these cells as effectors (producing an externally directed response), like muscles. Color changes may assist an organism in absorbing radiant energy from the sun, in hiding from enemies, or in communication.

DERMAL BONE: EVOLUTIONARY TRENDS

Normally, we do not think of bone as part of the skin, but in many of the earliest jawless fishes, heavy plates of bone, formed by the direct deposition of calcium salts within the dermis of the skin, provided an external armor. In contrast, bones forming the internal skeleton (the backbone or vertebral column and base of the skull) first developed as cartilage which was replaced from within by bone during maturation. Thus, two types of bone were found in early vertebrates: **dermal bone** formed in the skin and **endochondral,** or **replacement, bone** formed by the replacement of cartilage to provide an endoskeleton.

In the earliest jawed fishes, the dermis produced **dermal scales** composed of layers of dermal bone overlaid by a layer of dentine-like material (also derived from the dermis) and an enamel-like layer applied to the surface, possibly by the epidermis. As fish evolved, several of these layers became modified or reduced. In sharks, dermal bone is largely absent and only the outer layers of dentine and enamel persist in the placoid scales. The scales of bony fishes are thin disks of bone that develop in overlapping dermal folds of the skin. In bony fishes, dermal bone also becomes incorporated into the skeleton as part of the skull and the pectoral girdle—the only remnants of dermal bone that persist in terrestrial animals. There are some exceptions, however; for example, in turtles, dermal bone forms part of the carapace (shell) and also becomes part of the skeleton.

As you examine the four representative vertebrates being studied by your group, be sure to notice the following adaptive trends associated with the phylogenetic spectrum from jawless fishes to endotherms: (1) a general reduction in the importance of bony elements of the dermis and (2) a corresponding increase in the diversity and importance of epidermal structures (Table 33B-1).

ⅠⅠⅠⅠⅠ **Objectives** ⅠⅠⅠⅠⅠⅠⅠⅠⅠⅠⅠⅠⅠⅠⅠⅠⅠⅠⅠⅠⅠⅠⅠⅠⅠⅠⅠⅠⅠⅠⅠⅠⅠⅠⅠⅠⅠⅠⅠ

☐ Describe the functions of skin.

☐ Trace the development of dermal and epidermal structures in the integument in the four representative vertebrates.

☐ Explain the adaptive roles of epidermal and dermal structures.

ⅠⅠⅠⅠⅠ **Procedure** ⅠⅠⅠⅠⅠⅠⅠⅠⅠⅠⅠⅠⅠⅠⅠⅠⅠⅠⅠⅠⅠⅠⅠⅠⅠⅠⅠⅠⅠⅠⅠⅠⅠⅠⅠⅠⅠⅠⅠ

Shark

1. If available, study a prepared whole-mount slide of shark skin using the 10× objective of your microscope. Alternatively, use a scalpel to make a small incision through the skin of your shark. Lift one edge of the cut and separate a small section of skin from the underlying muscle. Mount this skin on a microscope slide for study. (Scales can be seen more easily if the skin is soaked overnight in glycerine and then mounted—your instructor may provide material for you to study.)

2. Recall that the shark's epidermis is made up entirely of living cells with few specializations.

a. How is the shark's skin different from that in the other organisms being studied?

_____

b. Where do new epidermal cells originate? _____

Gland cells are present and produce mucus, but sharks do not develop a "mucous cuticle," a streamlining cover found over the epidermis in some other aquatic vertebrates.

3. The dark color of the skin of sharks is due to pigment cells, **melanophores,** containing a dark brown-black pigment, melanin. Can you find the melanophores?

4. Note the regularly spaced dermal scales, often called denticles or "little teeth" because their structure resembles that of a tooth. These are the **placoid scales** (Figure 33B-1).

5. If available, examine the slide of fish skin using the 10× objective.

c. Describe the origin of the scales and the relationship of the scales to the dermis and the epidermis.

_____

_____

d. How do the scales of other fishes differ from the placoid scales of the shark?

_____

_____

Figure 33B-1 *The integument of sharks. Cross section through the skin and placoid scales. Cells within the epidermis form an enamel organ which induces a thickening in the underlying dermis. Each scale has an inner dentine layer, produced by mesenchymal cells of the dermis, and an outer enamel layer. The origin of the enamel is unclear, but possibly is secreted by epidermal cells. Placoid scales develop similarly to the teeth of higher vertebrates; the enamel contains the same proteins found in mammalian teeth.*

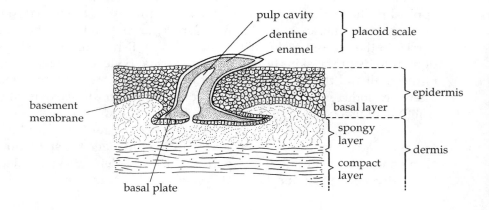

Frog

1. Obtain a prepared slide of a cross section through the skin of an adult frog or other amphibian. Study it under the 10× objective of your microscope. Identify the epidermis and the basal layer. In contrast to the epidermis of the shark, the outer layers of cells in the frog's epidermis are dead and provide a protective covering. The frog's epidermis is generally only five to eight cells thick (Figure 33B-2).

a. Describe the process of keratinization in the formation of the outer cornified layer of the epidermis.

_____

Figure 33B-2   *Section of the skin of a frog.*

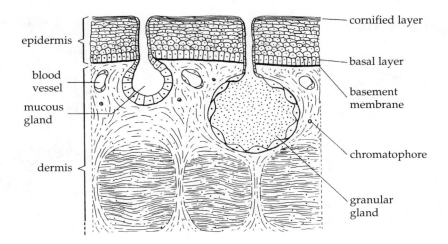

epidermis

blood vessel

mucous gland

dermis

cornified layer

basal layer

basement membrane

chromatophore

granular gland

2. Study the dermis in the prepared slide of frog skin. Identify the multicellular **mucous glands** and large **granular glands,** both epidermal structures that have grown into the dermis. (Curiously, in the Surinam toad, granular glands produce secretions that nourish developing tadpoles held in skin pockets on the backs of females.)

   b. *What are the functions of mucous glands in most adult amphibians?*

   _____

   Of granular glands? _____

3. The dermis also contains muscles, pigment cells, blood vessels, nerves and nerve endings, a variety of connective tissue cells and fibers, and other structures. Do you find any trace of dermal scales? (Remnants of dermal scales are present in some amphibians such as caecilians and a few anurans.)

   Because the skin of the adult frog is thin and richly supplied with capillaries, oxygen can diffuse through the body surface and into the blood supply. Thus a frog's skin serves as an accessory respiratory organ. (In the evolution of one line of salamanders, the plethodonts, lungs have been lost and all gas exchange must occur across the skin and the lining of the mouth cavity.) On the other hand, thin amphibian skin does little to protect the occupant from desiccation, and adult amphibians are generally restricted to humid environments. Some remain aquatic.

   c. *In what parts of the world would you expect to find the greatest amphibian diversity?*

   _____

Turtle

1. Recall that the large keratinized epidermal scales (**scutes**) that make up the outer covering of the carapace and plastron of the turtle are derived from the epidermis. Now observe the skin on the legs of the turtle.

   a. *How do the epidermal scales on the appendages differ from those covering the carapace?*

   _____

   b. *How does the epidermis of the turtle differ from that of the frog?* _____

   _____

   Increased keratinization of the epidermis in reptiles markedly retards water loss from the body and is a major factor in the adaptation of reptiles, birds, and mammals to land. In reptiles, the epidermis is

Figure 33B-3 *A turtle skeleton showing elements of the axial and dermal skeletons forming the carapace.*

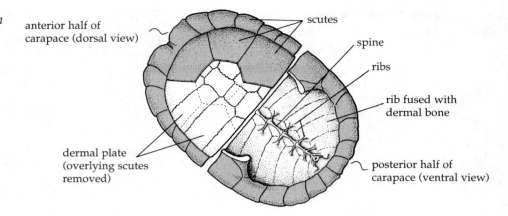

anterior half of carapace (dorsal view)

scutes

spine

ribs

rib fused with dermal bone

dermal plate (overlying scutes removed)

posterior half of carapace (ventral view)

generally thin and consists, as in amphibians, of a basal layer where cells multiply and successive layers where keratin is added. However, the outer layer of dead cells, the **cornified layer**, is much thicker and is less permeable to water than the outer layer of the skin in frogs. In reptiles this layer may be shed (molted) regularly (as in snakes and lizards) or retained as growth occurs (as in most turtles and alligators), giving the skin its characteristic "scaly" texture. The retention of the cornified layers results in a gradual increase in scute thickness and this may be used to estimate age in some turtles. Recall that terrestrial vertebrates have lost most of their skin glands in the course of evolution.

c. *Why has the move from an aquatic to a terrestrial environment caused a reduction in the number of skin glands?*

_____

_____

The epidermal scutes and underlying plates of dermal bone that form the carapace and plastron of the turtle both develop from the skin. Thus, the "shell" of the turtle is really a part of its skin; it is continuous with the skin covering the head, tail, and appendages. (Contrary to the animated liberties taken by cartoonists, turtles cannot leave their shells!) Dorsally, the dermal elements of the carapace fuse with the endochondral bone of the vertebral column and ribs of the endoskeleton.

2. Study the skeleton on demonstration to identify the bony dermal elements of the carapace and plastron (Figure 33B-3).

   d. *Do the shapes and locations of dermal bones forming the carapace match the edges of the epidermal scutes seen on your specimen?* _____

   e. *How are the development and structure of the bony dermal elements in the turtle different from the development and structure of scales in fish?* _____

   _____

   _____

As in sharks and frogs, **chromatophores** (pigment cells) are found in the turtle's dermis, where they produce the specific patterns and colors of the skin.

Rat

1. Observe a cross section of human skin (or other mammalian skin) using the 10× objective of your microscope. The outer epidermis is thick and cornified to reduce water loss, but does not form the keratinized scales present in reptiles.

Several epidermal specializations are found in mammals. Epidermal hairs, growing down into the dermis, are associated with smooth muscles that can change the orientation of the hair shaft, thus

Figure 33B-4 *Schematic drawing of a section of human skin. The integument of the rat is similar, but contains more hair follicles and lacks eccrine sweat glands. Apocrine sweat glands, not involved in evaporative cooling of the warm body surface, are associated with hair follicles in rats and are found in the axillary and pubic regions of humans.*

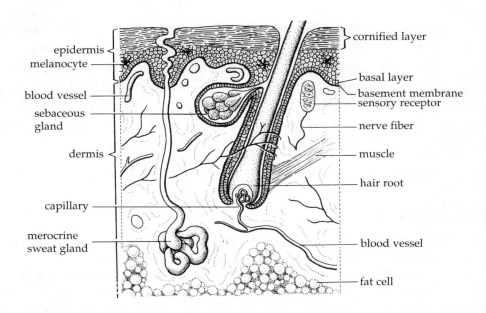

epidermis
melanocyte
blood vessel
sebaceous gland
dermis
capillary
merocrine sweat gland

cornified layer
basal layer
basement membrane
sensory receptor
nerve fiber
muscle
hair root
blood vessel
fat cell

changing the thickness of the layer of air trapped by the coat of hair. (In humans, "goose bumps" are a remnant of this thermoregulatory response—ineffectual because during the course of evolution we have become much less hairy than our ancestors.)

2. Epidermal glands are another specialization. Refer to Figure 33B-4 and identify the **sebaceous glands.**

   a. *What is the function of sebaceous glands?* _____

   _____

3. If you are examining a cross section of human skin, locate a **eccrine sweat gland** (Figure 33B-4).

   b. *What is the function of the eccrine sweat glands?* _____

   _____

   Recall that these glands are not found in rats. In rats, cooling is accomplished by evaporation of water from the mouth and respiratory tract. Urine may also be used to moisten the fur and legs to facilitate heat loss.

4. Find the external openings, the **nipples,** of the mammary glands of your rat. These specialized integumentary glands are found along each side of the body.

   c. *How many nipples are there on your rat?* _____ *Why do you suppose rats have more nipples than humans?* _____

   _____

Other keratinized, epidermal structures of mammals include the **horns** of cattle (but not the antlers of deer, which have a bony core covered with skin), **baleen** of some whales (plates in the mouth cavity that filter food—tiny plankton—from the water), horny **scales** (in the pangolin and armadillo), and **nails, claws,** and **hooves.**

The dermis of mammals contains connective tissue cells and fibers; muscles, nerves and sensors; blood vessels and capillaries; glands and hair follicles extending down from the epidermis; and, toward its base, a layer of fat cells that becomes quite thick in aquatic mammals such as seals and whales. In water, it is difficult for hair to retain a trapped layer of air to reduce heat loss, so, in the most advanced mammalian aquatic organisms, the coat of hair has been reduced and thick layers of fatty tissue have been substituted as an insulating layer.

  5. Study your slide. Identify the epidermal hair follicles and glands that invade the dermis. Locate muscles that attach to the hair follicles. Find connective tissue elements. Examine the innermost layers of dermis on your slide. Can you find fat cells?

  d. *Why is fat tissue generally a less efficient thermal insulator than hair? (Hint: Which tissue is living and vascularized?)* _____

_____

SUMMARY

Review Table 33B-1 on page 33-14. Be sure you understand the phylogenetic trends in the integument of the representative vertebrates dissected in this laboratory.

   Recall that the vertebrates studied in this lab show several specializations overlying general phylogenetic trends. Thus, dermal bone has been lost almost completely in the shark, but some other fishes retain some bone in their scales; it also reappears in the turtle. Nevertheless, the trend is toward restriction of dermal bone to specific sites—the skull and pectoral girdle—in the phylogenetic transition from aquatic to terrestrial organisms.

a. What phylogenetic trends are illustrated by epidermal structures in your representative vertebrates?

_____

_____

_____

Laboratory Review Questions and Problems

  1. Use anatomical terms for position or direction to describe the locations of the following vertebrate organs:

  Mouth

  Toes of a pig (in relation to the limb)

  Toes of a human (in relation to the limb)

  Tail

  2. List the organ systems of vertebrates. Given an example of at least one organ contained in each.

3. Compare and contrast integumentary structures in the shark, frog, turtle, and rat. What are the characteristics of the skin in each animal?

What are the major similarities of the skin in these animals?

What are the major differences in the skin of these animals?

4. How do the different structures in the skin of the four vertebrates studies in this lab adapt each animal to its major environmental factors?

Shark

Frog

Turtle

Rat

5. Why is it advantageous to terrestrial vertebrates to have a nonliving outer layer of epidermis in contact with the environment?

6. Birds and mammals use feathers and hair, respectively, to trap an insulating layer of air around the body, retarding heat loss and assisting in the maintenance of a relatively constant internal body temperature (endothermy). In other living vertebrates, body temperature is ultimately dependent upon that of the environment (ectothermy). What are the advantages of endothermy? Of ectothermy? What are the disadvantages of endothermy? Of ectothermy? How would these advantages and disadvantages be modified in animals of increasing or decreasing body size (mass)?

COMPARATIVE ANATOMY AND CONCEPTS OF VERTEBRATE EVOLUTION

Now that you have been introduced to the series of dissections of the four representative vertebrates, you should keep in mind that studies such as the one you are undertaking introduce many terms with which you may be unfamiliar. You are <u>NOT</u> expected to memorize these terms. Rather, they provide a "vocabulary" that enables you to locate various structures by using both written descriptions and diagrams. If you wish to develop your understanding of the vocabulary of biology, Appendix V—Key to Common Roots, Prefixes, and Suffixes—at the end of the laboratory manual will help you to understand the meanings of biological terms and relationships among terms. As you study the vertebrates, focus on the following concepts of vertebrate evolution rather than on details.

- Dermal structures become reduced and epidermal structures diversify.
- Major sensors remain similar in structure but, with the transition from water to land, the lateral line system for "distant-touch" disappears and some of the structures of the pharynx become modified for hearing.
- The forebrain increases in size and importance; in mammals, both optic and auditory sensory information are projected to the forebrain rather than to the midbrain.
- Muscles and bones work together to produce movement which becomes more complex with the development of limbs.
- The coelom becomes partitioned to separate the heart and lungs from the general body cavity.
- The digestive system is modified to accommodate increasing complexity in diet.
- Swim bladders and lungs replace gills as the major respiratory organs, facilitating the transition to land.
- The heart is modified to separate pulmonary circulation from general body circulation in terrestrial vertebrates.
- Posterior elements of the kidney become increasingly important in urine formation, and anterior elements become associated with the male reproductive system.
- Developing young tend to be retained within the mother or within an enclosed egg.

The Anatomy of Representative Vertebrates: Behavioral Systems

OVERVIEW

In this series of laboratories (33–36), we use four representative vertebrates to illustrate evolutionary or phylogenetic trends within the subphylum Vertebrata. To better understand these trends it is important to determine whether an anatomical feature that seems structurally and functionally similar in different animals can be traced to a common ancestral origin. Since the highest degree of similarity between structures is often observed during embryonic development, before adult specializations have obscured fundamental features, we can study the evolutionary origins of similar structures by comparing their embryological origins. Structures that show fundamental embryological or developmental similarities, suggesting a common evolutionary origin (or common genetic heritage), are said to be **homologous.**

In constructing phylogenies, it is helpful to determine the sequence in which homologous characteristics originated in the evolution of a particular group of organisms. When only two groups share a characteristic not present in other groups at the same level of classification, they are said to have "recently" diverged from a common ancestor. The shared characteristic is called a **derived** (or **advanced**) **character.** The characteristic that gave rise to the derived character evolved earlier in the lineage and is called an **ancestral** (or **primitive**) **character.*** In using these characters to compare groups, we must keep in mind that "ancestral" and "derived" are relative terms and must be thought of in relation to an organism's position on the phylogenetic tree (see Figure 27-II). For example, the tympanic membrane and middle ear of a frog are *derived* when compared with the open spiracle (first gill slit) of the dogfish shark, but *ancestral* when compared with the recessed tympanic membrane separating the outer and middle parts of the ear in the rat.

In this laboratory you will study anatomical homologies that help us to understand the evolution of **behavioral systems.** Organs and organ systems constituting the behavioral systems include the **sensors** (affectors) discussed in Exercise A, the **control systems** (the nervous and endocrine systems) described in Exercise B, and the **effectors,** systems that produce externally directed activities (behavior), outlined in Exercise C.

STUDENT PREPARATION

Prepare for this laboratory by completing Laboratory 33 and reading the text pages indicated by your instructor. Familiarizing yourself in advance with the information and procedures covered in this laboratory will give you a better understanding of the material and improve your efficiency.

**Since the terms "primitive" and "advanced" carry judgmental connotations that do not apply to their technical meanings, it is best to use the terms "ancestral" and "derived" when possible.*

If dissection tools are not provided, bring your dissecting kit to laboratory with you. Do not wear contact lenses to this laboratory.

You will *not* dissect all four representative vertebrates—the behavioral systems are complicated and some are covered in other laboratories. Instead, you will dissect one representative vertebrate selected to illustrate anatomical features of the behavioral systems. Two members of your group should work through Exercises A and B using the shark, while the two other members move on to Exercise C using the rat. Be sure to read the introductory material for each exercise and review all the material covered in the assigned exercises by sharing observations within your group.

✔ **EXERCISE A**   **Sensors (Affectors)**

Sensors continuously monitor the environment and are ready to produce signals in response to changes in the surroundings. There are a variety of sensors: those that sample the external environment, those that are in contact with the internal environment, and those that receive information from chemicals, light, and mechanical sources, including vibrations and the position of the organism and its parts. In early vertebrates, special sensors—the olfactory epithelia, the eyes, and the ears—developed in close association with the central nervous system. In fishes and aquatic amphibians, the lateral-line system is integrated with the ear (acoustico-lateralis system) and provides information about surrounding currents, body movements, low-frequency sounds, and, in some, the electromagnetic fields around the organism. The sensors of the lateral-line system are functional only in a dense, aqueous medium. In animals that made the transition to land, this system is not functional and is lost. Terrestrial animals evolved a variety of sensors distributed over the body surface and among the internal organs to monitor the position of body organs, touch, pressure, temperature, and other features of the changing environment. (In Laboratory 39, you will explore many of these sensors in your own body.)

Many of the types of vertebrate sensors that monitor the external environment are listed in Table 34A-1. Study this table to note some of the phylogenetic trends as you proceed with your dissections.

✔ **PART I**   **The Lateral-Line System**

The lateral-line system and inner ear form the **acoustico-lateralis system.*** Sensors of the acoustico-lateralis system, the **neuromasts** (Figure 34A-1a), are located over much of the body surface and are associated with the **lateral-line canal** (Figure 34A-1b) in all aquatic vertebrates. This canal is located between the dorsal and ventral muscles of the body and has several branches that extend into the head. The neuromasts of the lateral line are composed of organs containing "hair cells" with small sensory cilia that are stimulated by bending. These cilia are embedded in a covering cap which is deflected by contact or by currents produced by movements in the surrounding water. The neuromasts act as **mechanoreceptors,** providing information about mechanical forces acting on the surface of the body. The neuromasts also provide information about the position of various parts of the body, serving the function of **proprioceptors,** specialized sensors in tetrapods that signal the position of individual bones and muscles. These same structures may also sense low-frequency vibrations or "sounds" in the water, acting as **phonoreceptors.** Specialized neuromasts found on the head (ampullary organs) also act as **electroreceptors,** recording the electrical patterns produced by the muscles of the fish and by surrounding objects.

Acoustico-lateralis" implies an association with hearing (acoustico-). Actually, the ancestral inner ear structure of aquatic vertebrates senses the position and change in position of the head and has only a limited ability to detect very low-frequency (sound) vibrations transmitted through the water. Hearing becomes important in some bony fishes but is best developed in terrestrial vertebrates in association with accessory structures of the middle (and outer) ear. It is, therefore, preferable to refer to this system as the octavo-lateralis system because of the involvement of the eighth cranial nerve (octavo-) in both static and acoustic sensory systems. Sensory input from the lateral-line sensors is carried by fibers in adjacent cranial nerves (seven, nine, and ten), so the compound name "octavo-lateralis" remains appropriate. We have elected, however, to retain the more familiar name, acoustico-lateralis, for this system.

Table 34A-1   Vertebrate Sensors

Receptors	Sense	Organ	Location	Phylogenetic Trends
Chemoreceptors	Smell (distant sources)	Olfactory epithelia	Nares	Sensory epithelia, associated with shallow, surface pits in ancestral fishes, extend inward to connect with the mouth.
	Taste (contact sources)	Taste buds	Body surface, mouth	Concentrated on the head and mouth in fishes; confined to the mouth cavity in terrestrial vertebrates and to the tongue in mammals.
Photoreceptors	Vision	Eyes	Head (lateral surface)	Protective lids develop in fishes; glands moisten and lubricate the surface in terrestrial vertebrates.
	Time	Median eyes (pineal, parapineal glands)	Head (dorsal surface)	Dorsal "eyes" develop in the earliest jawless fishes; lose sensory function in mammals.
Mechanoreceptors				
Phonoreceptors	Hearing	Neuromasts; organ of Corti (lagena)	Inner ear	Hearing becomes much more important in terrestrial forms; organ of Corti becomes elongated.
Statoreceptors	Position	Neuromasts	Inner ear	Several different bands of neuromasts are found in the representative vertebrates.
	Acceleration	Semicircular canals (and neuromasts)	Inner ear	All vertebrates have three canals (except lampreys and hagfishes).
	Distant-touch	Neuromasts	Lateral line	Present only in aquatic vertebrates.
Pressure and touch receptors	Contact stimuli	Dermal sensors	Skin	Develop in terrestrial animals; replace lateral-line sensors
Electroreceptors	Electromagnetic fields	Ampullary organs	Lateral line	Well developed in fishes with electric organs.
		Dermal sensors	Mandible	Found in one mammal, the platypus.
Thermoreceptors	Temperature	Free nerve endings	Skin, brain	Free nerve endings in terrestrial vertebrates.
Proprioceptors	Organ position	Tendon organs, muscle spindles, other encapsulated sensors	Muscles, joints	Appear in fishes; replace lateral-line sensors in terrestrial vertebrates.
Nociceptors	Pain	Free nerve endings	Most organs	Absent in nervous tissues; nature of pain obscure in fishes.

Figure 34A-1 *(a) A neuromast organ of the lateral-line system. (b) The lateral-line canal and neuromasts in a bony fish.*

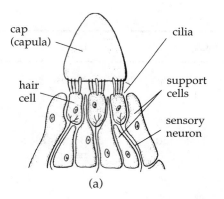

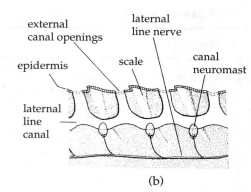

(a)   (b)

||||| **Objectives** ||||||||||||||||||||||||||||||||||||||

☐   Locate components of the lateral-line system and understand their sensory roles.

||||| **Procedure** ||||||||||||||||||||||||||||||||||||||

1. Find the light-colored, fine line along the side of the body of your shark. This line indicates the position of the lateral-line canal which lies beneath it. This canal, plus a complex of canals on the head, contains sensors of the lateral-line system, the neuromasts. (Individual neuromasts are also widely distributed over the body but cannot be seen without a microscope.)

    a. *The lateral-line system has been described as giving an animal a sense of "distant-touch." What does this mean?* _____

    _____

2. Push on the skin of the head. You should be able to find patches of pores from which a shiny, jellylike substance exudes under pressure. These are the openings of the **ampullary organs of Lorenzini,** specialized sensors of the lateral-line system that are sensitive to mechanical stimulation and weak electrical fields. They are used in the detection of prey.

    b. *Can you think of other uses for electroreceptors besides the detection of nearby objects?* _____

    _____

✔   **PART 2**   **The Inner Ear**

The inner ear is a second component of the acoustico-lateralis system. In fishes, the inner ear is not specialized as an organ of hearing (phonoreception), but is primarily an organ associated with sensing changes in position of the organism (or its head)—a form of mechanoreception called **statoreception** (the ancestral function of the inner ear). Only in terrestrial vertebrates, which receive sound waves propagated in the air, does the inner ear develop specializations for hearing.

Recall that the ear of a mammal consists of three parts: the outer, middle, and inner ears. Embryologically, the outer and middle ears of mammals are derived from the first gill slit, the spiracle, of their ancestors. In sharks, the ancestral pattern persists: the spiracle still connects the pharynx with the exterior. Frogs and turtles have only a middle and an inner ear—the tympanic membrane bounding the middle ear is flush with the surface of the head. In mammals, the tympanic membrane is recessed from the surface, thus forming an outer ear, which channels sound waves to the tympanic membrane. Sound waves in the air are transduced by the tympanic membrane of terrestrial vertebrates into mechanical vibrations that are transmitted to the inner ear via a bone (or bones) of the middle ear.

The inner ear is a membranous sac, the **membranous labyrinth,** filled with a fluid called **endolymph.** The membranous labyrinth is embedded in the base of the skull surrounded by a **bony labyrinth**—the membranous labyrinth is like a cast within a mold.

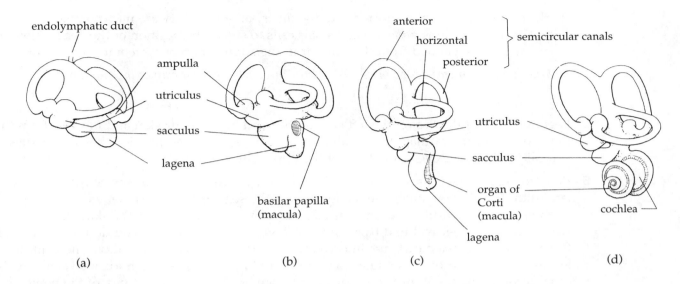

Figure 34A-2 *Membranous labyrinths of (a) shark, (b) turtle, (c) bird, and (d) mammal. The inner ears of all of these vertebrates have three semicircular canals as part of the utriculus. All also possess one or more clusters of sensory hair cells within the utriculus, within the sacculus proper, and within the lagena, a saclike extension of the sacculus. In birds and mammals, the macula within the lengthening lagena is called the organ of Corti and is associated with hearing. Note that the lagena becomes coiled in mammals, increasing the total length of the duct and, possibly, the auditory acuity of the ear.*

The membranous labyrinth is divided into two parts: the vestigial endolymphatic sac which, in sharks, remains connected to the surface of the body by a duct, and the **utriculus** and the **sacculus** (Figure 34A-2a). The most obvious parts of the utriculus are three **semicircular canals** or **ducts**, which lie at right angles to each other and at an angle of 45° to the body axis. Each canal has an expanded region, the **ampulla,** near its base, and projecting into each ampulla is a typical neuromast organ (**crista**) which can be deflected by movements of the endolymph within the canal. When the head moves, fluid flows within the labyrinth (*why?*) and differentially bends the hair cells within each of the semicircular canals, thus providing the organism with a perception of the direction of its movements.

The utriculus and sacculus also contain several neuromast organs in which the hair cells are arranged in small clusters (**cristae**) or broader bands (**maculae**). The caps of these neuromasts are impregnated with calcium salts, forming "ear stones" or **statoliths.** These statoreceptors respond to the force of gravity and provide the organism with a sense of "up" and "down." (Statoliths can also be affected by low-frequency vibrations and may serve as crude sound receptors, or **phonoreceptors.**)

In most fishes, there is a short ventral extension (**lagena**) of the utriculus; with the transition to land, this elongated to form the **cochlear duct** found in birds and mammals. Within the cochlear duct, hair cells form the **organ of Corti,** an elongate macula, (Figure 34A-2c, d), the principal sensory structure involved in hearing.

||||| **Objectives** |||||||||||||||||||||||||||||||||||||||

☐ Dissect the inner ear of the shark and learn its role in determining position.

||||| **Procedure** ||||||||||||||||||||||||||||||||||||||

1. Examine the dorsal surface of the head of your shark along the midline between the two spiracles (use a magnifying lens if necessary). Locate the small pair of **endolymphatic pores** along the midline. These pores open into the endolymphatic ducts of the inner ear and enable

sharks to take small grains of sand into the inner ear, where they are incorporated into statoliths. The statoliths are coupled to hair cells to signal the position of the head. (Note that the inner ear is separated from the exterior in the adults of other vertebrates.)

   a. What material increases the mass of statoliths in vertebrates that cannot take up sand?

   _____

2. Beginning *behind* the eye on the left side of the head, remove the skin and muscles from the top and side of the cartilaginous skull to behind the spiracle on that side. Remove tissues ventrally behind the eye to the level of the upper jaw (Figure 34A-3).

3. Refer to Figure 34A-3 (or to an embedded skull in which dye has been injected into the bony labyrinth) to help you anticipate the location of the structures of the inner ear. Using your one-piece scalpel (cartilage knife), *carefully* shave away cartilage from the skull. (*Do not* use a two-piece scalpel—it will fold back on your finger!) Use care so that you do not accidentally flip a piece of the fixed cartilage into your eye (wear glasses as a precaution—no contact lenses). As you approach the inner ear, you should be able to see the horizontal semicircular canal through the semitranslucent cartilage before breaking into the cavity of the bony labyrinth in which it lies. Be careful—remove little slivers at a time! Once you break through the cartilage into the bony labyrinth, continue to shave cartilage away from the membranous labyrinth, exposing the length of the first semicircular canal and the remaining structures of the inner ear. Use Figure 34A-3 to help you follow the position of the parts of the membranous labyrinth as you work.

4. Remove the membranous labyrinth from the skull, place it in a small finger bowl with water, and study it. (If prepared specimens are provided, observe them.) Identify the anterior and horizontal **semicircular canals** with their common duct, the anterior **utriculus** (Figure 34A-2a). Find the posterior semicircular duct. At the base of each semicircular canal, find an expanded segment, the **ampulla.** Find the ventral **sacculus** with its extension, the **lagena.**

   b. What is the function of the three semicircular canals? _____

   _____

   c. What structure does the lagena form in birds and mammals? _____

Figure 34A-3 *The head of the shark, showing the region to be skinned and cleared of muscles. The membranous labyrinth of the inner ear is shown in place within the chondrocranium.*

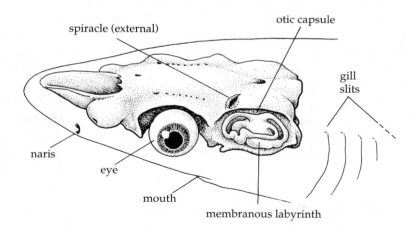

PART 3   Other Sensors of the Head

The nasal epithelia form a primitive sensory system in all vertebrates. The sensory elements (hair cells) are **chemoreceptors** stimulated by molecules carried in water or air from distant sources. In most sensors, specialized epithelial cells receive stimuli and transduce them into electrical events that are relayed to separate sensory neurons. These sensory neurons then carry the nerve impulses to the central nervous system. In the olfactory system, however, the hair cells themselves function as neurons that carry information directly to the brain. Unlike most nerve cells, olfactory neurons that are renewed throughout the life of the organism.

The eyes are also part of a sensory system in all vertebrate groups. As **photoreceptors** they are stimulated by light. The relative shapes of elements within the eye and the way in which they accommodate to objects situated at different distances may vary in different vertebrates, but the basic structure of the eye is remarkably constant in all vertebrates.

✔ ### EXERCISE B   Control Systems

Rapid responses to environmental stimuli are usually mediated by the **nervous system;** slower, longer-lasting responses may involve components of the **endocrine system.** Working together, these two systems allow the vertebrate to make adaptive adjustments to input from the environment (behavior) and to maintain a relatively constant internal environment (homeostasis).

✔ ### PART I   The Nervous System

The nervous system is composed of two parts: the **central nervous system (CNS)** and the **peripheral nervous system (PNS).** The central nervous system includes the dorsal **spinal cord** and its anterior expansion, the **brain.** Within the CNS, sensory input and past experience are evaluated against the genetically determined range of potential responses, and behavioral and physiological actions are initiated. The PNS constitutes the "wiring" that brings information in (from the sensors) and takes command signals out (to the effectors).

CENTRAL NERVOUS SYSTEM

During embryonic development, the anterior portion of the neural tube enlarges and forms three primary divisions; two of these subdivide later to form a total of five regions in the adult brain. These regions are:

Prosencephalon (Forebrain)	1. Telencephalon
	2. Diencephalon
Mesencephalon (Midbrain)	3. Mesencephalon
Rhombencephalon (Hindbrain)	4. Metencephalon
	5. Myelencephalon

Functionally, however, the brain is organized into only two major areas, the **brainstem,** which is a continuation of the spinal cord, and three dorsal expansions (hemispheres or lobes) associated with the primary sensors of the head—the nose, eyes, and acoustico-lateralis system, including the ears (Figure 34B-1).

Much of the brainstem is made up of nerve-fiber tracts, including an ancestral coordinating system of neurons participating in motor and other control activities, called the **reticular system,** and groups of cell bodies (**nuclei**) that act as "relay stations."

Cranial nerves of the peripheral nervous system connect to the brainstem, which functions as a "visceral brain," playing a major role in many of the homeostatic adjustments of the body, including regulation of blood pressure, heart rate, sleep/wake cycles, reproductive cycles, the intake of food and water, and the secretory activity of the pituitary gland, which, in turn, regulates many other body functions.

In contrast, the dorsal expansions of the brain function as a "somatic brain," initiating and coordinating behavioral events and integrating them with functions of the brainstem.

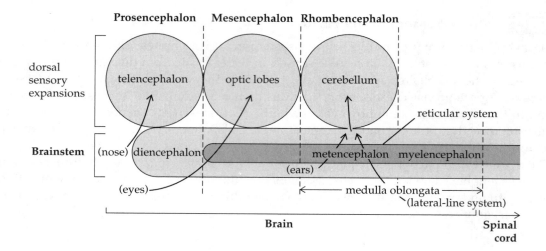

Figure 34B-1 *The brain of vertebrates is functionally divided into the brainstem (which includes an ancient control area, the reticular system), and three dorsal hemispheres or lobes. The most anterior of these dorsal lobes (the telencephalon) is associated with the sense of smell; the second (the optic lobes), with vision; and the third (the cerebellar hemispheres) with input from the inner ear and lateral-line system. The brainstem is continuous with the spinal cord.*

The first of the dorsal expansions of the brain includes the **telencephalon,** part of the forebrain, connected to the sensors of the nose through **olfactory lobes** in all vertebrates. In the earliest vertebrates, the highest proportion of sensory input was provided by nasal epithelia. Phylogenetically, there is a trend toward the enlargement of the telencephalon, leading to the formation of the prominent **cerebral hemispheres.** As the telencephalon became larger, certain motor activities came to be initiated within this region.

The second expansion, the **optic lobes** of the midbrain (**mesencephalon**), receives visual information from the eyes. As vision increased in importance to vertebrates, particularly terrestrial vertebrates (which live in surroundings where light is not attenuated by water nor vision obscured by sediment), this region increased in size and many more behavioral activities began to be initiated in this area of the brain. Information can pass from this area to the cerebral hemispheres of the forebrain, where conflicts between visual and olfactory input may be resolved. In the evolution of mammals, the increasingly important task of coordinating visual information was taken over by the cerebral hemispheres, and the optic lobes themselves decreased in size (Figure 34B-2).

The third expansion forms the **cerebellum,** a pair of hemispheres derived from the upper portion of the metencephalon. This area is associated with sensory input from the acoustico-lateralis system (ear and lateral line) in aquatic vertebrates and from the ear and various proprioceptors in terrestrial vertebrates. The cerebellum receives information about body position (from statoreceptors), changes in body position (from the semicircular canals of the inner ear), movements in the surrounding medium in fishes (from the lateral-line neuromasts), and the position of major muscles and bones in terrestrial vertebrates (from proprioceptors). The cerebellum does not initiate movements or behavior; instead, it integrates the command decisions made in the cerebral hemispheres and optic lobes with information about the position of the organism and its parts.

With the relative unimportance of hearing in fishes, no major expansion of brain tissue is associated with this function. With the origin of the organ of Corti in terrestrial vertebrates, however, important auditory information is projected to the dorsal midbrain.

This basic organization of the brain is found in all vertebrates. There are relative differences in the volume of the dorsal expansions based upon the relative importance of sensors (and muscular control). Follow these changes in Figure 34B-2. In the shark with its massive trunk muscles, the cerebral hemispheres are no larger than the olfactory lobes and the optic lobes are also relatively small, but the

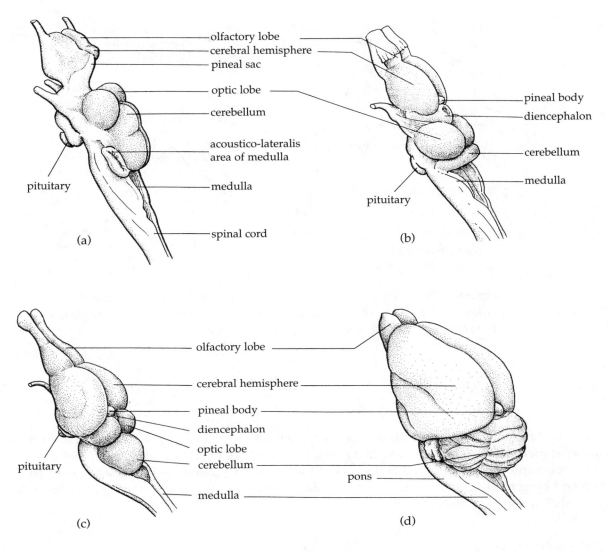

olfactory lobe
cerebral hemisphere
pineal sac

optic lobe
cerebellum

acoustico-lateralis
area of medulla

medulla

spinal cord

pituitary

(a)

pineal body
diencephalon

cerebellum

medulla

pituitary

(b)

olfactory lobe

cerebral hemisphere

pineal body
diencephalon
optic lobe
cerebellum

medulla

pituitary

(c)

pons

(d)

Figure 34B-2 *Brains of (a) shark, (b) frog, (c) turtle, and (d) rat. Note the progressive increase in the size of the cerebral hemispheres through the series and the reduction of the optic lobes to two of the corpora quadrigemina (located beneath the cerebral hemispheres, not visible here) in mammals.*

cerebellum is prominent. In the frog and turtle, the optic lobes are much larger than in the shark, reflecting the importance of vision in these terrestrial vertebrates. As the relative mass of the trunk muscles decreases, the cerebellum becomes somewhat reduced. Note that the most interesting trend is the increase in size of the cerebral hemispheres moving from the shark to the frog to the turtle and then to the rat. In mammals, the organization of the brain shows some innovations that go well beyond a basic increase in relative size of the cerebral hemispheres (Figure 34B-2).

Another part of the central nervous system, the spinal cord, begins at the back of the skull and extends into the tail. Like the brainstem, it includes two types of tissues: **white matter** (myelinated nerve cell fibers) and **gray matter** (nerve cell bodies). Within the spinal cord (and brainstem), the cell bodies are located centrally (Figure 34B-3) and are grouped functionally as sensory or motor and as somatic or visceral. **Sensory fibers** carry information from sensors to the central nervous system; **motor fibers** carry commands from the central nervous system to effectors. **Somatic fibers** innervate superficial parts of the body and generally mediate behavior at a conscious, voluntary level. **Visceral fibers** innervate deeper structures and mediate subconscious, involuntary responses.

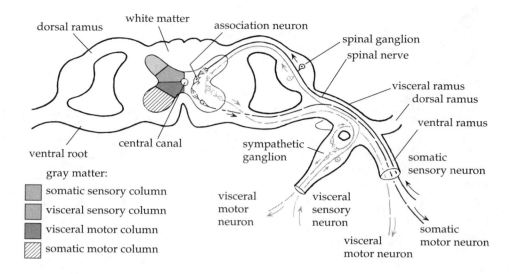

Figure 34B-3 *The mammalian spinal cord and spinal nerves. The organization of cell bodies in the gray matter of the spinal cord is suggested to the left. The pathway of somatic and visceral fibers and of sensory and motor fibers is shown on the right.*

Finally, remember that the central nervous system develops as a hollow tube. The central cavity or **neurocoel** includes the various ventricles of the brain and the neural canal in the spinal cord; it is filled with **cerebrospinal fluid** secreted by vessels of the connective tissue sheaths (**meninges**) surrounding the brain and by special vascular organs, the **choroid plexus** (*telae choroideae*), in the ventricles of the diencephalon and rhombencephalon.

PERIPHERAL NERVOUS SYSTEM

Directly associated with the central gray columns are fibers, comprising the **spinal nerves,** that join the spinal cord through two roots. These **dorsal** and **ventral roots** merge outside the spinal cord to form a complete spinal nerve. During the course of evolution, visceral motor fibers joined the somatic motor fibers in the ventral root and, in mammals, the dorsal root has finally become sensory and the ventral root motor.

As soon as the roots join to form the spinal nerve, they split into branches (**rami**). The dorsal ramus carries fibers to and from the dorsal musculature; the ventral ramus serves the ventral musculature. Visceral fibers travel to various internal organs. Visceral *motor* fibers form the **autonomic nervous system (ANS),** characterized by having two neurons in the motor "chain" or pathway; there is a synapse in the chain along its course, either in special visceral ganglia near the spinal cord or in the organ innervated. (Somatic and visceral sensory innervations have single-neuron pathways.) As we move through the vertebrate phyla, there is a trend toward the functional separation of this dual innervation such that the autonomic nervous system is divided into two opposing (antagonistic) systems: the sympathetic nervous system and the parasympathetic nervous system. Generally, physiological activities promoted by one system are retarded by the other.

In the cranial nerves associated with the brainstem, the roots remain unconnected. Dorsal root nerves contain all sensory and visceral motor fibers, and ventral root nerves contain only somatic motor nerve fibers. There are also three special somatic sensory nerves associated with the brain: the olfactory nerve (composed of fibers from the olfactory neurons), the optic tract (composed of neurons from the eye), and the stato-acoustic nerve (containing sensory neurons from the inner ear).

▒▒▒▒▒ **Objectives** ▒▒▒▒▒▒▒▒▒▒▒▒▒▒▒▒▒▒▒▒▒▒▒▒▒▒▒▒▒▒

- ☐ List the parts of the vertebrate brain and explain the association of each with sensory input and motor outflow.

- ☐ Describe the organization of spinal and cranial nerves in the shark.

▒▒▒▒▒ **Procedure** ▒▒▒▒▒▒▒▒▒▒▒▒▒▒▒▒▒▒▒▒▒▒▒▒▒▒▒▒▒▒▒▒▒

Central Nervous System

1. Carefully remove the skin and tissues that remain on the dorsal portion of the shark's head above, and lateral to, the entire cartilaginous skull, leaving the eyes in place in their orbits for now. Use Figure 34B-4 to guide you in your dissection. Once the dorsal and lateral aspects of the skull are exposed, shave away the cartilage of the skull using your one-piece scalpel. Begin in the area where the inner ear was removed and remove tissue medially and anteriorly. You will encounter the cerebellum, and, once it is exposed, you should be able to remove larger pieces of cartilage without damaging the brain. As you move laterally, you will encounter branches of several cranial nerves—they are white and relatively delicate. Try to preserve them.

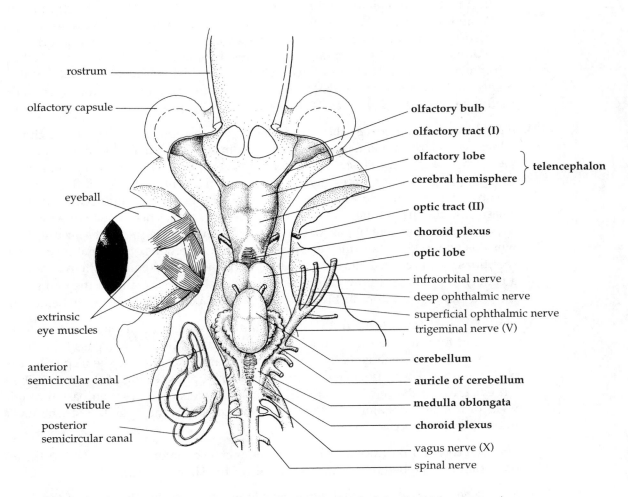

Figure 34B-4   *A frontal section through the head of the shark showing major sense organs, the brain, and several branches of cranial nerves.*

2. Note that the brain is covered by a tough outer membrane, the **dura mater** (or **dura**) and a thin, vascular **pia mater** (or **pia**). These membranes are connective tissue **meninges** that encase the brain. Use a dissecting needle to tear the dura apart. Vessels in the pia cover the brain (these vessels appear dark because hemoglobin in the blood loses its natural color on contact with formalin).

3. Beginning at the anterior end of the brain, locate the **olfactory lobes** and the **cerebral hemispheres** (the two parts of the telencephalon). The forebrain is mainly concerned with input from the olfactory sensors located in the external nares. The olfactory tract connects the sensors and olfactory bulb to the telencephalon.

   a. *What are the functions of the cerebral hemispheres in sharks?* _____

   _____

   b. *What added functions are assumed by the cerebral hemispheres in mammals?*

   _____

   c. *How do the cerebral hemispheres of sharks and mammals differ in relative size?*

   _____

   The second part of the forebrain, the diencephalon, is relatively undeveloped dorsally and is inconspicuous. It gives rise to the pineal body which extends through the roof of the skull. (In submammalian vertebrates, the pineal body or a similar structure is photoreceptive and may form an eyelike structure, the "third" or median eye. In humans, the pineal body is reduced to a small neuroendocrine organ—an endocrine organ of neural origin—that is embedded between the cerebral hemispheres. It produces melatonin, involved in the control of pigment cells and rhythms of activity, including sleep. The lateral walls of the diencephalon constitute the **thalamus,** a region of interconnections between fibers connecting the forebrain with other areas of the brain.

4. Find the dark area between the telencephalon and the optic lobes. This is the **choroid plexus,** a folded, vascularized membrane that produces the cerebrospinal fluid filling the cavity (neurocoel) of the central nervous system. It is also part of the diencephalon. If the brain is later removed, or if you have access to a model, study the ventral surface of the diencephalon. Find the optic nerves that enter this part of the forebrain through the optic chiasma. Behind the chiasma, locate the ventral extension of the diencephalon. This extends downward from the hypothalamus (the ventral part of the diencephalon) to form the **posterior pituitary gland,** which is wrapped by the **anterior pituitary gland.** Secretions of the anterior pituitary control the secretions produced by a number of other endocrine organs (gonads, thyroid, adrenal cortex) and a variety of processes (growth, reproduction, etc.). When you remove the brain from the skull, the pituitary gland will probably be torn off and remain in the skull. See whether you can find it.

5. Find the **optic lobes** located dorsally behind the cerebral hemispheres on each side of the brain and forming the roof of the midbrain. Ventrally, optic fibers pass from the eye through the optic nerve (which is really a tract of the brain) and into the diencephalon through the optic chiasma. There, fibers cross and pass to the opposite side of the brain (fibers from the left eye cross to the right side and vice versa). The nerve fibers then pass through the lateral walls of the diencephalon (the thalamus) and project to the optic lobes, where visual information is processed.

6. Find the **cerebellum** (the dorsal part of the metencephalon) located behind the optic lobes. Note that the cerebellum is the largest part of the shark's brain.

   d. *How does the size of the cerebellum relate to the structures and functions of the shark's body?*

   _____

7. Locate the second choroid plexus posterior to the cerebellum. It is formed from the roof of the myelencephalon. The myelencephalon and ventral part of the metencephalon form the **medulla oblongata.** This structure is continuous with the spinal cord and gives rise to several cranial nerves. Anteriorly, it connects with the midbrain and the diencephalon. Dissect away enough cartilage to follow the medulla to the back of the skull.

8. The spinal cord continues caudally from the brain. Remove the rear of the skull and trace the medulla oblongata to the spinal cord where it emerges from the brain case.

Peripheral Nervous System

9. Dissect away tissues lateral to the vertebral column in front of the first dorsal fin and locate several of the typical spinal nerves (see Figure 34B-4). Locate dorsal roots (with their spinal ganglia) and ventral roots.

   e. *What types of fibers travel through the dorsal root of the spinal nerve in the shark?*

   _____

   The ventral root? _____

10. The size of the different cranial nerves varies greatly, and branches of several may travel together for a part of their length. Several of these branches can be located without removing the brain from the brain case. Using Figure 34B-4, identify as many of the branches of the cranial nerves as you can. You should be able to find the large, **superficial ophthalmic nerve** (a branch of the fifth cranial nerve, the trigeminal), which passes through the orbit of the eye to the rostrum; the **optic "nerve"** (the second cranial nerve, actually a brain tract), which passes from the eye to the ventral diencephalon; the **vagus nerve** (cranial nerve ten), which innervates the last four pharyngeal arches and continues posteriorly to supply visceral (including autonomic) fibers to the heart and anterior abdominal organs.

11. If time permits, you may continue to cut away cartilage of the brain case ventrally and remove the brain intact. (As you do this, you will need to remove one eye—note the straplike extrinsic muscles that connect the eye with the skull. You might open the eye to see the lens and the sensory retina inside). As you cut ventrally, try to preserve the roots of the cranial nerves as they join the brain. Identify as many of the cranial nerves as you can. Try to find the vagus nerve as it emerges at several levels along the ventrolateral surface of the posterior medulla oblongata. Review the divisions of the brain ventrally. Can you find the pituitary gland?

✔ **PART 2**   **The Endocrine System**

All cells in our bodies affect adjacent cells through their metabolic activities and the release of waste materials. In the evolution of chemical controls, some of these products have come to assume a controlling role—first of neighboring cells and later of cells that are at a distance, with the product transported through the circulatory system. Those tissues or glands that have specialized in this direction form a functional group of ductless organs that make up the **endocrine system;** their secretions are known as **hormones.**

Chemical control mechanisms are ancient in vertebrates. Both lancelets and tunicates concentrate iodine in the endostyle at the base of their pharyngeal basket (Laboratory 27), which may be involved in the production of a hormone similar to that produced by the thyroid. The similar location of these glands and similar metabolic activities of their products suggest that the thyroid gland of vertebrates and the endostyle of protochordates may be homologous. This homology indicates a very early ancestry of endocrine control in the phylum.

Vertebrates have several different types of glands belonging to several different organ systems that produce hormones. In fact, since all cells influence adjacent cells, cells derived from any type of tissue may develop a controlling function. Thus, endocrine glands have developed within several systems including the nervous system, the digestive system, and the reproductive system.

We will make no attempt to systematically locate all of the endocrine organs, since many are associated with systems yet to be dissected. However, as you encounter them in later laboratories, keep in mind that it is the control systems, both nervous and endocrine, that enable the active animal to adjust rapidly to many environmental changes and to survive and reproduce.

✔ **EXERCISE C** | **Effectors: Muscles and Bones**

In fishes, paired limbs supply little of the power used in swimming—rather, they fine-tune the fish's position in space. In terrestrial vertebrates, however, the appendages assume a supporting role. They carry the weight of the organism and must, in turn, be securely anchored to the trunk. In tetrapods, the bones supporting the hindlimbs (pelvic girdle) are attached to the vertebral column to provide this support. In contrast, the bones supporting the forelimbs (pectoral girdle) are never directly fused to the spinal column. Muscles and connective tissue link the pectoral girdle to the rib cage. In some vertebrates, the collar bone (clavicle) connects the girdle to the breastbone (sternum), which is, in turn, tied to the vertebral column by the ribs. In many terrestrial vertebrates, the pelvic girdle and hindlimbs play the larger role in locomotion on land, while the forelimbs may be modified for other purposes, including other types of locomotion (flight in birds and bats, for example).

a. How are the structures of the hindlimbs related to their function? _____

_____

_____

_____

In the shark, trunk muscles flex the vertebral column, propelling the fish forward as the caudal fin pushes against the water. In terrestrial animals, muscles of the trunk remain important—the body bends from side to side and the appendages provide points of contact with the substrate, but little independent propelling force. As the limbs become more specialized and take over a major role in locomotion, muscles of the trunk are reduced and the muscles of the appendages become increasingly adapted for finer movements associated with individual skeletal elements. This principle of muscle and skeleton working together to provide externally directed movements is important throughout the vertebrate group.

PART I | **Muscles**

Muscles are organs composed of tissues specialized for contraction (see Laboratory 32). Usually acting with elements of the skeleton, they form the most obvious effector system in vertebrates. Other effectors include electric organs (modified muscle and nerve tissues), pigment cells, and glands.

Muscles are classified as **somatic,** the voluntary skeletal muscles, and **visceral,** mainly the involuntary muscles of visceral organ systems. Somatic muscles are divided into two major groups: **axial** muscles, those associated with the skull and vertebral column, and **appendicular** muscles, those associated with the paired appendages. Fibers of somatic muscles are striated; those of visceral muscles are usually smooth.

In this exercise, you will learn to dissect and identify some somatic skeletal muscles of the shark and the rat.

Skeletal muscles generally have an origin, a belly, and an insertion (Figure 34C-1a). The **origin** is on a bone (or connective tissue sheath of an organ) and is the relatively "fixed," proximal end of the muscle. The **belly** of the muscle, containing the majority of the muscle fibers, is interlaced with connective tissue that continues as a **tendon,** attaching the muscle to its proximal origin and distal **insertion** (Figure 34C-1c). Skeletal muscles usually occur as antagonistic pairs: typically, contraction in one member of the pair flexes a limb; contraction in the other member extends the limb (Figure 34C-1a, b).

Muscles should be named for their origin and insertion, but many are named for their position or carry older names given to human muscles before consistent naming conventions arose. Homologies of muscles in widely separated groups of vertebrates are difficult to establish and may not be reflected in their names.

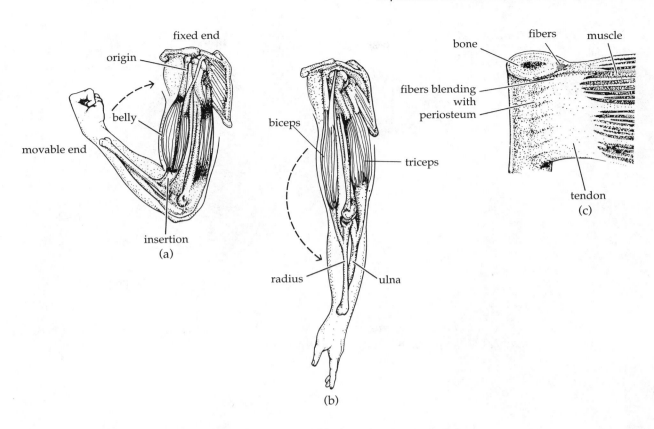

Figure 34C-1   *Structure of muscles. Skeletal muscles normally occur in antagonistic pairs: (a) the biceps flexes the human forearm and (b) its antagonist, the triceps, extends it. (c) A tendon attaches a muscle to bone.*

IIIII Objectives IIIIIIIIIIIIIIIIIIIIIIIIIIIIIIIIIIIIIIIIIII

☐ Study the general arrangement of the muscles of the trunk in the shark.

☐ Dissect the layers of muscles forming the body wall of the rat.

☐ Dissect some of the muscles of the forelimbs or hindlimbs of the rat.

☐ Examine the action of muscles working with bones to produce movements.

IIIII Procedure IIIIIIIIIIIIIIIIIIIIIIIIIIIIIIIIIIIIIIIIII

The dissection should be done by two members of your group while the other two are dissecting the ear and brain of the shark (Exercises A and B). Be sure to share your results.

Trunk Muscles of the Shark

1. Study a demonstration specimen in which the skin has been removed from a segment of the shark's tail. Find the **horizontal septum,** a connective tissue sheet lying below the lateral-line canal. This separates the upper **epaxial muscles** from the lower **hypaxial muscles.** Short longitudinally arranged muscle fibers form **myomeres** separated by connective tissue septa, which serve as the origin and insertion points of the muscles (Figure 34C-2). In fishes, epaxial muscles serve the upper surfaces of the appendages and the hypaxial series provides muscles for the lower surfaces.

Figure 34C-2 *Trunk muscles of the shark.*

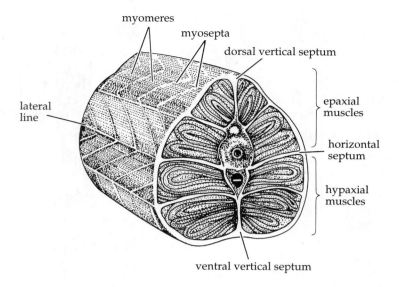

myomeres

myosepta

dorsal vertical septum

lateral line

epaxial muscles

horizontal septum

hypaxial muscles

ventral vertical septum

a. How do you think shark muscles work to cause movement? Use the space below to make one or more drawings of the patterns of muscle contraction.

Skinning the Rat

2. Place your rat in a dissecting pan. Obtain a piece of string. Tie it to one forelimb, loop it under the pan, and tie it to the other forelimb. Be sure that the string is tight enough to spread the forelimbs. Repeat this process with the hindlimbs.

3. If your specimen has had its circulatory system injected with colored latex, there will be a cut in the ventral neck region. If no incision is present, pinch a fold of skin in the midline of the neck, lifting it from the underlying tissues, and start a cut with your scissors. Separate the skin from the underlying tissues with a probe and insert your scissors under the skin. Cut along the midline toward the tail, lifting the skin with the scissors to avoid damaging underlying structures. Divert the incision to avoid the penis of the male, cutting on each side to form a **Y** around it.

4. Extend the cut from the midline to the paw of each limb. Starting at the midline, peel the skin away from the underlying tissues using your fingers or a probe. Loosen the skin around the limbs and work around the body toward the back. Cut through the skin around the wrist and ankle and free the skin from the lateral surface of the leg (you will need to untie the animal to do this). Cut through the skin around the neck, behind the ears, and peel the skin back toward the hind part of the body. Extend the caudal end of your mid-ventral incision dorsally around the base of the tail and remove the skin. Place it to one side—use it to wrap the rat's carcass for storage at the end of the laboratory.

Muscles of the Trunk

5. Place your skinned rat on its side in your dissecting tray. The neck forms the **cervical region** of the body. The **thoracic region** of the trunk includes that part of the trunk containing the rib cage and the pectoral girdle supporting the forelimbs. The **lumbar region** is between the rib cage and the pelvic girdle. The **pelvic region,** containing the pelvic girdle which supports the hindlimbs, is the most caudal region of the trunk. It is followed by the **caudal region** or tail.

6. A tough sheet of connective tissue covers the lumbar region of the back. Lateral to this, the **external oblique** muscle fibers of the trunk form a broad layer extending diagonally and caudally from the ribs to insert along a mid-ventral sheath of connective tissue. The **internal oblique** lies beneath and at right angles to the fibers of the external oblique. The internal oblique also inserts on the mid-ventral sheet of connective tissue. Using a dissecting needle, tease apart the fibers of the external oblique and the deeper fibers of the internal oblique.

7. Tease apart the fibers of the internal oblique to find the fibers of the **transversus abdominus,** a third layer of trunk muscles. If you separate these fibers, you will find the lining of the abdominal cavity, the peritoneum.

Superficial Muscles of the Pectoral and Pelvic Girdles

8. Use your probe to separate the dorsal muscles of the shoulder and hip regions into separate structures. Refer to Figure 34C-3 to locate the muscles listed in Table 34C-1. If possible, find the origin and insertion of each muscle and determine the action performed by the muscle. For reference, several additional muscles are labeled in the figure. If time permits, locate as many of these muscles as possible.

Note the added complexity of muscles found in the appendages of the rat in comparison with the simple repeated arrangement of muscles in the trunk of the shark.

b. Where would you look for a muscle that antagonizes the acromiotrapezius?

_____

Table 34C-1   Superficial Appendicular Muscles of the Rat

Muscle	Pectoral Girdle (dorsal) Origin (O) and Insertion (I)	Function
Acromiotrapezius	O: cervical and anterior thoracic vertebrae I: scapula (dorsal part of pectoral girdle)	Draws scapular medially
Clavotrapezius (anterior to and below acromiotrapezius)	O: skull I: clavicle (collar bone)	Pulls clavicle and scapula anteriorly
Spinotrapezius (posterior to acromiotrapezius)	O: posterior thoracic and anterior lumbar vertebrae I: scapula	Pulls scapula posteriorly
	Pelvic Girdle and Leg (dorsal)	
Gluteus superficialis	O: dorsal border of ilium (pectoral girdle) I: femur (thighbone)	Moves thigh away from the body
Biceps femoris (posterior to gluteus superficialis)	O: sacral and caudal vertebrae shank of leg I: distal femur and proximal tibia (shank bone)	Moves thigh away from the body, flexes
Semitendinosus (posterior to biceps femoris)	O: sacral and caudal vertebrae I: tibia	Flexes shank
Gastrocnemius (and Achilles tendon) (medial edge of leg)	O: femur I: bores of the foot	Extends the foot

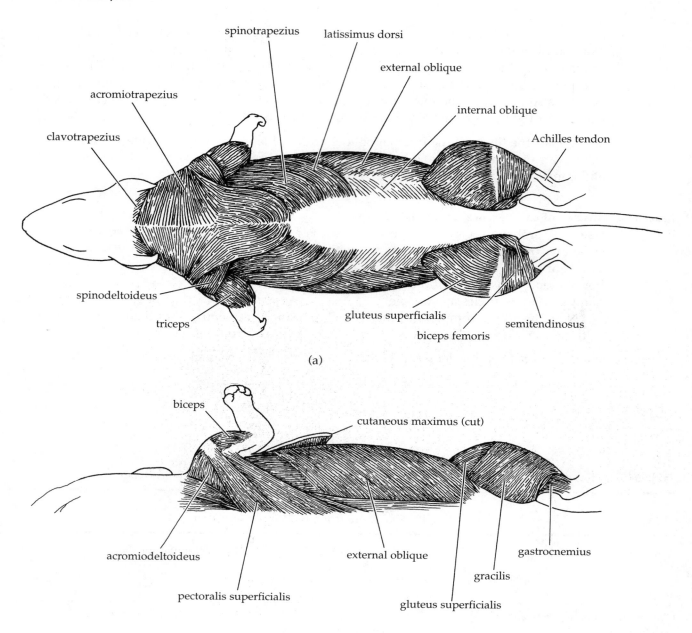

Figure 34C-3   *Muscles of the rat. (a) Dorsal view showing several superficial muscles of the trunk and appendages. The acromiotrapezius has been removed on the right side to show underlying muscles. (b) Ventral view of the pectoral and the pelvic musculature of the rat.*

c. *In dissecting muscles, did you find blood vessels (they are tough structures and are probably injected with colored latex)? Did you find nerves (tough, tendon-like white structures)?* _____

d. *Where are blood vessels and nerves located with respect to the muscles you dissected?*

_____

e. *Describe your technique for separating muscles. How do you know where to tear with the probe? How can you tell when you are within a muscle rather than between muscles?* _____

_____

  f. How do the trunk muscles of the rat differ from those of the shark? _____

  g. How do you think these differences relate to the structures and functions of the animals?

  _____

✔ **PART 2**   **Bones**

The skeleton of vertebrates protects some of the internal organs, supports the body, and provides the mechanical levers moved by muscles for procuring food, providing power for locomotion, and producing behavioral responses. It is an internal skeleton, or **endoskeleton.**

  The vertebrate skeleton is composed of the following parts:

Axial skeleton	Skull
	Vertebral column
Appendicular skeleton	Pectoral girdle and forelimbs
	Pelvic girdle and hindlimbs
Visceral (pharyngeal) skeleton	In ancestral animals, the bony structure that supports the pharyngeal arches and to which gills are attached; parts of the pharyngeal skeleton later evolved into jaws, the base of the tongue, and the bones of the inner ear

Bones of the skeleton are derived from three sources: (1) bone preformed in cartilage, which gives rise to the basic endoskeleton; (2) dermal bone associated with the skin, which becomes incorporated into the skull and pectoral girdle in more advanced vertebrates; and (3) bones, also preformed in cartilage, associated with the visceral (pharyngeal) skeleton. The skull is the most complex element of the skeleton, containing bones from all three sources.

  In this exercise, you will study the basic elements of the cartilaginous skeleton of the shark. (You can consult Laboratory 27, pages 27-12–27-14, for more detail about the appendicular skeletons of the frog and the turtle. Pages 27-14–27-18 cover the mammalian skeleton.) The skeleton of the shark shows clearly the basic arrangement of the skeletal system of vertebrates.

ııııı **Objectives** ıııııııııııııııııııııııııııııııııı

☐ Examine the axial skeleton of the shark.

☐ Study the relationship of the visceral skeleton to the chondrocranium of the shark.

☐ Identify the major elements of the appendicular skeleton.

ııııı **Procedure** ııııııııııııııııııııııııııııııııııııııı

Axial Skeleton—Skull

In the shark, the skull is made up of the **chondrocranium,** formed from cartilaginous elements of the axial skeleton, and the cartilaginous **visceral skeleton,** which forms the jaws and pharyngeal arches.

  1. If a separate chondrocranium is available as a wet preparation or embedded in plastic, use it to study features of this part of the skull. Compare the structures that you can locate with those shown in Figure 34C-4. The location of the sense organs (olfactory capsules, orbits of the eye, and optic capsules) and brain should be familiar to you from the dissection in Exercise B. Note the anterior **rostrum,** a cartilaginous support for the snout. The caudal occipital region of the skull surrounds the **foramen magnum,** a large opening where the spinal cord exits the cranium.

  2. Examine the visceral skeleton in a complete skull or in a mounted skeleton. It is composed of seven pharyngeal arches (Figure 34C-5). The first or **mandibular arch** forms the upper and lower jaws (note that the upper jaw is not attached to the chondrocranium). The second arch is the **hyoid arch.** The first gill slit, the **spiracle,** lies between the hyoid arch and the upper

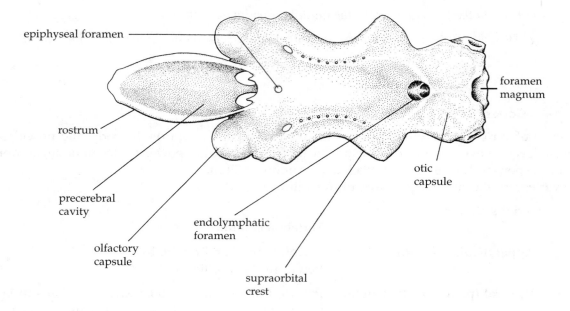

Figure 34C-4 *Dorsal view of the chondrocranium of the shark. Note that the photoreceptive pineal gland is located beneath the epiphyseal foramen.*

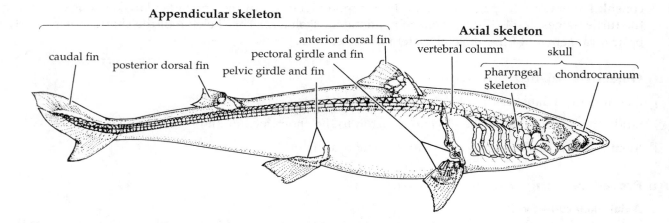

Figure 34C-5 *Skeleton of the shark.*

jaw. The last five arches form the skeletal elements that separate the remaining gill slits. With the transition to land, many elements forming the visceral skeleton have been reduced or lost, but some persist in the jaws, bones of the hyoid apparatus that supports the tongue, and bones of the middle ear.

a. *Why do you suppose the spiracle has not developed into a complete gill slit and has been displaced dorsally in the shark?* _____

_____

Axial Skeleton—Spinal Cord

3. Study the remainder of the axial skeleton in the wet preparation (Figure 34C-5). The vertebral column is composed of cartilaginous vertebrae. Vertebrae of the trunk are composed of two parts, the solid, ventral **centrum** and the dorsal **neural arch** and **intercalary plate** which surround the spinal cord (Figure 34C-6). The face of each centrum is biconcave and is filled with gelatinous remnants of the embryonic notochord, which is continuous along the length of the spinal column, coursing through small openings in the centra. Vertebrae in the more anterior part of the trunk bear short rib cartilages attached to each centrum. Caudal vertebrae of the tail each have a **hemal arch** below the centrum enclosing the caudal artery and vein. In most vertebrates, remnants of the notochord exist only as the disks that separate the adjoining bones of the vertebral column. The persistence of the notochord in sharks probably represents an ancestral characteristic (which also characterizes the embryonic development of vertebrates).

Figure 34C-6 *Trunk vertebrae of the shark.*

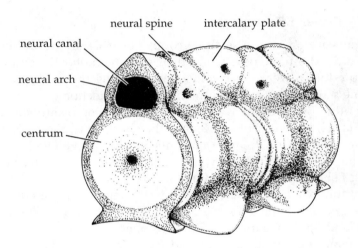

neural spine   intercalary plate

neural canal

neural arch

centrum

4. On demonstration is a sagittal (dorsal to ventral) section through the midline of the vertebral column. Note the **centra,** the spaces filled with persistent notochord, and the neural canal filled with the spinal cord.

5. In the dissection on demonstration, locate the skeletal elements of the **dorsal fins** above the vertebral column. Compare their structure with that of the **caudal fin.** These structures are also part of the axial skeleton.

   b. *Which of the fins is most important in propelling the organism through the water?* _____

   c. *Can you suggest how the myomeric muscles work to move this fin? (It might help to make a drawing in the space below.)* _____

   _____

   _____

Appendicular Skeleton

6. The appendicular skeleton consists of the pectoral girdle and forelimbs and pelvic girdle and hindlimbs. Locate the pectoral girdle behind the last of the pharyngeal arches. It consists of a **U**-shaped bar and several cartilages that support the fibrous elements of the fin. Dermal elements are missing in the pectoral girdle of the shark.

7. Find the pelvic girdle. In the live shark it is embedded in the ventral body wall anterior to the cloacal opening. Like the pectoral girdle, it is made up of several endochondral elements that support fibrous elements in the fins.

Cartilages of the appendages associated with the pectoral and pelvic girdles of sharks bear no clear relation to the skeleton supporting the limbs of tetrapods. However, in the lobe-finned bony fishes, which gave rise to amphibians, bones of the appendages are arranged as they are in our limbs. (See Laboratory 27, pages 27-12–27-14.)

✔ **PART 3**   **Other Effectors**

Electric organs are present in several groups of cartilaginous and bony fishes. These organs are composed of highly modified muscle tissue, and their somatic motor fibers are arranged in series (like the cells of a battery) to produce a weak electric field surrounding the fish. Deviations in the electric field can be sensed by lateral-line sensors and may aid the fish in the detection of nearby objects. Electric discharges may also be detected by other fishes in the water and, like bird song or human speech, may be used in communication. In the electric eel, the electric catfish, and the electric ray, these electric organs have become highly developed to produce much larger currents and higher voltages. In the electric ray, currents may reach 50 amperes at 50 volts, yielding a power of 2,500 watts! Electric discharges of this magnitude paralyze surrounding fishes on which the animal can feed.

Chromatophores, pigment cells found in the integument, are also considered effectors in some vertebrates—remember that effectors are organs that produce externally directed responses. In some fishes, amphibians, and reptiles, chromatophores respond directly to light, contracting in bright light to concentrate their pigment and lighten the animal, and vice versa in dim light. In many of these organisms, they also respond to a hormone (melanophore-stimulating hormone or MSH) from the pituitary gland and to another hormone, melatonin (melanophore-concentrating hormone or MCH), from the pineal gland. In certain species such as flounders that can rapidly camouflage themselves to match changing background patterns, the chromatophores are innervated and the nervous system responds quickly to input from the eyes.

Integumentary glands can respond to local, hormonal, and nervous stimulation in various situations. In some fishes, mucous glands have become specialized as bioluminescent organs (**photophores**), producing light in the dark waters of the deep ocean to help in finding prey and mates. In some glands, light is produced when the gland is activated by motor nerves. Others contain cultures of luminescent bacteria which provide a constant source of light (emissions may be controlled by lidlike shutters).

Laboratory Review Questions and Problems

1. Complete the following table for the three major sense organs of the vertebrate head.

Sense Organ	Stimulus	Cranial Nerve	Brain Projection

2. How does the structure of the membranous labyrinth in our inner ear differ from that of the shark?

3. You place your hand on a hot burner on the stove. Describe what happens. (Outline sensory events, control events, and the effectors involved in your response.)

4. Explain how muscles and bones normally work together as effectors.

5. List the effectors found in vertebrates.

The Anatomy of Representative Vertebrates: Digestive and Respiratory Systems

OVERVIEW

The digestive and respiratory systems, integrated by the control systems, participate in the procurement and metabolism of energy-containing materials. Food is taken in through the mouth and digested in the digestive tract, and nutrients are transported to all parts of the body by the circulatory system. Molecules obtained as food are stored at a variety of sites in the body: glycogen is stored in the liver and muscle tissues; fats are stored in the liver and muscles in lower vertebrates and in adipose tissues in more advanced vertebrates. The components of proteins (amino acids), however, are not stored—they are used to build new proteins or are deaminated in the liver to make simple sugars or fatty acids, which can be converted to storage forms.

When needed for the production of ATP, stored molecules are metabolized to yield water and carbon dioxide. This requires oxygen, which is supplied by the respiratory system and transported by the circulatory system. Carbon dioxide is removed from the body by the respiratory system. Other waste materials are carried to the liver where they are metabolized or repackaged and to the kidneys where they are eliminated from the body.

In this laboratory, we will concentrate on those systems responsible for providing food and carrying out the exchange of gases necessary to metabolize energy-containing molecules—the digestive and respiratory systems. Most of the organs of these systems lie within the body cavity (**coelom**), which will be examined first.

STUDENT PREPARATION

Prepare for this laboratory by completing Laboratory 34 and reading the text pages indicated by your instructor. Familiarizing yourself in advance with the information and procedures covered in this laboratory will give you a better understanding of the material and improve your efficiency.

If dissection tools are not provided, bring your dissecting kit to laboratory with you.

✔ **EXERCISE A** **Coelom and Mesenteries**

The coelom is a body cavity lined by mesoderm. In many invertebrates, a body cavity filled with fluid (*hydrocoel*) functions as a supporting skeleton for the body. However, in vertebrates the endoskeleton assumes this role, and the major role of the coelom in the course of evolution has been to allow the internal organs to lengthen, coil, and move independently of the outer body wall.

During evolution, the vertebrate coelom (originally a single cavity) was subdivided into several body cavities containing different organs (Figure 35A-1). Early in vertebrate phylogeny, the heart came to occupy the **pericardial cavity** and was separated from the abdominal cavity by the **transverse septum.** As swim

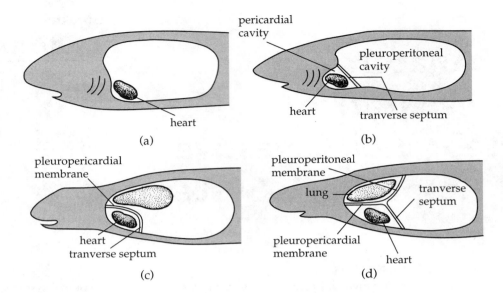

Figure 35A-I *Divisions of the coelom in vertebrates. In the vertebrae ancestors (a), there was a single body cavity. In fishes (b), the pericardial cavity is separated from the main body cavity. In amphibians and reptiles (c), lungs lie in the anterior portion of the pleuroperitoneal cavity above the heart, which moved caudally as the neck developed. In the mammals (d), lungs occupy a separate pleural cavity. The transverse septum and pleuroperitoneal membranes together form the diaphragm.*

bladders or lungs developed during the course of vertebrate phylogeny, they began to protrude into the abdominal cavity (**pleuroperitoneal cavity**) at the level of the heart and behind it (Figure 35A-1c). As the development of the neck separated the head from the body, a lateral fold of the body wall joined the transverse septum, forming a **pleuropericardial membrane** separating the heart from the lungs (Figure 35A-1c). In higher vertebrates, the pleural cavities encasing the lungs became completely separated from the abdominal cavity by the **pleuroperitoneal membranes** (and transverse septum), which in mammals form the **diaphragm,** a muscular structure assisting the muscles of the body wall in ventilating the lungs (Figure 35A-1d).

Coelomic body cavities are lined with a thin, permeable epithelium. On the inside surface of the body wall, this lining is called the **parietal peritoneum.** Visceral organs in the body cavity are suspended by extensions of the peritoneum called **mesenteries** (Table 35A-1). Mesenteries contain small amounts of

Table 35A-1   Mesenteries

Mesentery	Organ(s) Supported
Dorsal mesenteries	
Mesogaster (greater omentum)	Stomach
Mesentery proper (mesointestine)	Small intestines
Mesocolon	Large intestine
Ventral mesenteries	
Gastrohepatic ligament	
(lesser omentum)	Liver (from stomach)
Falciform ligament	Liver (to ventral body wall)
Median ligament	Bladder

connective tissue, blood vessels, and nerves, and in higher vertebrates may also contain adipose (fat) tissue. Body organs are covered by the **visceral peritoneum,** also part of the living of the coelom.

The heart tends to be free of suspensory mesenteries or ligaments which could constrain its activity. The pericardial cavity containing the heart is lined by a **parietal pericardium** and the heart is covered by the **visceral pericardium.**

◫◫◫ Objectives ◫◫

☐ Open the coelomic cavities of representative vertebrates and describe how the coelom of each is subdivided.

☐ Study the divisions of the coelomic cavities of the vertebrates and their peritoneal linings.

☐ Distinguish among the parietal peritoneum, the visceral peritoneum, and the parietal and visceral pericardia and be able to locate each.

☐ Describe the structure and the major functions of the mesenteries.

◫◫◫ Procedure ◫◫◫

It is customary to give dissection directions (right and left) in terms of the animal's right and left. When the animal's ventral surface is toward you, the animal's right is on your left and vice versa. We will follow this convention.

Work in groups of four. Before the end of the laboratory, show others in your group the features that you have found in your specimen and inspect theirs—you are responsible for knowing the general anatomical features of all four representative vertebrates.

Obtain your specimen (shark, frog, turtle, or rat). Rinse it in the sink and place it on a tray or in a dissecting pan.

Shark

1. Place your specimen with its ventral side up. Locate the pelvic and pectoral girdles by feeling them with your fingers (they are at the base of the paired fins). Beginning in front of the cloacal aperture, puncture the relatively thin body wall with the sharp blade of your scissors and make a shallow cut along the mid-ventral line to the level of the pectoral girdle. Cut through the pelvic (posterior) girdle in making this incision, but be careful to cut only the body wall—do not damage the underlying visceral organs. See Figure 35A-2a.

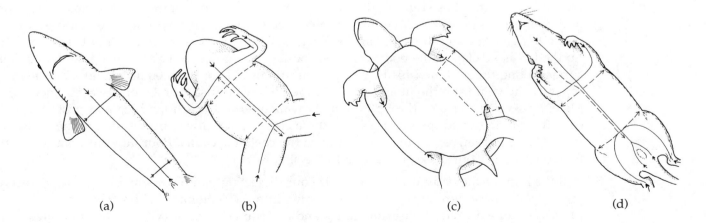

(a)   (b)   (c)   (d)

Figure 35A-2 *Schematic diagrams of (a) shark, (b) frog, (c) turtle, (d) rat. Solid lines indicate incisions through the skin and dashed lines show incisions through the body wall.*

2. Make lateral cuts from the mid-ventral incision through the body wall just behind the pectoral girdle and in front of the pelvic girdle. Extend the cut to about the level of the lateral line. This will give you access to the body cavity without damaging any essential structures. Do *not* remove these flaps—they will protect the internal organs during storage. Gently rinse the preservative and dried blood from the coelomic cavity. (Pour any preservative that has collected in your pan into the sink.)

In the shark, the coelomic cavity is incompletely divided into two parts: the **pericardial cavity** containing the heart and the **pleuroperitoneal** (abdominal or visceral) **cavity** containing the abdominal visceral organs. These cavities are *almost* completely separated by a thick **transverse septum.** (As in many fishes, these cavities connect through a small canal, the pericardioperitoneal canal.)

3. Pull apart the body wall to expose the pleuroperitoneal cavity and find the transverse septum—look in front of the liver, the large dark organ occupying much of the cavity. (*Use care not to break or abrade the liver; it contains oils which stain clothes and have an unpleasant odor.*) Find the **parietal** and **visceral peritonea.**

   a. *What structure forms the anterior end of the pleuroperitoneal cavity?* _____

   _____

4. Carefully move the liver and abdominal viscera to the shark's right to see the **mesenteries.** Find as many of the mesenteries listed in Table 35A-1 as you can without tearing them or damaging the visceral organs. Note their structure—recall that they are translucent membranes composed of two layers of peritonea and contain nerves and blood vessels.

5. The pleuroperitoneal cavity opens to the exterior through two **abdominal pores** on each side of the cloacal opening. Find them and use your probe to verify that they pass through the body wall. In lampreys and hagfishes, similar pores (which may not be homologous to those of the shark) carry gametes to the exterior, but in other vertebrates, including sharks, their function is unknown.

Frog

1. Turn your frog ventral side up. See Figure 35A-2b. Using scissors, make a *shallow* longitudinal cut through the thin skin of the ventral surface of the body (the skin will separate easily from the underlying muscles). Extend the cut from just in front of the cloacal aperture to the pectoral girdle. Make transverse cuts along the ventral surface of the forelimbs and hindlimbs as if making a large letter **I**. Free the skin from the underlying trunk and limbs and pull it back.

2. Now cut through the muscle layers of the ventral body wall slightly to the *right of the mid-ventral line* (your left as you face the frog). Extend the cut from in front of the cloacal aperture to the hind margin of the forelimbs (you may need to veer left as you approach the pectoral girdle to avoid the sternum). Be careful not to damage the visceral organs and the heart. Lift up the left-hand flap of the body wall and locate the large ventral abdominal vein on its inner surface along the midline (this is why you cut to one side of the median plane). Carefully separate the vein from the inner surface of the body wall (use a probe or dissecting needle). Make transverse incisions through the body wall anterior to each hind leg and posterior to each front limb (avoid the vein). Extend these cuts about halfway up the side of the body. Rinse the preservative and dried blood from the coelomic cavity. Pour any preservative that has collected in your dissecting tray into the sink.

3. Pull the flaps of the body wall outward. If your dissecting pan has a wax lining, use dissecting pins to hold these flaps away from the opening into the coelomic cavities. The large **pleuroperitoneal cavity** houses the lungs and abdominal viscera. With the head pointed away from you, push the viscera to the frog's right to see the nature of the peritoneal wall and mesenteries.

   a. *Describe the appearance of the parietal peritoneum.* _____

   _____

b. Describe the appearance of the mesenteries. _____

What other organs or tissues are contained within them in the frog? _____

_____

4. Find the **transverse septum** that forms the anterior wall of the pleuroperitoneal cavity. The **liver,** a large dark organ filling much of the anterior pleuroperitoneal cavity, is attached to it by the **coronary ligament.**

5. Find the **heart,** nestled between the lobes of the liver behind the pectoral girdle. It is located in the **pericardial cavity.**

   c. What structure separates the pericardial cavity from the pleuroperitoneal cavity?

   _____

   Does this separation appear to be complete in the frog? _____

Turtle

1. Place your turtle ventral side up. If the bony bridges uniting the carapace and plastron have not been cut for you, obtain a pair of bone shears and cut through these bony structures. Using a scalpel, separate the muscles of the pectoral and pelvic girdles and the underlying membrane, the **parietal peritoneum,** from the inner surface of the plastron (skin). See Figure 35A-2c. Two large veins, the **ventral abdominal veins,** form a letter **H** within the ventral peritoneum. The veins appear posteriorly in the pelvic girdle; anteriorly they enter the right and left lobes of the liver. They are connected in the middle by the **transverse abdominal vein.** Cut away the ventral peritoneum covering the body cavities, but leave the membrane intact around the veins. Now cut the veins posterior to their junction and reflect them forward (you will need to trace them in Laboratory 36). Rinse the preservative and dried blood from the coelomic cavity. Pour any preservative that has collected in your dissecting pan into the sink.

   a. The ventral body wall consists only of skin, which forms the plastron, and the peritoneum. The body

   wall has no muscle layers. Why not? _____

   _____

2. Posterior to the pectoral girdle, find the membranous pericardial sac. The **lungs** lie above the heart in an anterior extension of the **pleuroperitoneal cavity.** Cut open the pericardial sac, exposing the **pericardial cavity** and **heart.** Note that the sac is an isolated chamber formed by membranes, whereas in the shark and frog the chamber is formed ventrally by the body wall.

3. Using bone shears, cut into the carapace between the limbs and remove its margin on the left side (your right). You may need to remove adipose tissue as you proceed. Examine the peritoneal lining of the coelom and locate the most obvious mesenteries, but save your detailed study for the next exercise. The membrane between the heart and the liver is the **transverse septum,** and the liver attaches to it by the **coronary ligament.**

   b. Describe the appearance of the parietal peritoneum. _____

   _____

   c. Describe the appearance of the mesenteries. _____

   What other organs or tissues are contained within the mesenteries in the turtle?

   _____

Rat

1. Remove the skin from your rat and place the animal in a dissecting pan. Keep the animal covered with a wet paper towel when you are not working with it, to prevent the exposed body wall from drying out. Open the abdominal and pleural (thoracic) cavities by cutting

through the ventral body wall beginning in the abdominal region and extending the incision anteriorly (Figure 35A-2d). When you reach the rib cage, cut slightly to the right of the mid-ventral line to avoid the sternum, and continue cutting just beyond the level of the forelimb. Extend the cut posteriorly, stopping in front of the external genitalia.

2. Place your thumbs in the edge of the cut in the region of the thorax and bend the rib cage laterally and dorsally on each side, breaking the ribs near the spinal column. Cut through the side of the body wall just behind the front legs and in front of the hind legs on each side, extending these cuts about two-thirds of the way up the body wall. Pull the lateral body walls back to expose the visceral organs. Rinse the preservative and dried blood from the coelomic cavity. Pour any preservative that has collected in your dissecting pan into the sink.

3. Note that the **pericardial cavity** is embedded anteriorly in a septum derived from the ventral mesentery. As in the turtle, the cavity is not bounded by the ventral body wall. Cut open the membrane (the **pericardium**) surrounding the **heart.** Note the dark, spongy **lungs** on each side of the heart. They are separated from the heart by the **pleuropericardial membranes** lying above the heart and from the body cavity by the **transverse septum** and **pleuroperitoneal membranes** to form separate **pleural cavities.** In mammals, the transverse septum and pleuroperitoneal membranes are invaded by skeletal muscle to form the muscular **diaphragm.** You may have torn this structure as you opened the body cavity, but you should be able to locate it. What color is muscle tissue in the preserved rat?

4. Examine the abdominal or **peritoneal cavity.** Gently push the organs to one side to see the mesenteries—these will be studied in more detail in the next exercise.

   a. Describe the appearance of the parietal peritoneum. _____

   _____

   _____

   b. Describe the appearance of the mesenteries. _____

   _____

   _____

   c. What other organs or tissues are contained within the mesenteries in the rat?

   _____

   _____

✔ **EXERCISE B**   **The Digestive System**

In primitive vertebrates, the digestive system consisted of little more than an anterior opening, the **mouth;** the **pharynx;** a **foregut** and a **hindgut** separated by a constriction (the **pylorus**); and a posterior opening, the **anus.** The first vertebrates were filter-feeders, feeding continuously on small particles of food suspended in the water. With the development of jaws, larger food items were taken at less regular intervals ("meals"), and a temporary storage area, the **stomach,** developed in the anterior part of the system. The remainder of the digestive system changed little during the course of evolution. See Figure 35B-1.

Oral Cavity   The oral cavity was formed as jaws evolved to enclose a chamber between the anterior opening of the digestive tract (mouth) and the pharynx.

Pharynx   Behind the oral cavity is the pharynx. Its primary role is associated with gas exchange, so we will postpone our exploration of this area.

Foregut   The foregut extends from the pharynx to the pyloric constriction and often expands near the pylorus to form a storage organ, the stomach. The anterior portion of the foregut forms a connecting tube, the esophagus.

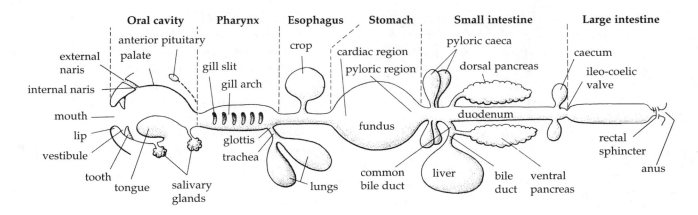

Figure 35B-1   *The complete gut. A generalized schematic diagram of structures and organs associated with the digestive tract in various vertebrates.* Note that no single vertebrate possesses all of these structures.

*When jaws developed, they enclosed a space in front of the pharynx, the **oral cavity. Teeth** lined the jaws and, with the change from gill to pulmonary respiration, the pharyngeal arches were modified to support a **tongue** used to gather and manipulate food. In terrestrial vertebrates, teeth became embedded in the bone of the jaw, anterior flaps of skin (**lips**) formed, and the angle of the jaw became closed (forming **cheeks**). In the evolutionary line leading to lobe-finned fishes and terrestrial vertebrates, **internal nares** arose and connected the roof of the oral cavity to the outside through nasal passages. In terrestrial vertebrates, **salivary glands** evolved to provide lubricating moisture for dry food and to protect the epithelial lining of the digestive tract.*

The oral cavity evolved into the most complicated part of the digestive system, adapting to the wide range of diets in vertebrates.

*The boundary between the oral cavity and **pharynx** lies behind the region where the anterior pituitary forms from the tissues of the roof of the oral cavity. In aquatic fishes, **gills,** derived from feeding structures, became adapted for gas exchange. **Lungs** (swim bladders) developed in ancestral bony fishes as outpocketings of the pharyngeal floor and assumed the major role in respiration in terrestrial vertebrates. Gills are lost in terrestrial vertebrates, and the pharynx serves only as a passageway for food.*

*The **foregut** extends from the pharynx to the **pyloric constriction.** In filter-feeders and jawless vertebrates, this portion of the digestive system was little more than a straight tube leading to the hindgut. However, with the development of jaws, which could provide large amounts of food at one time, the region near the pylorus expanded to form a storage organ, the **stomach,** with the anterior portion forming the esophagus. In advanced vertebrates, enzymes and an acid environment combine to initiate the chemical digestion of proteins in the stomach, but few substances are assimilated (absorbed) in this region.*

*The digestive tract behind the pylorus is the **hindgut,** which is responsible for the chemical digestion of food and the uptake of the products formed by this process. The hindgut shows few phylogenetic changes except for an increase in the area of the absorptive surface, as seen in the cigar-shaped spiral intestine of the shark and some other fishes or the coils of intestine found in most vertebrates. The size of the **small intestine** is generally correlated with diet—it is shortest in those forms that feed on microscopic food particles and easily digested foods such as nectar; it is longer in animals with a high-protein diet of insects or other animals; and it is longest in herbivores that feed on masses of grass or foliage. In early bony* *fishes and most higher bony fishes, extra surface area is also added to the intestinal surface by pyloric **caeca**—blind sacs of the duodenum located near the pylorus.*

*The first section of the small intestine, the **duodenum**, receives ducts from the **liver** and **pancreas**, large exocrine glands (glands with ducts) in the abdominal cavity. The liver produces bile, a fluid containing bile pigments (the end product of metabolized blood pigments which color the bile brown to green) and bile salts (which break up fats into small droplets that can be digested or absorbed directly). Enzymes secreted by the pancreas and the wall of the duodenum digest food materials into simple chemical compounds that can be assimilated as food passes through the remainder of the small intestine. The posterior part of the intestine is called the **large intestine** or **colon**. In more advanced vertebrates, one or two sacs or caeca mark the junction of the small and large intestines. The large intestine is primarily a site where water is removed from the intestinal contents to make the semisolid feces.*

Hindgut   The digestive tube behind the pylorus is the hindgut. It shows few phylogenetic changes in the vertebrate series except for an increase in internal surface area correlated with a more varied diet. The first section of the hindgut, or **duodenum**, receives ducts from large visceral organs, the **liver** and **pancreas**. The posterior part of the hindgut is the **large intestine** or **colon**. In many vertebrates (bony fishes and mammals are among the exceptions), the colon opens into the **cloaca**, a common chamber receiving fecal material from the digestive system, urine from the excretory system, and gametes from the reproductive system. You will study this chamber further in the next laboratory.

Anus   The anus is the terminal opening of the digestive system. The term may be used for the exterior opening through which feces are voided, regardless of whether or not they pass through a cloaca en route. The term "cloacal aperture" is also used for the opening of the cloaca.

ııııı **Objectives** ıı

☐   Observe the digestive tracts of representative vertebrates and note differences in their structures.

☐   Describe the role of the organs of the digestive system and their functional relationships to each other in the process of digestion.

ııııı **Procedure** ıı

Shark

1.   Make a cut through the angle of the jaws on the left side of the shark (your right), cutting through the cartilages of the jaw and pharyngeal skeleton. Continue this incision through the center of the external gill slits. Cut through the pectoral girdle on the left side, near the base of the fins, and continue cutting medially to meet the mid-ventral incision already made. Deepen the cut in the region of the gills to open the oral and pharyngeal cavities and the esophagus. Keep your cut on the side at the level of the digestive organs: the heart is located beneath the area just in front of the pectoral girdle and should not be disturbed at this stage of your dissection.

2.   Pull the bottom jaw and lower part of the oral cavity and pharynx to one side and study the structures of the mouth (Figure 35B-2a). Note the rows of teeth on the upper and lower jaws. Does the shark have a tongue? The boundary between the **oral cavity** and the **pharynx** lies in front of the **spiracle;** find the entrance to the spiracle behind the angle of the jaw. Locate the five **internal gill slits** behind the spiracle—note **gill rakers** on the arches; these protect the gills from damage by food in the pharynx. Behind the last slit, the lining of the cavity becomes folded as the pharynx grades into the esophagus.

a. In the shark, is the oral cavity larger or smaller than the pharynx?

_____

Figure 35B-2 *(a) Oral and pharyngeal cavities of the shark. (b) Organs of the digestive system of the shark.*

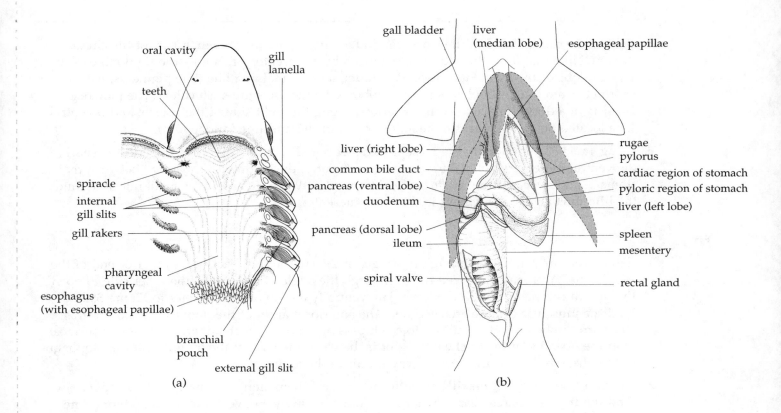

(a)

oral cavity
gill lamella
teeth
spiracle
internal gill slits
gill rakers
pharyngeal cavity
esophagus (with esophageal papillae)
branchial pouch
external gill slit

(b)

gall bladder
liver (median lobe)
esophageal papillae
liver (right lobe)
common bile duct
pancreas (ventral lobe)
duodenum
pancreas (dorsal lobe)
ileum
spiral valve
rugae
pylorus
cardiac region of stomach
pyloric region of stomach
liver (left lobe)
spleen
mesentery
rectal gland

3. Open the visceral cavity and study the organs associated with the digestive system (Figure 35B-2b). The largest organ lying at the anterior end of the body cavity, with lobes extending through most of the cavity, is the **liver.** Lift the left lobe and find the **esophagus** and **stomach** under it. Caudally, the stomach curves to the shark's left, forming a **J** and ends with a constriction, the **pylorus,** at the point where the digestive tract turns caudally. Using your scissors, make an incision about 3 to 4 inches long on the left side of the stomach and open it. Remove the contents of the stomach—it may be packed—and wash out the cavity. The wall of the stomach contains folds (**rugae**) which allow the stomach to expand, increasing its capacity to hold prey. These folds grade into smaller fingerlike **esophageal papillae** in the esophagus; there is no sharp demarcation between these portions of the foregut.

   b. *Can you identify what your shark had been eating?* _____

   _____

4. The intestine is a long, cigar-shaped structure extending from the pylorus to the region where the **rectal gland** (an osmoregulatory structure) enters the digestive tract. The **rectum** (not homologous with the rectum of mammals) extends from the intestine to the **anus,** which projects into the cloacal chamber. Locate all of these structures. Using your scissors, cut through the ventral wall of the intestine. Note the transverse partitions. Like a spiral staircase, these form a helical passageway leading through the intestine, lengthening the path that food

must follow and increasing the relative surface area of the intestine for the absorption of digested products. This type of small intestine is called a **spiral** (or **valvular**) **intestine.**

   c. *How do the other representative vertebrates increase the absorptive surface area in their small intestine?* _____

_____

5. Behind the caudal bend in the stomach, find a dark organ, the **spleen,** a part of the circulatory system. Note that it is attached to the stomach by a mesentery. Examine the right edge of the median lobe of the liver. Find the **gall bladder,** a sac containing bile (bile pigments, waste products from the metabolism of hemoglobin, give the sac a greenish color). The **pancreas** consists of a dorsal lobe extending anteriorly along the right side of the spleen, and a ventral lobe located below (ventral to) the anterior end of the intestine (duodenum).

6. Using Table 35A-1, identify the mesenteries associated with the digestive tract. In the shark, the ventral mesentery splits into two parts anteriorly: the *gastrohepatic ligament* between the stomach and liver and the *hepatoduodenal ligament* between the liver and small intestine (find the **bile duct** and blood vessels in the latter mesentery).

Frog

1. Extend the cut you made through the angle of the jaw until it reaches the anterior end of the esophagus. (You may also need to cut through the angle of the jaw on the other side to open the mouth completely.) Pull the lower jaw ventrally and to one side. Refer to Figure 35B-3a. Find the muscular **tongue** attached near the anterior margin of the floor of the mouth (this structure can be extended quickly to catch passing insects, which adhere to its sticky surface). Find the opening of the trachea, the **glottis,** between the base of the tongue and the esophagus in the pharynx. This is the passageway for air to the lungs.

2. Did you find the small **maxillary teeth** earlier? Feel them along the margin of the upper jaw. Find the **internal nares** near the margin of the **oral cavity.** Between the nares are **vomerine teeth,** extensions of the bones in the palate of the skull. These structures help the frog to hold prey after it is captured. Locate the openings of the **Eustachian tubes** near the angle of the jaw. Between the Eustachian tubes and the internal nares, the roof of the oral cavity bulges downward because of the eyes, located above the cavity.

   a. *Where is the boundary between the oral cavity and the pharynx in the frog?*

   _____

   b. *Which is larger in the frog: the oral cavity or the pharynx?* _____

3. In opening the body cavity of the frog, you exposed and located the heart anteriorly (Figure 35B-3b). The large dark organ behind and beside it is the **liver**—it consists of three lobes in the frog. The liver produces bile which is stored in the **gall bladder,** a small greenish sac embedded in the middle (left posterior) lobe of the liver.

   c. *What is the function of bile in digestion?* _____

      In excretion? _____

4. Lift the left lobes of the liver. Find the **esophagus** and the **stomach,** an enlarged part of the foregut that arcs to the left. Make an incision through the outer curvature of the stomach; remove any food and examine the wall of the organ. It contains folds or **rugae,** much like those of the shark, which allow the stomach to expand to hold food.

   d. *Identify any food remains in the stomach.* _____

Figure 35B-3 *(a) Oral and pharyngeal cavities of the frog. (b) Visceral organs of the frog.*

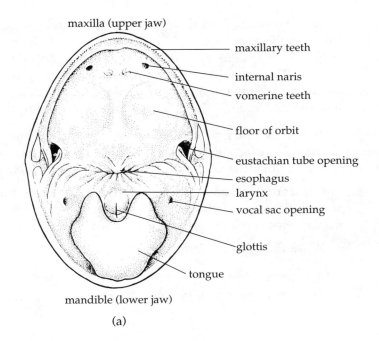

maxilla (upper jaw)

maxillary teeth

internal naris

vomerine teeth

floor of orbit

eustachian tube opening

esophagus

larynx

vocal sac opening

glottis

tongue

mandible (lower jaw)

(a)

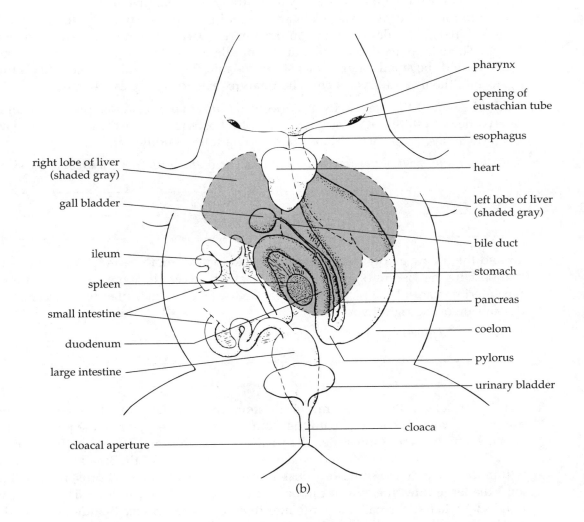

pharynx

opening of eustachian tube

esophagus

heart

left lobe of liver (shaded gray)

bile duct

stomach

pancreas

coelom

pylorus

urinary bladder

cloaca

right lobe of liver (shaded gray)

gall bladder

ileum

spleen

small intestine

duodenum

large intestine

cloacal aperture

(b)

5. Follow the stomach to the right. Locate the **pylorus** where the stomach joins the **small intestine.** The first section of the small intestine is the **duodenum;** ducts from the liver and pancreas enter here and many of the enzymes of the intestine are produced in this section of the gut. The remainder of the small intestine is coiled within the body cavity. Trace it to its junction with the large intestine (**colon**) in the dorsal area of the coelom. The colon empties into the cloaca. Find the **urinary bladder** ventral to the large intestine.

In reptiles, birds, and mammals, the bladder develops from one of the extraembryonic membranes (the allantois) formed during development of the embryo. However, frogs lack this membrane, so it is likely that the urinary bladder of the frog developed independently from the cloacal wall.

6. The **spleen** is a rounded, dark-colored organ located in the coils of the intestine. The **pancreas** is a whitish organ between the stomach and duodenum. Find and name as many of the mesenteries supporting the abdominal visceral organs as you can (Table 35A-1).

   e. Which belong to the dorsal mesentery system? _____

   _____

   The ventral mesentery system? _____

Turtle

1. Extend the cut made earlier through the jaw angle and make a second cut on the other side to open the **oral cavity.** Pull the lower jaw ventrally. The margins of the jaws are lined with keratin to form a horny beak—teeth are absent in turtles. Study the lower jaw, the mandible (Figure 35B-4a). The fleshy **tongue** in the anterior part of the mouth is attached along its length. There is a small raised area in the middle of the floor of the oral cavity—use your probe to find the small longitudinal slit in the middle of this raised area. This is the **glottis;** it opens into the respiratory system. The **pharynx** extends to the esophagus.

2. Study the roof of the oral cavity. Anteriorly, the **internal nares** open into the vault of the cavity; they are still located very near its front. Posteriorly, near the angle of the jaw, find the small openings of the **eustachian tubes** leading to the middle ear.

   a. Is the arrangement of the oral cavity and pharynx of the turtle more like that of the shark or that of the frog? _____

   _____

   _____

3. To find the following organs, you may need to remove more peritoneum and masses of fat that obscure the viscera (Figure 35B-4b). Locate the **liver,** the large dark organ filling much of the anterior space of the pleuroperitoneal cavity. Find the **gall bladder,** a greenish sac on the dorsal side of the right liver lobe.

   b. What is the function of the gall bladder? _____

   _____

   _____

4. Lift the left lobe of the liver to reveal the **stomach** to the left side of the body cavity. Find the **esophagus** where it connects with the stomach. Locate the **pylorus,** the junction of the stomach and the **small intestine.** The first section of the small intestine, the **duodenum,** receives ducts from the liver and pancreas. The remainder of the small intestine forms coils within the body cavity. Anteriorly, just above and caudal to the pylorus, the small intestine joins the **large intestine,** which can be traced to the cloaca posteriorly. There is a **caecum** (a blind sac) where the small and large intestines join. The **urinary bladder,** a bilobed structure, can be seen ventral to the large intestine.

Figure 35B-4 *(a) Oral and pharyngeal cavities of the turtle. (b) Visceral organs of the turtle.*

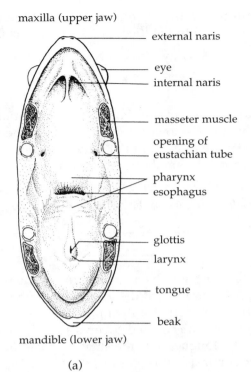

maxilla (upper jaw)

external naris

eye

internal naris

masseter muscle

opening of eustachian tube

pharynx

esophagus

glottis

larynx

tongue

beak

mandible (lower jaw)

(a)

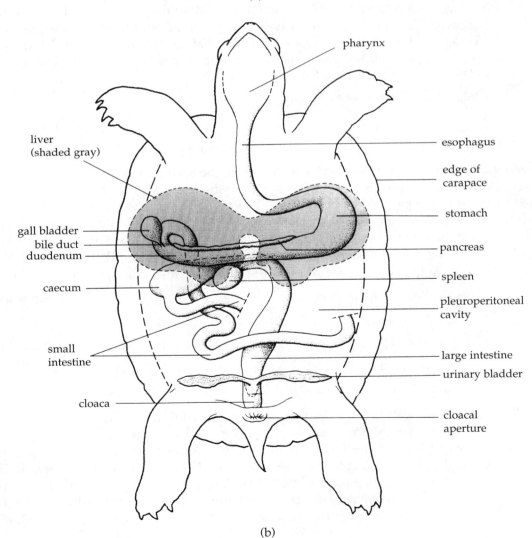

pharynx

liver (shaded gray)

gall bladder
bile duct
duodenum

caecum

small intestine

cloaca

esophagus

edge of carapace

stomach

pancreas

spleen

pleuroperitoneal cavity

large intestine

urinary bladder

cloacal aperture

(b)

5. Find the **spleen,** a dark-colored organ, located in the mesentery supporting the large intestine anterior and dorsal to the caecum. The **pancreas** lies along the duodenum.

6. Find and name as many of the mesenteries listed in Table 35A-1 as possible.

Rat

1. Make a mid-ventral incision from the throat region to the tip of the lower mandible through the skin covering the lower jaw, and separate the skin from the head on the side of the jaw that was not cut earlier. Find the salivary glands. Three pairs—the parotid, the mandibular, and the sublingual—are located behind the ear and extend ventrolaterally into the neck region.

   a. What are the functions of the salivary glands? _____

   _____

   _____

2. Using scissors (or bone shears), cut through the tissues forming the angle of the jaw on the intact side of the head. Pull the lower jaw ventrally. Identify the lips; note that they are separated from the bones of the upper and lower jaw by a space, the **vestibule.** Laterally, the angles of the jaws are covered by muscle and skin (which you have cut) forming the **cheeks.** Find the fleshy **tongue;** note the small papillae covering the surface of the tongue; these papillae contain taste buds. Basally, the tongue is supported by elements of the pharyngeal skeleton. Behind the tongue is the **epiglottis,** a flap that covers the slitlike **glottis** when food is swallowed. The roof of the **oral cavity** is separated from the nasal passages by the **hard palate** anteriorly and a posterior extension, the **soft palate,** which separates the roof of the **pharynx** from the nasal passages. The **internal nares** open into the pharynx behind the soft palate at the level of the glottis. Find these structures.

   b. What is the function of the epiglottis? _____

   _____

   _____

As vertebrates developed a more active metabolism, the regular replacement of air in the lungs became increasingly important. In reptiles and birds, the roof of the oral cavity is vaulted (raised) and partially separated from the lower part of the cavity. This allows food to be handled in a space largely separate from that carrying the respiratory air supply. In mammals, the separation of oral and nasal cavities provides an added function—it allows infants to develop the suction required to suckle or nurse, which was a prerequisite to the provision of milk to the young in placental mammals. However, in all vertebrates with lungs, there remains an incomplete separation of digestive and respiratory systems within the pharynx.

   c. What are some of the advantages and disadvantages of the incomplete separation of these two systems?

   _____

   _____

   _____

   _____

3. Pin the body wall flaps to the wax in your dissecting tray. Lift and push the heart and lungs to the animal's right. Find the **esophagus** embedded in the median septum above the heart (Figure 35B-5). Trace it posteriorly and follow it through the muscular diaphragm. Lift the **liver** and push the lobes to the right. Find the **U**-shaped **stomach** with the dark-colored **spleen** attached to its greater (outer) curvature. Holding the liver up, find the **pylorus,** the region where the stomach joins the **duodenum** of the **small intestine.** Part of the **pancreas** is located ventral to the pylorus; another part extends in the mesentery along the loop of the

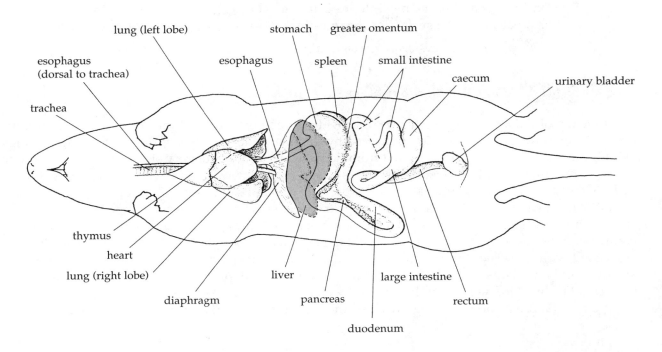

Figure 35B-5 *Visceral organs of the rat.*

duodenum. The small intestine is composed of a short segment, the **jejunum,** followed by the **ileum,** which coils to fill the posterior abdominal cavity. Ventrally the intestines are covered by an expanded part of the dorsal mesentery, the greater omentum, which forms a sac, or *omental bursa.* Lift this structure from the intestines. Find the junction of the jejunum with the **large intestine.** Projecting caudally from this junction is the large **caecum.** Find the **rectum,** the straight terminal portion of the large intestine passing to the **anus.**

4. The **liver** is the large dark organ lying just behind the diaphragm, which you moved to see the stomach and anterior intestines. It is divided into four lobes. Unlike the other representative vertebrates, *the rat has no gall bladder;* bile collects in the ducts and passes directly to the duodenum. Cut open the stomach along its greater curvature (left side). Rinse out any food that is inside and spread the stomach.

   d. *Describe the difference between the inner lining of the anterior part of the stomach and that of the*
      caudal portion of the stomach. _____

      _____

      _____

   In the rat, the anterior part of the stomach is expanded and has a cornified lining that includes striated muscle in its wall, much like the esophagus. The most posterior glandular part of the stomach produces mucus, enzymes, and acid, and its wall contains smooth muscle fibers.

   e. *What is a possible function of the anterior part of the stomach in rats?* _____

      _____

      _____

5. Identify as many of the mesenteries listed in Table 35A-1 as you can. Spread the omental bursa back over the surface of the intestines to see its extent and study its structure.

f. Can you see blood vessels in the omental bursa? _____ *Does it contain fat?* _____

✔ | **EXERCISE C** | **The Respiratory System**

The function of the respiratory system is to provide oxygen that can be transported to the tissues of the body and to remove carbon dioxide produced by the metabolic activities of the tissues. This requires a large surface area that must be protected from damage and osmoregulatory stress or drying. In vertebrates, there are two major respiratory surfaces: the gills and the lung/swim bladder. Accessory structures include the skin, the oropharyngeal cavity, and cloacal "bladders" in several groups.

Gills originated in the first chordates as pharyngeal slits used in filter-feeding. The organisms were small and unprotected by scales, so gas exchange occurred through the body surface or skin. In the earliest vertebrates, the jawless fishes, the gill bars (pharyngeal arches between pharyngeal slits), associated with vessels of the vascular arches, assumed the major role in gas exchange. Among the ancestors of bony fishes, a saclike outpocketing of the pharynx provided an accessory respiratory organ, the **lung.** Freshwater fishes living in stagnant water could simply "swallow" air and hold it in the lung (Figure 35C-1a). This ancestral lung provided the basis for two different specializations among vertebrates.

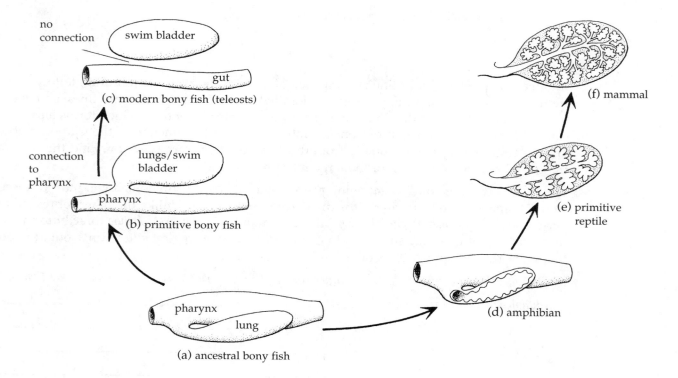

Figure 35C-1 *Evolution of swim bladders and lungs. (a) A bilobed outpocketing of the pharynx serves as an accessory respiratory organ. (b) The lung/swim bladder assumes a more dorsal position to help the fish float upright, and the lobes of the lung/swim bladder fuse. (c) The swim bladder has lost its connection to the pharynx in adult teleost fishes and no longer serves as a lung, only as a swim bladder or "ballast" organ. (d) The lung of an amphibian is a relatively simple vascular sac. (e) In primitive reptiles, the surface area is increased by folds of vascularized tissue. (f) Further partitioning in mammals leads to the formation of a copious number of microscopic sacs, the alveoli.*

First, in primitive bony fishes that live in well-aerated waters, the gills continue to function as the major respiratory organs, and gases in the lung reduce the density of the fish (like a ballast tank in a submarine). In this context, this structure functions as a **lung/swim bladder** and remains connected to the pharynx (Figure 35C-1b). In modern bony fishes (teleosts), this structure loses its connection to the pharynx during embryonic development and functions *only* as a swim bladder, not as a lung. Its contents are regulated by the circulatory system and a gland, which "secretes" oxygen into the swim bladder (Figure 35C-1c). The second line of specialization of this structure is as a true lung. In frogs, gas exchange is supplemented by the skin and the lung remains a relatively simple sac (Figure 35C-1d). In the turtle, and particularly in other reptiles, there is an increasing division of the lining of the lung; in mammals, this culminates in extremely small alveoli where air comes in close contact with the vascular system (Figure 35C-1e, f).

ııııı Objectives ııııııııııııııııııııııııııııııııııı

☐ Trace the evolution of respiratory mechanisms in fishes and terrestrial vertebrates.

☐ Explore mechanisms that ventilate respiratory surfaces in vertebrates.

ııııı Procedure ıııııııııııııııııııııııııııııııııııı

Shark

Set the lower jaw to one side. Find the **internal gill slits** protected by **gill rakers** (review Figure 35B-2a). Each internal slit opens into a large **branchial** (gill) **pouch,** then opens to the exterior through an **external gill slit.** The tissues between each pouch form a pharyngeal (gill) arch. Externally, extensions of the pharyngeal arches can close the external gill slits. The internal surfaces of most of the arches bear vascular **gill lamellae**, where gas exchange occurs between blood and water.

Note that the lack of a swim bladder and lungs is an ancestral trait in the shark. These accessory respiratory structures developed along the line leading to early bony fishes.

In the dogfish, gills are ventilated as muscles close the external gill slits and the floor of the oral cavity is depressed to draw water in through the mouth and spiracle. This is *inspiration.* During *expiration,* the mouth and spiracle are closed and the gill slits open as the floor of the oral cavity relaxes and is elevated, forcing water out over the gills.

Frog

1. If live tadpoles are available in the laboratory, study their **external gills.** Early in development, four pharyngeal (gill) slits appear but their outer openings are covered by a fold of skin, the **operculum.** This fold covers the undersurface of the neck and anterior part of the trunk, including the forelimbs as they develop. The chamber between the body wall and the operculum becomes filled with a mass of vascularized filaments. Water entering the mouth is passed through the pharyngeal slits to the filaments, where gas exchange occurs, and then out through an off-center opening, the **ostium.** Operculum, gills, and pharyngeal slits all disappear at metamorphosis. The appearance of the tadpoles will depend on their developmental stage. Some may already have legs protruding through the structure of the operculum.

    a. *What similarities exist between the respiratory systems of the shark and the tadpole?*

    _____

    _____

    _____

    b. *What are the differences between the respiratory systems of these animals?*

    _____

    _____

    _____

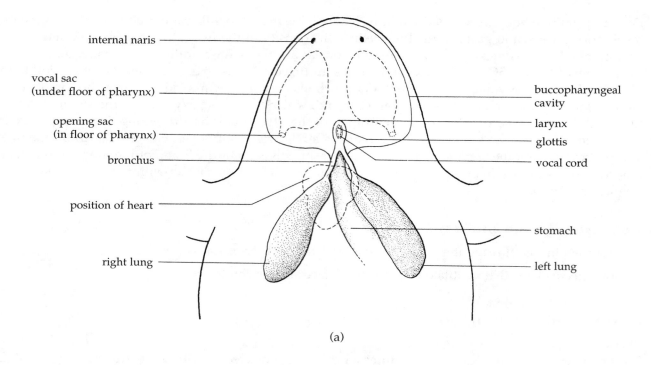

(a)

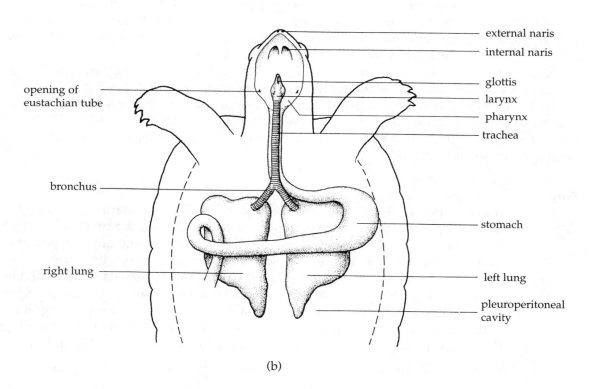

(b)

Figure 35C-2   *(a) Respiratory system of the frog (shown with tongue removed). (b) Respiratory system of the turtle.*

2. Open the oral cavity of your specimen and trace the passage of air from the outside through the nares and into the mouth to the **glottis.** This opening is surrounded by several cartilages forming the **larynx,** which contains the vocal cords (Figure 35C-2a). Push the abdominal viscera to the right and find the left **lung,** a saclike structure extending into the pleuroperitoneal cavity. Find the right lung. Anteriorly each lung attaches to the larynx by a short connecting tube, the **bronchus.**

3. Cut open one of the lungs and observe its inner surface.

    c. *Describe the surface of the lung.* _____

    _____

When a frog inhales, air first enters the oral cavity through open external nares as the floor of the mouth drops with the glottis closed. Then the external nares close, the glottis opens, and the floor of the oral cavity rises to force air into the lungs. Exhalation is essentially the reverse: with external nares closed and the glottis open, the floor of the oral cavity drops and then, nares open and glottis closed, the floor is raised and air is exhaled through the external nares. This respiratory mechanism can be described as an "oral pump." If live frogs are available, watch them breathe.

4. You have already noted that the skin is thin, moist, and richly vascularized. Study the inner surface of the skin. Note the large number of blood vessels. Gas exchange across this surface is of major importance in many amphibians (in the plethodont salamanders, lungs have been lost completely and gas exchange is confined to the skin and oral cavity).

Turtle

1. Trace the course of air through the nostrils and oral cavity to the **glottis,** an opening into the **larynx** (Figure 35C-2b). The larynx is composed of cartilages that support flaps regulating the size of the opening. Do turtles make sounds with their larynx? Cut open the glottis, cutting through the larynx and the anterior part of the **trachea.** Note the cartilaginous rings of the trachea that prevent it from collapsing when air pressure inside the tube is less than that of the surrounding air.

2. Use your scissors to cut the skin of the neck along the ventral midline and trace the trachea posteriorly (do not follow it all the way to the lungs—you would damage the circulatory system.) Near the lungs the trachea splits into two **bronchi.** The **lungs** are the dark structures lying against the carapace just posterior to the pectoral girdle. Lift the left lung and locate its bronchus. Make a lateral cut into the left lung using your scissors. Note the pockets of tissue separated by connective tisue strands.

    a. *Compare the internal structure of the turtle's lung with that of the frog (Figure 35C-2). How do they differ?* _____

    _____

    _____

The method of ventilation in the turtle differs from that in the frog. In the turtle, paired muscular membranes enclose the viscera and expel air by compressing the visceral organs against the lungs, an effect augmented by pulling in the neck and legs. Another pair of muscular membranes that enclose the flanks enlarges the body cavity, reducing pressure in the lungs and causing air to flow into them. The glottis is normally closed except during inhalation or exhalation.

Accessory respiratory organs in turtles include the vascularized skin, pharynx, and cloacal bladders. Pharyngeal respiration may also be important in some terrestrial reptiles. However, the principal organ for gas exchange in most turtles and in all other reptiles, birds, and mammals is the lung.

Rat

In mammals, the evolution of a secondary palate separated the oral cavity into a ventral food-handling area and a dorsal respiratory passageway. The paired **nasal cavities** above the palate contain three folds of bone, which increase the surface area of the mucosa exposed to air entering and leaving the body. Entering air is warmed, moistened, and filtered; the warmth of departing air helps to warm these same surfaces, thereby conserving heat essential to maintaining body temperature in an endotherm such as the rat.

1. Study special preparations of the skull showing the nasal cavities. Trace the course of air from the nostrils through the nasal passages to the internal nares.

2. In your organism, find the **glottis,** the opening through the **larynx,** a cartilaginous structure bearing vocal cords on its lateral walls. Insert the tip of your scissors into the glottis and cut through the larynx into the **trachea.** Note that the wall of the trachea is supported by cartilaginous rings that keep its passageway open during inspiration.

3. Open the thoracic cavity. Cut through the ventral body wall around the sternum, remove the sternum, and push back the wall of the thorax, breaking ribs as necessary. Trace the trachea caudally without damaging the blood vessels around the heart; near the lungs it branches, giving rise to two **bronchi.** Follow one bronchus into the **lung.** Use your probe to tease away lung tissue around the bronchus.

   a. *What happens to the cartilaginous supports as the bronchus becomes divided into finer branches?*

   _____

4. Using the scalpel, make a cut into the tissue of the lung and separate the edges.

   b. *Describe the tissue you see.* _____

Ventilation of the lungs in mammals depends on the muscles of the rib cage and the muscular diaphragm. On inspiration, the diaphragm is tightened, depressing the viscera and increasing the volume of the thoracic (pleural) cavity; during more vigorous breathing, the rib cage is also elevated, further increasing the space in this cavity. Because the lungs are mechanically coupled to the thoracic cavity, these actions draw in air. On expiration, the diaphragm relaxes and the rib cage is depressed, which "pushes" on the lungs so that air flows out.

Laboratory Review Questions and Problems

1. Define homeostasis. Give as many examples as you can of controlled physiological processes that contribute to homeostasis. (Add to this answer as you complete Laboratory 36.)

2. The heart develops beneath the digestive tract and, in reptiles, birds, and mammals, is supported during development by the ventral mesentery (dorsal mesocardium above and ventral mesocardium below). In the adult, however, the heart lies free in the pericardial sac. Why are the suspending mesenteries lost?

3. Why is it important that the *surface area* of the intestine be increased relative to the *volume* of the intestine? Why does the length of the intestine correlate with the diet of the animal?

4. List the functions of the liver. How does this organ contribute to homeostasis in vertebrates?

5. The first type of lung to arise in fishes was a simple vascularized sac. Describe the changes to the structure of the lung as it evolved.

6. Using information from this laboratory and from your lectures and textbook, construct a chart listing the various combinations of respiratory organs that may make up the respiratory systems of vertebrates. Give examples of animals that have each type of system, and describe the advantages and disadvantages of each system.

The Anatomy of Representative Vertebrates: Circulatory and Urogenital Systems

OVERVIEW

The **circulatory system** integrates many of the body's physiological responses. It provides nutrients and oxygen to, and removes wastes and carbon dioxide from, all living tissues of the body. It also carries information in the form of hormones and metabolites and plays a major role in immune responses and disease prevention.

The **urinary system,** which produces urine, and the **reproductive system,** which produces gametes, are closely associated during development. Except in jawless fishes and bony fishes, there is a general tendency for the male reproductive system to take over ducts of the urinary system, further intertwining their developmental history. For this reason, these systems are considered together as the **urogenital system.**

STUDENT PREPARATION

Prepare for this laboratory by completing Laboratory 35 and reading the text pages indicated by your instructor. Familiarizing yourself in advance with the information and procedures covered in this laboratory will give you a better understanding of the material and improve your efficiency.

If dissection tools are not provided, bring your dissecting kit to the laboratory with you.

EXERCISE A | **The Circulatory System**

The circulatory system includes a pump, the **heart,** and associated vessels: **arteries** that carry blood away from the heart and **veins** that carry blood back to the heart.

In vertebrates, the system is closed: blood is contained in vessels in almost all organs and tissues. **Capillaries** link arteries and veins and penetrate virtually every tissue of the body; these small-diameter, thin-walled vessels permit exchange of solutes and gases. In some tissues such as the liver, blood bathes tissues directly; these open channels, called **sinusoids,** also function in the exchange of materials.

Pressure supplied by contractions of the heart forces water, dissolved materials, and low-molecular-weight molecules through the single cell layer that makes up the capillary wall into the surrounding tissue spaces. As the vessels lose water, the high-molecular-weight molecules in the blood (albumins and other blood proteins) become more concentrated, providing an osmotic gradient that causes most of the fluid filtered from the system to return to the vessels. In terrestrial vertebrates, much of the remaining fluid is eventually returned to the venous system by vessels of the **lymphatic system.** Within the heart, veins, and lymphatic vessels, one-way **valves** ensure the proper circulation of blood.

PART I | **Arteries**

Arteries are thick-walled vessels carrying blood from the heart to the gills in fishes and to the body tissues in all vertebrates. During the course of vertebrate evolution, systems of arteries appear to have been derived from a basic ancestral plan. In this plan, a **ventral aorta** carries blood anteriorly from the heart to a series of **aortic arches** located within the pharyngeal arches. (The pharyngeal arches make up the wall of the pharynx; some of these arches bear gills.) Blood passes upward through the aortic arches, where capillaries associated with the gills provide for gas exchange. Blood is then collected by the **dorsal aorta.**

Evolutionary changes in the arrangement of these aortic arches, including redirection or loss of some and maintenance of others, occurred in response to a redistribution of gas exchange surfaces from gills to lungs, the development and partitioning of the heart, and the formation of paired appendages (Figure 36A-1). Derivatives of these arches carrying blood to the lungs are part of the **pulmonary circuit,** and those delivering blood to the body are part of the **systemic circuit.**

Posteriorly, the dorsal aorta branches to form unpaired vessels, including both **coeliac** and **mesenteric arteries,** which carry blood to visceral organs. Paired **segmental arteries** generally supply the "segmented" dorsal body wall. Paired **renal** and **gonadal arteries** supply the kidneys and gonads, respectively. A large pair of arteries, the **iliac arteries,** provides blood to the pelvic appendages. If a tail is present, the dorsal aorta continues into it as the **caudal artery.**

IIIII **Objectives** IIIIIIIIIIIIIIIIIIIIIIIIIIIIIIIIIIIIIIIIIIIIIIII

☐ Trace the evolutionary fates of the ancestral aortic arches as illustrated by the anatomical differences observed in the shark, frog, turtle, and rat.

☐ Locate the major arteries in the four representative vertebrates and learn how to dissect them so as to trace them to the periphery.

IIIII **Procedure** IIIIIIIIIIIIIIIIIIIIIIIIIIIIIIIIIIIIIIIIIIIIIIIIIII

Work in groups of four.

1. Obtain your specimen (shark, frog, turtle, or rat). Rinse it in the sink and place it on a tray or in a dissecting pan.

2. Locate Figures 36A-2 through 36A-5 (pages 36-20, 36-21, at the end of this laboratory). One of these diagrams illustrates the arterial system of your organism. Remove these pages from your manual and place them where you can refer to them easily during your dissection.

3. Follow the general dissection directions given below for each vertebrate. Remember that the terms "right" or "left" designate the animal's orientation.

   • **Shark** Orient your shark ventral side up. Remove the skin covering the area between the lower jaw and the pectoral girdle. Use your scalpel to remove the muscle tissue over the area of the pectoral girdle. With scissors, continue the mid-ventral incision from the abdomen through the middle of the pectoral girdle. Remove additional muscle tissue in front of the girdle until you reach a gray membrane—the **pericardial membrane.** Cut through this membrane to expose the **pericardial cavity.** Leave the heart in place. Refer to Figure 36A-2, page 36-20.

   • **Frog** Be sure that the heart is exposed—cut away the pericardium if necessary. Use your probe to separate the arteries leaving the heart from surrounding connective tissues; trace each branch. Refer to Figure 36A-3, page 36-20.

   • **Turtle** Be sure that the heart is exposed. Use your probe to separate the major arteries leaving the heart from surrounding connective tissues. Trace each branch as far as you are able. Refer to Figure 36A-4, page 36-21.

   • **Rat** Spread the body walls of your rat so that the heart is exposed. Identify the large **left aortic arch** (systemic aorta) arising from the heart and carefully clean away the tissue surrounding it. Refer to Figure 36A-5, page 36-21.

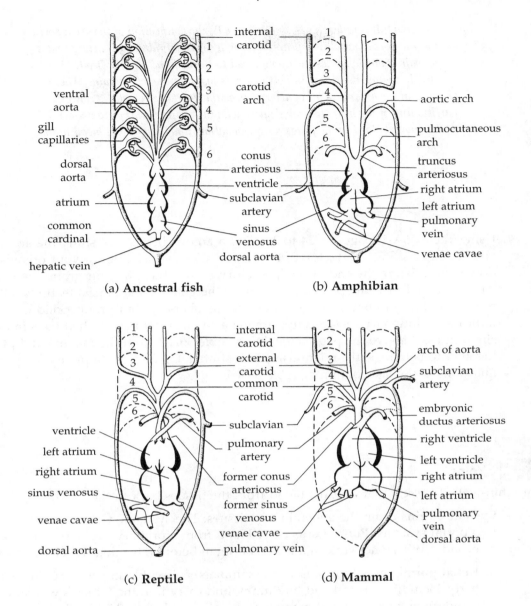

(a) **Ancestral fish**

(b) **Amphibian**

(c) **Reptile**

(d) **Mammal**

Figure 36A-1 *As you study the evolution of the aortic arches in vertebrates, observe how these structures of blood supply and drainage became modified as other structures and functions of the animals changed.*

(a) In ancestral fishes, the ventral aorta gave rise to six afferent branchial arteries carrying blood to the gills; six efferent branchial arteries picked up the blood from the gills and joined on each side to form a branch of the dorsal aorta.

With the evolution of jaws, the first arch became reduced. With the origin of the swim bladder (lung) in bony fishes, the sixth aortic arch became associated with the pulmonary circulation to the lungs. As the second pharyngeal arch became modified to form the operculum covering the gills in teleost fishes, the second aortic arch became reduced and was eventually lost.

(b) In adult frogs, the portions of the dorsal aorta between the third and fourth arches are lost, separating the general systemic circulation of the head from that of the body. The third aortic arch now carries blood only to the head. The fourth aortic arch serves as the major systemic trunk to the body. The sixth arch is primarily devoted to the pulmocutaneous circulation, providing blood to the skin and lungs, the primary organs of gas exchange in frogs. In some amphibians, the sixth aortic arch remains attached to the dorsal aorta.

(c) In adult reptiles, birds, and mammals, the connection of the sixth aortic arch to the dorsal aorta disappears, leaving a separate pulmonary artery and a single pair of aortic arches to supply blood to the body behind the head.

(d) In mammals, the right aortic arch is lost and the left remains. These changes in the arrangement of arteries near the heart are correlated with the partitioning of the heart to provide parallel, but separate, pumps to deliver deoxygenated blood to the lungs for oxygenation, and oxygenated blood to the body.

4. Using Table 36A-1, pages 36-24 to 36-27 as a guide, dissect and study the arterial system of your organism. Start at the top of the chart under the heading for your organism and work downward. Before the end of the laboratory, show others in your group the features that you have observed in your specimen and study the general anatomical features of the other three representative vertebrates dissected by the members of your group. Solid lines in Table 36A-1 indicate evolutionary relationships among arterial structures; dashed lines indicate functional similarities. Remove the pages containing Table 36A-1 from the manual and place them close together to better see the relationships illustrated. On a separate piece of paper, answer the questions in the last column of Table 36A-1.

PART 2   Veins

Veins are thin-walled vessels that conduct blood from the tissues back to the heart.

• **Systemic veins** drain the general body tissues. The systemic venous system includes one or two portal systems. **Portal veins** carry blood from a primary capillary network in one organ to a second capillary network in a different organ before returning the blood to the heart.

    a. **Renal portal veins** (not present in mammals) collect blood from capillaries in the posterior body, including the tail and hindlimbs, and carry it to the kidneys where, within a second capillary network, filtered materials are collected before the blood is returned to the heart.

    b. **Hepatic portal veins** carry blood containing many of the products of digestion from the intestines to a second capillary network in the liver where these materials can be metabolized or synthesized into storage products.

• **Pulmonary veins** (found only in vertebrates with lungs) return oxygenated blood from the lungs to the heart.

Phylogenetically and developmentally, veins arise as a series of paired vessels along each side of the body. Anteriorly, two major vessels, the **anterior venae cavae,** tend to persist. Posteriorly, there is an early trend toward the formation of a single **posterior vena cava,** derived by the fusion of several embryonic veins. This vein serves only the liver and anterior intestinal organs, and the kidneys in vertebrates with a well-developed renal portal system. In mammals, the renal portal system is absent and the posterior vena cava carries blood from *all* posterior regions of the body back to the heart.

Branches of veins tend to parallel those of arteries, and many have similar names. Because veins are more variable than arteries among vertebrate groups, and also have thin walls that make them more difficult to locate in preserved specimens, you will study evolutionary trends by comparing schematic diagrams of the circulatory systems of the four representative vertebrate groups. You will dissect a few of the major veins in the shark and rat as examples.

Objectives

☐ Describe the basic structural pattern of veins in vertebrates.

☐ Describe the functional differences that characterize the systemic, portal, and pulmonary systems of veins.

Procedure

1. In the schematic diagrams of Figure 36A-6, write the name of each vessel (in boldface type in figure legend) in the appropriate blank(s). *Note:* The diagrams in Figure 36A-6 are schematic and do not show the correct positions of arteries and veins relative to dorsal and ventral aspects of each organism.

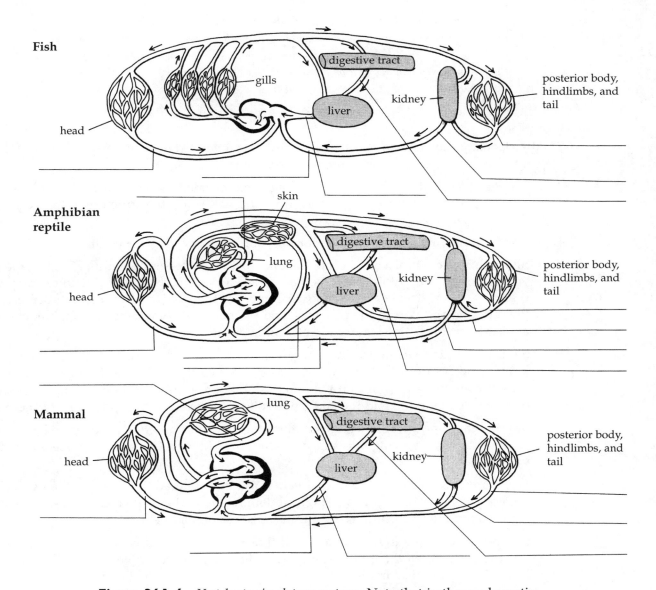

Figure 36A-6 *Vertebrate circulatory systems.* Note that in these schematic diagrams, only one side of the body is represented, so only one of each pair of vessels is shown. *Paired veins drain the head region in all vertebrates. In the shark, two anterior cardinal veins carry blood from the head to the first chamber of the heart. In the frog, turtle, and rat, these veins are called the anterior venae cavae. In the shark, two veins, the posterior cardinal veins, carry blood from*

the liver, kidney, gonads, and body wall back to the heart. In more advanced vertebrates, such as the frog, turtle, and rat, embryonic veins, including the posterior cardinal veins, fuse to form a single posterior vena cava. A hepatic portal system drains the digestive tract, carrying blood to the liver by way of the hepatic portal vein.

In all vertebrates except mammals, a renal portal system drains the posterior parts of the body, carrying blood to the kidneys by way of the renal portal veins. Blood from the kidneys and liver is returned to the systemic circulation through the renal and hepatic veins that join the posterior cardinal veins (in sharks) or the posterior vena cava (in the other representative vertebrates). In the frog and turtle, these portal systems are interconnected by the ventral abdominal vein, allowing blood returning from the posterior portions of the body to pass through either the kidneys or the liver.

In the rat, the renal portal system is absent and blood from all posterior regions of the body returns directly through the posterior vena cava. Blood from the limbs is returned to the posterior vena cava by the paired iliac veins and from the tail by the caudal vein. In vertebrates with lungs, pulmonary veins return oxygenated blood from the lungs to the heart.

2. After labeling the veins in Figure 36A-6, use your knowledge from Part 1 of this exercise to identify the major arteries shown in each diagram.

3. Use a pencil to shade those vessels carrying deoxygenated blood in each diagram. Do not shade the chambers of the heart at this time. You will study the structure of the heart in Part 3 of this exercise.

4. If time permits, find the following veins in your representative vertebrate.

   • **Shark: hepatic portal vein** Locate the bile duct in the lesser omentum. The hepatic portal vein is dorsal to it within the mesentery. Trace it posteriorly and find its three major branches: the gastric vein to the posterior part of the stomach; the splenomesenteric vein to the intestine and part of the pancreas; and the pancreaticomesenteric vein, branching to the anterior part of the intestine, pancreas, and spleen.

   • **Shark: posterior cardinal veins** These veins run anteriorly between the medial edges of the kidney and the dorsal aorta. Uninjected veins are easily located in the shark as thin-walled vessels containing the dark brown remains of blood. Run your probe along the peritoneum on one side along the midline and try to follow the fixed blood within one of these vessels.

   • **Shark: renal portal veins** The renal portal veins course anteriorly along the lateral margins of the kidneys, emptying into sinuses within the kidneys. Using your probe, locate these vessels.

   • **Frog and turtle: ventral abdominal veins** You encountered these veins when you opened the body cavity. The frog has a single abdominal vein; the turtle has two ventral abdominal veins connected to each other by a transverse abdominal vein.

   • **Rat: posterior and anterior venae cavae** Push the abdominal viscera to the organism's left. Look to the right of the midline and locate the single posterior vena cava. Follow it to the heart. Locate the two anterior venae cavae where they join the heart to return blood from the anterior parts of the body.

   • **Rat: pulmonary veins** Lift the heart to the right. Using a probe, try to find one of the four short, thin pulmonary veins returning blood from the lungs to the heart.

5. If prepared slides of arteries and veins are on display, observe them first at low power (10×) then in more detail at high power (40×).

a. Describe the visible differences between these two types of vessels. _____

_____

_____

_____

✔ **PART 3** **Heart**

In ancestral vertebrates (without lungs), including living representatives such as the shark, the heart is composed of four chambers arranged in a row—it is a **serial heart.** Blood returning from body tissues flows into the most posterior chamber, the **sinus venosus,** and then to the **atrium,** which injects blood into the third chamber, the **ventricle.** The thick muscular walls of the ventricle supply the major force for moving blood through the vascular system. Finally, blood enters the most anterior chamber, the **conus arteriosus,** which connects to the ventral aorta.

All four of these chambers are readily recognizable in the shark (Figure 36A-7a, page 36-22), although the heart tube has folded upon itself.

- Label the four chambers of the shark's heart on Figure 36A-6 (page 36-5). Use the following abbreviations: **SV** = sinus venosus; **A** = atrium; **V** = ventricle; and **CA** = conus arteriosus.

In other vertebrates the embryonic heart is tubular but, during development, it folds and twists, and two of the chambers—the conus arteriosus and sinus venosus—lose their independent existence as they become incorporated into the tissues of the atrium and ventricle. The rhythm of contraction in the heart originates within the cardiac muscle tissue of the sinus venosus. If the heart is cut into separate parts, the rhythm of the sinus venosus is fastest and, in the intact heart, this rhythm initiates the wave of contraction passing through the heart. In other words, the sinus venosus acts as a **pacemaker.** In mammals, the sinus venosus becomes incorporated into the wall of the chamber anterior to it, the atrium, where it remains as a small patch of tissue that serves as the pacemaker.

With the appearance of lungs during the course of evolution, a septum developed that separated the atrium into two parts: oxygenated blood returning to the atrium from the lungs is separated from deoxygenated blood returning from the body. In the frog, the ventricle remains unpartitioned, allowing some oxyenated blood and deoxygenated blood to mix before being pumped to the body organs. In turtles, the ventricle is partially divided into two chambers by an interventricular septum that is nearly complete, so that little mixing of oxygenated and deoxygenated blood occurs. Oxygenated blood is shunted first to the head and body, while deoxygenated blood is pumped to the lungs.

- Label the right and left atria and ventricle of the frog/turtle heart in Figure 36A-6. Use the following abbreviations: **LA** = left atrium; **RA** = right atrium, **V** = ventricle. The sinus venosus, emptying into the right atrium, is reduced and not shown in this diagram. The conus arteriosus is divided and forms the basal portion of the two arterial trunks—each called a truncus arteriosus—leaving the heart. Label this area **TA** in Figure 36A-6.

In birds and mammals (and the alligator), the partitioning of the ventricle is complete, and systemic and pulmonary blood occupy separate circulatory channels in the heart.

- Label the right and left atria and ventricles of the rat heart in Figure 36A-6. Use the following abbreviations: **LA** = left atrium; **RA** = right atrium; **LV** = left ventricle; **RV** = right ventricle. The sinus venosus persists as the pacemaker in the wall of the right atrium.

Each chamber of the heart is separated from the next by flaps in the wall of the heart, forming valves. The sinus venosus and atrium are separated by the **sinoatrial valve;** the atrium and ventricle by the **atrioventricular valve;** and the ventricle and conus arteriosus by a series of **semilunar valves.** In mammals, the semilunar valves are a remnant of the conus arteriosus. Valves prevent backflow and maintain a directed and effective circulation.

Thus, during the course of evolution, we see the circulatory system evolve from a **serial** arrangement (heart → gills → systemic circulation → heart) to one in which systemic and respiratory flows are **parallel** (heart → systemic circulation → heart) and (heart → pulmonary circulation → heart).

ııııı **Objectives** ıııııııııııııııııııııııııııııııııııııı

☐ Describe structural changes in the hearts of vertebrates as respiration shifted from gills to lungs and as tetrapods moved onto land.

ııııı **Procedure** ıı

Use the following directions to dissect and study the structure of the heart in your specimen. Remove Figure 36A-7, page 36-22, and place it where you can refer to it easily. Share what you learn with others in your group. Compare anatomical structures in the four animals to observe evolutionary adaptations to the presence of gills and of lungs.

Shark

1. Refer to Figure 37A-7a to identify the structures of the shark's heart. The thick muscular wall of the heart lying just in front of the transverse septum is the **ventricle.** Anteriorly, the ventricle connects with the **conus arteriosus** and then with the ventral aorta. Dorsal to the ventricle is the thin-walled **atrium.** Lift the ventricle to see the atrium. It connects to a broad triangular sac, the **sinus venosus,** applied to the anterior face of the transverse septum. Note that the heart is folded in an **S** shape, with the atrium above and somewhat anterior to the ventricle.

   a. *List the four chambers of the shark's heart.* _____

   _____

   b. *Return to Figure 36A-6. Use your pencil to shade the portions of the shark's heart and other vessels that carry deoxygenated blood from the heart to the gills.*

   c. *We humans also have four chambers in our heart. How does their arrangement differ from that in*

   the shark? _____

   _____

   _____

   _____

Frog

1. Refer to Figure 36A-7b to identify the structures of the frog's heart. Lift the muscular **ventricle** and follow the posterior vena cava beneath it to the **sinus venosus** (reduced in size from the sinus seen in the shark). Cut the posterior vena cava and lift the heart forward if necessary. Find the point at which the paired anterior venae cavae enter the sinus venosus, which empties into the **right atrium.**

2. In the adult frog, the atrium is completely divided by an **interatrial septum** into right and left chambers, the **right atrium** receiving the deoxygenated blood from the systemic circulation and the **left atrium** receiving the oxygenated blood from the pulmonary and cutaneous veins. Blood from both chambers passes into the undivided muscular **ventricle.** However, deoxygenated systemic blood is injected into the ventricle first and occupies the bottom of the ventricle, and oxygenated pulmonary blood, injected last, occupies the upper portion nearest the **conus arteriosus.** The conus is highly specialized to direct blood selectively into the three major branches of the ventral aorta on each side: the most oxygenated blood from the pulmonary circulation passes to the head, oxygenated blood returned from the cutaneous circulation passes to the body, and the least oxygenated blood, collected by the sinus venosus, is moved to the respiratory surfaces.

   a. *Return to Figure 36A-6. Use a pencil to shade the parts of the frog/turtle heart that carry deoxygenated blood.*

b. *The frog heart retains the four parts found in the shark: sinus venosus, atrium, ventricle, and conus arteriosus, although the first and last of these regions are reduced. However, the atrium is completely divided. How many chambers does the frog heart possess?* _____

c. *What difference do you see in the thickness of the atrial and ventricular walls?*

_____

d. *What differences in the functions of the atrium and ventricle might explain the difference in the thickness of their walls?* _____

_____

_____

_____

Turtle

1. Refer to Figure 36A-7c to identify the structures of the turtle's heart. The large muscular **ventricle** is located ventrally and gives rise to the major arterial trunks that you have already identified. The bases of the systemic and pulmonary aortae contain remnants of the **conus arteriosus** (the conus arteriosus contributes to the formation of the semilunar valves that prevent backflow of blood in each trunk). The **right atrium** is located to the right of the arterial trunks and receives blood from the anterior and posterior venae cavae; the smaller **left atrium,** receiving the **pulmonary veins,** is largely covered ventrally by the arterial trunks. The atria are completely separated by an interatrial septum. Lift the ventricle and find the **sinus venosus.** It is further reduced from that of the frog and is represented by the base of the anterior venae cavae, posterior vena cava, and hepatic vein as they enter the right atrium.

2. In turtles, unlike frogs, an **interventricular septum** partially divides the ventricles into dorsal and ventral chambers. The smaller ventral chamber is associated with the pulmonary trunk; the dorsal chamber gives rise to the systemic aortic arches. *Both* atria open into the dorsal chamber, but the mechanics of circulation are such that deoxygenated blood passes to the right side of the heart and into the pulmonary circulation, while oxygenated blood passes to the left and travels to the body. These shunts may be important in redirecting blood flow depending upon environmental circumstances—whether the turtle is afloat or submerged.

   a. *Which part of the heart of the turtle contains the pacemaker?* _____

Rat

1. Refer to Figure 36A-7d to identify the structures of the rat's heart. The mammalian heart has developed a four-chambered construction consisting of two atria and two ventricles supporting parallel circulatory pathways, one to the lungs and one to the body. Find the **ventricles** either in your rat, or in a larger sheep heart that may be available on demonstration, or in a model. They are thick, muscular structures forming the "pointed" end of the heart and are completely divided internally by an **interventricular septum.** The **atria** are the two small sacs located anteriorly to each side of the heart. As in the frog and turtle, they are divided by an **interatrial septum.**

2. Examine the interior of the heart. If you are studying the rat's heart, use a razor blade to make a cross section through the ventricle. Remove any latex or blood in the chambers.

In mammals, deoxygenated blood is returned by the three venae cavae to the right atrium. The sinus venosus has been incorporated into the wall of this structure and is represented by the **sino-atrial node,** a patch of modified cardiac muscle tissue that can spontaneously initiate its own impulse and contract, thus serving as the pacemaker of the heart. Upon contraction of the right atrium, blood passes into the right ventricle and is forced into the pulmonary trunk as the contraction proceeds. The pulmonary veins bring blood back to the left atrium. This oxygenated blood then passes into the left ventricle and is pumped to body tissues.

Contraction of the ventricles is speeded by a second node of specialized tissue, the **atrio-ventricular node,** which gives rise to conductile fibers, also composed of modified cardiac muscle tissue. As the wave of contraction initiated by the pacemaker (sino-atrial node) reaches this structure, these fibers carry action potentials through the interventricular septum to all areas of the ventricle, leading to a rapid and forceful contraction. Backflow to the atria is prevented by atrio-ventricular valves, the **tricuspid** ("three-toothed") **valve** on the right side and **bicuspid** ("two-toothed") or **mitral valve** on the left. Backflow into the heart is prevented by **semilunar valves,** remnants of the conus arteriosus.

a. Explain how the forms of the valves are related to their function of preventing the backflow of blood.

_____

_____

_____

    3. Remove the heart from your rat and identify the systemic and pulmonary arteries and the venae cavae and pulmonary veins. Make a frontal section of the remainder of the ventricle, passing through the atria. Remove any latex or blood and try to find the atrio-ventricular valves and large tendons (**chordae tendinae**), the "heartstrings," that support the valves, preventing them from being pushed back into the atria as the ventricles contact.

EXERCISE B | Urogenital System

In Laboratory 34 we mentioned that aspects of an organism's evolutionary history and its phylogenetic relationships with other organisms are sometimes revealed in the study of its embryology. The development of the vertebrate kidney illustrates this principle.

In early development, the urogenital system of vertebrates, including the kidneys, gonads, and their associated ducts, develops within two ridges of mesoderm extending along the anterior–posterior axis of the embryo on either side of its midline. Within each ridge, a longitudinal series of kidney tubules drains into a duct that carries urine to the outside. As development proceeds, the general pattern among vertebrates is that most anterior kidney tubules (**pronephric** tubules) degenerate, while some of those in the mid-region (**mesonephric** tubules) are appropriated by the male reproductive system to carry sperm. Kidney tubules in the more posterior region of the embryonic ridge (**metanephric** tubules) become part of the functional adult kidney (Figure 36B-1).

From a phylogenetic perspective, as we proceed from ancestral animals to those derived more recently, the adult kidney tends to become more compact, and the functional tubules, which increase in number, are increasingly derived from the more posterior portion of the embryonic kidney, while the more anterior areas of the embryonic kidney are increasingly taken over by the male reproductive system (Figure 36B-2).

These trends can be observed in the representative vertebrates you are studying. For example, in adult sharks, the kidney extends along the entire length of the coelomic cavity, much as in the vertebrate embryo. However, the most anterior of the embryonic tubules have degenerated and some of the mesonephric tubules connect to the testes. The remaining mesonephric and metanephric tubules form the functional kidney. A duct extending the length of the kidney carries urine and sperm to the outside. The same pattern is seen in amphibians, but the kidney does not extend the length of the coelomic cavity. In reptiles, birds, and mammals, only the most posterior metanephric tubules produce urine; the kidney is a **metanephric kidney.** The mesonephric tubules are associated with the testes, and the embryonic kidney duct is completely taken over by the gonads in adult males and carries only sperm to the outside. Another duct, the ureter, drains the kidney.

a. Briefly describe how the embryonic development of vertebrates reflects some aspects of evolutionary development.

_____

_____

_____

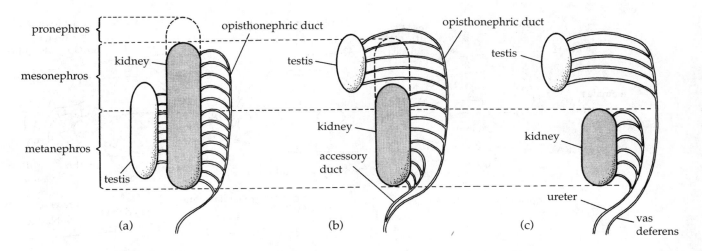

Figure 36B-1   *The urogenital systems of representative male vertebrates. During embryonic development of the kidney, three regions can be identified: the anterior region (pronephros), the mid-region (mesonephros), and the posterior region (metanephros). As development proceeds, kidney tubules of the anterior region degenerate.*

(a) As development continues in primitive jawed fishes and many amphibians, the remaining mesonephric tubules and metanephric tubules multiply to form the kidney. The opisthonephric duct in the adult carries both sperm and urine.

(b) In sharks and some amphibians, the more anterior mesonephric tubules degenerate while the posterior mesonephric tubules and metanephric tubules form a more compact kidney. The anterior segment of the embryonic duct now carries sperm, but posteriorly, reproductive cells mix with urine from part of the kidney.

(c) In reptiles, birds, and mammals the kidney tubules of the mesonephros (functional in the embryo) are modified to carry sperm in the male adult, becoming the sperm duct, or vas deferens; these tubules degenerate in the female. The adult kidney is now derived exclusively from metanephric tubules and is drained by another duct, the ureter.

In the development of most vertebrates, a second pair of ducts, the Müllerian ducts, also forms parallel to the original kidney duct. As development proceeds, the Müllerian ducts disappear if the individual is male, but persist as the oviducts if the individual is female.

In many fishes, amphibians, and reptiles, ducts carrying urine and reproductive products open into the urogenital sinus which joins the **cloaca** (Figure 36B-2a, b). Digestive wastes are also received by the cloaca. In mammals, a urinary bladder receives urine from the kidneys via the ureters, and urine passes to the outside through a **urethra.** In females, reproductive ducts form **uteri** (or join to form a single **uterus**) connected to the outside by a second opening, the **vagina.** In males, gametes travel to the outside through the same urethra used to carry urine. Digestive wastes are excreted directly through the anus (Figure 36B-2c).

ıɪɪɪɪ **Objectives** ıɪɪɪɪɪɪɪɪɪɪɪɪɪɪɪɪɪɪɪɪɪɪɪɪɪɪɪɪɪɪɪɪɪɪɪɪɪɪ

☐ Locate the kidneys and their ducts in representative vertebrates and relate their structure to their function.

☐ Describe the embryonic and phylogenetic development of kidneys in vertebrates.

☐ Locate the gonads and their ducts in representative vertebrates and relate their structure to their function.

☐ Describe how the ducts of the gonads and kidneys are related both embryologically and phylogenetically.

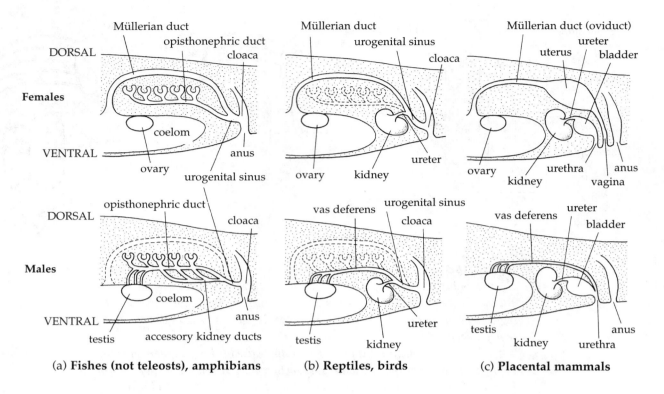

DORSAL

Females

VENTRAL

Müllerian duct
opisthonephric duct
cloaca
coelom
anus
ovary
urogenital sinus

DORSAL

Males

VENTRAL

opisthonephric duct
cloaca
coelom
anus
testis
accessory kidney ducts

(a) **Fishes (not teleosts), amphibians**

Müllerian duct
urogenital sinus
cloaca
ureter
ovary   kidney

vas deferens   urogenital sinus
cloaca
ureter
testis   kidney

(b) **Reptiles, birds**

Müllerian duct (oviduct)
ureter
uterus   bladder
urethra
ovary   kidney   anus
vagina

vas deferens   ureter
bladder
testis   kidney   anus
urethra

(c) **Placental mammals**

Figure 36B-2   *The urogenital systems and cloacae of vertebrates. The schematic diagrams at the top show structures found in females; the drawings below show structures found in males.*

In females, the embryonic Müllerian duct connects the coelom to the outside. These ducts develop into the oviducts and uteri (if present) in adults. (a) In fishes and amphibia, eggs are shed into the coelom, picked up by the oviduct, and carried to the outside. Fertilization is usually external. (b) In reptiles, birds, and mammals, fertilization is internal and, in the first two groups, the oviduct is modified to add materials, including a shell, around the egg before it is laid. (c) In mammals, the fertilized egg is retained in a modified portion of the oviduct, the uterus, where the embryo grows and develops before being born.

The fate of the reproductive structures and urinary structures is intertwined. Among fishes and amphibians, modified mesonephric tubules carry sperm into the opisthonephric duct and, to some extent, sperm and urine are mixed in this common duct. In reptiles, birds, and mammals, mesonephric elements, which function in the embryonic kidney, degenerate in the adult, leaving that part of the embryonic duct free to be only a sperm duct, the vas deferens in the adult. Rudiments of the Müllerian duct degenerate in the adult male.

In most nonmammalian vertebrates, digestive, urinary, and reproductive products pass through a common chamber, the cloaca, on their way to the exterior. In adult mammals, however, a partition separates the digestive system from the urogenital ducts and, in female mammals, the common urogenital chamber may be further reduced as the vagina and urethra become largely separated. In mammals, a bladder separates the ureter from the urethra, which carries urine to the outside.

Procedure ||||||||||||||||||||||||||||||||||||||||||||

Shark

1. Remove Figure 36B-3, page 36-28, and refer to it to identify and study the urogenital structures of the shark. Push the viscera of your specimen to one side, or the organs can be removed if your instructor approves. In the shark, the paired **kidneys** lie lateral to the mid-dorsal line of the body cavity and above the parietal peritoneum. They extend throughout much of the body cavity. The anterior portion is less well developed than the caudal region, and much of the production of urine takes place in the caudal portion of the kidney. Make a 1-to 2-inch cut through the peritoneum on one side along the lateral edge of the kidney and remove the peritoneum to the midline. Locate the **opisthonephric duct** on the ventral surface of the kidney. This duct carries sperm in males and urine in both sexes. The adult kidney of the shark is derived from both mesonephric and metanephric kidney elements formed during development.

2. If you have a male, locate the paired **testes** at the anterior end of the body cavity (Figure 36B-3a). Each is attached to the dorsal wall of the body by a mesentery. Several small **efferent ductules** can be seen passing through this mesentery to connect the tubules of the testis to the opisthonephric duct in the anterior part of the kidney.

    Caudally, the opisthonephric duct expands to form a **seminal vesicle.** Posterior outpocketings from these vesicles form the **sperm sac** on each side of the body. The union of the sperm sacs along the midline forms the **urogenital sinus** opening into the cloaca. Find these structures.

3. If you have a female, locate the **ovaries** at the anterior end of the pleuroperitoneal cavity (Figure 36B-3b). They may contain eggs up to an inch (or more) in diameter. The ovaries are also suspended by a mesentery. The **oviducts** originate with a mid-ventral opening into the body cavity located just behind the transverse septum. Trace the oviducts dorsally and caudally. They are suspended by a mesentery. Anteriorly, there is a slight swelling, the **shell gland,** and posteriorly, the **uterus** may appear as a slightly enlarged section of the oviduct or it may be greatly distended with young if the shark is pregnant.

    Examine the **cloaca.** Find the **urinary papilla,** a conical projection enclosing the urinary sinus formed by the distal ends of the opisthonephric ducts.

4. If your shark is a pregnant female, make an incision through the wall of the uterus. Note the structure of the uterine wall. Extend your cut until you can remove a pup, but leave the yolk sac inside the uterus. Carefully fold back the cut edge of the uterus to examine its relation to the yolk sac.

    b. Describe any modification of the uterine lining in the vicinity of the yolk sac.

    _____

    c. Describe how mother and infant are bound to each other. _____

    _____

    _____

    _____

Frog

1. Remove Figure 36B-4, page 36-29, and refer to it to identify and study the urogenital structures of the frog. Locate the **kidneys** in your specimen. They are dark, elongate organs found in the mid-dorsal body wall of the body cavity on each side of the midline. As in the shark, they lie outside the peritoneum. Make a lateral incision through the peritoneum and pull it toward the midline. The frog's kidneys, like those of the shark, are composed of tubules from the middle and posterior parts of the embryonic kidney. They are drained by **opisthonephric ducts** that originate from the convex outer margins of the kidney and pass to the cloaca. In males, each

expands near the cloaca to form a **seminal vesicle.** In males, the opisthonephric ducts carry both urine and gametes; in females, they carry only products from the kidney. A ventral diverticulum of the cloaca forms a sac, the **urinary bladder,** that stores urine. Cut through the ventral wall of the cloaca and locate the entry of the opisthonephric ducts and urinary bladder.

2. If your frog is a male, locate the **testis** on one side (Figure 36B-4a). It is an ovoid, cream-colored body ventral to the anterior end of the kidney. The testes is attached to the body wall by a mesentery through which the **vasa efferentia** (derived from mesonephric kidney elements) carry sperm from the testis to the opisthonephric duct in the kidney. Lying over the anterior end of the testis is a yellowish **gonadal fat body.** Lipids and hormones stored in this body are important for reproductive activity in the frog.

3. If your frog is a female, find the **ovaries** (Figure 36B-4b). They may occupy much of the body cavity in the spring; in the fall they are small and located ventral to the anterior end of the kidney. They are attached to the wall of the body cavity by a mesentery. Coiled oviducts with **ostia** on each side carry eggs to the exterior. Eggs pass through an expanded portion of each duct, the **uterus,** near its entry into the cloaca. Find the entry of the two uteri into the cloaca. Females also have yellowish **gonadal fat bodies.** Find them.

   a. *Where are the eggs of the frog fertilized?* _____

      How is this reflected in the structure of the reproductive system? _____

      _____

   b. *If you look carefully, you may observe an oviduct in the male frog—a "leftover" from embryonic development. How could this be possible?* _____

      _____

4. Note the **adrenal glands,** thin bands of whitish tissue lying along the ventral surface of the kidneys.

   c. *What is the function of the adrenal glands?* _____

      _____   _____

Turtle

1. Remove Figure 36B-5, page 36-30, and refer to it to identify and study the urogenital structures of the turtle. Cut through the large intestine of your turtle about an inch from the **cloaca** and lift the abdominal viscera forward (cut or tear the mesenteries). The digestive organs may be removed completely by cutting through the esophagus if your instructor approves. Expose the cloaca by cutting through the pelvic girdle on each side of the mid-ventral line with bone shears. Remove tissues around the cloaca and expose it from its aperture to the base of the urinary bladder. Separate the large bilobed **urinary bladder** from underlying tissues. Make an incision through the wall of the bladder and study the texture of its surface.

2. Push the reproductive organs to one side if necessary and locate the **kidney** on that side. It is a lobed structure closely applied to the dorsal wall of the pleuroperitoneal cavity, from which it is separated by a layer of peritoneum. Remove the peritoneum covering the kidney if you have not already done so in following the renal portal system. Remove the renal portal vein carefully. The whitish tube lying beneath (dorsal to it) is the **ureter.** Make an incision through the cloacal wall to one side of the mid-ventral line (avoid the penis if your specimen is a male) and spread the cloacal walls. Try to pass a dissecting needle through the ureter to the cloaca. Using your probe, find the entry of the urinary bladder into the cloaca. Above it, find the opening of the rectum into the cloacal chamber.

3. Tease away tissue from the sides of the cloaca at the level of the bladder. Find two thin-walled sacs, the **accessory urinary bladders,** on each side. Open one of these sacs and probe its

connection with the cloaca. These sacs may be used by females to carry water to soften the earth as they dig nests; they may also be used for respiratory gas exchange in some turtles.

4. If your specimen is a male, locate the **penis,** a mass of tissue in the ventral cloacal wall (Figure 36B-5a). This structure has two cavernous bodies which can be engorged by blood, causing it to become erect and protrude from the cloacal aperture. It is inserted into a female's cloacal aperture and sperm, carried along the dorsal **urethral groove** by cilia, enter the female's cloaca and fertilize her eggs. Because the turtle does not have a closed sperm duct in the penis, copulation may take a relatively long time. The erect penis must be relatively long to enable it to reach the female's body, given that both partners are encumbered by rigid shells. (Ogden Nash: "The turtle lives 'twixt plated decks which practically conceal its sex; I think it wonderful of the turtle, in such a fix, to be so fertile.")

5. If your specimen is a male, locate the **testes,** ovoid whitish organs attached to the body wall ventral to the anterior end of the kidney by a mesentery. The mesentery carries **efferent ductules** (mesonephric kidney elements) which carry sperm to the **epididymis,** an elongated mass of coiled tubules, in which sperm can be stored (Figure 36B-5a). Distally, the sperm travel from the epididymis through the **vas deferens** to the cloaca. Trace these ducts to their union with the cloaca near the base of the retracted penis.

6. If your specimen is a female, find the **ovaries**—large, egg-filled organs ventral to the kidney on each side of the body cavity (Figure 36B-5b). The spherical bulges in the ovary are the eggs in various stages of development. At ovulation, eggs are shed into the coelom and pass into the oviduct through its opening, the **ostium.** You can find this opening in the mesentery of the ovary near the cephalic end of the oviduct. Follow the oviduct to the cloaca. Examine the mid-ventral wall of the cloaca and locate the **clitoris,** a series of dark thickenings that are homologous to the male's penis. As eggs pass down the oviduct, various secondary membranes are added, including albumin, shell membranes, and a shell, often leathery in turtles.

Rat

1. Remove Figure 36B-6, page 36-31, and refer to it to identify and study the urogenital structures of the rat. Locate the **kidneys,** embedded in a cushion of fat against the mid-dorsal wall of the body cavity above the peritoneum. Using your probe, remove the peritoneum and fat around one of the kidneys, preserving the blood vessels serving the organ. The medial surface of the organ is concave, and renal vessels and the **ureter,** the duct carrying urine from the kidney, enter the organ in this region. Using your probe, follow the ureter posteriorly until it joins the base of the **urinary bladder.** If your specimen is male, be careful not to damage the sperm ducts that loop around the ureter near the bladder.

2. Using your scalpel or a razor blade, section the kidney in the plane of the body wall. The outer tissue is the **cortex.** It contains most of the microscopic glomeruli within the capsules of the kidney tubules (**nephrons**), where blood is filtered to start the process of urine formation. The inner **medulla** is darker in color than the cortex and contains the long loops of Henle (loops extending from the capsule of the kidney tubule), where water is removed from the filtrate and the urine is concentrated. Near the point where the ureter enters the kidney is the **renal pelvis,** a cavity that collects urine before it leaves the kidney.

   As in the turtle, the kidney of the rat is metanephric, composed of elements developing from the posterior portion of the embryonic kidney ridge. In the adult male, its duct, the ureter, carries only urine, and the vas deferens carries only sperm.

3. Using bone shears or heavy scissors, cut through the pelvic girdle on each side of the midline. Remove muscle and bony tissue from this area. Lift the **urinary bladder** and follow the **urethra** from its base inside the pelvic girdle (free the urethra from surrounding tissues using your probe). In females, the urethra opens directly to the outside. Find this opening. In males, it extends through the **penis,** the male's copulatory organ, as the **penile urethra.**

MALES (FIGURE 36B-6A)

4. Locate the **scrotum.** This structure protrudes from the body wall and contains two coelomic compartments into which the testes move at maturity. Using scissors, cut through the skin on the ventral surface on one side. Cut through the muscle and connective tissue surrounding the white oval **testis.** The testis is surrounded by the wall of the pouch, which is lined by parietal peritoneum. The testis is attached to the posterior wall of the scrotum by a ligament.

5. Using your probe, locate the **vas deferens,** and free it from the surrounding tissue between the body wall and testis. Cut the ligament and remove the testis from the scrotum, using care not to break the sperm cord. Find the vas deferens on the dorsal surface of the testis. Note the many coils of this duct that form (along with some elements derived from the embryonic kidney) the **epididymis,** a sperm storage and maturation area.

6. Follow the vas deferens as it passes into the body cavity through the **inguinal canal.** Find where the duct loops over the ureters and, using your probe, follow it as it passes under (dorsal to) the urethra. Find its junction with the urethra. From this point to the exterior, the urethra is a **urogenital sinus,** carrying products of both the excretory and reproductive systems.

7. Find the large glands lying alongside the urethra at the level where the vasa deferentia enter. These are the **prostate glands.** Distal to the entrance of the vasa deferentia are large, lobed **vesicular glands, coagulating glands,** and **ampullary glands** that add various components to the sperm to form **seminal fluid,** the fluid medium carrying the sperm when they are expelled (ejaculated).

8. Follow the urethra to the penis. Cut through the skin covering the **shaft** of the penis from the base through the fold of skin enclosing the enlarged end of the penis, the **glans.** You may find glands at the base and tip of the penile shaft that lubricate the erect organ and assist in copulation. Locate the opening of the urethra at the tip of the glans. Make a cross section through the penis. There are two lateral cavernous bodies, as in the penis of the turtle, and a third, additional erectile body that surrounds the penile urethra and forms the glans. There is also a bone, the **os penis,** that stiffens the penis and assists in copulation.

FEMALES (FIGURE 36B-6B)

4. Find the **ovaries.** In the rat, they are enclosed by a fold of the oviduct so that eggs are not shed free into the coelom (this is *not* the typical vertebrate condition). Remove any fat and the membranes surrounding the ovaries and observe their surface. They contain many follicles in various stages of development. The ovaries are suspended by mesenteries.

5. Find the **oviduct** (Fallopian tube) leading to the horn of the **uterus.** In the rat, there are two separate uteri, each opening separately into the **vagina** through its own **cervix.**

6. Using your probe, free the base of the urinary bladder and uterus from surrounding tissues (keep the ureters intact). Find the barrel of the vagina dorsal to the ureter and follow it to the exterior. In the rat, the vagina and urethra open separately to the exterior. In humans and many mammals, however, these structures open into a shallow **urogenital sinus.** Locate the **clitoris,** a structure homologous to the male's penis, just ventral to the vaginal orifice. Cut into the vagina and study the nature of its wall. Cut open one of the uterine horns and examine its lining, the **endometrium.** If your rat is pregnant, observe the relation of the **placenta,** a structure derived from both maternal and fetal tissues, to the endometrium and the amnion surrounding the pup. Open the amnion surrounding the pup and study the umbilical cord carrying blood to and from the uterus.

After completing your dissections, review the circulatory and urogenital systems of the representatives dissected by your group. Be sure to observe both sexes for all organisms.

Laboratory Review Questions and Problems

1. What are the functions of the circulatory system? In what parts of the system are these functions performed?

2. Fill in the following table to show what has happened to each of the six ancestral aortic arches in the representative vertebrates.

Arch	Shark	Frog	Turtle	Rat
1				
2				
3				
4				
5				
6				

3. Trace changes in the role of the renal portal system through the vertebrate group using the representative vertebrates to illustrate your answer.

4. Diagram the heart of the shark, frog, turtle, and rat. How has the serial circulation through the heart of the shark been modified to provide the parallel circulation found in the rat?

5. Use the correct terms for the urogenital structures in your representative vertebrates to complete the following table.

Structure	Shark	Frog	Turtle	Rat
Kidney				
Kidney duct				
Accessory ducts				
Testis ducts				
Oviducts				
Method of fertilization				
Copulatory organ				
Site of development of young				
Maternal–fetal connection				

6. Based on the names of the arteries and veins studied in your organisms, label the human arteries and veins in the diagrams below.

Arteries

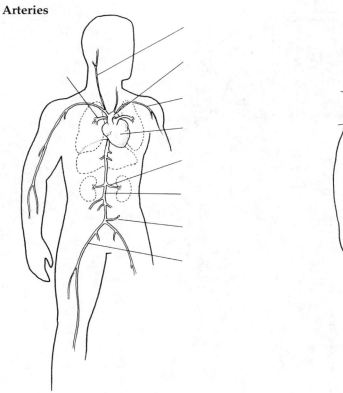

Veins

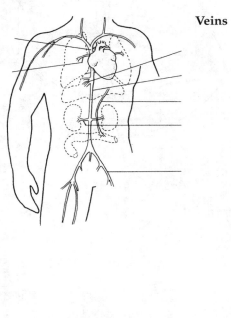

Human heart (ventral view)

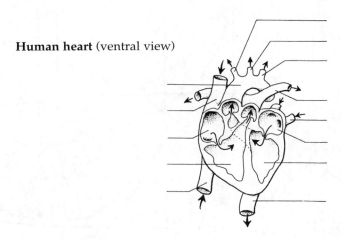

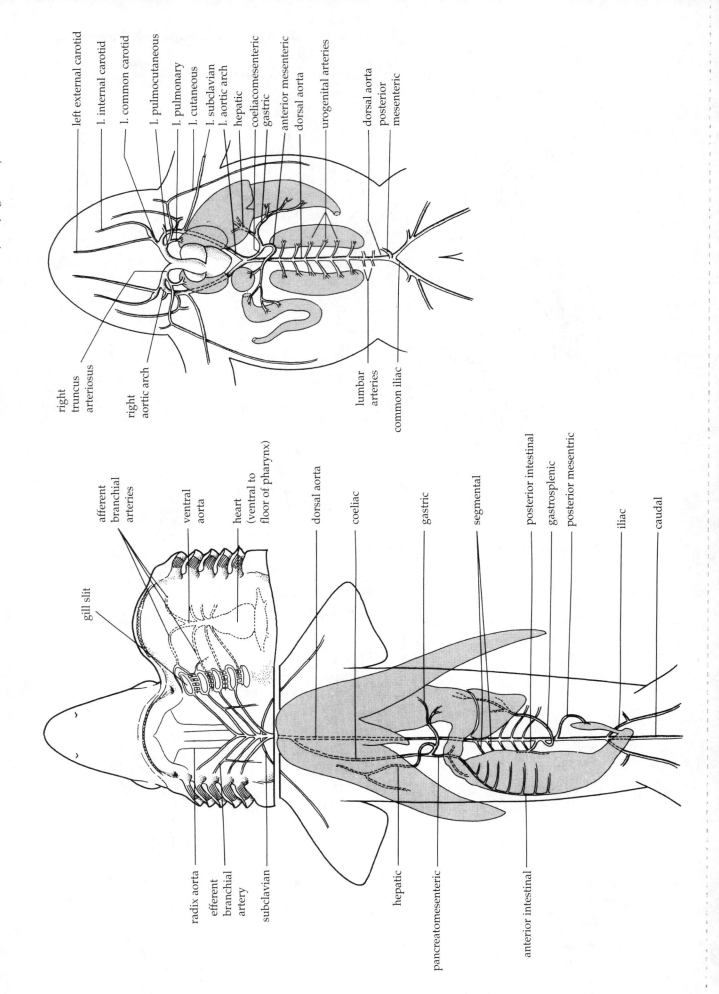

Figure 36A-2 *The aortic arches and arteries of the shark.*

Figure 36A-3 *The arteries of the frog (l = left).*

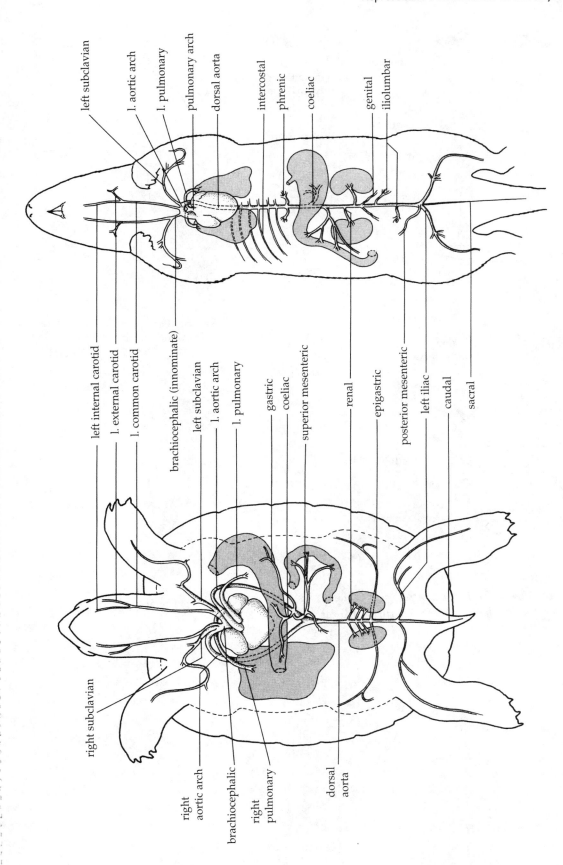

left subclavian

l. aortic arch

l. pulmonary

pulmonary arch

dorsal aorta

intercostal

phrenic

coeliac

genital

iliolumbar

Figure 36A-5   *The arteries of the rat.*

left internal carotid

l. external carotid

l. common carotid

brachiocephalic (innominate)

left subclavian

l. aortic arch

l. pulmonary

gastric

coeliac

superior mesenteric

renal

epigastric

posterior mesenteric

left iliac

caudal

sacral

right subclavian

right
aortic arch

brachiocephalic

right
pulmonary

dorsal
aorta

Figure 36A-4   *The arteries of the turtle.*

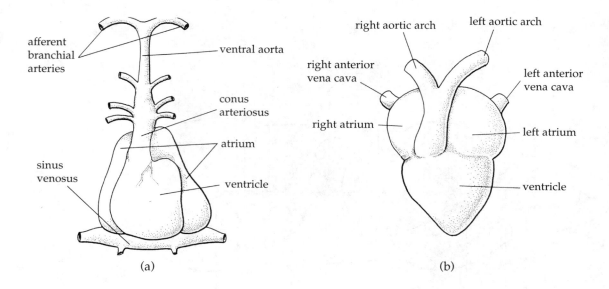

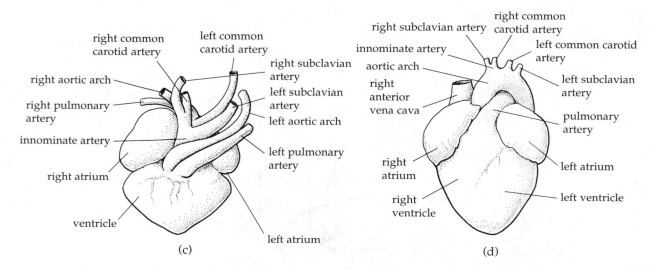

Figure 36A-7 *Interior ventral view of the hearts of the representative vertebrates: (a) shark, (b) frog, (c) turtle, (d) rat.*

NOTES

Table 36A-1   Descriptions and Dissection Directions for the Arterial Systems of the Four Representative Vertebrates*

	Shark	Frog
Aortic Arches and Associated Branches	**ventral aorta** Identify this thick vessel that leaves the heart anteriorly, carrying deoxygenated blood away from the heart toward the gills.	**truncus arteriosus** The ventral aorta has shortened and is divided into two large arteries, the right and left truncus arteriosus. Internally, each truncus arteriosus becomes separated into three channels: one leading to the body, one to the head, and one to the pulmocutaneous system supplying the lungs and skin:
	afferent branchial arteries Using your probe, work anteriorly along the ventral aorta to expose the afferent branchials. They branch from the ventral aorta laterally and pass to the gills. How many are there?_____	(1) **aortic arches** (to body) The left and right branches of the truncus arteriosus continue dorsally to form two aortic arches derived from the fourth arches of ancestral vertebrates. These give rise to branches to the larynx, esophagus, jaws, and vertebral column. A *subclavian* artery arises from each aortic arch to supply the forelimbs.
	efferent branchial arteries Open the mouth by cutting through the pharyngeal arches on the right side. Carefully make a longitudinal incision through the mucous lining on the dorsal side of the mouth. Peel back the membrane to expose the efferent branchial arteries. Try to follow one of the afferent branchials to the gill septum from which it originates. Dorsally, the efferent branchials join one of the two anterior branches (radices) of the dorsal aorta.	(2) **common carotids** (to head) The left and right common carotids are derived from the ventral aorta. Where they branch from the truncus there is a swelling, the **carotid body,** which houses sensors that determine pressure and pH. Each common carotid gives rise to an **external carotid artery** (an extension of the original ventral aorta), and an **internal carotid artery** derived from the third aortic arch of ancestral vertebrates goes to the external superficial ventral tissues of the head and jaw. The internal passes deeper into the head and skull.
		(3) **pulmocutaneous artery** (to lungs and skin) This is the third branch of the truncus arteriosus—it divides into the **cutaneous artery** supplying the skin and **pulmonary artery** to the lung. Trace these branches. This artery arises from the sixth aortic arch of the ancestral vertebrate.
	dorsal aorta The two radices of the dorsal aorta fuse in the roof of the pharynx, and the dorsal aorta extends into the body cavity. Open the pleuroperitoneal cavity, push the organs to the right, and identify the dorsal aorta.	**dorsal aorta** Follow the aortic arches on each side dorsally, above the viscera, to their junction to form the dorsal aorta (Figure 36A-3).

*Solid lines indicate evolutionary relationships; dashed lines indicate functional similarities.

Table 36A-1 *(continued)*

Turtle	Rat	Questions
ventral aorta This vessel is further reduced in the turtle so that three major trunks emerge from the heart:	**ventral aorta** In the rat, this vessel no longer exists—the muscular left ventricle of the heart gives rise ventrally to the systemic left aortic arch.	In the ancestral arterial pattern, both a dorsal and ventral aorta were present. How has this pattern changed during the course of evolution?
(1) **left aortic arch** The middle trunk arising from the heart. It is derived from the left side of the fourth aortic arch	**left aortic arch** Derived from the left fourth aortic arch of the ancestral vertebrate. It serves as the arch of the systemic aorta (dorsal aorta) which is not joined by the right arch (as in the shark, frog and turtle)—see *right subclavian* below. The aortic arch has three branches.	Distinguish the function of afferent branchial arteries from that of efferent branchial arteries. Are these found in the frog, turtle, or rat? Why or why not?
(2) **right aortic arch and brachiocephalic artery** The right side of the fourth aortic arch (originally branching from the ventral aorta) becomes the brachiocephalic, one of the three major trunks arising from the heart. The brachiocephalic is the right-hand trunk. In clearing away the tissue from this artery, you may notice the small, round **thyroid gland.** The brachiocephalic branches to form the right and left **subclavian arteries** to the forelegs and the **common carotid** arteries carrying blood to both sides of the head. The common carotids are actually extensions of the ancestral ventral aorta that continue as the **external carotids. Internal carotids** are formed from the third aortic arch of ancestral vertebrates.	(1) **innominate artery** (*brachiocephalic*) This is the first branch of the dorsal aorta and is derived from the right fourth aortic arch of ancestral vertebrates. It gives rise to the **right common carotid** and **right subclavian arteries.** Anteriorly, the right common carotid branches to form the **right internal carotid** to the brain and **right external carotid** to the ventral portions of the head. The subclavian branches to the neck, mammary region, and forearm. From which ancestral aortic arches are the external and internal carotids formed? _____ (2) **left common carotid** This is the second branch of the aortic arch; it supplies the head with external and internal branches. (3) **left subclavian artery** The third branch of the aortic arch; it supplies the left side of the body and left forelimb.	Which aortic arch present in the ancestral vertebrate plan gave rise to the common carotid? Arch of the systemic aorta? Pulmonary arch? Describe the relative contributions of the right and left aortic arches to the systemic circulation in the shark, frog, turtle, and rat. How are the fourth aortic arches of the ancestral vertebrate and the arch of the systemic aorta and the brachiocephalic artery related as evolutionary modifications?
(3) **pulmonary artery** The left-hand trunk from the heart. It is derived from the sixth aortic arch and carries blood to the lungs. The pulmonary circuit for gas exchange is now complete in terrestrial forms.	**pulmonary arch** Remove the surrounding tissue from the base of the aortic arch. As you work dorsally, you will find the pulmonary arch located to the left and below the aortic arch. It divides into right and left pulmonary arteries serving the lungs. Trace one of these branches to the lung. Try to separate the pulmonary and aortic arches as they leave the heart. A strong ligament connects these arches—this is a remnant of the **ductus arteriosus,** which connects the two arches during development, allowing blood to bypass the lungs before birth. Phylogenetically, it connected the sixth aortic arch to the dorsal aorta.	How does the pattern of arteries derived from the aortic arches change as vertebrates evolved paired appendages and moved onto land? To what general body area does each of the following arteries carry blood: carotid, subclavian, branchial, systemic aorta, subclavian, pulmonary?
dorsal aorta As the left aortic arch approaches the mid-dorsal wall of the body, it is joined by the smaller right aortic arch to form the dorsal aorta. (Do not try to follow this one now, you may destroy the vena cava.)	**dorsal aorta** The aortic arch curves sharply dorsad under the heart and continues as the dorsal aorta. Follow this with your probe and free it from connective tissue.	

(continued next page)

Table 36A-1 (*continued*)*

	Shark	Frog
Unpaired Branches of Dorsal Aorta (in order of appearance, anterior to posterior)	Follow the dorsal aorta posteriorly and identify the following unpaired branches: **coeliac artery** First unpaired branch of the dorsal aorta; gives rise to branches to gonads, esophagus, stomach, pancreas, and the ventral part of the small intestine. Find the three major branches. **posterior intestinal artery** Supplies the dorsal surface of the intestine. **gastrosplenic artery** Supplies the spleen and part of the stomach. **posterior mesenteric artery** Supplies the rectal gland. **caudal artery** Leaves the body cavity to supply the tail.	**coeliacomesenteric artery** Leaves the dorsal aorta near its formation from the two aortic arches; branches to the liver, stomach, spleen, and small and large intestines. It is the major artery supplying the digestive system. Find the following four branches: **gastric** To stomach. **coeliac** To stomach and spleen. **hepatic** To liver. **anterior mesenteric** To small intestine. **posterior mesenteric artery** Originates posteriorly and carries blood to the rectum. (*Note:* No caudal artery is present. Why?)
Paired Branches of Dorsal Aorta (in order of appearance, anterior to posterior)	As you follow the dorsal aorta posteriorly, be sure to identify the following paired branches: **subclavian arteries** Open the mouth and find the paired arteries leaving the dorsal aorta in front of the last pair of efferent branchial arteries. These serve the pectoral girdle and ventral body wall. **segmental arteries** Carry blood to each block of muscle tissue (myomere) along the length of the body. Some branches serve the gonads and kidneys. **iliac arteries** Posteriorly, these paired arteries arise just in front of the cloaca. Each divides into branches supplying the body wall and the fins.	(*Note:* In the frog, subclavian arteries branch from the two arches of the aorta: see *aortic arches* above.) **urogenital arteries** Six pairs supply kidneys, gonads, and gonadal fat bodies. **lumbar arteries** Paired arteries serving the dorsal body wall. **common iliac arteries** Paired arteries arising from the dorsal aorta near the posterior end of the body cavity; provide branches to ventral body wall and leg.

*Solid lines indicate evolutionary relationships; dashed lines indicate functional similarities.

Table 36A-1 (*continued*)

Turtle	Rat	Questions
gastric artery Unpaired artery branches from the left aortic arch and carries blood to the anterior part of the stomach.	**coeliac artery** Unpaired artery serving the cranial end of body cavity with branches to liver, stomach, small intestine, pancreas, and spleen.	In general, the unpaired arteries that serve the digestive organs have not changed much during the course of evolution from those seen in ancestral fishes. Why?
coeliac artery Branches from the left aortic arch; supplies remainder of the stomach, liver, and pancreas and first portions of the small intestine.	**superior (anterior) mesenteric** Provides blood to the small intestine.	To what organ(s) does each of the following arteries carry blood: coeliac, mesenteric, gastric, hepatic, caudal?
superior mesenteric artery Sends branches to remainder of the small intestine and large intestine.	**posterior (inferior) mesenteric** Supplies blood to the colon and rectum.	
caudal artery Continuation of dorsal aorta into the tail.	**sacral artery** Posterior extension of dorsal aorta, supplying blood to the tail.	
(*Note:* These branch from the *brachiocephalic;* see above.)	(*Note:* Subclavians branch from left aortic arch.)	To what general area of the body or to what body organ does each of the following arteries carry blood: iliac, renal, epigastric, lumbar (iliolumbar), genital, segmental?
	intercostal arteries Supply muscles between ribs.	
	phrenic arteries Supply muscles of diaphragm.	In the rat, phrenic arteries are present. Why are these not found in the other vertebrates you examined?
renal arteries To kidney.	**renal arteries** To kidneys.	
genital arteries To gonads.	**spermatic arteries** To testes.	
epigastric arteries To body wall.	**ovarian arteries** To ovaries.	
common iliac arteries Paired arteries arising at posterior end of the dorsal aorta; branch to form internal and external iliac arteries.	**iliolumbar arteries** Furnish blood to the dorsal body wall.	
	iliac arteries Supply blood to legs.	

Figure 36B-3   *The urogenital system of the shark. (a) Male. (b) Female.*

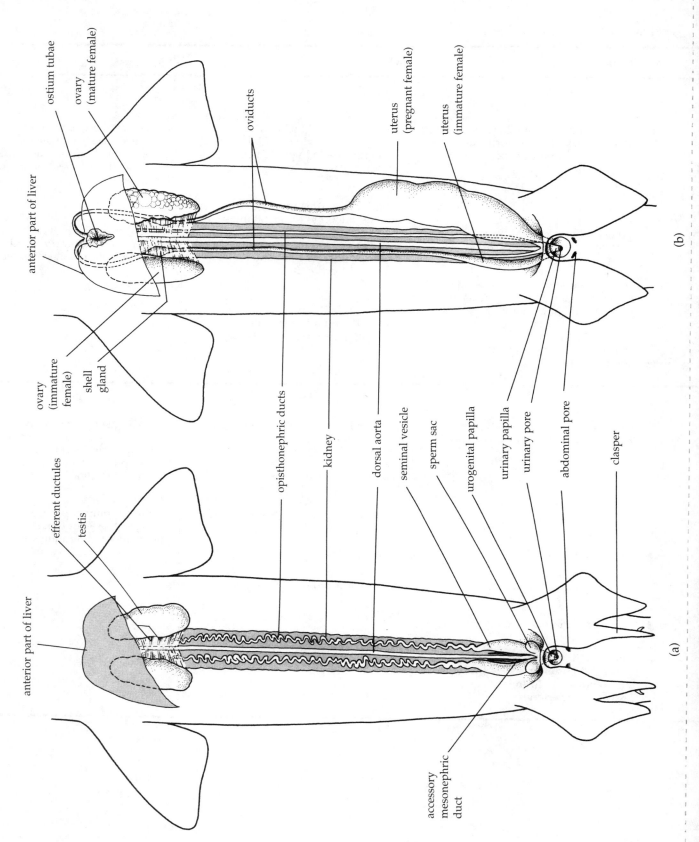

ostium tubae
ovary (mature female)
anterior part of liver
oviducts
uterus (pregnant female)
uterus (immature female)
ovary (immature female)
shell gland
opisthonephric ducts
kidney
dorsal aorta
seminal vesicle
sperm sac
urogenital papilla
urinary papilla
urinary pore
abdominal pore
clasper
efferent ductules
testis
anterior part of liver
accessory mesonephric duct

(a)

(b)

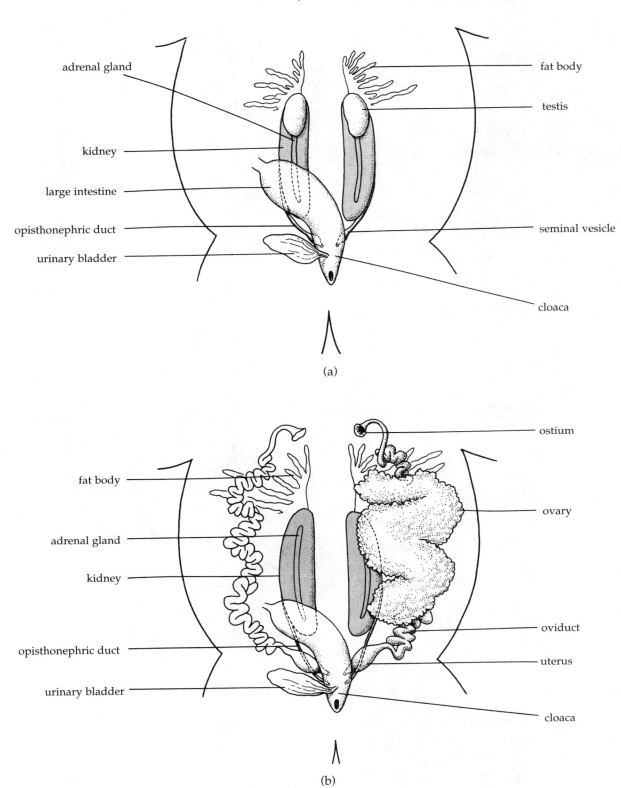

Figure 36B-4 *The urogenital system of the frog. (a) Male. (b) Female.*

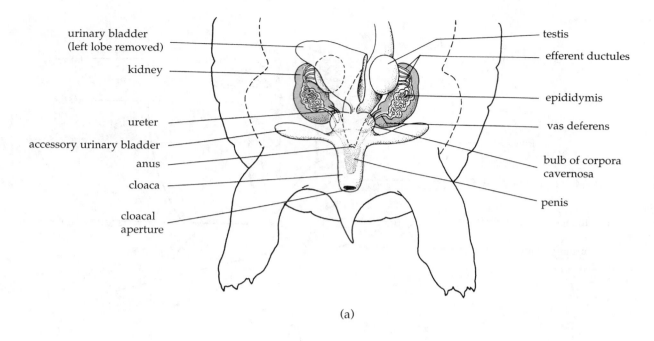

urinary bladder
(left lobe removed)

kidney

ureter

accessory urinary bladder

anus

cloaca

cloacal
aperture

testis

efferent ductules

epididymis

vas deferens

bulb of corpora
cavernosa

penis

(a)

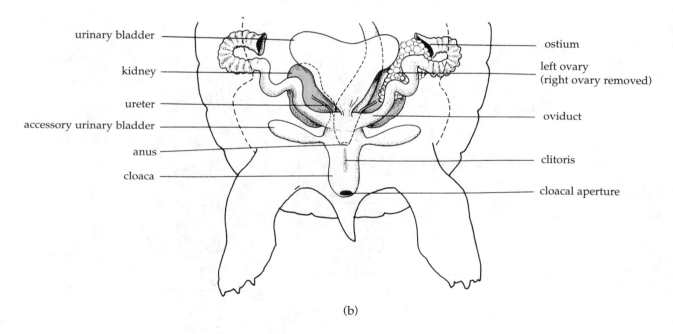

urinary bladder

kidney

ureter

accessory urinary bladder

anus

cloaca

ostium

left ovary
(right ovary removed)

oviduct

clitoris

cloacal aperture

(b)

Figure 36B-5 *The urogenital system of the turtle. (a) Male. (b) Female.*

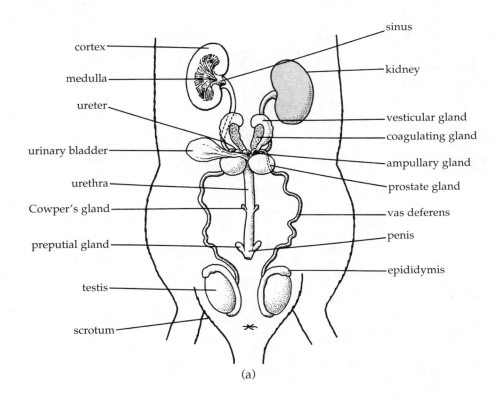

(a)

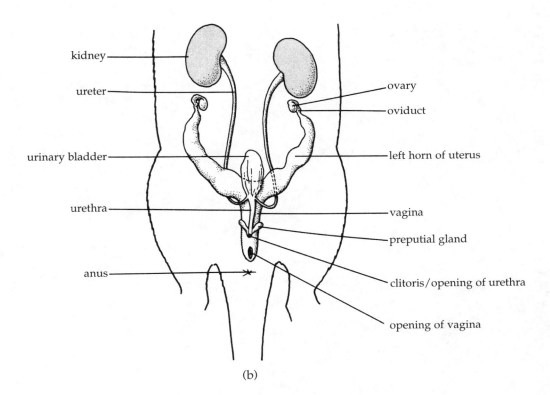

(b)

Figure 36B-6 *The urogenital system of the rat. (a) Male. (b) Female.*

The Basics of Animal Form: Skin, Bones, and Muscles

OVERVIEW

Among the great diversity of vertebrates living today, some are adapted to water and some are adapted to land. Some live in cold, dark places while others prefer hot, dry environments. Some vertebrates walk on two appendages and some on four. Some swim, others slither or fly. There is, however, a unity in this diversity of form and function; certain vertebrates share common anatomical, morphological, embryological, and physiological characteristics (Table 37-I).

During this laboratory, you will examine structural and functional systems that comprise the bulk of the vertebrate body: the integumentary, skeletal, and muscular systems. The skin, a relatively thin but collectively massive organ, provides an interface for living tissue that separates the body from the environment. Structures of the skin provide support and protection and mechanically link the body with the environment so that movements produced by muscles acting on an internal skeleton (effectors) are translated into externally directed activity (behavior).

STUDENT PREPARATION

Prepare for this laboratory by reading the text pages indicated by your instructor. Familiarizing yourself in advance with the information and procedures covered in this laboratory will give you a better understanding of the material and improve your efficiency.

If you have completed Laboratory 33, you can skip Exercises A and B. You may find it useful to review Tables 37A-1 and 37B-1 before proceeding to Exercise C.

✓ **EXERCISE A**   **The Language of the Body**

Multicellular organisms, including vertebrates, are composed of groups of interacting and cooperating *organs*, often with common developmental origins, which form *organ systems* such as the integument, the digestive system, and the respiratory system. For example, the digestive system includes the pharynx, esophagus, stomach, small intestine, liver, and other organs. Each organ is made up of one or more types of *tissue*.

To describe the position of body parts or organ systems and their location in relation to specified reference points, a variety of terms is used. It is important to understand these terms so you can discuss both form and function and can carry out the suggested experiments.

||||| **Objectives** ||||||||||||||||||||||||||||||||||||||||||||||||||||

☐ Describe and demonstrate anatomical positions for any representative vertebrate.

☐ Use proper anatomical terms to describe body directions and landmarks.

Table 37-I   Features of Representative Vertebrates

	Fish	Amphibian	Reptile	Bird	Mammal
Habitat	Aquatic	Transitional (aquatic larvae)	Terrestrial (some aquatic)	Terrestrial (flight)	Terrestrial (some aquatic)
Body Divisions	Head, trunk, tail	Head, trunk; tail in salamanders	Head with neck, trunk, tail; head, trunk, and tail in snakes	Head with neck, trunk, tail	Head with neck, trunk, tail
Digestive System	Mouth, cloaca; primitive fish (agnatha) have no jaws	Mouth, cloaca	Mouth, cloaca	Mouth, cloaca	Mouth, anus
Teeth	Homodont (all teeth are alike); none in primitive fish	Homodont	Homodont; none in turtles	None; muscular "gizzard" replaces teeth	Heterodont (having teeth of different types)
Respiratory System	Internal gills	External gills (larva); lungs, pharynx, skin (adult)	Lungs	Lungs (with air sacs)	Lungs
Sense Organs	External nares Eyes No middle ear Lateral-line system	Nares to mouth Eyes Tympanum (middle ear) in frogs; external and middle ear in salamanders	Nares to mouth Eyes External and middle ear (except snakes)	Nares to mouth or pharynx Eyes External and middle ear	Nares to pharynx Eyes External and middle ear
Locomotion	Swim	Swim (larva); swim, walk, or jump (adult)	Walk (in snakes, serpentine locomotion); swim	Fly, walk, swim	Walk, swim; bats fly
Integument	Dermal scales Unicellular glands	Moist, vascularized Multicellular glands	Epidermal scales (horny) Glands reduced (scent glands)	Epidermal scales Feathers Glands reduced (uropygial gland)	Hair Sweat and sebaceous glands Mammary glands
Reproduction	Oviparous (egg-laying); some viviparous (live birth)	Oviparous; external or internal fertilization	Oviparous; some viviparous; internal fertilization	Oviparous; internal fertilization (cloaca)	Most viviparous; internal fertilization

ⅠⅠⅠⅠⅠ Procedure ⅠⅠ

1. Study the relationships of anatomical locations by referring to Figure 37A-1 and the terms in Table 37A-1. Notice that there is some ambiguity in using terms such as "anterior" or "ventral" in humans because of our upright posture. Answer the following questions by inserting the appropriate term for the anatomical position.

   a. The head is _____ to the shoulders in a human.

   b. The back is _____ to the chest in a dog.

   c. The ears are _____ to the nose in a cat.

   d. The hips are _____ to the body, and the ankles are _____ to the trunk, in a human.

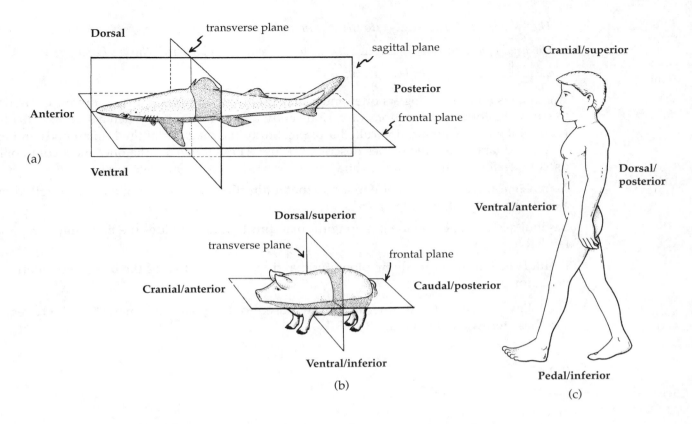

Figure 37A-1 *Lateral view of (a) shark, (b) pig, and (c) human. Refer to the terms in Table 37A-1.*

Table 37A-1 Definition of Terms Used to Describe Position or Direction

Position or Direction	Definition
Anterior	Pertaining to the head end (in humans, equivalent to *ventral*)
Posterior	Pertaining to the end opposite the head (in humans, equivalent to *dorsal*)
Dorsal	Toward the vertebral column; the upper surface of an animal or organ; the back
Ventral	Toward the chest and abdomen; the underside of an animal or organ
Cranial	Toward the head
Caudal*	Toward the tail
Pedal	Toward the feet
Superior	Toward the upper parts of the body
Inferior	Toward the lower parts of the body
Lateral	Away from the midline of the body
Medial	Toward the midline of the body
Superficial	At or near the surface
Distal	Away from the point of reference, usually the body trunk
Proximal	Near the point of reference, usually the body trunk

*The word *caudal* is sometimes used to mean "toward the feet" in the human. But caudal literally means "toward the tail," and the human vestigial tail (the coccyx) is in the region of the pelvis, not the feet. For this reason, we have defined caudal to mean "toward the tail" and pedal bones to mean "toward the feet."

e. The knee is _____ to the toes in a cat.

f. The head of a human is _____ to the body trunk, whereas the head of a dog is

_____ to the body trunk.

Anatomists and physiologists often must make sections (samples cut from organs or tissues) to study the form and functions of various body organs or tissues. When a section is made, it is cut along an imaginary line called a **plane.** Study the diagram of the human body in Figure 37A-2 and label the three planes described below. These terms are the ones most commonly used to designate the location of body parts or sections of organs and tissues.

Transverse section A cut made in a horizontal direction, at a right angle to the longitudinal axis of the body; often called a *cross section.*

Sagittal section A cut made in a longitudinal direction that divides the body into right and left halves.

Frontal section A cut made in a longitudinal direction that divides the body into anterior and posterior parts.

2. The following list includes other terms often used to designate "landmarks" for body regions. Use these terms to label Figure 37A-3.

cranial	head
thoracic	chest
axillary	armpit
brachial	arm
abdominal	anterior trunk below ribs
groin	where thighs meet the body trunk
femoral	thigh
flank	lateral body surface from lower rib cage to hip
pubic	genital region

✔ **EXERCISE B   Vertebrate Coverings: Skin, Scales, Feathers, and Hair**

The outer tissue layers of vertebrates form the skin or **integument,** which provides an interface between the internal and external environments. The integument functions in the following ways:

Protection

Mechanical Provides a barrier against abrasion, injury, attack, parasites, pathogens.

Chemical Forms a buffering barrier to passage of water and ions; can also include the production of mucus and other secretions.

Visual Provides pattern or color for camouflage or warning and also for attraction and communication.

Light Intercepts ultraviolet radiation and prevents its penetration into deeper layers of the body.

Locomotion Ciliary movement of aquatic larvae; skin transmits the physical force required for movement; scales, pads, or claws provide attachment to surfaces.

Respiration Allows gas exchange with blood vessels close to the surface in some species.

Secretion Glandular secretions modify the interface with the environment (e.g., a mucous coat "streamlines" certain fishes); secretions may provide mechanical protection or may be

Figure 37A-2 *Body planes.*

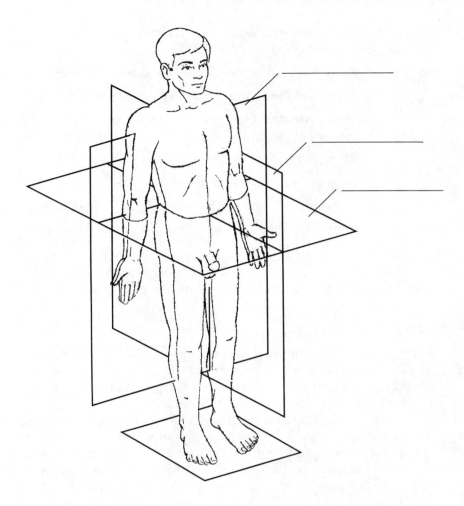

Figure 37A-3 *Body landmarks.*

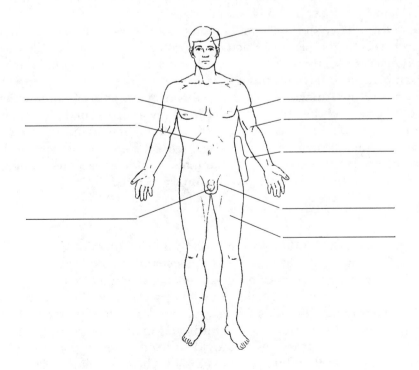

toxic; some secretions may provide species identification or function as behavioral cues (pheromones).

Excretion Ammonia and simple wastes diffuse across the integument in some vertebrates.

Water uptake Some amphibians can take up water through their skin.

Heat exchange Heat gain or loss can be affected by the presence of pigments and by other epidermal structures such as sweat glands, hair, and feathers.

Energy storage Fat deposits.

Monitoring the environment Sensors.

Behavior Effectors include muscles, glands, and pigment cells.

Synthesis Gives rise to dermal bone and the dentine of teeth; produces vitamin D (with sunlight).

The skin of vertebrates is composed of two layers: the outer **epidermis** and the deeper **dermis.** Scales, feathers, and hair that cover the body of many vertebrates are derived from these tissue layers.

EPIDERMIS

The outermost layer of the skin is the **epidermis.** The deepest layer of the epidermis, the basal layer, usually consists of a single layer of cells. It is this layer of the epidermis that is most mitotically active; new cells produced in the basal layer are pushed up through the other layers of the skin to replace cells that are worn away or shed from the surface.

In many aquatic organisms, the epidermis is relatively thin and all of its cells are living (Table 37B-1). In vertebrates that made the transition from water to land, the epidermis is thicker. As deeper cells within the epidermis die and move toward the surface, a horny proteinaceous material called **keratin** is deposited within the cells. This process, called **cornification,** reduces the permeability of the skin to water and protects the organism from desiccation. The outer layer of the skin of terrestrial animals is composed of dead, cornified cells. Dandruff is a good example of cornified cells that are continually shed.

Life on land led to further modification of the epidermis, including specialized, highly keratinized structures such as the epidermal scales of reptiles and birds, the feathers of birds, and the hair of mammals. Feathers and hair trap an insulating layer of air around the body and are important in maintaining a constant body temperature.

Within the epidermis, glands perform a variety of functions. In aquatic vertebrates, numerous **mucous glands** secrete the viscous mucus that protects the surface of the epidermis, reduces the friction of the body in water, and retards the loss of water to the environment. In addition to mucous glands, amphibians also have **granular glands** that produce a variety of watery secretions—some of them toxic. (Indians of the Amazon basin use toxins from amphibians in making their poison arrows.) Animals with a fully terrestrial life style generally have fewer epidermal glands—in part, to prevent the loss of water in secretions. Reptiles, for example, have only a few glands; these play a role in species recognition, reproduction, or defense. In birds, a single oil gland at the base of the tail is used to condition feathers. The major epidermal glands of mammals include **sebaceous glands,** associated with oil production in hair follicles, **sweat glands,** and **mammary glands.**

DERMIS

The **dermis,** beneath the epidermis, is composed of connective tissue, mainly collagen and elastin fibers in a gel-like matrix. In addition, the dermis contains blood vessels, nerves, sensors, and fat (adipose tissue).

Pigment cells, **chromatophores,** are a conspicuous element of the dermis. In most vertebrates, the most common type of pigment cell is a **melanophore** containing the brown pigment melanin. (Some vertebrates also have lipophores containing red, orange, or yellow pigments.) Chromatophores are concentrated in the upper layer of the dermis and, in birds and mammals, portions of these cells sometimes bud off and invade epidermal structures to provide a fixed pigment pattern to feathers and hair. However, in many vertebrates other than birds and mammals, chromatophores are capable of changing their size and shape

Table 37B-1 Comparison of the Skins of Representative Vertebrates

Trait	Fishes	Amphibians	Reptiles	Birds	Mammals
Epidermis	Living, thin	Dead, thin	Dead, thick	Dead, thin	Dead, thick
Specializations	None	Some keratin	Highly keratinized epidermal scales	Feathers and highly keratinized epidermal scales	Hair, composed of keratin
Glands	Unicellular mucous glands	Multicellular mucous and granular glands	Few glands	One gland, the uropygial (oil) gland	Sebaceous, sweat, and mammary glands
Dermis	Thin, tightly bound to body tissues	Thin	Thin	Thin, loosely bound to body tissues; stores fat	Thick, loosely bound to body tissues; stores fat
Chromatophores	Present	Present	Present	Present; melanophores invade epidermal structures	Present; melanophores invade epidermal structures
Dermal bone	Forms dermal scales; missing in cartilaginous fishes	Reduced, becomes associated with skull and shoulder girdle	Redeveloped as dermal plates of shell in turtles	Reduced	Reduced
Teeth*	Teeth; placoid scales	Small teeth	Teeth; reduced in turtles	No teeth	Teeth in sockets

*Dentine and pulp arise in the dermis; enamel is epidermal.

and thus the pattern of coloration. These changes may be mediated directly by light, indirectly by hormones, or rapidly through nerve impulses. Color changes may assist an organism in absorbing radiant energy from the sun, in hiding from enemies, or in communication.

Normally, we do not think of bone as part of the skin of vertebrates, but in many of the earliest (jawless) fishes, heavy plates of dermal bone, formed by the direct deposition of calcium salts within the dermis of the skin, provided an external armor. In primitive fishes with jaws, the dermis produced **dermal scales** composed of layers of dermal bone overlaid by a layer of dentine-like material (also derived from the dermis) and an enamel-like layer applied to the surface by the epidermis.

As fish evolved, several of these layers became modified or reduced. In cartilaginous fishes such as the shark, dermal bone is largely absent and only the outer layers of dentine and enamel persist in the placoid scales. The scales of bony fishes are thin disks of dermal bone that develop in overlapping folds in the skin.

In general, two adaptive trends in the vertebrate integument are associated with the phylogenetic spectrum from primitive fishes to mammals: (1) a general reduction in the importance of bony elements of the dermis and (2) a corresponding increase in the diversity and importance of epidermal structures (Table 37B-1).

ǀǀǀǀǀ **Objectives** ǀǀǀǀǀǀǀǀǀǀǀǀǀǀǀǀǀǀǀǀǀǀǀǀǀǀǀǀǀǀǀǀǀǀǀǀǀǀ

☐ Identify the layers and structures of the integument by microscopic examination.

☐ Describe how scales, feathers, and hair are formed.

ǀǀǀǀǀ **Procedure** ǀǀǀǀǀǀǀǀǀǀǀǀǀǀǀǀǀǀǀǀǀǀǀǀǀǀǀǀǀǀǀǀǀǀǀǀǀǀ

Fish

1. Study a prepared slide of shark (a cartilaginous fish) using the 10× objective of your microscope. Recall that the shark's epidermis is made up entirely of living cells with few specializations.

   a. Where do new epidermal cells originate? _____

Gland cells are present and produce mucus, but sharks do not develop a mucous covering like the streamlining layer found over the skin in some other aquatic vertebrates.

2. The dark color of the skin of sharks is due to dermal pigment cells, melanophores, containing the dark, brown-black pigment melanin. Can you find the melanophores?

3. Note the regularly spaced dermal scales, often called denticles or "little teeth" because their structure resembles that of a tooth. These are the **placoid scales** (Figure 37B-1).

4. Examine the slide of the skin of a bony fish using the 10× objective. Scale structure is one of the major differences between cartilaginous fishes such as the shark and bony fishes such as the perch.

  b. *How do the scales of other fishes differ from the placoid scales of the shark?*

  _____

  _____

  c. *The scales of the perch are described as* cycloid. *Explain why.* _____

  Cycloid scales have a pattern of growth rings.

Figure 37B-1 *Cross section through the skin and placoid scale of a shark. These scales develop similarly to the teeth of higher vertebrates; the enamel contains the same proteins found in mammalian teeth.*

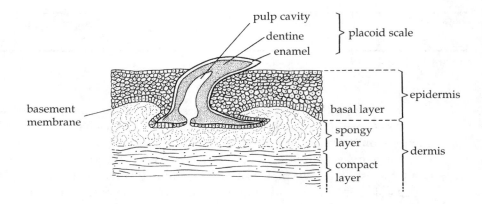

Amphibians

1. Obtain a prepared slide of a cross section through the skin of an adult frog or other amphibian. Study it under the 10× objective of your microscope. Identify the epidermis and the basal layer. In contrast to the epidermis of the shark, the outer layers of cells in the frog's epidermis are dead and provide a protective covering. The frog's epidermis is generally only five to eight cells thick (Figure 37B-2).

  a. *Does the frog have epidermal scales?* _____

2. Study the dermis in the prepared slide of frog skin. Identify the multicellular mucous glands and the large granular glands, both epidermal structures that have grown into the dermis. (Curiously, in the Surinam toad, granular glands produce secretions that nourish developing tadpoles held in skin pockets on the backs of females.)

  b. *What are the functions of mucous glands in most adult amphibians?*

  _____

  Of granular glands? _____

  _____

Figure 37B-2 *Section of the skin of a frog.*

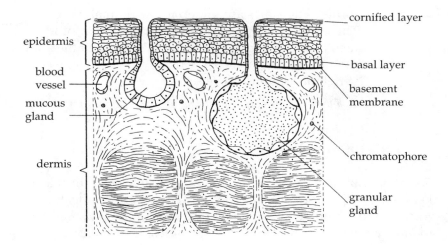

3. The dermis also contains muscles, pigment cells, blood vessels, nerves and nerve endings, a variety of connective tissue cells and fibers, and other structures. Do you find any trace of dermal scales? (Remnants of dermal scales are present in some limbless amphibians and a few frogs.)

Reptiles

In reptiles, the outer cornified layer formed by the epidermis is much thicker than that of amphibians. It is thickest where it forms epidermal scales, and then thins out between scales so that expansion or movement of the skin is possible. The thickness of the keratin layer and the scarcity of glands minimizes water loss through the skin. The cornified layer may be shed or molted regularly, as in snakes and lizards, or retained as growth occurs, as in most turtles and alligators.

1. Obtain a slide of snake skin showing the origin of the epidermal scales. Observe the skin using the 10× objective of your microscope. Identify the keratin layer, epidermis, and dermis.

   a. Are there some areas where the keratin layer appears to be thicker than in other areas? _____

   b. How does the epidermis of the snake differ from that of the frog? _____

   _____

2. Study the turtle skeleton on display. The large keratinized epidermal scales that make up the outer covering of the dorsal and ventral "shells" of the turtle are derived from epidermis. The retention of the cornified layers that make up the scales (turtles do not shed these scales) results in a gradual increase in their thickness. This may be used to estimate age in some

Figure 37B-3 *A turtle skeleton showing elements of the axial and dermal skeletons forming the carapace.*

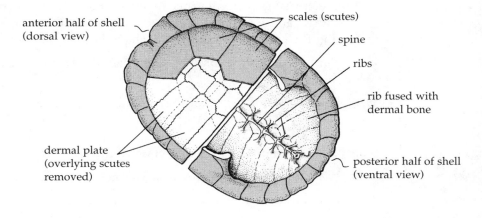

turtles. The scales are underlain by dermal bone—also a product of the dermis of the skin. Thus, the "shell" of the turtle is really a part of its skin; it is continuous with the skin covering the head, tail, and appendages. (Contrary to the animated liberties taken by cartoonists, turtles cannot leave their shells!) Dorsally, the dermal elements of the shell fuse with bones of the vertebral column and ribs.

> c. *Do the shapes and locations of dermal bones forming the shell match the shapes of the epidermal*
>
> *scales (Figure 37B-3)?* ————

Birds

The skin of birds is generally thin and is covered by a thick layer of feathers. The feathers, made of keratin, are a product of the epidermis. Flight feathers (feathers of the wing and tail) consists of an **axis** (rachis) and a **vane,** which is composed of laterally arranged **barbs** (Figure 37B-4). The barbs are attached to one another with small hooks or **barbules.** If the barbs of a feather become separated, a bird can reattach them by "preening"—pulling the feather through its bill. (Oil from a large sebaceous gland at the base of the tail conditions and waterproofs the feathers and may also be transferred to the feathers by the bill during preening.)

   Plumes include flight feathers and feathers that shape the body, while *plumules* are down feathers. Use the microscope to observe the feathers on demonstration.

> a. *How does the structure of plumules (down) differ from that of plumes (flight feathers)?*
>
> _____

Figure 37B-4   *The primary flight feather of a bird.*

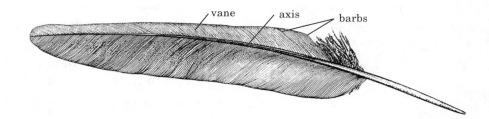

Mammals

1. Observe a cross section of human skin (or other mammalian skin) using the 10× objective of your microscope. The outer epidermis is thick and cornified to reduce water loss, but does not form the keratinized scales that are found in reptiles.

   Several epidermal specializations occur in mammals. Epidermal hairs, growing down into the dermis, are associated with smooth muscles that can change the orientation of the hair, thus increasing or decreasing the insulating layer of air trapped by the coat of hair. (In humans, "goose bumps" are a remnant of this thermoregulatory response—ineffectual because during the course of evolution we have become much less hairy than our ancestors.)

2. Epidermal glands are another specialization. Use Figure 37B-5 to identify the sebaceous glands in your cross section.

> a. *What is the function of sebaceous glands?* _____
>
> _____

3. If you are examining a cross section of human skin, locate a sweat gland (Figure 37B-5).

> b. *What is the function of the sweat glands?* _____
>
> _____

Figure 37B-5 *Schematic drawing of a section of human skin.*

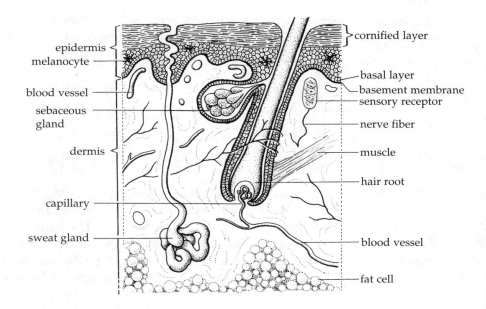

4. Using glycerol, make a wet-mount slide of a piece of your hair. Hair is composed of dead, cornified cells that line up behind one another and push the hair out of the follicle, much like toothpaste being squeezed out of a tube. A cross section of hair would show the cells arranged in concentric layers. The outer layer contains pigment (generally absent from this layer in blond hairs) and is covered by a cuticle. Draw a diagram of your hair in the space below and describe what you observe.

Differences in the structure of hair can be used to determine species, sex, and race. Hair samples can also be used for DNA fingerprinting.

Other keratinized, epidermal structures of mammals include the **horns** of cattle (but not the antlers of deer, which have a bony core covered with skin), **baleen** of some whales (plates in the mouth cavity that filter food—tiny plankton—from the water), horny **scales** (found in the pangolin and armadillo), and **nails, claws,** and **hooves.**

The dermis of mammals contains connective tissue cells and fibers; muscles, nerves, and sensors; blood vessels and capillaries; glands and hair follicles from the epidermis; and, toward its base, a layer of fat cells that becomes quite thick in aquatic mammals such as seals and whales. In an aquatic environment, hair is not effective in trapping a layer of air to reduce heat loss, so in the evolution of aquatic mammals, thick layers of fatty tissue took on the insulating function.

5. Study your slide. Identify the epidermal hair follicles and glands that invade the dermis. Locate muscles that attach to the hair follicles. Find connective tissue elements. Examine the innermost layers of dermis on your slide. Can you find fat cells?

   c. *Why is fat tissue generally a less efficient insulator than hair? (Hint: Which tissue is living and vascularized?)* _____

   _____

Summary

Review Table 37B-1. Be sure you understand the phylogenetic trends among the vertebrates. For instance, dermal bone has been lost almost completely in the shark, but many other fishes retain some bone in their scales; it also reappears in the turtle. Nevertheless, the trend is toward restriction of dermal bone to specific sites—the skull and pectoral girdle—as we move from aquatic to terrestrial organisms.

a. What phylogenetic trends are illustrated by epidermal structures in amphibians, reptiles, birds, and mammals?

_____

_____

✔ **EXERCISE C** | **Bones and Joints**

The skeleton is composed of bones that are connected at joints. The bones of the trunk and head region of the body are part of the **axial skeleton,** while those of the legs and arms make up the **appendicular skeleton.**

There are 206 bones in the adult human skeleton. The bones may be **long,** composed of a shaft with rounded heads at the two ends, or **short,** such as the tarsals of the upper foot and carpals of the upper hand. Long bones are generally smooth and **compact,** whereas short bones tend to contain more **spongy** bone. Bones may also be flat, composed of a layer of spongy bone sandwiched between two layers of compact bone—for example, the bones of the skull.

⁞⁞⁞⁞ **Objectives** ⁞⁞

☐ Identify the major anatomical features of a bone. Identify major types of joints.

✔ **PART I** | **Bones**

⁞⁞⁞⁞ **Procedure** ⁞⁞⁞

Obtain a specimen of long bone. The bone may be fresh or dry, cut longitudinally with a saw. On a separate sheet of paper, draw a picture of the bone. Use Figure 37C-1 to identify and label the following parts of the bone.

Diaphysis Shaft of the bone. Note the presence of any cartilage covering the bone, making a smooth surface to form joints.

a. Is the shaft composed of compact or spongy bone? _____

Periosteum If you are using a fresh specimen, carefully pull away the fibrous membrane covering the bone.

b. Does the membrane separate easily? _____ *c. Does it penetrate into the bone?* _____

The periosteum contains blood vessels and nerves that enter the bone.

Epiphysis Knoblike end of the long bone. Note the layer of compact bone on the surface.

d. What type of bone is present below the compact bone? _____

Epiphyseal line In a growing animal, a plate of thin hyaline cartilage is present in the epiphysis. This is an area of growth. When the bone stops growing, the area is replaced by bone, and only a thin line—the epiphyseal line—may be visible.

e. What material makes up the skeleton of embryonic vertebrates? _____

(Review Laboratory 32, Exercise B.)

f. How does the composition of the vertebrate skeleton change as the organism matures?

_____

Medullary cavity Also called the marrow cavity. In an adult, the interior cavity is usually filled with adipose tissue that makes the marrow appear yellow. **Red bone marrow** is found

Figure 37C-1 *The ends of long bones, such as this femur, consist of spongy bone, containing large spaces, surrounded by compact bone. The shaft, which is hollow, is composed of compact bone. A central cavity containing bone marrow extends through the center of the shaft. The marrow of long bones is yellow because of stored fat. The periosteum is a fibrous sheath, which contains the blood vessels that supply oxygen and nutrients to the bone tissues. Blood vessels emerge from bone through openings known as nutrient canals.*

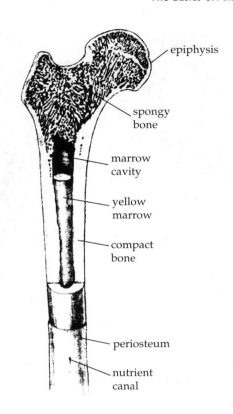

epiphysis

spongy bone

marrow cavity

yellow marrow

compact bone

periosteum

nutrient canal

in the spaces within the spongy bone of the epiphysis. Bone marrow is responsible for the formation of blood cells.

g. *In birds, much of the marrow cavity is filled by air sacs. Why is this an important adaptation for these organisms?* _____

_____

If you are using a fresh bone, can you distinguish the marrow within the bone? If you are using a chicken bone, can you find the air sacs?

✔ **PART 2**   Joints

Every bone in your body (except the hyoid bone of your jaw) is connected to other bones at **joints.** The joints hold the skeleton together; they also allow it to flex and move. There are three major types of joints:

Immovable joints Some bones may interlock to form **sutures** such as the ones between the bones of the skull. At other immovable joints, bones are connected by dense fibrous tissue.

Slightly movable joints Some bones are connected by a broad, flat disk of cartilage—for example, the spinal disks between vertebrae.

Freely movable joints It is the **synovial joints** that allow maximum movement of the skeleton, sometimes in a single direction, but often in multiple directions. The joint is encased by a sleeve of connective tissue that forms a "capsule" around the bones (Figure 37C-2a). The capsule is lined by a synovial membrane which produces synovial fluid, a liquid that acts as a lubricant in the joint. The ends of the articulating bones are covered with a smooth layer of hyaline cartilage (see Laboratory 32, Exercise B). **Ligaments** (sheets of dense connective tissue) often reinforce the fibrous capsule, attaching bones to bones (Figure 37C-2b). Additional fluid-filled sacs or **bursae** may also be present within the joint, where tendons that attach muscles to bones actually cross over the bones. Damage to the bursa of a joint can cause a painful

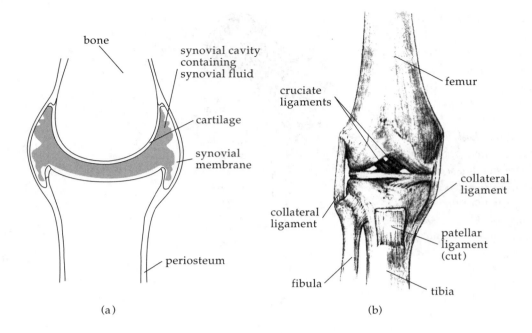

Figure 37C-2 *(a) A synovial joint is filled with fluid. Ligaments reinforce the joint as shown in (b), the right knee viewed from the front with the patella removed.*

condition called bursitis. Other types of damage such as sprains result from damage to the ligaments and sometimes to the tendons associated with the joint.

IIIII Procedure IIIIIIIIIIIIIIIIIIIIIIIIIIIIIIIIIIIIIIIIIIIIIII

Indicate the type of joint found in each of the following locations in the human skeleton.

Skull   _____

Knee   _____

Shoulder   _____

Spine   _____

Sternum   _____

Wrist   _____

Ankle   _____

Hip   _____

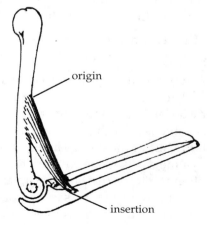

Figure 37C-3 *Muscles attach to bones at two points. The origin is stationary; the insertion point is attached to the bone, which will move when the muscle contracts.*

Body movement occurs when muscles contract across freely movable joints. Muscles attach to bones at two points: the **origin** (the stationary or less movable location) and the **insertion** (the movable point of attachment). When muscles contract, the insertion moves toward the origin (Figure 37C-3). The type of movement depends on the way in which the joint is constructed and on how the muscles are attached.

👁 **EXERCISE D**   **Muscles and Bones Working Together**

Muscles can only contract. Some other type of force must be used to return the muscle to its uncontracted form. For instance, when the biceps muscle in the upper arm (the one you contract to show off your muscles) contracts, the lower arm is pulled upward. To straighten the arm, the triceps on the back of the arm must contract (Figure 37D-1).

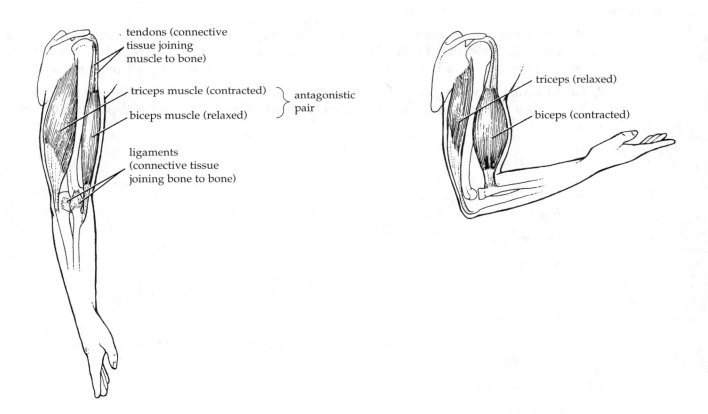

Figure 37D-1   *Muscles attached to bone move the vertebrate endoskeleton. They often work in antagonistic pairs, with one relaxing as the other contracts. Muscles cannot lengthen spontaneously; they lengthen only when the joint moves in the opposite direction, due to contraction of the anatagonistic muscles. For example, when you move your hand toward your shoulder, as shown here, the biceps contracts and the triceps relaxes. When you move your hand down again, the triceps contracts while the biceps relaxes. The muscles that move the skeleton, such as those diagrammed here, are known as skeletal muscles. They are striated, as shown in Figure 37E-1.*

When you move a heavy object, the force exerted on the object by the muscle is referred to as **muscle tension** and the force exerted on the muscle is the **load.** To move an object, muscle tension must be greater than the load. To hold an object steady, muscle tension must equal load.

Muscles can contract isometrically or isotonically. Muscles that produce movement contract **isotonically,** while those involved in static processes, such as maintaining posture, contract **isometrically.** During isometric contractions, muscles do not shorten; most of the cellular energy (ATP) is used to generate heat and tension. Even when you simply stand, the muscles of your back, lower legs, and thighs isometrically contract to support the weight of your body. If you lift weights, isotonic contractions (during which the muscle shortens) move the weight to above your head. Once there, the weights are held above your head by isometric contraction of the antagonistic muscles.

ııııı **Objectives** ıı

☐ Describe how the agonist and antagonist of antagonistic muscle pairs work to move a load.

☐ Differentiate between isotonic and isometric muscle exercise.

👁 **PART I**   Isotonic and Isometric Muscle Contractions

ııııı **Procedure** ıı

1. Work in pairs. Use a tape measure to determine the distance around the biceps muscle at its midpoint for both arms. During isotonic exercise, as the muscle shortens, it increases in circumference. This increase in muscle circumference is not seen during isometric exercise.

2. Note the four types of movements you will be asked to perform (see Table 37D-1). Form hypotheses that predict what will happen to muscle size during each of these movements.

    HYPOTHESIS 1:

    NULL HYPOTHESIS:

    HYPOTHESIS 2:

    NULL HYPOTHESIS:

    HYPOTHESIS 3:

    NULL HYPOTHESIS:

    HYPOTHESIS 4:

    NULL HYPOTHESIS:

    a. *What type of exercise, isotonic or isometric, do you* **predict** *will be demonstrated by each of these movements?*

    Movement 1 _____

    Movement 2 _____

    Movement 3 _____

    Movement 4 _____

    What is the **independent variable** *for these investigations?* _____

    What is the **dependent variable** *for these investigations?* _____

3. Place a small piece of tape or mark the bare arm with an **X** so that you can repeat the measurement at the same spot. Make measurements for each of the movements listed in Table 37D-1 and record all data. Perform tasks one limb at a time.

Table 37D-1

Movement	Left Arm	Right Arm	Isotonic or Isometric
1. Extend arm outward at shoulder and hold.			
2. Flex arm with palm upward.			
3. Flex arm while grasping a 20-pound barbell.			
4. Extend arm outward while holding a 10-pound barbell.			

b. *Do your results and conclusions recorded in Table 37D-1 agree with your hypotheses? Your null hypotheses?*

Movement 1 _____

Movement 2 _____

Movement 3 _____

Movement 4 _____

c. *Was there any different in the results for right and left sides of the body? If so, explain why this occurs.* _____

_____

4. Based on what you have observed, describe three exercises that would be isometric.

Exercise _____

Exercise _____

Exercise _____

👁 **PART 2**   **Muscle Fatigue**

Muscles that are continually stimulated eventually fatigue. You can't hold a barbell above your head forever!

▌▌▌▌▌ **Procedure** ▌▌

1. Work in pairs. Hold a soft rubber ball in one hand and squeeze it as rapidly as possible. Your partners should record the number of contractions for every 15-second interval for 3 minutes. Record the data in Table 37D-2 on the following page.

2. After resting for 1 minute, repeat the contractions for 3 minutes (trial 2).

3. Rest for 5 minutes and once again repeat the contractions (trial 3).

4. Change hands and repeat steps 1–3.

5. Each laboratory partner should carry out steps 1–4.

a. *What happens during the second trial?* _____

b. *Does the effect that you see in trial 2 also occur in trial 3?* _____

c. *Do your hands fatigue equally or does one hand fatigue faster than the other?* _____

Suggest why this happens. _____

Table 37D-2

	Minute 1 (in seconds)				Minute 2 (in seconds)				Minute 3 (in seconds)			
	0–15	16–30	31–45	46–60	0–15	16–30	31–45	46–60	0–15	16–30	31–45	46–60
Right Hand												
Trial 1												
Trial 2												
Trial 3												
Left Hand												
Trial 1												
Trial 2												
Trial 3												

EXTENDING YOUR INVESTIGATION: INTERACTIONS OF MUSCLES AND BONES

The types of movements an animal can make depend on the precise way that its bones and muscles fit together.

PROCEDURE

1. Obtain a raw chicken wing from your instructor or the local grocery store. Rinse it with tap water. (*Warning:* Do not touch your hands to your face or lips while working with raw chicken.) Lift the skin at the shoulder end (use forceps, if available, to make this task easier) and peel the skin back to the wing tip. (You may use scissors to cut the skin down the length of the wing, but be careful to tilt the scissors away from the muscle underneath.) As you peel the skin back, you may need to cut away some of the connective tissue. Rinse the wing often and dry it with a paper towel.

2. After removing the skin, examine the shape and texture of the muscles. On a separate sheet of paper, draw a diagram of the wing and label as many muscles as you can. Using the diagram provided by your instructor, separate the muscles using a probe so that you can locate the origin and insertion of each. From the names given in the muscle diagram, can you suggest what each muscle achieves when it contracts?

Choose three muscles and formulate a hypothesis about their function when made to contract.

HYPOTHESIS

Muscle 1

Muscle 2

Muscle 3

What do you **predict** will happen to the parts of the wing when each of the muscles contracts?

3. You can produce an imitation of muscle contraction. Locate the tendon that attaches the muscle to a bone and pull on this tendon with a pair of forceps to determine the action of the muscle.

RESULTS

Muscle 1

Muscle 2

Muscle 3

Does each muscle perform the task that you predicted? Do your results support your hypothesis?

4. Now remove a single muscle by cutting the tendons at both ends. Examine the shape of the muscle and notice how the whitish tendons are attached.

5. Remove the remaining muscles to expose the bones. Note how the different muscles are attached to the bones. Can you find any ligaments holding bones together at the joints?

6. Cut the ligaments between the upper arm and lower arm of the wing. Describe how the bones fit into each other at the joint. Describe the texture of the ends of the bones at the joint. Wash your hands carefully after handling raw chicken.

EXERCISE E | The Biochemistry of Muscle Contraction

A skeletal muscle is composed of a collection of muscle bundles. Each muscle bundle is, in turn, composed of a collection of **muscle fibers.** Muscle fibers are long, thin muscle cells (these can be as long as several centimeters). Each cell is covered by a cell membrane, the **sarcolemma.** Inside each muscle cell (muscle fiber) are thousands of structural elements called **myofibrils** (Figure 37E-1a, b).

Individual muscle cells also contain many of the same structures typically present in cells, including one or several nuclei (skeletal muscle cells are multinucleate), mitochondria, a highly developed endoplasmic reticulum (the sarcoplasmic reticulum), and cytoplasm (sarcoplasm).

a. *Would you expect a muscle cell to have very few or very many mitochondria?* _____

   Why? _____

Examined under the microscope, the myofibrils within skeletal muscle cells appear to be striated. Because of the arrangement of the myofibrils, the entire muscle fiber also appears to be striated. The striated appearance of the myofibrils is due to the presence of protein **filaments** arranged into subunits called **sarcomeres** (Figure 37E-1b). The sarcomeres are arranged side by side throughout the length of the myofibril.

A single sarcomere is composed of **thick filaments** and **thin filaments** arranged in a highly ordered manner (Figure 37E-1c, d). The thick filaments are composed of the protein **myosin,** and the thin filaments are composed of the protein **actin.** The arrangement of the actin and myosin in the myofibril accounts for the striated appearance or banding pattern of the sarcomere. The ends of a sarcomere are delineated by dark bands called Z lines which are attached to the actin filaments. The I band is composed of actin and spans two sarcomeres. The A band is composed of myosin alone in the H zone, and myosin plus actin outside the H zone.

Figure 37E-1d, e illustrates the arrangement of actin and myosin filaments in a relaxed muscle and in a muscle that has contracted. Notice that in the relaxed state the actin filaments partially overlap the myosin filaments in the A band. However, in the middle of the A band—the H zone—there are no actin filaments. The M line is a protein substructure that surrounds the myosin. In the contracted state, the actin filaments pull inward over the myosin filaments and completely overlap them.

b. *What does this do to the size of the H zone?* _____

c. *What does this do to the size of the A band?* _____

Notice that the Z lines have also been pulled by the actin filaments so that they move much closer to the end of the myosin filaments.

d. *What does this do to the size of the I band?* _____

Figure 37E-1 *The composition and action of skeletal muscle.*

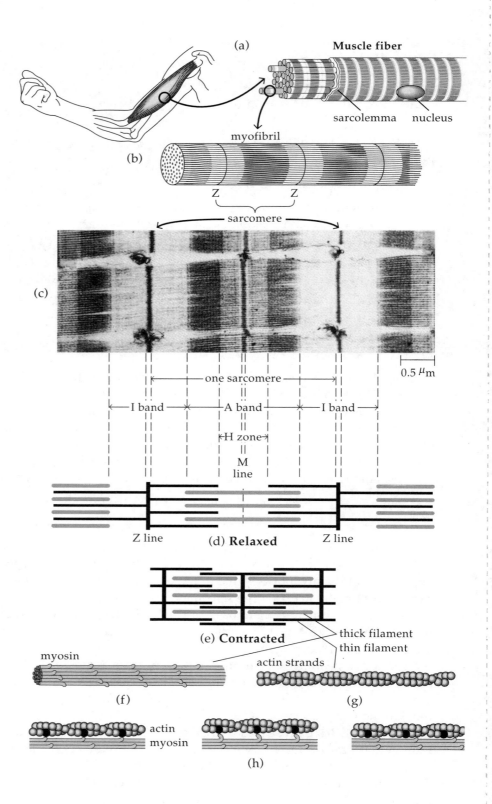

As the sarcomeres contract, so does the entire muscle. Muscle contraction occurs by a **sliding filament** mechanism. The myosin filaments have protuberances known as cross bridges. When ATP and calcium ions (Ca^{2+}) are present, the cross bridges of the myosin interact with active sites on the actin filaments, causing the actin to slide over the myosin filaments, which causes the sarcomere to shorten (Figure 37E-1h).

||||| **Procedure** ||||||||||||||||||||||||||||||||||||||||

1. Obtain a piece of glycerinated rabbit skeletal muscle which has been prepared by your instructor. Tease a segment of skeletal muscle into *very thin* groups of fibers (muscle cells). *Single* fibers will give the best results. Strands of muscle should not exceed 0.2 mm in width. Thick strands will curl when stimulated to contract and will be difficult to observe.

2. Mount one strand of muscle on a clean glass slide. (Do not stretch it!) Add a drop of glycerol solution and a coverslip. (Do not allow it to dry out!) Examine the preparation under the microscope at low and high power. Note the striations in the fibers and the smooth walls of the muscle cell.

   e. How many nuclei can you see? ———————

3. In the space below, draw the structure of a relaxed muscle fiber, as it appears under high power.

4. You will see a series of double, dark lines between two lighter areas. Each of the lighter areas also contains a *thin* dark line. These lines represent the various parts of a sarcomere as shown earlier (Figure 37E-1). Label the H, A, and Z areas on your drawing above.

5. Transfer a second muscle fiber group to a clean glass slide. Add a small drop of glycerol to make sure that the preparation does not dry out, but use the corner of a piece of lens paper to remove any excess glycerol. Do not use a coverslip.

6. Use a dissecting microscope and a plastic millimeter ruler to measure the length of the fiber. Record your data in the "Before" column of Table 37E-1.

7. Now, transfer the preparation to your light microscope and observe the fibers at high power while your lab partner floods the preparation (not all over the microscope!) with a solution containing ATP and Ca^{2+} ions.

8. Form a hypothesis that predicts what will happen when ATP and Ca^{2+} are added to the muscle preparation.

   HYPOTHESIS:

   NULL HYPOTHESIS:

   What changes do you predict in the band pattern of the myofibrils of the muscle fiber?

   What is the **independent variable**?

   What is the **dependent variable**?

9. Observe any changes that occur.

   f. What happens to the structure of the myofibril? ————————————————————————

   ——

10. After 30 seconds, use the dissecting microscope and once again measure the length of the fiber.

    g. Was there any change in length? ————————————————————————

Table 37E-1

	Myofiber Length		
	Before	After	Change in Length
ATP/Ca^{2+}			
KCl, MgCl$_2$			
ATP/CA^{2+} and KCl, MgCl$_2$			

    h. Considering what you observed using your light microscope, would you expect to see any change in myofibril length? _____ Why or why not? _____

_____

    i. Do your observations support your hypothesis? _____ Your null hypothesis? _____

11. Repeat steps 6, 7, 9, and 10 using a new fiber preparation that is flooded with a solution containing the mineral salts KCl and MgCl$_2$.

    j. Did the mineral salts (KCl and MgCl$_2$) cause the myofibril structure to change? _____

    Did the fiber length change? _____

12. Form a hypothesis about what will happen when KCl and MgCl$_2$ alone, or in combination with ATP and Ca^{2+}, are added to a myofibril.

    HYPOTHESIS (KCl and MgCl$_2$ alone):

    NULL HYPOTHESIS:

    HYPOTHESIS (KCl, MgCl$_2$, ATP, Ca^{2+}):

    NULL HYPOTHESIS:

What changes do you predict in the band pattern of the myofibril with each of these treatments?

What is the **independent variable**?

What is the **dependent variable**?

13. Observe any changes that occur following step 11.

14. Repeat steps 6, 7, 9, and 10 using a new fiber preparation flooded with a solution containing *both* ATP/Ca^{2+} and mineral salts (KCL and MgCl$_2$). What do you observe under the high power of your light microscope? Draw the myofibril in the space below.

15. Compare the structure of the myofibril with what you observed and drew in step 3.

k. From your observations and data, what do you conclude about the factors required for muscle contraction? _____

_____

ATP provides the energy for muscle fiber contraction. Ca^{2+} is needed to form cross bridges between actin and myosin. Mineral salts such as $MgCl_2$ and KCl are necessary cofactors for ATPase activity. ATPase catalyzes the hydrolysis reaction: $ATP \rightarrow ADP + P_i$.

Laboratory Review Questions and Problems

1. Use anatomical terms for position and direction to describe the location of the following body parts:

Mouth of a human in relation to ears

Toes of a human in relation to leg

Tail of a horse in relation to head

Tail of a horse in relation to hooves

2. Compare and contrast integumentary structures in fishes, amphibians, reptiles, birds, and mammals.
What are the characteristics of the skin in each type of animal?

What are the major similarities in the skin of these animals?

What are the major differences in the skin of these animals?

3. How do different structures in the skin of the vertebrate groups studied in this lab adapt each group to major environmental factors that they face?
Fishes

Amphibians

Reptiles

Birds

Mammals

4. Why is it advantageous to terrestrial vertebrates to have a nonliving outer layer of epidermis in contact with the environment?

5. If a young child breaks a bone, the doctor becomes very concerned if the fracture is in the "growth zone." Where is this growth zone, and why is it so important?

6. You read in the local paper that an "isometric exercise specialist" has joined the staff of the local health club. What types of exercises would you expect this specialist to teach?

7. In the space below, draw the pattern of actin and myosin filaments that you would expect to observe in a contracted muscle and a relaxed muscle, as viewed with a microscope.

8. What are the roles of ATP, Ca^{2+}, and Mg^{2+} in muscle contraction?

Physiology of Circulation

OVERVIEW

Animals too large to accomplish internal transport by diffusion are equipped with a system of branching vessels filled with blood, which is usually propelled through the system by the muscular contractions of the heart. Vertebrates have a closed circulatory system—a circuit of continuous vessels. Other animals, such as arthropods and annelids, have an open circulatory system: blood flows from vessels to open spaces in the tissues and then to vessels again. A pump is used to move the blood throughout the vessels of the circulatory system.

Many organisms couple the circulatory system with a respiratory surface such as lungs or gills, where gases can be exchanged between the blood and the environment. In vertebrates other than birds and mammals, a serial circuit delivers blood directly from the respiratory surface to the tissues. The parallel circuitry (pulmonary and systemic systems) present in birds and mammals is more efficient.

Blood traveling in the circulatory system is made up of a fluid matrix called plasma, which carries cells, oxygen, nutrients, wastes, and other materials from one region of the body to another. Blood often contains special respiratory pigments that deliver the oxygen throughout the body (see Laboratory 39). The rate at which the blood is pumped by the heart can be measured as one's pulse. Blood pressure is a measure of the force exerted by blood against the walls of the blood vessels. Both pulse and blood pressure can be influenced by a variety of factors, including diet, exercise, hormones, age, smoking, alcohol intake, and a number of other environmental factors.

During this laboratory, you will investigate the properties of blood cells, arteries, and veins. You will explore the structure of the vertebrate heart and determine the effects of hormones and environmental factors, including temperature, on heart rate and pulse.

STUDENT PREPARATION

Prepare for this laboratory by reading the text pages indicated by your instructor. Familiarizing yourself in advance with the information and procedures covered in this laboratory will give you a better understanding of the material and improve your efficiency.

✔ | **EXERCISE A** | **Microscopic Examination of Human Blood Cells**

Blood contains white cells (**leukocytes**) and red cells (**erythrocytes**). Mature mammalian erythrocytes are biconcave disks that lack a nucleus and contain hemoglobin for the transport of oxygen. Leukocytes are nucleated cells. **Granulocytes** and **monocytes,** types of leukocytes, transform into macrophages that migrate to infected areas, where they perform a clean-up function. **Lymphocytes,** another type of leukocyte, are responsible for immune reactions. Many infections are characterized by an increase in the white blood cell count.

⁞⁞⁞⁞ **Objectives** ⁞⁞⁞⁞⁞⁞⁞⁞⁞⁞⁞⁞⁞⁞⁞⁞⁞⁞⁞⁞⁞⁞⁞⁞⁞⁞⁞⁞⁞⁞⁞⁞⁞⁞⁞⁞

☐ Recognize and describe red blood cells.

☐ Recognize and describe granulocytes, monocytes, and lymphocytes.

☐ Describe the difference between normal and sickle-cell erythrocyte structure.

☐ Describe the characteristics of blood that can be used to identify infections such as mononucleosis.

⁞⁞⁞⁞ **Procedure** ⁞⁞⁞⁞⁞⁞⁞⁞⁞⁞⁞⁞⁞⁞⁞⁞⁞⁞⁞⁞⁞⁞⁞⁞⁞⁞⁞⁞⁞⁞⁞⁞⁞⁞⁞⁞⁞

1. Use high power (40×) to observe a prepared slide of human blood. Use Figure 38A-1 and Table 32B-1, page 32-10, to help identify cell types.

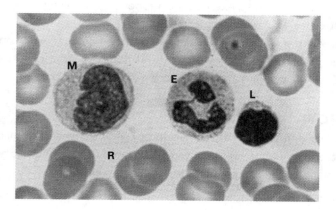
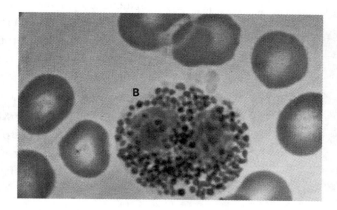
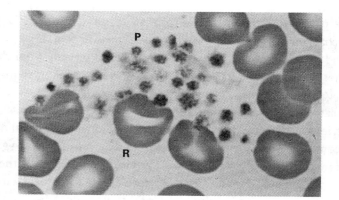
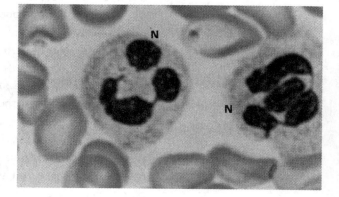

Figure 38A-1   *Blood cell types.* **R** Red blood cells *(erythrocytes);* white blood cells *(leukocytes):* **E** *eosinophil (stains dark pink to reddish orange),* **N** *neutrophil (stains light pink to lavender),* **B** *basophil,* **L** *lymphocyte,* **M** *monocyte;* **P** *platelet (thrombocyte).*

a. Which type of blood cells are stained pink on your slide? _____

   Do you see their nuclei? _____

b. What pigment is present in these cells? _____ What is its function?

   _____

c. What proportion of the cells on your slide appear to be leukocytes? _____

d. Granulocytes are characterized by nuclei of many different shapes and by the granules in their

   cytoplasm. What color are these granules? _____

e. Do you see white blood cells with no granules in the cytoplasm? These are either monocytes or
   lymphocytes. Monocytes are the largest leukocytes (approximately twice the diameter of
   lymphocytes).

   Which do you see? _____

f. Which type of leukocyte is most abundant? _____

2. Use high power (40✕) to examine the slide (on demonstration) of blood from a carrier of
   sickle-cell anemia in low-oxygen crisis.

   g. Describe your observations.

   _____

   The hemoglobin of an individual who carries the recessive sickle-cell allele in the
heterozygous condition is less soluble than normal hemoglobin. When the oxygen supply is
inadequate or when the carbon dioxide concentration increases, sickle-cell hemoglobin
molecules tend to crystallize to form hairlike rods that pile up and transform the cell into a
sickle shape. The cells then clump and clog the blood vessels and cannot carry out their
function of transporting oxygen. In order to determine whether a person is a carrier of the
sickle-cell allele, blood is subjected to a low-oxygen atmosphere and examined with a
microscope. In an individual who is homozygous for the sickle-cell allele (that is, has sickle-cell
anemia), hemoglobin is abnormal even at normal oxygen and carbon dioxide concentrations.

3. Use high power (40✕) to examine the slide (on demonstration) of blood from a person with
   mononucleosis. Mononucleosis is a disease characterized by fever, headache, scratchy throat,
   fatigue, and enlargement of the lymph glands.

   h. Describe your observations. _____

   _____

   i. How does the number of white blood cells in this preparation compare with the number of these cells

      in a normal blood smear? _____

   j. Is there a type of white blood cell that tends to be most common? If so, name the type.

   _____

4. Examine a prepared slide of frog blood.

   k. How do frog erythrocytes differ from those of humans? _____

✔ **EXERCISE B**   **Pumps—The Vertebrate Heart**

In fishes, the heart is tubular and is composed of a single atrium and ventricle. Deoxygenated blood
returning from the body enters a chamber at the rear of the heart and then flows into the atrium. From
there, blood flows into the ventricle, which pumps it to the gills (Figure 38B-1a). In amphibians,
deoxygenated blood returns to the right atrium, and oxygenated blood from the lungs moves into the

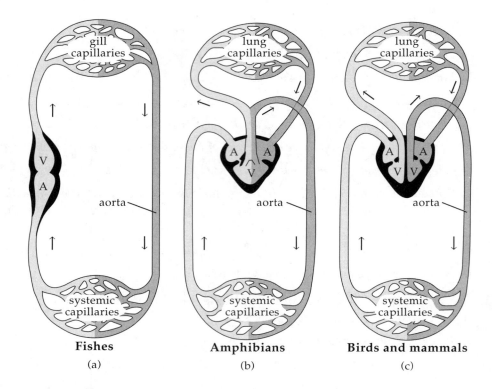

Figure 38B-I *Vertebrate circulatory systems. Oxygen-rich blood is shown as dark gray, and oxygen-poor blood as light gray.*

(a) In fishes, the heart has only one atrium (A) and one ventricle (V). Blood oxygenated in the gill capillaries goes straight to the capillaries of the systemic circulation without first returning to the heart.

(b) In amphibians, the single primitive atrium has been divided into two separate chambers. Oxygen-rich blood from the lungs enters one atrium, and oxygen-poor blood from the tissues enters the other. Little mixing of the blood occurs in the ventricle, despite its lack of a structural division. From the ventricle, oxygen-rich blood is pumped to the body tissues at the same time that oxygen-poor blood is pumped to the lungs. Some of the oxygen-poor blood is diverted from the lungs to the skin, a major respiratory organ in amphibians.

(c) In birds and mammals, both the atrium and the ventricle are divided into two separate chambers, so that there are, in effect, two hearts—one for pumping oxygen-poor blood through the lungs and one for pumping oxygen-rich blood through the body tissues.

left atrium. There is only one ventricle, and thus the amphibian heart is "three-chambered." The oxygenated blood entering the ventricle becomes layered on top of returning deoxygenated blood and is pumped to the head, heart, and body while the deoxygenated blood is pumped to the lungs (Figure 38B-1b).

In reptiles, there is a trend toward dividing the ventricle to form separate pulmonary (lung) and systemic (body) circulations. In birds and mammals, separation of the ventricle is complete (Figure 38B-1c). Systemic and pulmonary circulations are parallel: (heart → systemic circulation → heart) and (heart → lungs → heart).

In the mammalian heart, the atria and ventricles are separated by **atrioventricular valves:** on the right side, the **tricuspid valve,** and on the left, the **bicuspid valve.** Backflow of blood from the pulmonary arteries that carry blood to the lungs, and from the aorta that carries blood to body tissues, is prevented by **semilunar valves.** The rhythm of the heartbeat is maintained and controlled by a patch of tissue, the sinoatrial node (SA node) or **pacemaker** (Figure 38B-2b). A wave of excitation initiated in the pacemaker is

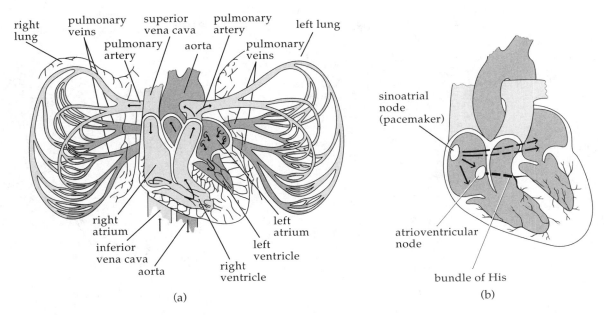

right lung
pulmonary veins
superior vena cava
pulmonary artery
pulmonary artery
aorta
left lung
pulmonary veins
right atrium
inferior vena cava
aorta
left atrium
left ventricle
right ventricle

(a)

sinoatrial node (pacemaker)
atrioventricular node
bundle of His

(b)

Figure 38B-2   *(a) The human heart. Blood returning from the systemic circulation through the superior and inferior venae cavae enters the right atrium and passes to the right ventricle, which propels blood through the pulmonary arteries to the lungs, where it is oxygenated. Blood from the lungs enters the left atrium through the pulmonary veins, passes to the left ventricle, and then is pumped through the aorta to the body tissues.*

(b) The beat of the mammalian heart is controlled by a region of specialized muscle tissue in the right atrium, the sinoatrial node, that functions as the heart's pacemaker. Some of the nerves regulating the heart have their endings in this region. (The tissue of the pacemaker is homologous to that of the chamber at the rear of the fish heart called the sinus venosus. *This tissue has an inherent ability to beat and sets the pace of the heartbeat for the fish.)*

Excitation spreads from the pacemaker through the atrial muscle cells, causing both atria to contract almost simultaneously. When the wave of excitation reaches the atrioventricular node, its conducting fibers pass the stimulation to the bundle of His, from which excitation spreads along specialized fibers of the ventricles. The result is an almost simultaneous contraction of the two ventricles. Because the fibers of the atrioventricular node conduct relatively slowly, the ventricles do not contract until after the atrial beat has been completed.

picked up by a special bundle of conducting fibers (the bundle of His) that originates in an area known as the atrioventricular node (AV node). These fibers conduct the impulse to the walls of the ventricle, causing it to contract.

Objectives

☐ Trace the pathway of blood through the mammalian heart.

☐ Trace the evolution of separate systemic and pulmonary circulations in the heart pump.

Procedure

1. Work in pairs. Wear gloves and safety glasses. If any preserved tissue gets into your eye, flush it with water immediately and notify your instructor. Place a fresh or preserved sheep heart in

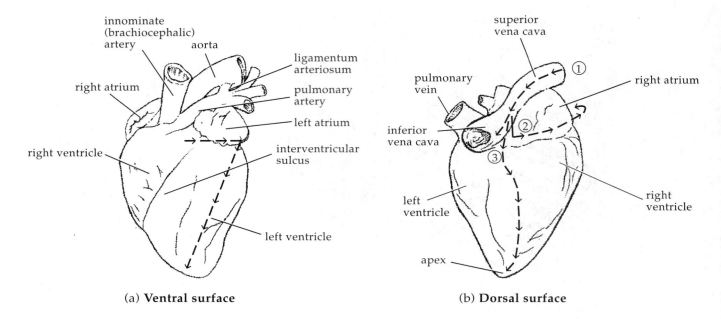

(a) **Ventral surface**　　　　　　　(b) **Dorsal surface**

Figure 38B-3　*Sheep heart. Dissection cuts are shown for each surface.*

a dissecting pan. Remove any remnants of the covering **pericardial membrane** surrounding the heart. Use a probe or forceps to scrape away excess fat from the heart and its vessels.

2. Position the heart so that you are looking at its ventral surface (locate the **interventricular sulcus** on the ventral surface, as shown in Figure 38B-3a). When viewing the heart from this aspect, note that structures to your right are actually located on the animal's left (and vice versa). Notice some twisting of the ventricles, with the larger left ventricle lying somewhat behind the right ventricle.

3. Locate the atria—smaller, thin-walled chambers anterior to the ventricles (Figure 38B-3a). Identify **right** and **left atria.** You should be able to identify the roots of three large arteries leaving the heart between the atria. The **pulmonary artery** carries deoxygenated blood from the right ventricle to the lungs (branches may or may not be found in your specimen).

a. Why is this vessel called an artery even though it carries deoxygenated blood?

_____

The pulmonary artery leaves the heart ventral to the aorta. The **aorta** carries oxygenated blood from the left ventricle. It branches immediately to give rise to the innominate artery, which carries blood to the right forelimb and head, and the aortic arch, which carries blood to the head and left forelimb and to the remainder of the body.

Can you find a ligament tightly joining the pulmonary artery and aortic arch (Figure 38B-3a)? This is the remnant of an embryonic open connection between these two vessels that allowed the fetus to circulate oxygen obtained from the umbilical vein (which entered the right atrium). If a diagram of fetal circulation is available, compare it with the circulatory system of an adult mammal. In the fetus, only a small amount of blood actually enters the pulmonary system. Instead, most blood empties directly into the systemic aorta through the **ductus arteriosus,** a shunt between the pulmonary trunk and the arch of the systemic aorta. This fetal structure constricts and closes after birth, becoming a tough strand of connective tissue, the **ligamentum arteriosum.**

b. Why is it not necessary for the fetus's blood to pass to its lungs?

_____

_____

4. Turn the heart over and examine the dorsal surface (Figure 38B-3b). Identify the **superior vena cava.** This vessel returns deoxygenated blood to the heart from the head and anterior body regions. Also identify the **inferior vena cava,** which returns deoxygenated blood from the remainder of the body. Try to find the **pulmonary veins.** These vessels return oxygenated blood from the lungs to the heart.

   c. *Can a blood vessel be defined as an artery or vein by the type of blood it carries (oxygenated or deoxygenated)?* _____

   d. *What determines whether a blood vessel is an artery or a vein?*

   _____

5. Looking at the dorsal surface of the heart and consulting Figure 38b-3b, begin cut ① by inserting your scissors into the superior vena cava. Cut through the dorsal wall of this vessel and extend the cut through the inferior vena cava. Begin cut ② by positioning your scissors in the inferior vena cava and cutting laterally through the right atrium to the base of the pulmonary artery. Make cut ③ through the right ventricle toward the apex of the heart. Separate the chambers and examine the internal structure (Figure 38B-4a).

6. Find the **tricuspid valve** (the right atrioventricular valve). As its name implies, the tricuspid valve is made up of three flaps of the heart wall and prevents backflow of blood during contraction. These flaps are prevented from everting into the atrium during contraction of the right ventricle by strands of tissue (**chordae tendinae**) connecting the ventricular wall to the valve flaps (Figure 38B-4a). Extend the opening of the ventricle into the base of the pulmonary artery. Locate the **pulmonary semilunar valves,** made up of three small flaps that prevent backflow of blood from the artery to the ventricle. Note the smooth layer of tissue, the endocardium, lining the cavity of the heart.

7. Now, turn the heart over to the ventral side. Starting about 2 cm from the left edge of the left atrium (Figure 38B-3a), cut laterally through the left atrium and then downward through the left ventricle to the apex of the heart.

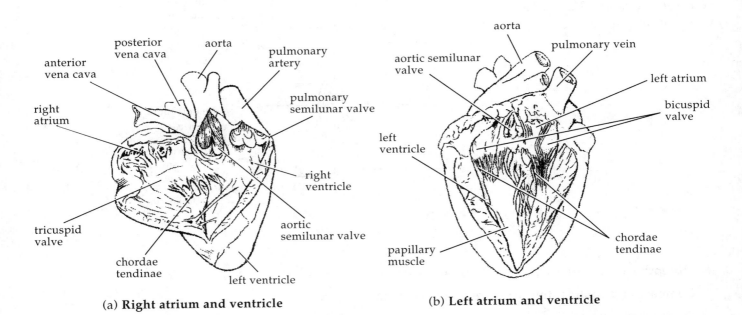

(a) **Right atrium and ventricle**

(b) **Left atrium and ventricle**

Figure 38B-4 *Internal structures of (a) the right atrium and ventricle and (b) the left atrium and ventricle.*

8. Open the chambers to examine interior structures. Find the pulmonary veins entering the atrium (Figure 38B-4b). Note that the left ventricular wall is thicker than the right.

   e. Why? _____

   The internal structures of both ventricles are similar. Find the **bicuspid** (mitral) **valve** (the left atrioventricular valve). This valve is made up of two flaps which prevent backflow into the left ventricle. Chordae tendinae provide support, as in the right ventricle. Extend your cut toward the aortic trunk and find the two **aortic semilunar valves** preventing backflow into the left ventricle.

9. On a separate piece of paper, diagram the flow of oxygenated blood from the lungs to the heart to the body and of deoxygenated blood from the body to the heart to the lungs.

   f. *Why would parallel circulation be an advantage to active organisms such as mammals? (Hint: Compare the mammalian circulatory system with that of fishes, as shown in Figure 38B-1.)*

   _____

   _____

✔ **EXERCISE C**   **The Structure of Arteries and Veins**

The blood of a closed circulatory system is confined within the blood vessels. Blood leaves the heart through large arteries—the pulmonary artery to the lungs and the aorta to the systemic or general body circulation. These arteries branch into smaller arteries, then into even smaller arterioles, and finally into very thin-walled capillaries. The capillaries join to form venules, which come together to form larger veins. Veins return blood to the heart. Label Figure 38C-1 to show the path of blood. Begin at the top with the artery.

Figure 38C-1 *Diagrammatic representation of the structure and arrangement of blood vessels in the circulatory system. The endothelium is only one cell thick.*

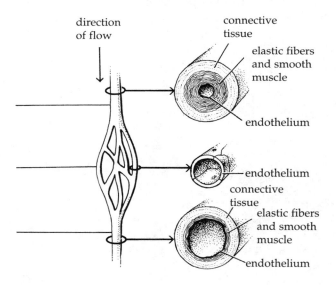

direction of flow

connective tissue

elastic fibers and smooth muscle

endothelium

endothelium

connective tissue

elastic fibers and smooth muscle

endothelium

ııııı **Objectives** ııııııııııııııııııııııııııııııııııııı

☐ Compare and contrast the structure of an artery and a vein.

☐ Describe the effects of atherosclerosis on the structure of blood vessels.

☐ Differentiate between atherosclerosis and arteriosclerosis.

Procedure ∎∎∎∎∎∎∎∎∎∎∎∎∎∎∎∎∎∎∎∎∎∎∎∎∎∎∎

1. Use low power (10×) to observe a prepared slide of an artery and a vein. Use high power (40×) to study both blood vessels in more detail. Identify the structures in the following description.

    An artery has thick walls with wavy-appearing elastic fibers and a relatively small central cavity (**lumen**). A vein has a thinner wall and a much larger lumen.

    a. *Why is the wall of the artery thicker, stronger, and more elastic than that of the corresponding vein?*

    _____

    _____

    Both arteries and veins consist of three layers that may vary somewhat in size and composition. The inner layer is the **endothelium**, found in all blood vessels, including capillaries. It is composed of an elastic membrane covered by a single layer of flat squamous epithelial cells. In an artery, this inner layer may also contain elastic connective tissue fibers. The middle layer is an area of smooth muscle fibers and elastic fibers. The middle layer of a vein contains a thinner layer of smooth muscle fibers and fewer elastic fibers than that of an artery. The outer layer of blood vessels consists of connective tissues, nerve fibers, and smaller blood vessels. In an artery, this area is composed chiefly of tough, nondistensible, white, fibrous connective tissue. The outer layer of a vein is even less distensible and elastic than that of an artery. The wall of a capillary consists of just a very thin layer of flat endothelial cells.

2. In the space beside Figure 38C-1, examine the cross sections of an artery and a vein. Note the three layers.

    b. *Veins contain* **valves**, *structures that are not found in arteries. What is their function?*

    _____

3. Observe a slide (on demonstration) showing the blood vessels of a person with atherosclerosis. Compare the artery in this slide with the normal artery.

    c. *What difference do you observe?* _____

    In **atherosclerosis**, principally a disease of the large arteries, lipid deposits, called **atheromatous plaques**, appear in the arteries. These plaques contain an especially large amount of cholesterol and are often associated with changes in the arterial wall. During later stages of the disease, the walls of the arteries become extremely hard due to the presence of fibroblasts in the affected area and the deposition of calcium and lipids. The disease is then called **arteriosclerosis**, or hardening of the arteries.

    d. *How might atherosclerosis interfere with the normal functioning of an artery?*

    _____

✔ **EXERCISE D**   **Movement of Blood in a Goldfish Tail (Demonstration)**

As blood moves from larger arteries into smaller capillaries, the speed of flow slows down and frictional resistance increases. In the smallest capillaries, blood cells must move in single file.

Objectives ∎∎∎∎∎∎∎∎∎∎∎∎∎∎∎∎∎∎∎∎∎∎∎∎∎∎∎∎∎∎∎∎∎∎∎∎

☐ Describe the movement of blood cells through large vessels and capillaries.

Procedure ∎∎∎∎∎∎∎∎∎∎∎∎∎∎∎∎∎∎∎∎∎∎∎∎∎∎∎∎∎∎∎∎∎

1. On demonstration you will find a goldfish wrapped in wet cotton with its tail placed between two slides.

2. Use scanning power (4×) and then low power (10×) to observe the movement of blood cells in the large vessels and then in the capillaries.

   a. In which vessels is blood moving fastest? _____

   b. How does the diameter of a red blood cell compare with that of a capillary? _____

   c. Why would it be advantageous for red blood cells to move more slowly in capillaries?

   (Hint: Think about the function of capillaries.) _____

   _____

3. Place the goldfish back in its tank.

👁 **EXERCISE E**   **Determining Blood Pressure**

A blood pressure reading is a measure of the force exerted by the blood against circulatory vessels, first during **systole,** contraction of the ventricles of the heart, and then at **diastole,** relaxation of the ventricles. When the ventricles contract, a greater volume of blood is forced through the open valves and into the arteries than can immediately exit through the narrowest of the arterioles. The result is a rise in pressure inside the arteries as blood pushes against the interior walls. Between ventricular contractions, the aortic valve closes, blood leaves the arteries, and there is a decrease in arterial pressure. The "rebound" of the elastic walls of the arteries helps to give the blood an additional push.

A sphygmomanometer (Figure 38E-1) is used to measure blood pressure. The cuff, designed to fit around the upper arm, can be expanded by pumping a rubber bulb connected to the cuff. The pressure gauge, scaled in millimeters of mercury (mm Hg), indicates the pressure inside the cuff. A stethoscope is used to listen to the subject's pulse.

Figure 38E-1 *Using the sphygmomanometer to measure blood pressure.*

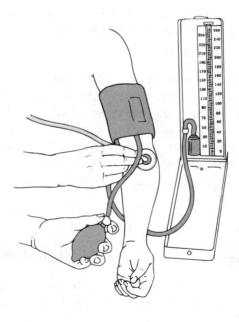

‖‖‖‖ **Objectives** ‖‖‖‖‖‖‖‖‖‖‖‖‖‖‖‖‖‖‖‖‖‖‖‖‖‖‖‖‖‖‖‖‖‖

☐ Determine systolic and diastolic blood pressure.

‖‖‖‖ **Procedure** ‖‖‖‖‖‖‖‖‖‖‖‖‖‖‖‖‖‖‖‖‖‖‖‖‖‖‖‖‖‖‖‖‖‖

1. Work in pairs. After your laboratory partner is seated, has rolled up his or her shirt sleeve, and is relaxed, attach the cuff of the sphygmomanometer snugly around the upper arm.

2. Place the stethoscope at a point directly below the cuff, preferably in the well of the elbow joint.

3. Close off the valve to the bulb by turning it clockwise and pump air into the cuff until the needle on the pressure gauge jumps just past the 200-mm mark.

4. Turn the valve on the bulb counterclockwise to slowly release air from the cuff at a rate of about 2 or 3 mm Hg per second. Listen for a pulse.

5. Just as you begin to hear the pulse, note the pressure on the gauge. This is the systolic pressure. As the cuff was inflated, the brachial artery in the arm collapsed. When the artery begins to expand as you release air from the cuff, you hear the pulse. As pressure in the artery is released, all sound ceases.

6. Continue to listen until the clear thumping sound of the heart becomes strong and then fades. When you hear the last part of the full heartbeat (lub *dub*), take note again of the pressure. This is the diastolic pressure.

7. Write down the complete blood pressure of your subject: systolic/diastolic = _____/_____

   Record your own blood pressure as measured by your partner: _____/_____

Table 38E-1   Normal Blood Pressure for Men and Women (Age 20)

Blood Pressure (mmHg)			
Men		Women	
Systolic	Diastolic	Systolic	Diastolic
105–140	62–86	100–130	60–85

a. *Does your subject's blood pressure lie within the normal range (Table 38E-1)?* _____
   Your own blood pressure? _____

b. *What circulatory problems may be indicated by high blood pressure (hypertension)?*

   _____

c. *Why is hypertension dangerous if it is allowed to continue indefinitely?*

   _____

d. *What condition exists when a person's blood pressure is well below normal?*

   _____

e. *How would blood pressure measured in the thigh compare with the reading you just obtained?*
   _____ *Why?* _____

   _____

8. The microscope slide on demonstration shows the cross sections of blood vessels from a person with atherosclerosis.

   f. *Explain how a narrowing of blood vessels affects blood pressure.*

   _____

👁 **EXERCISE F**   **Measuring Pulse Rate and Blood Pressure**

Heart rate, pulse rate, and blood pressure can all be affected by many factors, including fitness, activity, smoking, and drugs.

▌▌▌▌▌ **Objectives** ▌▌▌▌▌▌▌▌▌▌▌▌▌▌▌▌▌▌▌▌▌▌▌▌▌▌▌▌▌▌▌▌▌

☐ Determine the relationship between pulse rate and heartbeat rate.

☐ Determine the effects of both moderate and strenuous exercise on the heart rate and pulse rate.

☐ Explain how smoking can affect pulse rate.

👁 **PART 1**   Relation of the Heartbeat to Circulation

▌▌▌▌▌ **Procedure** ▌▌▌▌▌▌▌▌▌▌▌▌▌▌▌▌▌▌▌▌▌▌▌▌▌▌▌▌▌▌▌▌▌

Work in pairs. Use a stethoscope to listen to your partner's heart. Your partner should be sitting down. At the same time, take his or her pulse rate by placing the tips of the first two fingers of one of your hands over the radial artery at the base of the wrist on the palm side of your partner's hand. Pulse rate of subject while sitting: _____ Record your own pulse rate as measured by your partner: _____

   a. *What is the relation between heartbeat rate and pulse rate?* _____

   b. *Why does this relationship exist?*

   _____

   c. *What is a pulse?* _____

   d. *Why can a pulse be found only in certain parts of the human body?* _____

👁 **PART 2**   Variability of the Heart Rate and Blood Pressure

▌▌▌▌▌ **Procedure** ▌▌▌▌▌▌▌▌▌▌▌▌▌▌▌▌▌▌▌▌▌▌▌▌▌▌▌▌▌▌▌▌▌▌▌▌▌▌▌

1. Work in pairs. One lab partner should recline on a laboratory bench for 5 minutes. At the end of this time, the other partner should take the pulse rate of the reclining subject, who should continue to recline for another 2 minutes. The subject should then stand and the pulse rate should be taken immediately. Record all pulse rates in Table 38F-1.

2. After the subject has remained standing for 3 minutes, record the standing pulse rate.

3. Add the data from Part 1 for the subject's pulse rate in the sitting position. Note the data on your own pulse rates, as measured by your partner, in the space next to the table.

Table 38F-1   Pulse Rate (beats/minute)

Reclining	
Immediately upon standing	
Standing	
Sitting	

The change from a reclining or sitting to a standing position causes blood to "fall" from the upper body under the influence of gravity, resulting in a decrease in blood pressure. Pressure receptors (baroreceptors) in the aortic arch and the carotid arteries compensate by signaling the medulla of the brain to increase the heartbeat, and, consequently, the pulse rate. Did you observe this increase? _____

a. How might the valves of the blood vessels be involved in the increase in heart rate immediately upon standing and the decrease in heart rate after a period of standing?

_____

_____

4. Do you think a person who is more physically fit than another person would exhibit a larger or smaller difference in pulse rate upon standing? _____ Compare your subject's data or your own data with those obtained for others of the same sex in your class whom you might consider to be less or more physically fit. Does this comparison support your answer? _____

b. Explain the remaining data in Table 38F-1. _____

_____

PART 3   The Effect of Exercise on Heart Rate

Procedure

1. Work in pairs. Determine your partner's sitting pulse rate ("before exercise" pulse rate).

2. One partner should now exercise moderately by raising alternate knees to the chest for approximately 30 seconds. The other partner should exercise strenuously by running up several flights of stairs as quickly as possible.

3. Record the pulse rate and rate of heartbeat immediately after exercise, then measure the time required for the return of sitting pulse rate for both individuals.

4. Reverse roles, and record data for the other type of exercise for each partner. Record all data either for yourself or for your partner in Table 38F-2.

Table 38F-2   Effect of Exercise on Heart Rate

	Before Exercise	Moderate Exercise	Heavy Exercise
Pulse rate			
Rate of heartbeat			
Time required for the return of sitting pulse rate			

Table 38F-3 lists the expected pulse rate 2 minutes after vigorous exercise in individuals of various physical conditions.

Table 38F-3   Pulse Rates

Physical Condition	Pulse Rate (beats per minute)
Excellent	71–78
Very good	79–83
Average	84–99
Below average	100–107
Poor	108–118
Very poor	Above 119

The speed with which the sitting pulse rate is restored after exercise serves as one index of circulatory efficiency. Here is an example of a normal response: before exercise, pulse 80; immediately after exercise, pulse 120; 2 minutes after exercise, pulse 84.

   a. *How efficient is your circulatory system?* _____

   b. *Why should people with "weak" hearts avoid strenuous exercise such as climbing stairs?*

   _____

PART 4   The Effect of Smoking on Heart Rate (*Optional*)

ııııı Procedure ııııııııııııııııııııııııııııııııııııı

If you are a cigarette smoker, go without smoking for one to several hours. Take your sitting pulse rate before and then 3 minutes after smoking a cigarette.

a. What was the effect of smoking on your heart rate? _____

EXERCISE G | Factors Influencing Heart Rate in the Water Flea, *Daphnia*

The heartbeat rate can be influenced by temperature and by certain chemicals such as the neurotransmitters **acetylcholine** and **epinephrine.** (Recall that epinephrine is also a hormone.)

ııııı Objectives ıııııııııııııııııııııııııııııııııııııı

☐ Determine the effect of temperature on the heart rate in *Daphnia*.

☐ Determine the effect of acetylcholine and epinephrine on the heart rate of *Daphnia*.

☐ Describe Q_{10} and calculate Q_{10} for heart rate in *Daphnia*.

PART I   The Effect of Temperature Change

ııııı Procedure ıııııııııııııııııııııııııııııııııııııı

   1. Pick up a *Daphnia* with a large-bore pipette or eye dropper (a broken-off Pasteur pipette will also work).

   2. Place the *Daphnia* into the large end of a Pasteur pipette and allow the culture fluid containing the *Daphnia* to run down into the narrow tip of the pipette.

   3. Use a paper towel to draw some of the culture fluid out of the tip of the pipette until the *Daphnia* no longer moves down the tube and the fluid level is approximately 5 mm above the *Daphnia*.

   4. Seal the narrow end of the pipette with petroleum jelly.

   5. Score the pipette with a file and break it off about 2 cm above the *Daphnia*. Seal the broken end by keeping the pipette upright (narrow end down). and inserting the broken end into an inverted container of petroleum jelly. The tube is now sealed with a "plug" of jelly at both ends.

   6. Place the tube containing the *Daphnia* in a Petri dish or finger bowl of water that is the same temperature as the culture fluid. Refer to Figure 38G-1 to locate *Daphnia's* heart. Use a dissecting microscope to count the heartbeats for 10 seconds; multiply by 6 to obtain the heart rate per minute. Record the heart rate and temperature on a separate sheet of paper.

   7. Now place the tube into a Petri dish containing water at 0 to 5°C. Record the temperature and the heart rate after it has stabilized.

Figure 38G-1   Daphnia. *Note the position of the heart.*

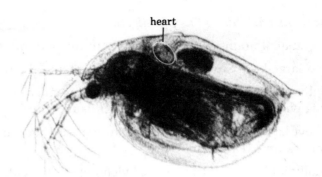

heart

8. Slowly add warm water to the dish and record the temperature and heart rate for every 5°C increase in temperature until you can no longer accurately record the beats.

9. Determine the Q_{10} for your *Daphnia*. Q_{10} is a strictly empirical temperature coefficient representing a ratio of the rate of a process (in this case heartbeat) at a given temperature to its rate at a temperature 10°C lower. Q_{10} may be calculated as

$$Q_{10} = \frac{R_1}{R_2} \times \frac{10}{T_1 - T_2}$$

where R_1 = rate (in this case, heart rate) at temperature T_1, and T_2 = rate at a lower temperature T_2. You can calculate a Q_{10} value for each 10°C change that you measured for your *Daphnia*.

   In most biological systems, Q_{10} decreases as the temperature increases. A value of 2 or 3 is normal at moderate temperatures.

a. *What is the* Q_{10} *for your* Daphnia *at low temperature?* _____ *At moderate temperature?*

   _____ *At high temperature?* _____

PART 2   The Effect of Chemicals

Acetylcholine and epinephrine are neurotransmitters that can affect heart rate. Acetylcholine is released by postganglionic fibers of the vagus nerve, a part of the parasympathetic nervous system. Epinephrine is released by postganglionic sympathetic fibers that originate in the thoraco-lumbar region of the spinal column.

A.   INFLUENCE OF ACETYLCHOLINE (A NEUROTRANSMITTER)

IIIII Procedure IIIIIIIIIIIIIIIIIIIIIIIIIIIIIIIIIIIIIIIIIII

Using your knowledge of the autonomic nervous system, form a hypothesis that predicts the effects of acetylcholine on heart rate.

   HYPOTHESIS:

   NULL HYPOTHESIS:

a. *What do you* **predict** *will happen to the heart rate under the influence of acetylcholine?*

_____

What is the **independent variable** *in this investigation?* _____

*What is the **dependent variable** in this investigation?* _____

1. Place a small amount of petroleum jelly in the bottom of the well of a depression slide.

2. Using a large-bore pipette, transfer a *Daphnia* into the well.

3. Carefully remove the water (use the dissecting microscope to watch your progress). The *Daphnia* will become attached to the petroleum jelly.

4. Place several drops of culture fluid into the well and determine the normal heart rate. Record this rate as time 0 in Table 38G-1.

5. Gradually add several drops of acetylcholine to the culture fluid.

6. Determine the heart rate each minute for 5 minutes and record these values in Table 38G-1.

Table 38G-1   Heart Rate

Time	Heart Rate
0 min	
1 min	
2 min	
3 min	
4 min	
5 min	

b. Do your results support your hypothesis? _____ *Your null hypothesis?* _____

c. How do you explain your results? _____

_____

d. What is the role of acetylcholine? _____

_____

B.   INFLUENCE OF EPINEPHRINE (A NEUROTRANSMITTER AND HORMONE)

ııııı **Procedure** ıı

Using your knowledge of the autonomic nervous system, form a hypothesis that predicts the effects of epinephrine on heart rate.

HYPOTHESIS:

NULL HYPOTHESIS:

*e. What do you **predict** will happen to the heart rate under the influence of epinephrine?*

_____

_____

*What is the **independent variable** in this investigation?* _____

*What is the **dependent variable** in this investigation?* _____

1. Immobilize a *Daphnia* as explained in steps 1 to 3 of Part A above. Add several drops of culture fluid.

2. Determine the normal heart rate and record it at time 0 in Table 38G-2.

3. Gradually add several drops of epinephrine to the culture fluid.

4. Determine the heart rate each minute for 5 minutes and record these values in Table 38G-2.

   f. Do your results support your hypothesis? _____ *Your null hypothesis?* _____

   g. How do you explain your results? _____
   _____

   h. Which of these substances, acetylcholine or epinephrine, would be produced in periods of stress?
   _____ *Why?* _____
   _____

Table 38G-2   Heart Rate

Time	Heart Rate
0 min	
1 min	
2 min	
3 min	
4 min	
5 min	

Laboratory Review Questions and Problems

1. Identify the type of circulatory system (open or closed) represented in each of the diagrams below. Label vein, artery, capillaries, heart, blood sinus, body cells as appropriate for each diagram.

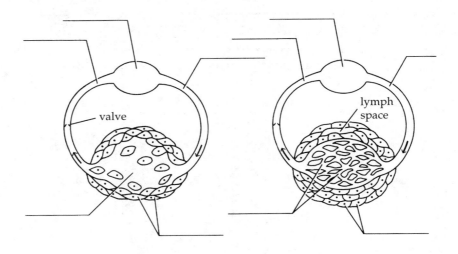

2. What major differences distinguish red blood cells from white blood cells?

3. Compare the structure of arteries and veins by completing the table below with descriptions of each tissue layer as thick, moderate, or thin.

	Connective Tissue	Elastic Fibers, Smooth Muscle	Endothelium
Artery			
Vein			

4. In the following graph of blood pressure changes over time, indicate systolic and diastolic pressure.

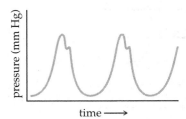

5. Using the following diagrams, explain why systolic pressure and diastolic pressure are represented as the "peaks" and "valleys" of the graph in question 4. (*Note:* the thickness of the arrows representing blood flow is significant.)

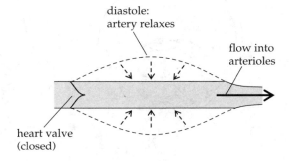

6. You determine that an organism has a Q_{10} of 2. What does this mean? Would you warm up or cool down the organism to lower its heart rate?

7. Compare the flow of blood in the fish, amphibian, and mammalian hearts by drawing arrows on the following diagrams using red and blue pencils to represent oxygenated and deoxygenated blood, respectively, in the atria and ventricles.

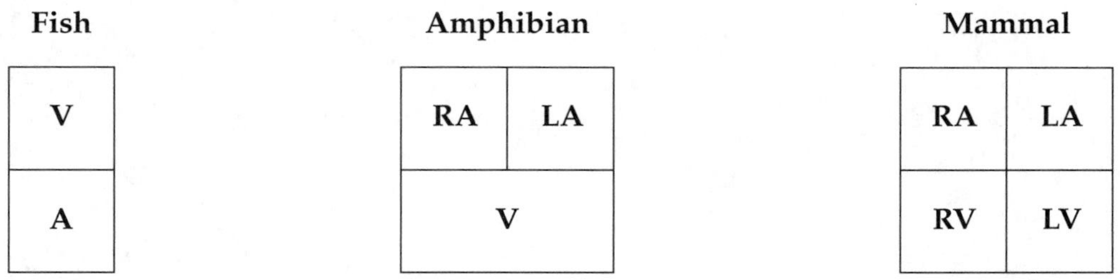

Fish

V
A

Amphibian

RA	LA
V	

Mammal

RA	LA
RV	LV

Alien

Alien is a computer simulation of cardiopulmonary (heart-lung) physiology that you can use as a data generator for simulated experiments.

- In one application, you assume the role of a physiologist hired by NASA to run experiments on the first extraterrestrials brought to Earth.

- To develop a basic understanding of physiological concepts, you first experiment on a human volunteer. You can investigate: the effects of exercise both at sea level and at high altitudes; the effects of increased blood levels of carbon dioxide as the carbon dioxide concentration of the atmosphere increases; the relationship between running speed, distance run, and exhaustion.

- You then have a choice of five aliens whose physiologies are well adapted to their exotic home planets. To make observations and collect data about an alien's heart rate, blood pressure, breathing, blood chemistry, and metabolic rate, you can perform experiments that involve increasing or decreasing levels of oxygen and carbon dioxide, using a treadmill, administering drugs, applying various stressors, and so on.

Alien allows you to make observations about both homeostasis and the breakdown in homeostasis when lethal limits are exceeded. Experimental subjects that are pushed beyond their physiological capacities may drop from fatigue, faint, have a heart attack, or die! (But you may be able to save them by CPR before brain damage becomes too severe.)

Gas Exchange and Respiratory Systems

OVERVIEW

In Laboratory 38, you explored the operations of the circulatory system—the heart, vessels, and blood. The circulatory system, in combination with the respiratory system, is also responsible for the transport and exchange of oxygen and carbon dioxide.

In vertebrates, the respiratory pigment, hemoglobin, carries O_2 from the gills or lungs to the tissues and organs and carries off excess CO_2 produced by metabolic activity of the body. Gases are exchanged across capillary walls by diffusion as a consequence of the partial pressures of oxygen (P_{O_2}) and carbon dioxide (P_{CO_2}) in the blood and tissue cells. Both the rate of breathing and ability to exchange O_2 and CO_2 can be affected by external factors such as exercise, smoking, or disease.

In this laboratory, you will study the functions of gills and lungs. You will examine how hemoglobin binds O_2 and CO_2 and learn how your breathing rate can be affected by exercise.

STUDENT PREPARATION

To prepare for this laboratory, read the text pages indicated by your instructor. Familiarizing yourself in advance with the information and procedures covered in this laboratory will give you a better understanding of the material and improve your efficiency.

EXERCISE A   The Vertebrate Respiratory System

The function of the respiratory system is to provide oxygen that can be transported to the tissues of the body and to remove carbon dioxide produced by the metabolic activities of the cells. This requires a large surface area that must be protected from damage, osmoregulatory stress, or drying. In vertebrates, there are two major respiratory surfaces: the gills and the lung/swim bladder. Accessory structures may include the skin and the cavities of the mouth and pharynx.

The first chordates were small and unprotected by scales, and gas exchange occurred through the body surface or skin. These organisms possessed slits in the pharynx, used in filter-feeding, that evolved into gills. In the earliest vertebrates, the jawless fishes, the pharynx tissue between the slits became vascularized, bringing more of the blood supply into closer contact with oxygenated water, thus increasing the amount of gas exchange; the **gills** assumed a major role in respiration. Gas exchange through gills increased in importance as fish developed thicker protective body coverings and increased their rates of activity in response to predation.

Among the earliest jawed fishes, a saclike outpocketing of the pharynx provided an accessory respiratory organ, the **lung** (Figure 39A-1a). This ancestral lung evolved into two different structural specializations among vertebrates. First, in primitive bony fishes (and their descendants that live in well-aerated waters), the gills function as the major respiratory organs, and gases in the lung are used mainly to reduce the density of the fish (like a ballast tank in a submarine). In these organisms, this structure functions as a **lung/swim bladder** and remains connected to the pharynx (Figure 39A-1b). In modern body fishes, this

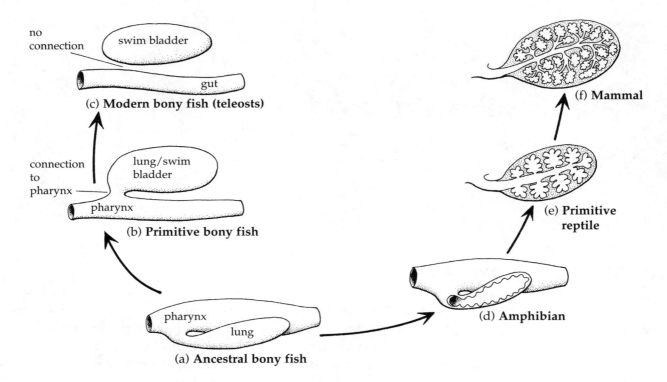

Figure 39A-1   *Evolution of swim bladders and lungs. (a) A bilobed outpocketing of the pharynx serves as an accessory respiratory organ. (b) The lung/swim bladder assumes a more dorsal position to help the fish float upright, and the lobes of the lung/swim bladder fuse. (c) The swim bladder has lost its connection to the pharynx in adult bony fishes and no longer serves as a lung, but only as a swim bladder or "ballast" organ.*
(d) The lung of an amphibian is a relatively simple vascular sac. (e) In primitive reptiles, the surface area is increased by folds of vascularized tissue. (f) Further partitioning in mammals leads to the formation of a copious number of microscopic sacs, the alveoli.

structure loses its connection to the pharynx during embryonic development and functions *only* as a swim bladder, not as a lung. Its contents are regulated by the circulatory system and a gland which "secretes" oxygen into the swim bladder (Figure 39A-1c).

In reptiles, there is an increasing division of the lining of the lung; in mammals, this culminates in extremely small alveoli where air comes in close contact with the vascular system (Figure 39A-1e, f).

a. *Examine Figure 39A-1d, e, f. What trend can be observed in the surface area of the lung, proceeding from amphibians to primitive reptiles to mammals?* _____

_____

b. *How would the process of gas exchange be enhanced by increasing the surface area of the lungs?*

_____

ⅢⅢ Objectives ⅢⅢⅢⅢⅢⅢⅢⅢⅢⅢⅢⅢⅢⅢⅢⅢⅢⅢⅢⅢⅢⅢⅢⅢⅢⅢⅢⅢ

☐ Trace the evolution of respiratory structures in vertebrates.

☐ Describe the circulatory pathway of blood and respiratory gases through the gills of a fish.

☐ Explain how oxygen is exchanged across the tissues of the circulatory system and lungs.

✔ **PART I**   **How a Fish Breathes—Gills as a Respiratory Surface**

In bony fishes, a series of gill arches support the gills. Attached to the arches are **gill filaments** (Figure 39A-2) with leaflike lobes (lamellae) that serve as the respiratory exchange surfaces. Blood vessels pass through the gill arches and send branches into the filaments and even smaller branches into the lamellae (Figure 39A-2c). Blood flows outward from the gill arch and through the lamella, moving across each lamella in one direction; oxygenated water flowing over the gills moves in the opposite direction (Figure 39A-2b). This **countercurrent flow** allows more oxygen to be picked up from the water by the capillaries in the lamellae (Figure 39A-3a). Exchange of oxygen from water to blood is dependent upon diffusion. The rate of diffusion increases as the difference in the concentration of the diffusing substance on the two sides of a membrane increases (see Laboratory 8, Exercises B and C). Countercurrent flow maximizes the difference between the oxygen content of the blood and that of the water at each point of contact, so the gills always pick up the maximum amount of oxygen from the water. The blood vessels going to the gills are called *afferent* arterioles; those that lead away from the gills are called *efferent* arterioles. From the gills, oxygenated blood is distributed to the body by the dorsal aorta and is returned to the heart by the cardinal and hepatic veins.

⁞⁞⁞⁞⁞ Procedure ⁞⁞⁞

1. Examine the mouth of a bony fish (perch). Feel the teeth on the upper jaw and on the roof and floor of the pharynx.

2. The pharynx wall is perforated by five pairs of gill slits. Four gill arches are present between the slits. The **operculum** (gill covering) extends over the entire grouping of arches and slits.

3. Place your finger into the mouth and extend it to emerge through one of the gill slits.

4. Cut through the opercular flap (Figure 39A-2a) and bend it backwards out of the way. Examine the gills of the fish. Each gill arch between the gill slits has two sets of filaments. Identify the two sets of gill filaments (Figure 39A-2b).

5. Observe a prepared slide of a gill filament if one is on demonstration. Otherwise, prepare a wet-mount slide of a gill filament and observe it under the microscope.

    c. *What is the advantage of the gill filaments having lamellae?* _____

    _____

6. Water passes into the mouth of the fish and then from the pharynx through the gill slits. It passes to the outside at the posterior edge of the opercular flap. As water passes over the gill filaments, oxygen is carried to the filament and diffuses into the blood that is flowing in the opposite direction within the capillaries.

    d. *Why does a fish continuously open and close its mouth?* _____

    _____

    e. *If water and blood flowed in the same direction, would the blood pick up as much as much oxygen as when blood and water flow in opposite directions?* _____

    f. *Why or why not? Examine Figure 39A-3a, b.* _____

    _____

    g. *Why are the blood vessels going both to and from the gills called arteries, or arterioles? What type of blood do they carry?* _____

    _____

At an early time in fetal development, humans have pharyngeal clefts (like gill slits but without four gills). We can still see evidence of these—the eustachian tube, middle ear, and external ear canal are derived from the first pharyngeal cleft.

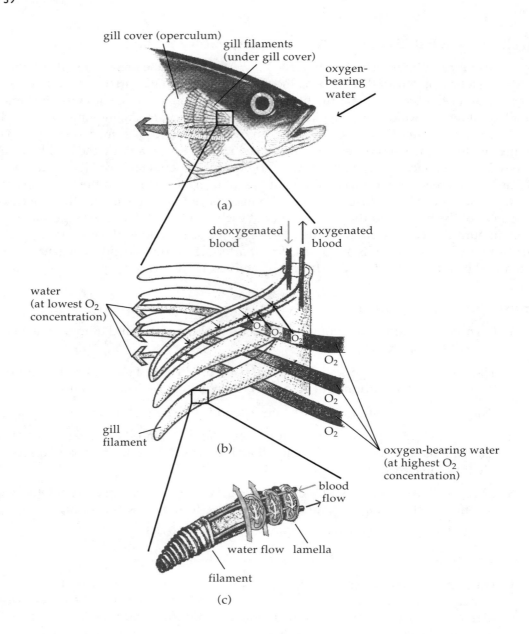

gill cover (operculum)
gill filaments
(under gill cover)
oxygen-
bearing
water

(a)

deoxygenated $\downarrow$ $\uparrow$ oxygenated
blood   blood

water
(at lowest O_2
concentration)

O_2 O_2 O_2

O_2

O_2

O_2

gill
filament
(b)

oxygen-bearing water
(at highest O_2
concentration)

blood
flow

water flow lamella

filament
(c)

Figure 39A-2 *(a) In fish, oxygen enters the blood by diffusion from water flowing through the gills. (b) The anatomical structure of the gills maximizes the rate of diffusion, which is proportional not only to the surface areas exposed but also to differences in concentration of the diffusing molecules. Water, carrying dissolved oxygen, flows between the gill filaments in one direction; blood flows through them in the opposite direction. Thus the blood carrying the most oxygen meets the water carrying the most oxygen, and the blood carrying the least oxygen meets the water carrying the least oxygen. The result is that the oxygen concentration of the blood in any region of the filament is less than the oxygen concentration of the water flowing over that region. In fact, the concentration gradient of oxygen between the blood in the gill filament and the water flowing over it is constant along the entire length of the filament. Thus, the transfer of oxygen to the blood by diffusion takes place across the entire surface of the filament. (c) Enlargement of a section of a gill filament showing the lamellae and the flow of blood within the capillaries.*

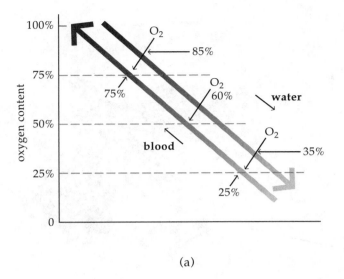

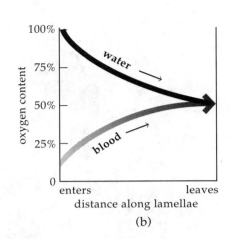

(a)

(b)

Figure 39A-3 *(a) When blood and water flow in opposite directions (countercurrent flow), blood becomes more highly saturated with O_2 than it would if blood and water flowed in the same direction (b). This is because in countercurrent flow, the difference in the O_2 content of the blood and the water is kept constant. If blood and water flow in the same direction, the difference in oxygen content disappears. Since it is the difference in the oxygen content of the two liquids that drives diffusion, without a difference no further change occurs.*

✓ **PART 2** ## How a Mammal Breathes—Lungs as a Respiratory Surface

In general, the lungs function as the center for gas exchange in mammals, but accessory structures such as the nasal cavity, surrounded by the sinuses, the pharynx, larynx, trachea, and the bronchus and its divisions, are also included when studying the respiratory system.

Air passing through the nasal cavity is warmed, moistened, and filtered by the mucosa of the nasal passages. Olfactory receptors in the nasal cavity provide us with a sense of smell, and the surrounding sinuses can act as resonance chambers for speech. Because the sheet of warm, moist mucosa that lines these areas is continuous, nasal infections often spread throughout these cavities and passages. The nasal passages are separated from the mouth (oral) cavity below by a partition, the palate (anteriorly, the hard palate, and posteriorly, the soft palate). In addition to entering the nasal passages, air can also enter the mouth.

Air moves from the nasal and oral cavities through the oral cavity into the **pharynx,** where it passes over the **larynx,** commonly referred to as the Adam's apple (Figure 39A-4a). The larynx consists of nine cartilages. One of these, the epiglottis, is located above (superior to) the larynx. The epiglottis forms a lid over the larynx when we swallow so that food is routed into the esophagus rather than into the trachea lying beyond the larynx. If anything other than air tries to pass, a cough reflex expels the material. The **trachea** is lined with a ciliated mucus-secreting epithelium. The cilia beat in unison to propel debris trapped in the mucus upwards to the oral cavity. The walls of the trachea are reinforced with C-shaped cartilage rings to provide support and flexibility.

After passing through the trachea, air moves into the right and left **bronchi** (singular, bronchus), which divide into smaller and smaller branches called **bronchioles** (Figure 39A-4b). These bronchioles are further subdivided to form alveolar ducts that end in alveolar sacs resembling clusters of grapes. The walls of the **alveoli** (terminal sacs) are composed of a single layer of epithelium that allows for gas exchange (by diffusion) with the weblike mass of capillaries in the lung tissue (Figure 39A-4b, c).

This gas exchange depends on differences in the concentrations of O_2 and CO_2 in the alveoli and in the lung capillaries (Figure 39A-5). Gas concentration is expressed in terms of partial pressure, which

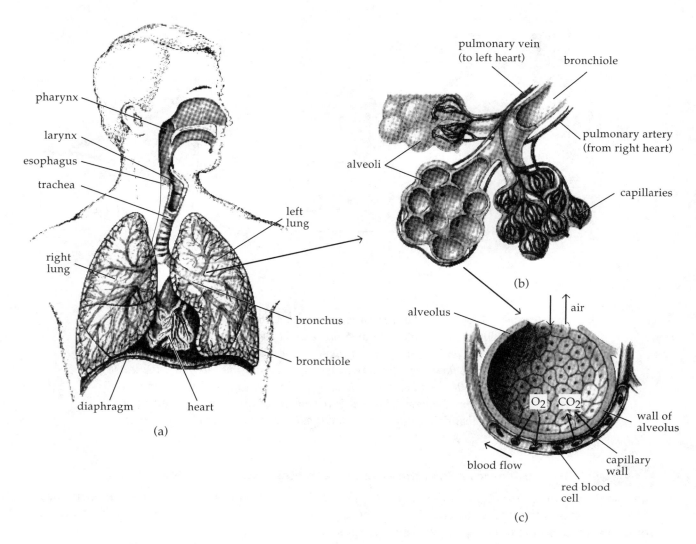

Figure 39A-4 *The human respiratory system. (a) Air enters through the nose or mouth and passes into the pharynx, past the larynx (voicebox), and down the trachea, bronchi, and bronchioles to the alveoli (b) in the lungs. The alveoli, of which there are some 300 million in a pair of lungs, are the sites of gas exchange. (c) Oxygen and carbon dioxide diffuse into and out of the bloodstream through the capillaries surrounding the walls of the alveoli.*

indicates the amount of atmospheric pressure exerted by one gas in the mixture of gases that make up the atmosphere.

The tissue of the lungs (other than respiratory passageways and capillaries) is mostly connective tissue. The lungs are covered by a thin membrane, a *pleural membrane* (visceral pleura). A second pleural membrane (parietal pleura) is attached to the walls of the thorax (chest cavity). The two pleural membranes lie close to one another, and each produces a lubricating fluid that fills the space between them. Pleurisy is a condition caused by inflammation of the pleural membranes and can be quite painful.

IIIII Objectives IIIIIIIIIIIIIIIIIIIIIIIIIIIIIIIIIIIIIIIIII

☐ Describe the anatomy of the mammalian respiratory system.

☐ Explain how the respiratory and circulatory systems are related anatomically and functionally.

Figure 39A-5 *Gases are exchanged by diffusion as a consequence of the different concentrations (partial pressures) of oxygen and carbon dioxide in the alveolus and the alveolar capillary. The numbers indicate partial pressures measured in millimeters of mercury.*

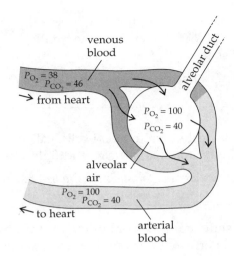

venous blood

$P_{O_2} = 38$ $P_{CO_2} = 46$

→ from heart

$P_{O_2} = 100$ $P_{CO_2} = 40$

alveolar duct

alveolar air

$P_{O_2} = 100$ $P_{CO_2} = 40$

← to heart

arterial blood

||||| **Procedure** ||||||||||||||||||||||||||||||||||||||||||

1. Examine the "sheep pluck" on demonstration. A sheep pluck includes the lungs and heart as well as the major blood vessels, the pulmonary artery and veins, that carry blood to and from the heart.

2. Identify the trachea and larynx. Try to feel the C-shaped cartilages in the trachea.

   a. *What happens to these cartilages as the trachea divides into bronchi and as bronchi divide into the smaller bronchioles?* _____

   _____

   b. *Why are these supporting elements necessary?* _____

3. Identify the lungs. c. *How many lobes are present in the right lung?* _____ *Left lung?* _____

4. Place a hose attached to an air compressor into the trachea and allow air to pass into each of the lungs. Notice how the lungs inflate.

   d. *Describe what you observe.* _____

   _____

5. Locate the pulmonary artery and veins. In the space below, diagram the passage of blood from the heart to the lungs and back to the heart. Indicate the names of the vessels and whether oxygenated or deoxygenated blood is found in each.

   e. *What kind of blood is carried by the pulmonary artery?* _____

   f. *What kind of blood is carried in other arteries?* _____

   g. *Why is the pulmonary called an artery?* _____

6. Obtain a cross section of cat lung tissue. Observe the specimen at 40×; identify the epithelium of the alveoli and red blood cells within the capillaries embedded in the lung tissue. The lung's connective tissue is filled with elastic fibers among the cells surrounding the alveoli.

   h. Suggest why there is an abundance of elastic fibers in this tissue. _____

   _____

   Look for the large nucleated white blood cells, called monocytes. These cells transform into macrophages that ingest foreign particles and fight infections.

   i. Why would it be advantageous for such cells to be present in the connective tissue of the lungs?

   _____

7. Obtain a prepared slide (cross section) of the cat trachea. Note the ciliated epithelium and hyaline cartilage support rings surrounding the trachea.

8. Use Figure 39A-5 to explain how oxygen and carbon dioxide are exchanged between lung and capillaries.

EXERCISE B Respiratory Pigments*

Whenever a circulatory system is responsible for the transport of oxygen, a respiratory pigment is usually present. All such pigments contain a metal ion capable of combining with oxygen; most contain iron, but some contain copper.

Hemoglobin, the respiratory pigment present in all vertebrates (and many invertebrates), consists of four subunits, each an iron porphyrin (*heme*) coupled with a *globin* protein molecule. In solution, hemoglobin appears pink. *Chlorocruorin,* present in polychaete worms, also contains iron and appears green in solution. *Hemerythrin,* a third iron-containing pigment, is violet in color and is found in polychaete worms and the brachiopod *Lingula.* In mollusks and some arthropods, the copper-containing respiratory pigment *hemocyanin* colors blood blue.

A molecule of hemoglobin is capable of carrying four molecules of oxygen. They are added one at a time—the binding of one increases hemoglobin's ability to bind the next. Whether oxygen combines with hemoglobin or is released from the oxygenated hemoglobin depends on the oxygen concentration (measured in terms of partial pressure of oxygen, P_{O_2}) in the surrounding blood plasma. When the concentration of oxygen is high, as in the capillaries of the lung, most oxygen is bound to hemoglobin. In the body tissues where the concentration of oxygen is low, oxygen is released from the hemoglobin molecule into the plasma and diffuses into the tissues (refer to Figure 39A-5). Carbon dioxide released from the tissues is picked up in the blood capillaries. Some CO_2 is dissolved in the plasma and some binds to the hemoglobin of the red blood cell, but most of the CO_2 picked up by red blood cells combines with water to form carbonic acid. When carbonic acid dissociates, it forms bicarbonate (HCO_3^-) and hydrogen ions (H^+).

a. How does an increase in the concentration of H^+ affect the pH of a solution?

_____

Most of the HCO_3^- diffuses out of the red blood cells into the plasma, and some of the H^+ ions, which could potentially reduce the pH of the blood, are bound to hemoglobin molecules within red blood cells. HCO_3^- can recombine with H^+ in the plasma or in red blood cells, helping to buffer the blood. (**Buffers** prevent large changes in the pH of the blood by accepting H^+ when its concentration rises above normal and giving up H^+ when its concentration falls below normal. See Laboratory 3, Exercise F.) Even though

*Professor James Colacino at Clemson University, Clemson, South Carolina, assisted in the preparation of this exercise.

blood is buffered, increasing the CO_2 concentration in plasma does decrease the pH of the blood. As acidity increases, hemoglobin's ability to bind oxygen decreases and the oxygenated hemoglobin gives up more oxygen to the tissues that need it.

IIIII Objectives IIIIIIIIIIIIIIIIIIIIIIIIIIIIIIIIIIIII

☐ Describe why respiratory pigments are an important component of the blood.

☐ Describe how differences between the oxygen concentrations (P_{O_2}) in the blood and in the body tissues cause oxygen to diffuse into the tissues.

IIIII Procedure IIIIIIIIIIIIIIIIIIIIIIIIIIIIIIIIIIIIIII

1. Obtain 1 ml of cow's blood in a conical centrifuge tube.

2. Add 5 ml of distilled water. Cover the tube with a piece of Parafilm and mix by inverting the tube. Remove the Parafilm.

   b. *What will happen to the red blood cells when they are suspended in distilled water? Where would you find their hemoglobin?* _____

   _____

3. Centrifuge the cell suspension for 1–2 minutes at low speed (use a setting of 7 on a clinical centrifuge).

4. Decant the supernatant and split it into two small test tubes.

5. Add enough activated yeast suspension to fill each of the tubes.

6. Fit a cork into the top of each tube such that oxygen is excluded from the tube. Mix by inverting the tubes.

7. Let the tubes sit for 5–10 minutes. Describe the color change that takes place.

   _____

8. After 10 minutes, pour a small amount of liquid out of one tube. Replace the cork and shake the tube vigorously. Describe the color change that takes place.

   _____

   Yeast respires, consuming oxygen and releasing carbon dioxide. After using up all of its own oxygen reserves, it begins to "steal" oxygen, as the cells of body tissues would do, from the hemoglobin molecules if no air is present.

   c. *How is this model similar to the dynamic relationship of blood flowing through capillaries in a body tissue?* _____

   _____

   d. *Where would you expect oxygen concentration to be higher—in the blood plasma or in the tissues of the body?* _____

   e. *In your experiment, as some carbon dioxide accumulates in the solution, what happens to the pH of the blood and yeast mixture?* _____

   f. *How does this change in pH favor the release of oxygen from the hemoglobin?*

   _____

   g. *Why do you see a color change in the pigment, hemoglobin?* _____

   _____

EXERCISE C | Lung Capacity

Normal breathing usually moves about 500 ml of air into and out of the lungs. This is called *tidal volume* (TV). After normal inspiration, you still have room for more; approximately 3,000 ml of air can be inspired forcibly. This forced inspiration is the *inspiratory reserve volume* (IRV). You can also forcibly expel air, about 1,100 ml. This is the *expiratory reserve volume* (ERV). **Vital capacity** of the lungs is calculated as follows:

Vital capacity = TV + IRV + ERV

Vital capacity falls within the range of about 3,000 to 5,500 ml. Variations in body size, age, and sex account for individual differences.

||||| Objectives |||||||||||||||||||||||||||||||||||||||||||

☐ Determine the vital capacity of your lungs.

☐ Understand the relationship between tidal and reserve respiratory volumes.

||||| Procedure |||||||||||||||||||||||||||||||||||||||||||

1. Stretch a balloon several times before blowing into it.

2. Take a normal breath and then exhale into the balloon, emptying your lungs normally.

3. Hold the end of the balloon tightly so that no gas can escape. Place the balloon next to a ruler and measure its diameter in centimeters. (Placing a piece of paper or cardboard on each side of the balloon may help you to read the diameter accurately.) Use the graph in Figure 39C-1 to convert the diameter of the balloon to liters of air. This value represents tidal volume (TV).

    TV = _____ liters

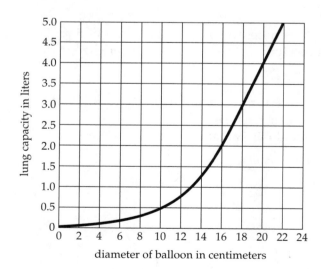

Figure 39C-1   *Measuring lung capacity (1,000 × capacity in liters = capacity in milliliters)*

4. Now, take a deep breath, forcing inspiration. Again, blow into the balloon, but do *not* forcibly exhale. Measure the balloon and convert this to volume of air. Subtract the value of TV from this measurement. The result will be the inspiratory reserve volume (IRV).

    IRV = _____ liters

5. Again, take a normal breath and exhale. Now place the balloon in your mouth and forcibly exhale any air that you can. Measure the balloon and convert this to volume of air. This represents the expiratory reserve volume (ERV).

ERV = _____ liters.

6. Calculate the vital capacity of your lungs (TV + IRV + ERV).

Vital capacity = _____ liters = _____ milliliters

a. *How does your result compare with those for 20-year-olds given in Table 39C-1?*

Table 39C-1   Normal Vital Capacity (in milliliters) for 20-Year-Olds*

Males		Females	
Height (in inches)	Normal Vital Capacity	Height (in inches)	Normal Vital Capacity
60	3,885	58	2,989
62	4,154	60	3,198
64	4,410	62	3,403
66	4,675	64	3,612
68	4,940	66	3,822
70	5,206	68	4,031
72	5,471	70	4,270
74	5,736	72	4,449

*Adapted from Donnesberger, Lesah, and Timmons, *A Manual of Anatomy and Physiology*, D. C. Heath, 1975.

EXERCISE D | How Does Smoking Affect Lung Capacity?

Hemoglobin binds carbon monoxide more quickly and strongly than it binds oxygen. This is why a person can die under conditions where oxygen is available but carbon monoxide concentrations are high: hemoglobin will preferentially bind carbon monoxide, which cannot be used in cellular respiration. By binding to many of the hemoglobin sites that would normally be occupied by oxygen, the carbon monoxide in cigarette smoke reduces the blood's ability to carry oxygen. In response, the body makes more red blood cells. As a result, the blood of smokers is usually thicker than that of nonsmokers, making the heart pump harder. Stress on the heart increases the incidence of heart attacks in smokers.

Smoking also affects the lungs by building up tars within the lung membranes. When enzymes in the cells begin to digest the tars, they may also digest the cell membranes themselves. This often results in the onset of emphysema, a disease characterized by deterioration of the walls of the alveoli and a loss of elasticity in the lung tissues.

What do you think smoking does to the vital capacity of the lungs? Is there a smoker in your class or among your friends? Formulate a hypothesis about how smoking might affect lung capacity.

HYPOTHESIS:

NULL HYPOTHESIS:

What do you **predict** will be the outcome of measuring lung capacity in smokers and nonsmokers?

What is the **independent variable**?

What is the **dependent variable**?

Follow the **procedure** outlined in Exercise C, and record your results below.

TV = _____

IRV = _____

ERV = _____

Vital capacity = _____

Do your results support your hypothesis? _____ *Your null hypothesis?* _____

*What do you **conclude** about the effects of smoking on respiration?* _____

Note: To obtain valid scientific data to support your conclusion, you would need to compare the lung capacities of large numbers of smokers and nonsmokers matched for many variables, including age, height, and weight.

EXERCISE E   The Effect of Exercise on Heart and Respiratory Rates

The effect of moderate and vigorous exercise on heart rate and respiratory rate may be observed in the following experiment.

Objectives

☐ Determine how exercise can affect respiratory rate.

☐ Explain how the elevation of pulse rate is related to increasing respiratory rate.

Procedure

1. Work in pairs. Determine your partner's sitting pulse rate.

2. Listen closely to your partner's respiration rate. In Table 39E-1, record the number of breaths per minute, and describe the depth of breathing in the "Before Exercise" column.

3. One partner should now exercise moderately by raising alternate knees to the chest for approximately 30 seconds. The other partner should exercise strenuously by running up several flights of stairs as quickly as possible.

4. In Table 39E-1, record the pulse rate, the rate and depth of respiration, the time required for the return of normal respiration, and the time required for the return of normal pulse rate for your partner in the appropriate column (moderate or heavy exercise).

5. Now repeat steps 3 and 4 with each partner performing the alternative type of exercise.

6. Record the data for your partner in Table 39E-1.

Table 39E-1

	Before Exercise	Moderate Exercise	Heavy Exercise
Pulse rate			
Respiration rate			
Depth of respiration			
Time required for the return of normal respiration	╳		
Time required for the return of normal pulse	╳		

The speed with which the heart returns to the sitting pulse rate after exercise serves as one index of circulatory efficiency. Here is an example of a normal response: before exercise, pulse 80; immediately after exercise, pulse 120; 2 minutes after exercise, pulse 84.

   a. *How is increased heart rate (measured as pulse rate) related to increased respiratory rate?*

   _____

   b. *As respiratory rates return to normal following exercise, what happens to the pulse rate?*

   _____

   c. *Can you explain why the observations you reported in* a *and* b *occur?*

   _____

Laboratory Review Questions and Problems

   1. The first type of lung to evolve in fishes was a simple vascularized sac. Describe the changes in the structure of the lung during the course of evolution.

   2. Why is it important that blood flows within gill capillaries in a direction opposite to that of water flowing over the gill filaments?

   3. How does the concentration of oxygen in the alveoli of the lungs compare with the concentration of oxygen in the pulmonary vein? The concentration of oxygen in the pulmonary artery?

4. Describe the structure of hemoglobin. How does it bind O_2? CO_2?

5. Iron-deficiency anemia can severely damage health. Why?

6. How is the pH of the blood affected by the presence of CO_2?

7. Describe the roles of H^+ and HCO_3^- in maintaining the pH of blood. How is hemoglobin involved in this pH maintenance?

8. How might the oxygen concentration of the blood passing from the lungs to the heart in the pulmonary vein be affected if a person just inhaled smoke from a cigarette?

9. How are pulse rate and respiratory rate affected during exercise?

10. Human respiration is often described as "negative pressure breathing." What does this mean? How is it accomplished?

11. What is meant by the "vital capacity" of the lungs? How is it determined?

Cycle

Cycle is a simulation of the menstrual cycle and human fertility. **Cycle** is a data generator that can be used in either a game or an experimental mode.

- In the **Cycle** game, you use data on blood hormone levels, follicle size, basal body temperature, and the thickness of the uterine lining to advise a woman about whether intercourse can result in pregnancy. Whenever you predict that pregnancy may occur, the couple refrains from intercourse (and you get no points). You receive points for correctly predicting when it is "safe" to have intercourse. However, if you predict safety at a time that is too close to ovulation, a pregnancy will occur and you will suffer a severe point penalty. The better your understanding of the menstrual cycle and fertility, the more accurate your predictions and the higher your score.

- The **Cycle** game offers four levels of difficulty. At the lowest level of difficulty, prediction of ovulation is easy—all cycles are the same length and all data on hormone levels and other factors are given. At the most challenging level, cycle lengths vary widely and only data about menstrual flow are available.

- In the experimental mode, you can use **Cycle** to change the secretion rates of estrogen, progesterone, LH, and FSH, and observe the effects on the menstrual cycle and pregnancy. You can also simulate the effects of birth control pills and RU-486 and conduct basic "research," for example, determining the combinations of hormones that induce ovulation most quickly.

The Digestive, Excretory, and Reproductive Systems

OVERVIEW

Food would not be of much use without the cooperative actions of the digestive, excretory, and circulatory systems. After food—carbohydrates, lipids, and nitrogen-containing molecules (proteins and nucleic acids)—is digested, blood delivers the products to the body's tissues and organs.

Once food has been metabolized, wastes and excess water must be excreted. A delicate balance between reabsorbing usable materials and excreting excess or unwanted breakdown products, as well as maintaining normal pH, is the job of the excretory system.

Closely associated with the excretory system is the reproductive system; based upon developmental events, excretory products and gametes may travel in the same ducts or may be emptied into the same regions of the body. Because of their relationship, the excretory and reproductive systems are often referred to collectively as the urogenital system.

STUDENT PREPARATION

Prepare for this laboratory by reading the text pages indicated by your instructor. Familiarizing yourself in advance with the information and procedures covered in this laboratory will give you a better understanding of the material and improve your efficiency.

EXERCISE A   Examining the Digestive System

In primitive vertebrates, the digestive system consisted of little more than an anterior opening, the **mouth;** the **pharynx;** a **foregut** and a **hindgut** separated by a constriction (the **pylorus**); and a posterior opening, the **anus** (Figure 40A-1). The first vertebrates were filter-feeders, feeding continuously on small particles of food suspended in the water. With the development of jaws, vertebrates could take in larger food items at intervals ("meals"), and a temporary storage area, the **stomach,** developed in the anterior part of the system.

Jaws enclosed a space in front of the pharynx, the **oral cavity.** This cavity evolved into the most complicated part of the digestive system, adapting to a wide range of diets. Lips, teeth, a muscular tongue, and internal nares (connecting the oral cavity to the outside through nasal passages) developed in association with the oral cavity. In terrestrial vertebrates, salivary glands evolved to add moisture to dry food.

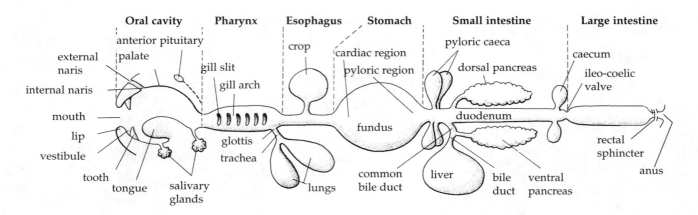

Figure 40A-1 *A generalized schematic diagram of structures and organs associated with the digestive tract in various vertebrates. Note that no single vertebrate possesses all of these structures.*

In aquatic vertebrates, gills (see Laboratory 39) and lungs became adapted for respiration, and the pharynx began to serve only as a passageway for food. Behind the pharynx, the foregut (extending from the pharynx to a point separating the stomach from the hindgut) served as a passageway for food, and the stomach, a major food storage organ, developed. In advanced vertebrates, enzymes and an acid environment combine to initiate the chemical breakdown of proteins in the stomach, but few substances are assimilated (absorbed) in this region.

The hindgut is responsible for most of the chemical digestion of food and uptake of products formed by this process. The hindgut shows few evolutionary changes except for an increase in the area of the absorptive surface. The length of the most anterior part of the hindgut, the **small intestine,** is generally correlated with diet—it is shortest in those forms that feed on microscopic food particles and easily digested foods; it is longer in animals with a high-protein diet; and it is longest in herbivores that feed on masses of grass or foliage.

The first section of the small intestine, the **duodenum,** receives ducts from the **liver** and **pancreas,** large exocrine (ducted) glands in the abdominal cavity. The liver produces **bile,** a fluid containing bile pigments (breakdown products of blood pigments, which color the bile brown to green) and bile salts, which break up fats into small droplets that can be digested or absorbed directly. Enzymes secreted by the pancreas and the wall of the duodenum break down food materials into simple chemical compounds that can be assimilated as food passes through the remainder of the small intestine.

The posterior part of the intestine is called the **large intestine** or colon. In more advanced vertebrates, one or two sacs (*caeca*) mark the junction of the small and large intestines. In humans, part of the single caecum forms an appendix. The large intestine is primarily a site where water is removed from the intestinal contents to make the semisolid feces. The **anus** is the terminal opening of the digestive system. In all vertebrates except bony fishes and mammals, the large intestine opens into the **cloaca,** a common chamber for receiving feces, urine, and gametes.

▐▐▐▐▐ Objectives ▐▐▐▐▐▐▐▐▐▐▐▐▐▐▐▐▐▐▐▐▐▐▐▐▐▐▐▐▐▐▐▐▐▐

☐ Trace the pathway of food from the mouth to the anus.

☐ Describe how food is moved through the digestive tract.

☐ Explain what types of special digestive structures occur in each part of the digestive tract.

👁 **PART I**   **Microscopic Anatomy of the Digestive System**

ııııı **Procedure** ıı

Obtain microscope slides of the following: salivary glands (sublingual), pancreas, liver, and small intestine (cross section).

1. Examine the tissue of the salivary glands using 10× power and then 40× power (Figure 40A-2a). Locate the ducts of these glands, which open into the oral cavity. The cells surrounding the ducts are triangular in shape. Do you see any difference in the appearance of the cells of the glandular tissue? _____ The cells filled with granules produce salivary **amylase,** an enzyme that digests starch. Mucus-secreting cells appear hollow or clear. On a separate piece of paper, draw a small portion of the salivary gland tissue and label its parts.

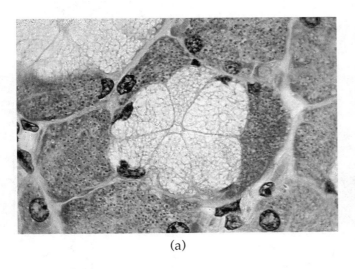
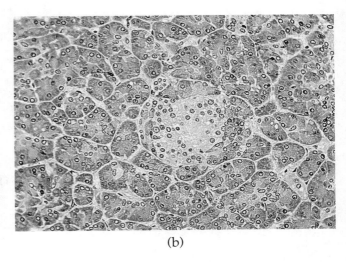

(a)   (b)

Figure 40A-2 *Photomicrographs of (a) salivary gland and (b) pancreatic tissue, with islet of Langerhans at center.*

   a. *Which of the foods that you eat might be partly digested by amylase within the oral cavity?*
   _____

2. Observe the pancreatic tissue using 10× and then 40× power (Figure 40A-2b). The clusters of cells that are lighter in color are the **islets of Langerhans,** endocrine cells that secrete the hormones glucagon and insulin directly into the bloodstream. These hormones are responsible for regulating blood sugar levels. The darker-staining cells are **acinar** cells, exocrine cells that produce digestive enzymes which travel to the small intestine (their products do not enter the bloodstream *directly,* as endocrine secretions do). Can you find any ducts in your preparation? _____ On a separate piece of paper, draw a portion of the pancreatic tissue and label its structures.

   b. *How do glucagon and insulin affect blood sugar levels?* _____
   _____

3. Observe a slide of liver tissue at 10× and 40× power. The liver produces bile, containing bile pigments (waste products) and bile salts, which emulsify fats in the small intestine. The liver is the site of many metabolic activities within the body. The small lobes (lobules) of liver cells

Figure 40A-3 *(a) Liver lobules (about 1 mm in diameter). (b) Bile duct, with arrows indicating the direction of blood flow.*

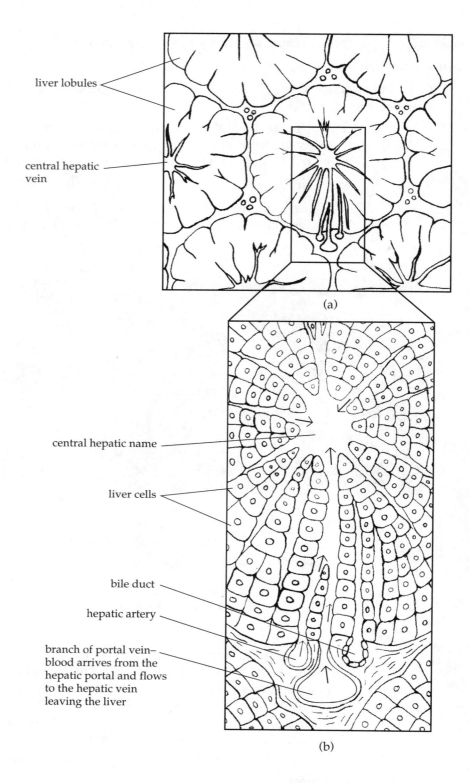

liver lobules

central hepatic vein

(a)

central hepatic name

liver cells

bile duct

hepatic artery

branch of portal vein– blood arrives from the hepatic portal and flows to the hepatic vein leaving the liver

(b)

are surrounded by branches of the hepatic artery, carrying arterial blood, and of the hepatic portal vein, carrying blood laden with digestive products from the intestine to the liver after a meal (Figure 40A-3). At the center of each lobule is a branch of the hepatic vein, which returns blood to the inferior vena cava and heart. Each lobule is also associated with small, thinwalled bile ducts, in the tissue between the lobules.

Using low power, can you identify individual liver lobules?_____ Can you see any of the circulatory vessels or branches of the bile duct? _____

Note that the liver cells are loosely packed, with blood sinuses between them. Blood passes through the liver tissue, depositing glucose to be made into glycogen and many other substances to be stored, metabolized, detoxified, etc. On a separate piece of paper, draw and label a small portion of liver tissue.

c. *Impairment of liver function caused by physical damage or disease can be extremely dangerous to health. Why?* _____

_____

4. At low power, observe a cross section of tissue from the small intestine. The intestinal wall is composed of four layers; from inside to outside, these include the mucosa, submucosa, muscle layer, and serosa (Figure 40A-4). Switch to 40× and focus on the fingerlike **villi**. These are extensions of the inner surface of the mucosa that increase the intestinal surface area for absorption. The epithelial cells that cover the surface of the villi are columnar. *d. What shape*

do they appear to have? _____

Figure 40A-4 *The layers of the digestive tract include (1) the mucosa; (2) the submucosa, which contains nerves and blood and lymph vessels; (3) the muscle layer; and (4) the serosa, also known as the visceral peritoneum, which covers the outer surfaces of all organs in the abdominal cavity. Mesenteries are folds of the peritoneum that suspend the intestines from the posterior abdominal wall. Glands outside the digestive tract, principally the pancreas and liver, discharge digestive enzymes and bile into the tract through various ducts.*

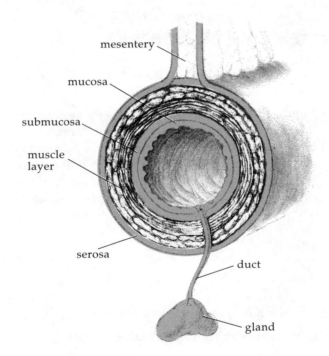

Locate an area where the epithelium extends into the intestinal wall and forms a pocket. These are glands called the crypts of Lieberkühn (Figure 40A-5); they secrete digestive enzymes that complete the breakdown of food materials into simple molecules that can be assimilated by the digestive system. Note the presence of many smaller mucous glands in the submucosa (Brunner's glands).

e. *Why would mucus secretion be of importance in the small intestine?* _____

_____

The muscle layer is composed of both circular and longitudinal muscles. Can you

identify these? _____ Locate the serosa—a layer of connective tissue covering the

outer surface of the intestine. On a separate sheet of paper, draw a section of small intestine showing the four layers. Label these.

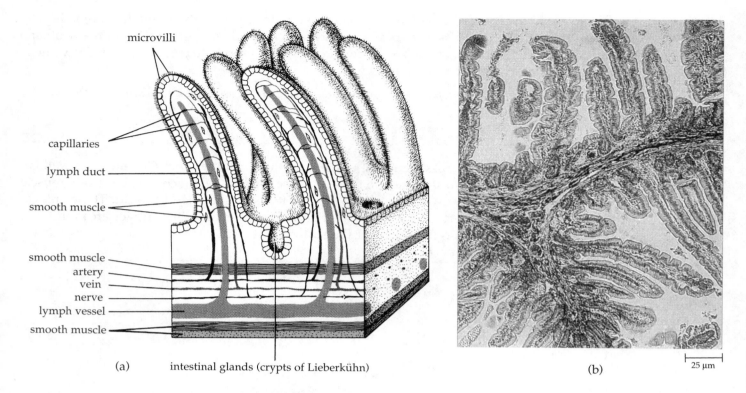

Figure 40A-5   *(a) Diagram and (b) photomicrograph of villi of the small intestine, as seen in longitudinal section. Nutrient molecules are absorbed through the walls of the villi and, with the exception of fat molecules, enter the bloodstream by means of the capillaries. Fats—hydrolyzed to fatty acids and glycerol, resynthesized into new fats, and packaged into particles known as chylomicrons—are taken up by the lymphatic system. The villi can move independently of one another; their motion increases after a meal.*

f. Why do you think organisms with a high-protein diet or a high-fiber diet would need a longer intestine than those feeding on simpler foods? _____

_____

g. Besides lengthening the intestine, how could its absorptive area be increased? _____

h. Can you give an example of an organism that uses this strategy? _____

✔ **PART 2   The Role of Peristalsis**

When food or fluid enters the mouth, swallowing occurs first by voluntary movements of the mouth and tongue. The food or fluid passes through the pharynx, then, on reaching the esophagus, it is moved by a series of involuntary contractions called **peristalsis.** Once food is partially digested in the stomach, peristalsis continues to move material through the small and large intestines.

Longitudinal and circular smooth muscles of the intestinal tract are responsible for peristalsis. Peristaltic movements occur in waves of contraction followed by relaxation. These pressure changes, caused by the muscle contractions, are controlled by the autonomic nervous system in response to the presence of food and fluid in the intestine.

||||| **Procedure** |||||||||||||||||||||||||||||||||||||||

1. Work in pairs. Obtain a large glass of water and a stethoscope.

2. Swallow a large gulp of water. Note the movement of your tongue.

    a. *Describe this movement.* _____

    b. *Would this movement be the same if solid food were swallowed?* _____

3. Watch your partner swallow and notice the movement of the larynx (Adam's apple).

    c. *Describe what you observe.*

    _____

    d. *Why does the larynx move in this way?* _____

    e. *What could happen if these movements did not occur?* _____

4. Place a stethoscope on your partner's upper abdomen, directly below the rib cage, a little to the left of center. Listen to the sounds made as your partner swallows a large gulp of water. As the water falls against the junction of the esophagus and stomach, you will hear a splashing sound. As water enters the stomach, you will hear a gurgling sound. Have your partner swallow several times until you are certain that you can distinguish the two sounds.

5. How much time passes between the two sounds? _____ This is a fairly good estimation of how long it takes a single wave of peristalsis to move along the esophagus.

6. Assuming the esophagus is 11 inches long, at what rate does peristalsis occur? Use the following equation:

    Rate = length/time = _____ inches per second

    f. *What factors or conditions do you think could affect the rate of peristalsis?*

    _____

👁 **EXERCISE B**   **The Chemistry of Digestion**

The digestion of carbohydrates, proteins, and fats takes place at various locations along the intestinal tract (Figure 40B-1). Polysaccharides are broken down into simple sugars such as glucose, galactose, and fructose (Figure 40B-1a). Proteins are catabolized into amino acids (Figure 40B-1b). Fats are emulsified by bile salts and then split into fatty acids and glycerol (or monoglycerides) (Figure 40B-1c).

Absorption of simple sugars, amino acids, glycerol, and smaller fatty acids occurs primarily in the small intestine. These are absorbed by the epithelial cells and transported by the blood. Within the epithelial cells, glycerol and fatty acids may also recombine to form fat. Fat is complexed with proteins to form small globules (chylomicrons) that get picked up by the lymphatic system rather than the blood capillaries. This system connects with the circulatory system near the heart. After a large fat-containing meal, your lymph turns cloudy with fat globules, but within a few hours these have been processed in the liver and your lymph is again clear.

Larger fats complexed with proteins (lipoproteins) are carried by the bloodstream. Lipoproteins carrying cholesterol occur in low-density (LDL) and high-density (HDL) forms. In general, elevated levels of LDL are associated with health problems, including arteriosclerosis and heart attack. Elevated HDL levels are usually associated with reducing the deposition of cholesterol in the arteries. Exercise tends to increase HDL levels, while smoking and using alcohol increase LDL levels.

||||| **Objectives** |||||||||||||||||||||||||||||||||||

☐ Describe how carbohydrates, proteins, and fats are broken down in the digestive tract.

☐ Explain how nutrients from digested foods can be absorbed.

☐ Trace the pathway of fat breakdown and absorption from the intestine to the liver and other body tissues.

(a) Carbohydrates

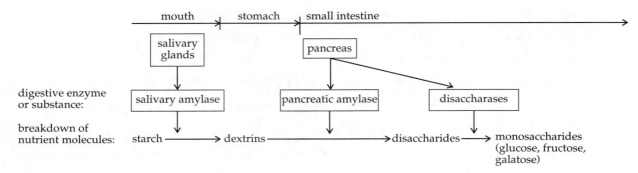

(b) Proteins

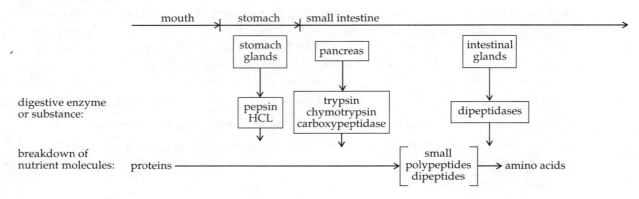

(c) Fats

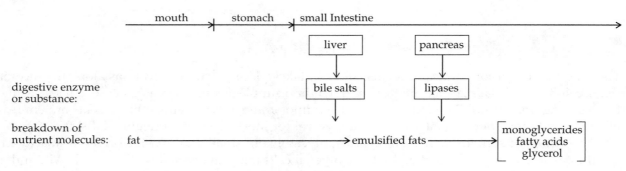

Figure 40B-I   *The breakdown of (a) carbohydrates, (b) proteins, and (c) fats in the digestive tract.*

👁 PART I   Carbohydrates

IIIII Procedure IIIIIIIIIIIIIIIIIIIIIIIIIIIIIIIIIIIIIIIIIIII

1. Obtain two test tubes and label them A_1 and A_2.

2. To both tubes, add 5 ml of starch. Add 5 ml of amylase to tube A_1, and 5 ml of water to tube A_2. Mix by covering the tubes with Parafilm and inverting them.

3. Using a Pasteur pipette, remove a small amount of solution from A_1, and dispense several drops into a porcelain spotting plate. Add a drop of I_2KI.

  a. What color does the solution turn? _____ Why? _____

  Record the color in the "Start" column in Table 40B-1.

4. Repeat the procedure in step 3 for tube A_2.

  b. What color does the solution turn? _____ Why? _____

  Record the color in the "Start" column of Table 40B-1.

5. Incubate the tubes for 10 minute at 37°C, then repeat steps 3 and 4 and record your results in the "Finish" column of Table 40B-1. As starch is broken down into dextrins, the color produced by I_2KI reacting with maltose (two glucose units bonded together as a disaccharide) changes to a red-blue or red-brown.

  c. What color did you observe?? _____ What does this indicate? _____

  _____

6. Remove 5 ml of solution from tube A_1 and place it into a tube marked B_1. At the same time, remove 5 ml of solution from tube A_2 and place it into a tube marked B_2. Save the remainder of the material in tubes A_1 and A_2.

7. To tube B_1, add 5 ml of the maltase solution, and to tube B_2, add 5 ml of water. Repeat step 3 (testing with I_2KI) for tubes B_1 and B_2. Record the color in the "Start" column of Table 40B-1.

8. Incubate tubes B_1 and B_2 at 37°C for 10 minutes. After incubation, repeat step 3 (testing with I_2KI). Record the color in the "Finish" column of Table 40B-1.

Table 40B-1

Tube	Enzyme	Substrate	Color with I_2KI	
			Start	Finish
A_1	5 ml amylase	5 ml starch		
A_2	None (5 ml H_2O)	5 ml starch		
B_1	5 ml maltase	5 ml A_1		
B_2	None (5 ml H_2O)	5 ml A_2		

9. Test for the presence of glucose in tubes A_1 and A_2 and B_1 and B_2 using Testape (as used by diabetics to test for the presence of sugar in urine). Compare the color change in the test tube with the scale on the Testape container.

  d. What do you observe? _____

  e. How do you explain your results? _____

  f. What do you conclude about the steps in digestion of carbohydrates? _____

  _____

  g. Why do you think carbohydrates do not undergo digestion in the stomach?

  _____

EXTENDING YOUR INVESTIGATION: WHERE IS IT DIGESTED? ENZYMES AND pH

When food enters the stomach, glands responsive to pressure secrete hydrochloric acid, thus lowering the pH of the digestive environment. Why do you think some enzymes are capable of digesting food in the stomach while others are not? Why would amylase and maltase not catalyze carbohydrate digestion in the stomach? Formulate a hypothesis to test why amylase and maltase do not have any digestive activity in the stomach.

HYPOTHESIS:

NULL HYPOTHESIS:

What do you **predict** will happen when you test your hypothesis?

What is the **independent variable** for this investigation?

What is the **dependent variable** for this investigation?

Design an experimental procedure to test your hypothesis. Base your procedures on the text in Exercise B, Part 1.

PROCEDURE

RESULTS

Do your observations support your hypothesis?

Your null hypothesis?

What do you **conclude?**

PART 2   Proteins

Procedure

1. Obtain four test tubes and label them P_1, P_2, P_3, and P_4.

2. Add the contents listed in Table 40B-2 to each tube.

3. Obtain a piece of exposed and developed (blackened) photographic film. Tape this to the bottom of a Petri dish. Use a grease pencil to mark "L" for left and "R" for right on the underside of the dish. Use a Pasteur pipette to place several drops of solution from each tube onto separate areas of the surface of the film. Cover the dish.

   In the space below, make a diagram to record the placement of the different solutions so that you will be able to identify their effects.

Table 40B-2   Action of Pepsin on Protein

Tube	Contents			Results		
	Enzyme	Substrate	Additive	Film	Biuret	Ninhydrin
P_1	5 ml pepsin	5 ml albumin solution	2 drops 2N HCl			
P_2	5 ml pepsin	5 ml albumin solution				
P_3	None (5 ml H_2O)	5 ml albumin solution				
P_4	None (5 ml H_2O)	5 ml albumin solution	2 drops 2N HCl			

4. Allow the film to set overnight. The next day, you (or your instructor) should wash the film, gently rubbing it with your thumb. Film is coated by gelatin (a protein). If the gelatin is removed (digested), the blackened silver particles will be removed and a clear spot will appear. Record your results in Table 40B-2.

   a. Was the protein, gelatin, digested by the pepsin in solution P_1? _____ P_2? _____

   b. Why did two of the mixtures (in tubes P_1 and P_4) use a lower pH? _____

5. If directed by your instructor, perform a biuret test (see Laboratory 5, Exercise C, Part 1) and a ninhydrin test (see Laboratory 5, Exercise C, Part 2) on a sample from each of your four tubes. Record the results in Table 40B-2. These tests allow you to detect undigested albumin protein (biuret) or amino acid breakdown products (ninhydrin).

   c. Did you detect the presence of albumin protein in any of the tubes? _____ Why or why not?

   _____

   d. Did you detect the presence of amino acids in any of the tubes? _____ Why or why not?

   _____

   e. What do you conclude about the digestion of proteins in the stomach? _____

   _____

PART 3   Fats

Procedure

1. Obtain two oil-emulsion agar Petri dishes. These were prepared by emulsifying a neutral fat (olive oil) with bile salts, then adding it to the agar. Use a grease pencil to mark a dividing line on the undersurface of the dish. Mark one side "L" and the other "W."

2. Using a Pasteur pipette, place three drops of lipase onto the surface of the agar on the side marked "L." Place three drops of water onto the surface of the agar on the side marked "W."

   a. What is the purpose of the water drops? _____

3. Incubate for 1 to 2 hours at 30°C in the incubator.

4. Pour a saturated $CuSO_4$ (copper sulfate) solution over the surface; let stand for 10 minutes, then rinse *gently* with water.

5. Lipase activity is indicated by the appearance of Cu^{2+} soaps that are blue-green in color. The fats have been saponified into glycerol and the fatty acids have formed Cu^{2+} salts, or soaps.

6. Record your results in Figure 40B-2.

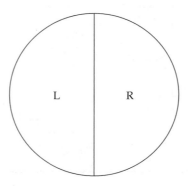

Figure 40B-2

b. In the process of digestion, what substance emulsifies fats? _____

c. How are glycerol and fatty acids, the products of fat digestion, carried to body tissues?

_____

d. How is the lymphatic system involved? _____

_____

7. Summarize your results from Exercise B in Table 40B-3.

Table 40B-3

Substance	Enzyme/Breakdown Products/Location of Digestion or Absorption
Carbohydrates	
Polysaccharides	
Proteins	
Dipeptides	
Fats	

✔ **EXERCISE C   The Mammalian Kidney**

Vertebrates take up a tremendous quantity of matter, both solid and liquid, from the environment. Some is used to provide energy and building blocks for new cellular materials, and the excess is excreted. An average adult human takes in approximately 5.05 pints of liquid each day and eliminates 5.0 pints of this in the form of urine, sweat, water vapor from the lungs, and water mixed with solid waste (feces). Of the approximately 4.25 lb of food taken in daily, about 0.33 lb is eliminated as feces. We should not forget that oxygen taken in is eliminated as CO_2.

In this exercise, you will concentrate on the work of one excretory organ, the mammalian kidney. Blood is cycled through the kidneys, which remove nitrogenous wastes and function in both osmoregulation and maintenance of body pH. Nitrogenous wastes are excreted as **urea** in a solution called **urine.** Nitrogenous waste products in other groups of organisms include **ammonia** and **uric acid.** Ammonia is the major excretory product of many aquatic vertebrates. It is highly soluble (and toxic), so it is excreted with copious quantities of water. Uric acid, excreted by reptiles and birds, is highly insoluble and relatively nontoxic, so it allows terrestrial vertebrates to conserve water. (The white material in bird droppings is composed of uric acid.) Urea, the major waste product in mammals and several other vertebrate groups, is intermediate in solubility and toxicity. In mammals, unique functional elements of the kidney form a highly concentrated urine and preserve body water. In fact, some desert mammals can live solely on water produced by the metabolism of food materials, without drinking free water.

a. *Where does the water produced by metabolism come from? (Hint: Write the summary equation for the oxidation of glucose.)* _____

_____

||||| **Objectives** |||||||||||||||||||||||||||||||||||||||||||

☐ Describe the structure of the mammalian kidney.

☐ Assign functions to the parts of the mammalian kidney.

☐ Understand how urine is formed.

✔ **PART I**   **Anatomy of the Mammalian Kidney**

||||| **Procedure** ||||||||||||||||||||||||||||||||||||||||||||

1. Obtain a sheep or beef kidney. Remove the covering of fatty material. This adipose tissue is structural fat used for cushioning the kidney against mechanical stresses. Be careful not to remove the renal artery or vein attached to the kidney at the **hilum,** an indentation on the concave side of the kidney. Note the overall shape of the organ.

2. Note the thick fibrous protective covering over the surface of the kidney. This is the **renal capsule.**

   b. *Why do you think protective layers such as the adipose tissue and fibrous connective tissue that cover the kidney are necessary?* _____

3. Carefully cut the kidney longitudinally, starting at the outer convex side and moving toward the hilum. Open the kidney and observe the structures listed below. Label these structures in Figure 40C-1 by matching the lettered descriptions with the diagram.

   a Ureter Carries urine from the kidney.

   b Renal pelvis The ureter expands within the kidney to form the renal pelvis. The kidney tissue converges into the renal pelvis, forming "projections" called **c renal papillae.**

   d Nephrons The remainder of the kidney tissue is divided into two regions, medulla and cortex, that differ in the parts of the functional units (nephrons) they contain. Each nephron consists of a **glomerulus** and **Bowman's capsule, proximal** and **distal tubules,** the **loop of Henle,** and **collecting ducts** (Figure 40C-2). The loop of Henle and the lower portion of the collecting ducts are in the **medulla e;** the other structures are in the **cortex f.**

   g The **renal artery** branches as it enters the kidney to supply arterial blood to the **glomerulus** within the renal corpuscle, where materials are filtered into the kidney tubule to form primary urine. Extensions of these arterial vessels surround the loops of Henle, and capillaries bathe the convoluted tubules where they reabsorb needed materials. Materials that are not reabsorbed are moved to the collecting duct and are eventually excreted. Venous blood leaves the kidney by the **h renal vein** (Figure 40C-2).

   c. *Does all uptake of materials from the kidney tubules occur by passive transport?* _____

   d. *If not, what other type of transport is used?* _____

Figure 40C-1 *In longitudinal section, the human kidney is seen to be made up of two regions. The outer region, the cortex, contains the fluid-filtering mechanisms. The inner region, the medulla, is traversed by long loops of renal tubules (loops of Henle) and by the collecting ducts carrying the urine. These ducts merge and empty into the funnel-shaped renal pelvis, which, in turn, empties into the ureter.*

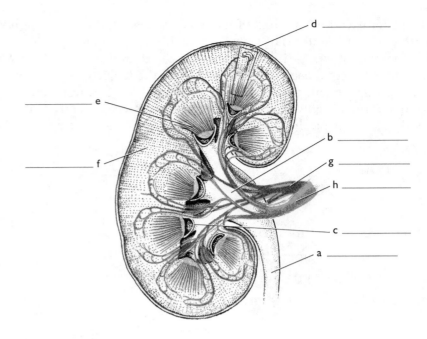

Figure 40C-2 *The nephron is the functional unit of the kidney. Blood enters the nephron through the afferent arteriole leading into the glomerulus. Fluid is forced out by the pressure of the blood through the thin capillary walls of the glomerulus into Bowman's capsule. The capsule connects with the long renal tubule, which has three regions: the proximal convoluted tubule; the loop of Henle, which extends into the medulla; and the distal convoluted tubule. As the fluid travels through the tubule, almost all the water, ions, and other useful substances are reabsorbed into the bloodstream through the peritubular capillaries. Other substances are secreted from the capillaries into the tubule. Waste materials and some water pass along the entire length of the tubule into the collecting duct and are excreted from the body as urine.*

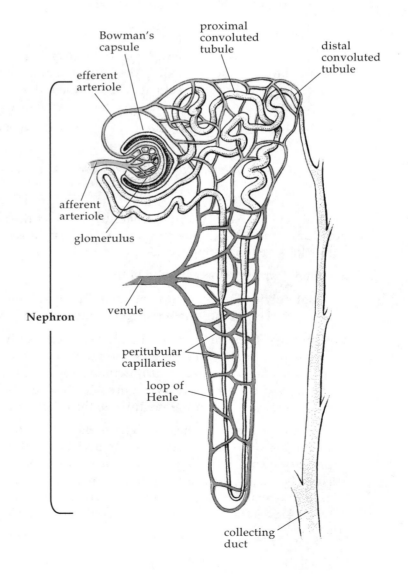

Countercurrent flow of water and blood in the fish gill (Laboratory 35) increases the efficiency of oxygen uptake. This process is a passive one. In the vertebrate kidney, active processes add a countercurrent **multiplier** to the exchange. The descending loop of Henle parallels the ascending segment of the loop (Figure 40C-2). Sodium is actively pumped into the surrounding tissues from the urine in the ascending loop and diffuses back into the urine in the descending loop. This results in a steep osmotic gradient—a large increase in the concentration of solutes from the peripheral portions of the kidney to the interior tissues. The collecting duct of the nephron passes through this gradient on its way to the cavity of the kidney (pelvis) and the ureter leading to the urinary bladder. The walls of the collecting duct are permeable to water; the degree of permeability is regulated by antidiuretic hormone (ADH) secreted by the pituitary gland in response to the organism's need to conserve or eliminate water. In the presence of ADH, the collecting duct is permeable to water, more water is drawn off from the duct into the surrounding tissues to be conserved, and the concentration of solutes in the urine increases. As the collecting duct passes through the zone of increasing solute concentration on its way to the ureter, water, moving down its concentration gradient, passes into these tissues, from which it is reabsorbed and conserved. As this physiological mechanism draws off water from the collecting duct, the urine within the duct becomes progressively more concentrated. In the absence of ADH, the walls of the collecting duct are not permeable to water, water is not conserved in this region, and the concentration of solutes in the urine is lower.

4. Examine Figure 40C-3 and study how the kidney nephron functions. Demonstrate your understanding by completing Table 40C-1.

e. *What effect does ADH (antidiuretic hormone) have on excretion?*

_____

f. *What part of the nephron does it influence?* _____

  How? _____

g. *From which portion(s) of the loop of Henle is the majority of salt and water reabsorbed?*

_____

h. *Do you think that an animal living in the desert would have a long or short loop of Henle?* _____

  Why? _____

Table 40C-1

Kidney Region	Structure	Function
Cortex	Glomerulus	
	Renal capsule	
	Proximal tubule	
	Distal tubule	
Medulla	Descending loop of Henle	
	Ascending loop of Henle	
	Collecting duct	

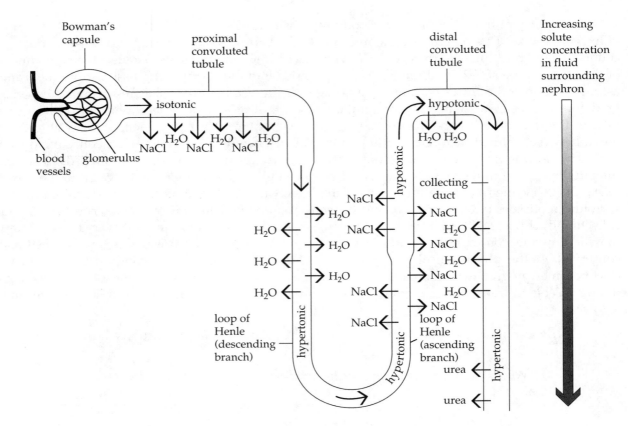

Figure 40C-3 *The formation of hypertonic urine in the human nephron. The filtrate entering the proximal convoluted tubule is isotonic with the blood plasma. Although sodium ions are pumped from the tubule here, with chloride ions following passively, the filtrate remains isotonic because water also moves out by osmosis. As the filtrate descends the loop of Henle, it becomes increasingly concentrated as water moves by osmosis into the surrounding zone of high solute concentration. This zone is created by the action of the wall of the ascending branch of the loop of Henle, which pumps out sodium and chloride ions, and by the diffusion of urea out of the lower portion of the collecting duct. Because the wall of the ascending branch of the loop is impermeable to water, the filtrate becomes less and less concentrated as sodium chloride is removed. By the time it reaches the distal convoluted tubule, it is hypotonic in relation to the blood plasma, and it remains hypotonic throughout the distal tubule. The filtrate then passes down the collecting duct, once more traversing the zone of high solute concentration.*

From this point onward, the urine concentration depends on antidiuretic hormone (ADH). If ADH is absent, the wall of the collecting duct is not permeable to water, no additional water is removed, and a less concentrated urine is excreted. If ADH is present, the cells of the collecting duct are permeable to water, which moves by osmosis into the surrounding fluid, as shown in the diagram. In this case a concentrated (hypertonic) urine is passed down the duct to the renal pelvis, the ureter, the bladder, and finally out through the urethra.

Based upon the structure of the kidney, mammals are the only vertebrates that regularly produce a urine more concentrated than their body fluids. Birds and reptiles conserve water by producing insoluble uric acid. Water is removed from the urine in the cloaca (and large intestine), and urine is a relatively dry paste in these organisms. Birds and reptiles that live in marine environments also possess accessory osmoregulatory structures located above the eye and draining into the nasal passageways. These organs enable them to secrete salt and retain water.

> *i. From your reading, how do you suppose marine fishes cope with the osmotic problems encountered in a saltwater environment?* _____
>
> _____

✔ **PART 2**   **Microscopic Anatomy of the Kidney**

||||| **Procedure** ||||||||||||||||||||||||||||||||||||||||||

1. Obtain a slide of monkey kidney tissue. Locate and identify the **glomerulus,** a tight knot of capillaries surrounded by **Bowman's capsule.**

2. Focus on the outer layer or **cortex** of the kidney and observe the walls of the proximal and distal convoluted tubules. Compare these with the collecting ducts.

   a. Do the walls of the collecting ducts appear to be of different thickness? _____ *Of different cell structure?* _____ *Describe.* _____

✔ **PART 3**   **The Urogenital System**

Within the vertebrates, special ducts have evolved to carry sperm and eggs separately from urine. In many vertebrates, the terminal portions of the digestive, urinary, and reproductive systems empty into a common chamber, the **cloaca.** In mammals, the cloaca is divided so that urogenital products (urine and gametes) exit separately from digestive wastes.

Reptiles, birds, and mammals reproduce on land. Mating requires a mechanism to ensure that sperm are introduced inside the body of the female, where internal fertilization occurs. Males may possess a special copulatory organ, the **penis,** to facilitate this event. In most reptiles and all birds, the fertilized, relatively yolky ovum is surrounded by nutritive albumin layers and a protective shell. The egg is laid and develops outside the female, with or without her care. In mammals, however, fertilized ova are retained inside the female and, in placental mammals, the embryo establishes an intimate connection with the material circulatory system through a special nutritive organ called the **placenta.**

✔ **EXERCISE D**   **Gamete Formation**

Gametes are formed within **gonads** by the process of meiosis. **Spermatogenesis** results in the formation of sperm within the **testes** and **oogenesis** results in the formation of eggs within the **ovary.** Maturation of gametes is controlled by hormones.

||||| **Objectives** |||||||||||||||||||||||||||||||||||||||||

☐ Relate the process of meiosis to the stages of oogenesis and spermatogenesis.

☐ Describe the structure of a seminiferous tubule (cross section) and relate this structure to the types of meiotic figures formed within it.

☐ Describe the structure of a Graafian follicle and relate the changes in this structure to the process of oogenesis and ovulation.

☐ Describe the structure of a mature spermatozoon (sperm cell).

☐ Identify the cellular organelles of a mature sperm cell and compare them with the structures of a typical animal cell.

✔ **PART 1**   **Spermatogenesis**

In males, the process of meiosis results in the formation of haploid sperm cells. Diploid cells, **spermatogonia**, embedded in the walls of the **seminiferous tubules** within the testes, at different time intervals change biochemically to form **primary spermatocytes.** During meiosis I, primary spermatocytes form **secondary spermatocytes;** during meiosis II, secondary spermatocytes form four haploid **spermatids.** Spermatids differentiate to form **spermatozoa** (sperm cells).

In Figure 40D-1, on the right-hand side, label primary and secondary spermatocytes, spermatids, and spermatozoa. On the left-hand side, indicate the processes required to produce each cell type. Within each circle, indicate whether the cell is haploid (*n*) or diploid (*2n*).

Figure 40D-1   *Schematic diagram of spermatogenesis.*

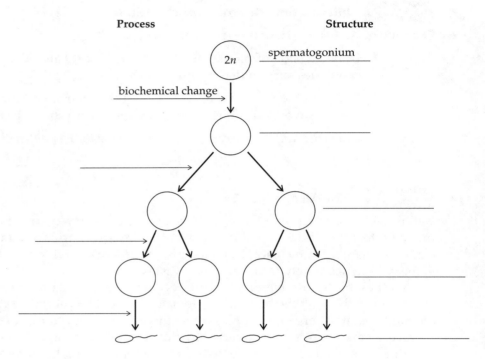

Process   **Structure**

2*n*   ___spermatogonium___

biochemical change

||||| Procedure |||||||||||||||||||||||||||||||||||||||||||||

1. Use high power (40×) to examine a prepared slide of a rat testis (cross section). Refer to Figure 40D-2 to identify spermatogonia, primary spermatocytes, secondary spermatocytes, spermatids, spermatozoa, Sertoli cells, and lumen of the seminiferous tubule.

   a. *Why is the adult male able to produce spermatozoa continually?* _____

   _____

   b. *After the sperm cell leaves the lumen of the seminiferous tubule, what path does it travel to get to*

   the outside? _____

   _____

2. Examine the microscope slide of rat or human spermatozoa on demonstration. Study the diagram of a typical sperm cell (Figure 40D-3). Remember, spermatozoa are single cells. Identify the head, midpiece, and tail of a sperm.

   c. *What shape is the head of the sperm?* _____

   d. *What typical animal cell structures are present in the sperm cell?*

   _____

Figure 40D-2 *Development of sperm within the seminiferous tubule. Spermatogonia are embedded in the wall of the tubule. As the cells mature, they progress toward the lumen (cavity) of the tubule, where mature sperm are found.*

Developing spermatids are connected to Sertoli cells, large "nurse" cells that provide nutrition during the process of differentiation, the transformation from the generalized spermatid cell to the highly specialized sperm cell.

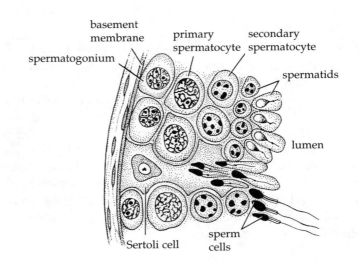

e. What typical animal cell structures are not present? _____

_____

f. What is the major structural component of the tail? _____

g. During sperm formation, the bulk of the cytoplasm and Golgi material is pinched off from the cell. What is the advantage of this process? _____

_____

h. The cap or acrosome is located on the head of the sperm. It is formed by a Golgi body and contains a "packet of enzymes." Why do you think a Golgi body is the appropriate organelle for acrosome formation? _____

_____

i. Can you see the acrosome on the rat or the human sperm? If you are observing a rat's sperm, you may see the acrosome as slightly detached. What shape is it? _____

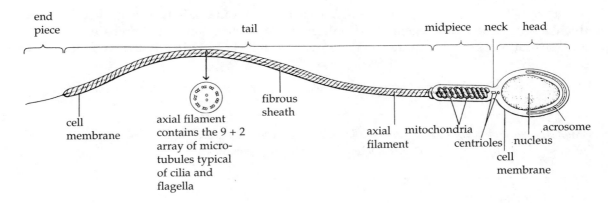

Figure 40D-3 *Diagrammatic representation of a typical sperm cell.*

PART 2   Oogenesis

The mammalian ovum develops within the outer tissue layer or cortex of the ovary. Each egg develops from the meiotic division of a diploid cell called an **oogonium.** Just as in the process of spermatogenesis, an oogonium will first become biochemically modified to form a **primary oocyte.** Meiosis I results in the formation of two cells—a large **secondary oocyte** and a very small **polar body.**

The larger cell or secondary oocyte divides again during meiosis II. This division is also unequal, giving rise to a much larger **ootid** and a very small polar body. The polar body formed during the first meiotic division may also divide into two smaller polar bodies. Thus meiosis during oogenesis results in the formation of one large ootid and two or three small polar bodies, which usually disintegrate. The ootid will differentiate biochemically to form the mature **ovum.**

In Figure 40D-4, label the primary oocyte, secondary oocyte, ootid, ovum, and polar bodies. Indicate where meiosis I, meiosis II, biochemical modification, and differentiation occur. Indicate whether each cell type is haploid (*n*) or diploid (2*n*).

Figure 40D-4   *Schematic diagram of oogenesis.*

Processes are designated by rules with arrows; **structures** are designated by rules.

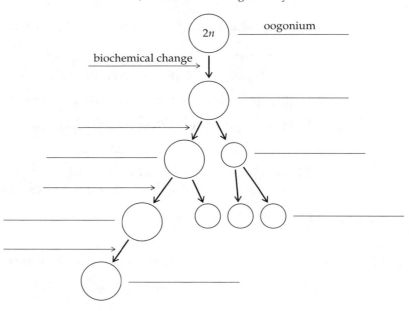

ııııı Procedure ıı

Label Figure 40D-5 as follows: **a** antrum, **b** cells of the **corona radiata, c egg, d follicle cells,** and **e zona pellucida.** Study the figure legend. Use high power (40✕) to examine the prepared slide of the cat ovary on demonstration. Study a mature Graafian (ovarian) follicle and identify the structures shown in Figure 40D-5.

a. *What structures surrounding the egg must a sperm penetrate before it can fertilize the egg?*

_____

b. *What sequence of events could lead to a fertilized egg developing within the body cavity rather than in the uterus? (This type of pregnancy is known as an ectopic pregnancy.)*

_____

c. *What is a tubal pregnancy?*

_____

Figure 40D-5 *Within the ovary, the maturing ovum is surrounded by follicle cells, which secrete a liquid into the space between the follicle cells and the ovum. This liquid gradually pushes the follicle cells away from the egg to form an enlarging fluid-filled cavity, the* antrum. *A few follicle cells adhere to the* zona pellucida, *a membrane surrounding the* plasma membrane *of the ovum. This group of follicle cells, the* corona radiata *(it gives the appearance of a radiating crown around the ovum), remains associated with the ovum when it is ovulated.*

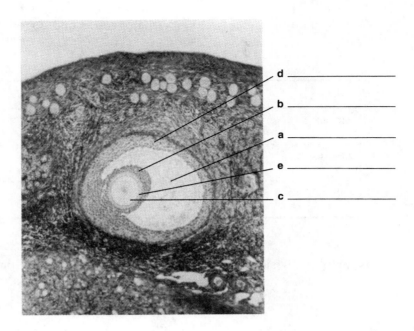

d _____

b _____

a _____

e _____

c _____

Laboratory Review Questions and Problems

1. Why is it important for the intestine to have a large surface area? How does the structure of the intestine increase its surface area?

2. List the functions of the liver. Why is this organ important in digestion and circulation?

3. You eat a chicken sandwich (chicken, bread, and mayonnaise). On a separate piece of paper, construct a chart showing where and how the parts of the sandwich will be digested.

4. How is the structure of a nephron related to the overall structure of the kidney itself?

5. How does the kidney work with the circulatory system, anatomically and chemically, to maintain water, salt, glucose, and pH balance?

6. What do we mean by the term "urogenital" system? Why do we often discuss the excretory system along with the reproductive system?

7. Compare the processes of spermatogenesis and oogenesis by completing the following table.

Process	Structures Formed
Meiosis I	
Meiosis II	
Differentiation	

8. Why is division of the cytoplasm uneven during the process of oogenesis?

Control—The Nervous System

OVERVIEW

To survive and reproduce, animals must maintain a relatively constant internal state in the midst of wide fluctuations of factors in the environment. This constancy, called **homeostasis,** is maintained by the nervous and endocrine systems, which interact to control an animal's internal functioning (**physiology**) and external activity (**behavior**). In the nervous system, sensors (**affectors**) monitor an animal's internal and external environments. When physiological adjustments are required, these are mediated by the nervous system and facilitated by chemical messengers (hormones) secreted by the endocrine system, under close control of the nervous system. When external action is called for, **effectors**—muscles and bones and glands—coordinate various behaviors. The activities of the nervous and endocrine systems can be directed or regulated by genetic information and by learning.

In this laboratory you will study three levels of components of the mammalian nervous system: the structural unit of the nervous system, the **neuron,** or nerve cell; several sensory receptors (sense organs); and the brain, the major component of the central nervous system.

STUDENT PREPARATION

Prepare for this laboratory by reviewing the text pages indicated by your instructor. Familiarizing yourself in advance with the information and procedures covered in this laboratory will give you a better understanding of the material and improve your efficiency.

PART 1. NEURONS

EXERCISE A   Examining Nerve Cells (Neurons)

Nerve tissue is composed of two types of cells: (1) neurons, which are highly specialized for irritability and conduction of nerve impulses, and (2) sheath cells and glial elements, which provide both nourishment and support; some also help to conduct impulses.

▐▐▐▐▐ **Objectives** ▐▐▐▐▐▐▐▐▐▐▐▐▐▐▐▐▐▐▐▐▐▐▐▐▐▐▐▐▐▐▐▐▐

☐ Diagram the neuron and state the function of each part.

▐▐▐▐▐ **Procedure** ▐▐▐▐▐▐▐▐▐▐▐▐▐▐▐▐▐▐▐▐▐▐▐▐▐▐▐▐▐▐▐▐▐

    1. On demonstration you will find a microscope slide of several neurons. Examine this slide and diagram a neuron on a separate sheet of paper. Label the neuron cell body, its axon, and any dendrites that may be visible.

2. Use your knowledge of nerve tissue to answer the following questions. (If you performed Laboratory 32, Animal Tissues, reviewing your notes on that lab will be helpful.)

a. *Identify the functions of each of the following parts of a neuron.*

Cell body _____

Axon _____

Dendrite _____

b. *Distinguish among the functions of the following types of neurons. Where in the nervous system is each type of neuron located?*

Sensory (afferent) neurons _____

Interneurons _____

Motor (efferent) neurons _____

c. *What is the difference between a neuron and a nerve?* _____

_____

d. *The functional junction between two neurons is the* **synapse.** *What happens at this junction and why are these events important? How does the synapse control the direction of information flow?*

_____

_____

PART II.   SENSORY RECEPTORS

Before an organism can respond to its external or internal environment, it must receive relevant information about its position within the environment and about the environment itself. Sensory receptors (sensors or affectors) transmit this information to the **central nervous system** (the brain and spinal cord), thus enabling the organism to coordinate its activities.

Prominent sensory receptors are sensitive to changes in particular forms of environmental energy such as heat, pressure, light, and vibration (sound). Others are sensitive to various ions or chemical molecules, some are sensitive to gravity, and others to the position of various skeletal elements and muscles. With the exception of receptors for smell and the free nerve endings occurring in various regions of the body, sensory cells are *not* derived from neurons. Sensors are specialized epithelial elements that synapse with one or more sensory neurons.

Some sensory receptors are functional parts of conspicuous sense organs (for example, the eyes, ears, and nose). However, diffuse receptors, such as those sensing touch, pain, temperature, and muscle or joint position, are equally important to an organism.

In Exercises B through J, work in pairs wherever necessary. Alternate the role of experimenter and experimental subject whenever appropriate.

EXERCISE B   Chemoreception: The Sense of Taste

In humans, the specialized receptor organs for the sense of taste (gustation) are clusters of cells called **taste buds** (Figure 41B-1), located on the tongue, especially on the tip, edges, and posterior third.

Taste sensations are traditionally divided into four basic groups: sweet, sour, salty, and bitter. Receptors for each group of substances are found in different areas of the tongue. Receptors are thought to register different gradations of intensity depending on the substance present.

ⅠⅠⅠⅠⅠ **Objectives** ⅠⅠⅠⅠⅠⅠⅠⅠⅠⅠⅠⅠⅠⅠⅠⅠⅠⅠⅠⅠⅠⅠⅠⅠⅠⅠⅠⅠⅠⅠⅠ

☐ Determine which areas of the tongue are sensitive to specific tastes.

Figure 41B-1 *Human taste bud. Cells of the taste bud may be classified as* receptor (hair) cells, *responsible for transducing sensory information, and* supporting (sustentacular) cells. *The external (distal) ends of the receptor cells project from a pore in the taste bud, enabling them to make direct contact with chemical substances. The internal (basal) ends of the receptor cells synapse with sensory neurons that enter the brain through one of two cranial nerves. Sustentacular cells can transform into taste cells.*

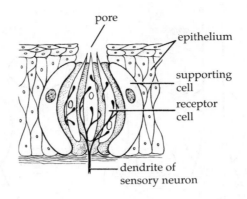

pore
epithelium
supporting cell
receptor cell
dendrite of sensory neuron

IIIII Procedure IIIIIIIIIIIIIIIIIIIIIIIIIIIIIIIIIIIIIIIIIIIIIIII

Five unknown solutions (A through E) are located at your lab station. Each student should obtain five cotton swabs for tasting the solutions and a paper cup for rinsing the mouth with water. Use only one end of each swab and discard it after use. *Do not place a used swab into the test solution.* Work in pairs, alternating the role of experimenter and subject. Record either your own or your partner's taste results.

1. The subject rinses his or her mouth with water.

2. The experimenter dips a cotton swab into solution A and touches it to the tip of the subject's tongue. (Do not use a drop large enough to spread over the tongue—the moist tip of the swab will be sufficient.)

3. The subject determines whether this substance tastes salty, bitter, sweet, or sour, or has no taste at all.

4. If the subject can taste solution A on the tip of the tongue, record a "+" on the tip of tongue A in Figure 41B-2. If the subject cannot taste this solution on the tip of the tongue, record a "−" on the tip of tongue A. The subject now rinses his or her mouth again.

5. Repeat this process, testing solution A on the tip, the sides, the upper back, and the upper from of the tongue, as indicated in Figure 41B-2. In each case, record the ability to taste or not taste solution A on the appropriate part of the diagram of tongue A. In the space below the diagram, indicate how solution A tastes to the subject (salty, sour, sweet, bitter, or no taste).

6. Repeat this entire process with each of the other solutions (B through E), with the subject rinsing his or her mouth between each taste test. Record your observations on the corresponding diagram of the tongue. Make sure that in each case you record the taste of the solution.

7. When you have finished collecting your data, discard any remaining solutions and throw cups and swabs into the trash container.

   The five solutions you tasted were water, sugar, salt, acetic acid (sour), and quinine (bitter). In the spaces below Figure 41B-2, write the names of the solutions.

   a. Why can you taste certain substances on one part of the tongue but not on other parts?

   _____

8. Now obtain a small amount of magnesium sulfate solution and a cotton swab. Test the solution on the upper tip of the tongue. *b. How did it taste?* _____

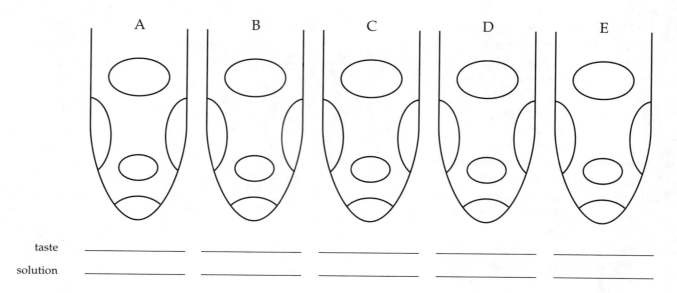

taste   _____   _____   _____   _____   _____

solution   _____   _____   _____   _____   _____

Figure 41B-2   *Localization of taste sensations on the tongue.*

9. Next, test the solution on the upper back of the tongue. *c. How did it taste?* _____

d. How do you explain these differences in the taste of the same substance?

_____

_____

◉ EXERCISE C   Chemoreception: Individual Differences in Taste

When you were a child, did your parents ever tell you it was "all in your head" if you didn't like the taste of some of the food they fixed? Do you think your likes and dislikes might be influenced by genetics? If so, would you expect to find differences in the way in which a particular food tastes to different people? Do chemoreceptors differ from person to person?

||||| Objectives |||||||||||||||||||||||||||||||||||||||

☐ Explain how individual differences in taste are related to food preferences.

During this exercise, you will have the opportunity to taste several specialized "taste papers" (PTC, thiourea, and sodium benzoate). Do you think these papers will taste the same to all members of your class? If you taste one of the taste papers as bitter while your laboratory partner tastes the same type of paper as sweet, do you think you would share the same food preferences?

Formulate a hypothesis relating chemoreception to individual differences in food preferences.

HYPOTHESIS:

NULL HYPOTHESIS:

Does your laboratory partner like the same types of foods as you do?

*a. What do you **predict** the two of you will find when you sample the taste papers?* _____

_____

What is the **independent variable**?

What is the **dependent variable**?

Use the following procedure to test your hypothesis.

||||| **Procedure** |||||||||||||||||||||||||||||||||||||||||

1. Each student should obtain from the demonstration bench one piece of each of the four taste papers (PTC, thiourea, sodium benzoate, and a control paper). Be sure to keep these papers in order.

2. Taste the control paper first and record the taste as sour, bitter, salty, sweet, no taste, or whatever terms may be appropriate. Touch each of the other papers to your tongue.

RESULTS: Record your sensation of taste for each paper.

Control (white) _____

PTC (blue) _____

Thiourea (yellow) _____

Sodium benzoate (pink) _____

3. Discard your papers in the trash container when you are finished.

b. Which chemical did you find most distasteful? _____

c. Did most of the students in the class have this same reaction to this particular chemical? _____

d. How did sodium benzoate taste to other members of your laboratory section? _____

Although sodium benzoate is tasteless to some individuals, to those who are capable of tasting it, it may seem sweet, sour, bitter, or salty. Sodium benzoate is sometimes used as a food preservative (check your favorite soft drink label).

e. How do you think the use of sodium benzoate an an ingredient in food processing affects the taste of the product

to different people? _____

_____

It is also interesting to note that those people who taste PTC as bitter and sodium benzoate as salty tend to like the taste of buttermilk, sauerkraut, turnips, and other sour foods more than average. Those who taste both PTC and sodium benzoate as bitter tend to dislike the taste of such foods.

f. How do your food preferences relate to the observations reported here? _____

g. Do your observations support your hypothesis? _____

Your null hypothesis? _____

h. What do you conclude about chemoreception and food preferences in different individuals?

_____

👁 **EXERCISE D**   **Chemoreception: Smell Discrimination and Its Influence on Taste**

Smell (olfaction) is another example of chemoreception. In this case, the receptors are part of a mucus-secreting membrane (hence, mucous membrane) in the upper part of the nasal cavity. The receptors, called **hair cells,** have hairlike cilia on one end extending into a layer of mucus (Figure 41D-1).

Before the odor of a substance can be detected, the substance must release molecules that diffuse into the air and pass into the region of the olfactory receptors. These molecules then dissolve in the mucus and

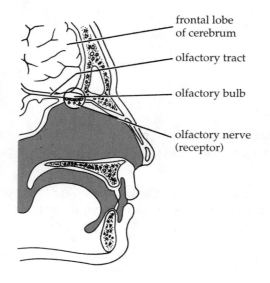

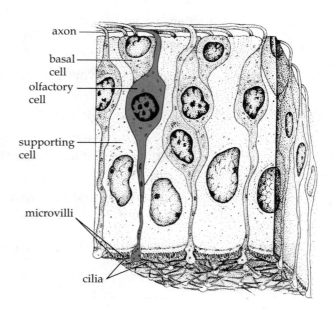

Figure 41D-1   *(a) A patch of special tissue, the olfactory epithelium, arching over the roof of each nasal cavity, is responsible for the sense of smell. (b) The olfactory epithelium is composed of three types of cells:* supporting cells, basal cells, *and* olfactory cells. *The olfactory cells are the sensory receptors. Cilia protruding from their upper surfaces are believed to be the odor receptors, although the way in which they function is not known. Note that the sensory cells are neurons that give rise to extensions (axons) that lead directly to the olfactory lobe of the brain.*

interact with receptor molecules in the cilia of the sensory cells. Exactly how these receptor cells and the brain distinguish odors is not understood. It is known, however, that we can distinguish a far greater number of odors than tastes. In fact, our enjoyment of food is actually the product of two sensations—the limited information conveyed by taste buds and the broader range provided by the olfactory receptors.

Do you think that information obtained by our olfactory receptors can influence how we think a particular food tastes? If you held your nose, would a cherry Life Saver® taste the same as a lemon Life Saver®? If you were blind-folded and given lemon Life Savers® to smell, but were fed cherry Life Savers®, what do you think you would report when asked how the candy tasted?

Formulate a hypothesis as a tentative explanation for how our sense of smell affects our sense of taste.

HYPOTHESIS:

NULL HYPOTHESIS:

a. What do you **predict** you will find if you test your hypothesis? _____

_____

What is the **independent variable**?

What is the **dependent variable**?

Use the following procedure to test your hypothesis.

ⅠⅠⅠⅠⅠ **Procedure** ⅠⅠⅠⅠⅠⅠⅠⅠⅠⅠⅠⅠⅠⅠⅠⅠⅠⅠⅠⅠⅠⅠⅠⅠⅠⅠⅠⅠⅠⅠⅠⅠ

1. The experimenter should obtain a Life Saver® from the various flavors available on the demonstration bench, without letting the subject know what the flavor is.

2. The subject closes his or her eyes and holds his or her nose.

3. The experimenter gives the Life Saver® to the subject and the subject places it on the tongue.

4. The subject, while still holding his or her nose, guesses the flavor of the candy. The experimenter records the guess in step 5.

5. The subject releases his or her nose and guesses the flavor again. The experimenter records this guess and the actual flavor of the Life Saver.®

   Flavor while holding nose _____

   Flavor without holding nose _____

   Actual flavor _____

   b. *From your results, how would you say that smell affects the taste of Life Savers®?*

   _____

   c. *Do your results support your hypothesis?* _____ *Your null hypothesis?* _____

   d. *What do you conclude about the effect of smell on your sense of taste?*

   _____

✔ **EXERCISE E** | **Photoreception: Vision—Structure of the Eye**

The human eye has often been compared to a camera. The **lens** focuses light onto the **retina** at the back of the eye just as the lens of a camera focuses light onto the film. The amount of light entering the eye is regulated by the **iris**, which controls the size of its opening, the pupil; the diaphragm of a camera regulates the amount of light that enters in much the same way. The retina consists of light-sensitive cells (**rods** and **cones**) that transduce light energy into nerve impulses that are partially processed in the retina and transmitted to the brain via the optic nerve. We "see" the picture as the brain processes the transmitted information. During development of the organism, the eye forms from an out-pocketing of the forebrain and associated elements. The arrangement of the functional elements of the eye reflects this origin; for example, the retina is really a part of the brain, and the optic nerve is, in fact, a tract in the brain.

ⅠⅠⅠⅠⅠ **Objectives** ⅠⅠⅠⅠⅠⅠⅠⅠⅠⅠⅠⅠⅠⅠⅠⅠⅠⅠⅠⅠⅠⅠⅠⅠⅠⅠⅠⅠⅠⅠⅠⅠ

☐ Identify the parts of the eye and state the function of each part.

☐ Trace the pathway of light through the eye and explain how the eye regulates the passage of light.

ⅠⅠⅠⅠⅠ **Procedure** ⅠⅠⅠⅠⅠⅠⅠⅠⅠⅠⅠⅠⅠⅠⅠⅠⅠⅠⅠⅠⅠⅠⅠⅠⅠⅠⅠⅠⅠⅠⅠⅠ

1. Match the boldface letters in the following list to the letters in Figure 41E-1, and write the name of the structure in the space provided on the figure.

   a **Optic nerve** Located opposite the pupil. The stub of the second cranial nerve (actually a fiber tract of the brain).

   Extrinsic eye muscle Six external muscles attached to the sclera, which appear as discrete masses of slightly darker tissue in the connective and fat tissue surrounding the eye. Responsible for movement of the eyeball within its socket (**orbit**).

   b **Sclera** Tough outer layer of the eye, continuous with the protective dura mater covering the brain. Attaches to extrinsic muscles.

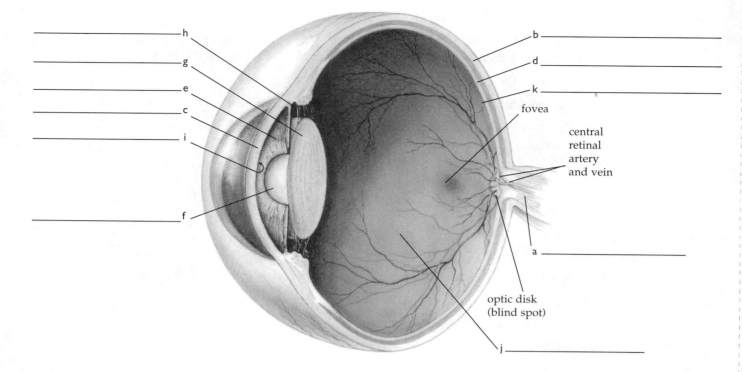

Figure 41E-1 *Structure of the eye.*

c Cornea Transparent anterior part of the sclera that admits light to the eye and focuses it on the retina. The transparent covering layer of epithelium is the **conjunctiva.** Asymmetries in the cornea (departures from a spherical shape) produce astigmatism.

d Choroid Middle layer of the eye providing a nutritive vascular blood supply. The anterior portion is modified into parts of the **iris, suspensory ligament,** and **ciliary body.** In nocturnal animals, the posterior portion behind the retina contains pigment that reflects light, causing "night shine."

e Iris A pigmented, diaphragm-like structure. Separates the aqueous chamber into anterior and posterior chambers (see **i**).

f Pupil Opening in the iris through which light passes.

g Lens Hard, transparent, rounded structure inside the eye; elastic in the living eye. The **ciliary body h,** a circle of smooth muscle, changes the shape of the lens to focus on near or distant objects. With age, the elasticity of the lens may be reduced, hence the progression to bifocal and trifocal glasses. Cataracts result from the loss of lens transparency.

i Aqueous chamber Chamber located in front of the lens; contains liquid **aqueous humor.** Divided into anterior and posterior chambers by the iris. Secretions of fluid maintain both the shape of the corneal surface and the proper pressure in the aqueous chamber. Excess pressure in the aqueous chamber causes glaucoma and can lead to blindness.

j Vitreous chamber Chamber of the eye behind the lens; contains jellylike **vitreous humor.** Transparent in the living eye.

k Retina Inner sensory layer of the eye. Contains sense elements (**rods** and **cones**), sensory neurons, and associative neurons. Technically, the retina is part of the brain. Fibers project from the retina to the brain through the optic tract or "nerve."

2. Examine a preserved sheep's eye. With a scalpel or razor blade, make a cut into the eye approximately 5 mm posterior to the edge of the cornea and continue the incision all the way around the eye. Remove the cornea and the lens, leaving the jellylike vitreous body in the back portion of the eye.

3. Referring to Figure 41E-1 and the list in step 1, identify as many of the parts as you can. Note that the retina appears black. This is due to a pigment layer behind the sensory cells. The iridescent appearance in this area is due to a reflecting pigment in the posterior portion of the choroid coat behind the retina, the *tapetum lucidum.*

4. The eye is actually composed of three layers of tissue that form a fluid-filled sphere. Organize your study of the structures of the eye by listing each of the following as part of the outer, middle, or inner layers of the eye: sclera, iris, cornea, lens, ciliary muscles, retina, choroid, sensory elements (rods and cones).

Outer Layer   Middle Layer   Inner Layer

_____   _____   _____

_____   _____   _____

_____   _____   _____

_____   _____   _____

_____   _____   _____

    a. *List in sequence those structures of the eye through which light must pass before arriving at the retina.* _____

_____

EXERCISE F | Photoreception: How We See

Light entering the eye passes through the chamber containing aqueous humor, then through the lens, and finally through the chamber containing vitreous humor before it falls on the retina. The retina is the sensitive layer of the eye and it is here that the energy of light is transduced into nerve impulses.

The light must pass through several layers of neurons on the retina before striking the photosensitive cells. In the human retina, there are two types of photoreceptor elements: cones and rods.

Cones are responsible for vision in bright light and for the perception of fine detail and color. Cones are concentrated in the **fovea centralis** at the center of the retina. Light-sensitive **rods** function in dim light and are insensitive to colors. Rods are more numerous in the periphery of the retina. Information from the cones and rods travels back toward the anterior of the retina through the layer of nerve cells. The first cell receiving information from the rods or cones is a bipolar cell, the sensory neuron (Figure 41F-1). Impulses from particular rods and cones may be modified by impulses from other rods or cones or sensory elements by means of transverse connections. Bipolar cells synapse with ganglion cells at the anterior of the retina. Axons of these neurons join together to form the optic nerve, which relays information to the brain.

a. *Which area would have the sharpest vision—the fovea or the periphery?* _____

b. *Why?* _____

c. *If you are trying to see an object in dim light, is it best to look at the object directly so that the image falls on the cones, or to look at it out of the side of your eye so that the image falls on the rods?* _____

d. *Why?* _____

_____

▪▪▪▪▪ Objectives ▪▪▪▪▪▪▪▪▪▪▪▪▪▪▪▪▪▪▪▪▪▪▪▪▪▪▪▪▪▪

☐ Relate the structure of the eye to its ability to perceive color, maintain peripheral vision, and compensate for the blind spot on the retina.

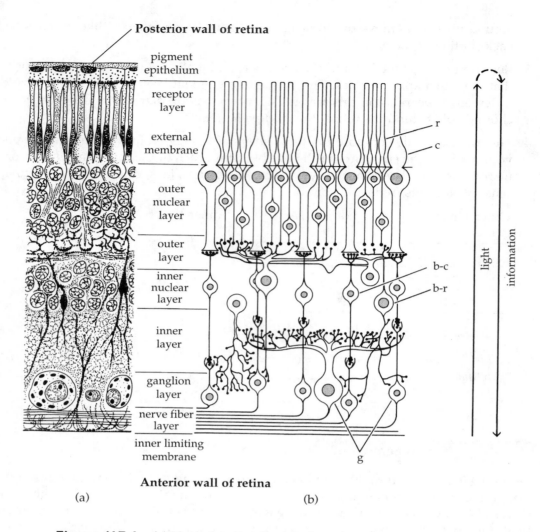

Posterior wall of retina

Anterior wall of retina

(a)   (b)

Figure 41F-1 *(a) Vertical section through the retina of a mammal. (b) Schematic drawing showing some of the connections among sensory and nervous elements in the retina: **r**, rods; **c**, cones; **b-r**, bipolar cells connecting with a series of rods; **b-c**, bipolar cells associated with a single cone; **g**, ganglion cells. Light enters at the anterior of the retina, traverses the nuclear layers, and arrives at the rods and cones at the posterior of the retina. Sensory information then travels from these sensory cells to the bipolar cells to the ganglion cells, which send it on its way to visual association areas in the brain. (After Alfred S. Romer and Thomas S. Parsons,* The Vertebrate Body, *6/e, p. 372, fig. 5B, CBS, New York, 1983.)*

PART 1   Peripheral Vision and Color Vision

||||| Procedure ||||||||||||||||||||||||||||||||||||||

1. The experimenter chooses a colored object from the demonstration bench, without letting the subject see what that object is.

2. The subject closes one eye and focuses the other eye on some point straight ahead.

3. The experimenter moves the object into view from the side. The subject should *not* look directly at the object.

a. What can be determined first, the shape of the object or its color? Explain.

_____

_____

 PART 2   **The Blind Spot**

The rods and cones make contact with neurons that conduct information from these receptor cells to the optic nerve fibers. Fibers from these cells located in all parts of the retina converge at a point on the retina known as the **optic disk.** This region has no rods or cones, thus producing a "**blind spot**": images falling on this area cannot be perceived. From the optic disk, nerve fibers proceed to the brain. Figure 41F-2 can be used to demonstrate the presence of the blind spot.

Figure 41F-2   *A test for the blind spot.*   ✚   ●

⁞⁞⁞⁞⁞ Procedure ⁞⁞⁞⁞⁞⁞⁞⁞⁞⁞⁞⁞⁞⁞⁞⁞⁞⁞⁞⁞⁞⁞⁞⁞⁞⁞⁞⁞⁞⁞⁞⁞⁞⁞⁞⁞⁞⁞⁞

1. Close your left eye and focus your right eye on the ✚ in Figure 41F-2.
2. Start with the figure about 12 cm (5 in) from your eye and gradually move the page away until the circle disappears.
3. Repeat this with your left eye, but focus on the circle and move the page until the ✚ disappears.

   a. Even though you have a blind spot in each eye, why do you not experience a blind spot in your vision when you look at an object with both eyes? _____

   _____

   b. When you look at an object with only one eye, you do not notice a blind spot in your vision. Why do you suppose this happens? ? _____

   _____

 EXERCISE G   **Mechanoreception: The Role of Sensory Receptors in Touch**

A large number of diverse sense organs in the skin contribute to tactile sensitivity. Receptors for each of the different sensations (pain, touch, deep pressure, cold, and warmth) are localized in different places in the skin (Figure 41G-1). The concentration of particular receptors also varies from place to place.

   One type of receptor, the **mechanoreceptor,** is sensitive to quantitative forces such as pressure, touch, stretching, sound, movement (acceleration), and gravity in the external and internal environments. Such receptors are abundant in the skin, muscles, tendons, various visceral organs, large blood vessels near the heart, and most connective tissues.

⁞⁞⁞⁞⁞ Objectives ⁞⁞⁞⁞⁞⁞⁞⁞⁞⁞⁞⁞⁞⁞⁞⁞⁞⁞⁞⁞⁞⁞⁞⁞⁞⁞⁞⁞⁞⁞⁞⁞⁞⁞⁞⁞⁞

☐ Describe the role of tactile sensory receptors in distinguishing the sensations of touch.

⁞⁞⁞⁞⁞ Procedure ⁞⁞⁞⁞⁞⁞⁞⁞⁞⁞⁞⁞⁞⁞⁞⁞⁞⁞⁞⁞⁞⁞⁞⁞⁞⁞⁞⁞⁞⁞⁞⁞⁞⁞⁞⁞⁞

This experiment will test your discrimination of various stimuli through receptors sensitive to light touch, the **Meissner's corpuscles.**

Figure 41G-1 *Sensory receptors present in the skin. Light touch is detected by Meissner's corpuscles and Merkel cells or by free nerve endings around hair follicles. Pacinian corpuscles are stimulated by deep pressure.*

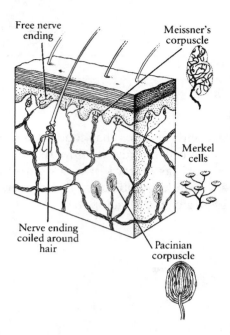

1. Obtain a ruler and a pair of compasses from the demonstration bench. Touch the tips of the compasses to the skin of the subject who, *with eyes closed,* should try to determine whether one or two compass tips are in contact with the skin.

2. Vary the distance between the compass tips (1, 3, 6, 12, and 24 mm) and try using one or two tips in a given area. Test the following areas: lips, fingertips, palm, inner forearm, inner wrist, outer wrist, back of the hand, and outer forearm. Record the smallest distance at which the subject can distinguish the compass tips as two points: lips _____, fingertips _____, palm _____, back of hand _____, inner wrist _____, outer wrist _____, inner forearm _____, outer forearm _____.

   a. *Explain why some areas of your body are better able than other parts to distinguish between two close stimuli.* _____

   _____

   b. *Referring to your data in step 2, determine which of the tested areas of your body are most sensitive to light touch. Suggest some possible advantages of greater sensitivity in these areas.*

   _____

👁 EXERCISE H | Thermoreception: Discriminating Temperature

Many types of receptors cease to become excited when they are subjected to a single, continuous stimulus. This phenomenon is called **habituation.** This exercise demonstrates habituation by heat and cold receptors in the skin.

▌▌▌▌▌ Objectives ▌▌▌▌▌▌▌▌▌▌▌▌▌▌▌▌▌▌▌▌▌▌▌▌▌▌▌▌▌▌

☐ Describe the role of tactile sensory receptors in distinguishing sensations of hot and cold.

☐ Define habituation as applied to the function of receptors for temperature.

||||| **Procedure** ||||||||||||||||||||||||||||||||||||||||||||

    1. Set up three beakers of water: one cold, one at room temperature, and one with tolerably hot water.

    2. Place one index finger in cold water and one in hot water. *a. How does each finger feel?*

    _____

    3. Leave your fingers in the water for several minutes. *b. How does each finger feel now?*

    _____

    4. Now move the two fingers directly from the cold and hot beakers into the beaker of room temperature water. *c. How does each finger feel now?* _____

    d. Is the stimulus the same for both fingers now? _____

    e. When you move your finger from cold water into warmer water, do the receptors in your finger respond to heat gain or heat loss? _____

    f. When you move your finger from hot water to cooler water, do the receptors in your finger respond to heat gain or heat loss? _____

👁 **EXERCISE I** | **Proprioception: The Role of Proprioceptors in Determining Position**

The activation of proprioceptors in the joints, associated ligaments, and muscles sends impulses through the nervous system, giving rise to the conscious awareness of the position and movement of joints. Input from these receptors is integrated with visual information and input from the inner ear to provide an awareness of the position of the body in space.

The various proprioceptors are activated by mechanical stimuli such as stretching, twisting, and compression, or by painful stimuli. This exercise demonstrates the sensitivity of these receptors.

||||| **Objectives** |||||||||||||||||||||||||||||||||||||||

☐ Describe the role of proprioceptors in maintaining the position of body members.

||||| **Procedure** |||||||||||||||||||||||||||||||||||||||

    1. Extend one of your arms and close your eyes. Extend your other arm.

    2. Bring the two arms together and touch the tips of your two index fingers.

    a. Did your index fingers actually touch when you brought them together or did they miss each other? _____

    3. Repeat this exercise by putting your hands behind your back and trying to touch the tips of the index fingers. *b. Could you get them to touch?* _____

    4. Obtain a piece of string. Close your eyes and tie a knot in the string. *c. Describe what happens.*

    _____

    d. How do you explain these results? _____

    _____

👁 **EXERCISE J** | **Mechanoreceptors of the Ear**

In humans, equilibrium depends upon stimuli from several different sources: (1) the eye, (2) proprioceptors, (3) mechanoreceptors sensitive to pressure in the soles of the feet, and (4) organs of the inner ear.

Figure 41J-1 *(a) The structure of the ear. (b) Within the interconnected, fluid-filled sacs of the* sacculus *and* utriculus *are sensory otoliths composed of bony material in which sensory "hairs" are embedded. Changes in the position of the otoliths relay information to the brain about changes in direction with respect to gravity.*

Each of the semicircular canals contains fluid and has a bulbous swelling at one end, the ampulla. These contain hair cells sensitive to displacement of fluid by acceleration and deceleration. As you accelerate in one direction during rotation, the fluid in the semicircular canal parallel to your motion pushes against the hair cells in the ampulla.

The body's position and balance are maintained by nerve impulses to the brain from the utriculus, sacculus, and all three semicircular canals of both ears.

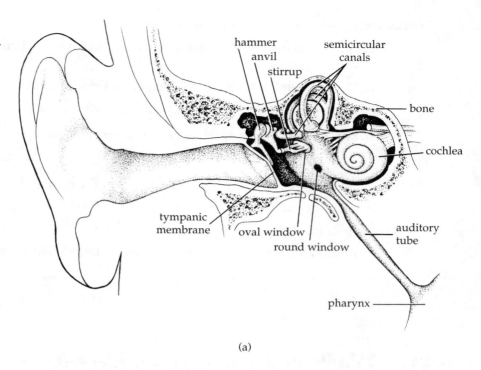

(a)

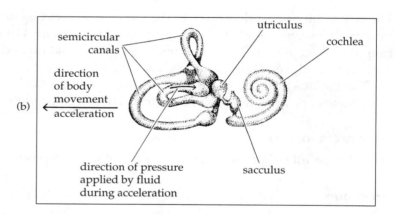

(b)

Sensors of the inner ear (Figure 41J-1) monitor the position of the head relative to gravity and detect changes in the motion of the head.

Certain reflex actions may occur in response to the stimulation of cells of the inner ear. When a person is seated on a revolving stool and rotated, the eyes move in a peculiar manner. The eyes fix upon, and keep in view, a certain object, resulting in a slow movement of the eyes in a direction opposite to that of the body. When the object can no longer be seen, the eyes shift very swiftly, moving in the direction of rotation, and fix on another object. The process is repeated over and over. Movement of the eyes in this manner is referred to as **nystagmus.** When rotation suddenly stops, a postrotational nystagmus occurs and the eyes move as they would have if rotation had begun in the opposite direction. These reflexes are dependent on information from the semicircular canals.

ⅢⅢ Objectives ⅢⅢⅢⅢⅢⅢⅢⅢⅢⅢⅢⅢⅢⅢⅢⅢⅢⅢⅢⅢⅢⅢⅢⅢⅢⅢ

☐ Identify the structures of the ear that determine how a body reacts to changes in position.

Procedure IIIIIIIIIIIIIIIIIIIIIIIIIIIIIIIIIIIII

1. If a rotating stool or chair is available in the laboratory, have your partner sit down and hold on tightly as you rotate the stool.

2. After several rotations, stop the stool and immediately observe the movement of your partner's eyes.

   a. *Which way do they move?* _____

   b. *What could be stimulating the eyes to move in the manner observed after the subject has stopped*

      rotating? _____

      _____

   c. *Many times a dancer will turn his or her head in the direction opposite to that in which he or she is*

      spinning. Why? _____

      _____

   d. *Why do you become dizzy if you spin around and around?*

      _____

PART III.   THE BRAIN

✔ **EXERCISE K** | **The Structure of the Mammalian Brain**

Sensory receptors relay information over sensory neurons to appropriate areas of the **brain.** Sensory input may be stored in memory, and motor impulses may be sent out to effectors (muscles or glands).

The brain and spinal cord make up the central nervous system. The major regions of the brain are the forebrain, the midbrain, and the hindbrain. Marked functional advances and structural changes in these regions have occurred among the vertebrates during the course of evolution. For example, in non-mammalian vertebrates, each of these regions of the brain receives input from one of the major sensory channels: the forebrain receives chemosensory information from the nose, the midbrain receives visual information from the eye, and the hindbrain receives information from the ear. However, in mammals, the portions of these inputs that are relevant to conscious and voluntary activities have been rerouted to the forebrain, which assumes a dominant role in behavioral coordination.

Objectives IIIIIIIIIIIIIIIIIIIIIIIIIIIIIIIIIIII

☐ Distinguish among the parts of the forebrain, midbrain, and hindbrain.

☐ State the functions of the major components of the forebrain, midbrain, and hindbrain.

☐ Describe how the brain processes information from sensory receptors.

Procedure IIIIIIIIIIIIIIIIIIIIIIIIIIIIIIIIIIIIIIIIIIII

1. On the demonstration bench you will find several containers of sheep brains. These specimens are extremely delicate and you should be very careful in handling them. Obtain one of the brains and study it with your partner.

2. Examine the parts of the brain while referring to Figure 41K-1 and Table 41K-1. The boldface letters in the table refer to lettered structures in the figure. Label the diagram as you identify each part.

   a. *Name the part of the brain concerned with the following functions or activities:*

      Controlling body temperature _____

      Coordinating equilibrium _____

Controlling breathing _____

Intelligence _____

Sensory perception _____

Heart rate _____

Relaying impulses between lower brain centers and the cerebrum _____

Figure 41K-1 *The sheep's brain.*   b _____

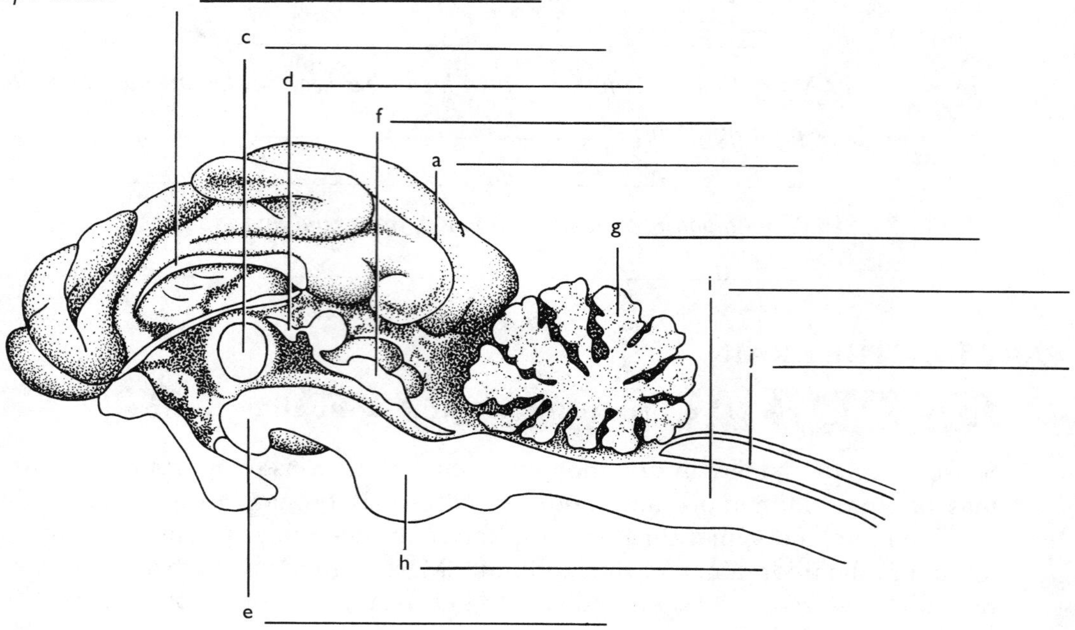

Table 41K-1 The Parts of the Mammalian Brain

Name	Location/Composition/Structure	Functions
Prosencephalon–Telencephalon Forebrain—Anterior Area **a** Cerebral hemispheres (olfactory lobes, basal nuclei, and cerebral cortex)	Most dorsal and anterior portion of brain. Most prominent and well-developed part of mammalian brain. Two lateral hemispheres. Convolutions increase surface area.	Highest integration center of the nervous system. Seat of psychic functions such as consciousness, sensation, perception, memory, and judgment (in humans).
Cortex (part of cerebral hemispheres)	Outer gray matter of cerebrum contains cell bodies; inner white matter composed of myelinated fibers and organized fiber tracts that connect cell masses in cerebral cortex to other parts of the brain.	Controls speech and voluntary movements. Functions may be mapped by studies of accidental injuries. Much of cortex is "silent," involved in associative activities. Visual and auditory information is projected to definite areas of the cortex.
b Corpus callosum	Located beneath the middle portion of cortex.	Fiber tracts interconnect the two halves of the cerebral hemispheres.

(continued next page)

Table 41K-1 *(continued)*

Name	Location/Composition/Structure	Functions
Olfactory bulbs (oldest part of the forebrain)	Much reduced in mammals, but they lead to an older, more primitive part of the telencephalon where sensory information from the nose is integrated.	Integrate olfactory sensory information.
Lateral ventricles	Cavities in the cerebral hemispheres.	Contain cerebrospinal fluid.
Prosencephalon–Diencephalon Forebrain—Posterior Area		
c Thalamus	Located posterior to cerebrum; forms the side walls of this more posterior region of the forebrain	"Way station" for information traveling from lower brain centers to cerebrum. Visual and auditory information relayed to cortex in this area in mammals.
d Epithalamus	Non-nervous choroid *plexus* (tissue layer) Site of pineal gland.	Produces the cerebrospinal fluid that fills the central cavities of the brain and spinal cord. Pineal gland is photoreceptive in some vertebrates. Produces melatonin, a hormone affecting rhythmic phenomena.
e Hypothalamus	Floor of the brain posterior to the cerebrum and below the thalamus. Median eminence Optic chiasma	Controls visceral nervous system, particularly emotions, temperature regulation, sleep, water balance, food and water intake, metabolic activity, and reproductive activity. Endocrine gland—produces neurohormones that regulate the anterior pituitary and other body functions. Optic nerve enters the brain through this area.
Mesencephalon Midbrain		
f Tectum	*Corpora quadrigemini*—four small bodies, two anterior and two posterior, constitute the midbrain roof. The first two correspond to a massive brain area, called the *optic lobes*, in nonmammalian vertebrates.	Anterior bodies associated with pupillary reflexes. Posterior bodies concerned with auditory reflexes (e.g., pricking of a dog's ears).
Tegmentum	Floor and sides of the midbrain.	Relays motor fibers and information traversing the ventral brainstem, including pyramidal tracts which relay information directly from the cerebral cortex to the spinal column. Controls a variety of involuntary processes.

(continued next page)

Table 41K-1   *(continued)*

Name	Location/Composition/Structure	Functions
Rhombencephalon–Metencephalon Hindbrain **g** Cerebellum	Anterior part of hindbrain; relatively large, rounded structure with a convoluted surface. In mammals the brain is folded so that the cerebellum lies directly behind the cerebrum. Surface contains cell bodies forming gray matter. White matter under the surface contains nerve fibers connecting the medulla and other areas of the brain.	Concerned with equilibrium, posture, and movement. Receives input from most sense organs, including proprioceptors in joints and the inner ear. Keeps track of the orientation of the body in space and the degree of contraction of skeletal muscles. Fine-tunes movements initiated in other areas of the brain. Injury to cerebellum results in impairment of muscular coordination but not in paralysis.
h Pons	Anterior end of medulla. Composed of thick bundles of myelinated (white) nerve fibers that carry impulses from one side of the cerebellum to the other in mammals.	Fiber tracts coordinate movements on both sides of the body.
Rhombencephalon–Myelencephalon **i** Medulla oblongata (begins in the metencephalon)	Located beneath and posterior to cerebellum; connects the spinal cord and midbrain. Roof has a second choroid plexus **j**	Contains nerve centers that control vital, largely subconscious activities such as respiration, heart rate, constriction and dilation of blood vessels, swallowing, and vomiting. Choroid plexus secretes cerebrospinal fluid.

Laboratory Review Questions and Problems

1. Define sensors (affectors) and effectors. Describe their relations to control systems in animals.

2. The structural units of the nervous system are neurons (and associated cells), but the functional units are synapses. What does this statement mean?

3. How are the senses of taste and smell alike? How are they different?

4. How is information from the inner ear related to information supplied by proprioceptors? Where in the brain is information from these sources integrated?

5. Identify the part of the brain associated with each of the following functions:

Vision _____

Smell _____

Hearing _____

Speech _____

Thought and voluntary
behavior _____

Visceral functions and control
of anterior pituitary _____

Respiration, heart rate, vasoconstriction
and vasodilation _____

Equilibrium, posture, and fine control of
voluntary movements _____

Behavior

OVERVIEW

Much of your behavior is externally directed activity, often elicited by specific changes in the environment. All cells are sensitive to external changes. However, the range of responses available to single cells (including single-celled organisms) is relatively small. In multicellular animals, an array of organs—the **sensors**—are specialized for transducing external stimuli into nerve impulses. These impulses are carried to a central nervous control system where they are evaluated against a spectrum of genetically determined response possibilities, some of which may be modified by learning acquired through experience. The outcome of this evaluation may lead to a series of nerve impulses carried through the motor fibers to **effectors**—the response organs. The simplest form of this type of behavior is a **reflex** such as the "knee jerk," a fixed or stereotyped response that is not modified by experience. Such behaviors are "instinctive." In more complex situations, the possible responses to a stimulus can range from pursuit to toleration to escape and avoidance. The choice and strength of the response are modified to a greater or lesser extent by learning and by immediate experience.

Instead of being tied directly to external stimuli, some behaviors are related to internal needs—the **motivation** or drive state of the organism. Hunger, for instance, is a drive state with no direct relation to external conditions. It appears at physiologically determined and, sometimes, socially conditioned times and results in searching behavior ("appetitive behavior"). When appropriate food is found, it serves as an effective stimulus leading to eating ("consummatory behavior"). While many consummatory activities are fixed ("instinctive"), the situations and places where the appropriate stimulus can be found, what constitutes an appropriate stimulus, and what constitutes appropriate responses are modified by experience ("learning").

The study of behavior has developed along two major lines. In Europe, a group of scientists, often called ethologists, first studied animals in the field in relatively natural surroundings and came to emphasize the relatively fixed or innate components of behavior. In the United States, behavioral studies by comparative psychologists focused on experiments with laboratory animals (primarily the rat) and emphasized the relatively flexible or learned components of behavior. In recent years, the contributions of both field and laboratory approaches have been synthesized into the field of animal behavior.

STUDENT PREPARATION

Prepare for this laboratory by reading the text pages indicated by your instructor. Familiarizing yourself in advance with the information and procedures covered in this laboratory will give you a better understanding of the material and improve your efficiency. Review the discussion of behavioral response systems in Laboratory 34.

EXERCISE A | Reactions of Isopods to Light and Humidity

Pill bugs, "rolly pollies," or slaters (isopods) are terrestrial crustaceans (arthropods) that you have likely found under rocks or boards—moist places with little light. They locate themselves in their preferred dark, moist habitat by two "instinctive" behaviors: a *kinesis* and a *taxis*.

A **kinesis** is a relatively fixed locomotor behavior in which the organism does *not* orient its body with respect to the stimulus. Instead, a combination of random turning and adjusting the speed of locomotion according to the intensity of a stimulus can result in an organism moving toward or away from the stimulus. For example, for pill bugs, a dry environment is a stimulus for increased activity, which continues until the animal reaches the damper environment it prefers. The pill bug's diminished activity in a moist habitat tends to keep it from wandering away from optimum conditions.

A **taxis** is a relatively fixed locomotor behavior in which movement is oriented—the animal heads either toward or away from the stimulus and steers according to its location. Pill bugs, assisted by photoreceptors (ocelli) sensitive to the level of illumination, make directed movements away from light— they exhibit **negative phototaxis.** Thus, individuals move away from light toward a dark environment.

In this experiment, you will study the interaction of a kinesis and a taxis in the behavior of the pill bug.

Objectives ||||||||||||||||||||||||||||||||||||||||||||

☐ Observe the locomotor behavior of isopods in relation to environmental gradients in light and humidity.

☐ Determine whether light or humidity is the stronger stimulus for maintaining the pill bug in its chosen habitat.

Procedure ||||||||||||||||||||||||||||||||||||||||||||

1. Work in pairs. Obtain two opaque paper tops from pint ice cream containers. Cut several ⋀-shaped notches in the rim and place the tops at the ends of an aluminum tray or clean dissecting pan. Place a moist piece of paper towel under one box top and a dry piece of towel under the other. Locate the tray so that it is illuminated evenly (you may use a light source, but be careful not to overheat the organisms). Place 10 pill bugs in the center of the tray and allow them to wander for 30 minutes. Form a hypothesis that predicts how the pill bugs will behave.

   HYPOTHESIS:

   NULL HYPOTHESIS:

   Where do you **predict** you will find the larger number of pill bugs after 30 minutes? Why?

   What is the **independent variable**?

   What is the **dependent variable**?

   Observe the pill bugs and record the number of exits from each box top. At the end of 30 minutes, count the number of animals that are wandering and the number under each box.

Number wandering _____ Number in moist habitat _____ Number in dry

habitat _____

Do your results support your hypothesis? _____ *Your null hypothesis?* _____

2. Repeat this experiment, but cover the tray with aluminum foil so that the animals are in darkness. Form a hypothesis that predicts how the pill bugs will behave.

    HYPOTHESIS:

    NULL HYPOTHESIS:

    Where do you **predict** you will find the larger number of pill bugs after 30 minutes? Why?

    What is the **independent variable**?

    What is the **dependent variable**?

    Record your results at the end of 30 minutes.

    Number wandering _____ Number in moist habitat _____ Number in dry

    habitat _____

    Do your results support your hypothesis? _____ *Your null hypothesis?* _____

    a. *From your observations in steps 1 and 2, do isopods demonstrate behavioral responses that lead*

       them to moist or dry environments? _____

    b. *Does movement toward or away from a moist environment appear to be affected by light?* _____

       If so, how? _____

3. Place a dry piece of paper under both a clear plastic box top and an opaque paper box top (both notched). Form a hypothesis that predicts how the pill bugs will behave.

    HYPOTHESIS:

    NULL HYPOTHESIS:

    Where do you **predict** you will find the larger number of pill bugs after 30 minutes? Why?

    What is the **independent variable**?

    What is the **dependent variable**?

    Record your results after 30 minutes.

    Number wandering _____ Number in light habitat _____ Number in dark

    habitat _____

    Do your results support your hypothesis? _____ *Your null hypothesis?* _____

4. Now place a piece of a wet paper towel under both the plastic and paper box tops and observe the results at the end of 30 minutes.

   Number wandering _____ Number in light habitat _____ Number in dark habitat _____

   c. *From your observations in steps 3 and 4, do isopods show behavioral responses that demonstrate a preference for dark or light environments?* _____

   d. *Do your results from this experiment support your hypothesis from experiment 3?*

   e. *Does movement toward or away from a dark environment appear to be affected by humidity?* _____ *If so, how?* _____

5. Place a piece of moist paper towel under the clear plastic box top and a piece of dry paper under one of the paper box tops. Form a hypothesis that predicts how the pill bugs will behave.

   HYPOTHESIS:

   NULL HYPOTHESIS:

   Where do you **predict** you will find the larger number of pill bugs after 30 minutes? Why?

   What is the **independent variable**?
   What is the **dependent variable**?

   Record your results at the end of 30 minutes.
   Number wandering _____ Number in moist-light habitat _____ Number in dry-dark habitat _____

   Do your results support your hypothesis? _____ *Your null hypothesis?* _____

6. Return the isopods to the container from which you obtained them at the beginning of the experiment.

   f. *From your observations would you conclude that pill bugs show a taxis or a kinesis with respect to light?* _____ *To humidity?* _____

   g. *Which combination of experimental results (steps 1–5) influenced your conclusion about pill bug behavioral responses to light?* _____

   _____

   To humidity? _____

   h. *Under which conditions did you observe the most wandering?* _____

   _____

   i. *Which sensors are you testing in these experiments?* _____

   _____

EXERCISE B | **Courtship Behavior in Fruit Flies**

In Exercise A, you studied responses to the physical environment. In this exercise you will observe the behavioral interactions of male and female fruit flies, *Drosophila melanogaster,* during courtship and mating. Reproductive behavior is based on a reproductive "drive" in adult flies. It includes searching or appetitive behavior and, once another fly is located, a sequence of consummatory events with specific responses from both animals. These events form a "chain" in which the performance of one activity elicits the next until it is completed by copulation or broken off for some reason (a male attempts to court another male or the female is unreceptive). The events in this chain are fixed or stereotyped and the various motor patterns appear without practice, thus they constitute a largely instinctive series of events.

The first component of the behavioral chain is **orientation.** The male stands close to and often behind the female. He remains facing her as he circles around her body. The second pattern is **vibration.** The male extends the wing nearest the female's head until it is perpendicular to the axis of his body and vibrates it vertically; he may continue circling as he vibrates his wing. **Licking** is the third component. While continuing to circle the female and to vibrate his extended wing, the male touches the female's genitalia with his extended proboscis. Next the male tries to mount the female: he places his first pair of legs on the female's back and tries to bring the tip of his abdomen into contact with her genital organs in **attempted copulation.** If the female is unreceptive, she may show several behaviors, including **ignoring** his advances while continuing her current activities, **depressing** her abdomen to prevent licking of the genitalia by the male, while keeping her wings in place to prevent mounting, or **decamping**—running, jumping, or flying away from the male. If, however, she is ready to mate, she will spread her wings and extend her genitalia and copulation ensues. These events are shown diagrammatically in Figure 42B-1.

ⅠⅠⅠⅠⅠ **Objectives** ⅠⅠ

☐ Observe and describe the behaviors constituting courtship and mating in the fruit fly.

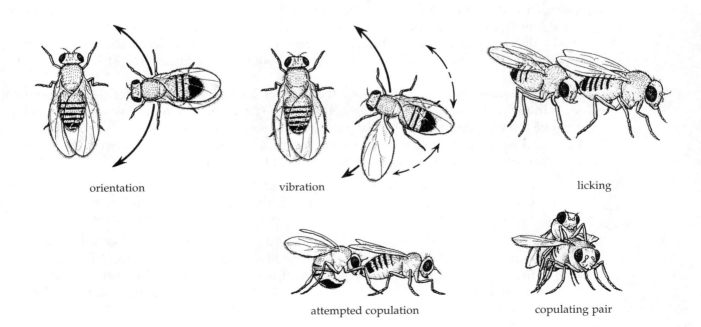

orientation   vibration   licking

attempted copulation   copulating pair

Figure 42B-1 *Movements during courtship behavior in fruit flies.*

ııııı Procedure ıııııııııııııııııııııııııııııııııııııı

Work in pairs; take turns making observations and recording results.

1. On a separate piece of notebook paper, construct a table resembling Table 42B-1, including enough intervals to continue recording observations of a pair of flies for 10 minutes.

Table 42B-1   Sample Table for Recording Observations of *Drosophila* Behavior

Minutes/Seconds		Pair A	Pair B	Pair C	Pair D	Pair E
01	10					
	20					
	30					
	40					
	50					
	60					
02	10					
	20					

2. Obtain a vial containing five male flies and one containing five virgin female flies. Using a hand lens or dissecting microscope (low power), study the flies in each vial to be sure you can distinguish males and females (males are somewhat smaller and have a rounded abdomen; females are larger with a pointed, black-tipped abdomen; see Figure 42B-2).

3. Obtain a mating chamber and remove the plunger. You must now transfer the flies from both vials to this chamber. Gently tap the bottom of the first vial on a notebook or pad. Quickly remove the plug and invert the vial over the mating chamber. Grasp the chamber and vial in one hand, using your fingers to surround and close the gap between the two tubes. Tap the chamber and vial on the pad to displace the flies from the vial.

4. Remove the first vial and, holding your hand over the mating chamber, repeat this process with the second vial. Reinsert the plunger to 1 inch below the screening on the opposite end of the chamber.

5. One student should observe the flies immediately to see the initial responses of a male encountering an unmated female; the second student records results in the data chart. Pick a single (focal) male and record his behavior at 10-second intervals, using a shorthand notation: **O** for orientation, **V** for vibration, **L** for licking, **AT** for attempted copulation, and **C** for copulation.

6. Describe any other behaviors in the pair of flies you observe and record these observations on your data sheet. These may include: *waving* (one wing spread outward from the body and

Figure 42B-2   *(a) Male and*
(b) female fruit flies.

Dorsal view   Ventral view of abdomen

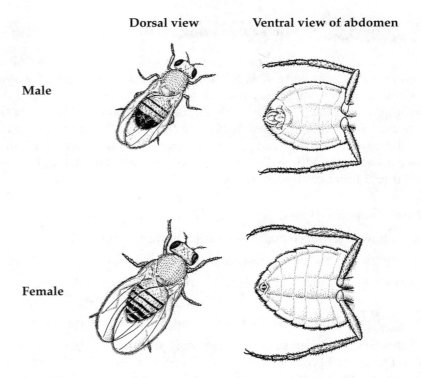

Male

Female

relaxed without vibration); *scissors* (opening and closing both wings); *fluttering* (a repelling movement of the wings made by females in which wings are opened slightly and vibrated rapidly); *tapping* (a foreleg motion motion by the male in which the limb is extended and then brought downward to touch the courted individual); *extruding* (an unreceptive female may extend her vaginal plates).

7. Describe the responses of unreceptive females and of males courted by another male. Make drawings to show the positions of legs, wing, head, and abdomen in each case.

8. Change roles as observer and recorder. Move on to other individuals or pairs if the male you first chose to observe is not cooperating or has completed the behavioral sequence. Note that many of the behaviors occur rapidly and together, so that careful observation is required. Continue to record data in your data table. Record observations of other behaviors as in steps 6 and 7. Be sure that both you and your partner observe all behaviors.

   a. *How long does copulation last? Time several pairs, collect class data, and record these data: shortest*

      period: _____ *minutes; longest period:* _____ *minutes; average (mean):* _____ *minutes.*

9. After completing your study, transfer the flies from the mating chamber to the morgue (a bottle of alcohol or oil). This will prevent them from populating the lab.

   b. *What evidence did you see of the exchange of signals or stimuli between the sexes?*

      _____

   c. *Which sex takes the more active role in courtship in* Drosophila? _____

   d. *Did you see any signs of competition among males for mates? Describe these signs.*

      _____

   e. *On a separate sheet of paper, outline an experiment to detect the role of various sensors (chemo-receptors, photoreceptors, mechanoreceptors) in courtship in the fruit fly.*

EXERCISE C | Social Behavior in Crickets

Studying crickets (*Acheta*) offers opportunities to observe many different behaviors, including aggression among males, courtship, and territorial behavior. Aggressive behavior, which is related to establishing dominance hierarchies, may include actions such as head-to-head contact or "fighting" with mandibles, chasing, and kicking with hind legs. The submissive male usually flees and avoids the more dominant male. Once dominance is established, a female will mate with the dominant male.

Territorial behavior can also be affected by dominance. Typically, a male in his home territory will have an advantage if challenged by another male. However, the home territory advantage can be affected by large differences in dominance.

⦙⦙⦙⦙⦙ Objectives ⦙⦙⦙⦙⦙⦙⦙⦙⦙⦙⦙⦙⦙⦙⦙⦙⦙⦙⦙⦙⦙⦙⦙⦙⦙⦙⦙⦙⦙⦙⦙⦙⦙⦙⦙⦙⦙⦙

☐ Learn how to observe and record behavioral phenomena.

☐ Discriminate between aggressive behavior and other types of behavior.

☐ Interpret observations and identify stimuli causing a behavior.

PART I   Aggressive Behavior and Social Dominance

In some species, aggressive behavior may spatially distribute individuals who maintain exclusive areas with recognized boundaries. In other species, aggressive behavior may distribute individuals to areas of dominance within which other individuals are tolerated if they behave submissively. At low densities, there may be few interactions, most occurring only when two animals come together by random movement. When density increases, encounters increase, and often a social hierarchy develops. Based on the outcomes of these encounters, animals arrange themselves in a more or less linear manner from the most dominant to the most subordinate. When resources (food, space, etc.) become scarce, it is the subordinates that must find relief and that frequently exhibit dispersal and searching (appetitive) behavior.

⦙⦙⦙⦙⦙ Procedure ⦙⦙⦙

Your instructor has isolated female crickets from male crickets. Males have been placed in individual containers for 1 week. Each male cricket has been marked with either red (R), blue (B), yellow (Y), or green (G) paint applied to its thorax so that individual males can be identified. Females can be distinguished from males by the presence of a long, narrow ovipositor at the rear of the abdomen (Figure 42C-1).

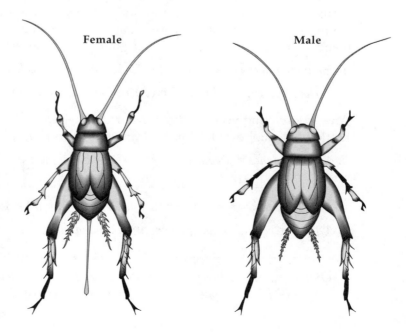

Figure 42C-1   (a) *Female cricket,* Acheta, *with a long ovipositor protruding from the abdomen;* (b) *male cricket.*

Note that containers and lids have also been marked with matching colors of paint so that each cricket's "home" can be identified; crickets should always be returned to the same home. Retrieve them by scooping them up in their container. Avoid trying to catch them in your hands. Handle the crickets gently and only when necessary. During observations, limit your movements in the vicinity of the crickets. Reduced room lighting is also helpful.

1. Work in groups of four. Obtain a plastic shoe box to serve as an observation cage. Obtain four male crickets, one of each color marking. At least two students should be assigned to evaluating and scoring aggressive behavior (steps 2–4). Another member of the group should be assigned to observe and record the use of "houses" by individual male crickets (step 5) as data on aggressive behavior is collected.

2. Place an open-end matchbox (the cricket "house") into the observation cage and then add two male crickets. Begin observing their aggressive and territorial behaviors. Score the level of intensity according to the criteria in Table 42C-1.

Table 42C-1   Levels of Aggression in Crickets

Level 1	Contact terminated without clear dominance, no apparent aggression, and no apparent retreat.
Level 2	Contact terminated by retreat with no apparent aggression.
Level 3	Contact terminated by retreat after mild to moderate aggression (one-sided) or mild reciprocal aggression.
Level 4	Contact terminated after moderate to intense reciprocal aggression.
Level 5	Contact terminated only after sustained aggression.

From R. D. Alexander, "Aggressiveness, territoriality and sexual behavior in field crickets (Orthoptera; Gyrillidae)," *Behavior*, vol. 17, p. 130, 1961.

As you can see, these descriptions of behavior are very generalized. You will find many gradations of aggressive behavior among individuals and will need to make decisions about how to evaluate the behavior of your crickets.

3. Each male cricket should be paired with every other male cricket (different colors) by rotation. Allow each pair of crickets to interact for **10 minutes.** Record all data in Tables 42C-2 and 42C-3. (See steps 4 and 5 before you begin your tests.)

4. Record data on aggressive encounters in Table 42C-2. An encounter should be recorded only when some part of the cricket touches a part of another. At least two students should identify (by agreement) the intensity of encounters according to the levels given in Table 42C-1. Record data for level 1 aggressive encounters by using tic marks (e.g., 卌 represents five

Table 42C-2   Aggressive Encounters

Pairing	Level 1	Level 2	Level 3	Level 4	Level 5
Y vs. G					
Y vs. B					
Y vs. R					
G vs. B					
G vs. R					
R vs. B					

encounters). For level 2 to level 5 encounters, indicate the "color letter" of the winner for each encounter in the appropriate box. For example, if you are pairing Y vs. G and there were four encounters at level 2, you might have "YGGY" in that box, indicating that each cricket "won" twice.

5. Matchboxes serve as "cricket houses" and are usually occupied by dominant crickets. One student in your group should be assigned the task of observing and recording the use of the "houses" by individual male crickets during successive trials. Record data on use of houses in Table 42C-3. Put the "color letter" for the individual that uses a house in the appropriate box for each use. For example, if you are pairing Y vs. G and the Y cricket used the house three times while the G cricket used it only twice, you might have YYGYG or YYYGG (or some other sequence of use) recorded.

Table 42C-3   Use of Cricket Houses

Pairing	Cricket Using House
Y vs. G	
Y vs. B	
Y vs. R	
G vs. B	
G vs. R	
R vs. B	

a. Did the dominant cricket in each trial (most wins) use the matchbox house more often?

_____

6. Determine the dominant cricket for your group by using the following scoring format:

A. **Each cricket** in an encounter (this means *both* crickets, not just the winner) should get points for encounters at particular levels for *all* pairings as follows:

Encounter Level	Points Scored
Level 1	1
Level 2	2
Level 3	3
Level 4	4
Level 5	5

B. **Each cricket** should get 1 point for using a house each time it is used as recorded in Table 42C-3.

C. For each aggressive encounter "**win**" (Table 42C-2), add 1 point to the total for each color of cricket as recorded in Table 42C-2.

D. Use Table 42C-4 to enter the total point scores from *all* pairings in *all* boxes in Tables 42C-2 and 42C-3. See example on the next page.

Table 42C-4   Cricket Dominance

Cricket	Total Point Score
Yellow	
Green	
Blue	
Red	

EXAMPLE. Just looking at a single pairing, add points from A, B, and C below.

A. Example (from Table 42C-2):

Give points to each cricket for each encounter.

Pairing	Level 1	Level 2	Level 3	Level 4	Level 5
Y vs. G	‖	Y	G, Y	G, G	

    Green = 18 points
    Yellow = 18 points

Do this for each pairing in the entire table to give a single cumulative score for each color of cricket.

B. Example (from Table 42C-3):

Give points to each cricket that used the house.

Pairing	Cricket Using House
Y vs. G	G G

    Green = 2 points
    Yellow = 0 points

Do this for each pairing in the entire table to give a single cumulative score for each color of cricket.

C. Winner points (from Table 42C-2) (see A above for this example):

    Green = 3 points
    Yellow = 2 points

D. Total points (A + B + C):

    Green = 23 points
    Yellow = 20 points

The dominant cricket, if this were the only pairing, would be the green cricket.

b. *What happens when two crickets encounter one another?*_____

c. *Do the crickets' songs show any changes in rate or loudness?* _____

d. *What part of the anatomy does the cricket use for chirping?* _____

e. *Do dominant males "call" more often than subordinate males?* _____

f. *Does the response of the cricket depend on the direction from which another cricket approaches?*

_____

g. *Was a hierarchy established?* _____

h. *How long does it take for a stable hierarchy to become established?* _____

i. *Which had more encounters—the dominant or the subordinate male?* _____

PART 2 **Aggressive Behavior, Dominance, and Courtship Among Male and Female Crickets**

Courtship and copulatory behaviors exhibited by crickets include the following sequence of events. A period of courtship which begins as agonistic behavior, including aggressive movements, is usually followed by three to five sound pulses (chirps) per second. The female mounts the male from behind and the male attaches a white or clear *spermatophore* to her genitalia. The spermatophore is a small sperm-containing capsule at the end of a short stalk. After the transfer has occurred, there is a short period of postcopulatory behavior. The drying of the spermatophore creates pressure that forces the sperm out of the capsule into the female's reproductive tract, where her eggs are fertilized. The female frequently bites off the spermatophore after a short period of time. Courtship and copulation behavior usually takes less than an hour.

 Procedure

Pair the dominant male from your group with a female from the holding tank. Observe which sex is dominant. Use M or F to record the results of aggressive behavior in Table 42C-5.

Table 42C-5   Courtship Dominance

Level 1	Level 2	Level 3	Level 4	Level 5

a. Which sex appears to be dominant? _____

b. Was any courtship or reproductive activity apparent during the experiment? _____

c. Do female crickets chirp? _____

d. How does the position of the wings help produce the cricket's sounds? _____

_____

e. Was there a difference in position of the wings during the different kinds of calls? _____

f. Does the male's song or chirp change when he encounters a female? _____

g. What stimuli are useful for the male to use in distinguishing females from males? _____

_____

h. Did you observe spermatophore transfer? _____

 EXERCISE D | **Learning in the Mealworm**

Learning is a process that "manifests itself by adaptive changes in individual behavior as a result of experience" (Thorpe, 1962). Several classifications of learning are possible; according to the scheme developed by Thorpe, three broad categories are recognized: habituation, conditioning, and insight learning.

In the process of **habituation,** an organism learns *not* to respond to stimuli that lack meaning. For example, a bird nesting near a sidewalk will cease fear or flight reactions to every person who walks past. **Conditioning** may be of at least two types: "classical" and "instrumental."

Classical conditioning involves the repeated, paired presentation of two unrelated stimuli: the **unconditioned stimulus** naturally elicits a particular response (the **unconditioned response**), and the **conditioned stimulus** does not normally elicit this response. Classical conditioning is said to occur when the conditioned stimulus alone elicits the response that had initially been elicited only by the unconditioned stimulus. This response to the conditioned stimulus is called the **conditioned response.** One of the first and most famous experiments in classical conditioning was described by Ivan Pavlov early in this century. When Pavlov presented dogs with food (the unconditioned stimulus), they salivated (the unconditioned response). When the sound of a ringing bell (the conditioned stimulus) was repeatedly paired with the presentation of the food, the dogs eventually began to salivate (the conditioned response) at the sound of the bell alone. This type of learning is commonly called *learning by association.*

Instrumental or **operant conditioning** occurs when a behavior not originally connected with a particular reward results in receiving that reward. In this case, the response (the behavior) *precedes* the stimulus (the reward). As an example, a hungry rat is placed in a box (called a "Skinner box") equipped with bars, which, when pressed, will result in the release of food. At first, the rat does not associate the act of pressing the bars with food. However, as the rat explores the box actively, searching for food, it may bump the bar; this action leads to a small pellet of food being delivered down a chute into the cage. The animal quickly learns that what was at first an accidental activity—pressing the bar—results in a reward of food, and it begins to press the bar to obtain food. This type of conditioning is found in *trial-and-error* behavior—and in *play* in young animals that depend upon learning for much of their behavioral repertoire.

Insight learning is an adaptive reorganization of experience—forming relationships without prior exploration or manipulation. For example, a hungry chimpanzee is presented with a bunch of bananas hanging out of reach, a short stick, and several boxes. If the chimp stacks the boxes to reach the food with the stick, the animal has had an "Aha!" experience in which it has determined the solution directly—it has used insight to organize useful relations of objects in its environment.

In this exercise you will use a "T-maze" to study conditioning in a trial-and-error behavior involving conditioning.

IIIII Objectives IIIIIIIIIIIIIIIIIIIIIIIIIIIIIIIIIIIIIIIIIIII

☐ Condition or "train" mealworms to make a consistent turning response using negative phototaxis.

IIIII Procedure IIIIIIIIIIIIIIIIIIIIIIIIIIIIIIIIIIIIIIIIIIII

1. Work in pairs. Obtain a shallow black box and T-maze. Place the box in a windowless or dimly lit room. Flip a coin to determine which arm of the maze will constitute a "correct" response: heads chooses the left arm, tails the right.

2. Place a mealworm larva into the alley of the maze. (The larvae may be marked with colored dots of model paint to allow individuals to be identified; record the colors used on your larva.) Shine a flashlight behind the animal to drive the larva up the alley toward the arms of the maze—mealworms are negatively phototactic and will move away from the light. When the animal reaches the choice point where the maze branches, it may turn either way. If the larva makes a "correct" choice, turn off the light and allow the animal to remain undisturbed for 1 minute. If the larva makes an incorrect choice, move the light to face the worm in the incorrect arm and keep the light on until the animal turns around. When the larva enters the correct arm, turn the light off and allow the worm to remain in the darkened arm for 1 minute.

3. Repeat this procedure 10 times, then return the worm to its home container for at least an hour. With closely spaced trial sequences, the larva may habituate to the light (learn not to respond), so series of no more than 10 trials must be separated by at least an hour. On a separate piece of paper, record the choice made by your larva during each trial. Include your data sheet in your laboratory notebook.

4. Repeat the training trials until the larva responds according to a preselected level—a correct response in 17 of 20 trials is reasonable (remember to keep a record of each trial). Do not be discouraged if it takes more than a hundred trials to reach this criterion level. You will need to return daily until your animal "passes." Alternatively, several sections may condition the same larva through the week, sharing data at the end.

a. *What other stimuli could be used to make the mealworm choose between the arms of the maze?*

_____

b. *Would you expect the larva to reach the criterion level more or less rapidly than a mammal such as the rat?* _____ *Why?* _____

_____

Laboratory Review Questions and Problems

1. Would a kinesis or a taxis be more effective in leading animals to congregate in optimal areas of their habitat? Why?

2. If an animal has been deprived of water for a period that is longer than the normal interval between drinking, what type of behavior would you expect to see?

   _____

   What would the animal do when presented with water? _____

   What type of behavior would this response constitute? _____

3. Give an example, based on your own experiences, of each of the following types of learning.
   Habituation

   Conditioning

   Insight learning

4. Most animals have a home range or a territory to which they confine their activities. What advantages would this offer in relation to learning?

5. Among animals that live in groups, individuals strive for dominance and aggressive behavior is common. Do you see this type of behavior in humans? Give examples.

6. Courtship between male and female animals usually takes the form of a ritual characterized by specific behaviors. Of what advantage is this for the individuals? For the species?

Communities and Ecosystems

OVERVIEW

Two of the most complex levels in the hierarchy of biological organization are the community and the ecosystem. A **community** consists of all the populations of species living within a particular locality. An **ecosystem** consists of one or more related communities plus the abiotic (nonliving) components that affect them, such as weather conditions and the type of soil present. Ecosystems may be large or small, with the boundaries variously defined. The earth can be considered a single ecosystem, as can a mountain valley or a fishtank in your laboratory. Ecosystems display characteristics other than those contributed by their separate components. Thus ecosystems are often studied in terms of their processes or products rather than by dissecting out the effects of particular organisms or other factors.

The nature of a geographically defined ecosystem is partly determined by its location and the biome with which it is associated. **Biomes** are geographic areas of the earth defined by dominant vegetation types that result from distinct patterns of rainfall and temperature.

During this laboratory period, you will study three factors important to understanding communities and ecosystems: **competition** among species sharing resources; **diversity,** the relative abundance of different types of organisms; and the relationship of **abiotic factors,** such as weather conditions, to particular types of biomes.

STUDENT PREPARATION

Prepare for this laboratory by reading the text pages indicated by your instructor. Familiarizing yourself in advance with the information and procedures covered in this laboratory will give you a better understanding of the material and improve your efficiency.

EXERCISE A | **Observing Competition Between Species Sharing Resources**

In a community, competition may take place among species sharing resources, particularly when these resources are in short supply relative to the demands of the organisms. In this exercise, you will observe competition between two species of *Paramecium* growing in mixed cultures.

If a species exhibits normal growth in a pure culture, but does not grow as well when another species is introduced, this negative effect is the result of competition. One possible outcome of competition is that both species will coexist, but at lowered densities; a second possible outcome is that one of the two species will become extinct. Extinction due to the effects of competition is called **competitive exclusion.**

ⅠⅠⅠⅠⅠ Objectives ⅠⅠⅠⅠⅠⅠⅠⅠⅠⅠⅠⅠⅠⅠⅠⅠⅠⅠⅠⅠⅠⅠⅠⅠⅠⅠⅠⅠⅠ

☐ Observe the effect of competition between two species of *Paramecium* by comparing their relative abundance over time in pure cultures and in mixed cultures.

ⅠⅠⅠⅠⅠ Procedure ⅠⅠⅠⅠⅠⅠⅠⅠⅠⅠⅠⅠⅠⅠⅠⅠⅠⅠⅠⅠⅠⅠⅠⅠⅠⅠⅠⅠⅠ

Three sets of *Paramecium* cultures have been prepared: the first set is one week old, the second set is two weeks old, and the third set is three weeks old. Within each set, cultures are labeled for the species of *Paramecium* they contain. Culture A contains only *Paramecium aurelia;* culture B, only *Paramecium caudatum;* culture C contains both species.

P. aurelia and P. caudatum can easily be distinguished using high power (40×) on your light microscope. *P. caudatum* is at least twice the size of *P. aurelia* and has a strikingly large macronucleus.

Work in groups of three students. Each student in the group should select one set of cultures of a certain age. Enter the age (number of weeks cultured) of your culture in Table 43A-1. Follow the steps below to take density estimates and record data.

Table 43A-1   Raw Data for _____ Week-Old Culture

	Number of Organisms in 0.01 ml of Culture Medium		
	(A) *P. aurelia*	(B) *P. caudatum*	(C) *P. caudatum/P. aurelia*
Sample 1			/
Sample 2			/
Average			/

1. To estimate the densities of the species in each culture flask, take two samples using a capillary tube (or micropipette, if available) at a point just above the bottom of the culture flask. The capillary tube will deliver a drop of culture (volume approximately 0.01 ml) that should *not* be larger than the microscope's field of view at high power (40×). Place a drop of your *P. caudatum* culture on a slide (without a coverslip). Count the number of *P. caudatum* within your field of view and record this number in Table 43A-1 as sample 1.

2. Take a fresh drop from the same culture and repeat this procedure. Record your count in Table 43A-1 as sample 2.

3. Repeat steps 1 and 2 first for the *P. aurelia* culture and then for the mixed species culture in your set. In the mixed culture, you will need to count the number of individuals of each species separately.

4. Record the average of your two samples from each culture in Table 43A-1.

5. Enter the starting densities (given to you by your instructor) for the three sets of cultures in Table 43A-2.

6. Enter your averages for each culture in the data table that your instructor has placed on the blackboard. Calculate grand averages for your cultures: add all the average values obtained for these cultures (of the same age) by all students, then divide the sum by the number of students supplying data. Enter the grand averages in Table 43A-2.

7. Obtain the grand averages for the cultures of the other two ages from the class data table and record them in Table 43A-2. When you have completed this table, analyze the results of the competition experiment by answering the following questions.

Table 43A-2   Density Estimates: Summary of Data for Entire Class

	Grand Average Number of Organisms in 0.01 ml of Culture Medium			
	Starting Density	One Week	Two Weeks	Three Weeks
P. aurelia (pure)				
P. caudatum (pure)				
Mixed culture: P. caudatum				
P. aurelia				

a. *Which species living in pure culture showed the greatest increase in numbers during the three weeks?* _____

b. *How was this growth rate (change in number per unit of time) affected by the presence of a competitor?* _____
_____

c. *Based on your experimental results, what do you predict will be the final outcome of competition in the mixed culture? (Or, if you have already seen extinction of one species, which one became extinct?)* _____
_____

8. On a separate sheet of paper, make a line graph to show the class data for the effects of species competition over time. Label the horizontal X-axis "time (weeks)"; label the vertical Y-axis "number of organisms." Draw one line representing class data for the pure *P. caudatum* culture over the three-week period; draw another for the pure *P. aurelia* culture. Draw one line each for *P. caudatum* and *P. aurelia* growing in a mixed culture.

👁 EXERCISE B | Measuring the Diversity of a Community

One way of gaining an understanding of a community's structure is to measure its diversity. Diversity is a measure of how many kinds of organisms (numbers of species) and how many of each of these kinds (numbers of individuals) are present in a community. You can calculate a numerical value called a **diversity index** (also known as the Shannon index) from information obtained by taking quantitative samples of organisms from a community. To specify community diversity, this index takes into account both the kinds and the numbers of organisms present. The diversity index can be used to compare the ways in which various communities are structured.

ᴵᴵᴵᴵᴵ Objectives ᴵᴵᴵᴵᴵᴵᴵᴵᴵᴵᴵᴵᴵᴵᴵᴵᴵᴵᴵᴵᴵᴵᴵᴵᴵᴵᴵᴵᴵᴵᴵᴵᴵᴵᴵᴵᴵᴵ

☐ Estimate the diversity of a community of organisms using a diversity index.

ᴵᴵᴵᴵᴵ Procedure ᴵᴵᴵᴵᴵᴵᴵᴵᴵᴵᴵᴵᴵᴵᴵᴵᴵᴵᴵᴵᴵᴵᴵᴵᴵᴵᴵᴵᴵᴵᴵᴵᴵᴵᴵᴵᴵᴵ

Leaf litter from a forest, a woodland, or a riparian environment (an area along a river or stream bank) contains a variety of decomposers feeding on decaying organic material. Although the communities represented by these samples include populations of soil bacteria, fungi, and protozoans, we will, for simplicity's sake, consider only the larger organisms found in leaf litter: the invertebrates.

At some time before this laboratory period the following procedure was performed. Plastic bags containing leaf litter are emptied into a collection apparatus. Litter invertebrates, which are negatively phototactic (that is, move away from light), are forced from the collection area into the funnel by light (and also by heat) from a 100-watt bulb. From the funnel they drop into the Erlenmeyer flask below. Alcohol in the flask both kills and preserves these invertebrates for viewing under the dissecting microscope.

Work in groups of four. Your instructor will supply each of you with a sample of invertebrates from a different environment.

1. Identify each organism in your sample according to its taxonomic class. Your instructor will provide you with sketches of representative organisms of the kinds you might expect to find (see Table 43B-1). The following characteristics should prove helpful in distinguishing invertebrates in leaf litter.

   1. Number of pairs of antennae = 0, 1, or 2
   2. Types of eyes = simple or compound
   3. Number of pairs of wings = 0, 1, or 2
   4. Number of pairs of legs = 0, 6, 8, or more than 8
   5. External organization of the body.

Table 43B-1   Representative Invertebrates Found in Leaf Litter

Phylum	Class	Representative Groups
Nematoda*		Roundworms (nematodes)
Annelida	Oligochaeta	Earthworms
Mollusca	Gastropoda	Snails and slugs
Arthropoda	Insecta	Beetles, bugs, roaches, flies, bees, lice, fleas, collembolans, mosquitoes, and termites
	Arachnida	Spiders, mites, ticks, pseudoscorpions, and harvestmen (daddy longlegs and their relatives)
	Crustacea	Pill bugs and sow bugs
	Diplopoda	Millipedes
	Chilopoda	Centipedes

*Nematodes are difficult to classify. Do not attempt to do so.

2. Record your identifications in Table 43B-2. Add the information recorded by your other group members.

3. Enter your group's data on the class data chart that your instructor has placed on the blackboard. Record class data in Table 43B-3.

4. Using these data, calculate the diversity index (Shannon index), D, of each community using the following formula:

$$D = -\sum \left(\frac{n_i}{N} \times \log_{10} \frac{n_i}{N} \right)$$

The variables in this formula are as follows:

n_i The number of organisms in a given class (n_i for each class of organisms in Table 43B-3).

Table 43B-2   Invertebrates Identified by Individual Students

Phylum	Class	Number of Organisms Collected		
		Woodland	Riparian	Forest
Nematoda				
Annelida	Oligochaeta			
Mollusca	Gastropoda			
Arthropoda	Insecta			
	Arachnida			
	Crustacea			
	Diplopoda			
	Chilopoda			
Other				

Table 43B-3   Class Data for All Invertebrates Collected

Phylum	Class	Number of Organisms Collected		
		Woodland (n_i)	Riparian (n_i)	Forest (n_i)
Nematoda				
Annelida	Oligochaeta			
Mollusca	Gastropoda			
Arthropoda	Insecta			
	Arachnida			
	Crustacea			
	Diplopoda			
	Chilopoda			
Other				
Total (N)				

N The total number of organisms in the sample for the community (N for each community in Table 43B-3).

Σ This symbol, sigma, indicates that you must sum (add) all the expressions in brackets. (If there are four classes of invertebrates in a sample, you will add four terms.) This sum will be a negative number. You must multiply by the minus sign that precedes the sigma to get a positive number for the value of diversity, D. The larger the value of D, the greater is the diversity of the community.

For example, if 20 organisms from leaf litter of the woodland habitat were sampled and found to belong to four classes—8 from one class, 3 from a second, 5 from a third, and 4 from the fourth—then the calculation for diversity of the woodland community would be set up as follows:

$$D = -\left[\left(\frac{8}{20} \times \log_{10} \frac{8}{20}\right) + \left(\frac{3}{20} \times \log_{10} \frac{3}{10}\right) + \left(\frac{5}{20} \times \log_{10} \frac{5}{20}\right) + \left(\frac{4}{20} \times \log_{10} \frac{4}{20}\right)\right]$$

5. Calculate a diversity index for each of the three environments in the space below. Rank the environments, from most diverse to least diverse, using the diversity index, D.

a. From the terms in your calculation of the diversity index for each community, which class of invertebrates contributed most to the diversity of the sample? Woodland _____;

riparian _____; forest _____.

Which contributed least? Woodland _____; riparian _____;

forest _____.

b. Which environment had the greatest number of classes? _____

The least number of classes? _____

c. Which environment had the greatest number of organisms? _____

The least number of organisms? _____

d. What are some differences in the physical conditions of the environments, for example, moisture, temperature, light, and humidity, that might have accounted for the differences in diversity among the leaf-litter communities? _____

_____

_____

During the process of succession, the diversity of an ecosystem and its communities increases as the ecosystem approaches its climax (equilibrium stage in which the community is typical of the biome rather than of a stage in biome development). More simply stated, as an ecosystem matures, it becomes more diverse.

e. Which of the sampled environments represents the most mature ecosystem? _____

👁 **EXERCISE C**   **Using Climate Data as an Index to Vegetation**

Climate is one of the major determinants of the kind of vegetation that grows in a particular region of the earth. The type of vegetation, in turn, determines what animal life is present. Two important climatic factors that govern the distribution of biomes are temperature and rainfall. Each biome has a characteristic pattern of temperature and rainfall. In fact, you can predict the type of vegetation found in a region from monthly temperature and rainfall data.

Sample data from a weather station are shown in Figure 43C-1a. Temperature and rainfall data are more adequately displayed in a graph called a *climatogram* (Figure 43C-1b), where each point represents average temperature and average rainfall for a given month.

▌▌▌▌▌ **Objectives** ▌▌▌

☐ Predict the biome within which a geographic location lies, based on weather data.

		Station: Kano, Nigeria (Africa)											
		Jan	Feb	Mar	Apr	May	Jun	Jul	Aug	Sep	Oct	Nov	Dec
Average monthly rainfall (cm)		1	1	1	3	5.5	12.5	19	29	14	1	1	1
Average monthly temperature (degrees Celsius)		20	22.5	27	31	30	29	28	27	27	27	26	24

(a)

Figure 43C-1 (a) *Sample weather station data. (b) Climatogram graphing weather station data.*

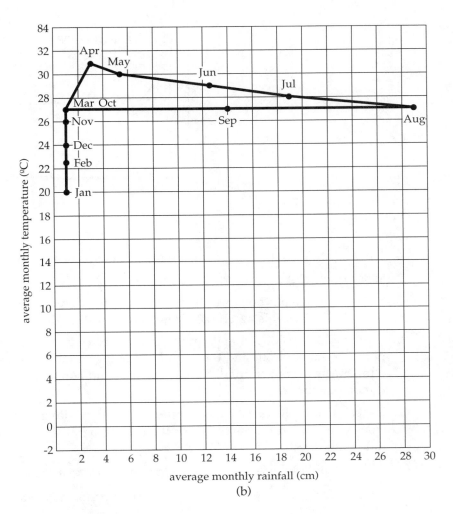

(b)

IIIII **Procedure** IIIIIIIIIIIIIIIIIIIIIIIIIIIIIIIIIIIIIIIIIIII

Your instructor will provide you with a handout of weather data for one or more cities.

1. Compute the total annual rainfall for the city by summing the 12 monthly averages given in the handout table.

2. Record the extremes of monthly average temperature that occur in the city.

3. Locate and mark the total annual rainfall of the city along the horizontal axis of the graph in Figure 43C-1b. Locate and mark the two temperature extremes along the vertical axis. Above the annual rainfall mark on the horizontal axis, draw a vertical line between the two temperature extremes. The area of the graph onto which this line falls indicates the biome in which

your city lies. Biome: _____

Figure 43C-2 *Composite
climatogram, showing biomes.*

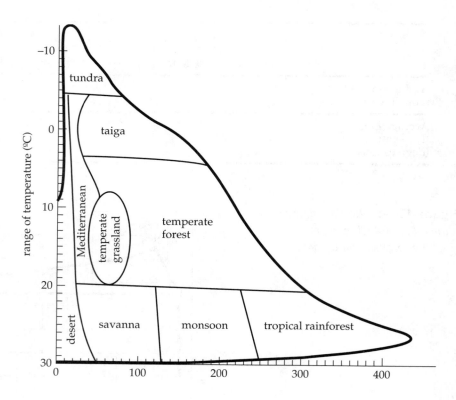

4. Now locate the city on a world map and note its longitude and latitude to the nearest degree:

   longitude _____ ; latitude _____ .

5. Locate the city on the world biome map on the demonstration table. Which biome is indicated? Does this agree with your result in step 3?

6. If time allows, plot a climatogram on a separate sheet of graph paper, using the weather data given in the handout and the sample climatogram of Figure 43C-1b as a guide. Plot the temperature versus rainfall for each month and connect the points on your graph in sequence (January, February, and so on).

7. Repeat steps 1–6 for each city on the lab handout. Referring to the weather data for all the cities you located on the maps, or their climatograms (if you have done them), answer the following questions.

   a. *Which of the cities reports the greatest annual extremes in average monthly rainfall?*

   _____ *b. What specific adaptations would plants in this biome exhibit?*
   What adaptations must animals make to such extreme annual conditions?

   _____

   c. *Which of the cities has both the hottest and driest conditions throughout the year?*

   _____ *d. What biological adaptations would be found in this biome?*

   _____

   e. *Which biome has the greatest temperature change from one month to another?*

   _____ *f. What changes in the life cycles of plants and animals would*
   you expect at this time of year? _____

   _____

Laboratory Review Questions and Problems

1. If the habitat occupied by two competing organisms is made more complex (in the case of the *Paramecium* cultures, capillary tubes, broken into small sections, could be added to the culture), what changes might occur in the competitive interactions of the species? Why?

2. If you were studying a community, would you be more likely to find competitive interactions among more closely related or less closely related species? Why?

3. The overall diversity of a community tends to increase as the community passes through stages of succession toward its stable climax state. Can you explain why this happens?

4. As you travel from the equator to the poles, community diversity declines. What happens to temperature, rainfall, light, and other environmental variables along this geographic transect? In relation to your answer to this question, why does diversity decline?

5. If you began at the base of a tall mountain situated near the equator, what changes would you see in community structure as you climbed that mountain? Can you predict what types of biomes you might traverse during your climb?

6. How might the following factors affect community structure and diversity?

   Photoperiod (length of the day)

   Exposure (north-facing or south-facing slope)

   Location in relation to mountains (on the windward or lee side)

Wind velocity

Partial pressure of gases along an altitudinal gradient

Permanently frozen substratum (permafrost)

Ultraviolet (UV) radiation

Ionizing radiation (cosmic rays, etc.), both natural and the product of human activities

Soil type and texture

Predator–Prey Relations

OVERVIEW

Predation, the killing and eating of one organism by another, is one of the basic mechanisms of energy flow in an ecosystem. It is also a potent selective agent, since success in obtaining prey can enhance the **fitness**—the ability to contribute genetically to the next generation—of the predator. Natural selection by predation also leads to the survival of prey individuals that possess more effective defenses such as camouflage, speed, armor, and defensive behavior. The pressures of selection lead predators to overcome the prey defenses, and thus sophisticated sensory systems, powerful attack weapons, pack hunting, and other specializations evolve. In fact, selection continues to refine the adaptations used by predators to pursue and catch prey at the same time that it leads to improved defense and escape mechanisms among the prey—a continuing evolutionary round-robin, or **coevolution.** When we examine the coadaptations of predator and prey, we expect to see adaptive solutions that hundreds of millennia of natural selection have produced.

Predation can benefit a prey population by the selective removal of excess young, older individuals less able to reproduce, and the injured and infirm—all of which require space, food, and other environmental resources. (All species possess the ability to produce far more individuals than can survive in the available space.) When humans remove natural predators such as large cats or wolves, prey populations increase to the point that habitat destruction by overgrazing reduces an area's overall **carrying capacity** (the number of individuals that can live in a given area), to the detriment of the prey population.

During this laboratory, you will use simulations to explore simple predator–prey interactions, the functional response of predators to changes in numbers of prey in a given area, and the effects of interspecific competition during predation.

STUDENT PREPARATION

Prepare for this laboratory by reading the text pages indicated by your instructor. Familiarizing yourself in advance with the information and procedures covered in this laboratory will give you a better understanding of the material and improve your efficiency.

EXERCISE A | Predation

Predation is an interspecific interaction in which one population uses another as its food (energy) source—and thereby benefits—while the prey population is reduced in number. This exercise involves a simulation of simple predator–prey interactions. A simulation is a model used to represent some real-life situation that allows you to learn about the dynamics of that situation without leaving the laboratory or immediate environment. Of course, direct observations taken in the field provide a basis for more realistic and general interpretations, but time, and access to the study organisms, among other factors, make this difficult. Therefore, simulations have become widely accepted by ecologists as learning tools, provided that the model adequately reflects the natural setting. Interpretations of results obtained from models are limited by the (over)simplified conditions of the simulation, yet such models offer insights and have proven their validity as an investigative tool. Even where models have proved to be quite inadequate in the real world, they have stimulated recognition of new concepts and have led to further research.

In this simulation, you will act as the predator, foraging on stationary, nondefensive prey—macaroni. As the simulation begins, it is the beginning of "winter," which lasts 12 days. Each "day" lasts 60 seconds, and you *must* capture five prey items per day to survive. After you have captured five prey in one day, you stop hunting. If you capture less than five prey, you starve and die.

‖‖‖ Objectives ‖‖‖‖‖‖‖‖‖‖‖‖‖‖‖‖‖‖‖‖‖‖‖‖‖‖‖‖‖‖‖‖‖‖‖‖‖‖

☐ Explore predation on prey inhabiting a limited area that are not replaced as they are consumed.

‖‖‖ Procedure ‖‖‖‖‖‖‖‖‖‖‖‖‖‖‖‖‖‖‖‖‖‖‖‖‖‖‖‖‖‖‖‖‖‖‖‖‖‖

This experiment allows everyone in the class to contribute data to a collective pool and gain experience with the general procedure employed throughout this laboratory. It also provides baseline data with which the results from other treatments can be compared.

1. Work in pairs, with one person acting as the predator and the other serving as the timer.

2. Each pair will be assigned a 10 foot × 10 foot territory. The predator sets up a "nest" (a dish or box lid) somewhere near the middle of one of the territory edges. While the predator turns his or her back and closes his or her eyes, the timer spreads 100 food items randomly throughout the territory and then takes up a position on the side of the territory opposite the nest.

3. The predator turns around and begins looking for prey when instructed to do so by the timer. Once released to hunt, the predator captures five (and only five) prey items and returns home to the nest. As soon as the predator places the captured prey in the nest, the timer records the amount of time it took for the predator to find all of the prey on day 1. Record the data in Table 44A-1. This task *must* be completed within 60 seconds if the predator is to survive.

4. The predator and timer than initiate day 2 by repeating step 3 above (release, hunt, return to the nest with five added prey). Again, the process is timed and must be completed within 60 seconds—if the predator fails to obtain five prey items and return home within the 60-second period, it dies. In this case, record "dead" in Table 44A-1 and the simulation is over.

5. Repeat the simulation for the 12 days of winter or until the predator expires.

6. When all predators have completed this simulation, you will be assigned a new territory and you and your lab partner will switch roles.

7. Enter all class data in Table 44A-1 and compute means for each day. Use a value of 60 seconds for all dead predators. Using a separate piece of graph paper, construct a graph of your class results for predation time. Also prepare a frequency histogram (see Appendix I, Part A) of the number of surviving predators over time.

   a. *What happens as prey nearer the nest are depleted?* _____

      As the total prey numbers are reduced? _____

   b. *What strategy or strategies were adopted by the predator as the prey numbers declined? Does experience help make a more efficient predator?* _____

   _____

Table 44A-1   Time Needed to Find Five Prey Items

Predator	Time (seconds) for Each Day											
	1	2	3	4	5	6	7	8	9	10	11	12
Mean												

c. *Relate your results to the natural world. Propose a natural situation that might be similar to this simulation.* _____

_____

d. *What do you expect to happen to predator and prey populations as the winter progresses?*

_____

e. *What could be added to this model to make it more realistic?* _____

_____

EXERCISE B   Functional Responses by Predators

This simulation measures a predator's functional response: the way a predator's capture rate varies with changes in prey density.

IIIII Objectives IIIIIIIIIIIIIIIIIIIIIIIIIIIIIIIIIIIII

☐ Describe how predators may be affected by the density of prey.

IIIII Procedure IIIIIIIIIIIIIIIIIIIIIIIIIIIIIIIIIIIIII

1. Work in pairs. This simulation is conducted just like the preceding one except that the number of prey is lower. Your instructor will give you two bags, each containing either 80, 60, or 40 prey items.

2. Move to a new 10 foot × 10 foot area. The timer should distribute prey items as in Exercise A.

3. Measure the time it takes to capture five prey. Record the data in Table 44B-1, in the appropriate section (80, 60, or 40 prey items). Continue the simulation for 12 60-second days or until the predator dies.

4. Switch roles and repeat the simulation, moving to a new 10 foot × 10 foot area.

5. Obtain data from other student pairs using the same number of prey items. Record these in the same section of Table 44B-1 and calculate means.

6. Obtain class data for other prey densities and enter these in Table 44B-1. Using a separate piece of graph paper, graph the results for each prey density on the same graph.

a. Combining the results of Exercises A and B, estimate *the critical prey density for your macaroni populations.*

   What is the critical, minimum *number of prey needed to maintain a predator through the winter?* _____

   How did you determine this number? _____

   _____

b. *In a natural ecosystem, what factors would increase the critical prey density?*

   _____

   _____

   Decrease the critical prey density? _____

   _____

It is often argued that predators control the population size of their prey species. Others argue that the prey control the size of the predator population.

c. *Using your data from Exercise B, defend one of these statements.* _____

   _____

d. *Does your argument agree with the trend seen in Exercise A as prey are captured and removed from the*

   population? _____

Table 44B-1   Time Needed to Find Five Prey Items at Various Prey Densities

80 Prey Items

Predator	Time (seconds) for Each Day											
	1	2	3	4	5	6	7	8	9	10	11	12
Mean												

60 Prey Items

Predator	Time (seconds) for Each Day											
	1	2	3	4	5	6	7	8	9	10	11	12
Mean												

40 Prey Items

Predator	Time (seconds) for Each Day											
	1	2	3	4	5	6	7	8	9	10	11	12
Mean												

| EXERCISE C | Competition |

Competition results when shared resources become limited. **Interspecific competition** involves populations of different species vying for the same food, habitat, nesting structure, or some other shared environmental resource. Interspecific competition can narrow some feature of the **niche** (the role of a species in its community) of one or more competing species. These interactions may also provide a selective force that can result in a niche shift and the evolution of distinctive morphological characteristics among competing, closely related species. **Intraspecific competition** involves individuals of the same species competing for shared resources such as access to mates or status within a social group. Intraspecific competition is usually more intense than interspecific competition because all competitors share an identical or very similar niche. (How might the niche of individuals of the same species differ? *Hint:* Consider sexual dimorphism.)

In this experiment, you will examine the potential effects of intraspecific competition on predator success, again using human predators and macaroni prey.

ⅠⅠⅠⅠ Objectives ⅠⅠⅠⅠⅠⅠⅠⅠⅠⅠⅠⅠⅠⅠⅠⅠⅠⅠⅠⅠⅠⅠⅠⅠⅠⅠⅠⅠⅠⅠⅠⅠⅠⅠⅠⅠⅠⅠ

☐ Investigate the potential effects of competition on predator success.

ⅠⅠⅠⅠ Procedure ⅠⅠⅠⅠⅠⅠⅠⅠⅠⅠⅠⅠⅠⅠⅠⅠⅠⅠⅠⅠⅠⅠⅠⅠⅠⅠⅠⅠⅠⅠⅠⅠⅠⅠⅠⅠⅠⅠ

1. Work in groups of four student pairs (eight students per group), so that there are four predators and four timers. Increase territory size to a 20 foot × 20 foot area. Timers scatter 400 prey items (100 prey items per timer) in the territory.

2. All other procedures are the same as in the previous simulations: five prey items must be captured and returned to the nest each day, a day is 60 seconds long, and winter lasts for 12 days. Failure to find five prey items causes immediate death. However, there is one additional feature to this simulation: *mild* harassment between foragers will be tolerated (four of you will be hunting at the same time and may get in each other's way or attempt to sequester particular resources). Record the times for each predator in Table 44C-1.

3. Obtain and record all class data and calculate means. Graph your results on a separate piece of graph paper.

   a. *If a predator had exclusive use of a feeding territory, would the individual be more or less likely to survive? To reproduce? Discuss.* _____

   _____

   b. *In what ways could a competitor affect an individual predator?* _____

   _____

   c. *What would be the effects of reproduction within interacting populations of prey and predators? For instance, what would happen if prey reproduced but predators did not, or vice versa? If both populations reproduced? How might these activities be interrelated and related to population numbers over time?* _____

   _____

   _____

Table 44C-1 Time Needed to Find Five Prey Items When Competitors Are Present

Predator	Time (seconds) for Each Day											
	1	2	3	4	5	6	7	8	9	10	11	12
Mean												

Laboratory Review Questions and Problems

1. Compare predation and parasitism. How are they alike? How do they differ?

2. Compare predation and herbivory. How are they alike? How do they differ?

3. Develop simulations to illustrate and explore the following situations.

    a. Crypsis (camouflage provided by pattern, color, shape, behavior, etc.)

    b. Toxic prey with warning (aposematic) coloration

    c. Mimicry (fully edible prey adopt the warning coloration of a toxic model species)

    d. Alternative prey (a predator is capable of using more than one species)

    e. Prey of different value (energy "packaged" in prey of different size)

    f. Effect of prey distribution (dispersion) on hunting success (prey can be distributed randomly, evenly, or in clumped patterns)

    g. Effect of habitat complexity (one-dimensional substrate, bare, grass, forest, etc.)

4. Suppose that prey are distributed within three different territories as follows.

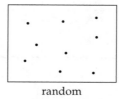
random

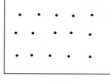
even

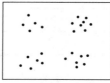
clumped

    a. Which type of distribution would probably allow for the highest survival of predators? Why?

b. Would it make a difference if the predator were blind (assuming the prey are stationary and give off no olfactory or auditory cues)? Why?

c. Would "learning" a pattern be advantageous? If so, which distribution(s) would allow for the highest survival of predators? Of prey?

5. Define the following terms.

Fitness

Coevolution

Niche

Interspecific competition

Intraspecific competition

Productivity in an Aquatic Ecosystem

OVERVIEW

One of the most important characteristics of an ecosystem is its **productivity,** or the rate at which it synthesizes biomass. Productivity, which can be measured in grams of biomass synthesized per square meter per year, ranges from over 2,000 $g/m^2/yr$ for tropical rainforests to less than 5 $g/m^2/yr$ in extreme deserts. One of the advantages of the aquatic environment you will study in this laboratory is that a sample of the ecosystem can conveniently be collected in a bottle. Rates of photosynthesis and respiration can be measured by changes in the dissolved oxygen in the water. In addition, since a liter of lake or ocean water may contain millions of organisms representing dozens of different species, a bottle can contain a small-scale but complex and interesting community.

   In this laboratory you will learn how to measure dissolved oxygen in water and investigate the effects of temperature on dissolved oxygen concentration. You can then use measurements of dissolved oxygen to determine **primary productivity,** the rate at which the radiant energy of the sun is converted into the chemical-bond energy of organic material in simulated eutrophic, oligotrophic, and polluted lakes. Or, you may want to study primary productivity in a plankton column, simulated by exposing algal cultures to varying amounts of light. A demonstration of lake "turnover" will also provide an opportunity for you to study oxygen distribution during thermal stratification.

STUDENT PREPARATION

Prepare for this laboratory by reading the text pages indicated by your instructor. Familiarizing yourself in advance with the information and procedures covered in this laboratory will give you a better understanding of the material and improve your efficiency.

EXERCISE A **Measuring Dissolved Oxygen: Effects of Temperature and Salinity**

In the aquatic environment, diversity, productivity, and the maintenance of life depend upon the availability of oxygen dissolved in the water. The concentration and distribution of oxygen are dependent upon several physical, chemical, and biological factors.

   There are two sources of oxygen in the water: diffusion from the atmosphere and photosynthesis by aquatic plants. Air contains approximately 210 ml of oxygen per liter, while the amount of **dissolved oxygen (DO)** in a liter of water is never more than 5 to 10 ml. Without movement of air or water, oxygen diffuses into water very slowly, but diffusion is accelerated by wind and water currents. Water temperature, dissolved salts, and the production of oxygen by photosynthesis can affect the concentration of oxygen. Throughout a body of water, daily, seasonal, and even spatial variations in dissolved oxygen can occur.

Salinity, the content of dissolved salts, is usually expressed in parts per thousand (ppt). As salinity increases, solubility of oxygen in water decreases. Likewise, as temperature increases, solubility of oxygen in water decreases (Figure 45A-1). As water is warmed, the kinetic energy (energy of movement) of water molecules increases, which tends to expel oxygen molecules from the solution. As a consequence, cold water can hold more DO than warm water. You can easily observe this difference. Next time you make ice cubes, make one tray using hot water and the other using cold water. The ice cubes made with hot water will turn out crystal clear, while those made with cold water will be fairly opaque. This is due to the presence of gas inclusions (small pockets of O_2) within the ice. To produce an oxygen-free solution, what might you do?

Figure 45A-1   *As salinity increases, solubility of dissolved oxygen in water decreases. The salinity of seawater is 35 ppt. Note: 1 ml O_2/l = 1.43 mg O_2/l at 20°C.*

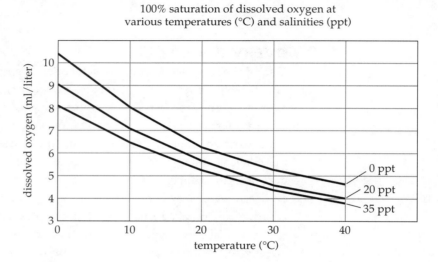

100% saturation of dissolved oxygen at various temperatures (°C) and salinities (ppt)

The amount of DO in water is of great consequence to aquatic animals and plants. In unpolluted lakes and streams, the greatest loss of oxygen occurs during the warm months of summer. Fish can die from lack of oxygen in warm waters, and you may see large numbers of dead fish floating on the surface of lakes or ponds during the hottest times of the year.

During the spring and summer months, increased temperature, light, and fertilizer run-off can cause algal "blooms" (rapid increases in population density) in lakes and ponds. When nutrients are exhausted, the algae die and are decomposed by microbial activity. Under such conditions, the amount of oxygen consumed by algal respiration and decomposition is greater than the amount of oxygen produced by photosynthesis. The result is a dramatic decrease in DO in the pond.

In high-altitude lakes, less DO is present due to decreased atmospheric pressure. In oceans, the presence of salts decreases the DO content. Solutes, including those that result from pollution, reduce the intermolecular space available for "housing" oxygen.

What factors, in addition to photosynthesis, increase the amount of O_2 dissolved in water? The rapid flow of water in streams and rivers or the presence of winds that disturb the water surface can increase the amount of DO in the water. Also, the inflow of oxygen-rich cold waters from tributaries, often from melting snows, may lower the temperature of rivers and lakes and increase their oxygen content.

How would thermal pollution—for example, warm effluent waters from nuclear reactors—affect the DO content of lake waters? Do you think that cutting trees in logging areas next to forest streams might also have an effect on the DO content of water? Why? Perhaps you can think of other factors that might increase or decrease the DO content of water. Discuss your ideas with the other students in your laboratory.

The **Winkler technique** is commonly used to measure DO. In this procedure, a series of reactions generates an amount of iodine (I_2) which is quantitatively identical to the amount of DO in the water sample. The quantity of iodine is then determined by **titration** with a sodium thiosulfate solution.

||||| **Objectives** |||||||||||||||||||||||||||||||||||||||||

☐ Describe the physical, chemical, and biological factors that affect the solubility of oxygen in aquatic ecosystems.

☐ Describe the Winkler technique for determining the dissolved oxygen content of water.

||||| **Procedure** |||||||||||||||||||||||||||||||||||||||

Work in groups of four. Freshwater samples are available at three temperatures: 0°C, 20°C, and 40°C. Your group will be asked to determine the dissolved oxygen (DO) content for a sample of fresh water at a specific temperature. In Table 45A-1, record the temperature of your sample.

Table 45A-1

Temperature	Titrant Used (ml)	DO (mg O_2/l)	Percent O_2 Saturation

Formulate a hypothesis that relates water temperature to its oxygen content.

HYPOTHESIS:

NULL HYPOTHESIS:

What do you **predict** will be the outcome of your experiment?

Identify the **independent variable** in this experiment.

Identify the **dependent variable** in this experiment.

Use the following procedure to test your hypothesis.

1. Using the water analysis kit provided by your instructor, test for the amount of dissolved oxygen in your sample of water.*

   a. Submerge each of two sample bottles in the container of warm or cold water (Figure 45A-2).

   b. Tap the sides of the bottles to release any air bubbles.

   c. While the bottles are submerged, replace the caps, then remove the bottles from the water. Be sure to label the bottles with appropriate temperature labels.

   d. Uncap one of the bottles and add 8 drops of manganous sulfate from the dropper bottle provided in your kit.

Directions are for the LaMotte Dissolved Oxygen Test Kit. The directions for other kits may vary.

Figure 45A-2 *Filling a sample bottle.*

e. Add 8 drops of alkaline potassium iodide azide solution.

f. Replace the cap and shake. A golden-brown precipitate will form. Let the precipitate settle to the bottom of the bottle.

g. Uncap the bottle and use the scoop provided in your kit to add 1 g of sulfamic acid powder to the sample bottle.

h. Cap the bottle and shake gently until all of the precipitate dissolves. The sample will appear straw-colored and is now "fixed." The addition of sulfamic acid, in combination with potassium iodide, reacts to bind or "fix" oxygen to manganous sulfate and release iodine (I_2). The iodine released corresponds (1:1 ratio) to the amount of oxygen that has been bound. (We can't test directly for oxygen, but we can test for the amount of iodine released by this reaction and thereby determine the oxygen content of the water.)

i. To determine the amount of O_2 fixed in the sample, fill the titration tube to the 20-ml line.

j. Fill the titrator syringe with sodium thiosulfate. If a bubble appears, overfill the syringe and then expel the air bubble. No bubbles allowed! See Figure 45A-3a, b.

k. Place the titrator syringe into the hole in the cap of the titration tube (it should fit snugly) and add one drop at a time of the sodium thiosulfate until the solution turns a pale yellow (Figure 45A-3c). Sodium thiosulfate reacts with the free I_2 to form sodium iodide.

l. Remove the titrator syringe and cap it so it does not leak. Remove the cap on the titration tube and add 8 drops of starch solution. Starch reacts with I_2 but not with sodium iodide. If any I_2 remains, the solution turns blue. This will allow you to see how much I_2 remains in your solution—the darker the blue, the more I_2 remains.

m. Recap the titration tube and gently mix the solution, making sure that you do not lose any solution from the hole in the cap.

n. Uncap the titrator syringe and reattach it to the titration tube. Continue to add sodium thiosulfate drop by drop until the blue color is gone. *When the blue color disappears, how much I_2 is left in the solution?* _____

o. Immediately remove the titrator syringe and cap it. Read the titrator syringe scale for results in parts per million (ppm) of dissolved oxygen (Figure 45A-3d). A certain amount of sodium thiosulfate was necessary to react with I_2 to produce sodium iodide. This amount allows you to calculate the amount of I_2 that was contained in the water. Since I_2 is proportional to O_2 (1:1 relationship), you can use this as a measure of O_2.

2. How much oxygen was present in your sample? Record your result as mg O_2/l in Table 45A-1 (*Note:* ppm = mg O_2/l or mg O_2/l × 0.698 = ml O_2/l.)

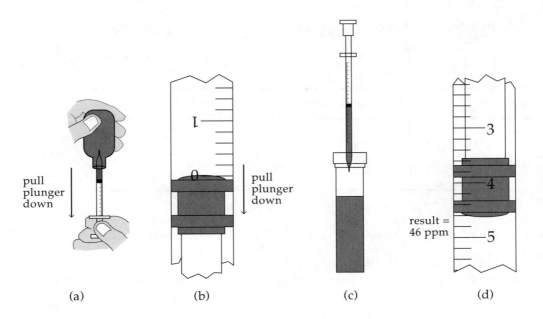

(a)   (b)   (c)   (d)

Figure 45A-3 *First, be sure the plunger of the titrating syringe is pushed all the way into the body of the syringe. Insert the needle of the syringe into the upright bottle of sodium thiosulfate. Then (a) fill the titrating syringe by inverting the bottle of sodium thiosulfate and slowly withdrawing the plunger until the end of the plunger is at the 0 mark on the scale (b). (c) Insert the tip of the syringe into the opening of the titrator cap. Slowly depress the plunger, gently swirling to mix, until the solution in the tube turns pale yellow. (d) Interpret the titration scale by reading the marks 3, 4, 5, etc., as 30, 40, 50 parts per million (ppm). The tic marks between the numbers each represent 2 ppm.*

3. Use Figure 45A-4 to estimate the percentage oxygen saturation (percent O_2 saturation) for your water sample. Position a straight-edge or ruler so that it intersects water temperature and oxygen (mg/liter) at the points that agree with your data. The point at which the ruler crosses the "% saturation" line indicates the percentage oxygen saturation of your sample. Record this value in Table 45A-1.

Figure 45A-4 *Nomograph of oxygen saturation.*

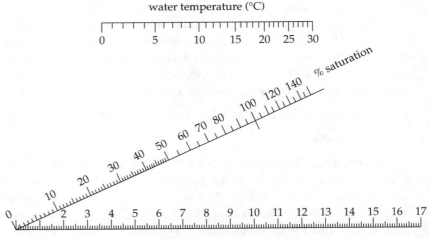

4. In Table 45A-2, record class data for water samples at all temperatures.

Table 45A-2

Temperature	DO (mg O_2/l)	Percent O_2 Saturation
0°C		
20°C		
40°C		

Do your results support your hypothesis? _____ *Your null hypothesis?* _____

Was your prediction correct? _____

What do you **conclude** *about how temperature might affect the solubility of oxygen in water?*

EXERCISE B   Measuring Production in an Aquatic Ecosystem

Productivity is the rate at which an ecosystem synthesizes organic matter (usually through photosynthesis by primary producer organisms). Productivity is one of the most important characteristics of an ecosystem because it determines how many herbivores and other organisms an ecosystem can support. Also, the rate of productivity gives us some indication of the metabolic vigor of the whole system.

Productivity is high whenever sunlight, temperature, nutrient availability, and especially water availability allow a brisk rate of photosynthesis (for example, in coral reefs, salt marshes, and rainforests). If one or more of these conditions is wrong for photosynthesis (lack of water in deserts, lack of nutrients in the open ocean, cold temperatures on the tundra), productivity is low and life is not as abundant.

Several different measures of productivity can be made. These include:

> **Gross primary productivity** (GPP) The total amount of matter photosynthesized by the primary producers of a community during some period of time, with no deduction for respiration.
>
> **Net primary productivity** (NPP) Gross primary productivity minus *plant* respiration.
>
> **Net community productivity** (NCP) Gross primary productivity minus the respiration of *all* organisms in the community, heterotrophs included.

Because net primary productivity measures the amount of biomass that is available for use by herbivores and higher trophic levels, it is a useful way to compare the productivity of various ecosystems (Table 45B-1).

Note: The relationship between net primary productivity and gross primary productivity can be compared to a series of bank transactions. Say that in the course of a week, you make several deposits to and several withdrawals from your account. If you add up all your deposits, the total will be your *gross* deposits. To find your *net* deposits, you must subtract all your withdrawals from your gross deposits.

a. Which amount—gross or net deposits—would provide a more accurate understanding of your account?

_____

In aquatic ecosystems, organisms, including green plants and photosynthetic algae, use light energy and CO_2 dissolved in the water to produce carbohydrates and oxygen. The oxygen produced may then be used up by other organisms such as microbes, animals, or the plants themselves during the process of respiration. (Keep in mind that all photosynthetic organisms carry on the processes of photosynthesis *and* respiration and thus are consumers as well as producers.)

Rates of CO_2 utilization, carbohydrate formation, or O_2 production can all be used to determine primary productivity values. Because there is a constant relationship between the amount of oxygen

Table 45B-1   Ecosystem Net Primary Productivity

Ecosystem	NPP (g/m^2/yr)
Sandy deserts	3
Oligotrophic lakes	100
Open ocean	125
Tundra	140
Eutrophic lakes	600
Temperate grasslands	700
Temperate deciduous forests	1,200
Estuaries	1,250
Marshes	2,000
Tropical rainforests	2,200
Algal turfs, reefs	2,500

produced and the amount of carbon dioxide converted into organic material, productivity can be expressed in terms of either oxygen produced or carbon fixed.

Whether oxygen production exceeds its consumption, and by how much, depends on factors such as temperature and the availability of light and nutrients. By measuring oxygen production over time, it is possible to determine productivity (the rate at which carbon is "fixed" into organic compounds). For each milligram of oxygen produced, approximately 0.375 mg of carbon is assimilated or "fixed" into chemical bonds.

It is easy to measure productivity (both gross and net) in aquatic ecosystems because oxygen dissolves in water and can readily be measured. To understand how productivity causes dissolved oxygen changes, recall that both photosynthesis and respiration are going on at the same time, and that photosynthesis produces oxygen while respiration consumes it. Figure 45B-1 shows what would happen if we could separate these two processes, and what happens when the two are combined.

Figure 45B-1   *In an aquatic ecosystem, dissolved oxygen increases due to photosynthesis (PS) and simultaneously decreases due to respiration (R). When photosynthesis exceeds respiration, the two processes add together to produce a small increase in dissolved oxygen.*

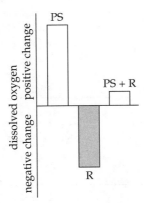

b. *In Figure 45B-1, which bar represents gross primary productivity?* _____

c. *If Figure 45B-1 represents an algal culture (autotrophs only), which bar represents net primary productivity?*

_____ *If it represents a mixture of autotrophs and heterotrophs (bacteria and animals), which bar*

represents net community productivity? _____

Figure 45B-2 *If photosynthesis (PS) is the same as in Figure 45B-1 but respiration (R) is higher, photosynthesis and respiration may combine to produce a net decrease in dissolved oxygen.*

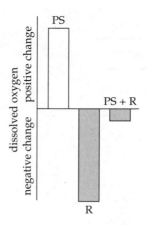

Of course, respiration may exceed photosynthesis, as shown in Figure 45B-2.

d. *Is the statement "gross primary productivity is always positive" correct? Why or why not?*

_____

_____

e. *Is the statement "net primary productivity and net community productivity can be either negative, positive, or zero" correct? Why or why not?* _____

_____

These principles are used to measure productivity by a simple technique called the "light bottle–dark bottle" method. By using the light bottle–dark bottle method in conjunction with Winkler titrations, you will measure O_2 production by comparing DO values for cultures in bottles incubated in the dark versus those exposed to light. In bottles exposed to light, both photosynthesis and respiration occur and the change in DO concentration from its initial concentration is a measure of **net primary** or **net community productivity** (defined on p. 45-6). In bottles kept in the dark, any change in DO concentration from the initial reading is a measure of the rate of respiration. The difference that develops over time between the DO concentrations in the light and the dark bottles is a measure of **gross primary productivity** (p. 45-6). See Figure 45B-3.

Figure 45B-3 *Net primary or community productivity (NPP or NPC) can be determined as the rate of storage of organic matter in plant tissue in excess of that consumed in respiration. Gross primary productivity (GPP) can be determined as the total formation of organic matter, including that consumed in respiration.*

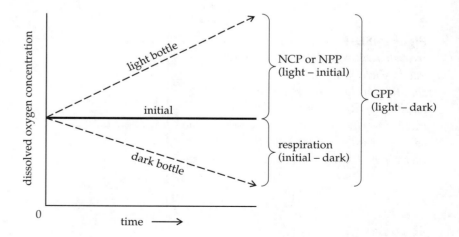

We can express these relationships in a series of equations:

$$NCP = L - I$$
$$R = I - D$$
$$GPP = L - D$$

where NCP is net community productivity; R is respiration; GPP is gross primary productivity; I is initial dissolved oxygen; L is dissolved oxygen in the light bottle after the measurement period; and D is dissolved oxygen in the dark bottle after the measurement period.

Consider the following example. For two light bottles and two dark bottles, the DO changed over two hours as recorded below, with the accompanying explanations.

Bottle	Initial DO	Final DO	Explanation
Light	7.3 mg/l	7.8 mg/l	Net community productivity (photosynthesis − respiration) was 0.5 mg/l/2 hr, (0.25 mg/l/hr)
Dark	7.3 mg/l	6.9 mg/l	Respiration was 0.4 mg/l/2 hr (0.2 mg/l/hr)
			Therefore, photosynthesis alone (GPP) was 0.5 + 0.4 = 0.9 mg/l/2 hr (0.45 mg/l/hr)

Now, consider a somewhat different example. Suppose you sample some polluted water and the dissolved oxygen concentration is 6.8 mg/l. After 6 hr, the light bottle has a DO of 5.7 mg/l, and the dark bottle has a DO of 3.6 mg/l. Compute net community productivity, respiration, and gross primary productivity, all in $mgO_2/l/hr$.

NCP = _____ mg O_2/l/hr

R = _____ mg O_2/l/hr

GPP = _____ mg O_2/l/hr

f. *Oxygen in the light bottle decreased. If photosynthesis was occurring, why didn't oxygen in the light bottle*

increase? _____

_____

Since NCP is photosynthesis minus respiration, anything that affects either of these processes will affect NCP.

g. *Will net community productivity become more positive or more negative if photosynthesis speeds up due to an*

influx of nutrients? _____ *An overcast day dims the light in the water column?* _____

An influx of organic matter causes an upsurge in the number of heterotrophic bacteria? _____

A critical fact about net community productivity is that it is a *difference* between photosynthesis and respiration. The magnitude of NCP tells us less than the magnitude of photosynthesis and respiration considered separately. Consider the three cases shown in Figure 45B-4. In the first two cases, NCP is

Figure 45B-4 *Photosynthesis (PS), respiration (R), and net community productivity (NCP) (calculated as P − R) in three cases: (a) photosynthesis and respiration both small; (b) photosynthesis and respiration both large; and (c) photosynthesis slightly reduced from (b), resulting in a negative NCP.*

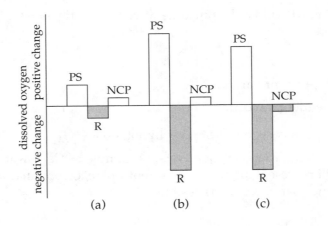

identical, but for different reasons. In (a), respiration and photosynthesis are small, possibly due to sparse populations of photoplankton and bacteria. In (b), there are high rates of both respiration and photosynthesis. The practical difference between the two cases is that at night, the water in (b) will suffer a sharp decline in oxygen content, but the water in (a) will not. We might assume that (c) is fundamentally different from the other two cases because its NCP is negative. However, at least in the short run, (b) and (c) are very similar. Small shifts in respiration and photosynthesis may change the NCP from negative to positive.

Again, what is important here is the *magnitudes* of photosynthesis and respiration, not the *difference* between photosynthesis and respiration. In this exercise, you will have the chance to measure the NCP, respiration, and GPP in phytoplankton cultures with (1) a small amount of nutrients, (2) a large amount of nutrients, and (3) enrichment by bacterial growth medium. These simulate conditions in three types of lakes:

1. **Oligotrophic** A clear lake with a low concentration of algal nutrients such as nitrate and phosphate and a low gross primary productivity (often deep, with cold water that remains oxygenated year round).

2. **Eutrophic** Turbid, with high concentrations of algal nutrients supporting a high gross primary productivity (often relatively shallow; light does not penetrate deeply because of algal populations; layers below the level of effective light penetration may lack oxygen at some times of the year).

3. **Polluted** Contains chemical or bacterial pollutants which have harmful or toxic effects on a variety of organisms (stagnant waters lacking oxygen below the upper surface, due to bacterial growth).

‖‖‖‖‖ Objectives ‖‖

☐ Define and measure net and gross primary productivity.

☐ Describe how photosynthesis and respiration affect dissolved oxygen concentration.

☐ Design an experiment to determine the effect of nutrient levels on primary productivity.

‖‖‖‖‖ Procedure ‖‖

Formulate a hypothesis that relates DO (dissolved oxygen) levels to conditions in oligotrophic, eutrophic, and polluted lakes.

HYPOTHESIS:

NULL HYPOTHESIS:

What do you **predict** will be the outcome of your investigation?

What is the **independent variable?**

What is the **dependent variable?**

Use the following procedure to test your hypothesis.

1. Work in groups of four. Obtain a sample of "lake water" from your instructor. The sample will represent an oligotrophic, eutrophic, or polluted lake. In the title of Table 45B-2, record which sample you are testing.

Table 45B-2   Dissolved Oxygen Data for

_____ **Sample**

Bottle	DO (mg O$_2$/l)
Initial	
Dark	
Light	

Start time: _____

Finish time: _____

Duration: _____ hr

2. Obtain three BOD (Biological Oxygen Determination) bottles and label them "I" for initial, "L" for light, and "D" for dark.

3. Fill all three bottles with the lake sample. Use a graduated cylinder or beaker to transfer the water. Record your starting time in Table 45B-2.

4. Immediately "fix" the initial bottle I by following steps 1d–h in Exercise A.

5. Cover the dark (D) bottle _completely_ with aluminum foil.

6. Place both the light (L) and dark (D) bottles beneath a fluorescent light fixture for 2 hours.

7. After 2 hours, "fix" both the light (L) and dark (D) bottles by following steps 1d–h in Exercise A. Record your finishing time in Table 45B-2 and calculate the duration of your incubation to the nearest 0.1 hour.

8. After fixation, samples can sit until the following laboratory period (if so, refrigerate), or they can be titrated immediately to determine their dissolved oxygen (DO) content. Titrate the contents of each of your three bottles—initial (I), light (L), and dark (D)—by following steps 9–15 in Exercise A.

9. Calculate the DO content for each bottle (ppm = mg O$_2$/l) and record this in Table 45B-2.

10. Calculate respiration rate (R) for the light bottle (L) using the formula below. Enter this value in Table 45B-3.

$$\text{Respiration rate (mg O}_2\text{/l/day)} = \left[\frac{\text{initial bottle (mg O}_2\text{/l)} - \text{dark bottle (mg O}_2\text{/l)}}{\text{hours incubated}} \right] \times 24 \text{ hr}$$

11. Use the formula below to calculate NCP (net community productivity) (in terms of oxygen produced) for the light bottle. Record the value in Table 45B-3.

Net community productivity (mg O$_2$/l/day)

$$= \left[\frac{\text{light bottle (mg O}_2\text{/l)} - \text{initial bottle (mg O}_2\text{/l)}}{\text{hours incubated}} \right] \times 24 \text{ hr}$$

12. Use the formula below to calculate the gross primary productivity, in terms of oxygen produced, for the light bottle.

Gross primary productivity (mg O$_2$/l/day)

$$= \left[\frac{\text{light bottle (mg O}_2\text{/l)} - \text{dark bottle (mg O}_2\text{/l)}}{\text{hours incubated}} \right] \times 24 \text{ hr}$$

Table 45B-3   Calculation of NCP and GPP

Initial DO (mg O$_2$/l) _____

Dark DO (mg O$_2$/l) _____

Respiration rate (mg O$_2$/l/day) _____

Sample	DO (mg O$_2$/l) ppm	Net Community Productivity NCP (light − initial) (mg O$_2$/l/day)	Gross Primary Productivity GPP (light − dark) mg O$_2$/l/day

13. Collect net community productivity and gross primary productivity data for all other groups using the same lake sample. Determine the mean. Enter these data in Table 45B-4.

Table 45B-4   Class Data for _____ **Sample**

Group	NCP	GPP
1		
2		
3		
4		
5		
Mean		

14. Collect class average NCP and GPP values for each type of lake sample. Enter these data in Table 45B-5.

Table 45B-5   Class Averages

Lake Type	NCP	GPP
Oligotrophic		
Eutrophic		
Polluted		

Do your results support your hypothesis? _____ *Your null hypothesis?* _____

Was your prediction correct? _____

From your data, what do you **conclude** *about the roles of photosynthesis and respiration in oligotrophic, eutrophic, and polluted lakes?* _____

h. Based on your results, which of the following conditions (A–D) occurred in each type of lake? **A,** *high rate of photosynthesis and low rate of respiration;* **B,** *high rate of photosynthesis and equally high rate of respiration;* **C,** *very little photosynthesis and high rate of respiration;* **D,** *very little photosynthesis and very little respiration.*

Oligotrophic _____

Eutrophic   _____

Polluted   _____

EXERCISE C   Productivity of Phytoplankton in a Water Column

Phytoplankton are photosynthetic algal cells that float or swim in lakes, in the ocean, and in other bodies of water. Small lakes may contain several quadrillion phytoplanktonic organisms belonging to hundreds of different species. Different environments within a body of water offer many challenges to survival. Limiting factors—available light, nutrients, grazing by zooplankton—that control phytoplankton growth may change rapidly over time and over a few meters of depth. You will study productivity in a vertical column of water. Pieces of wire screen will be wrapped around the light bottles, increasing the layers to simulate decreasing light at increasing depth. Temperature, which usually varies with depth in natural ecosystems, will be held constant.

Objectives

☐ Define net and gross primary productivity.

☐ Describe how photosynthesis and respiration affect dissolved oxygen concentration.

☐ Explain how photosynthesis and respiration are related to primary productivity.

☐ Design an experiment to determine the effect of light on primary productivity.

☐ Design an experiment to investigate the effects of the nutrients phosphorus and nitrogen on primary productivity.

Procedure

Formulate a hypothesis about the effects of light on productivity in a vertical column of a lake.

HYPOTHESIS:

NULL HYPOTHESIS:

What do you **predict** will be the outcome of your experiment?

What is the **independent variable?**

What is the **dependent variable?**

Use the following procedure to test your hypothesis.

1. Work in groups of four. Obtain five BOD bottles. Rinse them with distilled water.

2. Obtain 3 l of water sample (your instructor will tell you whether this is a natural water sample or an algal culture).

3. Label your bottles 1 to 5 as follows.

Bottle	Treatment
1	Initial
2	Dark
3	Light
4	Light + N(itrogen)
5	Light + P(hosphorus)

4. Your group will be assigned to work at a given light intensity (percentage of light) that will require a certain number of layers of screen to cover bottles 3, 4, and 5 (light, light + N, and light + P). The number of screens your group is to use can be determined from the following table.

Percentage of Light	Number of Screens
100%	0
65%	1
25%	3
10%	5
2%	8

5. Fill all five bottles. Carefully siphon water from the bottom of the sample into the bottom of a BOD bottle. To avoid introducing extra oxygen into your sample, keep both ends of the siphon beneath the water at all times. If a siphon is not available, follow directions in Exercise A, step 3.

6. Bottle 1 will serve as the initial bottle. Wrap bottle 2 in aluminum foil to serve as a dark bottle. Wrap the appropriate number of screens around the light bottle (bottle 3). Add 1 ml of nitrogen enrichment solution to bottle 4 and 1 ml of phosphorus enrichment solution to bottle 5. Wrap bottles 4 and 5 with the same number of screens used for bottle 3.

7. Place the dark and light bottles in front of or below a fluorescent light source as indicated by your instructor.

8. Perform steps 1d–h in Exercise A on the "initial" bottle (bottle 1). Refrigerate until the other bottles are processed.

9. Let the bottles incubate for 2 hours. (Alternatively, your instructor may request that you incubate bottles overnight. In this case, you may be asked to fix and process samples from a class that met the previous day. Your instructor will distribute class data.) Record total incubation time in Table 45C-1.

10. After incubation, "fix" bottles 1 to 4 belonging to your group. (The initial bottle has already been fixed.) Do this by performing steps 1d–h in Exercise A.

11. Determine concentration (mg O_2/l) for all five bottles by following steps 9–15 of Exercise A. Enter these data in Table 45C-1.

12. Calculate the respiration rate by using the formula below. Enter this value in Table 45C-1.

Table 45C-1 Calculation of NCP and GPP

Initial DO mg O_2/l _____

Dark DO mg O_2/l _____

Respiration rate mg O_2/l/day _____

Start time: _____

Finish time: _____

Duration: _____ hr

Sample	DO (mg O_2/l)	NPP or NCP (light − initial) (mg O_2/l/day)	GPP (light − dark) mg O_2/l/day
Light			
Light + N			
Light + P			

$$\text{Respiration rate (mg } O_2\text{/l/day)} = \left[\frac{\text{initial bottle (mg } O_2\text{/l)} - \text{dark bottle (mg } O_2\text{/l)}}{\text{hours incubated}} \right] \times 24 \text{ hr}$$

13. Use the formula to calculate net primary productivity (in terms of oxygen produced) for the three light bottles (light, light + N, and light + P). Record your data in Table 45C-1.

Net primary productivity (mg O_2/l/day)

$$= \left[\frac{\text{light bottle (mg } O_2\text{/l)} - \text{initial bottle (mg } O_2\text{/l)}}{\text{hours incubated}} \right] \times 24 \text{ hr}$$

14. Use the formula below to calculate the gross primary productivity.

Gross primary productivity (mg/l/day)

$$= \left[\frac{\text{light bottle (mg } O_2\text{/l)} - \text{dark bottle (mg } O_2\text{/l)}}{\text{hours incubated}} \right] \times 24 \text{ hr}$$

15. Collect average class productivity data (in terms of oxygen produced) for each of the different light intensities and record these data in Table 45C-2.

Table 45C-2 Class Averages for Productivity Data

Light	NPP (mg O_2/l/day)			Gross primary productivity (mg O_2/l/day)		
	Light	Light + N	Light + P	Light	Light + N	Light + P
100%						
65%						
25%						
10%						
2%						

16. On a piece of graph paper, plot net primary productivity (*Y*-axis) versus percentage light intensity (*X*-axis) for all three samples.

a. Was light a limiting factor for any of the three samples? _____

b. What was the apparent effect of phosphorus enrichment? _____

_____

Of nitrogen enrichment? _____

_____

Do your results support your hypothesis? _____ *Your null hypothesis?* _____
*What do you **conclude** about the rates of primary production as you move to greater depths within a lake?*

_____

17. Convert both the net and gross productivity data (in mg O_2/l/day) from your group's light bottle (not enriched) to milligrams of carbon per cubic meter per day (mg C/m³/day) using the following formula (remember that each mg O_2 produced is equivalent to 0.375 mg C fixed):

$$\frac{mg\ O_2}{l/day} \times 0.375\ \frac{mg\ C}{mg\ O_2} \times \frac{1{,}000\ l}{m^3} = \frac{mg\ C}{m^3/day}$$

Light bottle net primary productivity = _____ mg C/m³/day

Light bottle gross primary productivity = _____ mg C/m³/day

18. Record class productivity data (mg C/m³/day) for all light regimens in Table 45C-3.

Table 45C-3   Class Data on Productivity

Percentage of Incident Light Available	Net Primary Productivity (mg C/m³/day)	Gross Primary Productivity (mg C/m³/day)
100%		
65%		
25%		
10%		
2%		

19. Plot the net primary productivity data at depths corresponding to the percentages of incident light in a *turbid lake,* as listed in Table 45C-4. Construct your graph on a separate piece of graph paper using the format shown in Figure 45C-1. Note that negative net productivity indicates that organic materials are being used up by respiration at a greater rate than they are produced by photosynthesis.

Table 45C-4

Percentage of Incident Light Available	Depth (m) (Turbid Lake)	Clear Lake
100%	0.0	0.0
65%	0.5	2.0
25%	1.5	6.4
10%	2.5	10.7
2%	4.2	18.2

Figure 45C-1 *This example graph plots net primary productivity against depth. Positive NPP— photosynthesis greater than respiration—appears to the right of the dashed vertical line, and negative NPP—respiration greater than photosynthesis—appears to the left of the dashed line.*

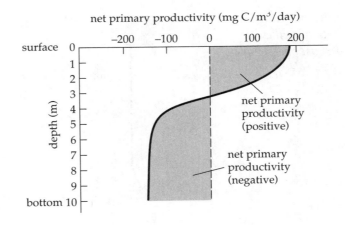

20. On a different graph, plot the net productivity data for depths corresponding to the percentages of incident light in a *clear lake* (Table 45C-4).

c. *Why do the surface waters of a lake (such as that for which data were graphed in Figure 45C-1) show positive net primary productivity, whereas the bottom waters show negative net primary productivity?* _____

_____

d. *At some depth, the light becomes sufficiently dim that photosynthesis just balances respiration. This depth is called the* compensation depth. *What is the compensation depth in Figure 45C-1?*

_____

e. *Would the compensation depth of the lake be closer to the surface or farther away from the surface if sunlight were reduced by a cloudy day?* _____

Why? _____

_____

EXERCISE D | **Thermal Stratification and Dissolved Oxygen Patterns in Lakes**

Photosynthesis and respiration produce distinctive patterns of dissolved oxygen (DO) in lakes and ponds, and these patterns can tell us much about ecosystem function. To understand these patterns, we must understand the physical properties of water that lead to **thermal stratification** of lakes and ponds, and then consider what will happen to DO concentrations in the surface and bottom waters.

In environments with winters cold enough to form an ice cover on lakes, when the ice melts in the spring, a lake is one temperature from the surface to the bottom and is freely mixed by the wind because the whole water column has the same temperature and density. This is "spring overturn."

As the sun continues to warm the surface waters, these waters become less dense than the underlying, colder water. If the wind is strong enough to mix the surface and bottom waters, the whole lake gradually warms. However, in warm, sunny weather, the surface layers are so much warmer and less dense than the layers underneath that the wind cannot mix the lake all the way to the bottom. It can mix the surface layers, though, and soon there is a sharp temperature discontinuity between a warm **epilimnion** (or mixed layer) and a cooler **hypolimnion.** The upper epilimnion is warm, lighted, turbulent, and nutrient-poor (because algal growth has stripped the nutrients from the water). Meanwhile, the lower hypolimnion is cold, dark, calm, and nutrient-rich. The transitional zone of rapid temperature change between these layers is the **thermocline.** In very large lakes such as Great Bear Lake in northern Canada (maximum depth 450 m), the thermocline may be at a depth of 200 m; in a sheltered, muddy pond it may be as shallow as 0.5 m. Figure 45D-1 shows the temperature profiles that might be expected in a 12-m-deep lake

Figure 45D-1 *Temperature profiles of a 12-m-deep northern lake during spring overturn and summer stratification.*

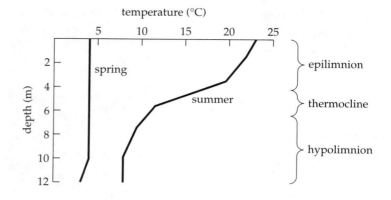

Figure 45D-2 *The density differences caused by the temperature profile in Figure 45D-1. The density differences are very slight (ranging from 1.00000 g/ml for the 4°C water to 0.99707 g/ml for the least dense water), but have been exaggerated here to show their pattern and relative size.*

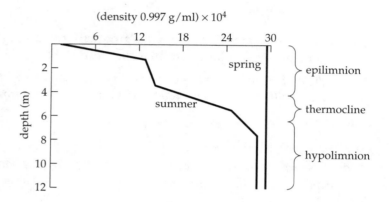

during spring overturn, and then again during thermal stratification. These temperature differences cause the density differences shown in Figure 45D-2.

When fall arrives, the surface waters gradually cool. The wind can mix the lake more deeply every day because the density difference between epilimnion and hypolimnion is decreasing. Then, probably on a cold day with strong wind, the wind can mix the lake all the way to the bottom. This "fall overturn" brings the lake to one temperature and one density.

One of the most important consequences of thermal stratification is its effect on dissolved oxygen. Oxygen distribution can vary dramatically with depth during thermal stratification, but the outcome depends on whether the lake is *oligotrophic* or *eutrophic*. An oligotrophic lake has a low supply of nutrients, a sparse phytoplankton crop, and clear water. There is only a slight rain of dead organic matter into the hypolimnion (reducing hypolimnion respiration), and light may penetrate all the way to the bottom and allow photosynthesis even at great depth. Therefore, oxygen will persist in an oligotrophic hypolimnion. In fact, because the water is colder in the hypolimnion, the DO concentration may be higher there than in the epilimnion.

A eutrophic lake, on the other hand, has a dense phytoplankton crop that shades the deep waters, and the rain of dead organic matter into the hypolimnion is heavy. The hypolimnion has intense respiration but no photosynthesis, and its oxygen may disappear. This may kill many kinds of sediment organisms and cause changes in the fish life of the lake. Because organic matter does not decay as well in the anaerobic deep waters, it may even accelerate the rate at which the lake fills in. Figure 45D-3 shows oxygen distributions that might be seen in the lake in Figure 45D-1 if the lake were either oligotrophic or eutrophic.

In this exercise, you will have the opportunity to observe thermal stratification and overturn with a model system. Using a colored 2 percent salt solution, your instructor will place a layer of dense solution under a layer of tap water in a container. Then, using an electric fan placed at various distances from the container, the class will experiment with how much "wind" is necessary to mix the "epilimnion" and "hypolimnion." Because of the dye in the salt solution, you will also observe the slow-motion undulations

Figure 45D-3 *Oxygen concentrations during thermal stratification in a lake with the temperature profile in Figure 45D-1.*

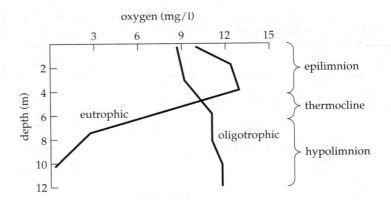

and shredding of the hypolimnion as the circulation of the epilimnion begins to affect it. You will use a spectrophotometer to monitor the changing color (absorbance) of the surface and bottom solutions.

a. *If the overlying layers of distilled water are all at the same temperature, how can they simulate the behavior of a hypolimnion and an epilimnion?* _____

_____

b. *The salt solution and the tap water will remain mostly unmixed for 12 hours or so if there is no "wind." Why will even gentle wind cause them to mix in a much shorter amount of time?*

_____

c. *What do you think will happen to the absorbance of the surface solution in our demonstration over time? How will this process be different with slow versus rapid wind speeds?*

_____

d. *What will happen to the absorbance* difference *between the surface and bottom solutions over time?*

_____

e. *If you wanted to find a thermally stratified lake that was* very resistant *to mixing, what kind of temperature difference between epilimnion and hypolimnion would you look for?*

_____

||||| **Procedure** |||||||||||||||||||||||||||||||||||||||

Your instructor has set up four simulations of thermally stratified lakes using tap water and a colored salt solution. The difference in density between the solutions results in stratification.

1. Obtain five spectrophotometer cuvettes (or tubes). Label them "TW" (tap water), "FS" (fan, surface), "FB" (fan, bottom), "NFS" (no fan, surface) and "NFB" (no fan, bottom). The tap water tube will serve as a blank.

2. Set the spectrophotometer to 510 nm.

3. At time zero, collect samples from the surface and bottom of each container and determine their absorbance at 510 nm. Use tap water as a blank. Use a Pasteur pipette and be careful not to disturb the layers. Record your data in Table 45D-1.

4. At 10-minute intervals, sample the surface in each container and record absorbance data in Table 45D-1, in the correct column for your fan distance and the no-fan column.

5. At the end of 60 minutes, once again sample both the surface and bottom in each container and record the data in Table 45D-1.

Table 45D-1 Absorbances of the Surface and Bottom Solutions in Stratified Solutions Subjected to Different Degrees of Wind Stress

Time (min)	Fan 0 m	Fan 0.5 m	Fan 1 m	No Fan
0 (surface)				
0 (bottom)				
10				
20				
30				
40				
50				
60 (surface)				
60 (bottom)				

6. Using your collected data, graph the absorbance of the surface solution against time on a separate piece of graph paper. Make separate curves for the no-fan container and the fan container. Use class data to complete similar curves for the two fan distances your group did not use. You will have to add appropriate numbers and tic marks to the Y-axis.

   f. Do your data show mixing of the top and bottom layers? Explain. _____

   _____

   g. What is the relation between wind speed and mixing? _____

   h. Do the "lakes" ever become completely mixed (surface and bottom absorbances the same)?

   _____

   i. If you had used a more concentrated salt solution (say, 5% instead of 2%), how would the result have been different? _____

   _____

   j. Compare the stability of lake thermal stratification on a hot, calm, sunny day with that on a chilly, windy, overcast day. Why is the stability so different in each case?

   _____

   _____

   k. In a eutrophic lake, why is oxygen stress for bottom animals greatest during the summer and least during the spring and fall? _____

   _____

Laboratory Review Questions and Problems

1. Describe how the dissolved oxygen of the light bottle and dark bottle would change from the initial bottle for a water sample with the following properties.

a. Very little photosynthesis and very little respiration.

   Light bottle _____

   Dark bottle _____

b. A high rate of photosynthesis and an equally high rate of respiration.

   Light bottle _____

   Dark bottle _____

c. A high rate of photosynthesis and a low rate of respiration.

   Light bottle _____

   Dark bottle _____

d. Very little photosynthesis and a high rate of respiration.

   Light bottle _____

   Dark bottle _____

2. For water that holds 7 mg O_2/l, use the nomograph in Figure 45A-4 to determine the percentage oxygen saturation for water at (a) 5°C _____ (b) 10°C _____ (c) 20°C _____.

3. How could the photosynthesis and respiration patterns in a eutrophic and an oligotrophic lake lead to the dissolved oxygen profiles depicted in the graph below?

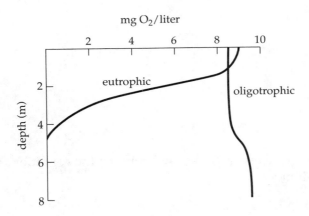

4. An oligotrophic lake is a clear lake with a relatively low gross primary productivity rate. A eutrophic lake has a high rate of gross primary productivity, but because of the high algal content, light does not penetrate very deeply. Oxygen is often absent from the hypolimnion of a eutrophic lake, but this is not the case for an oligotrophic lake. Suggest two reasons why these oxygen patterns occur.

5. The hypolimnion of a eutrophic lake is said to be nutrient-rich, cold, dark, and calm. Why does each of these conditions occur?

6. What factors would limit the phytoplankton population in the surface waters of a eutrophic lake in summer? What factors would limit populations in the hypolimnion?

7. Gross productivity in the water column of a lake is shown by curve A in the graph below. Explain what conditions might give rise to these data. How would you explain the conditions that lead to curve B?

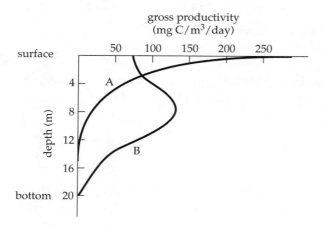

8. Explain why thermal stratification occurs in a lake, and contrast the conditions that characterize the epilimnion and hypolimnion.

Presenting and Analyzing Experimental Data: Tables, Graphs, and Statistical Methods

In many of the experimental laboratory exercises in this manual, observations result in data that can be recorded. In most cases, the data recorded are not all the same—measurements vary. Characteristics that show variability are called **variables.**

Discrete variables include those in which observations fall into one of several mutually exclusive categories (for example, red, yellow, or white flowers) or in which observations are not observed on a continuous scale (for example, the number of eggs in a bird's nest or the number of legs on an insect). These variables can take on only a limited number of values—a bird's nest cannot contain 1.23 eggs.

Continuous variables are derived from quantitative observations in which the data can assume any value in a continuous interval of measurement. Thus, one of the eggs in a bird's nest may weigh 1.20 or 1.23 or 1.25 grams—all values within a given range are possible. Continuous variables are often distributed within their possible range according to a frequency distribution—more observations tend to fall toward the middle of the range than toward the limits of that range. (One type of distribution is the normal distribution, and many statistical analyses assume that this distribution exists. You will see what a normal distribution looks like below.)

Familiarity with the number and types of variables as well as attention to the number of data points to be reported allows us to make appropriate decisions about the way in which experimental data are to be presented. Data are usually reported in the form of **tables** or **graphs** (see Part A).

After recording and presenting data from an experiment, it is important to determine how much the experimental results differ from the expected. (In fact, all of science could be considered a comparison of experimental results with what is expected.) This comparison is usually done by one or more **statistical tests.** In this manual we will be using the chi-square or chi-square median test (see Part B).

PART A   PRESENTING EXPERIMENTAL DATA: TABLES AND GRAPHS

Data from an experiment must always be presented in a form that allows the results to be readily communicated to other scientists. Tables or various types of graphs are most often used for this purpose. Scientists must be able to accurately report data collected from their experiments and interpret tables or graphs showing results from experiments done by others.

TABLES

Tables are the best choice of presentation format when all of the following criteria apply:

- The independent variable is discrete.
- Very few data points will be presented.
- Several dependent variables will be reported at one time.

Tables have a table number and a descriptive title, and all columns are labeled with units of measure. The independent variable (or variables) (month in Table AI-1) is placed on the left. The dependent variable (or variables) (number of tickets and fines in this case) should be placed to the right. The standardized variables are identified in the table title.

Table AI-1   Number of Speeding Citations Awarded and Fines Received in Hometown, USA, during the spring months, 1994

Month	Number of Tickets	Fines Received ($)
March	498	27,390
April	387	21,285

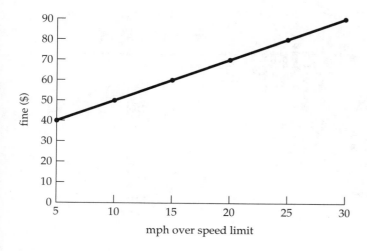

Figure AI-I   *Line graph showing fee schedule for speeding violations in Hometown, USA, in 1994.*

LINE GRAPHS

Line graphs (straight lines or curved) are the best choice of presentation format when all of the following criteria apply:

- The independent variable is continuous.

- More than three or four data points will be plotted.

- One or two dependent variables will be reported.

Line graphs have a figure number and a descriptive title, and both axes are clearly labeled, have units of measure and have equally spaced intervals. The independent variable is recorded on the X-axis. The dependent variable is recorded on the Y-axis. All data points are present and easy to see, and the curve takes up most of the area of the graph (Figure AI-1).
    Some general rules for line graph construction:

1. Always use graph paper to plot the points accurately.

2. Always use an accurate title for the graph.

3. Always record the independent variable on the X-(horizontal) axis and the dependent variable on the Y-(vertical) axis.

4. Space intervals evenly along each axis, and plot data points only at coordinates where data were collected.

5. Set the maximum values on both axes slightly above the maximum values of the data. The data should extend over most of the graph's area.

Use the following data to graph the growth of a plant on a sheet of graph paper.

Time (days)	Height (cm)
2	1.7
4	2.9
6	4.2
8	6.8
10	9.3
12	11.3
14	14.5

It should be clear that only the measurements actually made are indicated by dots (data points). However, because the information on both scales of the graph is assumed to be continuous, we can use the graph to find out what the height of the plant would be if measurements were made at some other time, at perhaps 3 days. Simply locate the position that corresponds to 3 days on the time scale, follow it up until it crosses the line graph, then locate the measurement that lies on the Y-axis, directly to the left of that point. To find on which day the plant reached a certain height, just reverse this process; start on the Y-axis and read the measurement on the X-axis.

a. *What was the height of the plant after 11 days of growth?* _____

b. *On what day was the plant 10 cm tall?* _____

    What would you do if you were asked to determine the height of the plant after 15 days? In making a line graph we are allowed to extend the line only between points representing actual measurements. Often scientists do extend the line, in the form of a broken line, predicting data the experimenter expects might appear beyond the given limits of the experiment. The extension is often the result of the experimenter's experience. The broken-line portion of the graph is an **extrapolation** because it goes beyond the experimenter's actual experience with this particular experiment. Between any two measured points it is possible to make an **interpolation** or estimate of size if one assumes that growth is occurring at a uniform

Time (days)	Plant height (cm)
2	0.8
4	1.5
6	2.1
8	3.4
10	4.7
12	5.7
14	7.3

rate between the two points. Interpolations can be made only between measured points on a graph; beyond the measured points we must extrapolate.

Suppose we study the growth of a second plant and graph the data on the same graph.

You will recognize that the slope of the line for the first set of data is steeper than that for this second set of data. The rate of growth can be calculated as

$$\text{Rate} = \frac{y_2 - y_1}{t_2 - t_1}$$

where y represents height (Y-axis variable) and t represents time. This is also the slope of a straight line between two points on the graph, and the slope is a measure of the speed of a reaction or event.

The type of graph examined above is a **progress graph** in which the amount of some quantity (in this case plant height) is shown on the Y-axis and time is shown on the X-axis. It is called a progress graph because it shows how the process (of plant growth) progresses over time. It may also be called a "time-course graph." A **rate graph** is constructed to indicate an amount per unit of time on the Y-axis. For example, if you were to graph an animal's weight gain per week, an appropriate label for the Y-axis might be "weight gain (grams)/week." Probably 80 percent of the graphs you encounter in biology will be progress or rate graphs.

How would you construct a rate graph for the data presented above for plant growth? Draw this graph below and label the axes correctly.

Scientists tend to use the words "curve" and "graph" interchangeably, but there are other types of graphs. One of these is the bar graph (histogram or frequency distribution).

BAR GRAPHS

Bar graphs are the best choice of presentation format when all of the following criteria apply:

- The independent variable is discrete.

- Three or more categories of independent variable will be reported.

- One dependent variable will be reported.

Bar graphs have a figure number and a descriptive title, and axes are clearly labeled, have units of measure, and have equally spaced intervals. The independent variable is recorded on the X-axis. The dependent variable is on the Y-axis. Bars are well separated and can be shaded differently. An example of a bar graph is given as Figure AI-2.

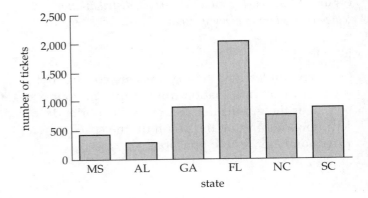

Figure AI-2   *Bar graph showing number of speeding tickets issued in southeastern states during March 1995. (Data have been fabricated and are being used solely for illustration.)*

In a **frequency distribution,** data are collected into classes and the number of observations in each class is shown on the Y-axis. This allows a large amount of "raw" information to be summarized.

For example, suppose we list the test scores of a group of biology students: 75, 86, 97, 53, 81, 75, 42, 83, 43, 75, 64, 72, 61, 83, 90, 73, 70, 91, 58, 84, 95, 71, 69, 87. The range of scores is 42 to 97. Now we choose classes (intervals), which must all be the same size. The choice of intervals depends on the data.

Next, we tally the number of observations (occurrences) in each class.

Class	Tally	Frequency
40–49	||	2
50–59	||	2
60–69	|||	3
70–79	⊞ ||	7
80–89	⊞ |	6
90–99	||||	4

Note that in Figure AI-3 we label the classes on the X-axis with the midpoint rather than the interval (class). Either way is acceptable. This type of graph is also called a **frequency histogram.**

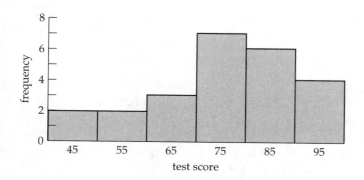

Figure AI-3 *Histogram or frequency distribution of test scores in a group of biology students.*

EXAMPLES

Once you understand how graphs are constructed, it is easier to get information from graphs in your textbook. For the graphs that follow, write a brief description interpreting the graph (in the space provided) and answer the questions.

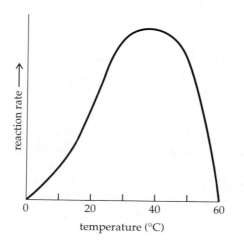

1. _____

_____

c. At what temperature is reaction rate highest? _____

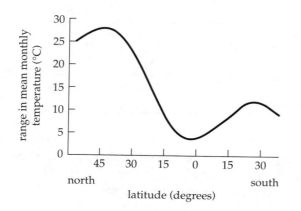

2. _____

_____

d. At what latitude does the least variation in temperature occur? _____

e. What is the range in mean monthly temperature in Clemson, SC, (approximately 34°N latitude)? _____

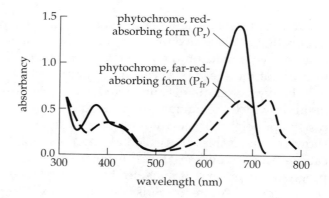

3. _____

_____

f. At what wavelengths does phytochrome (both forms) absorb the most light? _____

Here is an example you can use to practice graphing and evaluating data.

An ecologist observed that sugar maples seemed to dominate the understory of a forest while the larger (older) trees were mainly white oaks. Are the maples replacing the oaks?

Null Hypothesis: There is no significant difference between the age distributions of sugar maples and white oaks in this forest.

Alternative Hypothesis: Sugar maples and white oaks in this forest have different age distributions. (Other alternative hypotheses are also possible.)

The ecologist counts tree rings in a sample of trees throughout the forest, and the following data are collected on tree age (in years):

Sugar maple: 2, 6, 12, 16, 18, 12, 4, 2, 18, 2, 10, 2, 4, 12, 12, 2, 4, 16, 28, 22, 26, 40, 34, 36, 38, 24, 14, 11, 13, 17, 18, 19, 12, 32, 36, 26, 30, 24, 28, 27, 28, 44, 48, 31, 34

White oak: 4, 12, 16, 10, 36, 24, 40, 22, 38, 48, 49, 60, 54, 50, 45, 49, 54, 66, 66, 74, 80, 75, 66, 62, 92, 88, 94, 94, 82, 90, 84, 90, 88, 90, 88, 91, 91, 95, 94, 116, 108, 109, 104, 106, 111, 126, 140

Graph the data on a sheet of graph paper and evaluate the hypothesis.

PART B PRESENTING EXPERIMENTAL DATA: STATISTICAL ANALYSIS

In any experiment, when evaluating variables, we need to know something about the population from which observations are drawn. In **statistics,** a population contains all possible values of a variable. Thus the number of rose buds on one bush, the weight of all eggs in a robin's nest, or the colors of all kernels on an ear of corn can be determined, and because all values for the population are known, the measured population can be described unambiguously.

It is, however, often difficult to measure an entire population. The number of rose buds on one bush may not tell us much about a larger universe of rose bushes, and it would obviously be impossible to count the number of buds on all of the rose bushes everywhere. Thus, scientists often draw **samples** from larger populations in an attempt to make inferences about those larger populations. When this is done, it is crucial that the samples are representative of the entire population—this is assured by drawing the sample items randomly so that there is no bias in the selection. **Randomness** is often obtained by selecting samples using a table of random numbers.

MEASURES OF CENTRAL TENDENCY: THE MEAN

Once a sample is taken, several measures (**statistics**) are of interest. The first of these **central tendency**—an average that tells us something about the population. The most commonly used measure of central tendency is the arithmetic **mean** (others include the median and mode). The mean is calculated by summing all observations and dividing the total by the number of observations. Thus, if five rose bushes, sampled randomly from a larger population, had the following number of buds—63, 45, 86, 49, and 32—the mean would be (63 + 45 + 86 + 49 + 32) divided by 5 = 55. Thus 55 is a measure of central tendency for this population.

A general formula for the mean is

$$\bar{x} = \Sigma X_i / n$$

where

$\bar{x}$ ("x bar") is the mean

X_i ($X_1, X_2 \ldots X_i$) is any observation in the sample

ΣX_i is the total of all observations

n is the number of observations

MEASURES OF VARIATION

STANDARD DEVIATION

A second statistic measures **variation.** The most frequently used measure of variation is the **variance** (or its square root, the **standard deviation**). In the observations given above, the variance (s^2) is calculated by summing the squares of the difference between each observation and the mean. This sum is then divided by the sample size minus 1. Formally, this is expressed as

$$s^2 = \Sigma(X_i - \bar{x})^2 / (n - 1).$$

Using the numbers given in the rosebud example in which the mean ($\bar{x}$) is 55, we make the following calculations.

Observation	Deviation	Deviation Squared
(X_i)	$(X_i - \bar{x})$	$(X_i - \bar{x})^2$
63	+8	64
45	−10	100
86	+31	961
49	−6	36
32	−23	529
$\Sigma X_i = 275$	$\Sigma(X_i - \bar{x}) = 0$	$\Sigma(X_i - \bar{x})^2 = 1{,}690$
$n = 5$		$(n - 1) = 4$
$\bar{x} = 55$		$s^2 = \dfrac{1{,}690}{4}$
		$s^2 = 422.5$
		$s = \sqrt{422.5} = 20.6$

Thus, in this sample, the mean ($\bar{x}$) is 55 (as calculated previously), the variance (s^2) is 422.5, and the standard deviation (s) is 20.6. Notice that the sum of the deviations from the mean is always zero. Division of the squared deviations by one less than the number of observations ($n - 1$) gives an unbiased estimate of the variance.

Calculation Shortcut If a calculator is available that sums the total of the observations (X_i) and their square (X_i^2), it is not necessary to calculate the mean and determine each deviation from it. Thus

$$s^2 = \frac{[\Sigma X_i^2 - (\Sigma X_i)^2 / n]}{(n - 1)}$$

The **variance,** s^2, equals the sum of the **observations squared,** ΣX_i^2, minus the sum of the **observations squared divided by the number of observations,** $(\Sigma X_i^2)/n$, all divided by the **number of observations minus one,** $(n - 1)$. Using the data supplied above, verify this with your calculator.

If the sample is distributed normally, the distribution of observations will approximate a bell-shaped curve (Figure AI-4). In a normal curve, approximately two-thirds of the observations will fall within one standard deviation (s) on either side of the mean (between 55.0 ± 20.6, or 34.4 and 75.6 in the example given above—three-fifths or 60 percent of the observations fall in this range in our actual sample). In addition, 95 percent of all observations should fall within two standard deviations on each side of the mean, and 99 percent within three standard deviations of the mean.

The standard deviation is an unbiased estimate of the variability of the population and reveals the predicted limits within which one can make inferences about populations, regardless of their sample size. The mean and standard deviation are useful in characterizing a single sample. However, if many random samples were taken from the same population, the means of these samples would, themselves, be distributed normally with their mean equal to that of the population. From this distribution, a standard deviation of sample means can be estimated—this statistic is called the **standard error of the mean** or, simply, **the standard error ($s_{\bar{x}}$).** The best estimate of the standard error is

$$s_{\bar{x}} = \frac{s}{\sqrt{n}}$$

or, the variance equals the standard deviation divided by the square root of n. Verify that, in the example above, $s_{\bar{x}} = 9.2$. If the sample size increased, increasing the denominator, the standard error

would decrease. Thus, one effect of taking a larger sample is to lower the standard error—to increase the reliability of the estimates of the mean and standard deviation which would (should) not change with changes in sample size. One aspect of **experimental design** involves some estimate of variation to determine the optimum sample size to use—too few observations may not allow relationships to be determined at a suitable level of probability; too many observations increase the cost and may remove samples from the population needlessly.

STATISTICAL TESTS

Often we want to do more than describe a single population—we want to compare two populations. For example, do seedlings with adequate nutrients grow faster than those with low levels of phosphorus? We could conduct an experiment in which an experimental population is grown on soil deficient in phosphorus and a control population is grown on a completely adequate mixture of nutrients. If 10 seedlings are weighed on day 20 of the experiment, we can calculate their mean mass (weight), or we can measure their mean length, or some other variable that might reflect their growth. We can also determine the standard deviation for each sample. In this example, we would expect to see a difference between the groups, but how can we tell if this difference is likely to be real—what is the **probability** that it did not occur by chance or sampling error?

To answer this question, we could conduct a **statistical test.** First, we make the assumption that any differences we observe between the two samples are due to chance—they do not reflect true differences between the sample groups. Statistical tests allow us to generalize from the subset samples during experimentation to a larger population, based on the probability that chance alone caused the difference observed between treatment groups or samples.

Recall that the **null hypothesis** for any experiment is that there is no difference between the treatment groups or samples. The null hypothesis is useful because it is easily testable by statistics. Statistics give the probability that the results are due to chance and not some real treatment difference between the treatment groups. Arbitrarily, we can define the level at which chance is not affecting our decision. For biologists, this is usually a probability level of $p = 0.05$. This means that we have only five chances in 100 (a 5% chance) of drawing the wrong conclusion for the entire population based on our sample data.

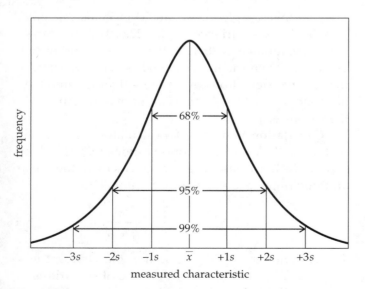

Figure AI-4 *A normal distribution. Two-thirds of all observations fall within one standard deviation* (s) *of the mean* ($\bar{x}$), *95 percent within two standard deviations, and 99 percent within three standard deviations.*

- If the computed probability is low (usually less than 5%) that the difference between two samples or treatment groups is caused by chance, we can *reject the null hypothesis* and accept the alternative hypothesis that there *is* a treatment effect.

- If the computed probability is *high* (usually greater than 5%), we *cannot reject the null hypothesis*; we accept the null hypothesis (no difference) because we do not have enough evidence to declare that there is a treatment effect.

In our plant example, the only difference between the two treatment groups was in the presence or absence of phosphorus. If the probability is less than 5%, we could conclude (but not prove), based on the results of statistical testing, that phosphorus has an important role in plant growth. (Recall that we can only *support* a hypothesis; there is a 5% chance that we are wrong in rejecting the null hypothesis in this case.) Conversely, if the probability is greater than 5% we could conclude with certainty that phosphorus does not affect plant growth (you can always disprove but never prove!)

These principles can be seen in the statistical test we will use in this laboratory manual, the **chi-square test.** The chi-square test gives us a χ^2 value which, when compared with critical values in a χ^2 table (see Table AI-2), allows us to determine the probability that our assumption of no difference or a difference due to chance only (null hypothesis) is a safe assumption. Depending on experimental design, we will use the chi-square statistic to perform an unpaired median test on a paired median test, or to analyze data from genetic crosses.

An **unpaired** test compares all observations in treatment group 1 with all observations in treatment group 2. An unpaired test is appropriate when there is no reason to link particular observations in treatment group 1 with particular observations in treatment group 2. An example would be a men-versus-women experiment. On the other hand, a **paired** test compares the first observation in treatment group 1 with the first observation in treatment group 2, the second observation in treatment group 1 with the second observation in treatment group 2, and so forth. It greatly increases the power of the test in "before-and-after" experiments where the effect of treatments *on the same individuals* are being examined. The following rules will work in most cases:

- For treatment groups of different individuals, use an unpaired test.

- For treatment groups of the same individuals, use a paired test.

THE CHI-SQUARE MEDIAN TEST FOR UNPAIRED SAMPLES

The chi-square (χ^2) median test is easy to understand and can be used in almost any situation in which two (or more) treatment groups are being compared. Its only drawback is that it cannot detect small treatment-group differences. Statisticians say it is "robust but insensitive."

Imagine that two people, treatment groups A and B, compare their reaction times (the time it takes to react to a stimulus; see Laboratory I). Each person does 10 tests. The results (in seconds) are as follows:

A: 0.24 0.28 .032 0.44 0.21 0.19 0.22 0.26 0.20 0.17

average = 0.253

B: 0.23 0.29 0.33 0.20 0.54 0.19 0.23 0.24 0.18 0.25

average = 0.268

We now pool these observations in one data set, order ("arrange") them from the fastest time to the slowest time (irrespective of treatment group; in this case, individual A is treatment 1 and individual B is treatment 2, if both are taking the same test), and code A observations as "A" and B observations as "B." The fastest time was 0.17 (from A), so the first letter on our list should be an A. The next fastest was 0.18 (from B), so the next letter will be a B. After ordering all the observations, we have the following coded list:

median
↓

fastest A B A B A B A A B B A B B A A B A B A B slowest

The median divides the data set into two parts. Half the observations will always be below the median and half above it. The number of As and Bs above and below the median are as follows:

	A	B
Below median	5	5
Above median	5	5

This is exactly the distribution above and below the median we would expect if the two people had the same reaction times.

On the other hand, let's say that person A now compares his or her reaction time with person C:

A: 0.24 0.28 0.32 0.44 0.21 0.19 0.22 0.26 0.20 0.17

average = 0.253

C: 0.15 0.19 0.19 0.16 0.17 0.17 0.15 0.16 0.40 0.25

average = 0.198

In this case the coded list is as follows:

median
↓

fastest C C C C C C A C A C A A A A C A A A C A slowest

and

	A	C
Below median	2	8
Above median	8	2

If there were no difference between treatment groups A and C, this very uneven distribution—with almost all A observations being slower than the median time and almost all C observations being faster—would be very improbable.

The statistic called chi-square (χ^2) can attach a probability to both arrangements around the median, above and below. Presented below are the χ^2 and probability values associated with increasingly uneven distributions of two treatment groups with 10 observations each. The probabilities shown are the probabilities that this distribution above and below the median could have originated due just to chance, not to any real difference between the treatment groups. Note how the distributions become more and more improbable as they become less even. The lower the probability, the less safe we are in accepting that the difference is due only to chance, so we reject this idea (the null hypothesis). Expressed another way, the probability is high that the difference is caused by the independent variable.

A-B Experiment I

	X	Y	
Below median	5	5	$\chi^2 = 0$
Above median	5	5	Probability > 0.9

A-B Experiment II

	X	Y	
Below median	6	4	$\chi^2 = 0.8$
Above median	4	6	Probability $= 0.25$–0.50

A-B Experiment III

	X	Y	
Below median	7	3	$\chi^2 = 3.2$
Above median	3	7	Probability $= 0.05$–0.10

A-C Experiment

	X	Y	
Below median	8	2	$\chi^2 = 7.2$
Above median	2	8	Probability $= 0.005$–0.01

In the first A-B experiment above, with a probability greater than 0.9, we would conclude that we cannot reject (we must accept) the null hypothesis and have no reason to declare that there is a difference between treatment groups. In the second and third A-B experiments, we also cannot reject the null hypothesis because the probability levels remain above $p = 0.05$. On the other hand, in the A-C experiment, with a probability less than or equal to 0.01 ($p = 0.005$–0.01, or one chance in 100 to 200 of being wrong if we reject the null hypothesis), we *do* have enough evidence to reject the null hypothesis, and we can accept the alternative hypothesis that there *is* a difference between treatment groups. We would say that the results are significant at the $p = 0.05$ level; a difference occurs 95 percent of the time (or no difference exists less than 5 percent of the time).

THE CHI-SQUARE MEDIAN TEST FOR PAIRED SAMPLES

A paired median test should be used when the *same individuals* make up both treatments, usually "before" and "after" treatment groups.

Let's say that two individuals (X and Y) test their reaction times before and after drinking a soft drink loaded with caffeine. Each person does 10 tests before the soft drink and 10 tests after the soft drink. The results appear below. The two treatments are before (B) and after (A) caffeine intake.

Individual X		Individual Y	
B	A	B	A
0.43	0.40	0.11	0.09
0.48	0.45	0.13	0.14
0.62	0.50	0.15	0.14
0.51	0.49	0.10	0.11
0.70	0.55	0.14	0.10
0.60	0.64	0.19	0.18
0.55	0.43	0.22	0.25
0.60	0.73	0.24	0.15
0.50	0.45	0.22	0.19
0.43	0.39	0.19	0.11

Did caffeine have an affect? The 20 fastest times all belong to individual Y, and they include an equal mix of "befores" and "afters." Likewise, the 20 slowest times all belong to individual X, and also are an even mix of "befores" and "afters."

	Before	After
Above median	10	10
Below median	10	10

$\chi^2 = 0$

The table makes it seem as if caffeine had no effect. However, note the tremendous variation between pitifully slow individual X and lightning fast individual Y. The test did not take this into account.

Instead of asking if *all* the "befores" are different from *all* the "afters," it would be more appropriate to ask if drinking the soft drink caused a change in reaction times *within the same individual.* In this way, variation between individuals won't matter. Let's re-code the data above as follows: if the "after" time is slower than the corresponding "before" time, code it +; if the "after" time is faster, code it −.

Individual X	Individual Y
−	−
−	+
−	−
−	+
−	−
+	−
−	+
+	−
−	−
−	−

If caffeine made no difference, we would expect 10 + signs and 10 − signs in the table. Instead, we see 5 + and 15 −. When we evaluate observed versus expected with χ^2, we find that $\chi^2 = 5$, and the probability of being wrong in rejecting the null hypothesis is less than 5% (approximately 2.5%). Thus, we reject the null hypothesis and accept the alternative hypothesis that there is likely to be a difference between treatment groups. Using a paired test when appropriate can make a big difference.

CHI-SQUARE CALCULATIONS

Now that you know the basics of unpaired and paired tests, you need to learn how to compute chi-square (χ^2) values. Chi-square compares the *observed number* of occurrences in a particular treatment group or "class" with the *expected number*. So, for example, χ^2 can be used to compare the num-

ber of observations above and below the median in two treatments, but it cannot directly work with treatment means, percentages, or the size of the difference between two treatment groups. It works only with the *numbers* of observations in different categories.

Chi-square is far more powerful with large numbers of observations and should not be used if the expected number of observations in a treatment group or class is less than five. In most cases, this means that you should plan to have at least 10 observations per treatment group.

Chi-square uses the formula

$$\chi^2 = \Sigma[(O - E)^2/E]$$

where O is the observed count, E is the expected or hypothetical count, and Σ means summation over all classes. Thus, we can write this as

$$\chi^2 = \Sigma\left[\frac{(\text{observed} - \text{hypothetical})^2}{\text{hypothetical}}\right]$$

If all the observed values are the same as the expected or hypothetical values, χ^2 is zero. On the other hand, the larger the deviation between observed and expected, the larger is χ^2 and the smaller its associated probability, making it more likely that you are safe to reject the null hypothesis—that is, the observed differences are due not to chance but to the treatment of the experimental groups or classes.

Chi-Square Calculations for Unpaired Median Tests

In the A-B experiment, observed and expected, if there is no reaction time difference between A and B, are identical. (The expected value represents an equal distribution of measurements on either side of the median for each sample group.)

		Observed	Expected
A	Above median	5	5
A	Below median	5	5
B	Above median	5	5
B	Below median	5	5

Therefore, for the A-B unpaired experiment,

$$\chi^2 = (5 - 5)^2/5 + (5 - 5)^2/5 + (5 - 5)^2/5 + (5 - 5)^2/5 = 0$$

Since a χ^2 of 0 is associated with $p > 0.9$, we accept the null hypothesis that there is no difference.

In the A-C unpaired experiment, observed and expected are different.

		Observed	Expected
A	Above median	8	5
A	Below median	2	5
C	Above median	2	5
C	Below median	8	5

For this experiment,

$$\chi^2 = (8 - 5)^2/5 + (2 - 5)^2/5 + (2 - 5)^2/5 + (8 - 5)^2/5 = 7.2$$

Since a χ^2 of 7.2 is associated with $p = 0.005$–0.001, we can reject the null hypothesis and conclude that the distribution observed is not due to chance.

Chi-Square Calculations for Paired Median Tests

For the paired test for individuals X and Y described above, there are only two categories, $+$ and $-$.

	Observed	Expected
+	5	10
−	15	10

The expected represents equal numbers in the $+$ and $-$ categories. In this case,

$$\chi^2 = (5 - 10)^2/10 + (15 - 10)^2/10 = 5.0$$

Since a χ^2 of 5 is associated with $p = 0.01$–0.05, we can reject the null hypothesis and conclude that the distribution observed is not due to chance.

Other Chi-Square Tests—Used for Genetics

One other area in which you will probably use statistics is genetics. Here, the treatment groups correspond to the numbers of offspring in different phe-notypic or genotypic classes. We test how the observed ratios of offspring conform to some expected ratio, such as 3:1 or 1:2:1. For example, consider a genetics problem in which we expect 60 mice to be 1/4 albino, 1/2 brown, and 1/4 gray. The null hypothesis proposes that there is no difference between observed and expected numbers. The observed and expected numbers are as follows:

Phenotype	Observed	Expected
Albino	21	15
Brown	29	30
Gray	10	15

The expected values are a 1:2:1 ratio of the 60 offspring produced, or 15:30:15. Thus we calculate

$$\chi^2 = (21 - 15)^2/15 + (29 - 30)^2/30 + (10 - 15)^2/15 = 4.1$$

The probability of this is less than 5%, so, at a critical value of $p = 0.05$, we have the evidence to reject the null hypothesis that there is no difference between observed and expected, and we must accept the alternative hypothesis that the ratio of mouse colors *is* different from 1:2:1.

DEGREES OF FREEDOM IN CHI-SQUARE

You have now learned to compute χ^2 values, to attach a probability to a χ^2 value, and to decide whether to reject or accept the null hypothesis. You still require one other value called **degrees of freedom,** *df*. This is related to the number of classes or treatment groups, each group being exposed to the independent variable (although *not* related to the number of observations).

Table AI-2   Critical Values of χ^2

Degrees of freedom	$p = 0.9$* (9 in 10)	$p = 0.5$ (1 in 2)	$p = 0.2$ (1 in 5)	$p = 0.05$ (1 in 20)	$p = 0.01$ (1 in 100)	$p = 0.001$ (1 in 1,000)
1	.016	.46	1.64	3.84	6.64	10.83
2	.21	1.39	3.22	5.99	9.21	13.82
3	.58	2.37	4.64	7.82	11.35	16.27
4	1.06	3.37	5.99	9.49	13.28	18.47
5	1.61	4.35	7.29	11.07	15.09	20.52
6	2.20	5.35	8.56	12.59	16.81	22.46
7	2.83	6.35	9.80	14.07	18.48	24.32
8	3.49	7.34	11.30	15.51	20.09	26.13
9	4.17	8.34	12.24	16.92	21.67	24.88
10	4.87	9.34	13.44	18.31	23.21	29.59

*p is the probability that results could be due to chance alone. The numbers in parentheses below each value of p restate p in terms of chance: 9 chances in 10 that results could be due to chance alone, and so on.

Degrees of freedom = number of classes
or treatment groups − 1

Thus, in the mouse (genetics) example (three pheno-typic categories or classes), $df = (3 − 1) = 2$. In both the unpaired A-B and A-C experiments (two treat-ment groups, A and B or A and C), $df = (2 − 1) = 1$. For the paired median test (two categories, + and −), $df = (2 − 1) = 1$.

(*Note:* Technically, the chi-square median test an-alyzes the distribution of data above and below a median and not actual student measurements. While the degrees of freedom will be the same as calcu-lated by the above method, df should be calculated by organizing data into "contingency tables.")

To determine the probability of a certain χ^2, com-pute the χ^2 then look it up in Table AI-2, on the line next to the appropriate degrees of freedom. Verify for yourself that the most-used χ^2 (5% significance level with 1 degree of freedom) is 3.84.

a. *If you computed a chi-square of 4.0 with df = 1, would you accept the null hypothesis?*

_____

b. *If you computed a chi-square of 4.0 with df = 3, would you accept the null hypothesis?*

_____

NON-SIGNIFICANT RESULTS

One final thing. Students are usually disappointed if their experiment does not show a significant treat-ment effect. This concern is misplaced. If a well-designed experiment does not disprove the null hy-pothesis, you have still found out reliable informa-tion about nature.

Non-significance *does not* imply insignificance.

Writing Laboratory Reports

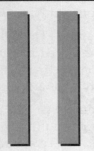

Researchers in biology communicate the results and significance of their work by publishing articles in professional journals where they will be seen and read by other investigators. Laboratory reports are written in the same general format as journal articles to communicate research in an effective and scientific manner.

All laboratory reports should include the following:

Title The title should indicate what the laboratory report is about. It should be brief, start with a key word, and indicate the nature of the investigation.

Abstract In a paragraph or two, explain the objectives of the research, how the experiment was conducted, the findings of the experiment, and finally, the implications and conclusions of the experiment.

Introduction All laboratory reports should begin with a short section describing (1) the purpose of the investigation; (2) justifications for performing the experiment; (3) the hypothesis to be tested; (4) predictions about possible outcomes; (5) how observations of others—for example, in references you have read—relate to your research; (6) the rationale for your experimental approach; and (7) how the information you gain can be used by others. Reference materials should be cited (paraphrase your citations, avoiding the use of direct quotations). Always write the introduction in the active voice.

Materials and Methods This section should contain a description of how you tested your hypothesis, including (1) experimental procedures followed, with a description of the different treatments you used; (2) the materials and equipment (you may identify these as they are discussed, rather than listing them separately); (3) the location of the study; and (4) any statistical techniques used. The reader should be able to reproduce your experiment after reading this section. This section is usually written in paragraph form and in the past tense.

Results In this part of the laboratory report, you organize and summarize the data generated by your experiment. General trends in the data should be discussed. Use tables and figures and graphs, as appropriate, to summarize the quantitative data. See Appendix I for graphing techniques. Data organized by statistical measures, such as means (a measure of central tendency) and standard deviations (a measure of variability) or standard errors (a measure of reliability), should be included in the form of tables and figures.

Also, you should describe the important aspects of your data in words. Make sure to label your tables and figures properly so they can be understood without having to read every word in the Results section. Remember to give all graphs and tables a title and make sure that all axes of graphs are clearly designated. Concisely state the results of any statistical analysis you perform. Interpretations are *not* made in this section!

Discussion In this section you should discuss and interpret your interpretation of the experimental results with regard to your hypothesis. Also discuss how the findings of other investigators relate to your results. It is helpful to note any limitations (e.g., small sample size) or problems (e.g., faulty equipment) that may have affected your results. Suggest how the investigation might have been improved. *Do not overstate this aspect of the discussion;* include it only if you think there is considerable room for improvement. In this section, you should also indicate the importance and possible applications of your findings and propose any new questions that occurred to you as the result of your study.

Bibliography   The bibliography references should be placed in alphabetical order by author in a special section at the end of your report. The correct style for a bibliographic entry is as follows:

Last name, first and middle initial (for first author) and first and middle initial and last name (of second author); extend series for additional authors. Date (year). Title of article. *Title of Journal* (italics or underlined) Volume (Issue number, optional): Pages.

For example:

Des Jarlais, D.C. and S.R. Friedman. 1994. "AIDS and the use of injected drugs." *Scientific American* 270(2): 82–88.

Appendix   Include raw data (all recorded observations), sample calculations, additional tables or figures not mentioned in the text, and any other relevant and *necessary* materials.

TO KEEP IN MIND WHEN WRITING LABORATORY REPORTS

1. Focus on carefully obtaining, analyzing, and interpreting *your* results. Do not concentrate on searching for the "right" answer or results that match those found by others. You should always consider the possibility that you have made an error; however, you should not reject your data just because they disagree with previously reported results. When drawing a conclusion, you must discuss and consider all of your data, not just those which support your views or hypothesis. Others may disagree with your interpretation of results, but you must consider all findings.

2. Remember, you can never be absolutely sure that your hypothesis is true no matter how good your results may appear. Your results will almost always contain some ambiguities (some data may support and some data may refute your hypothesis). Your conclusions do not have to be completely positive or negative—they may reflect a position in between.

3. Make sure that figures, tables, or diagrams are clearly labeled, an often-overlooked aspect of good communication in writing reports.

4. Another common mistake that interferes with communication is to write too much—without getting to the point. Before beginning to write, decide what is most important to say, then express yourself briefly and succinctly. Use the active voice and the first person ("I found that...").

5. Do *not* wait until the last minute to write your laboratory report. First, make a detailed outline and write a draft of your report. Let it sit for a day or so, re-read it for clarity and make corrections, then write the final copy. Have a friend read your first draft and make suggestions. Always keep an extra copy of your report (and a backup copy of your computer disk)—things do get lost! And turn in your report on time.

EXAMPLE OF A LAB REPORT

ROOT GROWTH AS A FUNCTION OF MOISTURE

Abstract   The purpose of this investigation is to determine whether the amount of moisture available to seeds affects the rate of seedling growth as measured by root length. Corn seeds (*Zea mays*) were subjected to three different moisture levels, and their roots were measured after one week. Seeds given 7.5 and 15.0 ml of water appeared to grow faster than those given 3.0 ml of water, but there was little difference between seeds given 7.5 ml and 15.0 ml of water. There may be an optimum level or range of moisture for seed growth.

Introduction

Research findings suggest that pea seeds germinate and grow at a faster rate (measured by root growth in terms of root length) when provided weekly with 15 ml of water as opposed to 5 ml of water (Smith, 1987). My hypothesis, constructed from previous observations and information, is that other cultivated seeds germinate and grow faster when moisture is provided in greater quantities. Corn, chosen as a typical, representative cultivated crop, will be used to test this hypothesis. Root growth is considered a reliable indicator of seedling growth. I predict that corn seedlings given more water will have longer roots. The results of this study could contribute to our knowledge about corn and its requirements for growth.

Materials and Methods

Materials required for this experiment [list is optional]:

    corn seeds
    metric ruler
    graduated cylinder (to measure water)
    Petri dishes
    filter paper

Corn seeds, chosen randomly, were treated with three different moisture levels in Petri dishes containing absorbent filter paper. Three groups of eight corn seeds each received total quantities of 3.0 ml, 7.5 ml, or 15.0 ml of water, measured using a graduated cylinder, over a period of one week. The Petri dishes were covered and placed in the same growth (environmental) chamber so that all conditions other than moisture level were identical for the three groups of seeds. All seeds were grown at 32°C. All seeds were watered at the same time. After one week, the length of the longest root of each seedling was measured using a metric ruler. Group means and standard deviations were calculated.

Results

The mean root lengths for the three groups of seedlings grown at different moisture levels are listed in Table 1. The mean root length of seedlings provided with 15.0 ml of water was greater than the mean for those given 3.0 ml of water. Similarly, the mean root length of seedlings that received 7.5 ml of water was greater than the mean for those that received 3.0 ml. However, there was little difference between the means of the 7.5 ml and 15.0 ml seed groups.

Table 1 Mean Root Lengths After One Week for Seeds Germinated at Three Different Mositure Levels

Moisture Level (mL of water)	Mean Root Length (cm)*
3.0	1.59 (0.394)
7.5	11.4 (1.61)
15.0	13.8 (2.87)

*Standard deviations in parentheses (see following page).

Discussion

Based on the length of roots at one week, seedlings given 7.5 or 15.0 ml of water grew faster than those given 3.0 ml of water. There was less difference between those given 7.5 ml and those given 15.0 ml. My conclusion is that corn seeds germinate and grow faster when moisture is provided in greater quantities, up to a point. Because the growth at 15.0 ml was not much greater than that at 7.5 ml, this study raises the question of whether there might be an optimum range of moisture for seed germination and seedling growth. On the basis of the results of this study, the hypothesis presented above can be neither accepted nor rejected. Additional experimentation with different kinds of seeds at more moisture levels is required. Also, the length of the longest root might not be a reliable indicator of growth; total root length would seem to be a better measure. Perhaps some indicator of growth in addition to root length should be used (root biomass, etc.).

Investigations of this sort have important applications. If it is found that there are optimum levels or ranges of moisture for the growth of corn at various stages, those geographic areas in which water is scarce might benefit from this information. In such areas, the amount of water supplemented could be limited to that necessary to achieve the known optimum degree of moisture for growth.

Bibliography

Smith, J.K. 1987. Peas need water. *J. Botany* 85(31): 53-59.

Appendix [additional or raw data may be included in this section]

Measured root lengths at the Three Moisture Levels

	Root	Length (cm)
3.0 ml	7.5 ml	15.0 ml
1.8	11.5	13.5
2.1	13.7	14.0
1.8	9.5	16.7
1.1	12.1	14.0
1.2	11.0	12.1
1.1	10.3	17.2
1.7	13.6	8.0
1.9	9.8	14.8

*Statistics are often used to analyze data in order to substantiate infer-
ences about the meaning of the observations. Standard deviation is an
unbiased statistical measure of sample variability. It is calculated by:

1. summing the squares of the deviations (differences) of each individual observation from the mean (arithmetic average) for the
   treatment group (population);
2. dividing this sum by one less than the number of observations in
   the sample (n − 1); and
3. taking the square root of the result.

Most scientific calculators produce this statistic (be sure that the result
is based on one less than the total number of observations rather than
on the total itself).

   Multiplying the standard deviation by the appropriate value of t
(found as a "Distribution of t Table" in statistical reference books) gives
a value for confidence interval (CI), which, when added to and subtracted from the sample mean, describes a range around the sample
mean in which the true mean values for a treatment are expected to be
found. (In this experiment, there were 8 observations in each treatment
group, so the appropriate value for t at a 95% confidence level with
n − 1 = 7 is 2.365.) Using this value for t, we can determine the range
around each mean in which 19 out of 20 (95% of) experimental means
for a particular treatment will lie. This is shown in the following table.

	Moisture Level (ml of water)		
	3.0 ml	7.5 ml	15.0 ml
Mean seedling root length (cm)	1.59	11.4	13.8
Standard deviation (SD)	0.394	1.61	2.87
Confidence Interval (CI = SD × t)	0.933	3.81	6.79
Range (mean ± CI)	0.7–2.5	7.6–15.2	7.0–20.6

Does the range for any treatment group include the mean of another
treatment group? If not, the compared treatments probably differ at the
5% level. Based on these statistics, the 3.0 ml treatment group differs
from the other two and the 7.5 and 15.0 ml groups do not differ from
each other.

   Note: The preferred method of analysis would be an analysis of variance, but this test would not tell you which of the means differed. A
t-test is not appropriate in this situation because there are more than
two treatment groups. See a statistics text for a discussion of other tests
and more information.

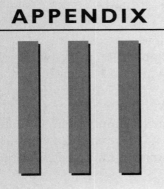

Preparing Solutions

Solutions are **homogeneous** mixtures of atoms, molecules, or ions of two or more different substances. The dissolved substance is called the **solute,** and the dissolving medium is called the **solvent.** At this point you should be familiar with measuring volume and mass. You will now learn how to use that knowledge to make solutions of **known composition.** Making solutions of known composition is one of the technically important aspects of science, since most chemical reactions occur in solution. To study these reactions it is necessary to know how to make solutions properly. You are now going to learn the types of solutions in common use, their properties and their uses. In particular, you will learn how to make molar, molal, normal, percent by mass, and percent by volume solutions.

MOLAR SOLUTIONS

A solution is 1 **molar** (M) when it contains 1 mole of solute in a liter of **solution.** One **mole** of a substance is the gram molecular weight or the molecular weight in grams of that substance. One mole of a substance contains Avogadro's number of molecules of that substance. For instance, the molecular weight (MW) of HCl (hydrogen chloride or hydrochloric acid) is 36.46. Thus, a mole of HCl has a mass of 36.46 g.

Example   Prepare a 1 M solution of NaCl. The molecular weight of NaCl = 58.5 g. To prepare a 1 M solution, add 58.5 g of NaCl to a 1000-ml volumetric flask. Add water to the 1000-ml mark. Notice that a molar solution is defined in moles of material *per liter of solution, not per liter of solvent.* Thus, the total of water and solute which make up the solution must equal 1000 ml. If you added 58.5 g of NaCl to 1000 ml of water you would increase the volume above the 1000 ml mark and you would have more

than a liter of solution. Remember when making a liter of a mol*ar* solution: Always make the solution **up to *1000 ml.***

How would you prepare the following?

0.03 M solution of NaCl

2 M solution of NaOH

1×10^{-4} M solution of Na_2CO_3

Hint: An easy way to determine the number of grams needed for a solution of a certain molarity is as follows:

grams required =
$$\text{mw (in g)} \times \text{volume (in liters)} \times \text{molarity}$$

MOLAL SOLUTIONS

A 1 molal solution (1 m) contains one mole of solute plus 1000 g of solvent. To make up a molal solution, add your solute to 1000 ml or 1000 g of solvent (water). (*Note:* It is generally assumed, when making molal solutions, that 1000 ml of water weighs 1000 g.) Also, be careful to distinguish between the symbols for molar (M) and molal (m).

The following experiment will illustrate the difference between molar and molal solutions.

Recall that mol*ar* solutions are defined by the final volume of solution. In contrast, as you have just read, mol*al* solutions are defined by the *mass* of solute per *mass* of solvent.

When dealing with very dilute solutions, for example, μM solutions, there is essentially no difference between molar and molal solutions. However, when more concentrated solutions are considered—for example, in the molar or molal range—distinct differences occur in the final volumes of the two types of solution. To illustrate this, prepare the following solutions and record the final volume of each solution.

Prepare 100 ml of molal solution for each of the following: (a) 1 m NaCl, (b) 0.1 m NaCl, (c) 0.5 m sucrose, (d) 0.05 m sucrose.

When you have prepared the solutions, measure the final volumes in a graduated cylinder and record the data. Compare the volumes of molal solutions with the volumes of similar molar solutions

NORMAL SOLUTIONS

Many solutions react with each other—for example, acids with bases. When dealing with solutions of this type it is convenient to use measures of solution that have **equal,** or **equivalent, numbers of reacting particles.** Solutions based on this criteria are referred to as **normal** solutions. A solution is 1 normal (1 N) when it contains 1 gram equivalent weight of solute per liter of solution. One gram equivalent weight is the weight of a substance that donates Avogadro's number (6.02×10^{23}) of reactive particles. In the neutralization of an acid with a base or vice versa, the gram equivalent weight of an acid contains 1 gram molecular weight of hydrogen ions (H^+) and the gram equivalent weight of a base contains one gram equivalent of hydroxide (OH^-) ions.

Example 1   Prepare a 1 N solution of HCl (MW = 36.46). Because it contains only one hydrogen, the gram equivalent weight is 36.46 g. A 1 N solution = 36.46 g HCl/liter of solution. Note that a 1 N solution of HCl is equivalent to a 1 M solution of HCl.

Example 2   Prepare a 1 N solution of H_2SO_4 (MW = 98). Because H_2SO_4 has two hydrogens that can dissociate in an acid-base reaction, it has a gram equivalent weight of 98/2 = 49 g.

A 1 N solution of H_2SO_4 = 49 g H_2SO_4/liter of solution. Note that this is equivalent to a 0.5 M solution of H_2SO_4.

How would you prepare the following?

1 N solution of NaOH

1 N solution of Ca(OH)$_2$

PERCENT BY VOLUME SOLUTIONS

Concentrations of a solution of two liquids are often prepared as percent by volume, rather than mass per unit volume, because volumes are easy to measure. This is common when preparing dilute solutions of alcohol, formaldehyde, etc. This is usually done volumetrically by using a graduated cylinder. To calculate the volumes needed to make a percent volume solution, use this equation:

$$C_i V_i = C_f V_f$$

where,

C_i is the percent concentration of the stock solution.

V_i is the volume needed of the stock solution.

C_f is the desired concentration of the final solution.

V_f is the desired volume of the final solution.

Example   Prepare 100 ml of a 70% solution of alcohol from a 95% stock solution. $C_i = 95\%$, $C_f = 70\%$, $V_f = 100$ ml; solve for V_i:

$$(0.95)\ V_i = (0.70)(100)$$

$$V_i = \frac{(0.70)(100)}{0.95} = 73.68$$

So you would fill a graduated cylinder up to the 73.7-ml mark with the 95% alcohol solution, and then add water up to the 100-ml mark to make 100 ml of 70% alcohol solution.

How would you prepare a 30% solution of alcohol using a 95% stock solution?

PERCENT BY MASS SOLUTION

A solution is 1% when it contains 1 g of solute per 100 ml of solution. How would you prepare the following?

4% solution of NaCl

0.1% solution of gelatin

10 ml of a 6% NaOH solution

OSMOLAR SOLUTIONS

The **osmolar** concentration of a solute is the molar concentration multiplied by the **total number of particles** produced per molecule in solution. For glucose, $C_6H_{12}O_6$, which does not dissociate in solution, a 1 M solution is also 1 osmolar; but for NaCl, which dissociates into two particles (Na^+ and Cl^-), a 1 M solution is 2 osmolar. Two solutions producing the same osmotic effect are said to be **isoösmolar (isotonic).** The osmolar concentration of plasma, which bathes the red blood cells, is 0.308. A glucose solution which is 0.308 M is also 0.308 osmolar and is, therefore, isotonic to plasma. Cells bathed in either plasma or 0.308 M glucose will not shrink or swell (osmotic effects). A solution of NaCl that is isotonic to plasma will _____ M.

PREPARING SERIAL DILUTIONS

Often the volume you wish to measure is too small to measure accurately with a pipette. For instance, suppose you wanted 100 ml of a 1×10^{-6} M solution of NaOH (MW = 40). To make this solution you would have to weigh out 0.000040 g (40 μg) of NaOH. This is beyond the capacity of most balances. You can, however, make this solution with precision by using the technique of **serial dilution.** The method of serial dilution is also widely used by microbiologists to estimate numbers of viable bacteria in various fluids and for preparing inocula for incubation tubes and broth. In some cases the technique of serial dilution is used for preparation of media used for cell culture; the media usually contain trace amounts of essential amino acids and vitamins.

Suppose you are making some culture media for an experiment and your recipe calls for μg/liter amounts of each constituent. Since your balance is only accurate to the second decimal place you decide to make stock solutions, containing 1 g/liter, of each component. This means that your stock solution is 10^6 more concentrated than you wish it to be. Since it would be almost impossible to make a 1:1,000,000 dilution accurately, you decide to use the technique of serial dilutions to make the solutions (Figure AIII-1).

1. Fill each of six tubes with 9 ml of water or buffer.

2. To the first tube add 1 ml of the concentrated stock solution, giving a final volume of 10 ml and a 1:10 dilution.

3. Mix the contents of the tube well, then transfer 1 ml to the second tube making a 1:100 dilution of the stock solution. Repeat this procedure six times until the original stock is diluted 1:1,000,000.

Suppose you wanted 10 ml of 1×10^{-7} M solution of NaOH. Describe how you would make such a solution.

The dilution described above is often called a 1 to 10 (or 1:10) dilution or a tenfold dilution. Note that you do not always have to use 1 ml and 9 ml. You could make a tenfold dilution by mixing 0.5 ml of sample with 4.5 ml of diluent just as easily. If you needed to make a hundredfold dilution (1 to 100) you could mix 0.05 ml with 4.95 ml of diluent. Note that you do not need to add 1 ml to 99 ml to make such a dilution. This would be wasteful if you did not need a large amount of solution. If you wished to make a 1 to 2 (1:2) dilution you would add 1 ml of the material to be diluted to 1 ml of the diluent to give a final volume of 2 ml. This could also be called a twofold dilution. If you needed 75 ml of a particular solution which required a twofold dilution of the material you had on hand, what would you do?

How would you prepare a 1:5 dilution of a broth solution containing 1000 bacteria per milliliter? How many bacteria per ml would be present in your final solution made by this fivefold dilution?

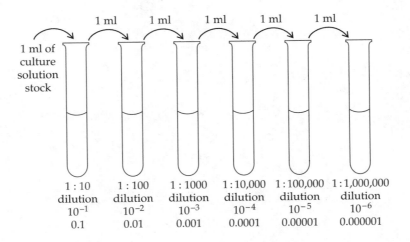

Figure AIII-I *Serial dilution technique and resultant concentrations.*

Calculate the amount of each reagent needed to make 100 ml of the following solutions. Prepare solutions 3, 4, and 7. *Have your calculations checked for accuracy before beginning!* Chemicals and balances are located on demonstration benches. **DO NOT USE YOUR FINGERS!** Always cover the chemical bottles when you are finished. Do not contaminate chemicals by using the same spatula in more than one bottle. **Keep the balances clean.**

Use graduated cylinders to make solutions. Though these are slightly less accurate than volumetric flasks, they are suitable for the present purpose and more convenient to use.

Ask your laboratory instructor for a demonstration. The small magnetic bars are easy to lose. Take care not to pour them out of the beaker and into the sink (they will go down the drain). Use the stirring bar retriever to remove the bars from solutions.

1. Prepare a 0.2 M solution of KH_2PO_4.

2. Prepare a 0.01 N solution of $Ca(OH)_2$.

3. Prepare a 0.1 M NaOH, 0.1 M KH_2PO_4 solution. What is the pH of this solution? _____ What is its H^+ concentration? _____

4. Prepare a 20% solution of red dye by using a 50% red-dye stock solution.

5. Prepare a 1% (w/v) solution of $NaHCO_3$.

6. Prepare a 0.1 m solution of KH_2PO_4.

7. Prepare a 1×10^{-4} solution of green culture medium using the stock solution (1 m) provided.

A Quick Guide to Using Scientific Notation

When using very large or very small numbers to report data, scientists write the numbers in **scientific notation.** A number is expressed in scientific notation when it is written as a product of a decimal number between 1 and 9 and the number 10 raised to the proper power. The power to which a number is raised is an indication of how many times that number is to be used as a factor in multiplication. For example, 10 raised to the power of 2, expressed 10^2, indicates that 10 is to be used as a factor twice: $10 \times 10 = 100$. So, in scientific notation, the number 100 can be written as 1×10^2. When a number is 10 or larger, the decimal point must be moved left to give a number between 1 and 9. The power of 10 is positive and equal to the number of decimal places moved; for example, $20 = 2.0 \times 10^1$; 34,698.5 can be expressed as 3.46985×10^4. Note the relationship between the movement of the decimal point and the power of 10. When the number is smaller than 1, the decimal point must be moved to the right to return to a whole number between 1 and 9. In this case, the power of 10 is negative and equal to the number of decimal places moved; for example, $0.01379 = 1.379 \times 10^{-2}$.

The most often-used measures in the laboratory are for mass (grams) and volume (liters). Note the relationships.

$$1\,0\,0\,0\,.\,0\,0\,0\,0\,0\,0\,0\,0\,0$$

(kg)	(g)	mg	μg	ng
	(l)	10^{-3} g	10^{-6} g	10^{-9} g
		ml	μl	

Moving the decimal point from right to left or left to right allows you to make simple changes from grams to milligrams and smaller units (*Note:* 1 g = 10^6 mg; the exponent is positive since there are many milligrams in a gram) or from milligrams to larger units (*Note:* 1 mg = 10^{-6} the exponent is negative since a milligram is only a very small fraction of a gram.)

Convert the following numbers to scientific notation and then convert to the metric units indicated in the second column.

	Scientific Notation	Metric Unit Conversion
1654 km	_____ km	_____ m
0.00013 g	_____ g	_____ mg
0.00000625 l	_____ l	_____ ml
2,323,000 m	_____ m	_____ μm
10 μg	_____ μg	_____ kg

A Quick Guide to Using the pH Meter

The standard laboratory pH meter has two electrodes: one electrode is sensitive to hydrogen ion concentration (usually expressed in moles/liter) and a second electrode, the reference electrode, completes the electrical circuit. On some pH meters, a combination electrode performs both functions. Other parts of the pH meter include the following:

Readout meter An analog meter with a pH scale for pH determinations and a millivolt (mV) scale for millivolt measurements

Mechanical meter zero An adjustment that mechanically zeros the meter pointer.

Function selector A switch that maintains the meter on standby when measurements are not being taken and that selects the measuring mode, either pH or millivolts.

Standardized control This control allows the meter to be set to the pH of the buffer solution used to standardize the instrument.

Temperature control This control compensates for the temperature of the solution being measured (it is active only when the function selector is in the pH mode).

DETERMINING pH

Before using the pH meter, you should always calibrate or "standardize" the instrument by adjusting the pH reading on the meter's scale to the known pH of a specific buffer solution (the standard). To obtain valid results when measuring the pH of a solution, the standardization buffer should have a pH close to that of the sample to be tested. In other words, if you are working with acidic solutions, you should standardize the pH meter with a known buffer solution of pH 4, *not* pH 13.

1. Examine the pH meter and identify each of the parts listed above.

2. Check to make sure that the function selector on the pH meter is on "standby." Raise the pH electrode out of the storage solution and rinse it with distilled water from a wash bottle (do this over an empty beaker).

3. Immerse the electrode in the solution to be analyzed. (If the solution is being stirred by a magnetic stirring bar, make sure that the stirring bar does not hit the electrode.)

4. Change the function selector from "standby" to "pH" and record your reading.

5. Raise the electrode out of the solution. Over a waste beaker, rinse the electrode with distilled water and place it into its storage beaker.

6. Repeat this procedure to measure the pH of solutions as needed.

A Quick Guide to Using the Spectrophotometer (Spectronic 20)

PRINCIPLES OF SPECTROPHOTOMETRY

Molecules either absorb or transmit energy in the form of electromagnetic radiation. White light (normal daylight) is made up of all the wavelengths of the electromagnetic radiation on the visible spectrum. How objects or chemical substances absorb and transmit the light that strikes them determines their color.

What we see as the color of an object, or a solution, is determined by what wavelengths of light are "left over" to be transmitted or reflected by the object after certain wavelengths are absorbed by its constituent molecules. For example, the pigment chlorophyll, present in the leaves of plants, absorbs a high percentage of the wavelengths of light in the red and violet-to-blue ranges. Green light, not absorbed by chlorophyll molecules, is reflected from the surface of the leaf—the reason why most plants appear green. A solution of chlorophyll extracted from a leaf would also appear green.

A spectrophotometer can be used to measure the amount of light absorbed or transmitted by molecules in a solution, which may depend on the concentration, reactions, or identity of the molecules. The spectrophotometer operates on the following principle: when a specific wavelength of light is transmitted through a solution, the radiant light energy absorbed, **absorbance** (A), is directly proportional to (1) the *absorptivity* of the solution—the ability of the solute molecules to absorb light of that wavelength; (2) the *concentration* of the solute, and (3) the *length of the path of light* (usually 1 cm) from its source, through the solution, to the point where the percentage of light energy transmitted or absorbed can be measured by a phototube.

Spectrophotometers that employ ultraviolet or visible light are the types most often used to study biological structures and reactions. The investigator selects a wavelength of light that will be maximally absorbed by a solute in a solution. (If visible light is used and the molecule of interest does not absorb, it is often possible to produce a chemical reaction that will yield a colored product.) After passing through the solution, the amount of light energy received at the phototube is expressed as **percent transmittance** ($\% T$). When this is compared with the intensity of light at the source, the amount of light absorbed (light that is not transmitted) can also be measured.

$$\frac{I_T}{I_0} \times 100 = \% \, T$$

where I_T is light that passes through the sample and I_0 is intensity of light at the source. And,

$$\log \frac{I_T}{I_0} = A \text{ (absorbance)}$$

By measuring the absorbance (or transmittance) it is possible to determine the concentration of the absorber (molecule) in solution. Concentration can be calculated directly if the *molar absorptivity* of the molecule (the amount of light at a specific wavelength absorbed by a specified concentration of solute in moles/liter) is known. Usually, however, molar absorptivity is not known, and absorbance readings indicate only relative concentrations—higher A resulting from a higher concentration.

USING THE SPECTROPHOTOMETER

The Bausch & Lomb Spectronic 20 Colorimeter (Figure AVI-1) is an extremely versatile instrument that is useful for the spectrophotometric, or colorimetric, determination of the concentration, reactivity, or identity of molecules in solution.

Within the optical system of the spectrophotometer, rotation of a prism (diffraction grating) perpendicular to the beam allows the investigator to select specific wavelengths of light in a range from 375 to 625 nm. Light of a selected wavelength is passed through the sample and is picked up by a measuring phototube, where the light energy is converted to a reading on the meter of the spectrophotometer.

Most spectrophotometers have two scales—one is a linear scale given as percent transmittance and the other is a logarithmic scale with the same gradations as the percent transmittance scale. (Note that 0.0 absorbance occurs at 100% transmittance, and infinite absorbance occurs at 0% transmitrance.) Examine the readout meter on the spectrophotometer you will be using. Note that the scales run in opposite directions.

1. Familiarize yourself with the parts of the spectrophotometer shown in Figure AVI-1.

2. To see how the wavelength control panel is responsible for selecting different wavelengths of light, cut a strip of white paper to just fit the diameter and length of a Spectronic 20 tube or cuvette. Slide the paper into the tube and insert the tube into the sample holder of the spectrophotometer.

3. Leave the sample holder open and place a cylinder of black paper around the opening.

4. Set the wavelength control to 620 nm and adjust the position of the tube containing the white paper until you see the maximum amount of red light on the right side of the paper.

5. Turn the wavelength dial in both directions and record the range of each wavelength at which you see a particular color. Range of wavelengths:

red _____   yellow _____

green _____   blue _____

violet _____

DETERMINING TRANSMITTANCE AND ABSORBANCE

To assure that spectrophotometer readings indicate only the concentration of the substance we wish to measure, a reading must first be obtained using a **blank,** a sample that contains all the components of the solution except the substance that will undergo (or cause) a change in absorbance. For instance, if you are using a reagent that changes color when mixed with a certain solute molecule, a blank should contain the reagent *minus* the solute molecule in question. With the blank inserted into the spectrophotometer, the instrument is adjusted to 100% transmittance. This step is similar to taring a balance: the transmittance of light through the blank is, of course, less than 100%, but the instrument can be adjusted to accept this reading as 100% transmittance with respect to the next reading, the sample to which the absorber molecule (or molecule that will cause the change in absorbance) has been added. The spectrophotometer readout for that sample will then indicate the absorbance for the substance of interest only.

1. Turn the power switch on, and allow a 5-minute warm-up period. The on-off switch is operated by the **zero control** knob.

2. Use the **wavelength control** knob to adjust the spectrophotometer to any wavelength between 500 and 600 nm. The selected wavelength is indicated on the **wavelength readout** in the window next to the knob.

3. When using the Spectronic 20, the meter must be adjusted to read its full scale—0% transmittance to 100% transmittance. With *no* sample tube in the machine, use the left-hand knob (zero control knob) to set the scale to 0% transmittance (infinite absorbance). (With no sample tube, the light path is automatically blocked and no light reaches the phototube; thus, 0% transmittance and infinite absorbance are simu-

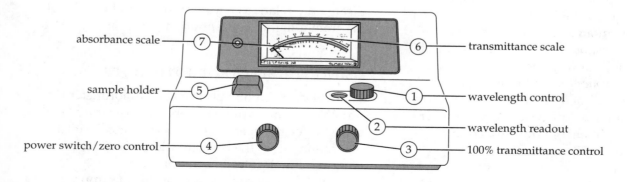

Figure AVI-1   *Features of the Bausch & Lomb Spectronic 20 Colorimeter. Note that the transmittance and absorbance scales run in opposite directions. Why?*

lated.) Be sure the cover on the sample holder is closed when you perform this step.

4. Insert your blank (be sure it is clean and dry on the outside) into the sample holder, and turn the right-hand control knob to set the **transmittance scale** to 100% transmittance, zero absorbance. This adjustment regulates the amount of light reaching the phototube in the absence of the absorber. *Whenever the wavelength is changed, the 100% transmittance adjustment must be reset.* Also, when operating at a fixed wavelength for an extended period of time, periodically check the 100% and 0% transmittance readouts and adjust if necessary.

5. Insert the sample to be tested into the chamber; read absorbance directly on the **absorbance scale** (lower scale). The reading on the absorbance scale is proportional to the concentration of your sample substance. *Note:* The absorbance scale reads from *right to left,* opposite to the direction for the transmittance scale.

6. Record your readings, making note of the wavelength used.

QUICK CHECKLIST FOR USING THE SPECTROPHOTOMETER

1. Turn the power on and allow a 5-minute warm-up period before taking sample readings.

2. Select the wavelength.

3. Check to make sure that the sample holder is empty and the cover is closed.

4. Use the zero control knob to set the meter to 0% transmittance.

5. Wipe off fingerprints from the reference-blank tube, insert it into the sample holder, and read the meter display in percent transmittance or absorbance.

For best results when using the spectrophotometer, always remember the following:

- All solutions *must* be free of bubbles.

- All sample holders must be at least one-half full.

- For best performance with test tube holders, the index mark on the tube or cuvette must align with the mark on the adapter.

- All sample tubes *must* be clean and free of scratches. Use lens paper to remove all fingerprints from the sample tubes and cuvettes.

- During extended operation at a fixed wavelength, you should make occasional checks of meter drift; use the blank to check for 100% transmittance.

a-	no, lacking, none
ab-	away from
ac-	to, toward
-aceus, -aceous	of, pertaining to
ad-	to, toward
adeno-	gland
agri-	field, soil
-al	having the character of
alb	white
-algia	pain
alto-	high
ameb-	change, alternation
amphibi-	leading a double life
ampho-, amb-	both
an-	not, without
ana-	up
andro-	male, man
anemo-	wind
angio-	vessel
ante-	before; ahead of time
antero-	front
antho-	flower
anti-	against
anthropo-	human
ap-	to, toward
aqu-	water
archaeo-	primitive, ancient
arthro-	jointed
aster-, astr-	stars
-ate	used in forming nouns from verbs
atom-	vapor
audi-	hear
auto-	self
bactr-	stick, club
barb-	beard
baro-	weight
bath-	depth, height
bene-	well, good
bi-	two, twice; double
bio-, bi-	life, living
-blast	sprout, germ

brachi-	having arms
branchi-	having fins
brev-	short
bronch-	windpipe
calor-	heat
carb-	coal, carbon
cardi-	heart
carn-	meat
carp-	fruit
-carpal, carpo-	wrist
card-	tail
cell-	storeroom; chamber
-cene	new, recent
centi-	hundredth
centr-	center
cephal-	head
cervic-	neck
chem-	referring to chemistry
chlor-	green
chrom-, chrome	color
chym-	juice
-cide	killing
circum-	around, about
cirro-	hairlike curls
co-	with, together
cocc-	seed
coel-	hollow
coll-	glue
com-	with, together
con-	against
coni-	cone
contra-	against
corp-	body
cosmo-	world; order; form
cotyl-	cup
counter-	against
crypt-	hidden, covered
-cule, -culus	added to nouns to form diminutive
cumul-	heaped
cuti-	skin
-cycle, cycl-	ring, circle
cyst-	bladder; pouch
cyt-, -cyte	cell; receptacle
dacty-	finger
deca-	ten

We thank Elizabeth Godrick, Boston College, for permission to use this material.

deci-	tenth	-graphy, graph	writing; record
deliquesce-	become fluid	-grav	heavy
demi-	half	gross	thick
dentri-	tree	gymno-	naked
dent-	tooth	gyn-	female, woman
derm-	skin	gry-	ring, circle, spiral
di-	two, double		
dia-	through, across	haem-, men-	blood
digit-	finger, toe	-hal, -hale	breathe, breath
din-	terrible	halo-	sea; salt
dis-	apart, out	hecto-	hundred
dorm-	sleep	-helminth	worm
dors-	back	hemi-	half
du-, duo-	two	hepat-	liver
-duct	lead	herb-	grass
dynam-	power	hetero-	different
dyan-	dark blue	hex-	six
dys-	ill, bad	hibern-	winter
		hipp-	horse
ec-	out of, outside	hist-	tissue
echin-	spiny, prickly	holo-	entire, whole
eco-	house	homo-	human
ect-	outside, without	hort-	garden
-ectomy	removal of	hybrid	mongrel
electro-	electric, electricity	hydr-	water
en-	in, into	hygr-	wet, moist
-em	made of	hypo-	beneath, under; less
-emia	blood	hyper-	above; beyond; over
encephal-	brain	hypho-	weaving
end-, ent-	within, in	hypno-	sleep
enter-	intestines		
-eous	nature of, like	-ia, -iasis	disease of; condition of
epi-	on, above	-ic	added to nouns to form adjectives
-err	wander; go astray	ichythy-	fish
erythro-	red	ign-	fire
ethno-	race, people	im-	not
eu-	well, good	in-	to, toward, into
extra-	beyond; outside of	in-	not
ex-	out of	-ine	of, pertaining to
		infra-	below, beneath
-fer	bear, carry; produce	inter-	between
ferro-	iron	intra	within, inside
fibr-	fiber, thread	-ion	go; come
-fid, fis-	divided into, split	-ism	a state or condition
-flect, flex-	bend	iso-	equal, same
flor-	flower	-itis	inflammation
fluor-	fluorine		
foli-	leaf	kilo-	thousand
fract-	break		
		lacry-	tear
galact-, galax-	milk, milky fluid	lact-	milk
gastro-	stomach	lat-	side, flank
geo-	land, earth	-less	without
-gen, -gine	producer, farmer	leuc-	white; bright; light
-gene, gnee-	origin	lign-	wood
-gest	carry; produce	lin-	line
glob-	all; round	lingu-	tongue
glottis	mouth of windpipe	liqu-	become fluid or liquid
glyco-	sweet	lip-	fat
-gon	angle, corner	lith-, -lite	stone; petrifying
-gony	offspring; generation coming into being	loc-	place
-grade	step; division	-logy	study
-gram	writing; record	lysis, -lyte, -lyst	dissolve, decompose

macro-	large
malle-	hammer
mamm-	breast
marg-	border, edge
mast-	breast
med-	middle
meg-	great; million
mela-, melan-	black; dark
mell-	soft
mes-	middle; half; intermediate
met-, meta-	between; along; after
meteor-	lofty, high; in air
-meter, -metry	way of measuring; instrument for measuring
micro-	small
milli-	thousandth
mis-	wrong, incorrect
mole-	mass
mono-	one, single
mont-	mountain
mort-	death
mov-, mot	move
morph-	shape, form
multi-	many
mycel-	threadlike
mycet-	fungus
myria-	many
nas-	nose
nemat-	thread
neo-	new, recent
nephro-	kidney
-ner	moist, liquid
neur-, nerv-	nerve; tendon
noct-, nox-	night
-node	knot
-nomy, -nome	distribute, arrange; law
non-,	not
not-	back
nuc-	center
ob-	against
ocul-	eye
oct-	eight
odent-	tooth
-oid	similar in form or shape
olf-	smell
-oma	tumor
omni-	all
oo-	egg
ophthal-	eye
opt-, -opay	eye; vision
orb-	circle, round, ring
orth-	straight; correct; right
oscu-	mouth
-osis	disease of
oste-	bone
-ous	full of, abounding in
ov-	egg
oxy-	sharp; acid; oxygen
pachy-	thick
paleo-	old, ancient

palm-	broad; flat
pan-	all
par-	beside, near; equal; bring forth
path-, pathy	disease, suffering
per-	through
peri-	around, on all sides
permea-	pass; go
phag-	eat
pheno-	show
phil-	loving, fond of
-phobia	excessive fear of
phon-, -phone	sound
-phore	bearer
photo-	light
phyc-	seaweed, algae
-phyll	leaf
physi-	nature, natural qualities
-phyte, phyt-	plant
plan-	roaming, wandering
plasm-, -plast	form, formed into
pleur-	rib; lung
pneumo-	lungs; air
pod-	foot
poly-	many; several
por-	opening
port-	carry
post-	after; behind
pom-	fruit
pro-	before; ahead of time; forward; favoring
proto-	first, primary
pseud-	false, deceptive
pter-	having wings or fins
pulmo-	lung
puls-	drive, push
pyr-	heat, fire
quadr-	four, fourfold
quin-	five
radi-	ray; spoke of wheel; energy in rays
re-	again; back
rect-	correct; right
ren-	kidney
ret-	net, made like a net
-rhage	flow
-rhaphy	suture
-rhea	flow
rhin-	nose
rhiz-	root
rubr-	red
sacchari-	sugar
sapr-	rotten
sour-	lizard
scler-	hard
sci-	know
-scope	look, observe
-scribe, -script	write
semi-	half, partly
sept-	partition; seven
septic	putrefaction, infection
sex-	six
-sis	condition, state

solv-	loosen, free		*therm-*	heat
somn-	sleep		*-tom*	cut, slice
son-	sound		*toxic-*	poison
spasm	tightening		*top-*	place
spec-, spic-	look at		*trans-*	across
sperm-	seed		*tri-*	three
spher-	ball		*trich-*	hair
-spire	breathe		*trop-*	turning; changing
spore	seed		*troph-*	one who feeds; well fed
stat-	standing, placed			
stell-	stars		*-ule*	diminutive
stereo-	solid, in three dimensions		*ultra-*	beyond
stern-	breast, chest		*ur-*	urine
stom-, stome	mouth			
strat-	layer		*vas-*	vessel
-stomy	to make an opening into		*vect-*	carry
strict-	drawn tight		*ven-, vent-*	come
styl-	pillar		*ventr-*	belly
sub-	under, below		*vice-*	in place of
super-	over, above, on top		*vig-*	strong
sur-	over, above, on top		*vit-, viv-*	life
sym-, syn-	together		*volv-*	roll; wander
			vor-	devour, eat
tachy-	quick, swift			
tarso-	ankle		*xanthin-*	yellow
tax-	arrangement			
tele-	far off, at a distance		*zo-, -zoa*	aminal
terr-	earth		*zyg-*	yolk
tetr-	four		*zym-*	yeast
thall-	young shoot			

ILLUSTRATION CREDITS